RESERVOIR EUTROPHICATION: PREVENTIVE MANAGEMENT

An applied example of Integrated Basin Management Interdisciplinary Research

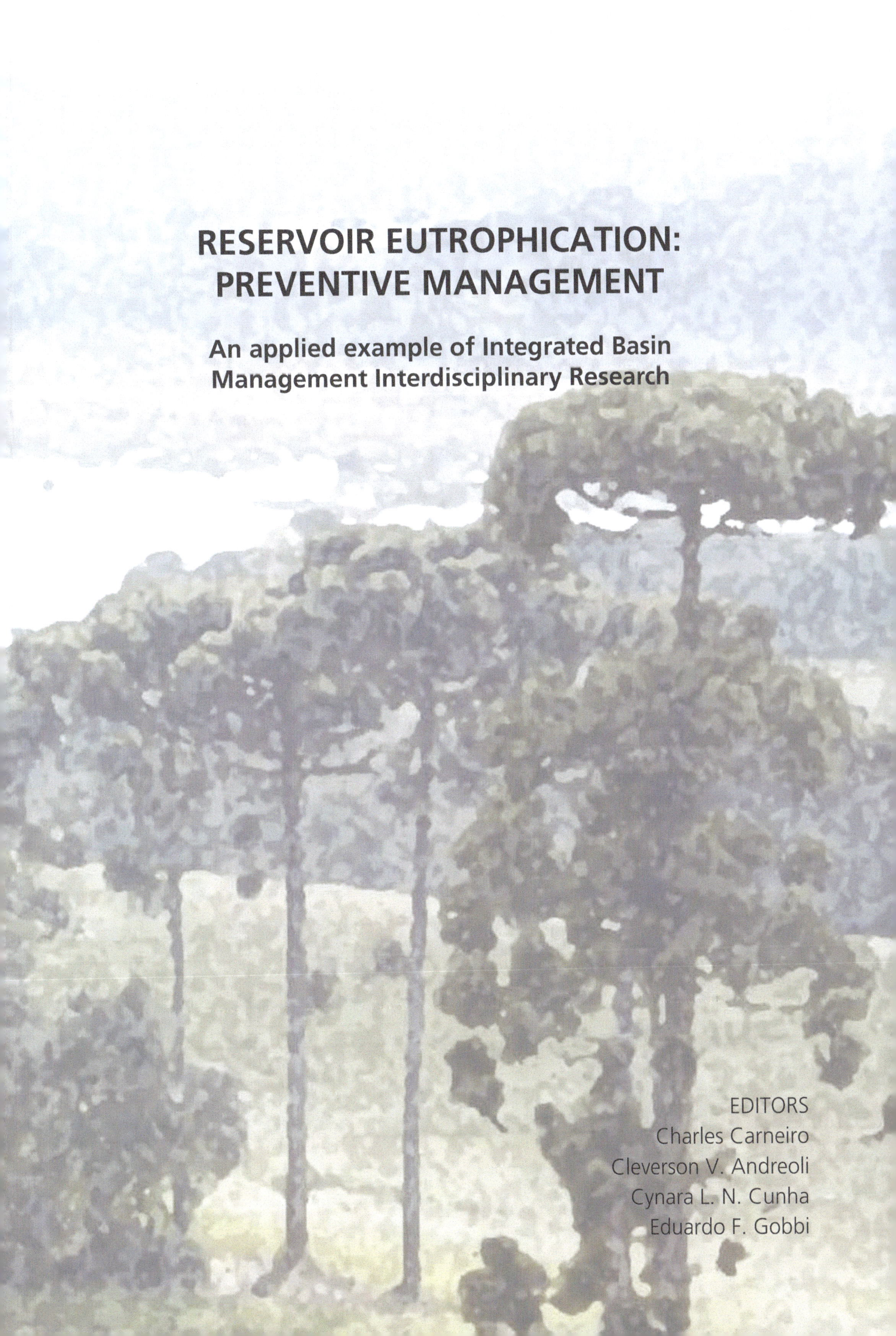

RESERVOIR EUTROPHICATION: PREVENTIVE MANAGEMENT

An applied example of Integrated Basin Management Interdisciplinary Research

EDITORS
Charles Carneiro
Cleverson V. Andreoli
Cynara L. N. Cunha
Eduardo F. Gobbi

Published by

IWA Publishing
Alliance House
12 Caxton Street
London SW1H 0QS, UK
Telephone: +44 (0)20 7654 5500
Fax: +44 (0)20 654 5555
Email: publications@iwap.co.uk
Web: www.iwapublishing.com

First published 2014

British Library Cataloguing in Publication Data
A CIP catalogue record for this book is available from the British Library

Library of Congress Cataloging- in-Publication Data
A catalog record for this book is available from the Library of Congress

ISBN: 9781780406473 (Hardback)
ISBN: 9781780406480 (eBook)

FOREWORD

RESERVOIR EUTROPHICATION: INTEGRATED STUDIES FOR PREVENTIVE MANAGEMENT IN THE RIO VERDE RESERVOIR, PARANÁ, BRAZIL

Sandra M. F. O. Azevedo
Bias Marçal de Faria
Francisco Barbosa

To be able to introduce this book is certainly a privilege and a pleasure. We do this with great pride and we wish to note that this opportunity became available to us through our participation as advisors of the projects that gave rise to this work. This participation, which encompassed bi-annual monitoring of the work developed by the teams of each sub--project over the past three years, was one of the most rewarding experiences of our professional careers.

This book provides an accurate summary of the results gained in the course of the 30-month study. The major challenge of harmoniously integrating a large number of studies related to human impacts on a drainage basin was boldly tackled by the coordinating team, resulting in this work, which certainly is a milestone in the history of Environmental Sciences in Brazil.

It is unlikely that another opportunity such as this will occur to study, in an integrated and preventive manner, the major environmental, social, and economic causes that lead to loss of water quality due to artificial eutrophication in our reservoirs. Therefore, the data presented throughout the 25 chapters, grouped into seven thematic sections, are of utmost relevance and represent a significant advancement in knowledge across all the areas incorporated into this study.

Section I clearly and precisely presents the general objectives of this study within the context of eutrophication and the general characterstics of the Rio Verde Basin. The physical environment of this drainage basin is discussed didatically but with necessary academic rigor, in Section II. Being able to systematically understand the geological and pedological data, vegetation cover and land-use, as well as an analysis of environmental sensitivity of an hydrographic basin through simultaneous and integrated studies using the most modern cartographic techniques, is undoubtedly a rare privilege.

The hydrodynamic aspects of the basin are thoroughly discussed in Section III and they represent a fundamental basis for understanding the potential impacts on the water quality of the reservoir in both spatial and temporal terms. The aquatic communities are reported in Section IV and are assessed through studies of phytoplankton, zooplankton, and fish population ecology. As is necessary in studies of this nature, the data interpretation of these studies is integrated with the physical and chemical environmental data.

The socioeconomic characteristics of the basin are well discussed in Section V, where the main economic and social aspects of the basin population are carefully analyzed.

Activities related to environmental education merited their own section (VI), for they certainly represent one of the greatest acheivements in the study. The ability to to build a collective environmental education program in a community believed to be lacking in social and economic resources is clear evidence that when one has real commitment to a mission, it becomes feasible even when the scenario suggests otherwise. The success of this sub-project is a milestone for the entire community and for all team members.

Finally, in section VII critical management tools are presented which were developed from the case studies using the collected preliminary data. Thereby, these tools can be used as a preventive action plan for the Rio Verde Basin, which in itself represents a major breakthrough for the region.

As we can see, this book presents not only the results of an extensive, complex, and very successful multidisciplinary study, but it is also a major source of data and information relevant to several areas of the environmental sciences, especially those that integrate the causes, consequences, and impacts of human activities on our aquatic ecosystems. This work represents unequivocal evidence that in this country we can indeed accomplish scientific studies at an international level that are fully dedicated to solving our local/regional problems. We hope that this study may serve as an example and motivate future generations.

Rio de Janeiro-RJ / Belo Horizonte-MG, May 25, 2011

TECHNICAL GROUP AND INSTITUTIONS

GENERAL MANAGEMENT

General Coordinators
Charles Carneiro (SANEPAR, ISAE)
Cleverson V. Andreoli (SANEPAR, ISAE, FAE)
Cynara de Lourdes da Nóbrega Cunha (UFPR)
Eduardo Felga Gobbi (UFPR)

Petrobras Coordinator
Ademar Kazuo Horiuchi (Petrobras)
Marcelo Bizzoni (Petrobras)

Technical Support
Helder Nocko (UFPR)
Kelly Cristiany Gutseit (SANEPAR, UFPR)
Patricia Esther Duarte Lagos E. D. Lagos (UFSCar)
Evelyn Nimmo (Translation Service)

COORDINATORS OF SUB-PROJECTS

Ana Maria Muratori (UFPR)
André Virmond Lima Bittencourt (UFPR)
Benno Doetzer (*Ambiental Florestas S.A.*)
Charles Carneiro (SANEPAR, ISAE)
Christine da Fonseca Xavier (IAP)
Cleverson V. Andreoli (SANEPAR, ISAE, FAE)
Cynara de Lourdes da Nóbrega Cunha (UFPR)
Eduardo Felga Gobbi (UFPR)
Everton Passos (UFPR)
Fabiano A. S. Dalto (UFPR)
Jose Edmilson de Souza-Lima (FAE, UFPR)
Lucia Izabel Czerwonka Sermann (*Sermanns– Assessoria e Treinamento*, SENAI, FAE)
Luciano Felício Fernandes (UFPR)
Mauricio Felga Gobbi (UFPR)
Miguel Mansur Aisse (UFPR)
Moacyr Serafim-Júnior (UFRB)
Patricio Peralta–Zamora (UFPR)
Ricardo Henrique Moreton Godoi (UFPR)
Sony Cortese Caneparo (UFPR)
Vinícius Abilhoa (Natural History Museum/PMC)

ADVISORS

Bias Marçal de Faria (CENPES/Petrobras)
Francisco Barbosa (UFMG)
Paulo Rubens de P. Faria Carvalho Cabral (CENPES/Petrobras)
Sandra M. F. O. Azevedo (UFRJ)

EDITORIAL ADVISORY BOARD

Arlindo Philippi Jr (USP)
Bias Marçal de Faria (CENPES/Petrobras)
Carlos A. C. Sampaio (FURB)
Fernando S. P. Sant'Anna (UFSC)
Francisco Barbosa (UFMG)
José Henrique de Faria (FAE and UFPR)
Maria do Carmo Sobral (UFPE)
Sandra M. F. O. Azevedo (UFRJ)
Selma S. Castro (UFG)
Severino S. Agra Filho (UFBA)
Tadeu F. Malheiros (USP)
Valdir Fernandes (UP, ISAE)

TECHNICAL PROJECT TEAM MEMBERS

CHAPTER 1: AN INTRODUCTION TO EUTROPHICATION AND THE STUDIES
Charles Carneiro – Researcher – SANEPAR, ISAE
Cleverson V. Andreoli – Researcher – SANEPAR, ISAE, FAE
Cynara de Lourdes da Nóbrega Cunha – Researcher – UFPR
Eduardo Felga Gobbi – Researcher – UFPR
Kelly C. Gutseit – Researcher and fellowship holder of the Rio Verde Project – SANEPAR, UFPR
Mariana Lagos – Fellowship holder – SANEPAR, FAE

CHAPTER 2: GENERAL CHARACTERIZATION OF THE BASIN
Charles Carneiro – Researcher – SANEPAR, ISAE
Christine da Fonseca Xavier – Researcher – IAP
Cleverson V. Andreoli – Researcher – SANEPAR, ISAE, FAE
Juliana Wojciechowski – Fellowship holder of the Rio Verde Project– UFPR
Kelly C. Gutseit – Researcher and fellowship holder of the Rio Verde Project – SANEPAR, UFPR
Rubia L. D. Lima – Fellowship holder – SANEPAR

CHAPTER 3: DEVELOPING A DATABASE FOR DIAGNOSING AND ASSESSING THE ENVIRONMENTAL CONDITIONS OF THE RIO VERDE BASIN – A DIGITAL ENVIRONMENTAL MODEL
Ana Maria Muratori – Sub-project vice-coordinator and collaborator – UFPR
Camila Strapassom dos Santos – Fellowship holder of the Rio Verde Project – UFPR
Elaine Cácia de Lima Frick – Sub-project vice-coordinator – UFPR
Everton Passos – Researcher – UFPR
Gabriel Henrique de Almeida Pereira – Fellowship holder of the Rio Verde Project – UFPR
Jean Jesus Ilsuk da Silva – Fellowship holder of the Rio Verde Project – UFPR
Lázaro Filipe de Souza – Fellowship holder of the Rio Verde Project – UFPR
Mariana Barbosa Druszcz – Fellowship holder of the Rio Verde Project – UFPR
Renan dos Santos Bastos Casagrande – Fellowship holder

of the Rio Verde Project – UFPR
Rodrigo Alves da Silva – Fellowship holder of the Rio Verde Project – UFPR
Sony Cortese Caneparo – Sub-project coordinator – UFPR

CHAPTER 4 – GEOLOGY
André Virmond Lima Bittencourt – Sub-project coordinator – UFPR.
Eduardo Chemas Hindi Geólo – Researcher – UFPR
Ernani Francisco da Rosa Filho – Researcher – UFPR.

CHAPTER 5 – PEDOLOGY AND LAND-USE SUITABILITY
Benno Henrique Weigert Doetzer – Sub-project coordinator– *Ambiental Paraná Florestas S.A.*
Carlos Henrique Zavelinski – Fellowship holder, Agronomy trainee UFPR
Oromar Bertol – Researcher – EMATER

CHAPTER 6 – NATURAL VEGETATION COVER AND LAND-USE MAPPING, CHARACTERIZATION AND DIAGNOSIS
Carlos Vellozo Roderjan – Sub-project coordinator and executor – UFPR
Pyramon Accioly – Researcher – UFPR/PUC
João Jailson Minuzzo – Fellowship holder – UFPR

CHAPTER 7 – POTENTIAL AND EMERGING FRAGILITY OF THE SOIL
Ana Maria Muratori – Sub-project coordinator – UFPR
Antônio Jorge dos Santos – Collaborator– UFPR
Camila Strapasson dos Santos – Fellowship holder – UFPR
Carla Gerhardt – Collaborator – COMEC
Elaine Cácia de Lima Frick – Collaborator – UFPR
Everton Passos – Sub-project coordinator– UFPR
Fabiana Moreno Casado – Collaborator – COMEC
Gabriel Henrique de A. Pereira – Fellowship holder – UFPR
Gustavo Roconi Nazareno –Fellowship holder – UFPR
Jean Jesus Ilsuk da Silva – Fellowship holder of the Rio Verde Project – UFPR
Jéssica Miara – Fellowship holder – UFPR
Lazaro Filipe de Souza – Fellowship holder of the Rio Verde Project – UFPR
Paulo Roberto Frick – Fellowship holder – UFPR
Renan Santos Bastos Casagrande – Fellowship holder – UFPR
Roberto Fulton Rivera – Fellowship holder – UFPR
Rodrigo Alves S. Martinez –Fellowship holder – UFPR
Sony Cortese Caneparo – Researcher – UFPR
Vânia Gomes Celso – Fellowship holder – UFPR

CHAPTER 8 – SUPPLY CAPACITY OF THE RIO VERDE RESERVOIR
Mauricio Felga Gobbi – Sub-project coordinator – UFPR
Helder Nocko – Researcher – UFPR
Osneri Andreoli – Collaborator – Governmental Water Institute

CHAPTER 9 – CHARACTERIZATION OF THE IONIC AND PARTICULATE SYSTEMS IN THE RESERVOIR
André Virmond Lima Bittencourt – Sub-project coordinator – UFPR
Jorge Eduardo de Souza Sarkis – Collaborator – IPEN
José Manoel dos Reis Neto – Collaborator – UFPR
Luiz Fabricio Zara – Collaborator – UnB
Miguel Jafelicci Junior – Collaborator – UNESP
Patricia Yassumoto Hirata – Fellowship holder of the Rio Verde Project – UFPR
Rene van Grieken – Collaborator – University of Antwerp, Belgium
Ricardo Henrique Moreton Godoi – Sub-project coordinator – UFPR

CHAPTER 10 – ASSESSING THE POLLUTION POTENTIAL OF LAKE BASINS: CONCENTRATIONS AND LOADS
Ana Flávia L. Godoi – Researcher – UFPR
Charles Carneiro – Sub-project coordinator – SANEPAR
Cleverson V. Andreoli – Researcher – SANEPAR, ISAE, FAE
Gustavo Kaminski –Fellowship holder – PUCPR
Gustavo R. Possetti – Collaborator – SANEPAR
Harry A.Bolmann – Researcher – PUCPR
Hugo H. Okumoto –Fellowship holder – UFPR
João Gomes – Collaborator – SANEPAR
Jonas H. Kondageski – Collaborator – SANEPAR
Josete de Fátima de Sá – Collaborator – SANEPAR
Juliana Wojciechowski – Fellowship holder of the Rio Verde Project – UFPR
Kelly C. Gutseit – Researcher and fellowship holder of Rio Verde Project – SANEPAR, UFPR
Maria Lúcia Ribeiro – Researcher – UNESP
Mary Marchi – Researcher – UNESP
Nicolas Lopardo – Collaborator – SANEPAR
Patricia E. D. Lagos – Researcher – UFSCar
Rafaela M. Clímaco – Fellowship holder – SANEPAR

CHAPTER 11 – SEDIMENT
Charles Carneiro – Subproject coordinator – SANEPAR, ISAE
Cleverson V. Andreoli – Researcher – SANEPAR, ISAE, FAE
Hugo H. Okumoto – Fellowship holder – UFPR
Juliana Wojciechowski – Fellowship holder of the Project – UFPR
Kelly C. Gutseit – Researcher and fellowship holder of the Rio Verde Project – SANEPAR, UFPR
Rubia L. D. Lima – Fellowship holder – SANEPAR

CAPÍTULO 12 – HYDRODYNAMICS AND TRANSPORTATION IN THE RIO VERDE RESERVOIR
Ana Paula Kelm Soares – Fellowship holder of the Project – UFPR.
Cynara de Lourdes da Nóbrega Cunha – Sub-project Coordinator – UFPR
Guilherme Augusto Stefanelo Franz – Fellowship holder of the Rio Verde Project – UFPR.
José Eduardo Gonçalves – Collaborator – SIMEPAR.
Paulo Cesar Colonna Rosman – Collaborator – UFRJ

CHAPTER 13 – PHYTOPLANKTON ECOLOGY IN THE RIO VERDE RESERVOIR
Ana Carolina Wosiack – Researcher - LECOTOX
Christine da Fonseca Xavier – Researcher – IAP

Juliana Wojciechowski – Fellowship holder of the Rio Verde Project – UFPR
Kelly C. Gutseit – Researcher and fellowship holder of the Rio Verde Project – SANEPAR, UFPR
Leda Neiva Dias – Collaboratora – IAP
Luciano Felício Fernandes – Subproject coordinator – UFPR
Patricia Esther Duarte Lagos – Collaborator – UFSCar
Renato Fernando Brunkow – Collaborator – IAP

CHAPTER 14 – ZOOPLANKTON
Alysson Troguer da Costa –Fellowship holder – PUC
Cláudia Fileto – Collaborator – University of São Paulo – São Carlos
Danilo Augusto de Oliveira Naliato – Collaborator – UNESP Botucatu
Gilmar Perbiche-Neves – Researcher – UNESP Botucatu
Letícia Fracaro – Fellowship holder – PUC
Moacyr Serafim Júnior – Sub-project coordinator– UFRB

CHAPTER 15 – ICHTHYOFAUNA
Hugo Bornatowski – Field assistant
Igor Kintopp Ribeiro – Field assistant
Jean Ricardo Simões Vitule – Collaborator – UFPR
Juliano Pilotto Abelardino da Silva – Fellowship holder
Letícia Lima – Fellowship holder – National History Museum/PMC
Letícia Michalik – Fellowship holder – PUCPR
Luis Fernando Favaro – Collaborator – UFPR
Vinícius Abilhoa – Sub-project coordinator– National History Museum/PMC
Wellen Laryessa Laynes – Fellowship holder – PUCPR

CAPÍTULO 16 – PARTICIPATIVE METHODOLOGIES APPLIED TO RIO VERDE BASIN, PR
Benno Doetzer – Researcher – *Ambiental Paraná Florestas S.A*
Daniel Larsen – Researcher – UFPR
Edna Batistella Lopes – Researcher – EMATER
Fabiano A. S. Dalto – Researcher – UFPR
Jose Edmilson de Souza-Lima – Researcher – FAE and UFPR
Miguel Mansur Aisse – Researcher – UFPR

CHAPTER 17 – SOCIAL AND ECONOMIC CHARACTERISTICS OF THE BASIN
Fabiano A. S. Dalto – Sub-project coordinator– UFPR
Felipe Almeida – Researcher – UNIPAMPA
Noam Goldstein – Researcher – UFPR

CHAPTER 18 – AN ASSESSMENT OF FISHING PRACTICES
Letícia Michalik – Fellowship holder – PUCPR
Vinícius Abilhoa – Sub-project coordinator – National History Museum/PMC
Wellen Laryessa Laynes – Fellowship holder – PUCPR

CHAPTER 19 – LIVESTOCK AND AGRICULTURAL ACTIVITIES
Benno Henrique Weigert Doetzer – Sub-project coordinator– *Ambiental Paraná Florestas S.A.*
Carlos Zavelinski – Fellowship holder, Agronomy trainee – UFPR
Edna Batistella Lopes – Collaborator – EMATER
Oromar Bertol – Researcher – EMATER

CHAPTER 20 – RURAL SANITATION
Daniel Larsen – Researcher – UFPR
Helio Luis Bzuneck – Researcher – SMMA of Municipality of Araucária
Marcio Roberto V. Canto – Researcher – Environment Surveillance Municipality of Araucária
Marco Antonio Archegas Ferreira – Researcher – SANEPAR
Miguel Mansur Aisse – Sub-project coordinator – UFPR
Ronaldo Barreto – Researcher – SANEPAR
Tânia Portela Costa – Researcher – Municipality of Campo Largo

CAPÍTULO 21 – THE VOICE OF THE POPULATION IN COLÔNIA CRISTINA: A STUDY OF RISK PERCEPTION
Claudia Cristina Machado – Researcher – UFPR
José Edmilson de Souza-Lima – Sub-project coordinator – FAE and UFPR
Lafaiete Santos Neves – Researcher – FAE
Priscila Cazarin Braga – Researcher – UFPR
Sandra Mara Maciel-Lima – Researcher – UFPR

CHAPTER 22 – ENVIRONMENTAL EDUCATION
Ana Maria Pereira Mendes – Researcher – FAE
Andreia Hoffman – Researcher – FAE
Claudia Cristina Machado – Collaborator – FAE
Edson José Veiga Silva – Collaborator – FAE
Janaina Kether Campos Piedade – Researcher, fellowship holder of Rio Verde Project – FAE
Jorge Gazal Carvalho – Fellowship holder – FAE
Lucia Izabel Czerwonka Sermann – Sub-project coordinator– *Sermanns– Assessoria e Treinamento,* SENAI, FAE
Osmar Ponchirolli – Researcher – FAE
Tânia Cristina Salomão Rodini – Researcher – FAE
Vânia Eloiza Cerutti– Fellowship holder – UP

CHAPTER 23 – ADVANCED OXIDATION PROCESSES
Loraine do Valle Jacobs – Research Student – UFPR
Marco A. Benedetti Durigan – Research Student – UFPR
Sérgio Renato Vaz – Researcher – UFPR
Patricio Peralta Zamora – Sub-project coordinator– UFPR

CAPITULO 24 – ENVIRONMENTAL ZONING OF THE PROTECTION AREAS IN THE RIO VERDE BASIN
Ednei Bueno do Nascimento – Researcher – EMATER
Eduardo Felga Gobbi – Sub-project coordinator– UFPR
Enéas Souza Machado – Researcher – *Instituto das Águas do Paraná*

CHAPTER 25 – PREVENTIVE ACTION PLAN OF THE RIO VERDE LAKE BASIN
Ana Sylvia Zeny – Collaborator – *Ingaia Engenharia Ambiental Ltda*
André Virmond Lima Bittencourt – Researcher – UFPR

Adenilson Ornelas – Researcher – PUCPR
Benno Henrique Weigert Doetzer – Researcher – *Environmental S.A*
Carlos Vellozo Roderjan – Researcher – UFPR
Charles Carneiro – Researcher – SANEPAR, ISAE
Cleverson V. Andreoli – Researcher – SANEPAR, ISAE, FAE
Cynara de Lourdes da Nóbrega Cunha – Researcher – UFPR
Daniel Larsen – Researcher – UFPR
Edilberto Nunes de Moura – Researcher – PUCPR
Eduardo Felga Gobbi – Researcher – UFPR
Fabiana de Nadai Andreoli – Researcher – PUCPR
Fabiano A. S. Dalto – Researcher – UFPR
Kelly C. Gutseit – Researcher and fellowship holder of the Rio Verde Project – SANEPAR, UFPR
Lucia Izabel Czerwonka Sermann – Researcher – *Sermanns*, SENAI, FAE
Luciano Felício Fernandes – Researcher – UFPR
Miguel Mansur Aisse – Researcher – UFPR
Patricio Peralta–Zamora – Researcher – UFPR
Ricardo Henrique Moreton Godoi – Researcher – UFPR.
Vilcimara Goncalves Quines – Collaborator – SANEPAR

COORDINATING INSTITUTION

UFPR – Federal University of Paraná State

SPONSORING INSTITUTION

PETROBRAS – Brazilian Petrol *S.A.*

RESOURCE MANAGEMENT INSTITUTION

FUNPAR – *Foundation of* Federal University of Paraná State

PARTICIPATING INSTITUTIONS

Environmental Paraná Forest S.A.
Bom Jesus Nossa Senhora de Lourdes High School
COMEC – Coordination of Curitiba Metropolitan Region
EMATER – Paraná Institute of Technical and Rural Assistance
FAE Administration and Economics University Center
FURB – Foundation University of Blumenau
IAP – Environmental Institute of Paraná State
IPAGUAS – Water Institute of Paraná State
IPEN – Nuclear and Energy Research Institute
ISAE – Institute of Business Administration and Economics
MHNC/PMC – Capão da Imbuia Natural History Museum / Municipality of Curitiba
PETROBRAS – Brazilian Petrol S/A
USP – University of São Paulo
PM de Campo Largo – Municipality of Campo Largo
PUCPR – Catholic University of Paraná State
SANEPAR – Water and Sanitation Company of Paraná State
SENAI – National Servive Agency for Industrial Learning
Sermanns Consulting
SIMEPAR – Technology and Environmental Data
SMMA PM Araucária – Environment Secretariat for the Municipality of Araucária City
UFBA – Federal University of Bahia
UFG – Federal University of Goiás
UFMG – Federal University of Minas Gerais
UFPE – Federal University of Pernambuco
UFPR – Federal University of Paraná
UFRB – Federal University of Recôncavo da Bahia
UFRJ – Federal University of Rio de Janeiro
UFSC – Federal University of Santa Catarina
UFSCar Federal University of São Carlos
UnB – University of Brasília
UNESP – Paulista State University
UNIPAMPA – Federal University of Pampa
University of Antwerp – Belgium
UP – Positivo University
Environmental Surveillance of the Municipality of Araucária City

ABOUT THE AUTHORS

ADEMAR KAZUO HORIUCHI
ademarh@petrobras.com.br
Degree in Chemical Engineering from UFPR and currently working as a Senior Processing Engineer at Petrobras. He has taken courses in Oil Processing Engineering and is specialized in Environmental Management. He works in the areas of design, monitoring and planning operations at REPAR, the President Getúlio Vargas Refinery, as well as a facilitator of the company's Internal Management System. He is the Environmental Sector Manager for REPAR.

ANA CAROLINA WOSIAK
carowosiak@bol.com.br
Full Bachelor Degree in Biology – Facultades Integradas Espírita (2001) and Master's in Botany from UFPR (2004). She works at the Laboratory of Environmental Analyses (LECOTOX) and has expertise in ecology, with an emphasis on physical, chemical and biological limnology. She is currently studying bacterioplankton, physical and chemical limnology, ecology, and reservoirs.

ANA MARIA MURATORI
amuratori@uol.com.br
She holds a PhD in Forest Engineering and a Master's in Geodesic Sciences, specializing in EAD. She is an accredited Geographer (CREA/PR-19779-D). Currently works as a teacher and an advisor in the Geography Graduate Program at UFPR and is a board member of AGB-Curitiba. She has authored several papers in the field of environmental sciences and books (in press) in the field of geography.

ANA PAULA KELM SOARES
ana_paula_ks@hotmail.com - UFPR
Environmental Engineer from UFPR. Currently a Master's student in the Graduate Studies Program in Numerical Methods in Engineering at UFPR.

ANDRÉ VIRMOND LIMA BITTENCOURT
andre@ufpr.br
Senior Professor at UFPR, where he was Head of the Geology Department and Director of the Earth Sciences Division. He is a consultant in Environmental Geochemistry, Hydro-geochemistry, Environmental Impact Assessment, and related areas. Author and co-author of more than 100 scientific papers and book chapters published in Brazil and abroad, and has developed dozens of technical papers. He has advised eight doctoral candidates and 12 Master's students and he is a member of the Life Institute Scientific Committee, Advisor of SPVS.

ANDREIA HOFFMANN BEZERRA FREITAS
andréia.hoffmann@bomjesus.br
Educator, Secondary Education Teaching Coordinator of Colégio Bom Jesus Nossa Senhora de Lourdes, specialist in Career Guidance, with a Master's in Organizations and Development from the FAE – University Center. Author and co-author of several scientific papers.

BIAS MARÇAL DE FARIA
biasfaria@petrobras.com.br
Currently works as a Consultant for Environmental Assessment and Monitoring Management at Petrobras Research and Development Center. Coordinates projects and research groups in recovering aquatic ecosystems and degraded wetlands, and assesses the impact in areas affected by oil spills. Author and co-author of 16 articles in scientific journals, two book chapters, and editor of a volume. Degree in Marine Biology from UFRJ, Master's in Geochemistry from the Universidade Federal Fluminense, and a Doctorate in Biophysics from UFRJ with a section of the thesis developed at the University of Lund, Sweden. Scientific and professional expertise in Limnology and Aquatic Micro-organism Ecology.

BENNO HENRIQUE WEIGERT DOETZER
benno@ambientalflorestas.com.br
Agronomist, with a Master's in Agronomy, focusing on soil, from UFPR. Rural Outreach for EMATER since 1991, state coordinator of the Environment (2003 to 2010), responsible for state irrigation programs, nocturnal poultry and swine production, soil fertility and agricultural recycling of sewage sludge. Technical Director of Ambiental Paraná Florestas S.A. beginning in January 2011.

CARLOS HENRIQUE ZAVELINSKI
UFPR Alumni, Agronomy (2006 – 2010), Fellowship holder of the Rio Verde Project.

CARLOS VELLOZO RODERJAN
roderjan@ufpr.br
Forest Engineer, Full Professor in the Department of Forest Sciences at UFPR and in the Forest Engineering Graduate Program, UFPR, in the disciplines of dendrology and phytogeography. He is an adjunct editor for the magazine Floresta and serves as coordinator and researcher on projects for the Fundação de Pesquisas Florestais do Paraná (FUPEF) and the FUNPAR Foundation at UFPR.

CHARLES CARNEIRO
charlesc@sanepar.com.br
Manager of Research & Development at SANEPAR - Water and Sanitation Company of Paraná State, and professor in Post-graduate programs, Federal University of Paraná State - UFPR and Institute of Business Administration and Economics – ISAE. Post-doctorate in Water Science and Engineering at UNESCO-IHE (Netherlands), Doctorate in Geology, focusing on water geochemistry and Master's in Soil Science at UFPR (Brazil). Improvement course in Lake Basin Management, from ILEC Shiga/Japan. Member of the Science and Technology Technical Board of the Water

Resources National Council. Author and co-author more than 50 scientific papers, 5 books and 11 book chapters. Has experience in Limnology, Lake Basin Management, Degraded Areas Recovery, Forest Planting, and Inventories of greenhouse gas emission.

CHRISTINE DA FONSECA XAVIER
christin@iap.pr.gov.br
Biologist in the Limnology Sector and a specialist in freshwater phytoplankton and limnology of reservoirs. Coordinator in the Research and Quality Department, IAP.

CLAUDIA FILETO
cfileto@hotmail.com
Completed the Post-graduate course on Environmental Engineering Sciences in the Department of Hydraulics and Sanitation, São Carlos School of Engineering. Post-doctorate at the Universidade Estadual Paulista (UNESP). Has experience in the field of Limnology with an emphasis on Experimental Limnology and Zooplankton. Has published articles in national and international journals.

CLÁUDIA CRISTINA MACHADO
claudiacristina.machado@hotmail.com
Doctoral candidate in Environment and Development (MADE-UFPR) and Professor at the SPEI Colleges. Researcher for the Public Policy Research and Development Group (CNPq/FAE) and a Sustainability Researcher at the FIEP System.

CLEVERSON V. ANDREOLI
c.andreoli@sanepar.com.br
Research Engineer at SANEPAR and Post-graduate Professor at Institute of Business Administration and Economics – ISAE, FAE – Administration and Economics University Center and Catholic Pontifical University - PUC. He has authored more than 150 articles, 20 environmental technical books and four children's books on environmental education. Received the MCT/FINEP Technological Innovation award in 1998 and 2008. He is a member of the ABES Editorial Board and the International Water Academy.

CYNARA DE LOURDES DA NÓBREGA CUNHA
cynara@ufpr.br
Doctorate in Civil Engineering from COPPE/UFRJ. Professor in Environmental Engineering, Graduate Program in Numerical Methods in Engineering, and the Graduate Program in Environmental Engineering, UFPR. Researcher at the Laboratory of Studies in Environmental Modeling and Monitoring.

DANIEL LARSEN
daniel.larsen@gmail.com
Environmental Engineer from PUC-PR, Master's Degree from the Water Resources and Environment Graduate Program, UFPR.

DANILO A. O. NALIATO
naliatobio@yahoo.com.br
Doctoral candidate in Biological Sciences at the Universidade Estadual Paulista with experience in the field of Aquatic Ecology/Limnology and an emphasis on zooplankton communities of continental waters. Has published in national and international journals, a book chapter and produced educational material on Aquatic Ecology.

EDNA BATISTELLA LOPES
edna@emater.pr.gov.br
Sociologist with a specialization in Adult Education and Higher Education Methodology. Has worked for 38 years at EMATER in the field of Outreach Methodology.

EDNEI BUENO DO NASCIMENTO
edneibueno@seab.pr.gov.br
Agronomist, employee of EMATER Paraná, has Specializations in the Formulation of Public Policies for Rural Development and in Environmental Analysis. Sustainability Coordinator for EMATER and National Negotiator for the network of organizations that operate with Rural Tourism in Family Agriculture (REDE TRAF). He is the author of several publications on Rural Tourism and was President of the Technical Support Board for the Passaúna River APA and for the Iraí River APA.

EDUARDO CHEMAS HINDI
híndi@ufpr.br
Adjunct Professor in Geology and Researcher for the Hydrogeological Research Laboratory at UFPR. Has developed applied geophysics projects in mining, hydrogeology and solid waste disposal. Published dozens of scientific papers in national and international journals with an emphasis on hydrogeology. He was editor of the magazine Águas Subterrâneas of ABAS and the ALHSUD magazine.

EDUARDO FELGA GOBBI
efgobbi@ufpr.br
Doctorate in Ocean Engineering with an emphasis on Environment from COPPE/UFRJ. Master's in Ocean Engineering from COPPE/UFRJ. Graduate degree in Economic Engineering and Industrial Administration from UFRJ. Civil Engineering degree from UFRJ. Professor in Environmental Engineering at UFPR. Researcher at the Laboratory of Studies in Monitoring and Environmental Modeling (LEMMA) at UFPR. Coordinator of Water Resources, Atmospheric and Basic Sanitation for the State Secretariat of the Environment and Water Resources for the Government of the State of Paraná.

ELAINE DE CACIA DE LIMA FRICK
elainecacia@ufpr.br
Bachelor's Degree in Geography from the Universidade Tuiuti do Paraná, with a Specialization in Environmental Analysis from the Department of Geography, UFPR. Master's in Forestry Science from the Department of Forestry, UFPR, and professor in Geography at UFPR.

ENÉAS SOUZA MACHADO
eneasmachado@aguasparana.pr.gov.br
Civil Engineering Degree from UFPR, Master's Degree in

Hydraulic and Sanitary Engineering from the Engineering School of São Carlos at USP, doctorate in Hydraulic and Sanitary Engineering from the Polytechnic School, USP. Has a graduate degree in Hydraulic Engineering from the Universidade Lomonosov, Moscow, Russia. He is currently director of Hydrographic Basin Management for the Instituto das Águas do Paraná.

ERNANI FRANCISCO DA ROSA FILHO
ernani@ufpr.br
Associate Professor, Department of Geology, UFPR. He was Coordinator of Geology and Advisor to the Dean of International Affairs. Was AHLSUD Vice-President. Author and co-author of 100 scientific papers and has advised several Master's Dissertations and Doctoral Theses. He is a CNPq research fellow, ad hoc consultant for CAPES, CNPq and other entities. He is leader of the Water Resources Group, UFPR, in collaboration with CNPq and is Coordinator of the LPH/UFPR.

EVERTON PASSOS
everton@ufpr.br or passosever@gmail.com
Specialization in Photointerpretation, Geo-processing, Master's in Agronomy, Soil Science, and Doctorate in Forestry, Forestry Management, from UFPR. Is Associate Professor in the Department of Geography, advisor in the Geography Master's and Doctorate Program, and teaches in the graduate disciplines of Geography, Environmental Engineering, Cartography and Oceanography. Author of various articles on the environment and books in the field of geosciences.

FABIANO ABRANCHES SILVA DALTO
dalto@ufpr.br
Doctorate in Economics from the University of Hertfordshire, England, Professor in the Economics Department and in the Public Policies Graduate Program, UFPR.

FELIPE ALMEIDA
felipealmeida@unipampa.edu.br
Doctorate in Economics from UFPR, currently an Adjunct Professor at Universidade Federal do Pampa. He has experience in the field of Economics with an emphasis on Institutional and Evolutionary Economics, focusing primarily on the topics of consumption, competition, collective learning and economic psychology.

FRANCISCO BARBOSA
barbosa@icb.ufmg.br
He has a Bachelor's Degree in Natural History from UFMG (1973), Master's in Ecology and Natural Resources from the Universidade Federal de São Carlos (1979), Doctorate in Ecology and Natural Resources from the Universidade Federal de São Carlos (1981) and Post-doctorate in algae ecophysiology from the Institute of Freshwater Ecology, England. Currently he is a CNPq researcher 1B, tenured professor at UFMG and coordinator of the Specialization course in Municipal Water Resources Management at ICB/UFMG. He has experience in the fields of Ecology/Limnology with an emphasis on ecology and aquatic systems conservation, focusing primarily on topics of primary productivity, phytoplankton ecology, aquatic biodiversity conservation, and water quality.

GABRIEL HENRIQUE DE ALMEIDA PEREIRA
Bachelor Degree in Environmental Engineering, UFPR.

GILMAR PERBICHE-NEVES
gilmarpneves@yahoo.com.br
Biology Degree, 2004, from PUC-PR. He has worked since 2002 with ecology and plankton systematics, limnology and aquaculture. He has more than 20 published articles and five book chapters. Currently he is a doctoral candidate in Biological Science (Zoology) at the Universidade Estadual Paulista.

GUILHERME AUGUSTO STEFANELO FRANZ
guifranz@gmail.com
Master's in Water Resources and Environment, UFPR. Holds a Specialization in Environmental Management and a Bachelor's Degree in Environmental Engineering, UFPR.

HELDER NOCKO
hel-nocko@ufpr.br
Environmental Engineer from UFPR with a Specialization in Field Engineering from PUC-PR. Currently a Master's Degree student in the Graduate Program in Numerical Methods for Engineering (PPGMNE), UFPR, and Director of EnvEx Engenharia e Consultoria Ambiental.

JANAINA KETHER CAMPO PIEDADE
janainakpiedade@yahoo.com,br
Psychologist and Occupational Therapist, Master's in Organization and Development from FAE – University Center, Specialization in Public Health and Community Action from FAFIJA, and Organizational Psychology Management from FAE – University Center. Instructor at SENAC and professor in the lato sensu graduate program at SPEI Colleges in Conflict Management, Interpersonal Communication and Cooperative Learning. Author and co-author of scientific papers and an environmental education booklet. Organizational consultant.

JEAN JESUS ILSUK DA SILVA
A trainee of the project who holds a Bachelor's degree in Geography, UFPR.

JEAN RICARDO SIMÕES VITULE
biovitule@gmail.com
Biologist, Master's and PhD in Zoology from UFPR. Professor in the Environmental Engineering Department at UFPR. Professor in the Zoology Graduate Program, Zoology Department, and in the Ecology and Conservation Graduate Program in the Biological Sciences Sector, UFPR.

JOSÉ EDMILSON DE SOUZA-LIMA
edmilson@fae.edu
Doctorate in Environment and Development (UFPR). Graduate Professor and researcher in the Organization and Development Master's degree program at FAE, and in the

Environment and Development Graduate Program (MADE-UFPR). Leader of the Public Policy Research and Development Group (CNPq/FAE). He develops interdisciplinary projects on concepts of nature and social development and environmental risk perception.

JOSÉ EDUARDO GONÇALVES
jgoncalv@simepar.br
Doctorate in Physical Oceanography from USP. Researcher from the Technological Institute SIMEPAR, in the fields of Research and Development (R&D) Project Management. Collaborating Professor in the Water Resources and Environmental Engineering Graduate Program at UFPR.

JOSÉ MANOEL DOS REIS NETO
jmreis@ufpr.br
Bachelor's Degree in Geology from UNB (1978), Master's in General and Applied Geology from USP (1983) and Doctorate in Geochemistry and Geotectonics, USP (1994). He is an Associate Professor I at UFPR in the field of Exploratory Geology. Conducts exploratory research on industrial minerals, limestone, geochemistry, karst and isotope geology. Responsible for research projects with Petrobras, CNPq, MINEROPAR and other companies. Coordinator of LAMIR—UFPR.

JULIANA WOJCIECHOWSKI
julianaw@ufpr.br
Degree in Biological Sciences from UFPR. Master's student in Botany (Taxonomy and Microalgae Ecology) at UFPR. Conducts research in the fields of limnology and toxic algae ecology in reservoirs.

KELLY CRISTIANY GUTSEIT
kellygutseit@hotmail.com,
Has a degree in Biological Sciences from UFPR (1997) and a doctorate in Aquatic Ecology from the University of Lund, Sweden (2007). She has experience in the fields of Limnology and Oceanography, with emphasis on plankton ecology and energy transfer in planktonic food chains.

LAFAIETE SANTOS NEVES
lafaiete.neves@gmail.com
Doctorate from the Economics Development, UFPR. Professor and researcher in the Organizations and Development Master's Program, FAE – University Center. IPEA/CAPES Fellowship holder. Leader of the Public Policy Research and Development Group (CNPq/FAE). Member of CEPAT/UNISINOS. He has experience in the field of Economics with an emphasis on Economic Development, Political Economics and Brazilian Economics.

LETÍCIA MICHALIK
leticiamichalik@hotmail.com
Bachelor's Degree in Biology, PUC-PR. Trainee of the Ichthyofauna Research Group, Natural History Museum, Municipality of Curitiba.

LETÍCIA LIMA
ledelima@hotmail.com
Biologist, technician of the Ichthyology Laboratory in the Natural History Museum, Municipality of Curitiba.

LUCIA IZABEL CZERWONKA SERMANN
lucia@sermann.com.br
Educator, Master's in Education from UFPR and Doctorate in Production Engineering from the UFSC. Professor at FAE – University Center Graduate Programs, SENAI-PR/SC, and SPEI. Director of Sermann's Assessoria e Treinamento. Organizational consultant. Institutional evaluator of INEP. She was a professor at PUC-PR in the Graduate course on Education. Coordinated the stricto sensu Program at FAE – University Center. Implemented and coordinated the CDE – Faculty Development Program at Universidade Positivo. She is author of scientific papers, book chapters, and material for distance education. Member of the Percurso magazine editorial board and the Uninter magazine.

LUCIANO FELÍCIO FERNANDES
lff@ufpr.br
Associate Professor III at UFPR, performing research on taxonomy and harmful algae ecology in reservoirs and marine waters, and studies on aquatic environmental impact. Master's in Botany and Doctorate in Zoology from UFPR, with a Post-Doctorate at the Woods Hole Oceanographic Institution, Massachusetts. Further international training courses in Harmful Microalgae and Phytoplankton Monitoring. Member of the Brazilian Society of Phicology, International Diatom Research and International Society of Harmful Algae.

LUIS FERNANDO FAVARO
lufavaro@ufpr.br
Biologist. Master's in Morphology (Cellular Biology) from UFPR and Doctorate in Ecology and Natural Resources from UFSCAR. Professor in the Cellular Biology Department at UFPR. Coordinator of the Research Group in marine and freshwater fish biology. Professor in the Zoology Graduate Program, Department of Zoology, and the Graduate Program in Ecology and Conservation in the Department of Biological Sciences, UFPR.

LUIZ FABRICIO ZARA
fabriciozara@gmail.com
Bachelor's Degree in Chemistry with technical assignments from the Universidade Federal de São Carlos (1993), Master's in Analytical Chemistry (Arar.) from the Universidade Estadual Paulista Júlio de Mesquita Filho (2001). Currently is an adjunct professor at the Universidade de Brasília and technical consultant of the civil society in the public interest organization – CAMARA-Health, Life Quality and Management. He has experience in the fields of Health and Environment with an emphasis on analysis of trace elements by atomic spectroscopy.

MARIANA BARBOSA DRUSZCZ
Bachelor's Degree in Environmental Engineering, UFPR.

MAURICIO FELGA GOBBI
gobbi@ufpr.br
Degree in Civil Engineering, UFRJ, Master's in Oceanic Engineering from COPPE and Doctorate in Coastal and Oceanographic Engineering from the University of Delaware. Currently he is an undergraduate and graduate professor in the Department of Environmental Engineering at UFPR. Has experience and written various papers in the fields of Fluid Mechanics, Coastal and Oceanographic Engineering, Water quality, Hydrology, Meteorology, and Numeric Modeling.

MIGUEL JAFELICCI JUNIOR
jafeli@iq.unesp.br
Research Productivity 1D Fellowship holder. Degree in Chemistry from UNESP (1976), Master's from the University of Bristol (1979), and Doctorate in Chemistry (Physicochemistry), USP (1983). Adjunct Professor at UNESP. Interested in Physicochemistry of Colloids and Interfaces, working mainly on materials chemistry, colloidal dispersions, magnetic materials and nanoparticles in self-organizing systems, nanostructured and biomedicine, metals and magnetocaloric alloys.

MIGUEL MANSUR AISSE
miguel.dhs@ufpr.br
Civil Engineer, Doctorate from the Polytechnic School, USP, Associate Professor of the PPGERHA/DHS, UFPR.

MOACYR SERAFIM-JUNIOR
moa.cwb@gmail.com
PhD. Professor in the Fishing Engineering Course at the Universidade Federal do Recôncavo da Bahia. He has published papers in national and international journals, is author of four technical book chapters and participated in the organization of three books in the environmental field. He was given the Ecology and Environmentalism award by the Curitiba City Council. He has experience in Limnology, Aquiculture and Environmental Management, focusing on the topics of ecology of continental and marine aquatic environments, aquatic organism production with an emphasis on mariculture.

NOAM GOLDSTEIN
ngolds@gmail.com
Economist, Bachelor's Degree from the Department of Economics, UFPR.

OROMAR JOÃO BERTOL
oromar@emater.pr.gov.br
Master's in Agricultural Engineering. Doctorate in Forest Engineering, with a focus on Nature Conservation. Area of expertise: integrated management of soils and water; Hydrographic Microbasin Diagnostics and Planning; Geo-processing.

PATRICIA ESTHER DUARTE LAGOS
estherlagos@yahoo.com.br
Biologist, Master's in Ecology and Conservation at UFPR and Doctoral candidate in Ecology and Natural Resources at the Universidade Federal de São Carlos. Author of papers in the field of limnology with an emphasis on eutrophication, phytoplankton ecology, and cyanobacteria.

PATRICIA YASSUMOTO HIRATA
paty.hirata@hotmail.com
Student in the Department of Environmental Engineering, UFPR.

PATRICIO PERALTA-ZAMORA
zamora@ufpr.br.
Chemistry Degree from the Universidad de Chile, with a Master's, Doctorate and Post-Doctorate from the Chemistry Institute, UNICAMP. Professor in the Chemistry Department at UFPR since 1997 and Coordinator of the Group for the Development of Advanced Techniques for Waste Treatment (TECNOTRATER).

PAULO CESAR COLONNA ROSMAN
pccrosman@ufrj.br
PhD from the Massachusetts Institute of Technology, M.I.T. Professor in the Oceanic Engineering Program in the area of Coastal & Oceanographic Engineering at COPPE/UFRJ.

PYRAMON ACCIOLY
pyramon@gmail.com
Forest Engineer, Master's degree, Doctoral candidate in the Forest Engineering Graduate Program at UFPR. Professor at the PUC-PR, in the disciplines of Forest Engineering and Biological Sciences. Conducts research in the field of remote sensing.

PRISCILA CAZARIN BRAGA
pri13luas@yahoo.com.br
Graduate student/fellowship holder in Environment and Development (MADE—UFPR). Researcher in the Public Policy Research and Development Group (CNPq/FAE). She has experience in the field of Environmental Sciences.

RENE VAN GRIEKEN
rene.vangrieken@ua.ac.be
Doctorate in Chemistry from the University of Gent, Belgium. Visiting professor (1972-1973) in the Department of Oceanography at Florida State University. Since 1973 has taught Analytical and Environmental Chemistry at the University of Antwerp, Belgium. He is chief Editor of the Journal X-Ray Spectrometry, web-editor of Spectroscopy NOW, associate editor of the Journal of Radioanalytical and Nuclear Chemistry. He was awarded several international commendations, is the author and co-author of more than 600 international papers, and is a member of the organizing committees for 240 international scientific conferences.

RICARDO HENRIQUE MORETON GODOI
rhmgodoi@ufpr.br
Holder of the CNPQ – Productivity in Research Fellowship 2. Doctorate in Physical Chemistry from UNESP (2000). Post-Doctorate at the University of Antwerp, Belgium

(2001-2005). Adjunct Professor in the discipline of Environmental Engineering, UFPR. Serves on the INCTs; PROANTAR/CNPq and DI FC Integrated Analysis of Environmental Risks. He has experience in the area of Physicochemistry, Chemistry of Colloids and Materials, using different techniques for Trace Analysis. Author and co-author of 38 published papers in international scientific journals and is author/co-author of nine book chapters in internationally published manuscripts.

SANDRA MARIA FELICIANO DE OLIVEIRA E AZEVEDO
sazevedo@biof.ufrj.br
Holder of Research Productivity 1B Fellowship. She has a degree in Biological Sciences from the Universidade Federal de São Carlos (1978), Master's in Ecology and Natural Resources from the Universidade Federal de São Carlos (1982) and Doctorate in Ecology and Natural Resources from the Universidade Federal de São Carlos (1988). Currently she is Associate Professor I at the Instituto de Biofísica Carlos Chagas Filho of UFRJ. She has experience in the area of Limnology and Ecotoxicology, with emphasis on Phytoplankton Ecology, operating primarily on the topics of cyanobacteria, cyanotoxins, bioaccumulation and effects on public health.

SANDRA MARA MACIEL-LIMA
sandralima@ufpr.br
Doctorate in Sociology (UFPR). Economics degree from UFPR. Professor at FATEC-PR. Researcher with the Research Group on Health Sociology (CNPq/UFPR) and the Public Policy Research and Development Group (CNPq/FAE). She conducts research in the areas of Health Sociology, Subjectivity, Medical Policies and Scientific Methodology.

SONY CORTESE CANEPARO
sony@ufpr.br
Bachelor's Degree in Geography. Doctorate in Environment and Development, Master's in Geodesic Sciences and Specialist in Geo-processing. Associate Professor in the Department of Geography and in the Geography Graduate Program, UFPR. Teaches graduate courses in Geography, Cartographical Engineering and Environment. Teaches and advises in the areas of Geographic Information Systems and Environmental Analysis. Coordinator for the Images and Geo-processing Processing Laboratory in the Department of Geography – UFPR.

TÂNIA CRISTINA SALOMÃO RODINI
direcaomc@cursoacesso.com.br / tanrodini@yahoo.com.br
Bachelor's degree in Biology, PUC-PR; Master's in Organization and Development from FAE. Biology Professor. Superintendent Director of Education and technical and pedagogical manager at Grupo Acesso. Sets policies and projects aimed at implementing and consolidating education programs; plans and supervises programs and actions inherent in the teaching process. Author and co-author of several scientific papers.

VÂNIA ELOIZA CERUTTI
van_cerutti@hotmail.com
Biology student and fellowship holder at Universidade Positivo. Author and co-author of scientific papers, book chapters (Parque Estadual de Vila Velha) and a booklet on environmental education.

VINÍCIUS ABILHOA
vabilhoa@uol.com.br
Biologist, Coordinator of the Ichthyofauna Research Group in the Natural History Museum, Municipality of Curitiba. Master's and Doctorate in Zoology (UFPR), with further training in Information Systems in Biodiversity and Statistics for Ecology and Conservation. Specialization in Environmental Analysis (UP). Instructor in the specialization course in Nature Conservation and Environmental Education, PUC-PR, and advisor in the Zoology Graduate Program at UFPR. Author of numerous scientific papers, book chapters and technical reports on ichthyofauna of streams, rivers and reservoirs. Member of the Brazilian Ichthyologic Society.

WELLEN LARYESSA LAYNES
wellen.bio@gmail.com
Biologist, Trainee in the Ichthyofauna Research Group of the Museum of Natural History, Municipality of Curitiba.

SUMMARY

SECTION I

INTRODUCTION AND PRESENTATION

CHAPTER 1

AN INTRODUCTION TO EUTROPHICATION AND THE STUDIES

Cleverson V. Andreoli
Charles Carneiro
Eduardo F. Gobbi
Cynara Cunha

AN INTRODUCTION TO EUTROPHICATION AND THE STUDIES

Cleverson V. Andreoli
Charles Carneiro
Eduardo F. Gobbi
Cynara Cunha

SUMMARY

This chapter discusses issues related to the demand and availability of water in Brazil, as well as problems associated with increased water consumption and human population growth. Currently more than 1 billion people worldwide live without adequate water for domestic consumption and this number grows considerably every decade. The way hydrographic or drainage basins are used, occupied and managed, along with the geomorphology, soils and environmental conditions of slopes, have a significant influence on water quantity and quality. The impact of human activities on continental ecosystems is reflected in water quality, which can lead to the eutrophication of water bodies thus limiting their potential uses. This chapter introduces the issue of water availability, highlights the processes that restrict water supply in the Metropolitan Region of Curitiba, and discusses the geomorphologic characteristics that determine the vulnerability of artificial reservoirs to eutrophication. The Rio Verde Reservoir located in Paraná State shows signs of deteriorating water quality. In 2005, a bloom of *Cylindrospermopsis raciborskii*, a potentially toxic cyanobacteria, was detected in the reservoir. To ensure the availability of good quality water in the reservoir, Petrobras supported a broad interdisciplinary research project in order to deepen our understanding of the structure and environmental dynamics within the Rio Verde Lake Basin. The project involved research on the physical environment, water quality, a quatic communities, environmental education and socioeconomic aspects. The entire project was designed to produce applicable results that would support the principles of preventive management of water sources at risk of eutrophication. At the end of this chapter, the research program structure is presented, together with the methodology used to integrate projects , which stimulated the interdisciplinary approach.

KEY WORDS

Water, water availability, reservoir, eutrophication.

1. INTRODUCTION

The growth of human populations, and the associated major increases in consumption, has created skyrocketing demand for natural resources and massively increased waste production. The subsequent significant growth in production has however not been able to meet the basic needs of most human populations. About one billion people around the world live in a state of food insecurity: they do not know if they will be able to eat properly today. On the other hand, wealthier societies, which have extremely high levels of consumption, make up about 20% of the population and are surpassing consumption levels that the planet can sustain. The balancing of environmental, social and economic disparity, therefore, depends on two major challenges: economic inclusion of the majority of the world's population and the reduction of consumption levels within developed countries.

To meet the needs of the poor majority and to support the high levels of consumption of the wealthy minority in society, humans intervene in nature, which in turn responds to these pressures through natural ecological processes in search of equilibrium. The extraction of natural resources, exploitation of nature through agricultural activities, environmental disturbances for the development of energy, transportation and urban infrastructure, coupled with the large amount of waste produced, generate environmental impacts that apparently have already exceeded the planet's capacity to recalibrate. The large amount of information available on the severity of environmental issues has not yet been successful in mobilizing humanity to adopt a more rational use of our planet's resources.

Water stands out as an essential resource for development, as it is not only important for human consumption, but also its availability is necessary in the production of food, energy and many industrial products. However, the way a hydrographic basin is used has a direct impact on the quality of water and waterways. It is a natural resource that depends on how other resources are handled: good quality water must be cultivated. The sepatation of environmental problems into the impacts on climate, resource depletion, water pollution and loss of biodiversity is merely didactic because all these crises are interconnected and are directly affect one another.

In this context, we must understand the water crisis as the result of an inadequate process of appropriation and use of natural resources, which has two major consequences: reduced volume due to increasing demand for water resources and the gradual depletion of water quality, which limits its availability.

2. WATER AVAILABILITY

The availability of water supply requires a guaranteed supply; hence different uses should be evaluated in terms of reduced availability. Public water supplies, electric power generation and irrigation systems should ensure a continued availability of water throughout the year and over the long-term. Therefore, in addition to seasonal variations, these systems should consider the different levels of availability over several years. Thus, the amount of water that is considered available is the amount of water that is critical over the long-term.

Rivers are complex systems which include the thalweg and geomorphologic areas of run-off or floodplains, which are riparian environments adjacent to thalwegs where water accumulates during periods of flooding. This excess water is returned to the river during dry periods generating an easement zone that regulates the natural flow of watercourses. Moreover, the river flow is related to several factors inherent to the basin, such as shape, slope, use and management, roughness coefficient and permeability.

River water originates from rainfall and from groundwater springs. In areas located downstream from snow-covered mountainous regions, water from thaw may also influence river flow. This precipitation can evaporate, infiltrate into the soil and flow along the surface. There is thus a direct relationship between the drainage area of the basin and the flow of the water body, which we call the discharge. Thus, the closer to the headwaters, the lesser the flow, while the size of the river increases proportionally to the increase of the basin area. In areas of heavy precipitation, the basin is more productive. Therefore, similar drainage areas may show significant differences in flow rates.

The way the basin is used, managed and occupied, as well as the geomorphologic and pedological features associated with the slope environment, has a major influence not only on the amount of water that flows, but also on the flow speed, and whether the water flows on the surface, subsurface or underground. For this reason, higher flood levels are more likely in environments in which the vegetation cover has been removed. Likewise, the use of floodplains for urban development or agriculture reduces the areas of flood easements, expanding the critical extent of river flow and exacerbating floods and droughts, directly resulting in a reduction in available water.

The loss of water quality is another important factor that limits the potential uses of water. River pollution can be caused by point-source releases of industrial and domestic pollutants as well as by the combination of several diffuse elements including agriculture, animal husbandry, and urbanization. Rain-water washes away the dirt and grime in the city, which ends up in the river. Similarly, in rural environments, water runoff carries particles with adsorbed elements and solubilizes chemical fertilizers and pesticides. As various different water uses have different requirements in terms of quality, compromising the natural characteristics of water may limit its use for in situations that demand high quality levels. Therefore, the concept of water availability is based on two variables: the amount of water and the demand for the resource. The United Nations World Conference on Water and Environment (ICWE *apud* Hespanhol, 2008) established basic criteria for water resource management in the XXI century. Also, it defined water as "[...] a finite and vulnerable resource, essential to life, development and environment" that "[...] has economic value for all its uses and should be considered as an economic good [...]" (HESPANHOL, 2008).

Another important milestone, Agenda 21, which was the result of the UN Conference on Environment and Development held shortly after the ICWE in Rio de Janeiro in 1992, proposed in chapter 18 actions to sustainably manage and redirect water resources (UNCED, 1992), something that was already urgent:

> "[...] to ensure maintenance of an adequate supply of good quality water for the entire population of the planet, while preserving the hydrological, chemical and biological functions of ecosystems, adapting human activities to the capacity limits of nature and fighting vectors of waterborn diseases."

Table 1 shows the Water Stress Indicator (WSI), associated with the Annual Water Supply (AWS) (m³/inhabitant/year), in a particular region or hydrographic basin.

Research has shown that WSI values above 1,700m³ per inhabitant per year correspond to circumstances in which there is adequate water. Values lower than 1,700m³ per inhabitant per year suggest conditions of significant water scarcity and values below 1,000m³ reflect situations of chronic water shortage (HESPANHOL, 2008).

It is estimated that the global water availability is approximately 40,000 km³/year but only 10% of this amount flows through rivers. It is important to note that approximately half of the water abstracted for public consumption returns to water courses with inferior quality to when it was extracted (SHIKLOMANOV, 1998).

According to Tundisi (2008), "Today, across the planet, the total volume of impounded water reaches

TABLE 1 – WATER STRESS INDICATOR (WSI) BASED ON WATER AVAILABILITY (AWS) AND RELATED MANAGEMENT ISSUES

WSI	ANNUAL WATER SUPPLY (m³/INHABITANT/YEAR)	WATER RESOURCE MANAGEMENT ISSUES
	DEA ≥ 10,000	No problems or limited problems
	10,000 > DEA ≥ 2,000	General management problems
	2,000 > DEA ≥ 1,000	Significant pressure on water resources
	1,000 > DEA ≥ 500	Chronic water shortage
	DEA < 500	Beyond availability limit

SOURCE: FALKENMARK *apud* HESPANHOL (2008)

over 10,000km³, occupying an area of approximately 650,000km²". The current tre nd is that there is a large increase in demand for water, while water resources remain relatively constant, which reduces overall water availability.

Currently more than one billion people live without adequate water availability for domestic consumption and it is estimated that in 30 years there will be 5.5 billion people living in areas with moderate to severe lack of water (DEMANBORO & MARION, 1999). In 1995, approximately 20% of the world's population, which was 5.7 billion at the time, already suffered from the lack of a reliable water supply and over 50% did not have an adequate sanitary system (LIMA, 2001).

The total amount of water on Earth is estimated to be 1386 million km³. Current research suggests that in the last 500 million years this volume has remained approximately the same. However, the quantities stored in different water reservoirs have varied substantially throughout this period (SHIKLOMANOV, 1998).

Despite the huge volume of water on our planet, the proportion actually available for human use and consumption is very small. The majority, 97.5% of the total volume, is salt water. The remaining 2.5% represents the total volume of fresh water on Earth but the majority of this percentage (68.7%) is stored in glaciers and ice caps or infiltrated into the subsoil (30.1%). Only 0.26% of the freshwater on Earth is concentrated in lakes, reservoirs and rivers (SHIKLOMANOV, 1998).

Fresh water found in rivers, lakes and reservoirs represents the means by which most water resources are available for humans and ecosystems, corresponding to 0.26% of total fresh water or about 93,100km³. However, due to the dynamics and variability of the hydrologic cycle, accessing fresh water stored in various different reserves on Earth is not a simple task.

The hydrologic cycle is responsible for displacing huge volumes of water around the world. Part of this displacement is rapid because water droplets are in a constant and continuous recycling process. A drop of water stays about eight days in the atmosphere, 16 days in a river, and a period of months to years in lakes. However, depending on the volume, this time may be longer and extend up to thousands of years if, for example, we consider water that moves slowly through a deep aquifer. Moreover, the overall circulation of water does not depend only on the stored volume of a given component of the cycle but largely on its period of renewal (LIMA, 2001).

Figure 1 presents data on the global dynamics of water use for a period of 125 years using data collected from 1900 until 1995 and estimated for the subsequent years.

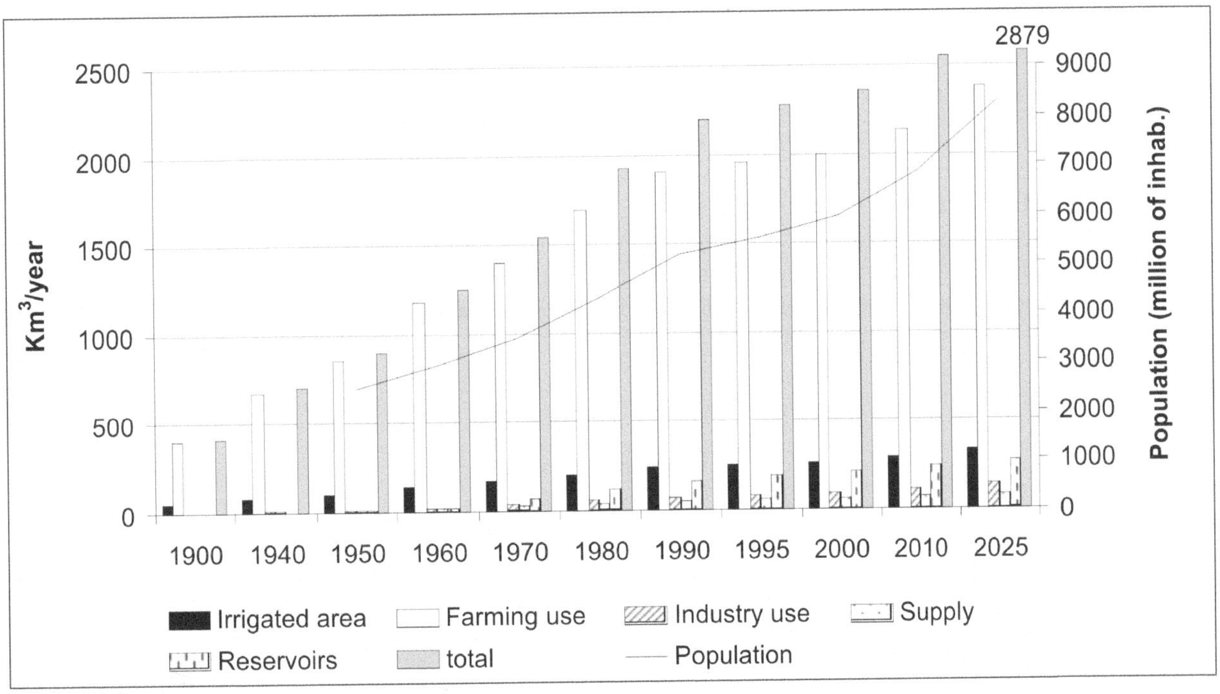

FIGURE 1 – GLOBAL WATER USE (km³/year).
SOURCE: Adapted from LIMA (2001).

According to data presented in 2002 by the National Water Agency (ANA, 2002), in general Brazil "presents a highly favorable condition" in relation to water availability as it has 33,944.73m³ per inhabitant per year. For the São Paulo State, the average value for 1996 was 3,014.4m³ per year and the forecast for 2010 was 2,339.6 m³ per year.

Figure 2 shows water demands in Brazil. The South and Southeast regions have higher water consumption for industrial purposes than other areas of the country which reflects greater development of the sector in these two regions.

The two standard ways of increasing water availability in a region are: importing water from distant basins, called basin transposition; and the regulation of flow through artificial dams that allow for the accumulation of excess water in periods of heavy precipitation for use in drier periods. A dam can greatly increase the abstraction availability, reaching more than five times the volume available in run-of-river abstractions. However, a more contemporary concept known as ecological flow should be taken into consideration as it allows for the development of water captures that are calculated to ensure the maintenance of a minimum flow, thus promoting the continuation of essential ecological processes.

According to Hespanhol (2008), basin transportation is not new:

> "The policy of importing water from increasingly distant basins to meet the growing demand goes back over two thousand years. The Romans, who practiced the intensive use of water for the supply of households and spas, tried, at first, to capture water from fountains available nearby, and as such became polluted by sewage disposed of without any treatment, or were unable to attend demand, they started using the second closest source, and so on. This practice gave rise to the construction of the great Roman aqueducts, of which some ruins still exist in various parts of the Old World".

According to Hofwegen & Svendsen (1999), "in many regions of the world, more than three quarters of the annual rainfall occurs in a period of less than six months." Therefore, the purpose of reservoirs is to better balance water availability throughout the year, i.e., water storage during wet periods and subsequent use in drier times.

3. EUTROPHICATION

A reservoir is a complex environment which creates interdependent relationships between water tributaries (rivers that contribute to the reservoir) and the environmental conditions of the basin, which are strongly influenced by its use. By reducing the speed of rivers with dams, the lentic environments are created which drastically alter the original conditions of rivers with greater flow (lotic environments).

The impact of human activities on continental ecosystems has resulted in the deterioration of water quality and changes in the hydrological cycle, biogeochemical cycles, and in biodiversity (TUNDISI, 2003). Among these outcomes, we can include the eutrophication of inland waters.

Artificial Eutrophication is the excessive increase in the concentration of nutrients, especially phosphorus and nitrogen, in aquatic ecosystems, normally caused by the discharge of agricultural, urban or industrial effluents. This enrichment of nutrients in the water leads to excessive proliferations of algae, breaking the natural balance of trophic chains and causing changes in biogeochemical cycles (ESTEVES, 1998; ROLAND, CESAR & MARINHO, 2005)

Natural Eutrophication of aquatic ecosystems also occurs; however, this is a process that can take hundreds of years as it depends on the inorganic load in the aquatic environment and on the influence of natural processes in hydrographic basins (TUNDISI, 2008).

The characterization of nutrient levels, trophic states, structure and composition of aquatic communities are essential tools to detect different types of influences on aquatic ecosystems and also to propose mitigation and conservation measures (PETRUCIO et al., 2005) related to these environments. According to Esteves (1998), "as a result of the process of eutrophication, the aquatic ecosystem goes from oligotrophic and mesotrophic to eutrophic or even hypereutrophic."

The overall classification of the trophic state of lakes and reservoirs is normally based on various indicators considered together, such as the concentration of nutrients, chlorophyll-a, and Secchi depth. Table 2 shows the criteria of several parameters for classifying the trophic state of lakes and reservoirs.

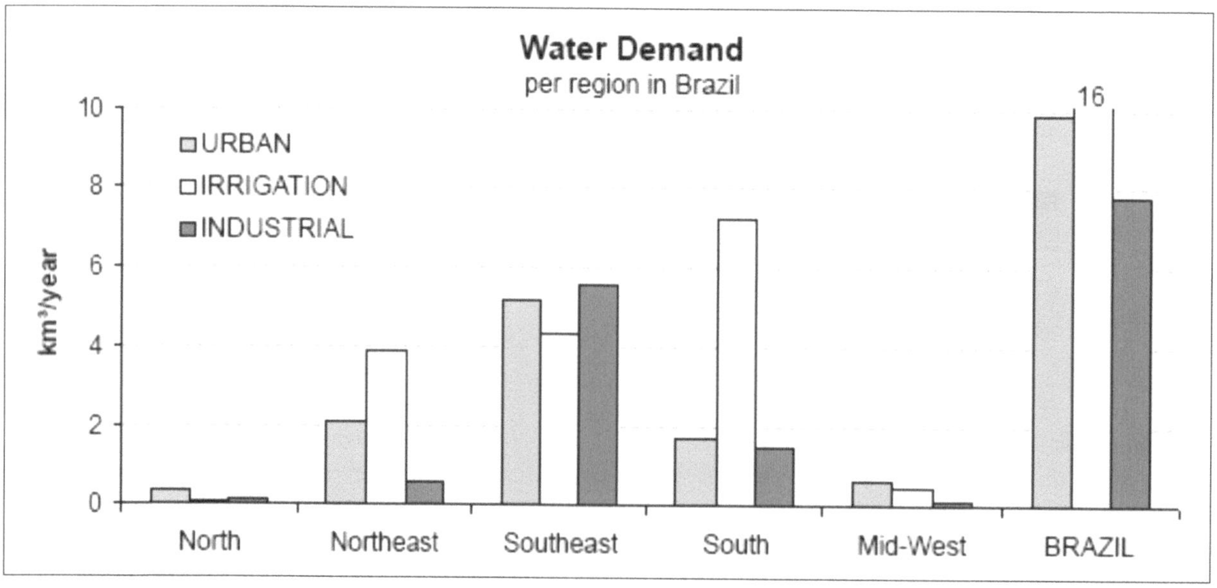

FIGURE 2 – WATER DEMAND PER BRAZILIAN REGION (km³/year)
SOURCE: Adapted from TUNDISI (2003)

The parameters typically used to classify lakes and reservoirs can be used to calculate indices of trophic states. One such index is the Carlson index (1977), which uses chlorophyll-a, Secchi depth and total phosphorus to estimate the algae biomass. The index can then be used to monitor changes in the studied aquatic ecosystem over time.

With regards to the sources of pollution of aquatic ecosystems, Andreoli & Carneiro (2005) note:

> "The organic discharge of treated and untreated effluents, as well as percolated waters from septic tanks and also illegal sewer connections to rainwater networks, in most cases, constitute the main nitrogen and phosphorus pollution sources in aquatic systems. The loads of these elements are always very significant in the effluents and eventually determine the level of trophism of the medium. Generally, releases are punctual and with high pollution potential".

The impact on water quality is amplified in a lentic environment as it responds to these changes with complex ecological dynamics. The factor that triggers the process of multiplication and growth of primary producers, mainly cyanobacteria, algae and aquatic weeds, is related to the availability of nutrients in the environment, particularly phosphorus and nitrogen.

As some algae are fixers of elementary nitrogen, which exists in abundance in the atmosphere, phosphorus is usually the element associated with the process of eutrophication in freshwater ecosystems. Other factors, such as light and temperature, can also play a role in controlling the proliferation of cyanobacteria.

The geomorphologic conditions of the region, climatic factors, land-use and occupation, as well as morphometry and management of the reservoir are some of the main interrelated factors that determine its susceptibility to eutrophication, which usually results in the proliferation of aquatic weeds and (or) algae growth. Among the various factors that affect the process of multiplication of algae cells, two important geomorphologic characteristics directly contribute to the process: the depth of the reservoir and length of time the water spends in the reservoir. In shallower reservoirs, the volume of water that receives solar radiation is much higher, which creates favorable conditions for the growth of phototrophic organisms such as cyanobacteria. Likewise, the longer the water remains in the reservoir, the greater the amount of cell divisions that can occur while the water is in the environment, thus increasing the density of such organisms in the water. As such, when other environmental conditions are stable, these factors strongly influence the potential for algal blooms.

The eutrophication process can have several consequences for the balance of the aquatic ecosystem under study and its various uses. The excessive proliferation of algae and the consequent increase in the decomposition of organic matter can dramatically decrease oxygen concentration (ESTEVES, 1998). This decrease in oxygen can cause the death of many aquatic organisms, including fish.

Another consequence of the eutrophication process related to increased decomposition of organic matter in the sediment is the release of malodorous gases such as hydrogen sulfide and methane. In addition, during algal blooms when the density of cyanobacteria is significantly high, a dense layer of cells can form on the surface, creating thick layers on the surface which may restrict recreational use.

However, according to Figueiredo *et al.* (2007), "one of the main problems related to eutrophication is the proliferation of cyanobacteria at the expense of other aquatic species." Cyanobacteria can produce different toxins such as hepatotoxins, neurotoxins, cytotoxins and dermatotoxins (AZEVEDO *et al.* 1994 and 2002, CARNEIRO *et al.* 2009), all of which can be fatal to humans and animals (AZEVEDO *et al.* 2002). In some cases, the removal of these toxins from the affected water is difficult and even after treatment the toxins may remain.

Besides the problems related to cyanotoxin removal, excessive proliferation of cyanobacteria in reservoirs used for public water supply may cause clogging of filters at water treatment plants, making the treatment process difficult and expensive.

From an ecological point of view, the cyanobacteria blooms can negatively affect energy transfer in aquatic food chains. Besides producing cyanotoxins that can adversely affect their consumers, most cyanobacteria are poor sources of the compounds necessary for zooplankton growth (BRETT & MÜLLER-NAVARRA, 1997). Furthermore, cyanobacteria can form colonies and be coated with mucilage, which can hinder ingestion by zooplankton.

According to Tundisi & Barbosa (1995), water use and socioeconomic aspects of the population in the drainage basin are crucial elements not only in understanding ongoing processes in aquatic ecosystems, but also in the development of strategies for conservation and public policies related to the management of these environments.

TABLE 2 – OVERALL CLASSIFICATION OF THE TROPHIC STATES OF LAKES AND RESERVOIRS REGARDING THE CONCENTRATION OF PHOSPHORUS, NITROGEN, CHLOROPHYLL-A AND SECCHI DEPTH.

PARAMETERS	ULTRAOLIGOTROPHIC	OLIGOTROPHIC	MESOTROPHIC	EUTROPHIC	Hypereutrophic
Total phosphorus (mg. m^{-3})	4	10	10 – 35	35 – 100	> 100
Total nitrogen (mg. m^{-3})	-	661	753	1875	-
Average chlorophyll-a (mg. m^{-3})	1	2.5	2.5 – 8.0	8.0 – 25	> 25
Maximum chlorophyll-a (mg. m^{-3})	2.5	8	8 – 25	25 – 75	> 75
Maximum Secchi depth (m)	12	6	3 – 6	1.5 – 3	1.5
Minimum Secchi depth (m)	6	3	1.5 – 3	1.5 – 0.7	0.7

SOURCE: Based on WETZEL (2005) and OECD apud TUNDISI (2008).

4. REGIONAL CONTEXT OF THE RIO VERDE BASIN

The Prata Basin in the South of Brazil is a unique feature in the region as it incorporates rivers that are both environmentally and geopolitically important, such as the Paranapanema and Iguaçu Rivers. These rivers originate in the high altitudes of the Serra do Mar near the coast, run from East to West, toward the countryside of the states of São Paulo, Paraná and Santa Catarina, and then flow into the Paraná River. The large urban centers of the Metropolitan Region of São Paulo and Curitiba therefore present major challenges for water supply as they are located near the headwaters of these rivers.

In São Paulo, in order to meet the huge demand generated by its Metropolitan Area (SPMA), reservoirs have been constructed; however, there is also the need to reserve large volumes of water from adjacent basins. In the Alto Tietê Basin, for example, where the SPMA is located, annual water supply in the year 2000 reached a low of 216m^3/inhabitant (IBGE, 2000), well below the chronic water shortage index (Table 1).

In the Metropolitan Region of Curitiba (RMC), the initial strategy has been to build reservoirs and transport water from near-by basins for public consumption and industrial use. The cities that form Greater Curitiba are surrounded by reservoirs for power generation, industrial use and public consumption, which impose significant restrictions on urbanization in these regions. To the east, limitations are imposed by the dams Iraí, Piraquara I and Piraquara II. To the south, we encounter the Miringuava dam (pending authorization) and the entire Várzea basin, which is already under consideration as a future water source. To the west, we have the dams Passaúna and Rio Verde and to the north, there is we are limited by the groundwater recharge basin of the Karst aquifer, which is also used sparingly to supply water.

The urban growth of RMC is, therefore, confined by the reservoirs constructed in order to meet its own water demands. However, due to the regional geomorphologic and morphometric characteristics, most of these reservoirs are susceptible to eutrophication. For the waters of these reservoirs to remain within acceptable quality levels, very careful management is required which guides the occupation and land-use of the surrounding areas, ensuring the maintenance of these reservoirs. The RMC therefore lies in a paradox: if the city expands into its water source regions without proper planning and care, there will be no water available for its development.

The Rio Verde Reservoir exists within this context. The Rio Verde dam was built between 1974 and 1976, by Petrobras, in order to ensure the availability of water for industrial processes of REPAR (President Getúlio Vargas Refinery). The waters of the Rio Verde basin are currently used mainly to meet the needs of REPAR and for public consumption, both of which are looking to expand the use of the reservoir.

In 2005 in the reservoir, there was an algal bloom of *Cylindrospermopsis raciborskii*, a potentially toxic cyanobacteria, which reached counts of 96,489 cells/mL (IAP, 2009). To ensure the availability of good quality water, Petrobras asked FUNPAR (Fundação da Universidade Federal do Paraná) to develop a study that would improve our understanding of the structure and environmental dynamics of the Rio Verde basin. Further, the study was tasked with proposing a preventive action plan for the management of the reservoir in collaboration with public institutions, municipalities and other social groups.

As the dynamics of the Rio Verde Basin are representative of the region, the results of this study are of particular importance because they can be used as a model to be applied across the RMC and as a methodology that can be replicated in other regions of Brazil.

5. ORGANIZATION OF THE INTERDISCIPLINARY RESEARCH

As the dynamics governing the phenomenon of eutrophication in reservoirs are relatively consistent, the preparation of this proposal was based on the experience of the Interdisciplinary Research Project on the Eutrophication of Public Water Supply in the Altíssimo Iguaçu Basin, which began in 2001 and focused on the Iraí Reservoir. This project resulted in the publication of the book "Integrated Management of Eutrophic Water Supply Sources" in 2005 (ANDREOLI & CARNEIRO, 2005).

Thus, a team was formed to coordinate the research with the goal of designing and developing a research program capable of providing the necessary technical guidance for the preventive management of the Rio Verde Reservoir and incorporating the experiences from Iraí. In the Rio Verde Project, a new themed group was created to address socio-environmental issues beyond the geological aspects of the basin, an aspect which was not addressed in the 2001 project. In addition, themes were adapted to the new project by creating new thematic clusters and defining the actions needed to stimulate collaboration between researchers and develop interdisciplinary outcomes. Another very important aspect of this project was the condition that all studies result in suggestions and actions that could be directly or indirectly implemented in the basin.

An important aspect that differentiates this new research program is the intention of developing a preventive management system, since the water quality in the Rio Verde Reservoir is currently in a mesotrophic state. The emphasis on preventive management is important and should be a constant practice in the management of basins as it allows actions to be taken before the problem intensifies. Such management planning greatly facilitates the conservation of the ecosystem and if performed responsibly enables multiple uses of the reservoir. Moreover, it is a significantly different dynamic from the activities undertaken during an emergency recovery. The project design included the following steps:

- Detailing of Petrobras' requirements;
- Creating a group to coordinate the project;
- Identifying the problems;
- Defining the variables to be studied;
- Developing a preliminary proposal for the interdisciplinary research project;

- Contacting prominent researchers from different institutions;
- Proposing specific research tracks and expected results;
- Preparing individual thematic research proposals;
- Adjusting project proposals to the objectives of the interdisciplinary research program;
- Tying up the individual proposals;
- Integrating individual proposals into thematic clusters;
- Developing an interdisciplinary approach;
- Defining the administrative system for program management;
- Contracting the research teams.

The studies were conducted according to following structure:

- Preliminary seminar (presentation of the research problem, overall structure of the research program, interdisciplinary approach and presentation of specific research projects).
- Meetings among thematic clusters to coordinate sampling schedules, standardize analytical methodologies and discuss ongoing projects and preliminary results.
- Meetings among thematic clusters for the development of the Preventive Action Plan for the reservoir.
- Seminars for the presentation of preliminary research results and modifications to project design based on the critical evaluation of the research team and external consultants. At least one seminar per semester
- Preparation of interim reports 9 with the following structure:
  Summary;
  Introduction;
  Literature Review;
  Methodology;
  Results;
  Conclusion;
  Next-steps.
- Preparation of the book structure by the project coordinators, distribution to external consultants and discussion to define the structure and content with the team of researchers in the last seminar.
- Arrangement of the chapter summaries for thematic adjustments, avoiding overlap and ensuring the most relevant topics were covered.
- Standardization of maps from the cartographic research.
- Critical analysis of the chapters prepared by the project coordinators and analyzed by the authors. Development of chapters that integrate the thematic clusters and prepare final integrating chapters. Review of chapters and final review by the editorial board.
- Final Seminar.
- Completion of research studies and proposal of implementation plan for monitoring.

Originally, the structure of the interdisciplinary program consisted of **19 projects grouped into five** thematic clusters:

THEMATIC CLUSTERS:
I – Physical Environment and Environmental Modeling;
II – Socioeconomics and Environmental Education;
III – Dynamics of Nutrients and Water Quality;
IV – Water Treatment;
V – Phytoplankton, Zooplankton and Ichthyofauna.

The original goals of each of the subprojects are presented below grouped by thematic cluster.

THEMATIC CLUSTER I - Physical Environment and Environmental Modeling

1 – **Vegetation cover mapping, characterization and diagnosis in the drainage basin**
Coordinator: Dr. Carlos V. Roderjan – UFPR
PURPOSE: Identify and quantify the natural vegetation cover, map different types of land use in order to identify places where restoration is needed.

2 – **Development of cartographic database**
Coordinator: Dra. Sony C. Caneparo – UFPR
PURPOSE: Generate a database to integrate geographic information systems aimed at developing an integrated study of the landscape.

3 – **Modeling of water quality**
Coordinator: Dra. Cynara Cunha – UFPR
PURPOSE: Development of an environmental model to study the hydrodynamics and transportation of pollutants in the reservoir which enables simulations/forecasts/studies of the eutrophication processes.

4 – **Identification of potential and emerging soil fragility**
Coordinator: Dr. Everton Passos – UFPR
PURPOSE: Definition of a classification system regarding land use and management that is consistent with the dynamics of the basin, identifying levels of potential and emerging fragility that may affect the water quality of the Rio Verde reservoir.

5 – **Water Resource Policy, and Land Use**
Coordinator: Dr. Eduardo F. Gobbi – UFPR
PURPOSE: Evaluate the relationship between land-use planning in the municipalities within the drainage area and the guidelines to be proposed under this project. Develop proposals to standardize regulations, adjust the Master Plan and approval of the Rio Verde APA.

THEMATIC CLUSTER II - Socioeconomics and Environmental Education

6 – **Agricultural Activities**
Coordinator: Msc. Benno Doetzer – EMATER
PURPOSE: Survey and diagnose livestock activities in

the areas of influence in the drainage basin, study proposals for the adjustment of properties, as well as conduct a cost analysis for the conversion of productive systems.

7 – **Environmental Risk Perception**
Coordinator: Dr. José Edimilson S. Lima – UFPR
PURPOSE: Study the perceptions of the basin's inhabitants related to environmental risks and local development resulting from the actions of the groups, their own actions and other groups in the area.

8 – **Rural Sanitation**
Coordinator: Dr. Miguel M. Aisse – UFPR
PURPOSE: Diagnose the water supply quality and develop proposals for wastewater treatment and final disposal upstream of the reservoir.

9 – **Socioeconomic Profile of the basin inhabitants**
Coordinator: Dr. Fabiano A. S. Dalto
PURPOSE: Identify the local productive system that delineates the conditions of use of resources and the development of the region to guide public policies related to development.

10 – **Environmental Education**
Coordinator: Dra. Lucia I. C. Sermann
PURPOSE: Develop an environmental education program aimed at the communities within the Rio Verde basin and use it to mitigate against the causes and consequences of environmental degradation present in this region.

THEMATIC CLUSTER III - Dynamics of Nutrients and Water Quality

11 – **Evaluation of Nutrient Loads and Quality of Sediment**
Coordinator: Dr. Charles Carneiro – SANEPAR
PURPOSE: Evaluate nutrient contribution, sediment and water quality of the reservoir..

12 – **Geochemical Controls for Water Quality**
Coordinator: Dr. André V. L. Bittencourt – UFPR
PURPOSE: Elucidate geochemical mechanisms that influence the quality of water with direct consequences to the reservoir.

13 – **Monitoring the Influence of Land Use Patterns in the Tributaries Water Quality**
Coordinator: Dr. Harry A. Bollmann – PUC-PR
PURPOSE: Monitor surface water quality in the main tributaries of the reservoir and develop an efficient, effective and low cost monitoring plan.

14 – **Water Balance**
Coordinator: Dr. Mauricio Felga Gobbi – UFPR
PURPOSE: Quantify the inflows to the reservoir, obtain the flows in real time and estimate the capacity of the reservoir to regulate flows with scenarios of water consumption and the probability of failure.

THEMATIC CLUSTER IV - Water Treatment

15 – **Potential of advanced oxidation processes in the degradation of cyanotoxins**
Coordinator: Dr. Patricio Guillermo Peralta Zamora – UFPR
PURPOSE: Develop procedures for the remediation of water contaminated by cyanotoxins.

THEMATIC CLUSTER V - Phytoplankton, Zooplankton and Ichthyofauna

16 – **Effects of weather events on phytoplankton and cyanobacteria**
Coordinator: Dr. Luciano F. Fernandes - UFPR
PURPOSE: Analyze the influence of meteorological and limnology factors in the composition and abundance of phytoplankton and cyanobacteria in the reservoir.

17 – **Evaluation of cyanotoxins**
Coordin ator: Msc. Christine da F. Xavier – IAP
PURPOSE: Evaluate the concentration of chlorophyll-a and cyanotoxins in "in natura" water of the reservoir.

18 – **Zooplankton**
Coordinator: Dr. Moacyr Serafim Junior – UFRB
PURPOSE: Study the zooplankton of the reservoir and evaluate the influence of environmental parametres on the composition and abundance of the zooplankton population.

19 – **Ichthyofauna**
Coordinator: Dr. Vinícius Abilhoa – MHNCI/PMC
PURPOSE: Identify changes in biological dynamics and distribution patterns of fish species by analyzing the process of accommodation of popultions resulting from the creation of the reservoir.

6. INTERDISCIPLINARY APPROACH

After detailing the program and based on the preferences of the researchers involved, an interdisciplinary strategy was defined. The goal of the strategy was to support collaboration among researchers within the thematic clusters where there is a natural convergence of research topics.

The main strategies used for the implementation of interdisciplinary and inter-institutional projects were integration seminars, involving all those involved in the research projects. The purpose of the first seminar was to present the projects, standardize the methodologies and coordinate data collection schedules, which allowed for a comparison of the data generated across different projects. The coordination of the sample collection schedule sought to encourage, wherever possible, researchers to go to the field together and preferably to work with the same samples, so that the data could be integrated. Subsequent seminars were held at least twice a year, when the preliminary results were presented for discussion.

The varied knowledge and expertise of the researchers allowed for the critical analysis of the research across a wide range of disciplines and the identification of strategies for the integration of results. Institutions that did not participate in research, but that are involved in the management of the area of influence of the reservoir, such as representatives of local government, local leaders, public institutions, etc., were also invited to attend the seminars. Independent researchers with an interest in the subject also participated in the integration seminars.

Whenever possible, assistance of external advisers should be sought as they can have a significant influence on research procedures. Beyond the assessment of the research program, and from the beginning of the program, external advisers attended all seminars in order assess the proposal in light of a qualified critical examination. The advisors selected to participate in the Rio Verde Project,Professor Dr. Sandra M. F. O. Azevedo (UFRJ), Professor Dr. Francisco A. R. Barbosa (UFMG), and researcher Dr. Bias de Marçal Faria (CENPES/PETROBRAS), attended the seminars, received reports and presented their recommendations for each subproject and overall.

In addition to seminars, technical meetings specific to the thematic clusters were held which addressed converging issues and depended required operational adjustments for the proper completion and integration of projects.

Another strategy used to support an interdisciplinary approach was the creation of a website with an exclusive area for researchers where databases of different projects were made available in order to facilitate the sharing of data from different researchers.

In the Rio Verde Project, the main research results showed the structure and functions of environmental components in the ecosystem and highlighted the interconnectedness between the appropriation and use of resources by society and their environmental outcomes. The study focused on understanding how interferences or disruptions to environmental components resulting from its use can have an impact on water quality and in turn measuring such influences. Beyond academic production, and through the study of these interrelated dynamics, the project enabled the development of proposals for the use and management of the basin to ensure the maintenance and improvement of water quality. These proposals led to the design of a preventative management plan.

The preparation of the management plan involved the following steps: based on the concepts of World Lake Vision (WLV) and International Lake Basin Management (ILBM), a preliminary structure outlining the main points of the proposed organization, was defined. Researchers responsible for specific projects produced a preliminary action plan within their areas of study and presented them in a dedicated seminar. These proposals were further developed by the project coordinators, ensuring the standardization of their layout and grouping together proposals on related topics. When a uniform structure was established, it was presented at meetings of specific thematic clusters in order to guide an action plan.

The research groups were organized in the same structure as this book, which shows the state of the art evolution of the defined themes and presents the main results of the research.

The book was reviewed by an external and independent editorial board, which was responsible for the final recommendations and adjustments to the material presented. Given the interdisciplinary nature of the project, external advisers were asked to assess the main thematic clusters and interdisciplinary approach.

7. REFERENCES

ANA – Agência Nacional de Águas. (2002) Regiões Hidrográficas do Brasil. Caracterização geral e aspectos prioritários. Brasília.

ANDREOLI, C. & CARNEIRO, C. (2005) Gestão integrada de mananciais de abastecimento eutrofizados. SANEPAR e FINEP. Curitiba, Editora Gráfica Capital, 500 p.

AZEVEDO, S. F. O., EVANS, W. R., CARMICHAEL, W. W. & NAMIKOSHI ,M. (1994) First report of microcystins from a Brazilian isolate of the cyanobacterium Microcystis aeruginosa. Journal of Applied Phycology 6: 261-265.

AZEVEDO, S.M.F.O., CARMICHAEL W.W., JOCKIMSEN E.M., RINEHART K.L, LAU S., SHAW G.R., EAGLESHAM, G. K. (2002) Human intoxication by microcystins during renal dialysis treatment in Caruaru-Brazil. Toxicology.

BRETT, M.T. & MÜLLER-NAVARRA, D.C. (1997) The role of highly unsaturated fatty acids in aquatic food web processes. Freshwater Biology, 38: 483-499.

CARNEIRO, R. L., SANTOS, M. E.V.., PACHECO, A. B. F & AZEVEDO, S. M. F. O. (2009) Effects of light intensity and light quality on growth and circadian rhythm of saxitoxins production in Cylindrospermopsis raciborskii (Cyanobacteria). Journal of Plankton Research, 31(5): 481-488.

CARLSON, R.E. (1977) A trophic state index for lakes. Limnology & Oceanography, 22(2): 361-369.

DEMANBORO, A. C.; MARIOTON, C. A. O conceito de escala e o desenvolvimento sustentável: implicações sobre os recursos hídricos e energéticos. Revista Brasileira de Energia, Rio de Janeiro, v. 7, n. 2, 1999. p. 71-88.

ESTEVES, F. A. (1998) Fundamentos de Limnologia. Rio de Janeiro. Interciência.

FALKENMARK, M. (1992) Water scarcity generates environmental stress and potential conflicts. Lewis Publishers, Inc.

FIGUEIRÊDO, M.C.B.; TEIXEIRA, A.S.; ARAÚJO, L.F.P.; ROSA, M.F.; PAULINO, W.D.; MOTA, S.; ARAÚJO, J.C. (2007) Avaliação da vulnerabilidade ambiental de reservatórios à eutrofização. Eng. Sanit. Ambient., v.12, n.4, p. 399-409.

HESPANHOL, I. (2008) Um Novo Paradigma para a Gestão de Recursos Hídricos. Estudos Avançados, v. 22, p. 131-158.

HOFWEGEN, P. van.; SVENDSEN, M. (1999) A vision of water for food and rural development. [S.l]: CEMAGREF; DVWK; FAO; HR Wallingford UK; ICID; ILRI; IPTRID; McGILL University Canada; Wageningen Agricultural University; World Bank, 77p.

IAP – Instituto Ambiental do Paraná. (2009) Monitoramento da qualidade das águas dos reservatórios do estado do Paraná, no período de 2005 a 2008. Curitiba, 120 p.

IBGE – Instituto Brasileiro de Geografia e Estatística. Atlas de Saneamento (2000) Available in: http://www.ibge.gov.br/home/estatistica/populacao/atlas_saneamento. Acessed in: December 15, 2010.

ICWE. (1992) The Dublin Statement and Report on the Conference and the International Conference on Water and the Environmental Development Issues for the Twenty-first Century. WMO. Genéve, Switzerland.

LIMA, J.E.F.W. (2001) Recursos Hídricos no Brasil e no mundo. Empresa Brasileira de Pesquisa Agropecuária - Embrapa Cerrados, Planaltina.

OECD. (1992) Eutrophication of Waters: Monitoring, assessment and control. Paris: OECD.

PETRUCIO, M.M., MEDEIROS, A.O., ROSA, C.A. e BARBOSA F.A.R. (2005) Trophic State and Microorganisms Community of Major Sub-Basins of the Middle Rio Doce Basin, Southeast Brazil. Vol.48, n. 4 : pp. 625-633, July.

ROLAND, F., CÉSAR, D., MARINHO, M. (2005) Lições de Limnologia. São Carlos.

SHIKLOMANOV, I. A. (1998) World water resources: a new appraisal and assessment for the 21st century. Paris: UNESCO, 76 p.

TUNDISI, J. G. (2003) Água no século XXI: enfrentando a escassez. São Carlos: Rima Artes e Textos, v. 1. 248 p.

TUNDISI, J.G. & MATSUMURA TUNDISI, T. (2008)Limnologia. São paulo: Oficina de Textos.

TUNDISI, J.G. and BARBOSA, F.A.R. (1995) Conservation of aquatic ecosystem: present status and perspectives. Limnology in Brazil. ABC/SBL. Rio de Janeiro: Graftex Comunicação Visual. pp.365-371.

UNCED. (1992) Agenda 21, United Nations Conference on Environment and Development. Rio de Janeiro.

WETZEL, R.G. (2001) Limnology – Lake and River Ecosystems. Academic Press, San Diego, CA.

CHAPTER 2

CHARACTERIZATION OF THE BASIN

Cleverson V. Andreoli
Charles Carneiro
Kelly C. Gutseit
Christine da Fonseca Xavier

CHARACTERIZATION OF THE BASIN

Cleverson V. Andreoli
Charles Carneiro
Kelly C. Gutseit
Christine da Fonseca Xavier

SUMMARY

This chapter presents a general overview of the Rio Verde Reservoir and its hydrographic basin and summarizes significant results from several subsequent chapters. The summary includes data on climatic conditions and soil to the socioeconomic status of the municipalities located within the basin. The objective is to give a general idea about briefly present the contents included in different chapters to facilitate access to the information presented in this study.

KEYWORDS

Rio Verde, hydrographic basin, reservoir, socioeconomic.

1. INTRODUCTION

The objective of this chapter is to present a general description of the Rio Verde Hydrographic Basin and its Reservoir as well as summarizing information detailed in the subsequent chapters. This chapter is structured based on six pillars of ILBM platform – Integrated Lake Basin Management and guidelines of WLV – World Lake Vision – International Lake Committee Foundation (ILEC) and United Nations Environment Programme-International Environmental Technology Centre (UNEP-IETC). The characterization of the basin includes climate, socioeconomic status of the municipalities located within the basin, land-use, aspects of the hydrology, pedology, and landscape, and the region's infrastructure. We then discuss the main characteristics of the Rio Verde Reservoir as well as the use of water resources and related impacts. At the end of the chapter, we summarize data on the implementation of an environmental education program in the region.

2. THE RIO VERDE BASIN

2.1 LOCATION

The Rio Verde Hydrographic Basin (25°31'S; 49°31'W) is located in the state of Paraná, in the municipalities of Campo Magro, Campo Largo and Araucária, and is part of the Alto Iguaçu basin. The basin is delimited to the east by the Passaúna river sub-basin and to the north by the Ribeira and Iguaçu sub-basins (Figure 1) and it is located in the Rio Verde Environmental Protection Area (APA). The Rio Verde APA spans across parts of the municipalities of Campo Largo and Araucária that are within the Metropolitan Region of Curitiba (RMC).

2.2 CLIMATE

The Rio Verde Hydrographic Basin is located in the climatic zone classified as humid temperate, with mild summers and a lack of a well-defined dry season. Climate is defined as type Cfb (according to KÖEPPEN), i.e., a subtropical climate with cool, humid summers and mild winters with the occasional occurrence of frost.

2.2.1 Air Temperature

The records from the last 30 years obtained from a meteorological station located near the study area show that the average annual temperature is 16.5°C. The average temperature of the coldest month (July) is 12.7°C and the warmest month (February) is 20.3°C, with minimum temperatures reaching values below -5°C and maximum temperatures above 33°C. Frosts are fairly common in the region, although the number of occurrences per year vary from two to twenty.

2.2.2 Relative air humidity and rainfall

Levels of air humidity fluctuates very little during the year, with an average of 80% (minimum of 17.3%). The annual average rainfall is approximately 1468mm. Historically, the months with the highest levels of rainfall occur during the summer months and the driest months occur in the winter. In general, the values recorded during the summer months (average above 70mm) are at least twice the values recorded during the driest months (July and August).

2.2.3 Prevailing wind direction

The prevailing wind direction in the region is from east to west and the most frequent intensities are between 2.0m/s and 4.0m/s (~38%) and between 4.0m/s and 6.0m/s (~26%). Average and maximum wind speeds are between 2.8m/s and 10.3m/s, respectively (Chapter 12).

2.3 INFORMATION ON THE MUNICIPALITIES LOCATED INTHE BASIN

2.3.1 General Information on the Municipalities

In total area, Campo Largo is the largest of the municipalities that are located in the Rio Verde Basin. It has a population of 112,548 people (Table 1, Figure 2) and it is the municipality with the largest percentage of area located within the basin (109.2km^2) and the greatest number

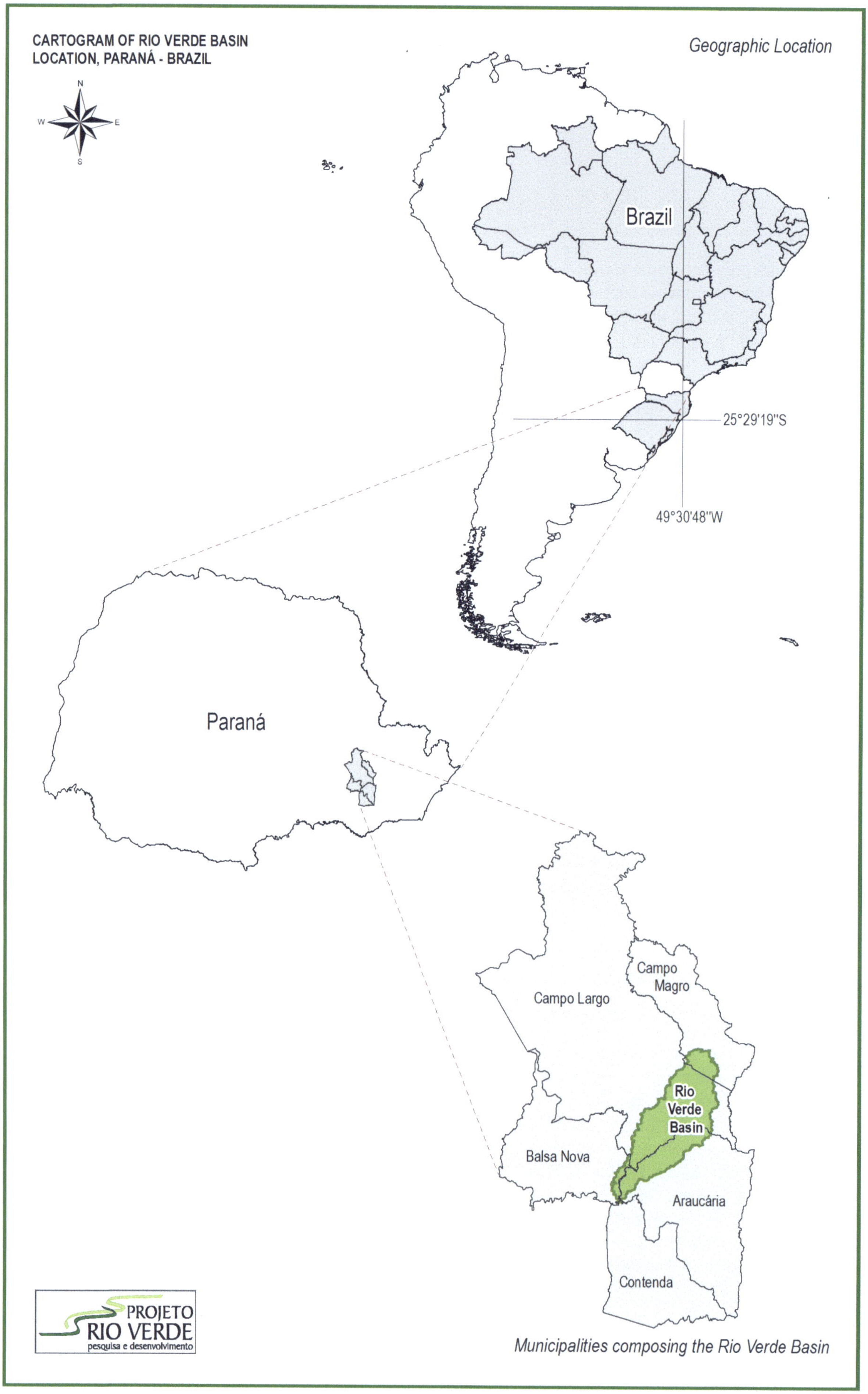

FIGURE 1 – LOCATION OF THE RIO VERDE BASIN

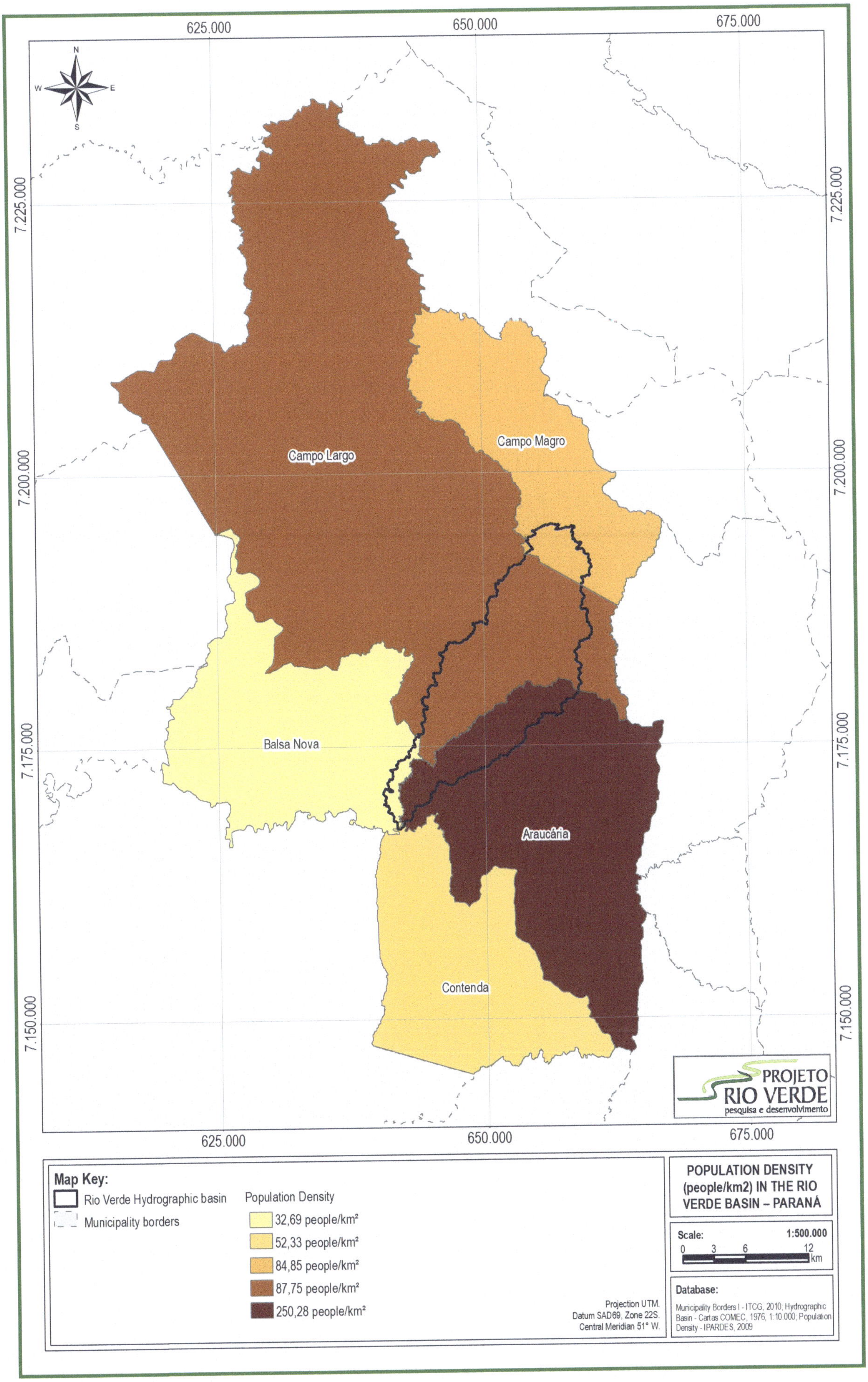

FIGURE 2 – MAP OF POPULATION DENSITY IN RIO VERDE BASIN

of inhabitants (close to 80%) residing within the drainage basin, followed by Araucária (38.4km²) and Campo Magro (18.4km²) (Table 2). The municipality of Campo Largo is known for the production of ceramics, furniture, metal industries and also for its wineries.

Araucária is the municipality with the greatest number of inhabitants, the highest population density (Figure 2), and the highest income per capita (Table 1). The installation of the President Getúlio Vargas Refinery in the 1970's contributed to the development of a strong industrial economy, which directly relates to the current high gross domestic product in Araucária (Table1)

Campo Magro is the municipality with the smallest total area but with a population density that is much higher than Campo Largo. The majority of the population of the three municipalities within the basin is located in urban areas (Figure 3). Campo Magro's economy is essentially agricultural and the most important crops are beans, corn and potatoes.

2.3.2 Education

The number of students enrolled in primary and secondary education in the municipalities within the study area is shown in Figure 4. Araucária was the municipality with the highest number of students enrolled in 2009, followed by Campo Largo, and Campo Magro. In the three municipalities located in the basin, the rate of illiteracy is higher among people aged 50 and over. The highest illiteracy rates, from any age group, were registered in Campo Magro (Table 3).

2.3.3 Culture

Most of the colonial communities located in the study area were founded by Italian and Polish immigrants and some residents still maintain the traditions of their country of origin. Most residents of these colonial communities are Catholics. The Church has an important role in rural communities as a place for prayer and religious practices, recreation, education, leisure, culture and political activities.

TABLE 1 – GENERAL INFORMATION ON THE MUNICIPALITIES

	CAMPO LARGO	CAMPO MAGRO	ARAUCÁRIA
Geographic Characteristics			
Area	1,249km²	276km²	469km²
Population	112,486 people IBGE/2010	24,836 people IBGE/2010	119,207 people IBGE/2010
Density	87.7people/km²	84.8people/km²	250.3people/km²
Altitude	956m	990.3m	897m
Bordering Municipalities	Castro, Campo Magro, Itaperuçu, Ponta Grossa, Araucária, Balsa Nova, Curitiba and Palmeira	Curitiba, Campo Largo, Itaperuçu and Almirante Tamandaré	Balsa Nova, Campo Largo, Contenda, Curitiba, Fazenda Rio Grande, Mandirituba and Quitandinha
Economic Indicators			
HDI	0.774 UNDP/2000	0.74 UNDP/2000	0.801 UNDP/2000
GDP	R$ 1,192,071 IBGE/2008	R$ 160,144 IBGE/2008	R$ 11,001,673 IBGE/2008
GDP per capita	R$ 94,965.63 IBGE/2008	R$ 10,759.15 IBGE/2008	R$ 6,864.87 IBGE/2008

TABLE 2 – POPULATION IN THE RIO VERDE DRAINAGE BASIN (2002)

MUNICIPALITY		CAMPO LARGO	CAMPO MAGRO	ARAUCÁRIA	TOTAL
Area within drainage basin (km²)	**km²**	109.2	18.4	38.4	165.9
	%	65.8	11.09	23.1	100
Population within drainage basin	**People**	14,966	3,059	764	18,789
	%	79.6	16.3	4.1	100

TABLE 3 – ILLITERACY RATE (%), ACCORDING TO AGE, IN THE MUNICIPALITIES OF CAMPO LARGO, CAMPO MAGRO AND ARAUCÁRIA

	CAMPO LARGO	CAMPO MAGRO	ARAUCÁRIA
Age (years)	Rate (%)		
Under 15	6.8	9.7	5.8
15 to 19	1.3	2.6	1.1
20 to 24	2.0	3.0	1.0
25 to 29	2.3	3.5	1.4
30 to 39	3.6	6.2	2.8
40 to 49	6.0	10.0	6.0
50 and over	19.0	29.0	21.0

SOURCE: IBGE – 2000 Demographic Census

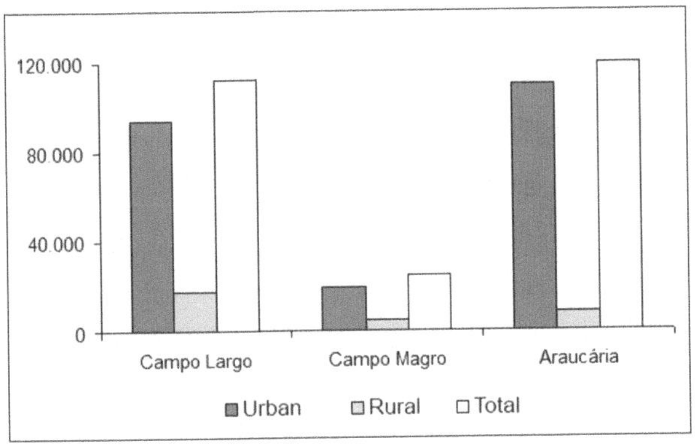

FIGURE 3 – URBAN, RURAL AND TOTAL POPULATION ACORDING TO MUNICIPALITY
SOURCE: IBGE – 2010 Demographic Census.

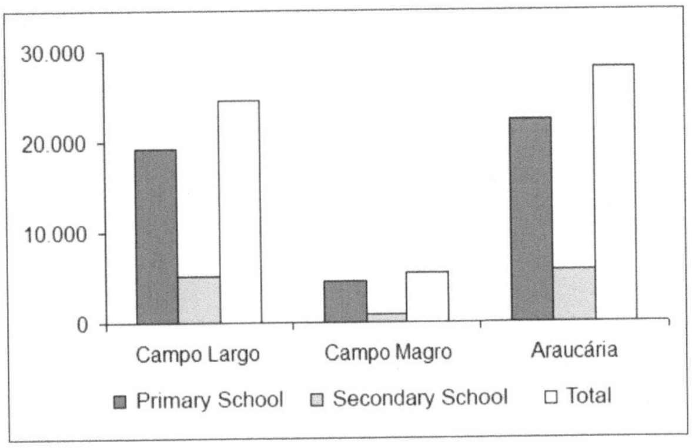

FIGURE 4 – STUDENTS ENROLLED IN PRIMARY AND SECONDARY EDUCATION IN 2009 IN CAMPO LARGO, CAMPO MAGRO AND ARAUCÁRIA

2.4 LAND-USE AND OCCUPANCY

2.4.1 Land-Use regulation

The Rio Verde APA is located to the west of the Curitiba Metropolitan Region (RMC) and it includes a portion of the municipalities of Araucária and Campo Largo, and areas upstream of the Rio Verde Reservoir, which are the drainage basin of this reservoir (Figure 5). The APA was established by State Decree 2.375, July 31, 2000, and it aims to protect and preserve the quality of the environment and its natural ecosystems with a particular focus on the quality and the quantity of water for public consumption. The APA establishes measures and instruments to manage conflicts arising from the various types of land-use that occur in the area.

The Campo Largo municipal zoning laws established by Municipal Decree 444/78, amended by Municipal Decree 97/81 and 123/82, by Municipal Law 1.236/96 and by Municipal Decree 061/00, are applied in the Rio Verde APA.

The laws of land-use and land-occupation in the macro-zoning of the Campo Magro Territorial Planning Unit (UTP), created by State Decree 1.611/99, are also applied in the APA because part of the municipality's territory is located within the Rio Verde Basin. This macro-zoning was designed based on the provisions of the Law for the Protection of Water Sources (State Law 12248/98) and considered areas under pressure due to occupation and environmental sustainability. The macro-zoning created six areas focused on targeted occupation, two on consolidated urbanization and one with high restrictions on occupation.

Part of the rural area that belongs to the Municipality of Araucária is located within the Rio Verde APA. The following legal provisions prevail: Federal Law 4.504/64, which provides for the Land Statute (*Estatuto da Terra*); and Federal Law 5.868/72, amended by Federal Law 10.267/01, which created the National System of Rural Registration (*Sistema Nacional de Cadastro Rural*).

2.4.2 Evolution of Land-use

Several anthropogenic activities have modified the Rio Verde Basin over the last two centuries, such as logging, increased agricultural and livestock activities and urban and industrial expansion. In Chapter 3, the evolution of land-use in the Rio Verde Basin is discussed in detail. In 1976, the areas with vegetation represented the majority of land-cover (~52%) and the built environment occupied a small portion in the eastern region of the basin (Campo Largo). In 2000 and 2009, both rural and urban areas increased. In addition to the population increase in the Campo Largo municipality, the built environment increased along highway BR 277 and the areas with vegetation decreased over time (from ~ 52% to ~ 41%; Chapter 3).

Another important fact that affected land-use in the basin was the development of Petrobras' President Getúlio Vargas Refinery in 1973 and the construction of the dam for water abstraction from the Rio Verde. Comparing area maps from 1976 with that of 2009, it is easy to observe an increase in the area covered by water (from 0.7% in 1976 to 3% in 2009) with the reservoir occupying approximately 8km^2 of the basin (Chapter 3).

2.4.3 Land-Use in Areas of Permanent Preservation

In Chapter 6, the vegetation cover and land-use in areas of permanent preservation (APPs) are assessed. The goal in creating areas of permanent preservation (Federal Law 4771/65) is to "preserve the water resources, landscape, geological stability, biodiversity, gene flow of fauna and flora, protect the soil and ensure the well-being of human populations."Approximately 50% of the surface area of the APP is occupied by natural vegetation, especially vegetation in the Middle Stages of Succession (35.5%) and Gallery Forests (4.7%). The categories of "Agricultural", "Reforestation", "Urbanized Areas" and "Exposed Soil" represent 47.4% of inappropriate land-use in the APP. The remaining surface area (1.2%) is occupied by "water bodies".

2.5 INFRASTRUTURE

2.5.1 Sewage system

The sewage collection system in the region has been expanded in recent years. In 2009 R$13.6 million were allocated for the expansion of the sewage collection system in the municipalities of Araucária, Campo Largo and Campo Magro, serving approximately 13,000 people. In the municipality of Campo Largo, a 4.2km sewage system was implemented and 210 homes were connected to the system. In

Araucária, the system for sewage collection and treatment was extended by 31km. In 2009, the first portion of the sewage system was implemented in the municipality of Campo Magro, benefitting 1,006 households in the town of Jardim Boa Vista, with more than 19km of piping.

In the project aiming to expand the Campo Largo Water Supply System (WSS), which began in 2011, the Cercadinho/São Luiz sub-system which captures water from the Rio Verde, will be available for local use and it will serve the areas to the east, towards Curitiba that are now under within the RMC Integrated System.

The construction of three sewage pumping stations and the extension of 33km of the collection system in São Luís, Cercadinho, Jardim Itália, Padre Anchieta and Vila Lurdes, in Campo Largo, as well as the construction of a sewage treatment station in Ithaqui, are among the projects scheduled to help clean up the Rio Verde Basin. These projects were scheduled to start in the first half of 2011 and should be completed within 12 months.

Most homes that are not connected to the sewage collection system have septic tanks to treat domestic sewage and some of them dispose of their wastewater directly into the water bodies without treatment.

2.5.2 Solid and hazardous waste management

Solid waste from about half of the population located within the APA is collected by municipalities. The other half of the population burns the solid waste that they generate. The Municipality of Campo Magro has the sole recycling program in the drainage basin. Even with regular collection, there are instances where garbage is disposed of in inappropriate places, such as water ways. All of the urban solid waste collected in the region of the APA is sent to the Cachimba landfill, located south of Curitiba.

2.6 TOURISM AND TOURISM INFRASTRUCTURE

Industries such as the production of ceramics and furniture attract tourists to Campo Largo as do wineries and properties that work with rural tourism. Rodeos (Colônia Mariana) are another attraction in the region. There is also a practical tourist program which includes specialized visits by farmers and professionals to organic producing properties and to the Chamel company that produces medicinal herbs and herb products.

In the Municipality of Araucária there are countless "fishing ponds" that attract tourists, mainly on weekends, as well as several properties that have the potential for rural tourism.

2.7 CHARACTERIZATION OF THE BASIN

2.7.1 Hydrologic aspects

The total area of the basin drainage is 242km^2 and represents about 9% of the entire drainage area of the Alto Iguaçu Basin. The drainage basin of the Rio Verde Reservoir was subdivided into 18 smaller hydrographic sub-basins, according to the natural divisions in the region (Figure 6). The pattern of the drainage network is dense, characterized as sub-dendritic to dendritic and it is little affected by tectonics. The main springs drain along straight steep slopes of geological formation of the Atuba Complex. In general, these drainage flows are fairly swift, because of the steep slopes. Several streams canals of 1st, 2nd and 3rd order begin in the subsequent relief. The main watercourse is the Rio Verde that flows to the south. In addition to the Rio Verde, 15 other tributaries supply the reservoir, including the Ribeirão dos Pessegueiros, Ribeirão Iguaçu, Cristal River, Arroio Rondinha and Arroio Formigueiro (Figure 6).

2.7.1.1 Underground sources

The underground water sources in the study area are characterized by five aquifers: 1) surface or unconfined aquifer, located at the base of the water table and present in the surface soil layers; 2) aquifer that is free from alluvial deposits and shows little consistency; 3) fractured aquifers of the cristaline formation (Atuba Complex), with concentrated movement of water in open fractures and flaws, highly variable permeability, average production volumes of 15 to 20L/s, and exceptional flow rates of 100L/s; 4) confined aquifer of sandy lenses of the Guabirotuba formation; 5) karst aquifer, with complex circulation of groundwater and a basic mechanism that consists of the dissolution of fissured carbonate rock by the water. This mechanism forms cavities in the rock through which the groundwater circulates.

2.7.2 Types of Soil

In the Rio Verde Basin Gleysols, Oxisols, Nitosols, Inceptisols and Histosols are the most common. The types of soils present in the basin are shown in Figure 7. The mapped soil types were grouped and identified as follows:

- LBd – Dystrophic Bruno Oxisol;
- LVd – Dystrophic Red Oxisol;
- NVd – Dystrophic Red Nitosol;
- CXbd – Dystrophic Tb Haplic Inceptisol;
- CXve – Eutrophic ta Haplic Inceptisol;
- GJo – Orthic Tiomorphic Gleysol;
- LVe – Association Eutrophic Red Oxisol + Dystrophic Bruno Oxisol;
- NBd – Association Distrophic Bruno Nitosol + Dystrophic Tb Haplic Inceptisol.

The Gleysols occupy the floodplain and these environments require special care in planning decision-making and environmental management. The areas in which Oxisols occur have the lowest restrictions on usage, while the portions of land with Nitosols and Inceptisols require an intermediate level of restrictions because they are mostly impermeable and have a high potential for hydric erosion. Clay soils, in turn, are quite porous, present good levels of water capacity and are naturally relatively impermeable.

2.7.3 Landscape characteristics of the basin

The most common types of natural vegetation cover and the most relevant types of land-use in the basin are shown in Figure 8. The basin has areas with Araucaria (Mixed Ombrophilous) Forest in different successional stages (*capoeirinhas, bracatinga, capoeirões,* forest and gallery forest), areas of pioneer species (floodplain), and areas of anthropogenic disturbance (reforestation, agriculture and livestock, urban areas, areas with exposed soil and water bodies).

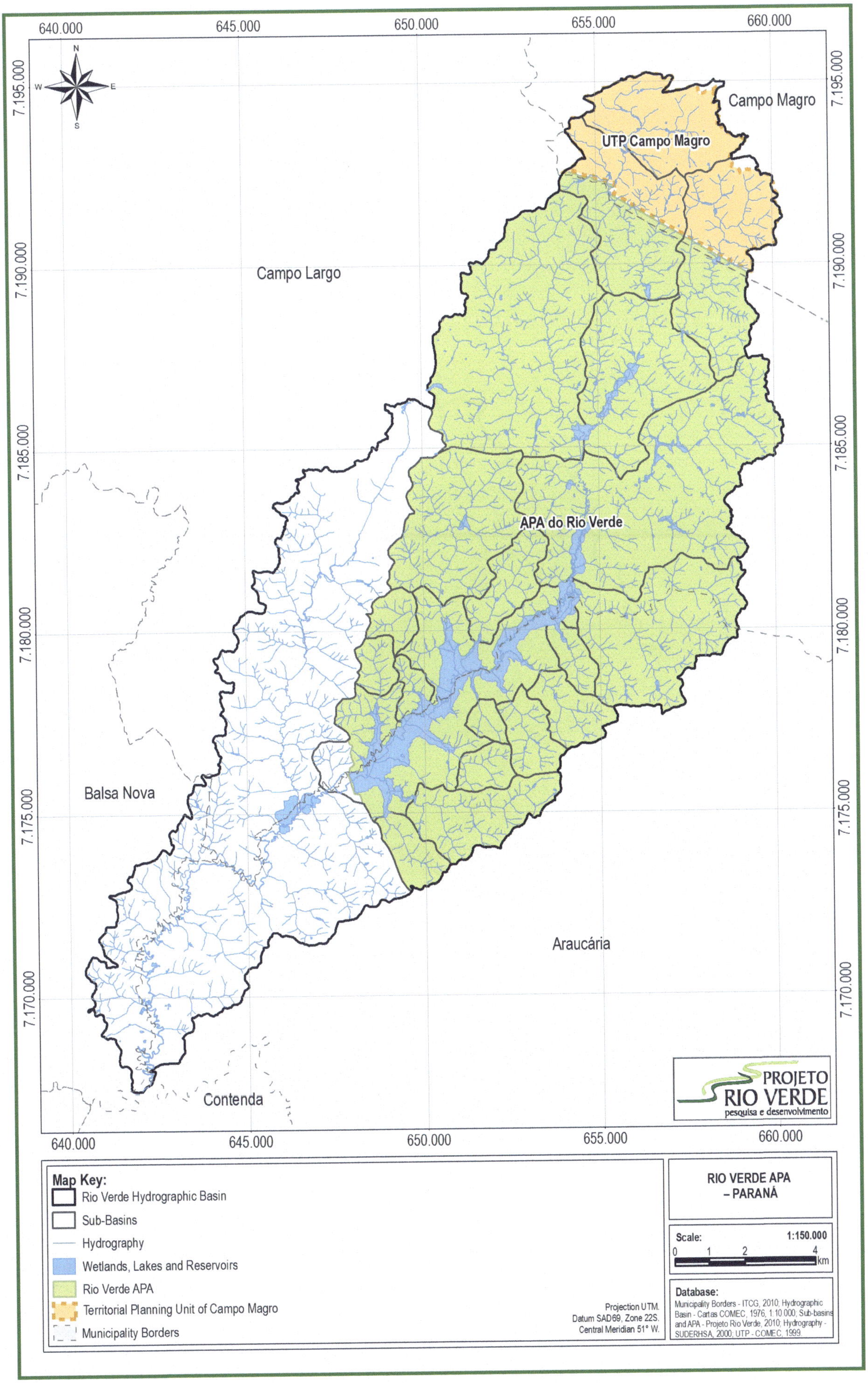

FIGURE 5 – MAP OF THE RIO VERDE ENVIRONMENTAL PROTECTION AREA (APA)

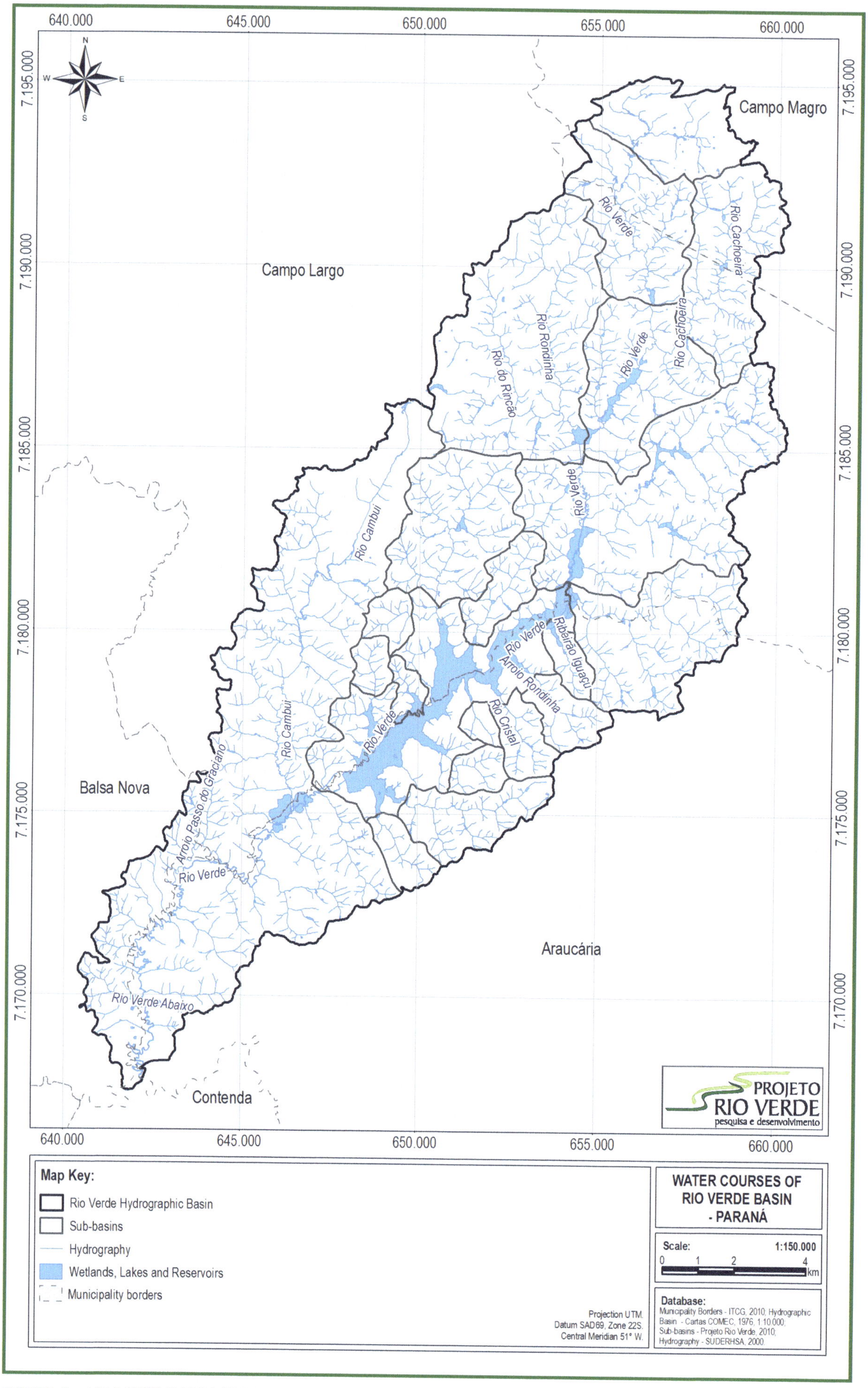

FIGURE 6 – MAP OF THE RIO VERDE HYDROGRAPHIC BASIN DRAINAGE AREA

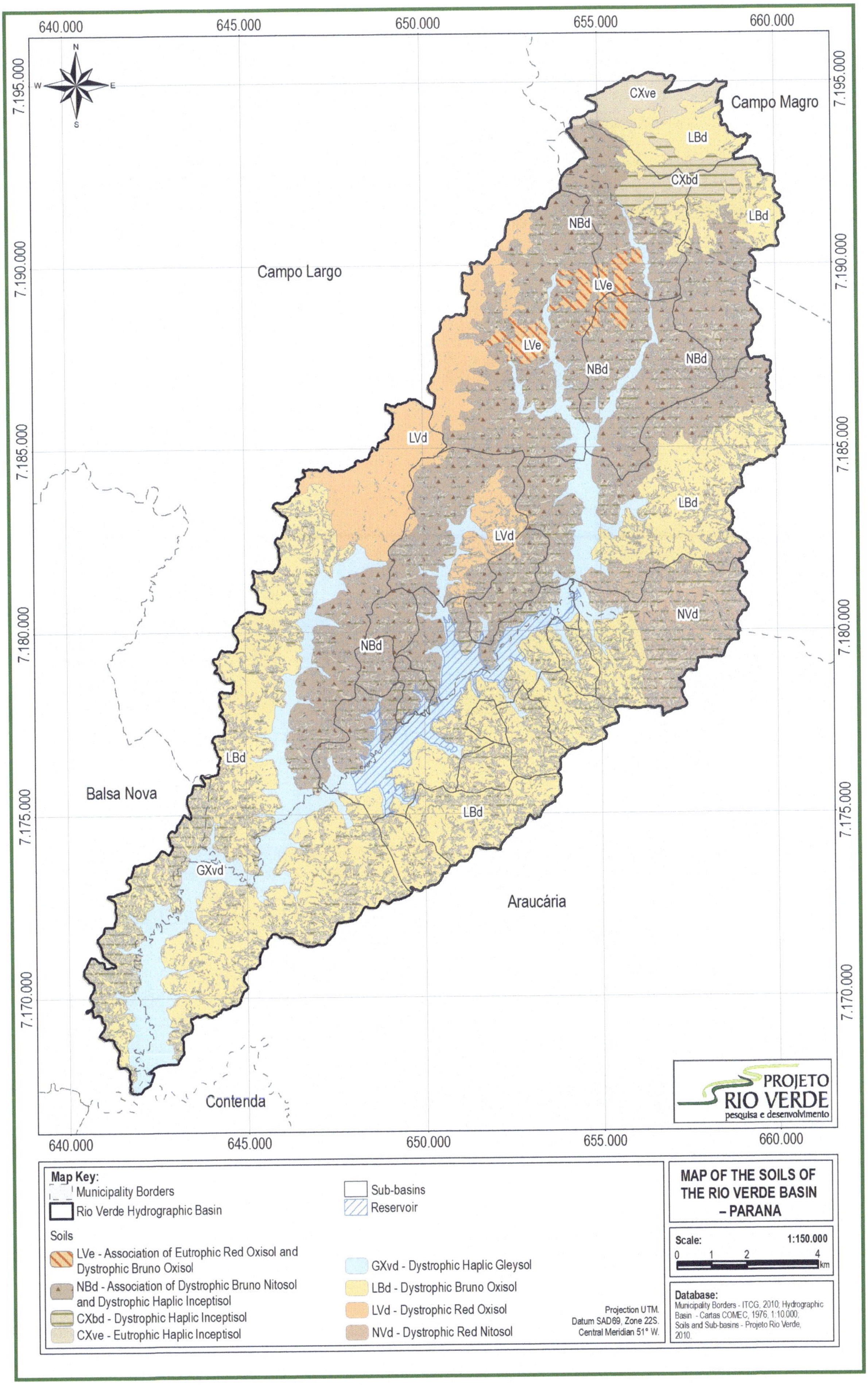

FIGURE 7 – MAP OF THE SOILS OF THE RIO VERDE HYDROGRAPHIC BASIN

FIGURE 8 – NATURAL VEGETATION COVER AND TYPES OF LAND-USE IN THE RIO VERDE BASIN. A: GALLERY FOREST; B: *CAPOEIRÃO*; C FLOODPLAIN; D: *BRACATINGA*; E REFORESTATION; F: AGRICULTURE; G: *CAPOEIRINHA*; H: LIVESTOCK; I: URBAN AREA

3. THE RIO VERDE RESERVOIR

The Rio Verde dam (25°31'30"S, 49°31'30"W; Figure 9) was constructed between 1974 and 1976 by Petrobras (Petróleo Brasileiro S.A.) with the initial purpose of serving the President Getúlio Vargas refinery. The region upstream from the dam is made up of about 165km² (~68% of the total basin area), around 147km² (~61%) makes up the Rio Verde APA, and 18.4km² (at north end of the basin – 7.6%) is located in the Municipality of Campo Magro, making up the Campo Magro UTP.

FIGURA 9 – AERIAL IMAGE OF RIO VERDE RESERVOIR – DAM AREA

3.1 GENERAL CHARACTERISTICS OF THE RESERVOIR

Table 4 contains general information about the Rio Verde Reservoir. The average and maximum depth shows that the reservoir is shallow with the majority of the reservoir being less than 8 m deep (Figure 10). The mean residence time of water is 218 days (Chapter 8).

The input flow is estimated at approximately 2.80m³/s. The sum of the output flows from ecological processes, licenses, irregular abstraction, evaporation, infiltration and percolation was estimated at 2.26m³/s (Table 5).

3.2 LIMNOLOGICAL CHARACTERIZATION OF THE RESERVOIR

The Rio Verde Reservoir has been monitored since 1987 by the Environmental Institute of Paraná (IAP), as part of the "Program for the Evaluation, Classification and Monitoring of Water Quality of the Reservoirs in the State of Paraná." This Program, developed with the technical cooperation of GTZ (German Agency of Technical Cooperation), aims at classifying the water quality of reservoirs

TABLE 4 – GENERAL CHARACTERISTICS OF THE RIO VERDE RESERVOIR

RESERVOIR DATA	
Reservoir area	7.9 km^2
Basin total area	239 km^2
Drainage basin	173 km^2
Mean volume	25,644,000 m^3
Reservoir volume at the spillway top	34,000,000 m^3
Surface	5,972 km^2
Length	9.06 km
Average width	0.645 km
Perimeter	36.4 km
Maximum depth	11 m
Average depth	5.6 m
Mean residence time	218 days

TABLE 5 – USE AND LOSS OF WATER IN THE RIO VERDE BASIN – 2010

DESCRIPTION	FLOW (m^3/s)
Instream flow	0.21
REPAR license	0.84
Fosfertil license	0.12
SANEPAR current license	0.30
Irregular abstraction	0.03
Loss through evaporation	0.25
Loss through percolation/infiltration	0.012
SANEPAR extra license	0.50

according to their degree of degradation. This method enables an understanding of the main limnological characteristics of each environment, particularly in relation to water quality and changes over time.

Through the Program a Water Quality Index is calculated for each Reservoir (IQAR). Each reservoir is monitored twice a year (winter and summer) and samples are taken from two depths of the water column: euphotic and aphotic zones, independent of the processes of thermal stratification. Physical, chemical and biological analyses are performed on each sample. In addition, thermal profiles, dissolved oxygen and saturation, pH, conductivity and water transparency using a Secchi disk are assessed.

Data collected up to 2009 show the following limnologic characteristics of the Rio Verde Reservoir:

- - thermal profile similar to warm monomitic lakes. In these environments, there is a complete circulation every year and thermal stratification during the warmest months;
- - dissolved oxygen distribution correlates with the thermal profile, with oxygen concentration decreasing with distance from the euphotic zone and anoxia occurring in the bottom layers, particularly during spring and summer months;
- - the average phosphorous concentration observed during the period (1987 to 2009) was 0.015mg/L which classifies the reservoir as mesotrophic, according to OECD (1982). Total phosphorous content varied from 0.004 to 0.027mg/L;
- the ratio of nitrogen/phosphorous indicates that in this environment phosphorous is the limiting nutrient for the primary production of phytoplankton;
- the reservoir was considered mesotrophic with regard to phytoplankton biomass, with an average chlorophyll-a concentration of 6.9µg/L. According to OECD (1982), in mesotrophic environments chorophyll-a concentration varies between 2.5 and 8µg/L;
- during the monitoring period, only one cyanobacterial bloom was observed in April 2005, and the prevalent species was *Cylindrospermopsis raciborskii*. This species is considered potentially toxic. Other cyanobacterium species were also detected in other samples but always in very low concentrations. It is important to highlight that the Rio Verde Reservoir presented the second largest diversity of phytoplankton species of all reservoirs monitored by IAP, with a total of 84 taxa;
- in the study of the zooplankton population, 40 taxa were registered and a group of Rotifers was prevalent. Among the identified Copepods, more Cyclopoida were present than Calanoida, indicating a mesoeutrophic trend in the environment.

Based on data collected throughout the study period, the Rio Verde Reservoir presented an water quality index of 3.5, which indicates a moderately degraded environment (IAP, 2009). The main characteristic of such environments is a significant deficit of dissolved oxygen in the water column, which may lead to anoxia in the lower layers during certain periods. These environments present an average supply of nutrients and organic matter, significant variety and/or density in algae (with some prevalent species), moderate tendency to eutrophication and significant water residence time. Water quality is considered to be fair/acceptable.

According to the index adopted by IAP, reservoirs classified as moderately degraded have sufficient water quality for the most demanding uses, such as domestic consumption, after conventional treatment and protection of aquatic communities. However, to maintain this level, conservation measures within the hydrographic basin, upstream of the reservoir, are recommended.

3.3 SAMPLING PLAN OF THE RIO VERDE PROJECT

In the Rio Verde project, the hydrographic basin was divided into 20 sub-basins and a sampling point was selected in each sub-basin (Figure 11). Each sub-basin has the same identification as the sampling point. In total, samples were taken from 25 points as follows:

- 4 points upstream of the reservoir in the main channel;
- 1 point downstream of the reservoir in the main channel;
- 5 points on the right bank;
- 8 points on the left bank;
- 2 points in indirect tributaries;
- 5 points in the reservoir

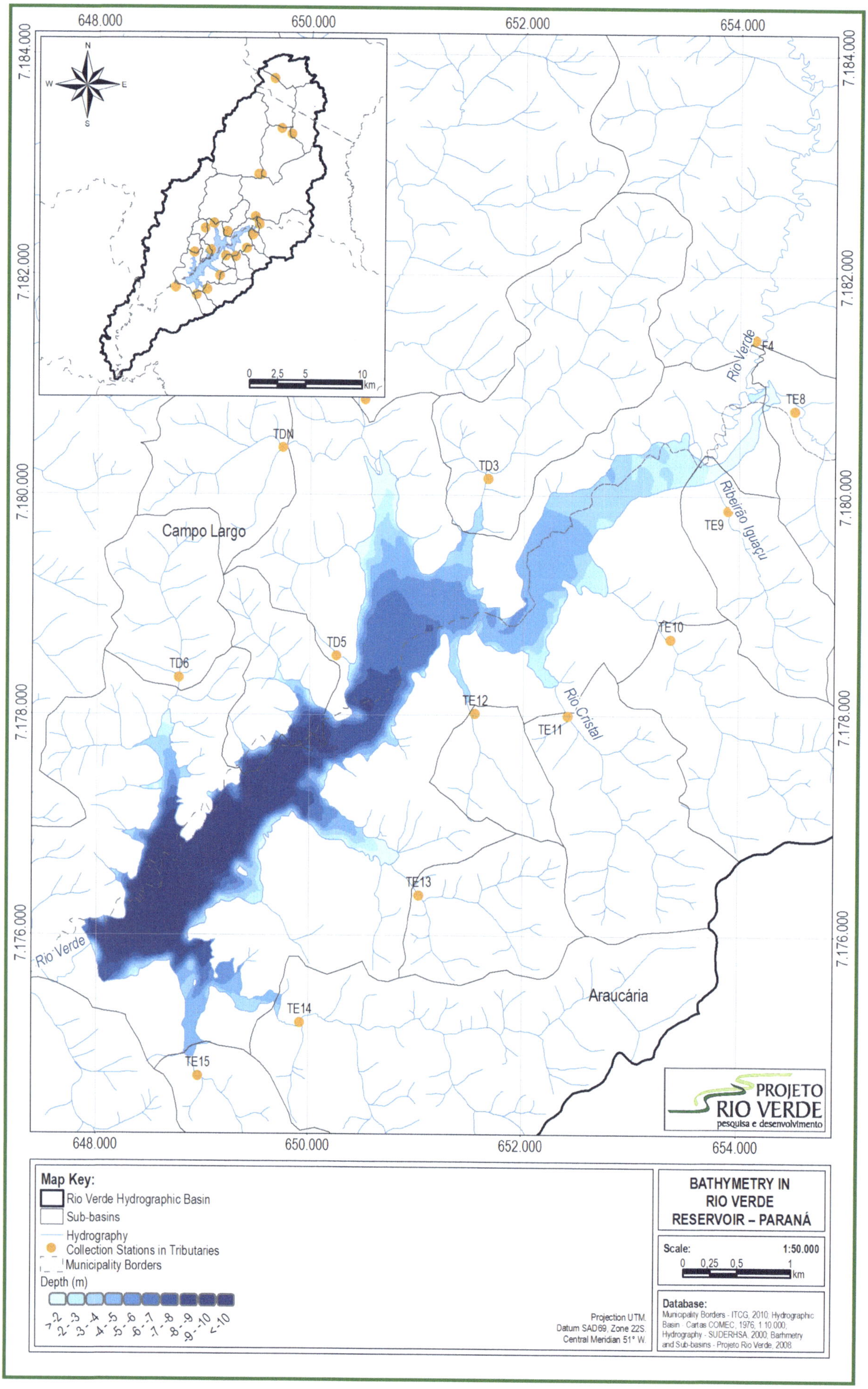

FIGURE 10 – BATHYMETRIC MAP OF RIO VERDE RESERVOIR

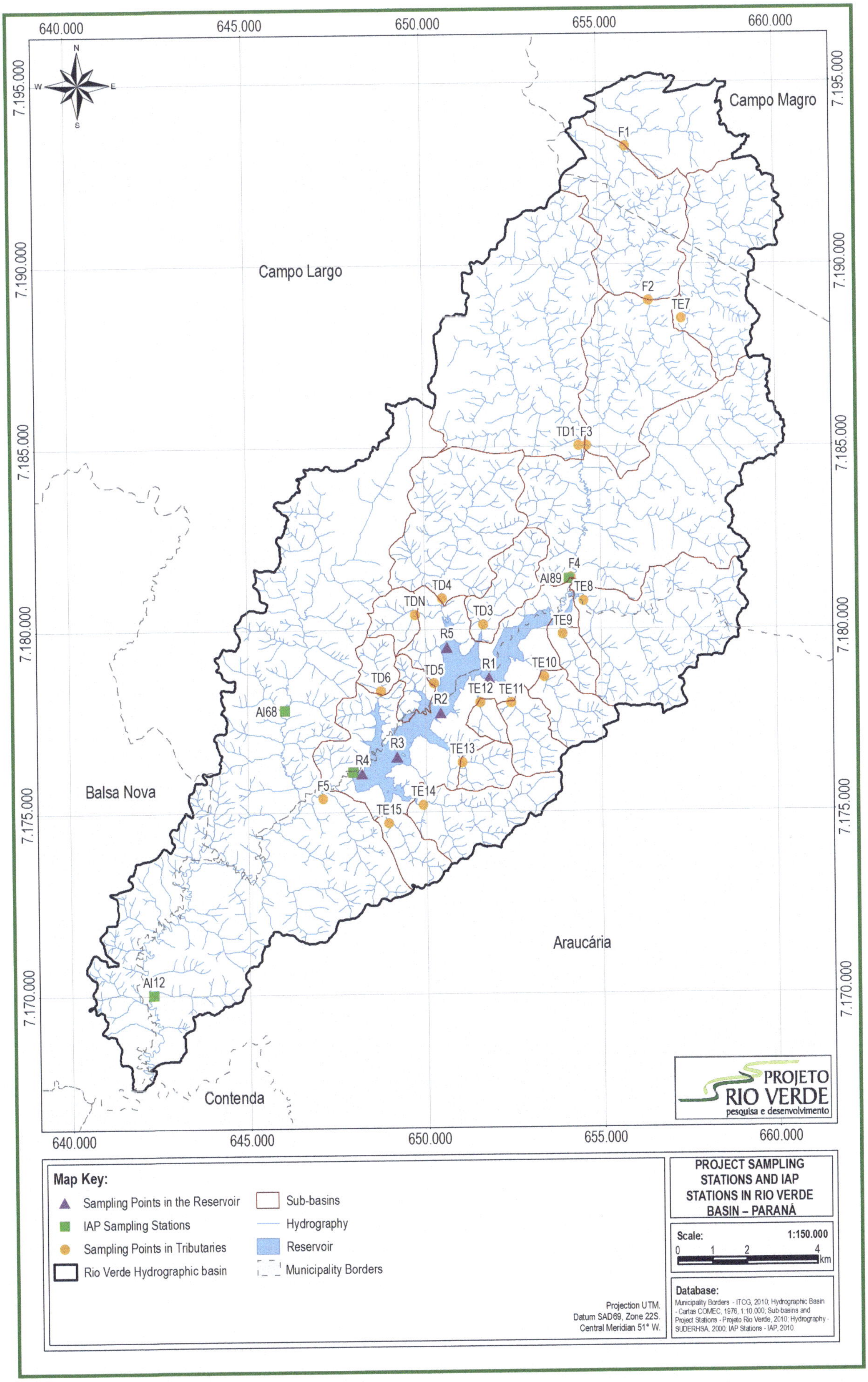

FIGURE 11 –LOCATION OF THE SAMPLING POINTS OF THE RIO VERDE PROJECT.

The physico-chemical and biologic parameters of the samples collected at those points were assessed. Data on the chemical analysis of water quality within the tributaries is presented in Chapter 10. Chapter 12 discusses the physico-chemical and meteorological parameters of the reservoir based on mathematical models. Data on the aquatic communities of the reservoir are presented in Chapters 13, 14, and 15.

Studies relating to socioeconomic conditions, rural sanitation and environmental education were conducted in the communities of Cercadinho, Colônia Antônio Rebouças, Colônia Rio Verde Faxinal do Tanque, Colônia Cristina, Colônia Figueiredo and Colônia Dom Pedro II.

3.4 USES OF THE RESERVOIR

3.4.1 Water

3.4.1.1 Drinking water abstraction from the reservoir

Currently SANEPAR (Water and Sanitation Company of Paraná State) does not make water abstraction system from the Reservoir for public consumption. However, water is collected from the main tributary, the Rio Verde, by the sub-system Cercadinho/São Luiz, which serves these communities (Figure 12). Abstraction of water from the reservoir is planned for a point 200 meters from the bridge that crosses the reservoir connecting Campo Largo to Araucária (Figure 12). This system will be comprised of: abstraction from the Rio Verde reservoir, water pump and pipeline, a water treatment station, a sludge treatment station, holding tanks, and the pumping and piping of treated water.

Data on water use and loss in the Rio Verde Basin are shown in Table 5. The Rio Verde dam has 1.87m^3/s of regularized flow. Two licenses for water withdrawal from the reservoir have been granted: one for Petrobras, with 0.84m^3/s (~ 45% of the total) and one for SANEPAR, with 0.3m^3/h. SANEPAR also has an extra license of 0.5m^3/s, which will be used for public supply after the enlargement of the Campo Largo water supply system. The total volume of reserved water after the enlargement of the system will be 9,600m^3, enough to meet the demand forecasted up to the end of the project horizon.

3.4.1.2 Water abstration for agricultural purposes

The water used for cleaning, irrigation, animal consumption and chemical spraying is withdrawn from private wells in Colônia Cristina and Formigueiro (municipality of Araucária) and in Colônia Rebouças and Dom Pedro II (municipality of Campo Largo). Despite the good water quality of the collective system, most producers prefer to use water from private wells for their primary consumption and in some cases they use the collective system for secondary consumption.

3.4.1.3 Water withdrawal for industrial purposes

REPAR (President Getúlio Vargas Refinery) uses water abstracted from the Rio Verde Reservoir for industrial purposes and distribution of drinking water in the refinery. The industrial uses include: 1) untreated water to pressurize the water system for emergency combat, for cleaning the area, and for testing tanks; 2) untreated water to replenish the cooling system of the refinery; 3) filtered and chlorinated water for industrial use in operations that require higher quality water; 4) drinking water (filtered and chlorinated); and 5) demineralized water for steam production. In addition to the water used at REPAR, untreated water, industrial water, drinking water and steam are also sent to Fosfértil (Fertilizantes Fosfatados S/A) in Araucária.

Water abstraction by REPAR occurs at the dam (Figure 13), the deepest part of the reservoir. At this location, there is a water intake with four manually activated gates. The depth of the water body at each gate is: Gate 01 = 2.60m; Gate 02 = 5.60m; Gate 03 = 8.65m; Gate 04 = 11.80m.

The water is piped through the pumping station to the water tower. From this tower, water is fed by gravity to the untreated water storage tanks (Figure 14) of the refinery, placed at an elevation of 915.5m. The purpose of the tanks is to ensure the continuity of untreated water supply and to allow solid particles to settle. To this end, tanks are arranged so that while one is being filled, another is in operation.

3.4.2 Fishing at the reservoir

Fishing around the reservoir is not commercialized, it is only done for leisure or sport (Chapter 18) and it is a practice that has been going on for at least 10 years. Fishing is practiced by residents of Curitiba, Araucária or Campo Largo and the targeted species are those of the *Astyanax* (*lambaris*) genus.

4. CONSERVATION STATUS OF THE ECOSYSTEM

The Rio Verde Basin is subject to pressure from urbanization, increasing industrial activities, development of the service sector and transport systems, creating conditions that can degrade the environment and significantly modify natural ecosystems.

The natural characteristics of a large proportion of the region are highly modified. Sections of the forest that are well preserved are mostly restricted to the areas under environmental protection. Still, the native forest cover in areas of permanent preservation (APP) does not meet the requirements of environmental legislation (Chapter 6). Beyond these areas, the remaining patches of Atlantic Forest have been replaced by pasture or reforestation with exotic species.

Nevertheless, the Rio Verde still has the highest levels of environmental integrity with discreet patterns of biodiversity in comparison to all other rivers located in hydrographic area of the Curitiba Metropolitan Region.

Aquatic vegetation is the common element throughout the watercourse. Gallery forests and foodplains occupy about 36% of the APP (Chapter 6). In the area surrounding the dam, the natural vegetation along the banks is deteriorated. In this area a large number of tree trunks are submerged.

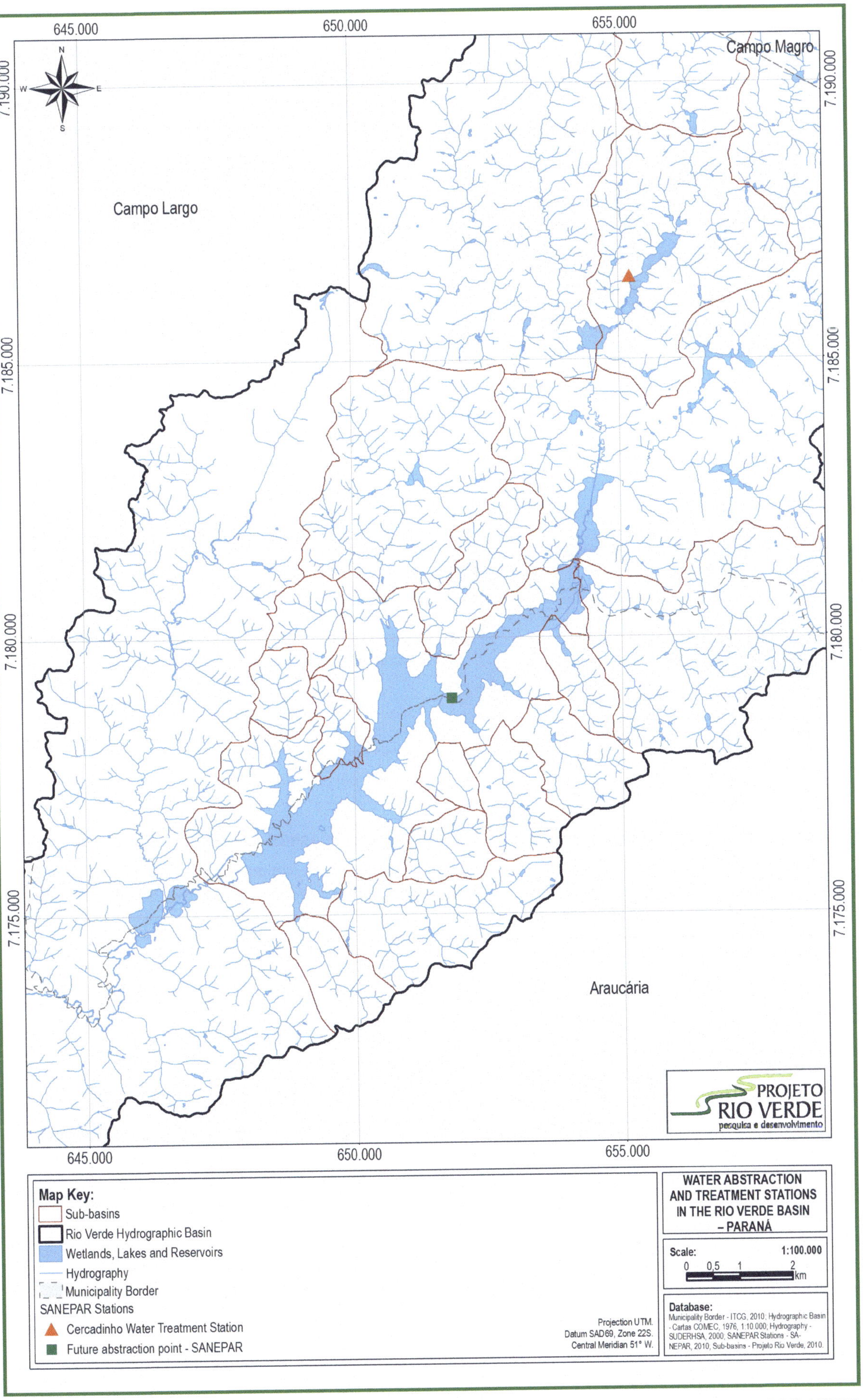

FIGURE 12 – LOCATION OF SANEPAR'S FUTURE INSTALLATIONS FOR WATER ABSTRACTION FROM THE RIO VERDE RESERVOIR AND THE CERCADINHO/SANEPAR WATER TREATMENT STATION

FIGURE 13 – PLACE OF WATER ABSTRACTION BY REPAR IN THE RIO VERDE RESERVOIR

FIGURE 14 – UNTREATED WATER STORAGE TANKS OF REPAR REFINERY

4.1 CAUSES OF ENVIRONMENTAL DEGRADATION

4.1.1 Agricultural and livestock uses

The production of livestock is not significant in the Rio Verde Basin (Chapter 19). The prevailing agricultural cultivation system is tillage, with significant and indiscriminate use of pesticides. Due to naturally low levels of soil fertility, farmers use strong fertilizers which are easily lost through erosion, due to conventional soil management practices. This type of management causes erosion through which agrochemical products, clay colloids and organic matter are carried away by runoff, causing siltation of the river beds and eutrophication. The impact on the river and reservoir produces significant challenges for water treatment. About 43% of the basin area presents fair to good suitability for agriculture, if soil control and management measures are used (Chapter 19). In the remainder of the basin, more drastic measures of control and management should be employed, due to the increased fragility of the soil.

4.1.2 Urban use

The urban or built areas total 1,511.97ha, or 6.32% of the basin surface, and are concentrated mainly within the municipality of Campo Largo, in the vicinity of highway BR 277, and at the north end of the basin, within the municipality of Campo Magro (Chapter 6).

4.2 IMPACTS CAUSED BY LAND-USE

Nutrient enrichment is one of the main causes of water degradation because it can boost the production of algae and aquatic plants, disturbing the balance of dissolved oxygen, reducing biological diversity and negatively affecting the water supply (FERNADES et al., 2005). Agricultural practices are the main contributor of the diffuse nutrient load into the country's rivers, while urban areas add concentrated point source loads into waterways with high pollution potential (ANDREOLI et al., 2003). Two aspects related to the misuse of water resources are addressed below.

4.2.1 Increase of the algae population

In 2005, a bloom of the cyanobacterium *Cylindrospermopsis raciborski* occurred in the Rio Verde Reservoir, reaching densities of 96,000 cells/mL. Cyanobacterial blooms can range from harmless to harmful or toxic to other organisms throughout the food chain, including humans. After the occurrence of the bloom in 2005, a significant increase in *C.raciborskii* was not observed. Currently, species of Chlorophyceae (green algae), diatoms and flagellates prevail in the microalgae population.

4.2.2 Decline of the fish population

The results of the ichthyofauna survey in the drainage area of the Rio Verde Basin produced a total of 27 fish species, of which 11were considered endemic to the Iguaçu River (Chapter 15). In comparison to the Rio Verde Reservoir, the Passaúna, Iraí and Piraquara reservoirs presented 15, 12 and 8 species, respectively, indicating that the fish population of the Rio Verde Reservoir is within the standards expected in the Curitiba Metropolitan Region, with few dominant species. According to the survey of fishing in the reservoir (Chapter 18), most fishermen (63%) believe that the amount of fish has not diminished over the years.

5. ENVIRONMENTAL EDUATION

Chapter 22 discusses the process of developing an Environmental Education program together with the Cercadinho (Campo Largo Municipality) community, located within the boundaries of the Rio Verde APA. The program was conceived and developed by the community group "Life for Rio Verde" (Grupo Vida ao Rio Verde) and rolled-out through activities designed to raise awareness in the community. The goal of the Environmental Education program is the continuous implementation of preservation actions of the River and its surroundings. The management team of the local Municipal School, members of the Parents, Teachers and Staff Association, leaders of the residents association, neighborhood merchants, residents and representatives of municipal agencies are part of this program.

The involvement of the community resulted in activities such as lectures and awareness workshops, ecological walks and a brochure on environmental education. The team also created an interactive blog (www.vidarioverde.blogspot.com) to communicate its activities and to share knowledge.

6. REFERENCES

ANDREOLI, C.V., HOPPEN, C., PEGORINI, E.S., DALARMI, O. (2003) A Crise da Água e os Mananciais de Abastecimento – A Disponibilidade de Água na RM Curitiba. In: Andreoli, C.V. (Ed.). *Mananciais de Abastecimento: Planejamento e Gestão* – Estudo de Caso do Altíssimo Iguaçu. SANEPAR / FINEP: Curitiba, p. 35-84.

FERNANDES, L.F., LAGOS, P.D., WOSIACK, A.C., PACHECO, C.V., DOMINGUES, L., ZENHDER-ALVES, L., COQUEMALA, V. (2005) Comunidades Fitoplanctônicas em Ambientes Lênticos. In: ANDREOLI, C.V.; CARNEIRO, C. (Eds.). *Gestão Integrada de Mananciais de Abastecimento Eutrofizados*. SANEPAR / FINEP: Curitiba, p. 303-366.

IAP – Instituto Ambiental do Paraná. (2009) *Monitoramento da qualidade das águas dos reservatórios do estado do Paraná, no período de 2005 a 2008*. Curitiba, p. 120.

IBGE – Instituto Brasileiro de Geografia e Estatística. *Censo Demográfico*, 2000. Available in: www.ibge.gov.br

IBGE – Instituto Brasileiro de Geografia e Estatística. *Censo Demográfico*, 2008. Available in: www.ibge.gov.br

IBGE – Instituto Brasileiro de Geografia e Estatística. *Censo Demográfico*, 2010. Available in: www.ibge.gov.br

IPARDES – Instituto Paranaense de Desenvolvimento Econômico e Social. *Perfil Municipal*, 2010. Available in: www.ipardes.gov.br

OECD – Organization for Economic Co-operation and Development. (1982) *Eutrophication of waters: monitoring, assessment and control*. OECD, Paris, 154 p.

PNDU – Programa das Nações Unidas para o Desenvolvimento. *Atlas do Desenvolvimento Humano*, 2000. Available in: www.pndu.org.br

SECTION II

THE PHYSICAL ENVIRONMENT

CHAPTER 3

DEVELOPING A DATABASE FOR DIAGNOSING AND ASSESSING ENVIRONMENTAL CONDITIONS OF THE RIO VERDE BASIN – A DIGITAL ENVIRONMENTAL MODEL

Sony Cortese Caneparo
Elaine de Cacia de Lima Frick
Gabriel Henrique de Almeida Pereira
Mariana Barbosa Druszcz
Jean Jesus Ilsuk da Silva

DEVELOPING A DATABASE FOR DIAGNOSING AND ASSESSING ENVIRONMENTAL CONDITIONS OF THE RIO VERDE BASIN – A DIGITAL ENVIRONMENTAL MODEL

Sony Cortese Caneparo
Elaine de Cacia de Lima Frick
Gabriel Henrique de Almeida Pereira
Mariana Barbosa Druszcz
Jean Jesus Ilsuk da Silva

ABSTRACT

The Rio Verde Basin, located in the Metropolitan Area of Curitiba (RMC) , Paraná, Brazil, is one of the region's main water supplies and it provides water to Petrobras' President Getúlio Vargas Refinery, the largest oil complex in the State of Paraná. The initial occupation of the basin, which was predominantly rural, was not planned or controlled. Similarly, urban development was not based on proper assessments of environmental impacts and weaknesses. This resulted in improper land-use practices, which resulted in significant changes to river environments in areas of human occupation. This situation has driven this research study, the primary purpose of which was to develop a cartographic database in order to study spatio-temporal dynamics and identify growth trends within the basin. The struggle between development and legal environmental requirements has also been identified through the use of geographic information systems (GIS). For the classification and analysis of mapping units, information layers were generated from topographic maps (scale 1:20,000 – 1976), aerial photographs (scale 1:30,000 - 2000), satellite images (2005) and field data. Our results show that anthropogenic pressures have had an impact on the basin and these pressures affect the quality of the environment and require mitigation measures. Because of human occupation in the basin, there is the possibility of eutrophication of the reservoir, which may affect the future use of this water source.

KEY WORDS

Environmental assessment, space-time dynamics, conflicts, environmental risk, geographic information systems.

1. INTRODUCTION

There are several common environmental problems associated with sustainable development in watersheds, including: siltation of waterways, altered flow regimes with larger and more frequent floods during rainy periods, erosion of canal banks and floodplains, levee breaks, land-use, and inadequate management of soils, slopes, and interfluves. These problems create conditions favorable to erosion by reducing the storage capacity of the soil as well as affecting the replenishment of the water table, which reduces flow during rainy periods. These are factors that, in severe cases, transform perennial streams into intermittent streams, or in relation to water pollution, make them unsuitable for consumption or irrigation. When added to other environmental factors resulting from un-planned use of the basin, they affect the quality of life of local and regional population.

These phenomena significantly compromise development and require managers to take immediate action especially in source-areas of the basin where these processes are usually triggered by the removal of vegetation, followed by the improper use and management of the land. Such actions can occur in both rural and urban areas; in urban areas the accumulation of inappropriately treated solid waste, along with the absence of a rainwater drainage system, and its use for excessive storm water runoff, can produce increased surface flow. The surface flow increase is sometimes the result of soil sealing related to intensified development and occupation within these basins (buildings, transportation routes and pavement).

These factors result in a decrease in the amount of rainwater that is infiltrated into the soil, which reduces the capacity of water storage in the soil. In addition, these factors can change river dynamics through modifications to surface and river flow, increased erosion and the frequency of flooding which in turn affect water quality.

In the Metropolitan Region of Curitiba (RMC), Paraná, where the Rio Verde Basin study area is located, human occupation and inappropriate land management on slopes and interfluves, as seen through the loss of topsoil due to sheet erosion, can cause significant changes in river environments (channels or low-flow channels) and alluvial plains (flood channel or flood plain). In these environments, even without human occupation, imbalances are often found that can trigger processes resulting in siltation and changes in water quality, thus limiting its use.

The Rio Verde Reservoir, which is owned by Petrobras, is located in the center of the hydrographic basin. The environment has been experiencing impacts from human occupation over time and is thus undergoing a process of eutrophication, a situation in which algae can proliferate with great speed and intensity.

The impact on water resources results in a significant social burden as the basin is an important part of the water supply of the RMC, where they often lack the information necessary to fully understand the problem presented herein.

Therefore, an interdisciplinary team was formed to conduct studies that investigate and indicate measures to prevent the eutrophication process, since algae-infested waters cannot be used for either human consumption or refinery operations.

A cartographic database was required to facilitate the analyses of the basin in order to not only understand the basin as a system in constant evolution, but also to develop scenarios for assessing levels of conservation and environmental protection.

To this end, we developed a methodology for the integrated study of the watershed landscapes using remote sensing tools coupled with a geographic information system (GIS). For the generation, classification and analysis of mapping units, layers were produced from topographic maps, aerial photographs, satellite images and field assessments.

With the digital cartographic database, several aspects of spatial analysis were developed for the study area, which will be presented in this chapter: spatio-temporal dynamics in the Rio Verde Basin (1976-2000-2009); simulation of the growth of areas of anthropogenic disturbance (AAD) in the Rio Verde Basin, using Markov Chain and Cellular Automata in a GIS environment (1976-2000-2009); and the identification of conflicts between land-use and legal restrictions in the Rio Verde Basin.

2. LOCATION AND GENERAL CHARACTERISTICS OF THE STUDY AREA

The Rio Verde Basin is located in the western region of the Metropolitan Region of Curitiba, Brazil, and covers parts of four different municipalities: Araucária, Campo Largo, Campo Magro and Balsa Nova. Its drainage area is 238.96 km^2 and it is located between Greenwich latitudes of 25°18'S and 25°40'S and longitudes 49°21'W and 49°49'W. The source of the basin is in Campo Magro and it flows into the Iguaçu River. The Rio Verde Reservoir is located along the course of the waterway and it is owned by Petrobras (Figure 1).

The climate is humid mesothermal subtropical, with average annual temperatures ranging from 16.5°C to 22.6°C and annual rainfall of 1450 millimeters.

The dominant vegetation consists of the *Montana* Mixed Ombrophilous Forest, as well as its initial stage of *Montana – Capoeira* Mixed Ombrophilous Forest, in different stages of succession. We also find vegetation specific to riparian environments, such as the Alluvial Mixed Ombrophilous Forest which is associated with steppe environments, known locally as Edaphic Fields, and occurs along the main drainage channels and flood plains (KLEIN & HATSHBACH, 1962).

The basin is economically diverse; the municipality of Araucaria has the highest Gross Domestic Product (GDP) at approximately R$ 7,023,744,000 (Brazilian Institute of Geography and Statistics, IBGE, 2005). This high GDP is mostly related to the Getúlio Vargas Refinery, built in 1972, which allowed the municipality to develop a predominantly industrial economy.

The Human Development Index (HDI) in the city of Campo Largo is 0.774 (United Nations Development Program - UNDP, 2000) and is considered average since the GDP is R$ 1,117,934,000 (IBGE, 2005). In relation to industry, Campo Largo is known for a variety of industries across several sectors, including furniture manufacturing, metal manufacturing, wineries and particularly ceramics (porcelain and tiles). The HDI of Campo Magro is 0.740 and is considered average (UNDP, 2000) and the GDP is R$ 105,368,000 (IBGE, 2005). The economy is based mainly on the agricultural production of beans, corn and potatoes. There also are many other economic activities, such as organic food production, crafts and tourism (CALADO, 2004 in SILVA, 2009).

The population of the municipalities in the basin have increased significantly since 2000 (Table 1); this growth can generate a series of conflicts between environmental pressures and land-use. Between 2000 and 2009, the total population of the municipalities of the Rio Verde Basin increased by 47,769 inhabitants (21.9%).

TABLE 1 – POPULATION GROWTH IN THE RIO VERDE BASIN

MUNICIPALITY	POPULATION 2000	POPULATION 2009
Araucária	94,258	117,964
Balsa Nova	10,153	11,252
Campo Magro	20,409	23,607
Campo Largo	92,782	112,548
TOTAL	217,602	265,371

SOURCE: http://www.ibge.gov.br/cidadesat Access on 05/20/2010

3. MATERIALS AND METHODS FOR THE CONSTRUCTION OF THE DATABASE

For the generation of the database, layers were prepared based on the cartographic sources listed below (Table 2).

TABLE 2 – CARTOGRAPHIC SOURCES USED TO GENERATE THE DATABASE

CARTOGRAPHIC SOURCE	SCALE/RESOLUTION	YEAR	PREPARED BY:
Topographical Maps (total of 21 maps)	1:10,000	1976	Coordination of the Metropolitan Region of Curitiba (Coordenação da Região Metropolitana de Curitiba - COMEC)
Panchromatic aerial photographs	1: 2 5,000	1980	Institute of Lands and Cartography (Instituto de Terras e Cartografia - ITC)
Color aerial photographs	1:30,000	2000	Superintendent of the Development Water Resources and Environmental Sanitation of Paraná (Superintendência de Recursos Hídricos do Estado do Paraná -SUDERHSA)
Satellite Images	10 meters	2005	Spot

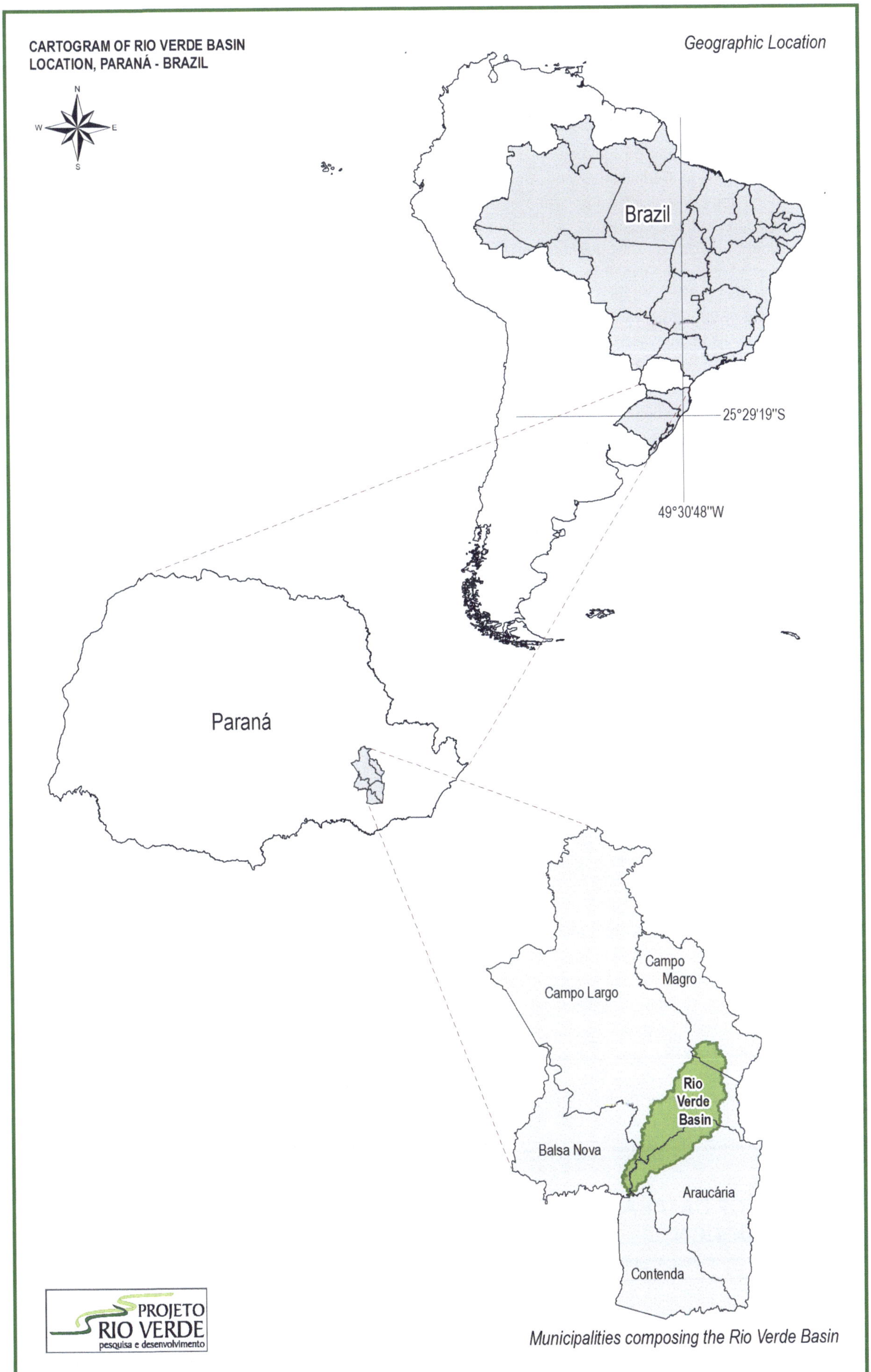

FIGURE 1 – MAP OF THE RIO VERDE BASIN, PARANÁ, BRAZIL

In addition to the cartographic sources mentioned above, bibliographic information was collected from the database available through IBGE, the Paraná Institute for Social Development (Instituto Paranaense de Desenvolvimento Social - IPARDES), COMEC, and SUDERHSA, among other sources.

The GIS software used in the generation of the database included Idrisi Taiga, CartaLinx 1.2 and Arc Gis 9.2.

The methodology was divided into five steps. The first four steps relate to the construction of the database itself and the final corresponds to the development of analytical procedures to assess the dynamics of growth in the basin, as well as the conflicts created between land-use and the appropriation of natural resources.

Step 1 – survey and analysis of cartographic and bibliographic sources and definition of software to be used in the research.

Step 2 – determine the final mapping scale or scales and resolution of layers. Based on the scale and resolution, 21 topographic maps were georeferenced for later integration. The final scale adopted was 1:50,000, using the Universal Transverse Mercator coordinate system and zone 22S. The resolution was set to 20 meters for information layers in raster format.

Step 3 – creation of a georeferenced database, in which the following layers were prepared (hereby defined as basic environmental data): perimeter of the Rio Verde Basin, definition of municipalities in the basin, drainage network, natural springs, contour lines, floodplains, the dam, delineation of sub-basins, historic land-use/land-cover (1976 and 2000), road systems and areas of preservation.

Step 4 – from the basic layers created in Step 3, derived information layers were generated, through spatial operators, intersections and reclassifications, including: hilltops, digital terrain elevation model, declivity, spatio-temporal dynamics, simulation of growth of areas of anthropogenic disturbance (AAD) of the basin, legal restrictions, conflicts between land-use and legal restrictions, geomorphology, environmental risks, emerging and potential issues.

Layers generated during the third and fourth steps allowed various spatial analyses within the Rio Verde Basin. Below we outline four studies, two that directly involve spatial analysis of human occupation in the basin and two that focus on environmental degradation.

Step 5: spatial analysis – in this stage, we analyzed the spatio-temporal dynamics of the basin, simulation of growth in areas of anthropogenic disturbance (AAD) in the basin, conflicts between land-use and legal restrictions, and environmental risks.

4. RESULTS OF DATABASE CONSTRUCTION

The database was used to support several sub-projects of the wider interdisciplinary project on the eutrophication of the Rio Verde Reservoir. We also supported studies that resulted not only in publication but also in graduate and master's degree dissertations, some complete and others under preparation.

Among the completed works, only three were selected to be part of this chapter.

4.1 SPATIO-TEMPORAL DYNAMICS IN THE RIO VERDE BASIN (1976-2000-2009)

4.1.1 Introduction

Because of the so-called rural exodus, the movement of populations from the countryside to the city, urban populations began to increase dramatically. However, this movement into urban areas created problems due to the lack of basic infrastructure required to provide a sufficient quality of life for its citizens (i.e. sanitation, electricity, hospitals, schools, industries and commerce to sustain the workforce). The absence of such infrastructure can lead to several physical, social and environmental issues.

A study by the Environmental Protection Plan and Territorial Reorganization in Watershed Protection Areas (Plano de Proteção Ambiental e Reordenamento Territorial em Áreas de Proteção aos Mananciais - PPART, 2002) in the Metropolitan Region of Curitiba (RMC) identified several problems in the Rio Verde Basin, including: illegal settlements in the central and upper portions of the basin, urban pressure from Campo Largo, lack of sufficient water quality monitoring, and significant impact from the highway transportation systems (BR-277 and Estrada do Cerne). The direct impact of urban expansion is an increase in demand for water that can exceed the projected local availability, which requires restrictions on development and limited occupation around natural springs (SUDERHSA, 2007).

The Rio Verde Basin is supplied sufficiently by rainfall, with a well-documented spatial distribution of the phenomenon. This abundance has enabled cities to grow, largely relying on the water supply from local sources (SUDERHSA, 2007).

Environmental management is normally used to solve problems related to urban development by restricting land-use in order to protect natural resources. As such, it uses environmental planning which prioritizes maintaining the support capacity of the affected ecosystems.

The purpose of this study was to analyze the temporal and spatial dynamics of human occupation in the Rio Verde Basin using GIS. This assessment was conducted by developing layers for the years 1976, 2000 and 2009. As a result, a tool was developed to identify environmental impacts caused by urban and rural growth in this region, as well as enable the planning and reorganization of land-use.

4.1.2 Materials and Methods

The results of the spatio-temporal dynamics of Human Occupation of the Rio Verde Basin (1:50,000) were obtained through a comparison of the land-use/land-cover (LULC), generated from: Topographical Maps from 1976, scale 1:10,000 (COMEC); LULC information from 2000 (SUDERHSA website); and land-cover layers (2009), prepared by Prof. Carlos Roderjan, PhD (Course of Forestry at UFPR) for the interdisciplinary project.

Given that the maps used come from various sources, it was necessary to develop an integrated land-use/land-cover (LULC) classification so that comparisons could be made across the study years (Table 3 and Figures 2, 3 and 4). The

TABLE 3 – integration of LULC classification systems FROM 1976-2000-2009

CLASSIFICATION - 1976		CLASSIFICATION - 2000		CLASSIFICATION - 2009	
Initial	**Reclassified**	**Initial**	**Reclassified**	**Initial**	**Reclassified**
Montana Mixed Ombrophilous Forest – FOMM	Montana Mixed Ombrophilous Forest – FOMM	Natural Tree Vegetation	Montana Mixed Ombrophilous Forest – FOMM	Shrub Initial – Initial Stage of Vegetation Succession – shrub (*capoerinha*)	FOMM Initial Succession Stage
	FOMM Initial Succession Stage	Natural Shrub Vegetation	FOMM Initial Succession Stage	Tree Initial – Initial Stage of Vegetation Succession – trees (*braçatinga*)	FOMM Initial Succession Stage
Culture	Rural Area	Fields	Rural Area	Intermediate – Medium State of Vegetation Succession (*capoeira*)	Montana Mixed Ombrophilous Forest – FOMM
Built Area	Built Area	Reforestation	Reforestation	Advance – Advanced Stage of Natural Succession (undergrowth/Forest)	Montana Mixed Ombrophilous Forest – FOMM
Flood plains in Edaphic Fields	Flood plains	Rivers, lakes, dams	Water bodies	Gallery – Gallery Forest	Montana Mixed Ombrophilous Forest – FOMM
Lakes	Water bodies	Exposed Soil	Rural Area	Floodplain	Flood plains
Rivers	Water bodies	Permanent Culture	Rural Area	Reforestation	Reforestation
BR	Built Areas	Temporary Culture	Rural Area	Agriculture and Livestock	Rural Area
Warehouses	Built Areas	Grange	Rural Area	Urban Area	Built Area
Reforestation	Reforestation	Warehouses/ Silos	Built Area	Exposed Soil	Rural Area
Urban Center	Built Area	Mining/Sand	Mining	Water bodies	Water bodies
		Mining/Others	Mining		
		Industrial Area	Built Area		
		Low Urban Area	Built Area		
		Village	Built Area		
		Division in Lots	Built Area		

software used was *Idrisi Taiga* from Clark University.

In order to obtain more general information about the spatio-temporal dynamics of human occupation in the Rio Verde Basin, we reclassified land-use/land-cover information from 1976, 2000 and 2009. Classifications corresponding to *Montana Mixed Ombrophilous Forest – FOMM*, *FOMM Initial Succession Stage*, *Flood plain* and *Reforestation* were reclassified and included in the category *Vegetation*. The *Mining* category of the 2000 land-use plan was reclassified as *Area of Anthropogenic Disturbance (AAD)*, since it was only mapped in this source and no further instances were recorded. Under the classification of *AAD*, we included the categories *Rural Area* and *Built Area* (Figures 5, 6 and 7).

The next step consisted of a cross-comparison of these maps. Thus, the 1976 LULC map (reclassified into three categories) was compared with the map from 2000, which resulted in a third map called 1976/2000 spatio-temporal dynamics.

In the next stage, we compared the 1976/2000 dynamics of LULC with the 2009 LULC map (reclassified into three categories), which resulted in the final map called 1976/2000/2009 Land-Use/Land-Cover Dynamics.

4.1.3 Results and Discussion

The resulting comparative measures (Table 7) were calculated from the 1976, 2000 and 2009 LULC maps (Figure 2).

In 1976, we can see that the predominant land-use in the basin was rural (44.50%), followed by the FOMM Initial Succession Stage, with 28.85%. This category significantly decreased during the period from 1976-2009 and by 2009, it was only found in 3.46% of the basin. The reforestation areas that reached a high of 5.72 km^2 in 2000 were reduced to 3.6 km^2 in 2009. This decrease can be explained by differences in the classification systems used in the databases for studying the spatio-temporal dynamics. The FOMM increased 17.37% from 1976 to 2000.

The AAD, including the categories of rural and built areas, also showed changes over time with the most significant change between 1976 and 2000, when the AAD grew 6 km^2 or 8.38%. We can see in Figure 2 that this increase mainly occurred in the regions adjacent to highway BR 277, which links Curitiba to Campo Largo. Analyzing the growth of these areas from Figure 2, we can see a increase in urban area to the north of the Rio Verde Basin, in the city of Campo Magro (where the new city hall will be located), indicating a trend towards greater human occupation in this area.

A summary of information related to vegetation cover, human occupation, and water bodies is presented in Table 8. Please note that these features are also represented in Figure 3.

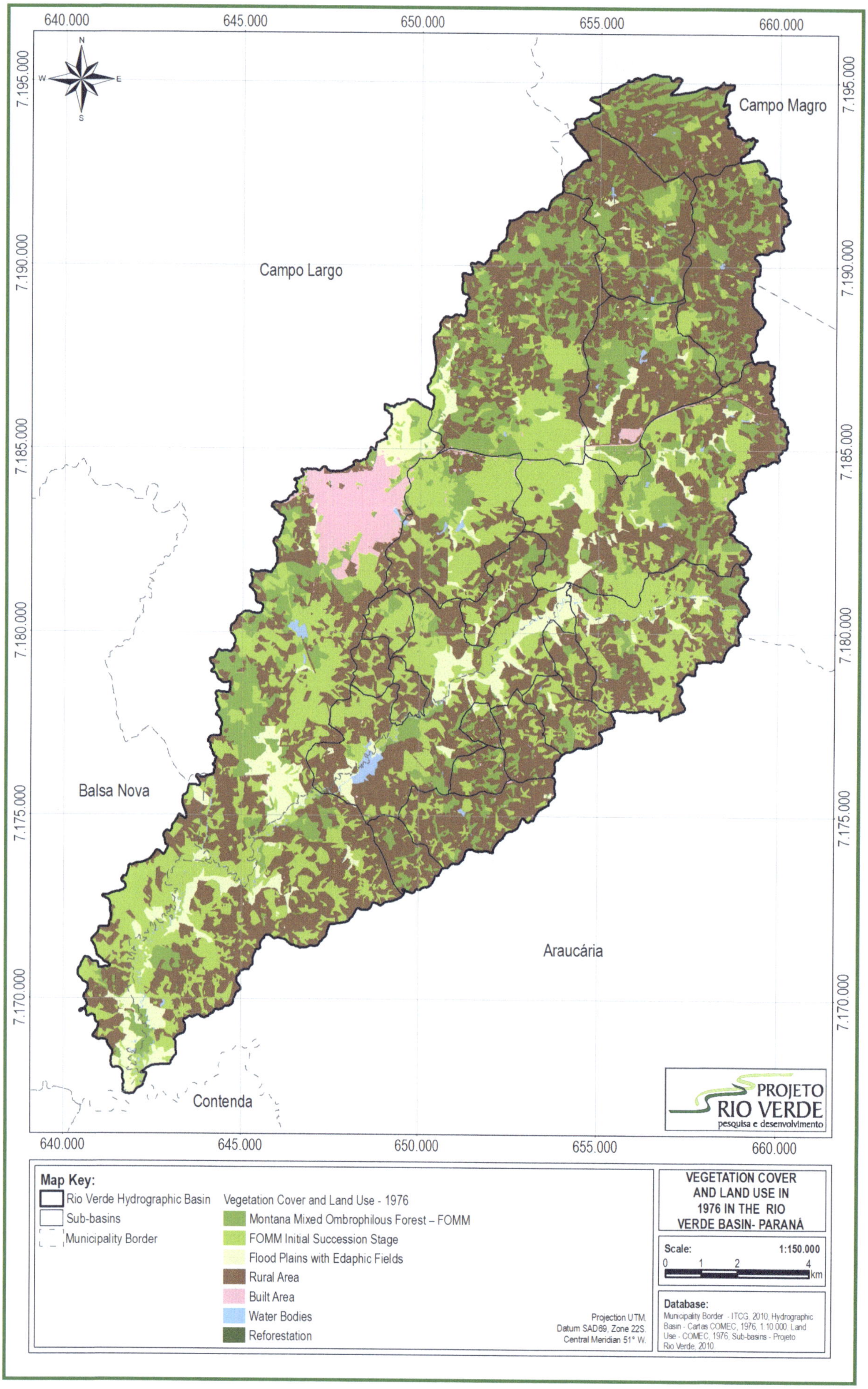

FIGURE 2 – MAP OF LULC IN THE RIO VERDE BASIN (1976)

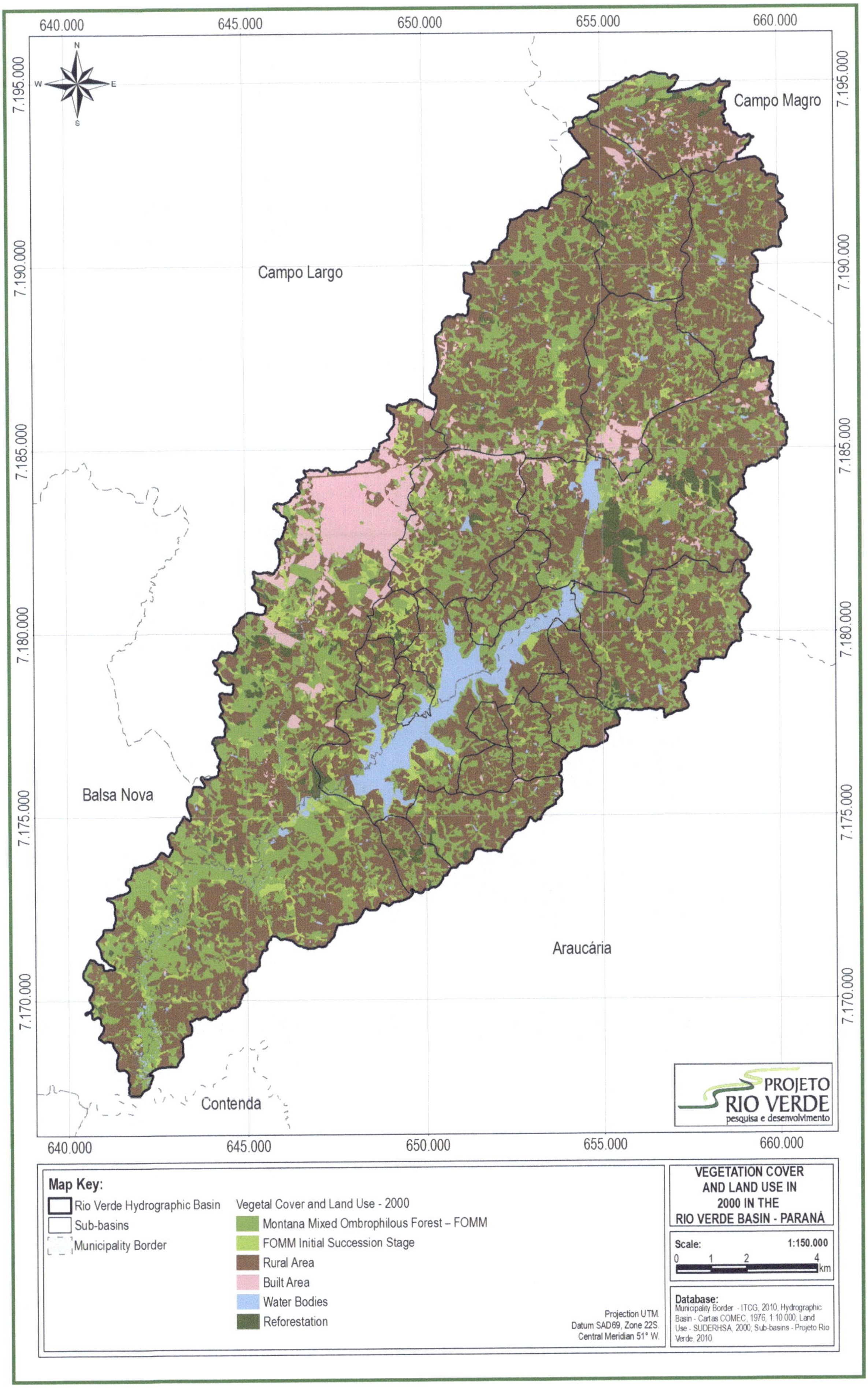

FIGURE 3 – MAP OF LULC IN THE RIO VERDE BASIN (2000)

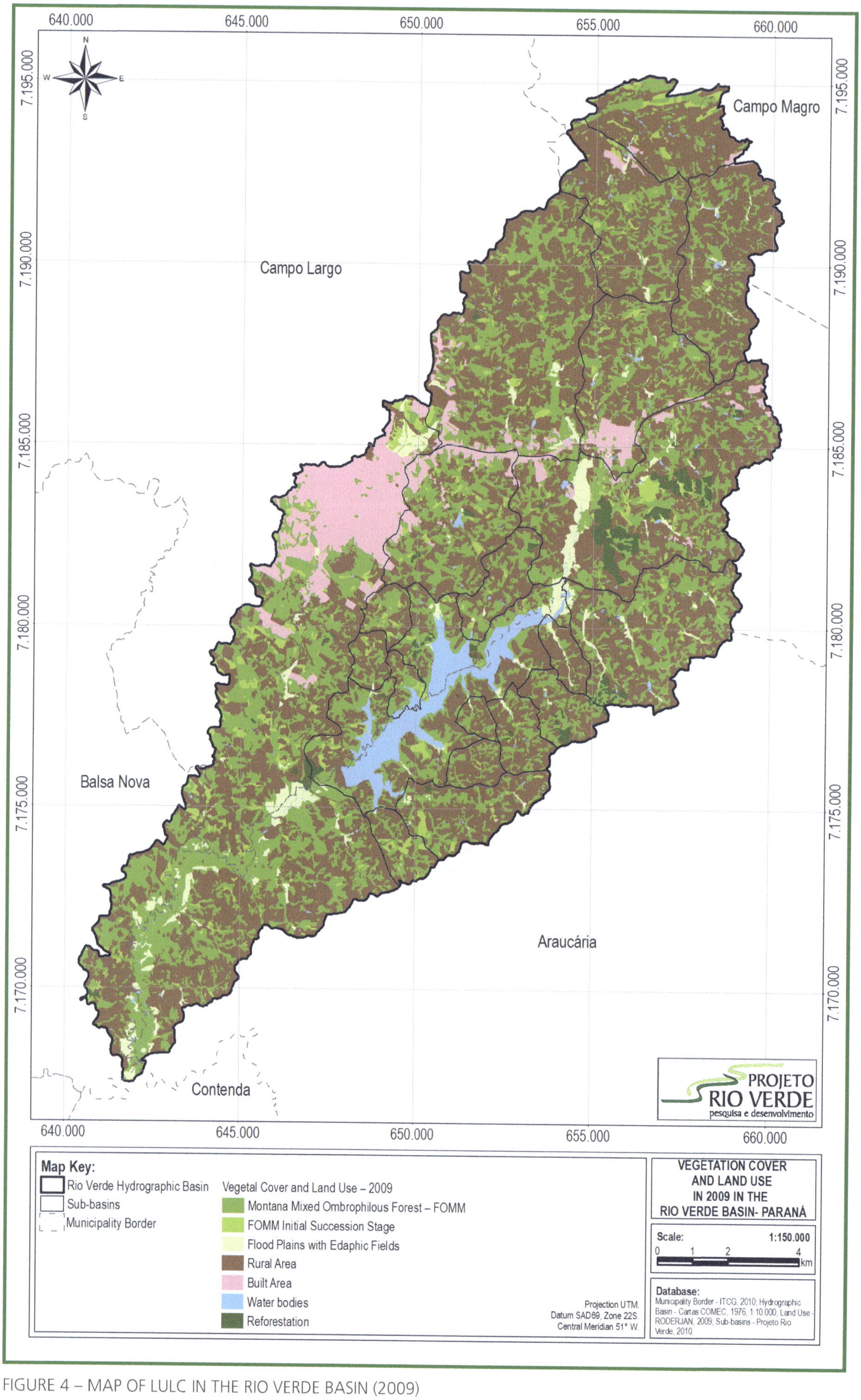

FIGURE 4 – MAP OF LULC IN THE RIO VERDE BASIN (2009)

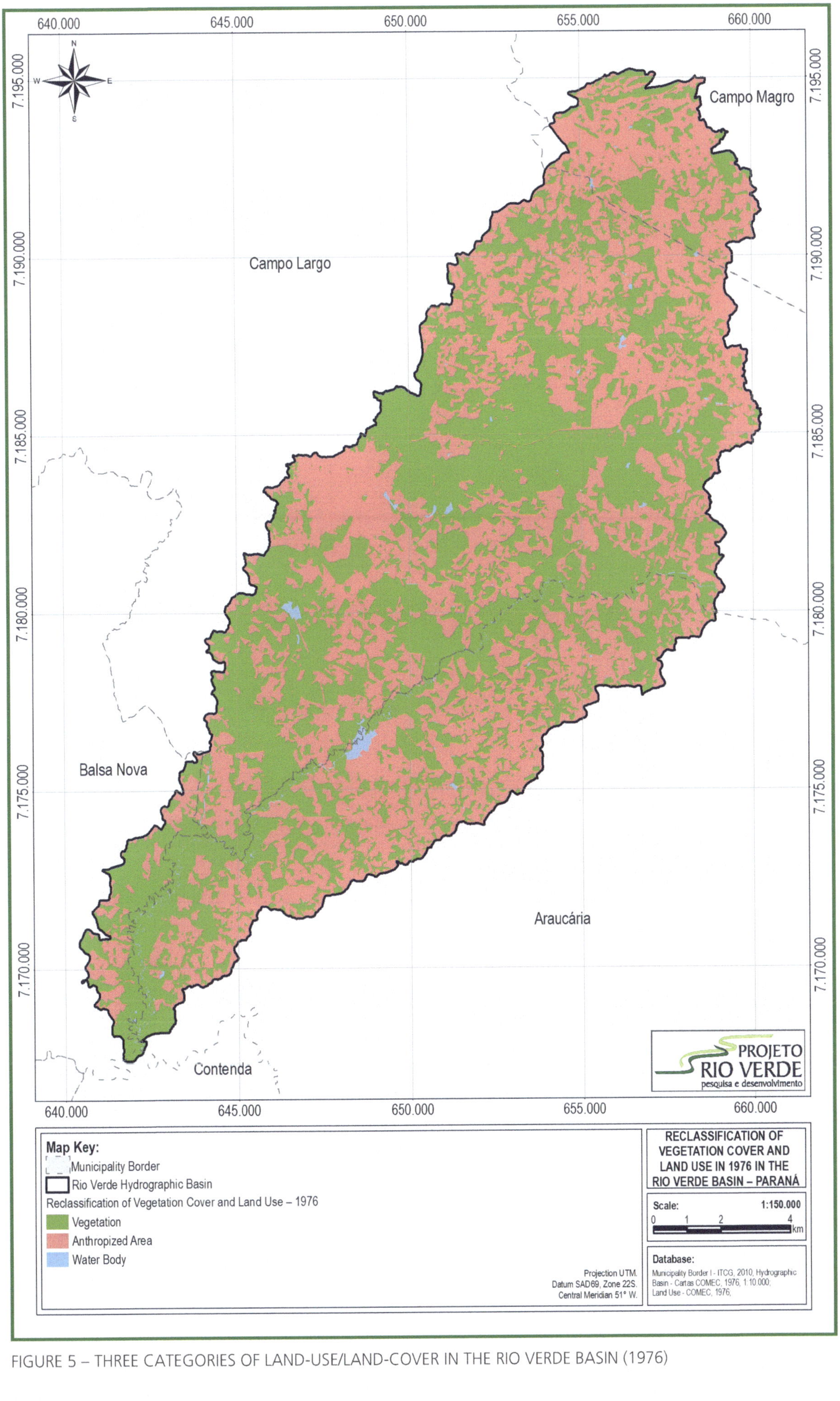

FIGURE 5 – THREE CATEGORIES OF LAND-USE/LAND-COVER IN THE RIO VERDE BASIN (1976)

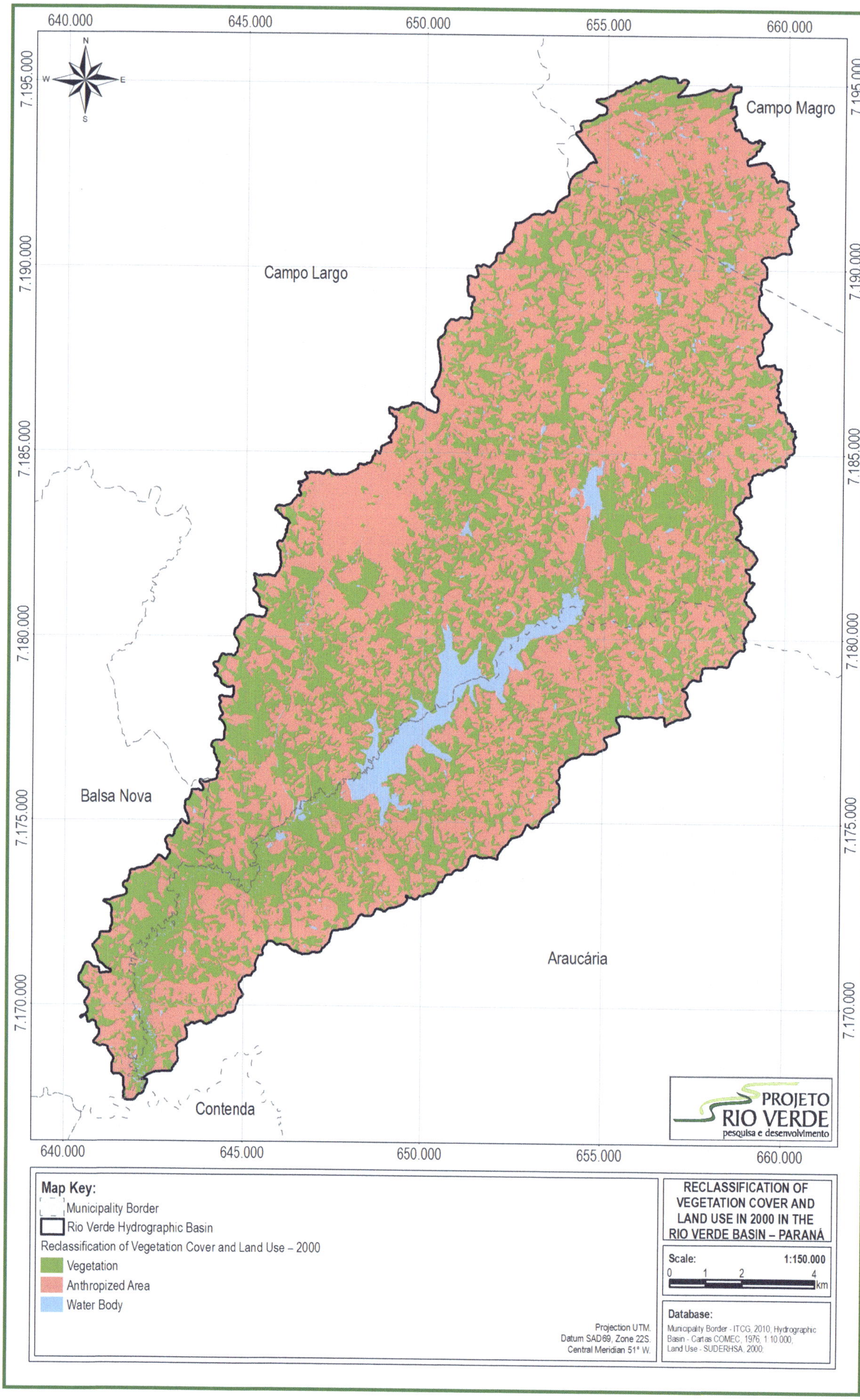

FIGURE 6 – THREE CATEGORIES OF LAND-USE/LAND-COVER IN THE RIO VERDE BASIN (2000)

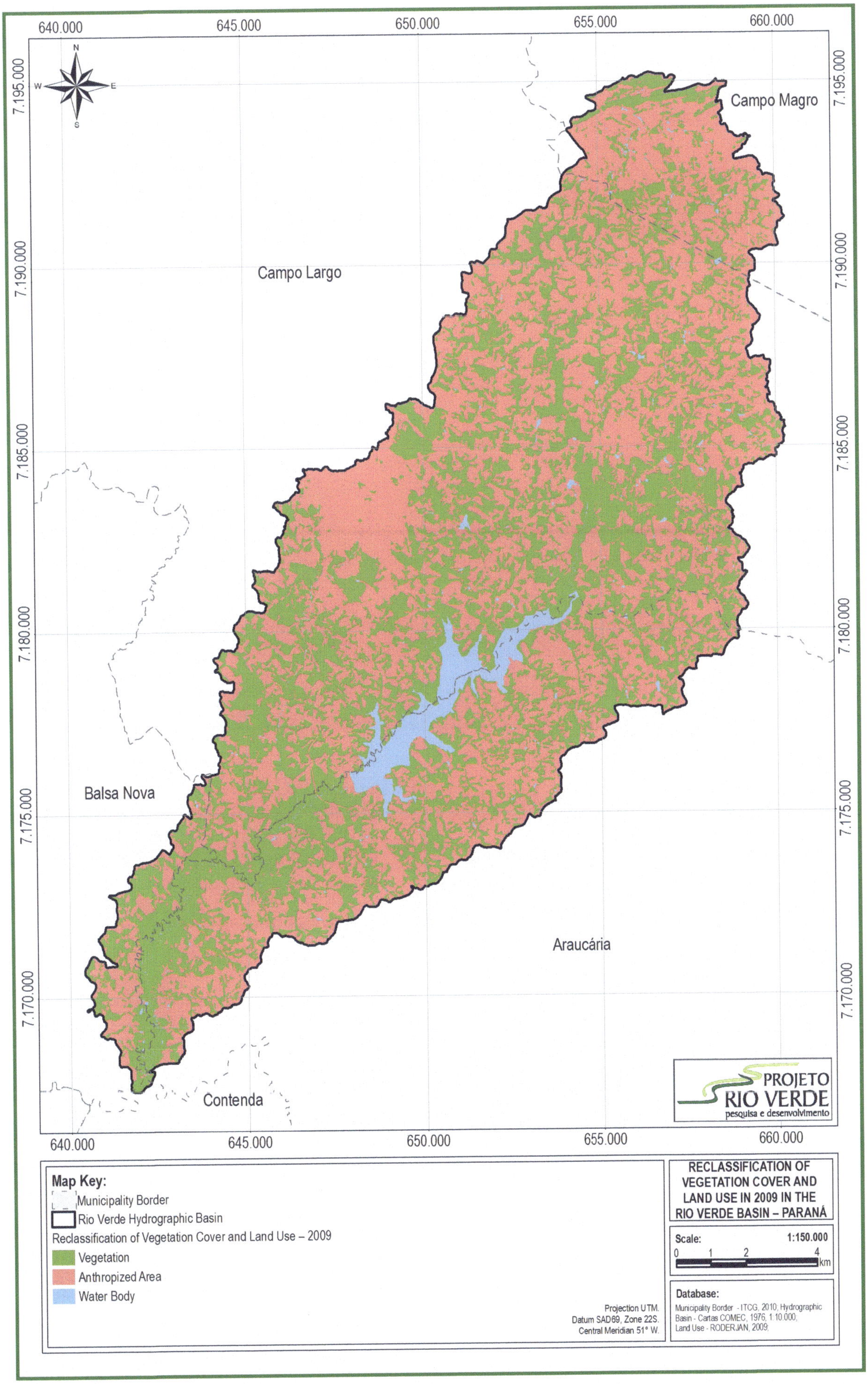

FIGURE 7 – THREE CATEGORIES OF LAND-USE/LAND-COVER IN THE RIO VERDE BASIN (2009)

TABLE 7 – COMPARISON OF LULC FROM 1976-2009

LULC	1976		2000		2009	
	Km²	%	Km²	%	Km²	%
Montana Mixed Ombrophilous Forest – FOMM	39.50	16.50	75.79	31.74	80.88	33.87
FOMM Initial Succession Stage	68.85	28.82	12.88	5.39	8.27	3.46
Flood plains	15.61	6.54	0.0	0.0	6.98	2.92
Rural Area	106.30	44.50	121.90	51.03	116.57	48.82
Built Area	6.98	2.93	13.00	5.45	14.99	6.28
Water bodies	1.69	0.70	9.57	4.00	7.56	3.14
Reforestation	0.02	0.01	5.72	2.39	3.61	1.51
TOTAL	238.86	100	238.86	100	238.86	100

TABLE 8 – AREAS WITH VEGETATION, AAD AND WATER BODIES (1976, 2000 AND 2009)

LULC	1976 AREA (km²)	%	2000 AREA (km²)	%	2009 AREA (km²)	%
Vegetation	123.89	51.87	94.39	39.52	99.74	41.76
AAD	113.28	47.43	134.90	56.48	131.56	55.10
Water bodies	1.69	0.70	9.57	4.00	7.56	3.14
TOTAL	238.86	100	238.86	100	238.86	100

In 1976, the areas with vegetation accounted for the majority of the basin land-cover (51.87%) and was concentrated mostly in its central area (east of the urban center of Campo Largo) and at the discharge point of the basin.

In 1973, Petrobras' Getúlio Vargas Refinery was created and later a dam was built to capture water from the Rio Verde River with an area of approximately 8 km². Comparing maps from before and after the construction of the dam, it is easy to see the water coverage area. Between 1976 and 2000, there was therefore an increase in the category of *Water bodies* to approximately 7.9 km².

We also noted that from 1976 to 2000 the area of anthropogenic disturbance grew 21.62 km²: in 1976, its percentage was 47.43% and in 2000, 56.48%. It is important to note that the central region of the basin, which used to have extensive vegetation cover, was almost completely disturbed by human occupation and only fragments of the vegetation cover are left. A significant amount of vegetation cover remains in the discharge area of the basin.

For the period 2000-2009, we observed that the AAD were reduced by 3.34 km². This can be explained by the fact that in 2000 flood plain areas were classified as *Rural Areas (Areas of Anthropogenic Disturbance*) and changed to *Vegetation* (*Flood Plain Areas*) in 2009. However, we cannot rule out the possibility of the regeneration of vegetation cover.

4.1.4 Final Comments

Through this study, it is possible to see the changes in land-use/land-cover over time. It is noteworthy that the areas with vegetation increased during the period from 2000-2009 and it now exceeds AAD (rural and urban).

It should be noted that the results presented herein show some variation due to differences in the cartographic databases used: i.e. topographic maps from 1976 (COMEC), 2000 (SUDERHSA) and 2009, and the Interdisciplinary Research Project on the Eutrophication of the Rio Verde Reservoir. Each of the cartographic sources used in this study was prepared for different purposes, which presented challenges for the integration of LULC classifications. Despite these challenges, we were able to produce a final result that represents well the current conditions in the Rio Verde Basin. In future, we recommend that cartographic databases be developed by the same team and with the same goals so that classification systems, as well as scale, are consistent across the entire process.

4.2 SIMULATING GROWTH OF AREAS OF ANTHROPOGENIC DISTURBANCE IN THE RIO VERDE BASIN USING MARKOV CHAIN AND CELLULAR AUTOMATA IN A GIS ENVIRONMENT (1976-2000-2009)

4.2.1 Introduction

The main advantage of using models is the ability to quickly study many different scenarios, many of which cannot be tested in actual experiments. Another important advantage of using simulation is their low cost. The greatest challenge in using models is working with the large amounts of data that describe the heterogeneity of natural systems. For these reasons, GIS is used to develop the database required to prepare and run models (MACHADO *et al.*, 2003).

GIS tools are able to acquire, store, process, integrate, retrieve, transform, manipulate, model, update, analyze and display digital georeferenced information, which is topologically structured and represents the relationships between entities, associated or not, in an alphanumeric database (ROCHA, 2000).

These systems allow for the manipulation of data from various sources such as maps, images and historic records, enabling the retrieval and combination of information and several different kinds of data analyses (ALVES, 1990). Therefore, GIS can be defined as a system that supports the automated processing of spatially referenced data.

In GIS, spatial data can divide large heterogeneous areas into small homogeneous units, which can then be used in modeling programs (MIT, 1994). Using GIS, some models and simulations have been developed based on spatial analysis including the Markov Chain Analysis and Cellular Automata.

According to the IDRISI Manual (2009), a Markov Chain process is one in which the status of a system at time n+1 can be found based on the status at time n, given a matrix of transition probabilities for each class of Land Use/ Land Cover. A Markov Chain Analysis is simply a model in which the future status of a system can be simulated based only on the immediately preceding status.

As for Cellular Automata (CA), it can be described as a grid of cells in which the status of each cell depends on the status of the neighboring cells. The grid is uniform and regular with a discrete variable and discrete time in each cell, with the value of the variable in each cell affected by the values in the neighboring cells (WOLFRAN, 1983). Thus, the Cellular Automata can be understood as a spatio-temporal dynamic and a relatively simple system in which the status of each cell depends on the previous status of the cells within a certain neighborhood, based on a set of transition rules (ROCHA, 2004).

According to Almeida *et al.* (2007), "it is wise to say that this type of model is also one of the best techniques currently available to meet the needs and interests of research on the dynamics of Vegetation Cover and urban and regional Land Use."

In using the Markov Chain Analysis and Cellular Automata together, the Markov chain is used as the transition matrix and a rule for changing the status of cells in Cellular Automata. Thus, the CA_MARKOV model combines the concept of fluctuating Cellular Automata cells based on the probability of the Markov chain (IDRISI, 2009).

The result of the Markov Chain and Cellular Automata process is that the simulated changes in land cover will occur based on the characteristics of change in the previous time period, the likelihood of change in LULC, and resemblance to neighboring regions resulting from spatial dependence of LULC in the immediate vicinity.

4.2.2 Materials and Methods

Analyses were performed with the Software IDRISI 32.2, with modules MARKOV and CA_MARKOV. We used the coordinate system "Universal Transverse Mercator" (UTM), Datum SAD 1969, Zone 22 S.

Analyses presented herein were performed using simplified maps of the Land-Use/Land-Cover for the years 1976, 2000 and 2009 (Figures 8, 9 and 10). Maps were simplified to only three categories of land-use: vegetation areas, areas of anthropogenic disturbance (AAD), and water. The areas of native vegetation, forest succession and reforestation were included in vegetation areas; AAD include urban, industrial and agricultural areas; and rivers, water bodies, and the reservoir are grouped as water areas.

Table 11 provides the total area in hectares and percentage in relation to the total area of the basin for each land-use category in each year based on the maps shown above.

4.2.3 Results

Table 12 shows the transition probability matrix resulting from the Markov chain, using the LULC maps for the years 1976 and 2000.

This matrix represents the probability that a particular area will become another. For instance, for areas with no anthropogenic disturbance (vegetation areas), the probability that they will continue as vegetation areas is 71.51%. The probability that these areas become AAD is 25.05% and the probability that they become areas covered with water is 3.44%.

From the matrix of transition probabilities and using the 2000 LULC map as a basis, we modeled LULC for year 2009 using the Markov Chain and Cellular Automata. The results obtained are shown in Figure 11.

Table 13 shows the simulated 2009 values for each land-use area in hectares and percentage of total basin area, as well as the actual values found for the existing 2009 LULC map.

There is a similarity between the simulated results and the actual data. The absolute value of each area in hectares and percentages of the total basin area were remarkably close; however, they are not identical. This variation in simulated results, and even in the total basin area, is due to differences in digitalization, in data used in the preparation of maps, and other methodological issues, as explained above.

For a more detailed analysis of the results, we used the CROSSTAB tool, which is a cross comparison, to check for changes between the year 2000 and the forecasted results for 2009. Using the data from 2000 and 2009, CROSSTAB calculated the area in hectares and percentage of total basin area. The results are shown in Table 14.

TABLE 11 – TOTAL AREA (IN HECTARES AND PERCENT OF BASIN) FOR EACH LULC IN EACH YEAR

	1976		2000		2009	
	AREA		AREA		AREA	
Category	(ha)	(%)	(ha)	(%)	(ha)	(%)
Vegetation	12,388.69	51.87	9,439.31	39.50	9,940.92	41.81
AAD	11,328.29	47.43	13,497.91	56.49	13,077.30	55.01
Water	169.27	0.71	957.14	4.01	756.07	3.18
TOTAL	23,886.25	100.00	23,894.35	100.00	23,774.29	100.00

SOURCE: Figures 8, 9 and 10

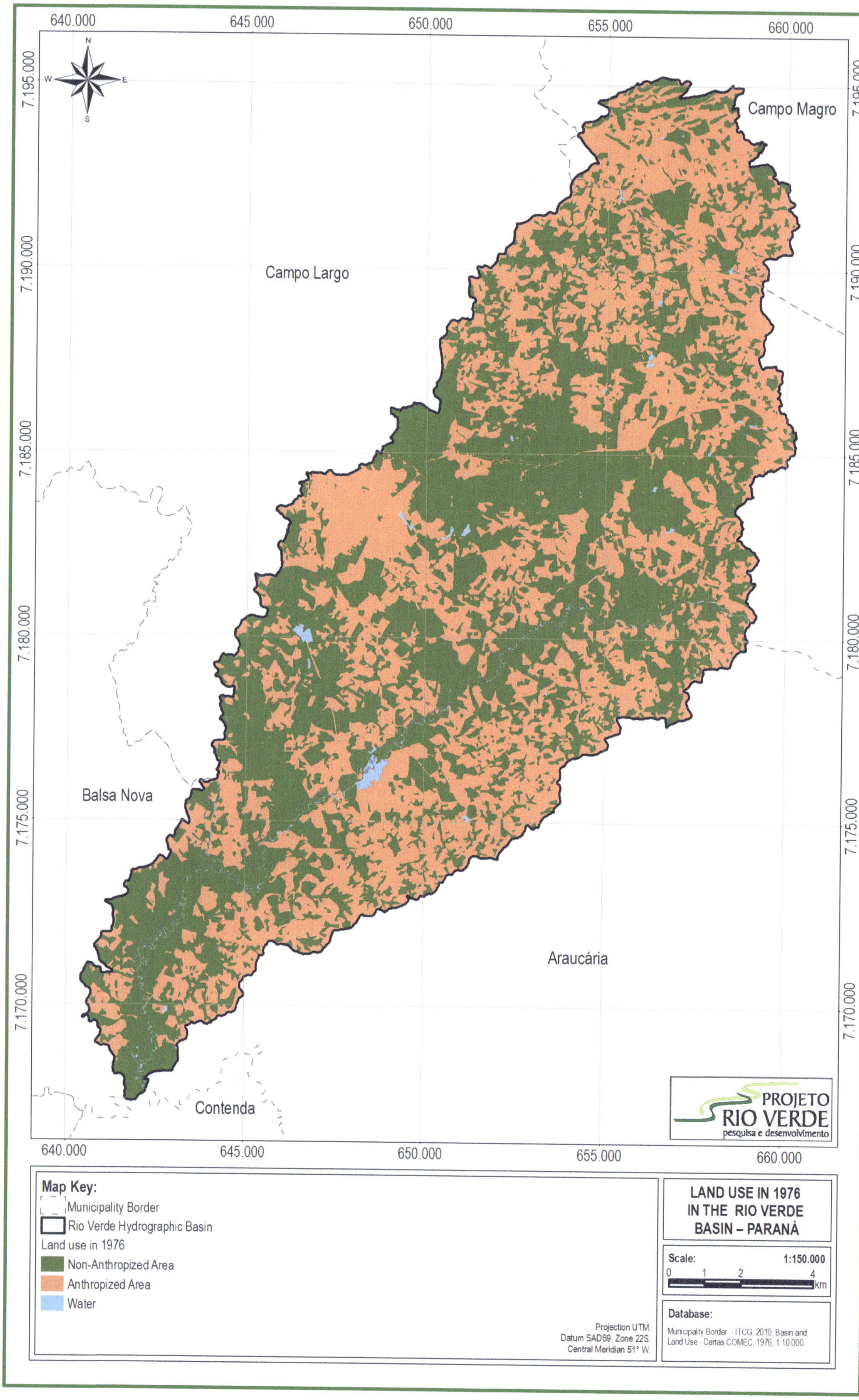

FIGURE 8 – THREE CATEGORIES OF LAND-USE/LAND-COVER IN THE RIO VERDE BASIN (1976)

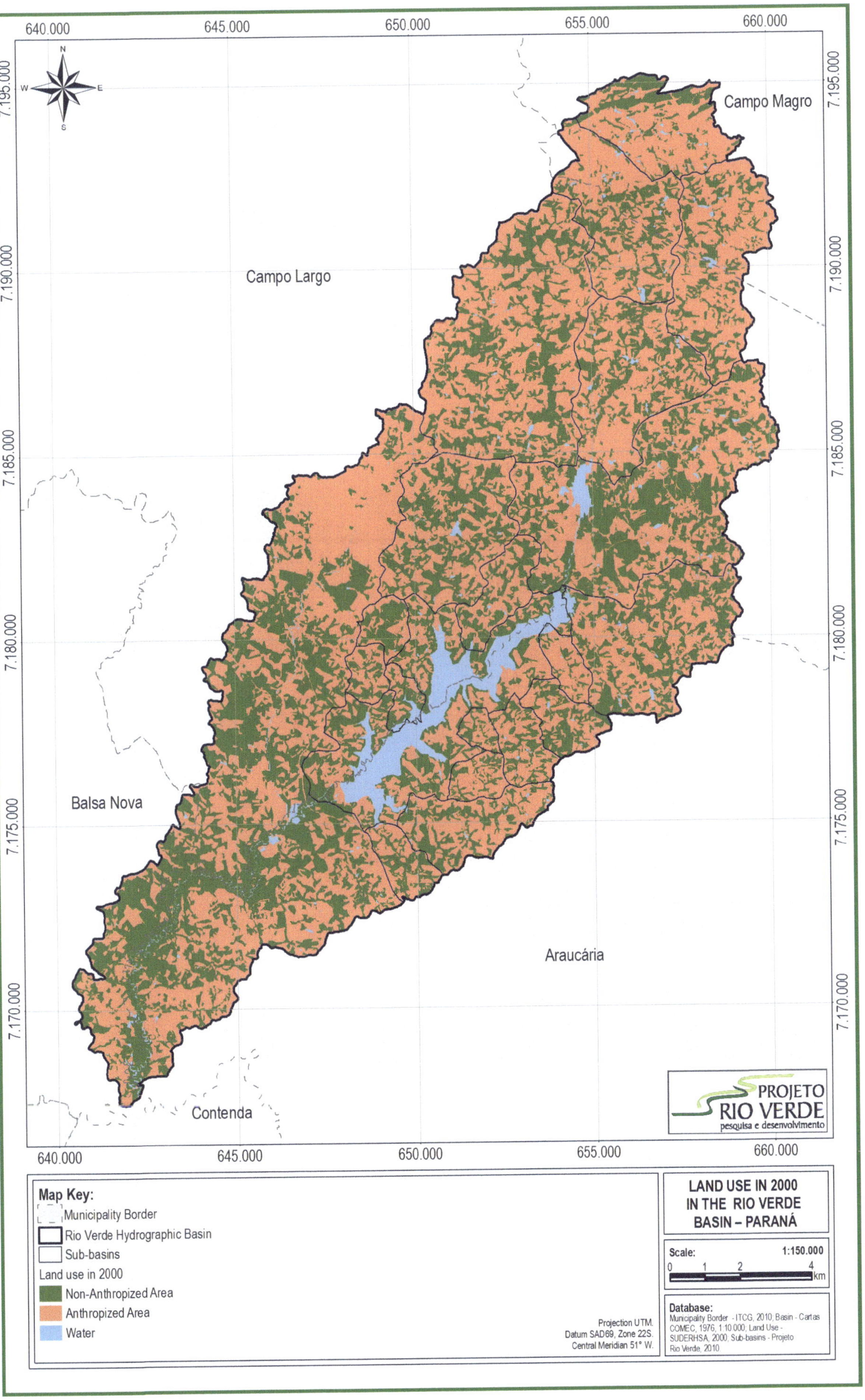

FIGURE 9 – THREE CATEGORIES OF LAND-USE/LAND-COVER IN THE RIO VERDE BASIN (2000)

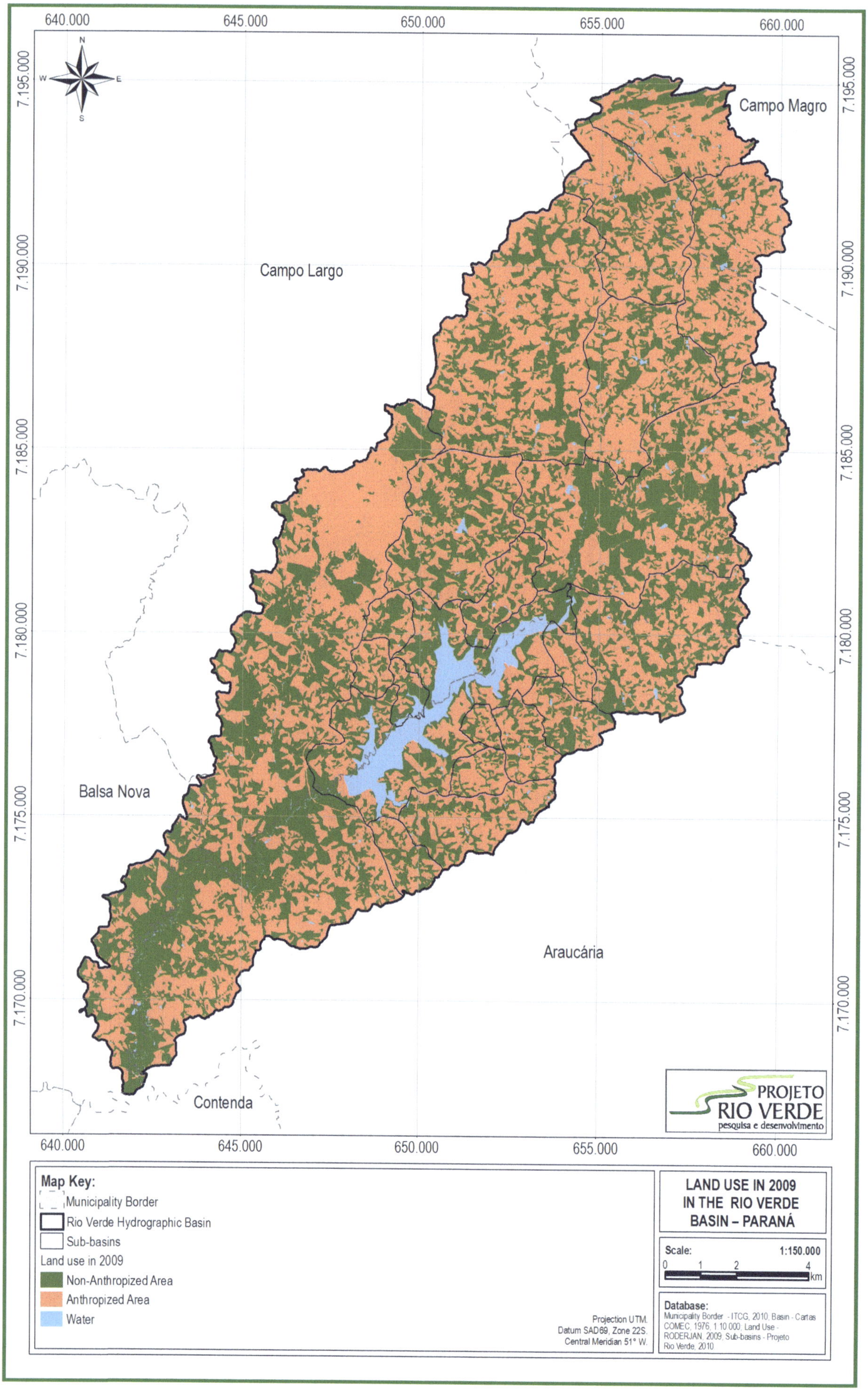

FIGURE 10 – THREE CATEGORIES OF LAND-USE/LAND-COVER IN THE RIO VERDE BASIN (2009)

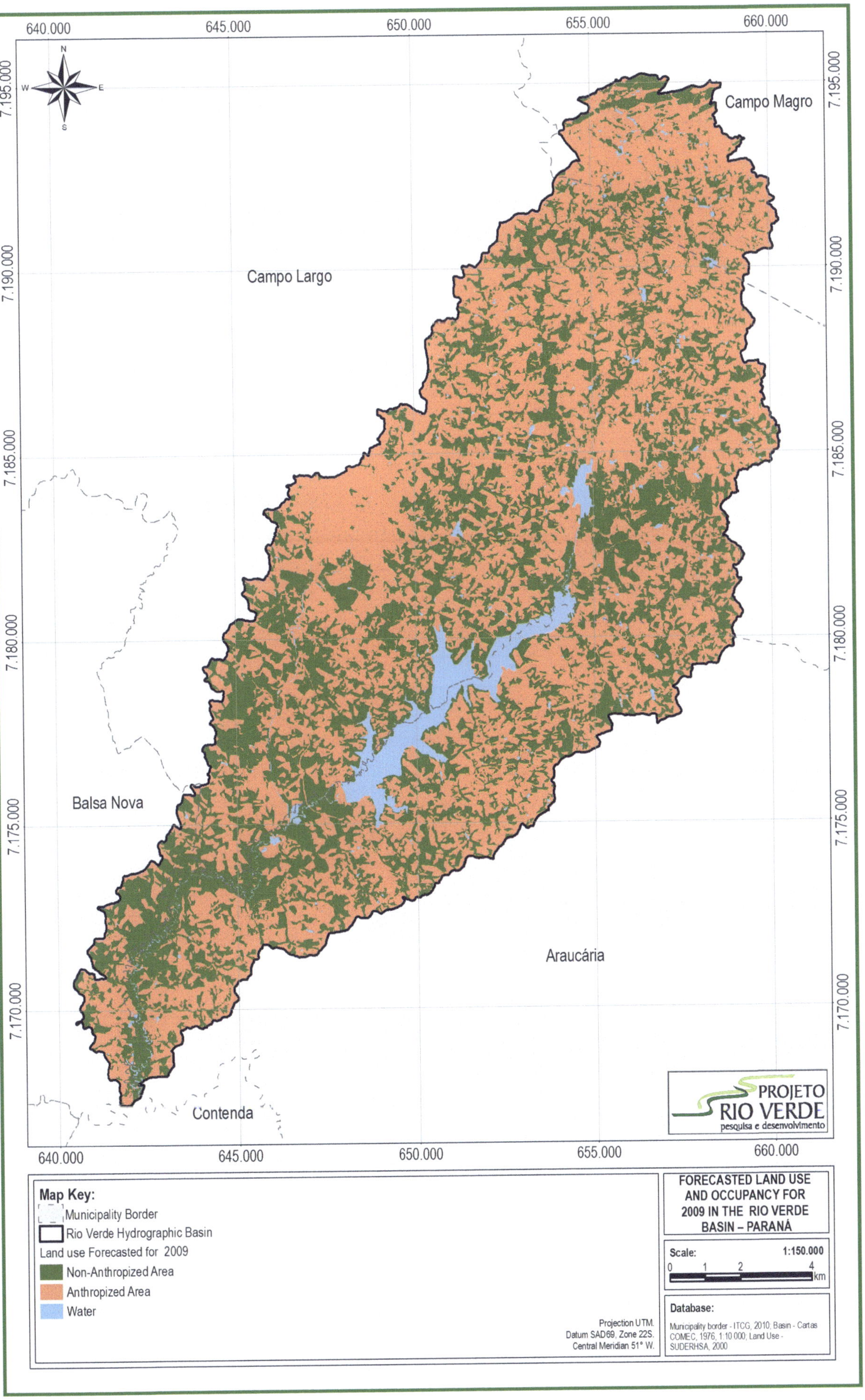

FIGURE 11 – RESULTS OF SIMULATED LULC FOR 2009

TABLE 12 – TRANSITION PROBABILITY MATRIX FOR 3 CATEGORIES OF LULC

	VEGETATION	AAD	WATER
Vegetation	0.7151	0.2505	0.0344
AAD	0.1395	0.8546	0.0060
Water	0.1369	0.0624	0.8007

SOURCE: Results of the Markov Chain analysis (Idrisi Taiga)

TABLE 13 – COMPARISON OF SIMULATED GROWTH OF AAD IN THE RIO VERDE BASIN FOR 2009 USING MARKOV CHAIN AND CELLULAR AUTOMATA AND THE ACTUAL 2009 LULC MAP

	2009 FORECAST		2009	
	AREA		AREA	
Category	(ha)	(%)	(ha)	(%)
Vegetation	9,380.974	39.26	9,940.92	41.81
AAD	13,554.86	56.73	13,077.30	55.01
Water	958.11	4.01	756.07	3.18
TOTAL	23,893.95	100.00	23,774.29	100.00

SOURCE: Results of the CA-Markov Chain analysis (Idrisi Taiga) and Figure 10

TABLE 14 – CHANGES IN LAND USE BETWEEN 2000 AND SIMULATED 2009 AND BETWEEN SIMULATED AND ACTUAL 2009 LULC MAP

	2000 X 2009 SIMULATED		2009 SIMULATED X 2009	
	AREA		AREA	
Category	(ha)	(%)	(ha)	(%)
Remained undisturbed vegetation	9,323.22	39.02	8,182.92	34.42
AAD x Vegetation	56.29	0.24	1,591.72	6.70
Water x Vegetation	1.47	0.01	166.28	0.70
Vegetation x AAD	114.00	0.48	1,175.52	4.94
Remained AAD	13,437.12	56.24	11,862.59	49.90
Water x AAD	3.74	0.02	39.19	0.16
Vegetation x Water	2,09	0.01	0.98	0.00
AAD x Water	4.50	0.02	2.67	0.01
Remained Water	951.52	3.98	752.42	3.16
TOTAL	23,893.95	100.00	23,774.29	100.00

One of the first observations that can be made, based on Table 14, is the change in each land use type from 2000 to that simulated for 2009. For example, only 0.48% of the total basin area transitioned from vegetation in 2000 to AAD in the 2009 simulation and the AAD identified in 2000 remained AAD in the 2009 forecast.

To check the validity of the results, we also compared the simulated 2009 restuls with the existing 2009 LULC map (Table 14). In this comparison, we expected the highest percentage values for areas that remain within the same category of LULC because the more similar the values between land uses, the lesser the variation between the simulation and the existing 2009 map.

In fact, we can see that the highest values obtained are from AAD that continue as disturbed areas, vegetation areas that remained undisturbed, and areas with water that remain water areas. However, the model predicted that 6.70% of the total area of the basin would suffer anthropogenic disturbance, when in fact these areas remain undisturbed in the actual 2009 LULC.

Despite this difference, the simulated values are similar to those expected, since the forecast changes are comparable to those that actually took place. The forecast changes occurred in areas where there actually was some variation in 2009, but without losing the connection with the 2000 LULC map on which the forecast was based.

4.2.4 Final Comments

The analyses presented herein were relatively simple compared to the capability of the parameters used. This simplicity was chosen in order to better understand the methodology and because such approaches are relatively novel in the region. Thus, we recommend that in future studies, analyses be made with different and possibly more complex parameters, in order to challenge these models and verify their simulations. We also suggest that, whenever possible, similar cartographic maps are used to ensure greater reliability of data and results.

Despite the differences in cartographic databases, the results obtained using Markov Chain and Cellular Automata were very consistent. In the simulations, the areas where certain land uses changed are very similar to the changes that actually took place when we compare the 2000 and 2009 LULC maps.

We can see that these models are an important tool in predicting changes in land use and human occupation.

4.3 IDENTIFYING CONFLICTS BETWEEN LAND-USE AND LEGAL RESTRICTIONS IN THE RIO VERDE BASIN

4.3.1 Introduction

The Rio Verde Basin is one of the main sources of water supply in the RMC. Around its waterways, we found unplanned and illegal land occupation, resulting in negative impacts on river environments and their surrounding areas. In light of this occupation and impact on the waterways, the purpose of this study was to identify and quantify the conflicts between land-use and legal restrictions in the Rio Verde Basin, using remote sensing and digital mapping tools combined with a geographic information system.

4.3.2 Materials and Methods

The materials used were the cartographic database on Land Use available on the website of SUDEHRSA and the topographic maps of COMEC, scale 1:10,000 (1976).

This analysis was divided into three basic steps.

First step – the proposed methodology allows for the identification and manipulation of causal associations between environmental variables, territorial dimensions and land-use conflicts, using the current legal requirements as a parameter. Inconsistencies between legislation and actual land-use patterns were found across the region; in the monitoring phase, changes over time are considered in relation to control over development and prospective development (CANEPARO & PASSOS, 2006).

Initially, an Environmental Survey was conducted and included the diagnoses of environmental situations that required further investigation in relation to land-use.

This stage corresponds to the creation of a geocoded database from the synthesized inventory developed for the LULC discussed above, which included basic environmental data from which the LULC layers were derived, and through spatial operators and reclassifications used in the GIS analysis.

Second step – layers relating to the perimeter of the basin, drainage network, springs, contour lines, terrain elevation and declivity model, were developed through manipulation and analysis (spatial operators), resulting in an information layer entitled Areas with Legal Restrictions.

The drainage and spring network was condensed into a single layer with buffer zones based on Article 2 of the Forest Code (item a, paragraph 1 and items b and c). Declivity was added to this intermediate layer, which was obtained through the contour line layer and terrain elevation model. The declivity layer was reclassified according to Article 2 of the Forest Code, item e. The other elements that were part of this layer were generated from the legislation included in Table 15.

TABLE 15 – COMPILATION OF ENVIRONMENTAL LAWS USED IN THE DEVELOPMENT OF THE MAP 'AREAS WITH LEGAL RESTRICTIONS'

AREAS OF PERMANENT PRESERVATION (APP)	LEGISLATION
River APPs – 30 m	1. Federal Law no. 4.771, dated 09/15/1965 – Art. 2, item "a", paragraph 1. 2. CONAMA Resolution no. 303, dated 03/20/2002, Art. 3°, Paragraph III, item "a".
Spring APPs – 50 m	1. Federal Law no. 4.771, dated 09/15/1965 – Art. 2, item "c". 2. CONAMA Resolution no. 303, dated 03/20/2002, Art. 3, Paragraph II.
Lake APPs – 30 m for those located in consolidated urban areas and 100 m for those in rural areas, except for water bodies with up to 20 ha of surface, which require APPs of 50 m	1. Federal Law no. 4.771, dated 09/15/1965 – Art. 2, item "b". 2. CONAMA Resolution no. 303, dated 03/20/2002, Art. 3°, Paragraph III, item "b". 3. IBAMA/SEMA/IAP Joint Resolution no. 05, dated 03/28/2008, Articles 3 and 5.
Dam APPs – 100 m	1. Federal Law no. 4.771, dated 09/15/1965 – Art. 2, item "b". 2. CONAMA Resolution no. 303, dated 03/20/2002, Art. 3°, Paragraph III, item "b". 3. IBAMA/SEMA/IAP Joint Resolution no. 05, dated 03/28/2008, Articles 3 and 5.
Hillside APPs – hillsides with declivity is higher than 45°.	1. Federal Law no. 4.771, dated 09/15/1965 – Art. 2, item "e". 2. CONAMA Resolution no. 303, dated 03/20/2002, Art. 3, Paragraph VII.
Flood plain APPs – 50 m	1. CONAMA Resolution no. 303, dated 03/20/2002, Art. 3, Paragraph IV. 3. IBAMA/SEMA/IAP Joint Resolution no. 05, dated 03/28/2008, Articles 3 and 5.
Hilltop APPs	1. Federal Law no. 4.771, dated 09/15/1965 – Art. 2, item "d". 2. CONAMA Resolution no. 303, dated 03/20/2002, Art. 3, Paragraph VI.

The layer called Land-Use/Land Cover was generated from shapes available on SUDERHSA's website (http://www.suderhsa.pr.gov.br/modules/conteudo/conteudo.php?conteudo=92). Information from SUDERHSA is based on data from the year 2000 and was prepared at a scale of 1:20,000. This layer was reclassified into three categories, areas of anthropogenic disturbance (AAD), areas with vegetation cover and water bodies, in order to synthesize the information.

Third step - through a cross-comparison and reclassification between the LULC information from 2000 and the Areas with Legal Restrictions, conflicts between land-use and the requirements of current legislation were identified. The map Conflicts between Land Use and Legal Requirements on the Rio Verde Basin was generated and compared to field data.

4.3.3 Results

Based on the LULC map (Figure 12), it was possible to determine the AAD, the areas of vegetation cover and the water bodies in the Rio Verde Basin in 2000. Table 16 shows these areas and the percentage of the total basin.

TABLE 16 – LAND USE/LAND COVER (2000)

CATEGORIES	AREA (km²)	% (IN RELATION TO TOTAL BASIN AREA)
Vegetation	125.94	52.71
AAD	86.89	36.36
Water bodies	26.12	10.93
TOTAL	238.95	100.00

According to Table 16 and Figure 12, we can see that the largest portion of the basin under study corresponds to vegetation cover, with 52.71% or 125.94 km^2, which is greater than the AAD, comprising 36.36% of the basin or 86.89 km^2. In Figure 9, we can see that the urbanized areas are found in the western portion of the basin, whereas the rural areas are distributed throughout the entire basin.

Figure 13 shows the areas with legal restrictions in the Rio Verde Basin generated from information provided in Table 16.

In Table 17, we can see the calculated area with legal restrictions and their respective percentages based on the categories used in Figure 10. Of the total 238.95 km^2 within the basin, 36.58% or 87.37 km^2 should be preserved. Thus, various land-use activities are permitted on 151.58 km^2; the Rio Verde Basin thus has a potential land-use equivalent of 63.42% of its total area.

TABLE 17 – AREAS WITH LEGAL RESTRICTIONS

TYPE OF RESTRICTION	AREA (km²)	% (IN RELATION TO TOTAL BASIN AREA)
River APPs (30 m)	30.59	12.80
Spring APPs (50 m)	15.69	6.57
Lake APPS (50 m – rural area)	5.40	2.26
Dam APP (100 m)	2.58	1.08
45 Degree APPs (100%)	0.01	0.00
Flood plain APPs (50 m)	5.21	2.18
Hilltops	0.45	0.19
Flood plains	5.68	2.38
Dam	6.13	2.57
Lakes	0.82	0.34
Rivers	14.05	5.88
Springs	0.76	0.32
Permitted Use	151.58	63.42
TOTAL	238.95	100.00

The map showing Conflicts between Land-Use and Legal Restrictions was generated from a cross-comparison between the two maps discussed above (LULC - 2000 and Legal Restrictions), and resulted *a priori* in 40 categories which were analyzed, grouped and reclassified into nine categories (Figure 14 and Table 18).

Table 18 shows that 196.04 km^2 (82.04%) of the basin area does not present conflicts. On the other hand, areas that should be preserved according to current legislation but are not relates to 16.79 km^2 or 7.03% of the total basin. Although this value seems insignificant, when we observe each type of conflict separately, we note that with a total area of 12.79 km^2 the particular degradation of riverbanks and springs is worrying. These areas, shown in Figure 14, are riparian areas, located along most first-order rivers as well as segments of main streams, that are in jeopardy or have completely disappeared. As for the 1,736 springs mapped during our analysis, we found that 1,389 (80.01%) throughout the basin area were deforested and occupied mainly by agricultural activities.

TABLE 18 – CONFLICTS BETWEEN LAND USE AND LEGAL RESTRICTIONS

TYPES OF AREAS	AREA (km²)	PERCENTAGE IN RELATION TO TOTAL BASIN AREA
AAD in River APPs (30 m)	6.80	2.85
AAD in Spring APPs (50 m)	5.99	2.51
AAD in Lake APPs (50 m)	1.42	0.59
AAD in Dam APPs (100 m)	0.39	0.16
AAD in APPs above 45°	0.01	0.00
AAD in Flood Plain APPs	2.07	0.87
AAD in Hilltop APPs	0.11	0.05
Areas without conflicts	196.04	82.04
Water bodies	26.12	10.93
Total basin area	238.95	100%

4.3.4 Final Comments

Our results show that, in general, the Rio Verde Basin is well preserved (82%); however, the analysis shows that the most affected areas are the springs (80%). The most troubling factor is that deforestation of springs is found throughout the basin and is mainly the result of agricultural activity.

Studies of this kind enable decision making by public officials in order to adopt public policies that are consistent both with the preservation and conservation of the basin and with the lessons learned from sustainable development. Such policies also help to maintain quality of life of the basin population in general.

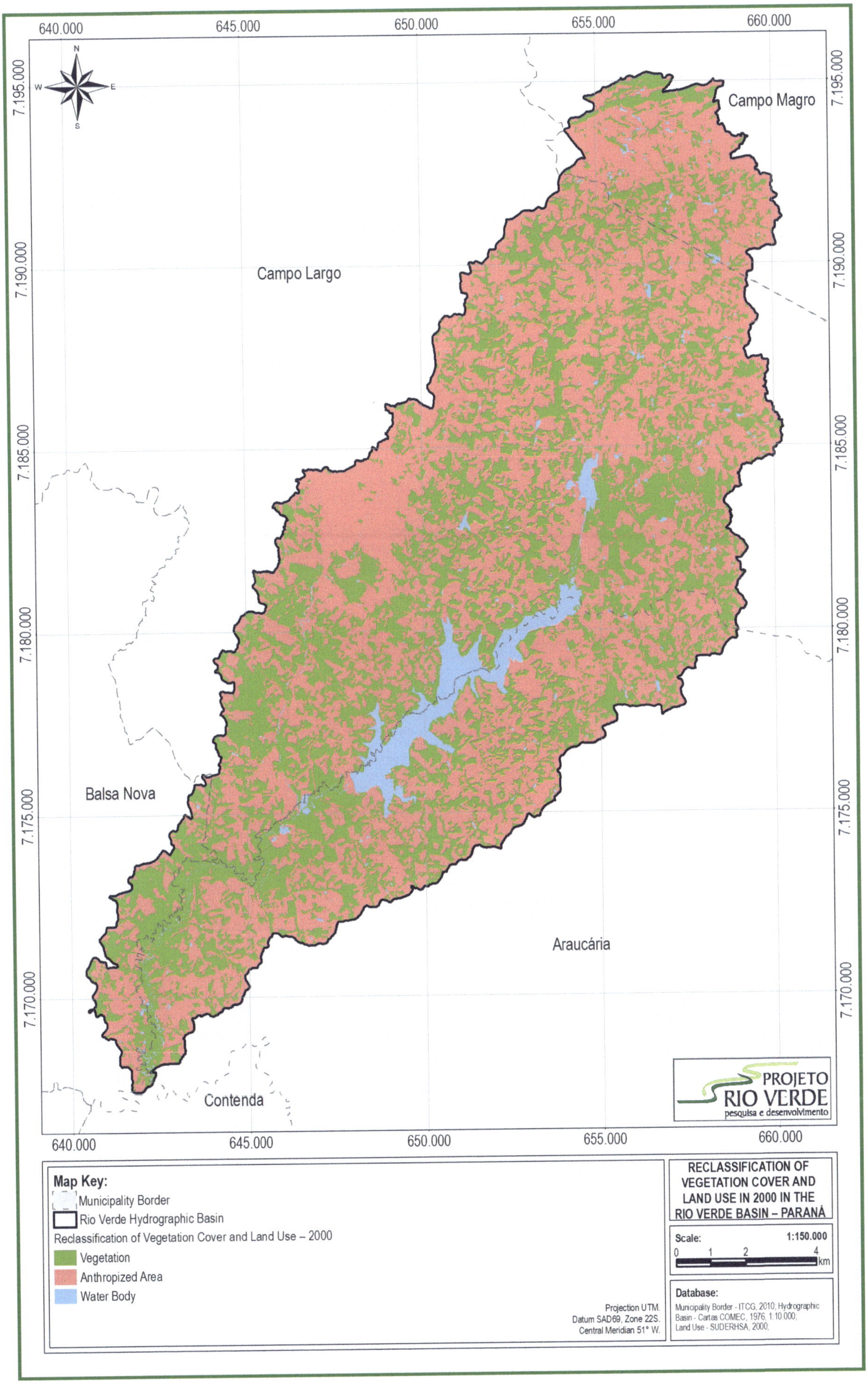

FIGURE 12 – LAND USE/LAND COVER (2000)

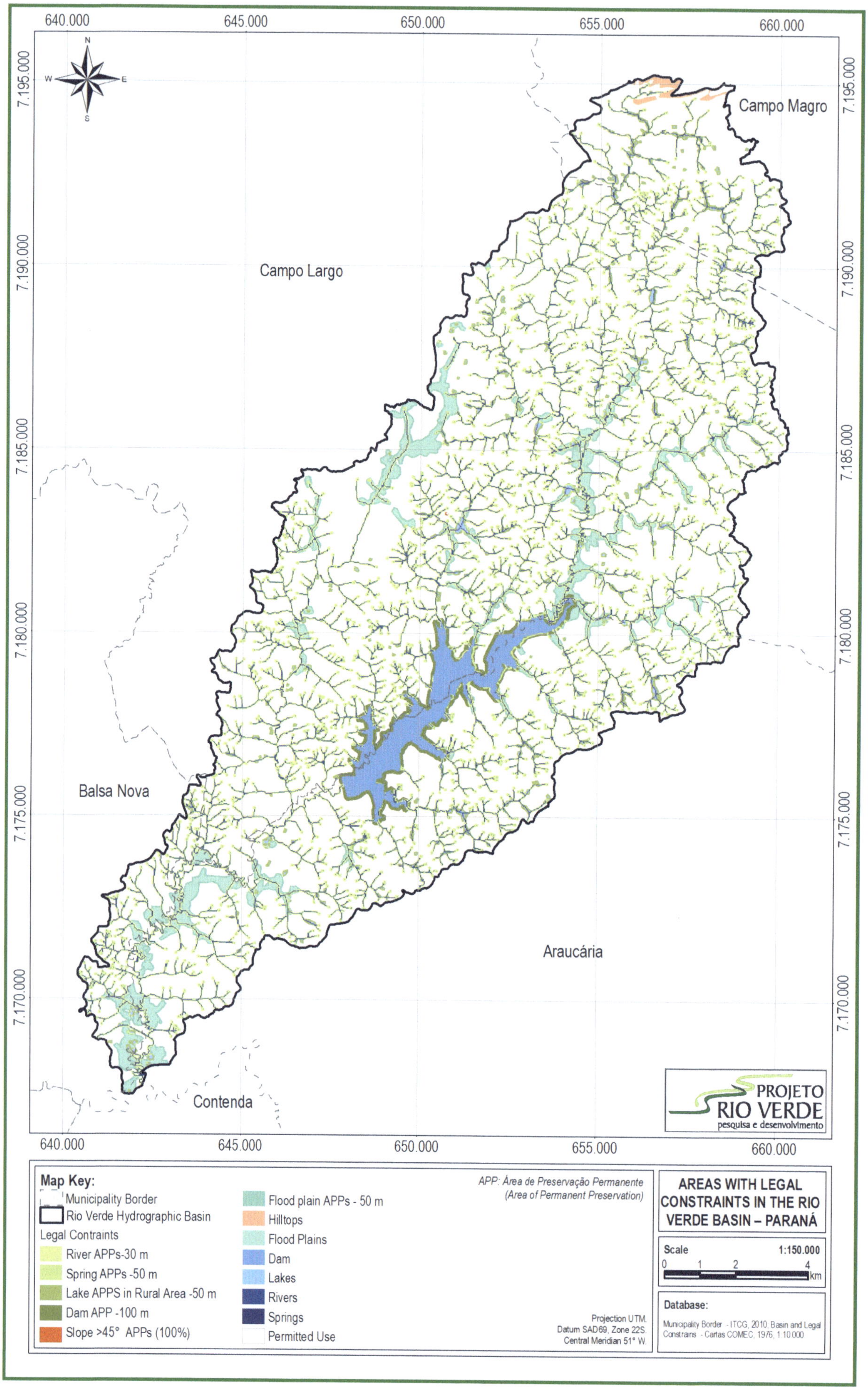

FIGURE 13 – LEGAL RESTRICTIONS ON THE RIO VERDE BASIN

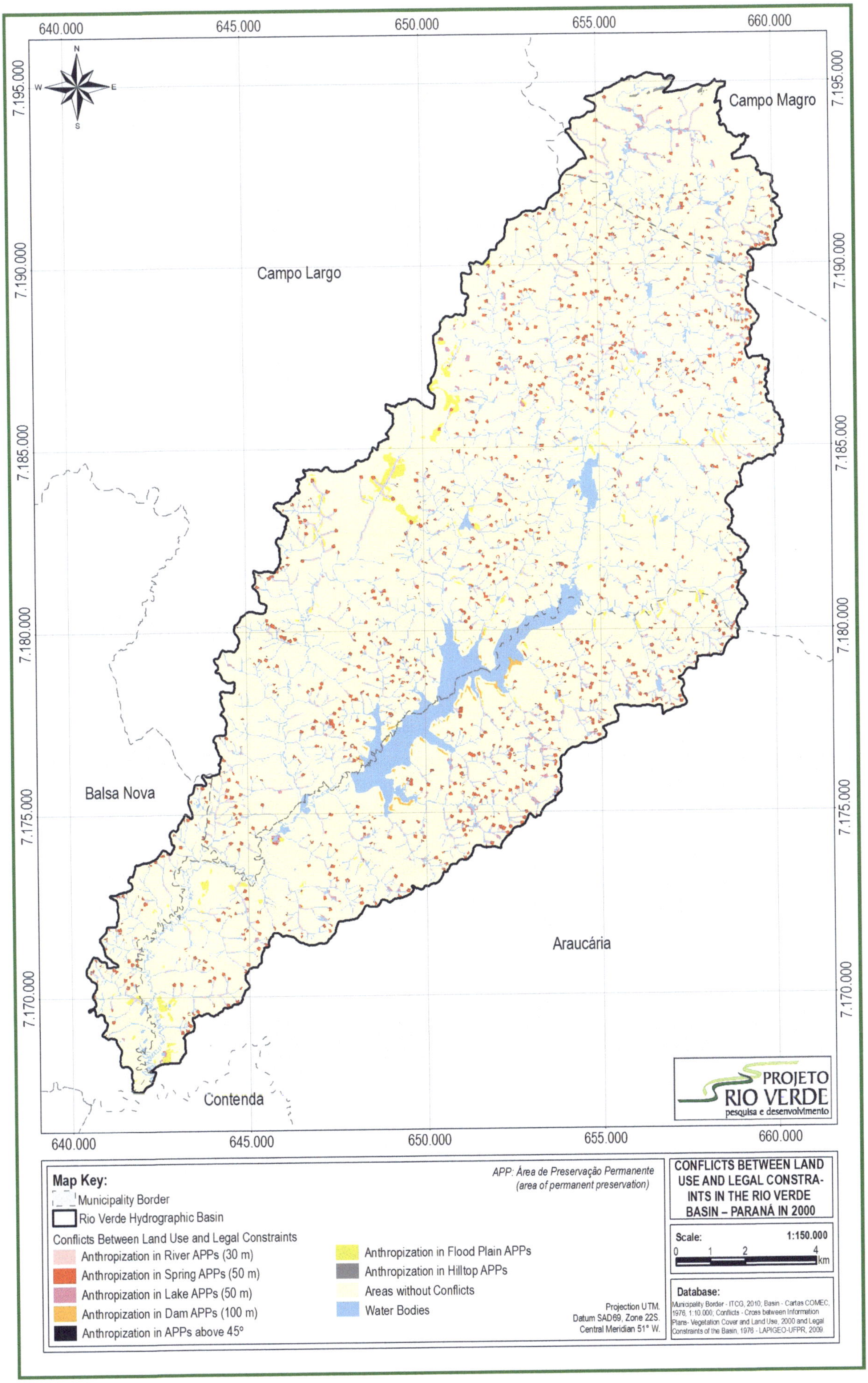

FIGURE 14 – CONFLICTS BETWEEN LAND USE AND LEGAL RESTRICTIONS

5. CONCLUSION

Geographic information systems are valuable and efficient tools in guiding environmental planning as they enable the manipulation and integration of differentiated data sets, facilitating consultation, analysis and the generation of new information, as well as the integration of digital image processing systems (aerial photographs and satellite images).

This GIS methodology is efficient and essential for studies of this nature, which involve the integration of cartographic information related to environmental problems in their spatial dimension (territorial), as well as to identify and analyze conflicts and spatio-temporal dynamics, allowing the precise analysis and identification of the way environmental phenomena evolve.

In this sense, the environmental analysis of hydrographic areas using GIS, provides important variables in decision making, including diagnosis and predicting possible outcomes, which enable preventive guidelines for both environmental management and the reorganization of land-use.

6. REFERENCES

ALMEIDA, C. M.; CÂMARA, G.; MONTEIRO, A. M. V.; BATTY, M. (2007) *Geoinformação em Urbanismo: Cidade Real x Cidade Virtual.* São Paulo: Oficina de Textos.

ALVES, D. S., Sistema de Informação Geográfica. (1990) In: SIMPÓSIO BRASILEIRO DE GEOPROCESSAMENTO,1., São Paulo, 1990. *Anais*. São Paulo: USP, Escola Politécnica, p.66-78.

ANDREOLI, C. V. (2003) *Mananciais de Abastecimento: Planejamento e Gestão: Estudo de Caso do Altíssimo Iguaçu.* Curitiba: SANEPAR.

ANDREOLI, C. V. et al. (1999) Os mananciais de abastecimento do sistema integrado da Região Metropolitana de Curitiba – RMC. In: *SANARE*, v.12, Curitiba: SANEPAR.

BRASIL, *Resolução CONAMA n.° 001*, dated January 23, 1986.

BRASIL, *Resolução CONAMA n.° 10*, dated December 14, 1988.

BRASIL, *Lei n.° 9.433*, de 8 de Janeiro de 1997. Institui a Política Nacional de Recursos Hídricos, cria o Sistema Nacional de Gerenciamento de Recursos Hídricos, regulamenta o inciso XIX do art. 21 da Constituição Federal, e altera o art. 1° da Lei n.° 8.001, de 13 de março de 1990, que modificou a Lei n.° 7.990, de 28 de dezembro de 1989.(Law n. 9.433, of January 8, 1997. National Policy of Water Resources, creates the National System of Water Resources Management, regulates paragraph XIX of art. 21 of the Federal Constitution, and changes the art. 1 of Law 8.001, of March 13, 1990, which amended Law 7.990, of December 28, 1989.)

CAMPO LARGO, (2009) Prefeitura Municipal. *Plano Local de Habitação de Interesse Social.*

CANEPARO, S. C.; PASSOS, E. (2006). Avaliação do estado de preservação ambiental a partir da espacialização de parâmetros geo-ambientais por meio de um sistema de informação geográfica – estudo de caso sub-bacia do rio Inspetor Carvalho – Paraná – Brasil. In: *IV Seminário Latinoamericano de Geografia Física* – Universidade Estadual de Maringá. Maringá.

COMEC, Available in: <www.comec.pr.gov.br> *Acessado em: 24 de maio de 2009.*

IBGE. (2005). *Produto Interno Bruto dos Municípios 2002-2005.* Available in:<<http://www.ibge.gov.br/home/estatistica/economia/pibmunicipios/2005/tab01.pdf>>. Acessed in June, 02 2009.

IBGE. (2008). *Estimativas das Populações Residentes, em 1° de julho de 2008, Segundo os Municípios.* Available in: <<http://www.ibge.gov.br/home /estatistica /populacao/estimativa2008/POP2008_DOU.pdf>>. Acessed in May12 , 2009.

KLEIN, R. M.; HATSHBACH, G. G. (1962). Fitofisionomia e notas sobre a vegetação para acompanhar a planta fitogeográfica do município de Curitiba e arredores. In: *Bol. Univ. Paraná;* 4:1-29.

MACHADO, R. E.; VETTORAZZI, C. A.; XAVIER, A. C., (2003) Simulação de Cenários Alternativos de Cobertura Vegetal e Uso do Solo em uma Microbacia Utilizando Técnicas de Modelagem e Geoprocessamento. In: *Revista Brasileira Cientifica Solo*, 27, p.727-733.

MANUAL IDRISI – EASTMAN, J. R. (2009) *IDRISI TAIGA: Guide to GIS and Image Processing.* Clark University. USA.

MOTA, S. (1999) *Urbanização e Meio Ambiente.* Rio de Janeiro: ABES.

PARANÁ, *Lei Estadual n.° 12.726*, de 26 de Novembro de 1999. Institui a Política Estadual de Recursos Hídricos, cria o Sistema Estadual de gerenciamento de Recursos Hídricos e dá outras providências. (State Law 12,726, of November 26, 1999. Establishing the State Policy on Water Resources, creates the State System of Management of Water Resources and other matters).

PARANÁ, *Decreto n.° 1.611*, de 03 de Dezembro de 1999. Declara para os fins de que trata a Lei Especial de Proteção dos Mananciais da RMC, como Unidade Territorial de Planejamento. (Decree n 1.611, of December 3, 1999. States for the purpose of dealing with the Special Law for the Protection of Water Sources of RMC, as Territorial Planning Unit)

PARANÁ, *Decreto n.° 2.375*, de 28 de Julho de 2000. Instituída a Área de Proteção Ambiental do Rio Verde, denominada APA do Rio Verde, localizada nos municípios de Araucária e Campo Largo, Estado do Paraná. (Decree 2375, 28 July 2000. Established the Environmental Protection Area of Rio Verde, called APA Rio Verde, located in the municipalities of Araucaria and Campo Largo, Paraná State.

PARANÁ, *Lei Complementar n.° 59*, de 1° de outubro de 1991. ICMS, a que alude o art. 2° da Lei 9.491/90, aos municípios com mananciais de abastecimento e unidades de conservação ambiental, assim como adota outras providências.

PARANÁ, *Decreto n.° 3.411*, de 10 de Setembro de 2008. Declara as Áreas de Interesse de Mananciais de Abastecimento Público para a Região Metropolitana de Curitiba e dá outras providências.

RIBEIRO, M. A. G; SANT'ANA, L. C. F., PASSOS, M. M., (2008) Aspectos Socioespaciais da Evolução Urbana da Rede Paranaense, Maringá: *Anais do I Simpósio sobre Pequenas Cidades e Desenvolvimento Local.*

ROCHA, C. H. B. (2000) *Geoprocessamento: Tecnologia Transdisciplinar.* Juiz de Fora.

ROCHA, J.; TENEDÓRIO, J. A. (2004) *GEOCOMPUTAÇÃO E ANÁLISE ESPACIAL: A utilização de Autómatos celulares com base em Redes Neuronais na previsão de alterações nos padrões de Cobertura Vegetal e Uso do Solo num contexto periurbano.* VIII Encontro de Utilizadores de Informação Geográfica. USIG. Oeiras, Portugal.

RODERJAN, C. V., (2009) *Projeto Interdisciplinar de Pesquisa sobre Eutrofização no Reservatório Rio Verde, Petrobras.* Departamento de Ciências Florestais da UFPR.

SILVA, J. J. I. (2009) *Identificação dos Conflitos Territoriais na Bacia Hidrográfica do Rio Verde – PR, através de Sistemas de Informações Geográficas.* Curitiba. 99p. Monography (Graduation in Geography), Universidade Federal do Paraná.

SUDERHSA, (2000) Disponível em: <http://www.suderhsa.pr.gov.br/modules/ conteudo/conteudo.php?conteudo=90> Acessed in May 23, 2009.

SUDERHSA, (2007) Relatório de Diagnóstico. Plano de Bacia do Alto Iguaçu e Afluentes do Alto Ribeira. Disponível em: <http://www.recursoshidricos.pr.gov.br/ modules/conteudo/conteudo.php?conteudo=47> Acessed in June 5, 2009.

SUDERHSA, Disponível em: http://www.suderhsa.pr.gov.br/meioambiente/arquivos /File/suderhsa/PLERH/Etapas.jpg. Acessed in: June 5, 2009.

TIM, U. S.; JOLLY, R. (1994) Evaluation Agricultural Nonpoint-Source Pollution Using Integrate Geographic Information Systems and Hydrology /Water Quality Model. *Journal I Environmental Quality*, v-23, p. 25-35.

WOLFRAN, S. (1983) The statistical mechanics of cellular automata. *Review of Modern Physics*, 55: 601-643.

CHAPTER 4

GEOLOGY

André Virmond Lima Bittencourt
Eduardo Chemas Hindi
Ernani Francisco da Rosa Filho

GEOLOGY

André Virmond Lima Bittencourt
Eduardo Chemas Hindi
Ernani Francisco da Rosa Filho

SUMMARY

Geology is a basic factor in understanding the chemical composition of water in a watershed, which is independent of, but influenced by, vegetation cover, fauna and human occupation. Geological mapping was not carried out in the Rio Verde Basin because existing and published data proved sufficient. The four geological features that make-up the substratum of the tributary basin of the Rio Verde Reservoir are: metamorphic units of the Atuba Complex, particularly migmatite and amphibolite bodies emerging in the middle portion of the basin; metamorphic units related to the Açungui Group containing carbonate rocks in the furthest upstream portion of the basln; mudstone-rich sedimentary rocks of the Guabirotuba and Tinguis formations containing 2:1 ratios of clay minerals and carbonate levels in the west and northwest of the basin; and recent alluvial deposits. In all these features, as is the case throughout the earth's crust, silicate minerals predominate. Differences can be observed through the types of silicates of each geological unit and other minerals that are added to the silicates, especially carbonates. Differences in the types of silicates and other minerals are fundamental determinants of the chemical composition of the water that feeds the Rio Verde Reservoir.

KEYWORDS

Rio Verde Basin, Atuba complex, Assungui group, Guabirotuba formation.

Hydrographic basins are formed on a rock substrate; climate and biota act on the substrate generating soils and together these aspects influence the quality of the water bodies they produce. Anthropogenic interferences only have a significant impact at a much later point in time. Considering these factors, knowledge of the main features of the geological structure and the lithology of the region are important aspects in understanding the composition of the waters of the Rio Verde Reservoir.

The Rio Verde Basin is located in the northwest portion of the Geological Basin of Curitiba. The Curitiba Basin was formed by a series of tectonic events that began during the formation of the Paraná Basin. Such events created geological faults that underwent several phases of reactivation until the Neogene (SALAMUNI, 1998 CHAVEZ-KUS & SALAMUNI, 2008). Both the lithological composition and the structural system of the basin were important factors in establishing the morphology of the basin under the prevailing climatic conditions.

Four geological features make-up the substratum of the basin: outcroppings of metamorphic units of the Atuba Complex in the middle portion of the basin; metamorphic units related to the Açungui Group that are present in the upstream portion of the basin; sedimentary rocks of Guabirotuba and Tinguis formations that occur to the west and northwest of the basin; and recent alluvial deposits (Figure 1). For all of these features, silicate minerals predominate, as is the case throughout the earth's crust; however, differences can be observed between the types of silicate that occur in each geological unit and through the inclusion of other minerals to the silicates. These differences in silicate type and the occurrence of minerals are fundamental determinants of the chemical composition of the water that feeds the Rio Verde Reservoir.

Accordingly, the following discussion outlines the geological characteristics deemed most important for each geological feature present in the basin.

1. ATUBA COMPLEX

The Atuba Complex is formed by paleoproterozoic rocks with high-grade metamorphism which are dominated by banded gneiss and partly migmatized granitoids. To a lesser degree, schists, granulites, amphibolites and metabasites also occur (BIGARELLA & SALAMUNI,1959; BATOLA Jr. *et al.*,1981; SIGA Jr. *et al.*, 1995).

The gneiss rocks are coarse to medium grained, have a banded structure, with dark bands of biotite and clear bands of quartz and calcium, sodium, and potassium feldspar.

The dark bands in the rocks indicate the significant proportion of silicates containing higher concentrations of iron and magnesium, as well as other elemental components of pyroxenes and biotite amphibole structures.

The migmatitic gneisses are irregular with granitic features and centimetric to decimetric dimensions interspersed in gneissic bands. The banded sections present features that are common to migmatitic gneisses, while the granitoid sections present a clear pink or white color, are rich in quartz, plagioclase and potassium feldspar, and have a granoblastic texture, ranging from fine to medium-grained.

Dikes of basic and intermediate rocks of the Jurassic to the Cretaceous period cut across the Proterozoic bodies, predominantly in a NW-SE direction. There are silicate rich rocks with a greater solubility than granitic rocks, which are dominated by potassium and sodium silicates and quartz itself, making them less soluble.

The largest portion of the Rio Verde Basin overlays banded migmatites of the Atuba Complex, while in the central and southern parts of the basin two zones of granitic gneiss and augen-gneiss occur. Between these two strips of high-grade metamorphism in the south-east portion of the basin, ultramafic rocks occur as either a large mass or in small lenses.

2. AÇUNGUI GROUP

In the northern and northwestern parts of the Rio Verde Basin, Capiru metamorphic rocks of the Capiru Formation belonging to the Upper Proterozoic Açungui Group occur. Preliminary studies on the litho-stratigraphic characterization of the Açungui Group were performed by Maack (1947) and sequenced by Bigarella (1948, 1953, 1956) and Bigarella & Salamuni (1956, 1958a, 1958b, 1959). Later Fiori (1992) proposed the subdivision of the Capiru Formation into the following clusters: Juruqui, Rio Branco and Morro Grande.

The lithology of the Capiru Formation that most influences the hydro-chemical composition is the dolomitic and metadolomitic marbles that occur and are spatially associated with quartzites and metacherts.

This portion of the Rio Verde Basin is part of the karst region of Paraná known for its mineral resources, water and the fragility of the environment. Much of the characteristics of this geological feature relate to the dolomite rocks, which are highly soluble due to the presence of magnesium and calcium carbonates.

3. GUABIROTUBA AND TINGUIS FORMATIONS

The Guabirotuba Formation (BIGARELLA & SALAMUNI, 1959 and 1962; SALAMUNI, 1998), of the Oligocene-Miocene period, is made up of pelitic sediments (clay and silt) with a light grey and greenish-grey color, interspersed with arcosean sand lenses of a light cream to yellow-orange color. The presence of caliches is common and it results from the precipitation of calcium carbonate. In some places, basal conglomerates occur with pebbles of varying compositions, particularly quartz. The maximum thickness of this formation, which was registered by drilling for water catchment in the vicinity of the Canguiri Farm of the Federal University of Paraná (UFPR), was 80m. In the Rio Verde hydrographic basin, the Guabirotuba Formation occurs along the western flank of the Basin with a maximum thickness of a few meters and occurring along only a few elevated slope sections.

4. TINGUIS FORMATION

Ranging in age from the Pliocene to Pleistocene (SALAMUNI, 1998), this formation, according to Becker (1982), is the result of the reworking of Guabirotuba Formation sediments that were deposited in floodplains.

This formation, with a centimetric to metric thicknesses, is comprised of clay and sandy pelitic material with a reddish color.

5. HOLOCENE DEPOSITS

These deposits correspond to recent alluvial deposits found in the riverbeds of the rivers flowing through Curitiba. They are composed predominantly of sandy and sandy-conglomeratic deposits and discontinuous layers of pelitic sediments and organic clay.

These deposits occur in the floodplain of the Rio Verde; upstream of the dam they are now mostly covered by the water body of the Reservoir.

6. HYDROGEOLOGY

In hydrogeological terms, the Rio Verde Basin has two deep aquifers: the fissured aquifer, established in the rocks of Atuba Complex, and the karst aquifer, formed through the dissolution of carbonate fractions of the Capiru Formation.

Because of sandy arcosean lenses, the Guabirotuba Formation has aquifer characteristics in some parts of the Curitiba Basin (ROSA FILHO et al., 1998); however, in the study region this formation does not behave as an aquifer as it is shallow and occupies only the upper layers.

The karst aquifer, one of the most productive sources of water in the Curitiba region (HINDI, 1999), is an important water source for the municipality of Campo Magro located in northern part of the Rio Verde Basin. A well with a depth of only 30m reached the karst aquifer and produced a flow rate of $72m^3.h^{-1}$; this well was used by SANEPAR to supply the municipality.

The Atuba Complex, which occurs across most of the Rio Verde Basin, has a fissure aquifer that dominates the Basin of the Metropolitan Region of Curitiba. In this aquifer, water inlets are generally located at intervals of 40 and 80m but intervals of up to 250m have been recorded (SALAMUNI, 1998).

Chavez-Kus (2008) recorded the average depth of deep wells at 112m and the median water intake depth of 84m. The average production of fractured aquifer wells is $6.2m^3.h^{-1}$ and they are generally used for gated communities, industries, hospitals and health service centers (ROSA FILHO et al, 2002).

Information about the wells that capture water from the fissure aquifer within the Rio Verde Basin is virtually non-existent. There is only one well on record with a depth of 36m located in community of Jardim Bom Pastor in the municipality of Campo Magro and the construction profile of the well shows 30m of regolith and 6m of migmatite. From these characteristics, it is believed that the flow rate of $1.3m^3.h^{-1}$ obtained from the well is the outcome of contact between the materials mentioned and not specifically from the fissure aquifer.

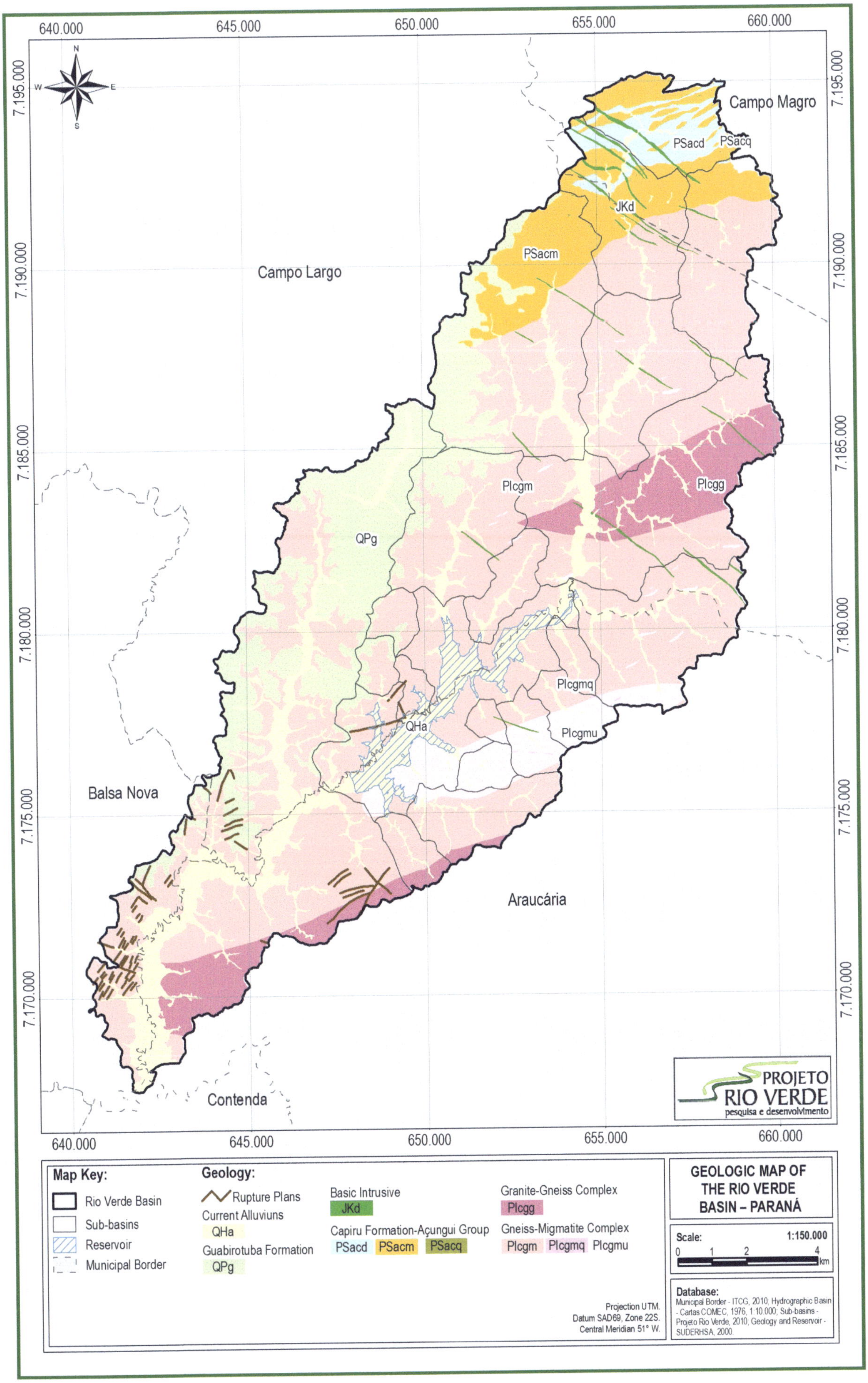

FIGURE 1 – LITHOLOGIC COMPARTMENTS OF THE RIVER VERDE BASIN

7. REFERENCES

BATOLA J.R., F., SILVA A.T.S.F., ALGARTE J.P. (1981) O pré-cambriano da região sul-sudeste do estado de São Paulo e Leste-nordeste de estado do Paraná. In: SBG, *Simpósio Regional de Geologia,* 3., Curitiba, Atas, p. 94-108.

BECKER, R.D. (1982) *Distribuição dos sedimentos cenozóicos na Região Metropolitana de Curitiba e sua relação com a estrutura geológica e geomorfológica regional.* Porto Alegre. Doctoral these. Universidade Federal do Rio Grande do Sul, 237 p.

BIGARELLA, J. J.; SALAMUNI R. (1959) Notas complementares à planta geológica da cidade de Curitiba e arredores. *Bol. do Inst. de Biologia e Pesquisas Tecnológicas.* Curitiba, n. 40 p.1-14.

BIGARELLA, J. J. (1956) Contribuição ao estudo dos calcáreos no Estado do Paraná. *Boletim do Instituto de Biologia e Pesquisas Tecnológicas.* Curitiba, n. 37, p. 1-79.

BIGARELLA, J. J. (1948) *Estudos preliminares na Série Açungui II – Rochas calcáreas.* Arquivos de Biologia e Tecnologia. Curitiba, v. III, p. 201-354.

BIGARELLA, J. J. (1953) *Estudos preliminares na Série Açungui III – Rochas calcáreas da faixa central e sua classificação.* Arquivos de Biologia e Tecnologia. Curitiba, v. VIII, p. 473-502.

BIGARELLA, J. J.; SALAMUNI, R. (1958a) Contribuição à geologia da região sul da série Açungui (Estado do Paraná). *Boletim Paulista de Geografia.* São Paulo, n. 29, p. 3-19.

BIGARELLA, J. J.; SALAMUNI, R. (1958b) Estudos preliminares na Série Açungui VIII – A Formação Votuverava. *Boletim do Instituto de História Natural,* Curitiba, v.2, p.1-6.

BIGARELLA, J. J.; SALAMUNI, R. (1956) *Estudos preliminares na Série Açungui VII – Estruturas singenéticas nos dolomitos da formação Capiru (PR).* Arquivos de Biologia e Tecnologia. Curitiba, v. XI, p. 197-205.

BIGARELLA, J. J.; SALAMUNI, R. (1959) *Planta geológica provisória de partes dos municípios de Rio Branco do Sul, Bocaiúva do Sul, Almirante Tamandaré e Colombo.* Curitiba: UFPR/Instituto de Geologia.

CHAVEZ-KUZ, L.; SALAMUNI, E. (2008) Evidência de tensão N-S intraplaca no neógeno, Complexo Atuba – região de Curitiba (PR). *Revista Brasileira de Geociências* n.38, v.3, p. 439-454.

FIORI, A. P. (1992) Tectônica e estratigrafia do Grupo Açungui – PR. *Boletim do Instituto de Geociências.* Série Científica, São Paulo, v. 23, p. 55-74

MAACK, R. (1947) *Breves notícias sobre a geologia dos Estados do Paraná e Santa Catarina.* Arquivos de Biologia e Tecnologia. Curitiba, v. II, p. 63-154.

MINERAIS DO PARANÁ S.A – MINEROPAR. (1997) *Projeto Geotecnia – Região Metroppolitana de Curitiba.* Mapeamento Geológico-Geotécnico nas Folhas Comec A100, A103 e A093 (parcial). 3 volumes.

SALAMUNI, E. (1998) *Tectônica da Bacia Sedimentar de Curitiba (PR).* Doctoral these, Inst. Geoc. e Ciências Exatas, UNESP, Rio Claro (SP), 214p.

SIGA Jr, O.; BASEI M.A.S.; REIS NETO, J.M.; MACHIAVELLI, A.; HARARA O.M. (1995) O Complexo Atuba: um cinturão Paleoproterozóico intensamente retrabalhado no Neoproterozóico. *Boletim IG-USP Série Científica,* 26: 69-98.

CHAPTER 5

PEDOLOGY AND LAND-USE SUITABILITY

Benno Henrique Weigert Doetzer
Oromar Berthol
Carlos Zavelinski

PEDOLOGY AND LAND-USE SUITABILITY

Benno Henrique Weigert Doetzer
Oromar Berthol
Carlos Zavelinski

SUMMARY

Soil degradation processes are a serious problem throughout the entire world, with significant environmental, social and economic implications. As the world population increases, so does the need to protect the soil as a vital resource, especially for increases in food production. To better use the resources offered by the soil (amount of nutrients, water availability, slopes, amount of exchangeable Al, etc.) it is necessary to classify it with the objective of establishing strengths and weaknesses and identifying where they occur. From this, a plan can be developed for the occupation and use of available resources. The aim of this study was to survey the soils in the Rio Verde Basin region, located in the Metropolitan Region of Curitiba (PR), which included classifying the soils that occur in the Basin, developing a soil map, identifying areas suitable for agriculture, and critical areas. The classification was based on field surveys, using geo-referencing as a tool for mapping, based on the methodology proposed by EMBRAPA. Within the area there is a wide variety of pedologic units, mainly due to the geomorphological characteristics of the area. The largest portion of the area is occupied by soils with good pedogenetic development (Oxisols and Alfisols), low natural fertility, and high levels of aluminum. However, due to intensive land-use, cultivated soils are generally quite degraded with significant accumulations of nutrients that can have a potential impact on the quality of the basin's water resources.

KEYWORDS

Sustainability, soils, suitability, management.

1. INTRODUCTION

Soil is a finite resource. Considering its potentially rapid degradation rates, which have been increasing in recent decades due to increasing pressure from human activities, and its extremely slow formation and rate of regeneration, it is a resource that is limited and non-renewable. The process of soil degradation is a serious problem worldwide, with significant environmental, social and economic implications. As the world population increases, the need to protect soil as a vital resource, especially for food production also increases. Soil degradation reduces the availability and long-term viability of the soil, reducing or altering its ability to perform functions that depend on it. Agriculture and forestry thus have a major impact on agricultural soil and may also have an impact on adjacent non-agricultural soils and groundwater.

To better use the resources offered by the soil (amount of nutrients, water availability, slope, amount of exchangeable Al, etc.) it is necessary to classify it, establish strengths and weaknesses, as well as identify where these soils occur. As such, plans can be prepared that manage occupation on the soil and the use of its resources.

Along with the classification of the soils within a region it is essential that maps are produced which incorporate the classification information and occurrence of soil types, cross-referenced with other types of information from the region (climate, geology, relief, rainfall, land-use, hydrography, etc.), resulting in a map of suitability for agricultural use. With such a map, a panoramic view of the region can be obtained, identifying areas suitable or unsuitable for a particular crop in a region. With this suitability map, we can identify the main strengths and weaknesses of a region and use the most well-suited techniques and crops adapted to these conditions.

From an environmental standpoint, the soil may have a direct influence on the maintenance of water quality. Soils with a high load (CEC - Cation Exchange Capacity) have high filtration power, retaining occasional soil contaminants, before reaching the water table or water body. Soils saturated with water have an almost nil load and therefore a low filtering capacity. This is one of the justifications for preserving the plains and low-lying levels near drainage channels.

The aim of this project was to conduct a soil survey of the area of the Rio Verde Basin, located in the Metropolitan Region of Curitiba (PR), which included the classification of the soils that occur in the region, the development of a soil map, and a map outlining agricultural suitability.

2. METHODOLOGICAL ASPECTS

This study was conducted in the Rio Verde Basin, located in the Metropolitan Region of Curitiba, within the municipalities of Campo Largo and Araucária. Predominant soils in the region are Ultisols, Oxisols, Inceptisols, Gleysols and Histosols (Soil Map of the State of Parana - updated Legend, 2008). The vegetation is classified as Mixed Ombrophilous Forest subdivided into Perennial Subtropical Forest, Subtropical Floodplain and Subtropical Grassland (SANTOS et al., 2006).

The climate of the region is characterized as Cfb, with an oceanic climate and abundant and well-distributed rainfall throughout the year, with fairly cool and humid summers. According to the normal climatologic measurements of INMET (1961-1990), the average annual temperature is 15.9 °C, with an annual temperature variation of 8 °C. The average temperature of the coldest month (July) is 12.5 °C and 19.9 °C of the hottest month (February). Rainfall reaches 1,500 mm on average per year because rain is a constant in the local climate.

The project was divided into several parts (Figure 1): diagnosis of the region/choosing the areas for sample collection; sampling; analysis and classification of soils; and finally drawing up maps.

2.1 STAGE 1 – DIAGNOSING THE REGION AND CHOOSING SAMPLE COLLECTION AREAS

This stage of the project consisted primarily of meetings held in communities within the Rio Verde Environmental Preservation Area (APA). At the meetings, we sought to explain the goals of the project to the local inhabitants, as well as discuss the possible implications of the project. In the meetings, residents participated in the diagnosis, during which they expressed their main concerns about the APA and reported their beliefs of how the region once was, what they currently observe, and what they expect for the future. The participants in the meeting were also asked to complete questionnaires regarding quality of life, income, etc. At the end of the meetings the soil collection team asked for permission to collect samples from participant's properties. A sample collection plan was defined based on consent.

This phase was very important to the project because through it we obtained permission to carry out the soil sampling within the region. From this, we could draw up a good sample collection plan for the area in which local land-holders living nearby were grouped and each sampling day we were able to give priority to a group, thus maximizing the efficiency of the collection team.

2.2 STAGE 2 – SOIL SAMPLE COLLECTION

This step basically consisted of field trips during which we collected samples at strategic points within a property, always seeking to sample the beginning, middle and end of a topographic sequence (relief), taking into account the size of the area, slope, vegetation, presence of obstacles (forest, rocks, streams, ravines etc.) and the available collection materials.

To collect soil, the following devices were used: a 1.2m Dutch auger, a handheld GPS, a camera, an inclinometer, a notebook and sample packaging. Each point was geo-referenced with GPS and the declivity, average elevation of the ground, and some soil features were recorded.

Among the observations of the soil and field characteristics, we recorded: soil color, presence or not of stones, roots, rock/source material, texture (only for reference), waxiness and plasticity.

Upon arriving at the property, our first step was to briefly observe the area to identify relief, size, the presence of drainage lines, visible variation of soil color, etc. With this first observation we were able to determine the amount of collection points to be used in the property. Thus, the number of points per area was variable, with a minimum number of 3 points per topographic sequence.

We always sought to start collection at the top of a topographic sequence. A collection team of two people visited the site and started the process of sampling. As one of the team members augered the ground and recorded the profile, the other recorded the point location using GPS, read the declivity with the clinometer, and photographed the sample location in relation to the landscape (Figures 2 and 3).

After the soil profile was recorded, we photographed the soil color characteristics and relative size compared to the size of the auger. In profile analysis we sought to separate horizons A and B because they presented different physicochemical properties and we recorded the individual characteristics of each profile (depth, waxiness, plasticity, texture, etc.). For soil samples, we took approximately 600 grams of each soil horizon (A and B), packaged them individually, sealed them with tape and sent the samples for analysis. In summary, each point generated two samples, one representing horizon A and the other horizon B.

It is important to mention that not all the points came from recorded properties. The team was dispersed throughout the region and a few places were observed that had soils with peculiar characteristics. Samples were therefore taken from these locations, along open roadside profiles, floodplains, etc., but without entering onto private property without permission.

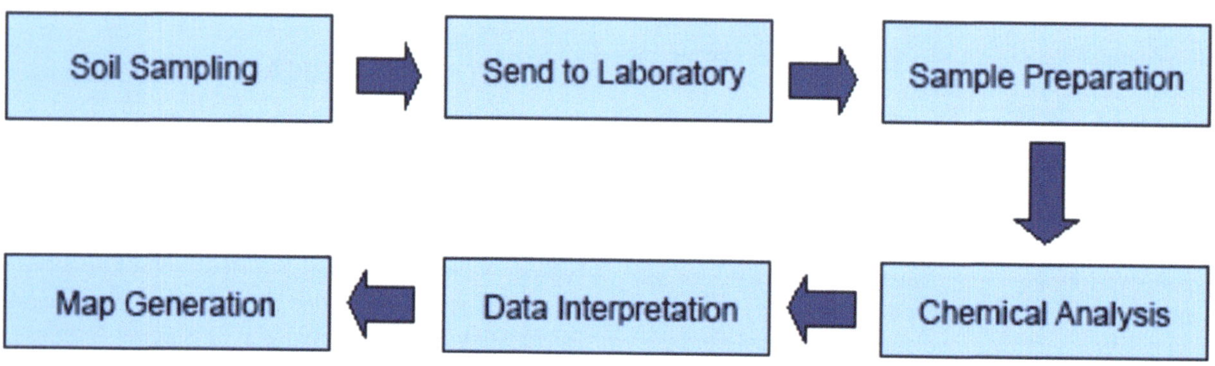

FIGURE 1 – FLOWCHART OF SOIL ANALYSIS ACTIVITIES

FIGURE 2 – DEMONSTRATION OF BOREHOLE OPERATIONS, SETTING UP OF PROFILE AND ANALYSIS OF SOIL CHARACTERISTICS

FIGURE 3 – DEMONSTRATION OF POINT RECORDING OPERATIONS BY PHOTOGRAPHY, GPS AND NOTES

2.3 STAGE 3: SOIL ANALYSIS AND CLASSIFICATION

2.3.1 Soil Analysis

Samples were passed on to the company SOLANALISE – Central de Analises Ltda. (located in the municipality of Cascavel), where physical and chemical analyses were carried out, according to the Manual of Soil Analysis Methods (*Manual de Métodos de Análise de Solo*; EMBRAPA, 1979).

2.3.1.1 Physical

Analyses were carried out to determine granulometric composition, fraction of sand, silt, and clay. The granulometric composition is determined by the dispersion of samples with NaOH, significant agitation, and settling: clay is determined by densimetry of the supernatant, course and fine sand are separated by sieving, and the silt is calculated by the difference (method 1.16.2).

2.3.1.2 Chemical

Analyses were conducted to determine SMPpH, CaClpH, exchangeable aluminum, as well as to determine macronutrients (Ca, Mg, K, P) and micro nutrients (Fe, Mn, Cu and Zn).

A Melich extractor was used to analyze nutrients K – P – Fe – Mn – Cu and Zn. For nutrients Ca – Mg and Al, a KCl extractor was used.

2.4 Soil Classification

The process of classification was based on the Brazilian System of Classification - SiBCS (EMBRAPA, 2006). According to the collection method and the variables analyzed, it was possible to classify the soil up to the fourth categorical level (orders, suborders, large groups, and subgroups).

Data processing of both field data and data obtained from laboratory analyses was done using MS Excel®; the data were stored and cross-referenced using formulas that facilitate classification.

2.5. Developing Maps

Spatial analyses were conducted using an Arc Info 9.2 environment, where we built a database with data provided by SUDERHSA. The data was from an aerial survey conducted in 2000. For this project we used the following data: Orthophotos (1:10,000) to verify areas of land-use; Sub-basins (1:10.000); Contour lines (1:10,000) spaced at five-meter intervals; Land-use (1:20.000); and Geology (1:20,000). Although data for the entire basin of the Upper Iguaçu are available, for this project only the Rio Verde Basin was included. All data were processed using the reference system UTM SAD 69 22J zone.

With the contour lines an altitude model was generated by TIN (Triangulated Irregular Network) and converted into raster format with a 3m resolution. From these, declivities were generated with a 3m resolution. With this database, areas were identified that likely had homogeneity in soil type and these areas were compared against the field analyses of profiles for classification of the region.

With the data available, along with the surveyed soil map, the agricultural suitability was classified (RAMALHO BEEK & SON, 1995). For this classification the shapes were converted to rasters with a 3m resolution with the help of the ArcGIS Spatial Analyst extension. Sorting operations were executed with a raster calculator using logical operators. After classification of agricultural suitability, our classification was compared to the current use of the basin, generating areas of land-use conflicts in the Rio Verde Basin. For this analysis the implementation of environmental legislation was not considered.

As the final product the rasters were converted back into shapes and used as the final product of the analysis.

3. CRITERIA USED FOR IDENTIFICATION OF MAPPING UNITS

With data from soil analysis in hand, the early phase of classification started with a basis in the Brazilian System of Soil Classification (SiBCS). All criteria of color, texture, depth, CEC, base saturation, etc. were calculated and plotted in Excel, which facilitated the calculation of some features that are critical in soil classification. As a result of the classification, we obtained four orders of main soils: Oxisols, Alfisols, Inceptisols and Gleysols. Within these order of soils, there are variations in classification level; with the resources available for the project a classification was made to the third categorical level (large groups).

Some features play a key role in soil classification up to the third categorical level, including the features described below.

Activity of the clay fraction

This feature refers to the capacity to exchange cations by soil clay fraction, calculated by the expression: T x $1000/g.kg^{-1}$ of clay. High activity (Ta) refers to a value equal to or exceeding 27 cmolc/kg clay, without correction for carbon, and low activity (Tb) corresponds to a value less than 27 cmolc/kg of clay, without correction for carbon. This criterion does not apply to soils which, by definition, have sand and loamy-sand textures.

Saturation of Bases

Refers to the proportion [percentage rate, V% = (100.S)/T] of exchangeable basic cations in relation to the exchange capacity determined at pH7. The term high saturation applies to soils with a base saturation equal to or higher than 50% (Eutrophic) and low saturation is for values below 50% (Dystrophic). A value of V greater than 65% is also used to identify the Chernozemic A horizon.

To distinguish between soil classes using this criterion, the base saturation in the subsurface diagnostic horizon (B or C) is considered. In the absence of these horizons, the criterion is defined for each specific class.

Acric character

Refers to the sum of exchangeable bases (Ca^{2+}, Mg^{2+}, K^+ and Na^+) plus extractable aluminum by KCl 1mol. L-1 (Al^{3+}) in a quantity equal to or less than 1.5 cmolc/kg of clay and which meets at least one of the following conditions:

- pH KCl 1 mol. $L^{-1} \geq 5.0$; or
- positive or zero pH (pH = pH KCl - pH H_2O).

Aluminic character

Refers to the condition where the soil is in a desaturated state and characterized by an extractable aluminum content that is greater than 4 $cmol_c$/kg soil, associated with clay activity <20 cmolc/kg clay, as well as aluminum saturation [100 Al^{3+} / (S + Al^{3+})] of > 50% and/or base saturation [V% = (100 . S)/T] of <50%.

To distinguish soil using this criterion, the content of extractable aluminum in the B horizon is considered, or the C horizon in the absence of B.

Abrupt Textural Change

Abrupt textural change refers to a considerable increase in clay content over a short distance in the transition zone between horizon A or E and the underlying B horizon. When horizon A or E has less than 200g clay/kg of soil, the clay content of the underlying B horizon (which is determined by a vertical distance greater than 7.5 cm) should be at least double the content of horizon A or E. When horizon A or E has a clay content equal to or greater than 200g/kg of soil, the increase of clay in the underlying B horizon (determined at a vertical distance <7.5 cm) must have an absolute fine soil fraction that is at least an additional 200g/kg (for example, from 300g/kg to 500g/kg and from 220g/kg to 420g/kg).

Plinthite

Plinthite is a formation consisting of a mixture of clay material that is low in organic carbon and rich in iron, or iron and aluminum, with grains of quartz and other minerals. It occurs commonly in the form of red, red/yellow and dark red mottles, with usually laminar, polygonal or cross-linked patterns. Plinthite is formed by the segregation of iron, important in mobilization, transport and final concentration of iron compounds that can be processed in any soil where the iron content is sufficient to enable the separation of the iron in the form of soft red spots.

Plinthic character

This criterion is used to distinguish soils with plinthite in insufficient quantity or thickness to characterize a plinthic horizon in one or more horizons, at some point in the control section that defines the class. A minimum of 5% plinthite per volume is required.

Silt/Clay Ratio

The ratio of silt to clay is calculated by dividing the silt content by clay content, based on particle size analysis. The silt/clay ratio serves as a basis for assessing the stage of weathering in tropical soils. It is used in soils with loamy-sand textures or finer and indicates low levels of silt and therefore a high degree of weathering. When present, and generally in the B horizon, values are below 0.7 in medium textured soils or below 0.6 in clayey or very clayey soils. This ratio is used to differentiate the oxic B horizon from incipient B when they exhibit similar morphological characteristics. This is used primarily for soils of which the original material comes from crystalline rocks such as granite and gneiss.

In soil classification there is a separate reading of soil horizon A (superficial) and B. The unique characteristics of each type of horizon are taken into account when classifying the soil in a particular order. The main types of surface diagnostic horizons were:

Chernozemic Horizon

This is a relatively thick mineral surface horizon that is dark in color, with high base saturation. Even after plowing the surface, the horizon meets the following characteristics:

a) soil structure is sufficiently developed, with aggregation and a moderate or strong degree of development, but not simultaneously a solid structure with hard or harder (very hard or extremely hard) consistency when dry. Prisms without secondary structure, larger than 30cm, similar to massive structure, are not included;

b) the color of the soil, in both undisturbed and crushed samples, has a chroma equal to or less than 3 when moist, and 5 when dry. If the surface horizon presents 400g/kg of soil or more of calcium carbonate equivalent, the chroma value limits for dry condition are not considered and when wet, the limit goes to 5 or less;

c) the base saturation (V%) is 65% or more, with a predominance of calcium and/or magnesium ions;

d) the organic carbon content of the soil is 6g/kg or more across the entire horizon, according to the criterion of thickness in the following item. If, due to the presence of calcium carbonate equivalent of 400g/kg of soil or more, color requirements are different from the norm. The organic carbon content is 25g/kg of soil or more in the surface 18cm. The upper limit of organic carbon content that characterizes the chernozemic A horizon, is the lower limit excluding the histic horizon;

e) thickness, including transitional horizons, such as AB, AE or AC, even when the soil material is upturned, must meet one of the following requirements:

1. 10 cm or more, if horizon A is followed by contact with rock; or
2. at least 18cm and more than a third of the thickness of the solum (A + B), if the solum is less than 75cm; or
3. for soils with no B horizon, at least 18cm and more than a third of the thickness of horizons A + C, if A + C is less than 75cm; or at least 18cm and more than a third of the thickness of the solum, or more than a third of the thickness of horizons A + C if B does not exist, if these are less than 75cm; or
4. at least 25cm, if the solum has a thickness of 75cm or more.

Prominent A Horizon

The outstanding characteristics of the prominent A horizon are similar to those of the chernozemic A in relation to color, organic carbon content, consistency, structure, and thickness, differing mainly by a base saturation (V%) of less than 65%. It differs from the humic A horizon in organic carbon content in conjunction with thickness and clay content.

Anthropogenic A Horizon

This horizon is formed or modified by the continued use of the soil, through human occupation or cultivation for prolonged periods, with additions of organic material, mixed or not with mineral material. In this layer, fragments of ceramics, lithic artifacts, remains of bones and shells can be found.

Together with the analysis of surface horizons the analysis of the subsurface horizons (B horizon) is also carried out. Among the types of B horizon, the most commonly found in the basin were:

Textural B Horizon

This is a mineral subsurface horizon with a loam-sandy texture or finer, where there was an increase of clay (fraction <0.002 mm), oriented or not, as long as it does not present discontinuity from the original material, resulting from the accumulation, absolute or relative concentration due to: processes of illuviation; and/or *in situ* formation; and/or inherited from the parent material; and/or infiltration of clay or clay plus silt with or without organic matter; and/or the destruction of clay in the A horizon; and/or the loss of clay in the A horizon by differential erosion. The clay content of the B horizon is greater than horizon A or E and may or may not be greater than in horizon C.

This horizon can be found on the surface if the soil was partially lost through erosion. The colloidal nature of clay makes it susceptible to mobility with water in the soil if percolation occurs. When it is deposited in an aqueous medium, the particles of mineral clay, usually lamellar, tend to lie flat at the place of support. When transported by water, the translocated clays tend to form films that are oriented parallel to coated surfaces, whereas clays formed *in situ* have a disordered orientation. However, other types of colloidal inorganic coating materials are also taken into account as characteristics of B horizons and recognized as waxy.

The waxiness considered when identifying textural B horizon includes mineral coatings of colloidal material which, if well developed, are easily identifiable by their glossy appearance and greasy shine in the form of pore fillings and coating of structural units (aggregates or peds).

In the field identification of most textural B horizons, waxiness is important. However, the mere occurrence of waxiness may not be enough to characterize the textural B horizons. Thus it is necessary to associate it with other auxiliary criteria because, due to the turbulent flow of water through cracks, filling the pores can occur during a single event of rain or flooding. For this reason, waxiness in a textural B horizon must be present in different aspects of the structural units and not solely on the vertical aspect.

Considered as textural horizon B is the occurrence

of lamellae, of loam-sandy texture or finer, which together make up 15cm or more of thickness, allowing that between them sandy and sand-loamy texture classes can occur.

It can be said that a textural B horizon is formed under a surface horizon or horizons and has a thickness that satisfies one of the following conditions:

a) has at least 10% of the sum of the thicknesses of overlying horizons and at least 7.5 cm; or
b) has 15cm or more, if horizons A and B add up to more than 150cm; or
c) has 15cm or more, if the texture of horizon E or A is loamy-sand or sand; or
d) if horizon B is entirely composed of lamellae, then together they will have a joint thickness exceeding 15 cm; or
e) if the texture is medium or clayey, the textural B horizon should have a thickness of at least 7.5cm. In addition to this, for the characterization of a B horizon, one or more of the following characteristics must occur:
f) the presence of the E horizon on the sequum above the identified B horizon, as long as B does not satisfy the requirements of a spodic, plinthic or planic B horizon;
g) a large increase of total clay from horizon A to B, enough to characterize an abrupt textural change; or
h) an increase in total clay from horizon A to B within a control section defined according to the thickness of the A horizon, that is sufficient in order for the textural relationship B/A8 to meet one of the following alternatives:
   h1) in soils with a clay content of more than 400 g/kg of soil in the A horizon, ratio greater than 1.50; or
   h2) in soils with a clay content of 150 to 400 g/kg of soil in the A horizon, with a ratio greater than 1.70; or
   h3) in soils with a clay content of less than 150g/kg of soil in the A horizon, with a ratio greater than 1.80.
i) when the increase of the total clay from horizon A to B is less than that specified in item h, the textural B horizon must meet one of the following conditions:
   i1) soils of medium or sandy/medium texture with no macro aggregates should present illuvial clay, represented by moderate waxiness in the form of coatings of individual grains of sand, oriented according to their surface, or forming bridges connecting the grains;
   i2) soils with medium texture B horizon and prismatic or moderate or more developed block structure must present at least moderate waxiness in one or more sub-horizons, from the upper part of B;
   i3) soils with clayey or very clayey B horizon textures and prismatic or block structures must have a waxiness that is at least common and weak or low and moderate in one or more sub-horizons, from the upper part of B;
   i4) soil with a textural relationship B/A equal to or greater than 1.4, combined with the presence of fragipan within 200cm of the surface as long as it does not meet the requirements for spodic B;
j) if the profile presents discontinuity of original material between the A or E horizons and textural B (mainly soils developed on recently added material, such as alluvial sediments) or if only one plowed layer is above textural B, it needs to satisfy one of the requirements specified in items h and/or i.

Oxic B Horizon

This is a subsurface mineral horizon, the elements of which show an advanced stage of weathering through the nearly complete transformation of alterable minerals, followed by intense desilting, leaching of bases, and residual concentration of sesquioxides and/or 1:1 clay-type minerals and minerals resistant to weathering. In general, it consists of variable amounts of iron and aluminum oxides, 1:1 type clay minerals, quartz and other minerals more resistant to weathering.

In the oxic B horizon there should be no more than 4% of changeable primary minerals (low resistance to weathering), or 6% in the case of muscovite, determined from the sandy fraction and recalculated for the fine earth fraction. The fraction smaller than 0.05 mm (silt + clay) can present small amounts of interstratified clay minerals or illites but should not contain more than mere traces of clay minerals of the smectite group. No more than 5% of the volume of the mass of the oxic B horizon can be made up of the original rock as thin stratifications, saprolite, or rock fragments with low resistance to weathering.

The oxic B horizon must have a minimum thickness of 50cm, loam-sandy texture or finer, and low silt content, so that the silt/clay ratio is less than 0.7 in medium textured soils and below 0.6 in clayey soils in most B sub-horizons to a depth of 200 cm (or 300 cm if the A horizon exceeds a thickness of 150 cm).

The oxic B horizon may have minimal and weak waxiness. It may contain more clay than the overlying horizon but the increase in clay with increasing depth is minimal. As such, comparisons made at intervals of 30cm or less between horizons A and B (or within the control section for calculating the textural relationship) present insufficient differences needed to characterize a textural B horizon.

The oxic B horizon has unclear differentiation between its sub-horizons with a generally diffuse transition. In some cases, the upper limit of the oxic B horizon is difficult to identify in the field as it presents minimal transition contrast with the horizon that precedes it, checking only a sharp contrast almost exclusively of color and structure between the bottom of the A horizon and the oxic B horizon.

The structure of this horizon may be strongly developed, when the elemental structures are granular and of a very small to small size, or weak and only moderately developed when dealing with sub-angular block structures. The consistency of the material of the B horizon, when dry, ranges from soft to very hard and from firm to very brittle when moist.

Variations in structure, consistency, and waxiness for retractable oxic horizons are accepted.

Oxic horizons usually have a high degree of flocculation in sub-horizons further away from the surface and with lower organic matter content, which shows the low mobility of clays and high resistance to dispersion.

Many medium textured soils may not have high flocculation, especially those with lower levels of clay and those that are very weathered with a balance of positive charges.

Incipient B Horizon

This is a sub-superficial horizon underlying the A, Ap or AB that has undergone physical and chemical change to a limited degree. The changes are, however, sufficient for the development of color or structural units in which over half of the volume of all sub-horizons should not consist of original rock structure.

Nitic B Horizon

The Nitic horizon is a mineral subsurface horizon that is not hydromorphic, with a clayey or very clayey texture, and without an increasing amount (or with a small increment) of clay from surface to subsurface horizon, which creates a textural relationship of B/A that is always less than 1.5. They ordinarily exhibit low activity clay or clay of an alitic character.

The structure, at a moderate or strong stage of development, is in sub-angular and/or angular blocks, or prismatic, which can be composed of blocks. Aggregates usually present a gleaming surface. This characteristic is usually described in the field as waxy in a quantity and degree of development that is described at least as common and moderate. It shows diffuse or gradual transition between the sub-horizons. This horizon can be found at the surface if the soil has been eroded.

Gley Horizon

This is a subsurface, or sometimes surface, mineral horizon with a thickness of 15cm or more, characterized by iron reduction and a prevalence of a reduced state, in whole or in part, mainly due to stagnant water, as evidenced by a neutral or close to neutral color in the matrix of the horizon, with or without mottling of more vivid colors. The horizon is strongly influenced by groundwater and reducing moisture regime. It is virtually free of dissolved oxygen because of water saturation throughout the year, or at least for long periods of time, associated with oxygen demand by biological activity.

This horizon may be comprised of any class of material texture and color with very low, neutral or near neutral chroma. However, the horizon turns more brown or yellowish when exposed to the air. When there is an aggregated structure, the faces of the structural elements have a continuous phase of gray, blue, green or neutral color and mottling may be in more vivid colors. The inside of the structural elements may have prominent mottling but usually there are a network of lineaments or low chroma bands surrounding the mottles. When structural elements are absent, the horizon matrix (bottom) generally has a chrome 1 or less, with or without mottling.

If periodically saturated with water or if the soil has been drained, the horizon should show some mottling of high chroma of yellowish or reddish colors resulting from iron segregation and precipitation in the form of oxides. Accumulations can occur in the form of black or black-red mottles, mild or semi-consolidated, or possibly nodules or concretions of manganese or manganese and iron.

4. GENERAL CHARACTERISTICS OF THE BASIN SOILS

The Basin is located on recent lithological formations typical of the sedimentary basin of Curitiba. On floodplains and in most significant drainage areas, recent alluvial sediments predominate. On slopes and elevated areas, wide and flattened features dominate that are typical of the clayey sediments of the Guabirotuba Formation.

In the Iguaçu River Basin, the predominant soils are characterized by the pluvial deposition of sediments, high levels of organic matter and hydromorphism. At the headwater, soils are shallow and rocky outcrops can occur. In both cases the soils may be classified as Entisols (fluvic and litholic, respectively). In the middle portions of the basin, at higher elevations, one can see the predominance of developed soils, such as oxisols and Alfisols, and soils that are minimally developed, such as inceptisols. Near the drainage channels gleysols may be present (hydromorphism).

The geomorphology of the Basin is characterized by short slopes and valleys in the shape of a "V", with a mountainous relief in areas of higher elevation in the upper reaches of the basin. These elevated areas have greater declivity, typically greater than 45°, and altitudes above 1000 meters mainly in the western slopes of the Serra do Mar, which is home to the sources of the tributaries of the Iguaçu River.

The central part of the Basin has a more even relief with large, rounded, half dome-shaped hills, and lower declivity. The presence of alluvial deposits is more significant than in the upper reaches of the basin (GUIMARÃES, 2000).

In the lower third, the southwest is characterized by the floodplains of the Iguaçu River with long slopes and low declivity. Outcrops of the Guabirotuba Formation occur as softly rounded hills and limit the floodplain of the Iguaçu River with flattened fluvial terraces, broad interfluves and alluvial deposits influenced by the confluence with the river.

The soils encountered in the Rio Verde Basin are: Inceptisol, Oxisol, Alfisol, Gleysol and Histosols. Their morphological characteristics are important determining factors that must be considered in an integrated analysis of an environment because the conventions used to establish the use and/or parceling of the land, among others, is related to soil genesis and formation, which will enable parameters to be established that limit or not their use.

A description of the main types of soil found, as well as a description of some key characteristics in the process of soil classification, is given below.

Oxisol:

A grouping of soils of the oxic B horizon with: very advanced evolution with significant oxisol formation processes (ferralitization or laterization); intense weathering of primary mineral elements and even weathering of less resistant secondary elements; relative concentration of resistant clay minerals and/or oxides and hydroxides of iron and aluminum; negligible mobilization or migration of clay, ferrolysis, gleyzation or plinthitization.

Gleysol:

A group of soils with significant gleyzation. The hydromorphy is seen through strong gleyzation resulting from processes of marked reduction of iron compounds in the presence of organic matter, with or without alternating of oxidation due to fluctuations in groundwater level under permanent or periodic conditions of excessive moisture. The preponderance and depth of gleyzation attributes are coupled with the characterization of the gley horizon.

Inceptisol:

A group of poorly developed soils of the incipient B horizon. Pedogenesis is little advanced, as shown by: soil structure development; alteration in source material shown by the near absence of rock structures or layering of sediments; stronger chroma; redder hue or higher clay content than the underlying horizons. Development of incipient B horizon following the surface horizon of any kind, including the chernozemic A horizon, the incipient B should present low activity clay and/or low base saturation.

Alfisol:

A grouping of soils of the nitic B horizon with low activity clay or alitic characteristics. Advanced pedogenetic evolution through ferralitization with intense hydrolysis, yielding kaolinitic-oxidic or virtually kaolinite composition, or with hydroxy-Al interlayers. The development (expression) of diagnostic nitic B horizon following any type A horizon, shows little textural gradient, but moderate to heavy sub-angular or angular block or prismatic structures, with significant waxiness in the structural units.

5. MAPPING UNITS

The relationships between the different types of soils and the geomorphology (landscape) in which they occur were determined using GIS and field survey. During this phase, variations in soil morphological features such as texture (surface and subsurface), drainage, type and thickness of the A and B horizons, hydromorphism and anthropogenic impacts, etc., were assessed and recorded along with the geomorphological characteristics, such as slope, length, shape and position of the incline, etc. Thus, it is possible to relate the different soil classes to the landscape in which they occur.

In assessing mapping units and soil types, it is necessary to create clusters with similar geomorphology and conditions of soil hydromorphism. Detailing topography, slope, type of incline and length, allows us to prepare a preliminary map of the mapping units from which we could define observation points.

The density and frequency of observation points were determined based on the geomorphologic heterogeneity of the area (slope, shape, length and type of incline, hydromorphism, anthropogenic disturbance, etc.) and their correlations with the various types of soils.

The soil classes identified in the landscape were associated with simple and compound mapping units. Whenever possible we tried to create simple mapping units, isolating different soil classes. However, due to the relative morphological complexity, there is a very heterogeneous soil distribution across the Basin. In general, the occurrence of soils is correlated, forming mosaics that hinder the precise delimitation of the different classes.

In delimiting composite units, also called associations, we sought primarily to group units with the most relevant characteristics which were similar from the point of view of the potential of the land for agricultural use. This allowed for the development of recommendations for use, occupancy, and management favorable for all units.

Based on morphological characteristics, soil analyzes and landscape studies, the mapped units can be grouped and identified in the landscape as follows (Figure 4):

LBd – Dystrophic Haplohumox Oxisol;
LVd – Dystrophic Red Oxisol;
NVd – Dystrophic Red Alfisol;
CXbd – Dystrophic Tb Haplic Inceptisol;
CXve – Eutrophic ta Haplic Inceptisol;
GJo – Orthic tiomorphic Gleysol
LVe – Association of eutrophic Red Oxisol + dystrophic Haplohumox
NBd – Association of dystrophic o alfisol + dystrophic Tb haplic Inceptisol

It should be noted that the transition from one class to another is gradual, i.e., there is no abrupt change from one unit to another.

6. DESCRIPTION OF MAPPING UNITS

Alfisols

In Figure 5 we can see a **Alfisol** profile. This class is made up of deep to very deep mineral soils, that are non-hydromorphic, clayey to very clayey, red or reddish in color, developed from basic and ultrabasic rocks. They have a low textural gradient and a B horizon with low activity clay, a moderate to strongly developed structure, and shiny brightness due to moderate or strong waxiness and/or compression surfaces. They are therefore considered pedologically well-developed soils. In addition to these characteristics, they typically have relatively high contents of Fe_2O_3 (³ 150g.kg-1) and TiO_2 (³ 15g.kg-1).

The B horizon is considered a textural B horizon despite the low textural gradient (<1.5), given the small increase in clay content in comparison to the A horizon. However, in light of this and other features, this type of soil has recently been incorporated into the category of the nitic B horizon (EMBRAPA, 1999).

These soils are deep to very deep with a horizon sequence of type A, Bt (or nitic B) and C, usually with poorly differentiated horizons, except for the transition from A to B which is usually clear and gradual.

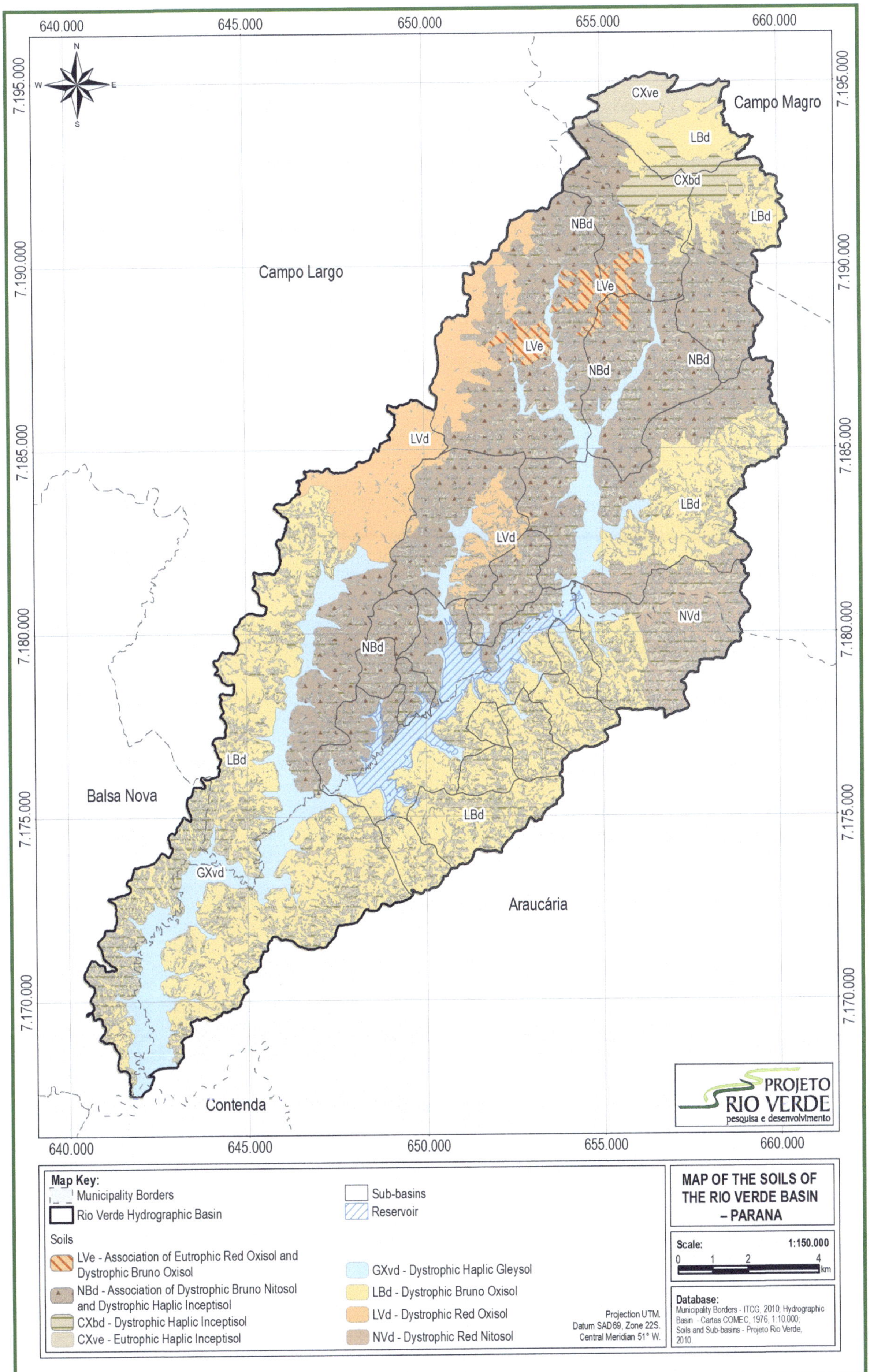

FIGURE 4 – SOIL MAP OF THE RIO VERDE BASIN

They present an A horizon of the moderate type with a thickness around 20 cm, dark reddish-brown color (2.5 YR 3/3 and 5YR 3/3) of clay or very clayey texture, with moderate to strong structure, and small to medium granular and small sub-angular blocks. The consistency of the soil in the dry state is usually hard, changing to brittle when humid and becoming sticky and plastic when the soil is wet. The transition to the B horizon occurs clearly and gradually.

The B horizon, 2-3 meters thick with a clayey to very clayey texture may have sub-horizons of BA, B1, B2 (or BT1, BT2) and BC types. It appears with dark red to dark reddish-brown color, in the 2.5YR hue, with value 3, and a chroma of 4 to 8. It is comprised of moderate to strong structures with a small size in the shape of sub-angular or angular blocks, or of moderately medium prismatic structures composed of blocks. In both cases the horizon always appears with a gleaming brightness due to the presence of waxiness and/or compression surfaces. The amount of waxiness usually varies from low to high with a weak to moderate level of development. The consistency of the soil ranges from hard to very hard when dry, from brittle to firm when humid, and becomes plastic and sticky to very sticky when wet. In some restricted locations, a line of quartz edged pebbles and small nodules were observed at the top of the B horizon.

Horizon C is 1 to 2 m thick, with subdivisions of type C1, C2, etc., and was observed with brownish colors and/or was reddish in hue (2.5YR), with values of around 3 and more commonly chroma of 4 to 5. The texture was observed to be silty or loam-silty. Where it was possible to observe the C horizon, the structures appeared with a weak to moderate degree of development, of small size and sub-angular-block shape. Consistency of the soil when dry varies from slightly hard to hard, brittle when moist, and becomes slightly sticky and plastic when the soil is wet.

As far as the physical properties, as a function of the types and degrees of structural development, soils are well drained, porous, and are considered to be under optimal conditions. There was virtually no occurrence of particles with a fraction greater than 2mm average diameter. In the fine earth fraction (<2 mm), clay was predominant, with values in the range of 55-66% in the A horizon, with a slight increase in the clay content in the B horizon from 70 to 80%, which then decreases with depth at the transition to the C horizon.

The percentage of silt is 23 and 33% in the A horizon, decreases at the top of B and then increases again, reaching values in the range of 33-83%, from the BC to C horizon.

The percentage of coarse and fine sand, which is around 12% in the Ap horizon, decreases from the top of the B horizon. In the B horizon it ranges from 2-6%, increasing again at the transition to C. Coarse sand predominates over fine sand in the superficial horizons, they occur in equal proportions in the B horizon, and course sand is reduced to values lower than those of fine sand in the underlying horizons.

The silt/clay ratio, which does not exceed 0.6 in most A and B horizons, undergoes an increase from B to C, and reaches values of up to 11 in the C horizon.

The clay dispersed in water varies between 28 to 45% in the A horizon and is nil in the other horizons, resulting in flocculation ranging from 32 to 49% at the surface and increasing to 100% in other horizons.

Regarding chemical characteristics, soils are strongly to moderately acidic, with pH varying in the range of 4.9 to 5.6. Generally they present a base saturation of less than 35% and aluminum saturation above 70% in most of the subsurface layers. The base desaturation is very low, with sum values from 0.6 to 3.6 cmolc.kg^{-1} soil in the surface and from 0.5 to 1.1 cmolc.kg^{-1} soil in the subsurface layers. With the contribution of exchangeable and/or extractable aluminum and hydrogen, the cation exchange capacity produces values of around 10 cmolc. kg^{-1} of soil in the surface and varies in the range of 4.2 to 13.3 cmolc.kg^{-1} soil in subsurface horizons. The organic carbon content ranges from 17.3 to 21.6 g.kg^{-1} soil in the A horizon.

The Ki ratio, which is indicative of the mineralogy of clays, frequently shows values between 1.85 and 1.98 in the Bt horizon. This suggests relatively weathered soils. As the Kr ratio is greater than 0.75, even with high concentrations of Fe203, the soils are considered kaolin. Results of mineralogical analyses also show the presence, and an increase with depth, of ferruginous nodules.

The dominant relief of these soils corresponds to rounded hills and hillocks with slightly convex slopes, valleys in a V shape, and declivity of usually 15 to 20%.

Soils are of average agricultural potential, with excellent physical conditions resulting in good drainage, among other qualities. The main constraints are low natural fertility and minimally active to active relief, with the consequent risk of erosion.

FIGURE 5 – ALFISOL PROFILE

Inceptisols

This group comprises mineral, non-hydromorphic soils with an incipient B horizon that is very heterogeneous in terms of color, thickness and texture as well as chemical activity of the clay fraction and base saturation (Figure 6). This horizon occurs immediately below any A horizon, except the weak or under a peat H horizon, producing a sequence of A, Bi, C or H, Bi, C. The group is derived from materials related to rocks of variable nature and composition, from the oldest which constitutes the foundation of the Brazilian Complex, to those of more recent origin, including metamorphic, intrusive granitic referred to as the Eo-Paleozoic, Paleozoic sedimentary, Botucatu sandstone, and extrusions of the Serra Geral Formation.

They are soils with a certain degree of progression but have not progressed enough to completely meteorize primary minerals that are more easily weathered, such as feldspar, mica, hornblende, and augite. They have no significant accumulations of iron oxide, clay and humus, which allows them to be identifying as having textural B or spodic B.

Due to the small difference in horizons and low textural gradient, many Inceptisols, especially the deeper ones, are mistaken for Oxisols. They differ from Oxisols because of: their lower pedogenetic development, reflected in the presence of a higher percentage of primary minerals less resistant to weathering (> 4%); or in the clay activity that is higher than the Oxisols (> 13 meq/100 g clay); or in the higher silt content and silt/clay ratio; or in the molecular relation of Si02/Al203 which is generally higher (> 2.2); or the paler color of the soil.

They are well to moderately drained, not very deep to deep, although both shallow profiles (<50cm) or very deep profiles (> 200cm) can occur. The thickness of the A horizon varies greatly, as a rule, from 15 to 80cm. As a result of the heterogeneity of the source material and the direct or indirect influence of climate, the soils in question present an uneven color.

The texture and other characteristics related to it also vary widely depending on the nature of the source material. Regardless whether it is derived from claystone or siltstone, the texture throughout the soil profile is usually uniform, with a small decrease or a small increase of clay from A to B, assuming a marked increase in Inceptisols developed from alluvial sediments or other cases of lithological discontinuity.

The class covers soils with high aluminum saturation (predominant) and eutrophic and dystrophic soils with low (dominant) and high clay activity. These soils are found throughout almost the entire state and occur from 20 to 1,200 meters of altitude, both in generally flat relief, such as those developed in alluvial deposits, and in mountainous landscapes, predominantly strongly undulated, undulated or smoothly undulated relief. The vegetation they support is related, among other causes, to climatic, edaphic and topographic variations and is the reason why Inceptisols have been identified as occurring under subtropical forest, transitional tropical/subtropical forest, tropical forest, subtropical grasslands and subtropical floodplain grasslands.

In the Rio Verde APA, they predominantly occupy the southern portion of the area, in the municipality of Araucária, usually associated with Xanthic Kandiudox . They also occur in the northwest border of the Basin, associated with Haplohumox usually in undulated relief.

FIGURE 6 – INCEPTISOL PROFILE

Red Oxisol and Haplohumox

Soils in this class develop from various source materials because, with the exception of materials from basic effusive rocks, they occur in virtually all other types of rocks found in the state. They are formed in areas of smoothly undulated to strongly undulated relief, under changing climatic conditions from tropical to subtropical, dominated by tropical, subtropical and intermediate tropical/subtropical forest and grasslands.

This class consists of non-hydromorphic mineral soils, with a dark reddish oxisolic B horizon with levels of iron oxides (Fe2O3) between 9 and 18% (Figure 7). They are clayey, very deep, well drained and derived from acidic effusive rocks and fine texture sedimentary rocks from the Paleozoic. They present an A, B, C, horizon sequence with usually clear transitions between A and B and gradual transitions among B sub-horizons.

They are typically more than 3 meters deep and the thickness of the A horizon generally varies from 25 to 40 centimeters, although it can reach 80 centimeters or more in humic varieties. The color of the A horizon varies greatly depending on the organic matter content but in general, it is more red in hue than 4YR with a value of 3 and chroma between 2 and 4. The B horizon varies from dark reddish--brown to dark red, with hues of 3.5 YR or redder in the upper section, with a tendency to become redder with depth. The value is 3 or very rarely 4 and chroma varies from 4 to 8. The distribution of clay is quite uniform throughout the profile, usually with a clay content greater than 60%. While the surface horizon has a granular structure from weak to moderately developed and soft or slightly hard consistency, friable, slightly plastic to plastic, and slightly sticky to sticky, the B horizon shows sub-angular blocky, weak to moderately developed structure, and variable consistency from slightly hard to hard with dry soil, from brittle to firm when moist, and slightly plastic to plastic and slightly sticky to sticky when wet.

Striking features of these soils are: a low silt/clay ratio, relatively uniform distribution of clay in the solum, and low clay mobility. Another important feature relates to the relatively low magnetic susceptibility as compared to the Purple Oxisol. This property is used in the field to differentiate Dark Red Oxisol clay texture from Purple Oxisol, which it resembles.

The cation exchange capacity is low, although at the soil surface, due to the contribution of organic matter, the T value is in general higher. The low percentages of base saturation (V% value) demonstrate the severe desaturation that the soil has undergone.

This group occurs in parts of the APA landscape with less sharp relief (smoothly undulated to undulated), such as the eastern part of the reservoir in the municipality of Araucária and the western portion of the basin with a more flat relief. They also occur in the northwest border of the basin in association with Oxisols.

FIGURE 7 – OXISOL PROFILE

Gleysols

Gleysols include hydromorphic soils consisting of mineral material which present a gley horizon within the first 150 cm of the soil surface, immediately below horizons A or E (with or without gleyzation) or of histic horizon with less than 40 cm thickness. They do not present exclusively sand or loamy-sand texture for all horizons within the first 150 cm of the soil surface or up to lithic contact. They also do not present a vertical horizon, nor a horizontal textural B horizon with abrupt textural change above, or coincident with the gley horizon, or any other type of B horizon identified above the gley horizon. A Plinthic horizon, if present, must be at a depth greater than 200cm of the soil surface.

The soils of this class are permanently or periodically saturated by water, unless artificially drained. The water stagnates within or saturation is caused by a lateral flow in the soil. In any event, ground water may reach the surface by capillarity.

They are characterized by strong gleyzation due to the moisture characteristics and virtually free from dissolved oxygen because of water saturation throughout the year, or for a long period of time, associated with the oxygen demand of biological activity.

The gleyzation process produces colors of gray, bluish or greenish, due to the reduction and solubility of iron which allows for the expression of neutral colors of clay minerals, or precipitation of ferrous compounds.

Under natural conditions, soils are poorly or very poorly drained and present the following sequence of horizons: A-Cg, A-Big-Cg, A-Btg-Cg, A-E Btg-Cg, A-Eg-Bt-Cg, Ag-Cg, H-Cg. The surface horizon has colors from gray to black (sometimes reaching 18), the A or E horizon itself can be concurrently a gley horizon, usually between 10 and 50cm thickness, and with medium to high levels of organic carbon.

The gley horizon, which may be a C, B, E or an A horizon, has more dominant blue color than 10Y, a very low, nearly neutral, chroma.

Occasionally, these soils may have a sandy texture (sand or loam-sandy) only on the surface horizons, as long as they are followed by a gley horizon of loam-sandy texture or finer.

Aside from the horizons A, H or E that might be present, the B horizon presents blocks or prismatic structures composed (or not) of angular and sub-angular blocks.

In the case of the C horizon, the structure is generally solid, but may have cracks and resemble prismatic structures when dry or after exposure for several days on the trench wall. They may present horizons that are sulfuric, calcic, solodic, sodic, salic or plinthitic in a quantity, or position insufficient to place or diagnose them as Plinthosols.

They are soils formed by original materials, stratified or not, and subject to periodic or constant excess of water, which can occur in many situations.

They commonly develop in recently formed sediments in the vicinity of watercourses and in colluvial-alluvial materials subject to hydromorphic conditions. They may also be formed in areas of flat relief along rivers, lakes and marine terraces, as well as in residual materials in basin areas and depressions. They are sometimes formed in sloped areas under the influence of emerging groundwater. They are soils that occur beneath hydrophilic or herbaceous hygrophilous vegetation, shrubs or trees.

7. UNIT SUITABILITY AND POTENTIAL CONSTRAINTS

Due to the characteristics discussed above, the region presents typical geotechnical, hydrological, pedological/agricultural, environmental and mineral characteristics.

The alluvial plains are occupied by Gleysols. For a variety of reasons, alluvial plains are the environments that require special care in planning, decision making, and environmental management. In the majority of cases, this type of soil is not taken into account in decision making related to land-use and occupancy. This results in a number of environmental problems, loss of life, and large amounts of money wasted on recovery projects, which in most cases are either ineffective or generate other larger problems rather than solving them.

It is curious to note that although this is known, inappropriate occupation in alluvial plains continues. In urban areas they are usually the first to be occupied; these regions are often invaded by low-income sectors of the population, creating a multitude of difficulties for public authorities, who end up having to provide them with urban infrastructure. As discussed below, this is quite complicated to do in these areas.

In planning, it is important to take into account that alluvial plains are very problematic for any kind of infrastructure, for the following reasons:

- In particular, the plains are likely to experience flooding which is often very rapid and long-lasting. Although the frequency of flooding is greater in areas closer to the riverbed, it is noted that even the most distant plains will at some point be flooded. They are areas of the river domain and as such they are inappropriate for urbanization. It is important to take into account that infrastructure to prevent floods are complex, expensive, and in most cases inefficient, or they can lead to more serious problems;
- for almost the entire expanse of the alluvial plain area, the ability to drain the surface and subsurface is very limited, i.e. the water barely circulates or circulate very slowly and in the majority of alluvial plains the groundwater emerges on the surface or is present at depths generally less than 2 meters;
- infrastructure that is placed below the surface of alluvial plains very quickly corrode and become damaged. Thus, special care must be taken with the quality of the materials used, especially in projects for the movement and storage of pollutants, such as gas pipelines, oil pipelines, fuel tanks, etc. We emphasize that if a leak occurs there is a strong possibility that the pollutants will directly contaminate the groundwater. If this happens, it will cause major, long-lasting, negative impacts, and much money will be spent on recovery. If these types of works exist on floodplains, it is important for them to be constantly monitored. These are environments that are naturally quite unhealthy for humans because soil moisture remains high most of the year and in the summer becomes very hot, supporting the proliferation of various types of insects, fungi and bacteria;
- it is a transitional land between terrestrial and aquatic environments, with fairly typical vegetation that survives well in both dry and aqueous environments. This vegetation has an important environmental role and as such it should be preserved. It acts as a barrier to retain debris washed down from the highlands by the rushing waters, preventing them from reaching the watercourses. Furthermore, they are the habitat of many predators of insects and rats, such as frogs, toads and snakes. Among the typical vegetation of these areas, there is the cattail, which can be used for making various items such as baskets, bags, mats, etc.

The areas occupied by Oxisols are those with the least constraints on usage. Because of the geology, the relief is characterized as undulating to smoothly undulating, although very active, i.e. a softened relief but with moderate to high density of elevated areas and drainage channels. The elevations are low, ranging between 30 and 50 meters, with relatively large flat tops, short convex slopes, and declivity between 5 and 15%, which in some cases may vary between 20 and 25%. The main drainage system presents open valleys and relatively wide plains.

These are lands where runoff occurs quite differently depending on the location in the relief. At the top, because the relief is smooth and covered with a thick layer of very permeable black soil, runoff is relatively slow. But on the slopes, with steep declivity, runoff is quite rapid. Thus, when laying draining networks and storm water management systems, especially at the lower levels, it is important to plan for a large volume of water and strong, erosive torrents that can occur suddenly even during rain of moderate intensity.

It is important to note that due to good hydrodynamic characteristics of surface soils, these terrains are important for local replenishment of groundwater, which in most of the basin is very low. They are also areas that contain many patches of natural forest with large populations of the Paraná Pine. Considering the forest cover, coupled with the contrast of its relief with the surrounding mountainous areas, it makes for beautiful rural landscapes.

The portions of the Basin in which Alfisols and Inceptisols occur have an intermediate level of restrictions in relation to usage. Because of the predominance of lithologies and soils that are poorly permeable, and because most lithology is folded, the entire length of this Subdomain is characterized as an environment of extremely active relief, that is diverse and mountainous. Even the more level areas contain a high-density of elevations intersected by a dense system of drainage channels of short and compact valleys. These characteristics indicate that the land is naturally of very low permeability with a high potential for hydric erosion. Thus, special care must be taken in order to preserve the natural vegetation and not increase impermeability.

Clay soils, independent from other variables, are quite porous and have the ability to retain and secure elements. As such, they respond well to fertilization, have good assimilation of organic matter, and during the dry season retain water availability for plants over a long period (high water capacity soils).

Soils of low fertility that are very rich in aluminum should predominate, so they should be quite acidic. As a consequence they often need to be fertilized and modified by applying dolomitic limestone.

Clayed soils are naturally of low permeability, indicating that drip irrigation methods should be used. If using other methods, rather than penetrating the soil, most of the water runs off creating highly erosive torrents.

These soils become compacted and impermeable if heavy mechanized equipment is continuously used or if the soils are trampled by cattle. Continuous high loads cause the formation of a very hardened and impermeable subsurface layer. The difference in permeability between this layer and the more friable and permeable upper layer generates a situation in which during rainfall the upper layer becomes saturated in water and is easily removed by sheet erosion. Thus, if handled improperly, a clayey soil that is naturally resistant to erosion becomes as erosive as or more than a sandy soil.

Much of this land, although it is quite rugged, was cleared and occupied by pastureland and particularly reforestation. There are also many abandoned pastures that are today covered by ferns, a type of vegetation highly susceptible to fire.

Because of their more adequate topographical and pedologic features, it is the granitic terrain that is most densely occupied in the region, mainly by agriculture and pastureland. For this reason they are also the most devoid of their natural features, especially because of deforestation.

They have many water springs that have good year--round flow, thus they are important for the maintenance of the regular river flow during dry periods

A positive aspect of this soil is the fact that in almost the entire area the alteration mantle is clayey and consequently of low natural erosion. In most parts of the region it can be easily excavated with only with tools and cutting machinery at depths greater than 10 meters.

As a downside, we note the predominance of the following factors:

- it is common to have soils and rocks with the more contrasting geotechnical characteristics side by side or within a few meters of each other. Thus, in the case of linear construction projects, very detailed geotechnical studies should be conducted and supported by sampling and testing of materials collected at various depths and across a tightly laid grid, which means high costs, both in the planning phase and during the execution of the work. Localized Geotechnical testing points have little lateral and vertical representation in most of the region;
- where the relief has significant and accelerated wear, it presents favorable topographic features for the natural movement of large amounts of water and erosion processes, both sheet and concentrated;
- when excavating a little deeper, the slopes of the cut are more likely to have exposed materials that are highly susceptible to erosion when subjected to concentrations of rainwater. In most of the area, the C horizon was very erosive and unstable along cut slopes, outcrops or was present at low depths;
- the lands tend to have very fast-running runoff. Therefore, drainage networks and storm water management should be planned in order to support the large volume of water and strong torrents with high erosive potential that are suddenly formed with stronger rain.

Because runoff is very fast when it rains, most of the water flows quickly to the drainage channels that, for the most part, are devoid of floodplains, thus creating turbulent waters with a high potential for transporting sediment. It is therefore an environment with high potential for water erosion, that can transport a high load of debris leading to the silting of rivers, with dispersed, rather than concentrated hydrogeological characteristics. Therefore, if a pollutant reaches a watercourse it can be quickly carried over long distances.

8. PHYSICAL AND CHEMICAL CHARACTERIZATION OF IDENTIFIED UNITS

Table 1 shows a summary of the main physical and chemical characteristics of soil units mapped in the Basin, which served as the basis for their classification with crite-

TABLE 1 – RESULTS OF SOIL ANALYSIS FOR THE CHARACTERISTIC PROFILES OF MAPPING UNITS

IDENTIFICATION			HORIZON A		Cmolc/dm³					g/dm³	g/dm³					%												
Soil Taype	Coord UTM SAD 69		Alt (m)	Decl %	Ca	Mg	k	Al	HAl	C	P	Fe	Mn	Cu	Zn	pH SMP	pH CaCl	Sand	Silt	Clay	Depth (cm)	Color	Texture	Plasticity	Waxiness	Rock	Stones	Roots
Gleysol	650470	7180886	892	0	7,82	3,7	0,3	0	3,69	19,37	4,5	189	64,1	2,43	3,84	6,4	5,9	25	37,5	37,5	20	Dark brown	A	High	B	Absent	Few	Few
Alfisol	65470	7189853	947	16	7,31	3,77	0,3	0	2,36	11,96	11,8	20,5	28,1	1	1,46	7	6,2	45	17,5	37,5	5	Dark brown	A	Low	B	Absent	Many	Absent
Oxisol	656813	7182018	952	5	1,7	0,99	0,04	0,8	6,69	14,04	14,5	49,2	3,35	1,45	0,8	5,6	4,5	52,5	15	32,5	30	Dark brown	A	ME	B	Absent	Absent	Few
Inceptisol	654096	7181401	896	8	3,87	3,57	0,17	0	4,96	17,81	3,45	48,9	30,6	1,67	3,89	6	5	55	27,5	17,5	50	Dark brown	ME	Low	B	Present	Present	Many
IDENTIFICATION			HORIZON B		Cmolc/dm³					g/dm³	g/dm³					%												
Soil Taype	Coord UTM SAD 69		Alt (m)	Decl %	Ca	Mg	k	Al	HAl	C	P	Fe	Mn	Cu	Zn	pH SMP	pH CaCl	Sand	Silt	Clay	Depth (cm)	Color	Texture	Plasticity	Waxiness	Rock	Stones	Roots
Gleysol	650470	7180886	892	0	3,13	1,05	0,1	0,9	7,2	9,23	9,65	127	138	2,97	3,4	5,5	4,4	40	30	30	40	Greyish brown	MA	High	B	Likeli	Absent	Absent
Alfisol	65470	7189853	947	16	3,34	1,7	0,04	0	3,97	3,51	1	19,7	7,25	0,28	0,26	6,3	5,3	52,5	17,5	30	0	Red	A	ME	E	Absent	Many	Absent
Oxisol	656813	7182018	952	5	0,8	0,45	0,04	0,9	7,2	2,86	0,5	22,3	4,63	0,38	0,41	5,5	4,5	47,5	15	37,5	0	Light reddish brown	A	High	MO	Absent	Absent	Absent
Inceptisol	654096	7181401	896	8	1,35	2,7	0,04	0,3	4,96	3,51	1	39,2	9,22	1,81	0,41	6	4,8	40	37,5	22,5	60	Yellowish brown	A	ME	E	Present	Present	Few

ria established by EMBRAPA.

9. CONCLUSION

The study of the pedology of the region of the Rio Verde Basin allows us to conclude that the factor that most influences the soil typology is geomorphology.

The geological base of the Basin differs little across the study area and the determinant factors in the characterization of the type of soil are declivity, length of the slope, position in the relief, and altitude.

In terms of fertility, in their natural state soils normally exhibit dystrophic characteristics (V% <50%) and average saturations of aluminum, producing a medium acidity in the soil. However, in this same aspect, there are soils that when corrected and managed, show good production potential. We observed that most of these cultivated soils are modified for normal fertility and the main problem identified was excess and not a lack of nutrients in the soil, especially phosphorus.

Regarding the environment, the cultivated soils are quite degraded due to intensive use combined with the lack of soil conservation practices.

The intersection of these factors (high level of fertilization + intensive use + inappropriate management) results in agricultural practices that have a high potential impact on the environment, especially in relation to water resources.

Given this reality, fertility management, adoption of conservationist cultural and mechanical practices, and proper land-use according to suitability, are of vital importance for both maintaining existing production systems in the Basin, as well as maintaining the quality of water resources.

10. REFERENCES

CHUEH, A.M. (2005) *Análise do Potencial de Degradação Ambiental na Bacia Hidrográfica do Rio Pequeno em São José dos Pinhais/PR, por meio do DFC – Diagnóstico Físico Conservacionista.* R. RA'E GA, Curitiba, n. 10, p. 61-71, Editora UFPR.

COORDENADORIA DA REGIÃO METROPOLITANA DE CURITIBA – COMEC. (1988) *Plano Diretor da Bacia hidrográfica: estudos geológicos.* CURITIBA: COMEC, 32 p.

COORDENADORIA DA REGIÃO METROPOLITANA DE CURITIBA – COMEC. (2002) *Proposta de Macrozoneamento da Bacia do Rio Verde.* CURITIBA: COMEC.

EMPRESA BRASILEIRA DE PESQUISA AGROPECUÁRIA / CENTRO NACIONAL DE PESQUISA DE SOLOS – EMBRAPA/CNPF. (1999) *Sistema brasileiro de classificação de solos.* Brasília: Embrapa/SPI e Embrapa Solos, 412 p.

EMPRESA BRASILEIRA DE PESQUISA AGROPECUÁRIA / SERVIÇO NACIONAL DE LEVANTAMENTO E CONSERVAÇÃO DE SOLOS – EMBRAPA/SNLCS. (1984) *Levantamento de Reconhecimento de Solos do Estado do Paraná.* Curitiba: Embrapa/ SNLCS/SUDESUL/ IAPAR, 791 p.

FONTES, J.L. (2006) Programa de Microbacias Hidrográficas: justificativa, histórico, estratégia e resultados. In: RODRIGUES, V.A.; BUCCI, L.A. *Manejo de microbacias hidrográficas: experiências nacionais e internacionais* (Org.). Botucatu, FEPAF, p. 200-215.

FREISCHFRESSER, V. (1999) Políticas públicas e a formação de redes conservacionistas em microbacias hidrográficas: o exemplo do Paraná Rural. *Revista Paranaense de Desenvolvimento.* Curitiba, n. 95, jan./abr., p. 61-77.

HESPANHOL, A.N.; HESPANHOL; R.A.M. (2004) Desenvolvimento rural e poder local: a situação dos CMDRS na região de presidente prudente – SP. In: *I Seminário Internacional, 2004, Rio Claro.* CD do I Seminário Internacional. Rio Claro: IGCE-UNESP, p. 1-11.

LEPSCH, I.F. (1983) *Manual para Levantamento utilitário de meio físico e classificação de terras no sistema de capacidade de uso.* 4ª aproximação. Campinas: Sociedade Brasileira de Ciência do Solo.

MACHADO, M. L. da S. & RODRIGUES, A. dos S. (1997) *A geração de tecnologia e o desenvolvimento da agricultura familiar.* Londrina, IAPAR, 80 p. (IAPAR. Boletim Técnico, 57).

MINEROPAR – (2000) *Proposição de Expansão Urbana na Região Metropolitana de Curitiba, com Base em Levantamentos e Cadastramentos Geotécnicos.* Curitiba.

MIRANDA, M. & MIRANDA, G.M. (1995) A experiência do Instituto Agronômico do Paraná – IAPAR, na busca do desenvolvimento da agricultura familiar. In: WORKSHOP O DESENVOLVIMENTO DE UMA OUTRA AGRICULTURA: ACESSO À TERRA E A MEIOS DE PRODUÇÃO, A QUESTÃO DA FOME E A 3 Sistemas de produção equilibrados e coerentes, em harmonia com seu meio ambiente, adaptados à diversidade das situações locais, reproduzíveis por um grande número de agricultores, rentáveis e sustentáveis. INTEGRAÇÃO SOCIAL (1995: Curitiba). *Anais...* Curitiba: Cooperação Inter-Universitária Franco-Brasileira, sp.

NAVARRO, Z. (2001) *Manejo de recursos naturais e desenvolvimento rural.* Porto Alegre: Programa de Pós-graduação em Desenvolvimento Rural. (Relatório preliminar).

PARANÁ. (1999) Secretaria de Estado da Agricultura e do Abastecimento. *Manual Operativo do Projeto Paraná 12 Meses.* Curitiba, 241 p.

PASSINI, J.J. *Geração e comunicação de inovações tecnológicas para a agricultura familiar.* Curitiba, 1999. Dissertation (Master's degree) – Graduation Program in Technology, CEFET/PR, 154p.

PASSINI, J.J. (1997) Redes de propriedades de referência. In: IAPAR. *Enfoque Sistêmico em P&D: a experiência metodológica do IAPAR.* Londrina, 152 p. Ilust. (IAPAR. Circular, 97).

PAYÉS, M. A. M. (1989) *Sistemas de produção predominantes no município de Rio Azul* – Paraná – uma proposta teórico-metodológica. Londrina: IAPAR (IAPAR. Boletim técnico, 27).

PRADO, H. (1993) *Manual de Classificação de Solos do Brasil.* Jaboticabal: FUNEP, 218 p.

RASCHIATORE R.A.; MOREIRA, D.A. (2006) Inovação na implementação do Programa Estadual de Microbacias Hidrográficas do Estado de São Paulo. *Gestão & Produção*, v. 13, n. 3, p. 517-529, Set.-Dez.

SABANES, L. (2002) *Manejo sócio-ambiental de recursos naturais e políticas públicas: um estudo comparativo dos Projetos Paraná Rural e Microbacias.* Programa de Pós-Graduação em Desenvolvimento Rural da UFRGS. (Master Dissertation)

VIEIRA, L.S.; VIEIRA, M.N. F. *Manual de Morfologia e Classificação de Solos.* 2. ed. São Paulo: Editora Agronômica Ceres Ltda. 313 p.

WORLD BANK. (1997) *Land Management III – Project São Paulo, – Environmentally and socially sustanaible development* – Sector Management Unit – Latin America and the Caribbean region.

CHAPTER 6

VEGETATION COVER AND LAND USE MAPPING, CHARACTERIZATION AND DIAGNOSIS

Carlos Vellozo Roderjan
Pyramon Accioly

VEGETATION COVER AND LAND USE MAPPING, CHARACTERIZATION AND DIAGNOSIS

Carlos Vellozo Roderjan
Pyramon Accioly

SUMMARY

The Rio Verde Basin was mapped in order to identify and quantify the natural vegetation cover. Different types of land-use were also mapped to identify places where the there is a need to modify land-use and promote vegetation restoration based on the region's natural characteristics. To meet the purposes of this study, a map legend was established that incorporated the most characteristic and significant types. Considering Middle and Advanced Vegetation Succession stages, Gallery Forests and Floodplains as complex environments with high levels of biodiversity that maintain environmental quality, together they represent a protected area of 36.78% of the total mapped area. Agriculture involving crops and pastureland are the most significant form of human modified land-use in the Rio Verde Basin, occupying almost half of its surface (48.56%). Productive stands or Reforestations, using *Pinus* and Eucalyptus species, are not very significant in the Rio Verde Basin, occupying only 1.51% of the surface area (360.85ha). The Urban and/ or Built areas make-up 6.34% of the basin surface and are concentrated mainly in the municipality of Campo Largo, within the vicinity of highway BR 277, and at the north end of the basin within the municipality of Campo Magro around the town center.

KEYWORDS

Rio Verde, vegetation cover, land use, mapping, diagnosis.

1. INTRODUCTION

Among the environmental impacts caused by the expansion of human occupation of the land, the impact on water resources is undoubtedly the most serious. Of more concern, however, is the protective function of vegetation around natural and artificial water bodies that has been poorly understood throughout human history. Hitherto, humans have ignored the facts and used the soil at their own convenience. Now, they seek for solutions to reverse this situation, starting with the obvious: the restoration of protective vegetation around springs and water sources, river banks, lakes and reservoirs.

To this end, the Rio Verde Basin was mapped and its natural vegetation cover and different forms of land-use were identified and quantified, aiming to detect areas where modifications in land-use or vegetation restoration are needed based on the region's natural characteristics.

2. NATURAL VEGETATION COVER IN THE RIO VERDE BASIN

The region of the Rio Verde Basin was originally covered by predominantly forest vegetation, where Araucária (Mixed Ombrophilous Forest) thrived. These forests are characterized by the dominant presence of *Araucaria (Araucaria angustifolia)* and a diverse set of other tree species, some of which are of high commercial value, such as the Brazilian walnut (Ocotea porosa), cedar (Cedrela fissilis), and cinnamon-sassafras (*Ocotea odorifera*).

In the plains along the basin's rivers, there was a different vegetation complex that was adapted to the conditions of water saturation (hydromorphic soils), popularly known as floodplain or swamps (Pioneer Formations with Fluvio-Lacustrine Influence). Depending on its development, the vegetation in floodplains could have an exclusive herbaceous physiognomy or it could be interspersed with isolated trees (corticeiras – *Erythrina crista-galli, branquilhos – Sebastiania* spp. and *gerivás – Syagrus romanzoffiana*). Usually associated with floodplains and in specific pedologic conditions, gallery forests or riparian forests (Alluvial Mixed Ombrophilous Forest) occurred in places where the substrate presented better drainage conditions.

3. CURRENT VEGETATION COVER OF THE RIO VERDE BASIN

Over the past two centuries, the region that includes the Rio Verde Basin has been heavily modified by a wide range of human interventions, beginning with logging and followed by the conversion of forests into crops and pastureland. In recent decades it was strongly influenced by urban and industrial expansion and currently there is little remaining natural vegetation.

To diagnose and quantify this transformation, a map legend was established that included the most characteristic and typical scenarios of natural vegetation cover and the most significant land-use activities, aiming at meeting the purposes of the present study. Thus, the following classification was defined:

Mixed Ombrophilous Forest

1. Initial Stage of Vegetation Succession – shrub (*capoeirinhas*).
2. Initial Stage of Vegetation Succession – arboreal (*bracatingais* or *capoeiras*).
3. Middle Stage of Vegetation Succession (*capoeirão*)

4. Advanced Stage of Vegetation Succession (forest)
5. Gallery forest

Pioneer Formation Areas
6. Floodplain

Areas of Anthropogenic Intervention
7. Reforestation
8. Agriculture and Livestock
9. Urban and/or Built Areas
10. Exposed soil
11. Water bodies

Table 1 presents the Rio Verde Basin surface occupation, in hectares and percent, for the entire mapped area (Figure 1) according to the proposed typology.

3.1 CHARACTERIZATION OF NATURAL VEGETATION COVER AND LAND-USE

3.1.1 Mixed Ombrophilous Forest (forest with Araucaria)

Initial Stage of Vegetation Succession – shrub (capoeirinhas)

This category included the initial phases of vegetation occupation after abandonment of an area, popularly called *capoeirinhas* or *quiçaças*. These are sparsely distributed over the study area, accounting for only 2.12% of the total mapped area. They are usually the result of recent interventions (agricultural areas or pastureland that has been abandoned or is temporary left fallow), on smoothly-undulated to undulated lands, free from hydromorphy. The vegetation varies from herbaceous-graminoid formations to the establishment of pioneer shrub species and the type of vegetation is directly related to the amount of time the land has been left undisturbed.

Among the herbaceous plants, the presence of pampas grass (*Cortaderia selloana)* is striking, with beautiful inflorescences, interspersed with *Eupatorium* (*E. laevigatum*), *Senecio* (*S. brasiliensis*), *Solidago* and *Mikania* (*M. cordifolia*) genera of Asteraceae, Verbenaceae *Lantana câmara*, *Phytolacca thirsiflora,* species of Phytolaccaceae, *Eryngium* sp, species of Apiaceae, and *Pteridium aracnoideae*, among others.

In subsequent stages of the natural succession process, these formations are occupied by shrub species of the *Baccharis* genus of Asteraceae, particularly represented by *B. uncinella*, forming dense bunches called *"vassourais"*, associated with other species of the same genus (Figure 2).

Initial Stage of Vegetation Succession – arboreal (bracatingais or capoeiras)

This stage was identified by tree formations where *bracatinga* (*Mimosa scabrella*) prevails. *Mimosa scabrellaia* is a pioneer leguminous species that grows quickly and has a short life-span (15 – 20 years). It spontaneously grows in abandoned or unused areas. It is widely cultivated in the Southern highlands as a source of energy (charcoal) and the production of stanchions used in civil construction. They form dense populations of a single stratification and can reach 10 to 15m in height at maturity. If these areas are not exploited, they provide a suitable environment for the establishment of another stage of tree growth enabling the succession process to occur naturally. This type of land cover occupies only 1.32% of the surface (316.45ha) of the study area and it is rapidly being replaced by plantations of fast-growing tree species (reforestations of *Pinus* and *Eucalyptus spp.*) (Figure 3).

Middle Stage of Vegetation Succession (capoeirão)

The vegetation that composes this stage of vegetation succession is formed by relatively homogenous and dense tree communities, with a canopy height varying between 10 and 15m, under which a second layer of trees may occur (understory). It consists of pioneer tree species, mainly established due to ornithologic dispersal (by birds), such as the Brazilian pepper (*Schinus terebinthifolius*), the *pinho-bravo* (*Podocarpus lambertii*), the *vacum* (*Alophyllus edulis*), the *canjica* (*Rhamnus sphaerosperma*), the *miguel-pintado* (*Matayba elaeagnoides*), the *guabiroba* (*Campomanesia xanthocarpa*), *mamica-de-porca* (*Zanthoxyllum kleinii* and *Z. rhoifolia) and the pessegueiro-bravo* (*Prunus brasiliensis*); and by anemochory (dispersal by the wind), trees such as the *vassourão-branco* (*Piptocarpha angustifolia*), the *vassourão-preto* (*Vernonia discolor*), and the *cambará (Gochnatia polymorpha*).

TABLE 1 – VEGETATION COVER IN HECTARES AND PERCENT

TYPE	AREA (ha)	%
1. Initial Stage of Vegetation Succession – shrub (*capoeirinhas*)	506.16	2.12
2.Initial Stage of Vegetation Succession – arboreal (*bracatingais*)capoeiras)capoeira)	316.45	1.32
3. Middle Stage of Vegetation Succession (*capoeirão*)	6,313.35	26.44
4. Advanced Stage of Vegetation Succession (forest)	973.23	4.08
5. Gallery forest	801.58	3.36
6. Floodplain	691.82	2.90
7. Reforestation	360.85	1.51
8. Agriculture and Livestock	11,595.74	48.56
9. Urban and/or Built Areas	1,514.97	6.34
10. Exposed soil	49.35	0.21
11. Water bodies	754.80	3.16
TOTAL	23,878.31	100.00

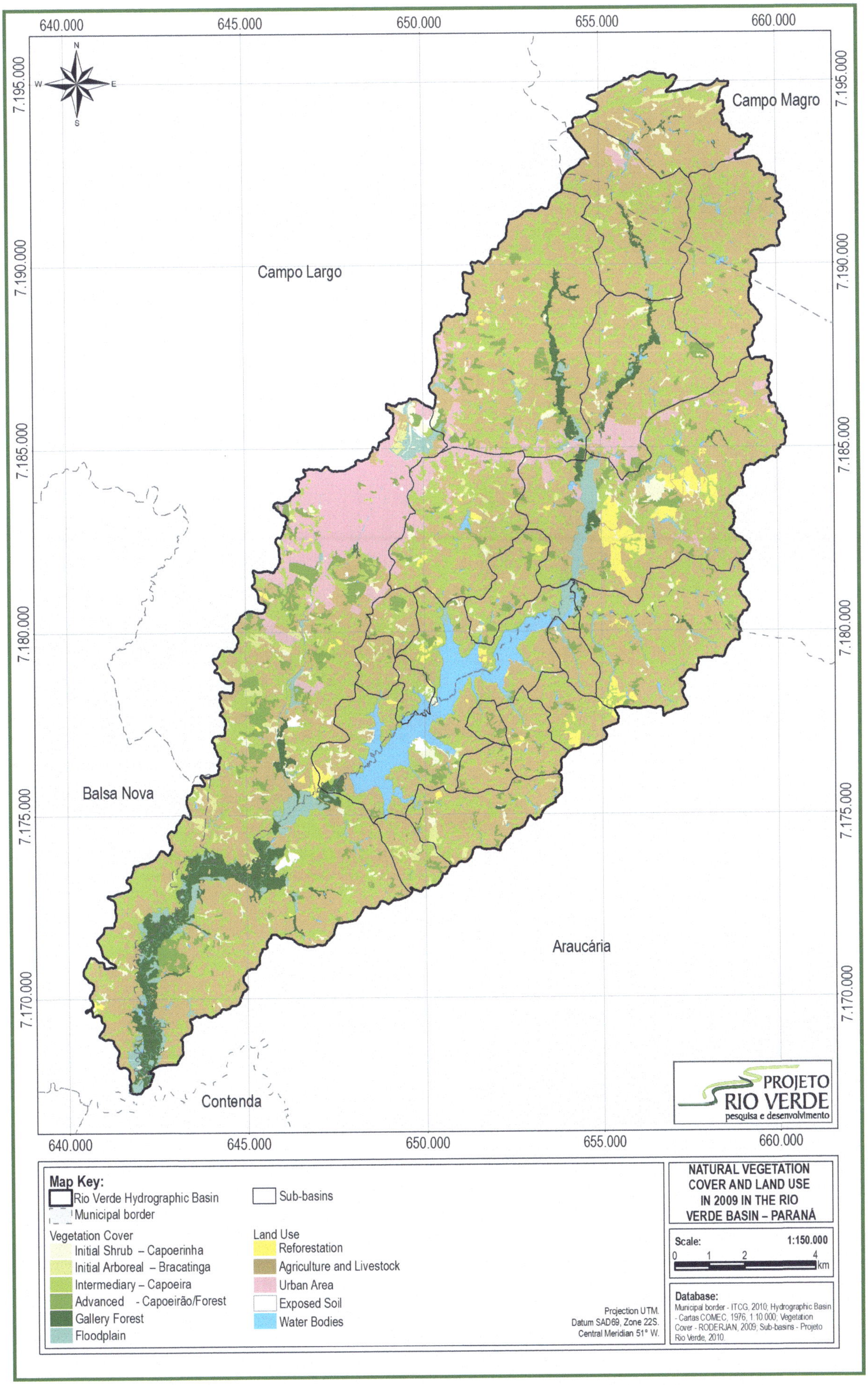

FIGURE 1 – MAP OF NATURAL VEGETATION COVER AND LAND USE IN THE RIO VERDE BASIN

FIGURE 2 – AN INITIAL STAGE OF VEGETATION SUCCESSION, SHRUB PHASE

FIGURE 3 – THE INITIAL STAGE OF VEGETATION SUCCESSION – ARBOREAL (*BRACATINGA*)

These areas constitute a more significant type of natural vegetation in the mapped area, occupying 26.44% of its surface (6,313.35 ha) (Figure 4). According to current environmental legislation, they are areas in which harvesting or land conversion is prohibited, ensuring the preservation of about one third of the surface of the drainage basin.

FIGURE 4 – THE MIDDLE STAGE OF VEGETATION SUCCESSION (*CAPOEIRA*)

Advanced Stage of Vegetation Succession (capoeirão/forest)

This type of land cover consists of the most evolved, diversified and stratified forests, that demonstrate more advanced phases of secondary succession (due to anthropogenic interventions) or have remnant individuals in forests that have been deeply altered by selective harvesting of the best wood. This stage of succession is usually characterized by the presence of Araucaria (*Araucaria angustifolia*). These areas are not significant in the Basin, constituting only 4.08% of the surface or 973.23 hectares. However, these areas have the greatest diversity and richness as seen through the presence of several life forms (trees, shrubs, herbs, epiphytes and lianas) (Figure 5) and the associated fauna.

Dominating the canopy, between 15 and 20m in height or below the tops of the remnant pine trees, it is common to find *miguel-pintado* (*Matayba elaeagnoides*), cedar (*Cedrela fissilis*), *guabiroba* (*Campomanesia xanthocarpa*), *pinho-bravo* (*Podocarpus lambertii*) and *cuvatã* (*Cupania vernalis*), among other tree species. In clearings opened by selective harvesting, some pioneer species such as *canela-guaicá* (*Ocotea puberula*), *vassourão-branco*, *vassourão-preto* and even *bracatinga* can be found.

FIGURE 5 – THE ADVANCED STAGE OF VEGETATION SUCCESSION (*CAPOEIRÃO*/FOREST), CHARACTERIZED BY THE PRESENCE OF ARAUCARIA (*ARAUCARIA ANGUSTIFOLIA*)

Gallery Forest or riparian forest (Alluvial mixed ombrophilous Forest)

This type of land cover is composed of forest formations established in alluvial plains adjacent to watercourses that are subject to the fluctuations of the water table as the result of rainfall and may present variable size and diversity depending on the degree of soil development that supports them (Fluvic Entisol and Gleysol). At any time, however, they are characterized by the striking presence and dominance of the *branquilho (Sebastiania commersoniana*), a pioneer tree species that colonizes these environments. This type of land cover can include dense groupings of trees of small stature (5 – 7 m in height) at early succession stages (*branquilha)* and varies to diversified, stratified, taller (up to 15 – 18 m in height) forests in more advanced stages of succession when the *branquilho* shares the area with the *aroeira* (*Schinus terebinthifolius*), *murta* (*Blepharocalyx salicifolius*), *veludo* (*Myrrhynium loranthoides*), *guabiroba*, *pessegueiro-bravo* (*Prunus brasiliensis*) and *jerivá* (*Syagrus romanzoffianum*), among others. In the understory, a second layer may be found characterized by the presence of *vacum* (*Allophylus edulis*), *embiras* (*Daphnopsis* spp.) and *cambuí* (*Myrciaria tenella*), among others. Araucaria (*A. angustifolia*) occurs sometimes forming round clusters in dykes and shoulders that are free from hydromorphy. This type of land cover constitutes 3.34% of the mapped area (801.58 ha) (Figure 6).

FIGURE 6 – THE GALLERY FOREST (ALLUVIAL MIXED OMBROPHILOUS FOREST), USUALLY ASSOCIATED WITH FLOODPLAINS

3.1.2 Pioneer Formations Areas with Floodplain influence (floodplains)

This type of land cover corresponds to herbaceous vegetation occupying the floodplain, subject to the fluctuations of the water table, depending on the local water conditions. Generally on Histosols and Gleysols, which are both hydromorphic soils, the floodplains are restrictive environments for the establishment of arboreal vegetation. On the other hand, these soils are efficient regulators of the water flow, mainly during periods of heavy rainfall and flooding. Their degree of development and their geographic coverage are variable depending on the characteristics of the location.

These areas are characterized by the prevalence of caespitose and rhizomatous herbaceous species along with some herbaceous and erect sub-shrubs, creating a vegetation formation up to 2 meters in height. Many of the species typical of this formation create clusters of varying sizes depending on the geomorphic design of the site. The caespitose species that physiognomically stand out are: *Panicum urticans* and *Echinochloa crusgalli* (Poaceae), *Carex brasiliensis* and *Cyperus ferax* (Cyperaceae), besides some other less conspicuous species of the same families. Among the rhizomatous, *Thalia geniculata* (Marantaceae), *Scirpus californicus* (Cyperaceae), *Echinodorus grandiflorus* (Alismataceae) and *Typha domingensis* (Typhaceae) occur more frequently. In the mapped area, 691.82 hectares of floodplains were recorded corresponding to 2.90% of the area (Figure 7).

3.1.3 Areas of anthropogenic interventions

Reforestation

These areas include all the commercial plantations of

FIGURE 7 – VIEW OF A FLOODPLAIN AREA (PIONEER FORMATION WITH FLUVIO-LACUSTRINE INFLUENCE)

fast-growing tree species, mainly of *Pinus* and *Eucalyptus* genera, with varying ages, shapes and sizes. The surface occupied by this activity is not significant, corresponding to only 1.51% (360.85ha) of the mapped area (Figure 8).

Agriculture and Livestock

This land-use classification includes the diverse forms of agricultural land uses, particularly crops and pastureland. This classification makes up the most significant surface area representing 48.56% of the mapped area, equaling 11,595.74 ha (Figure 9).

Exposed soil

Areas of exposed soil consist of areas where mining activities were detected, or areas that were void of vegetation cover when flying over the region in 2000. It composes only 49.34 ha, or 0.21% of the mapped surface.

Urban and/or built areas

This classification consists of all urban spaces, whether or not they are amalgamated. This land-use makes-up 6.34% of the mapped area, or 1,514.97 ha.

FIGURE 8 – A REFORESTATION (STAND) OF *EUCALYPTUS* SP. IN THE CENTRE AND A MIDDLE STAGE OF VEGETATION SUCCESSION IN THE BACKGROUND

FIGURE 9 – OTHER FORMS OF LAND-USE CONSIDERED IN THE MAPPING (A: AGRICULTURAL CROPS; B: PASTURES; C: EXPOSED SOIL; D: URBAN AND/OR BUILT AREAS)

3.2 DIAGNOSIS OF VEGETATION COVER AND LAND-USE IN AREAS OF PERMANENT PRESERVATION (APP)

3.2.1 Categories of Areas of Permanent Preservation (APPs)

Considering the Brazilian Forest Code (Federal Law 4.771/65 from September 15, 1965), the CONAMA resolutions complementary to the Forest Code, and the Joint Resolution IBAMA/SEMA/IAP 5/2008 that defines the criteria for the assessment of Wetlands and their Protective Environs in the State of Paraná, the following areas were considered as areas of environmental sensitivity that are protected by law:

- 30 meter marginal strip along the water courses up to 10 meters in width;
- 50 meter radius around springs;
- 100 meter strip around the Rio V erde Reservoir (comprised as rural area);
- 50 meter strip around ponds and lakes;
- 50 meter marginal strip around floodplains*, basically represented by alluvial plains;
- areas with slopes greater than 45°;
- areas located on hilltops;
- crests lines;
- middle and advanced stages of vegetation succession.

* Minimum value adopted in this diagnosis is based on the scale adopted for the Protective Environs established by the Joint Resolution IBAMA/SEMA/IAP 5/2008, that varies from 50 to 90 m depending on the slope and the texture of the soil.

In order to identify inconsistencies between land-use practices observed in the Rio Verde Basin and the areas legally established as APPs, the legal constraints mentioned above were spatialized using GIS (Geographic Information System) and subsequently compared with the vegetation cover and land-use maps. From this, we assessed the level of environmental preservation of the APPs by cross-referencing the selected geo-environmental parameters which allowed us to determine the different forms of land-use in the APPs in terms of area and representativeness (%). The method of analysis using raster models gives flexibility to the data cross-referencing; however, it does require a small amount of rounding-off but this represents a variation not greater than 2% of the calculated areas.

3.2.2 Vegetation cover and land-use in areas of permanent preservation

Federal Law 4771/65 establishes that the function of the areas of permanent preservation is to "preserve the water resources, landscape, geological stability, biodiversity, gene flow of fauna and flora, protect the soil and ensure the well-being of human populations." The objective of Municipal Law 1.963/2007 of Land Use and Occupancy is to "protect the valley floors, springs, public parks, and other areas of environmental interest." From this, it was possible to define, identify and quantify the inappropriate uses of APPs. Thus, the classifications of Agriculture and Livestock, Reforestation, Urbanized Areas and Exposed Soil represented 36.6% (2,705.34 ha) of inappropriate land-use in the APPs, revealing significant disagreement between what is defined by the legislation and what is practiced in an area of interest, such as a source for public water supply for the Metropolitan Region of Curitiba (DE 3411/2008).

On the other hand, it is important to note that 59.22% (4,377.29 ha) or practically half of the surface of the APPs is occupied by natural forms of vegetation, mainly the Middle Stage of Vegetation Succession with 40.93% and Gallery Forests with 6.75%.

Table 2 shows the total values of surface area for each land-use classification in the APPs, in both hectares and percent.

Table 3 contains the classes of vegetation cover and land-use for each legal constraint on land-use included in the present analysis.

TABLE 2 – NATURAL VEGETATION COVER AND LAND-USE IN AREAS OF PERMANENT PRESERVATION (APP) IN THE RIO VERDE BASIN

TYPEC	AREA (ha)	%
APP Shrub Initial Stage	184.31	2.49
APP Arboreal Initial Stage	79.99	1.08
APP Middle Stage	3,025.92	40.93
APP Advanced Stage (Forest)	280.53	3.79
APP Gallery Forest	499.11	6.75
APP Floodplain	307.43	4.16
APP Reforestation	74.07	1.00
APP Agriculture and Livestock	2,398.98	32.45
APP Urbanized and/or Built Areas	223.47	3.02
APP Exposed Soil	8.82	0.12
APP Water bodies	309.38	4.18
TOTAL	7,392.01	100.00

TABLE 3 – CLASSES OF NATURAL VEGETATION COVER AND LAND-USE IN AREAS OF PERMANENT PRESERVATION IN THE RIO VERDE BASIN

100M AROUND THE RIO VERDE DAM		
TYPE	**AREA (ha)**	**%**
100m dam + shrub initial stage	14.66	4.16
100m dam + arboreal initial stage	1.01	0.29
100m dam + middle stage	189.60	53.82
100m dam + advanced stage	5.19	1.47
100m dam + gallery forest	1.98	0.56
100m dam + floodplain	25.01	7.10
100m dam + reforestation	5.82	1.65
100m dam + agriculture and livestock	103.94	29.51
100m dam + exposed soil	4.03	1.14
100m dam + water bodies	1.01	0.29
TOTAL	352.25	100.00
30M FROM THE RIVER BANKS		
TYPE	**AREA (ha)**	**%**
30m rivers + shrub initial stage	103.76	2.49
30m rivers + arboreal initial stage	39.95	0.96
30m rivers + middle stage	1940.13	46.48
30m rivers + advanced stage	165.20	3.96
30m rivers + gallery	285.69	6.84
30m rivers + floodplain	232.08	5.56
30m rivers + reforestation	39.03	0.93
30m rivers + agriculture and livestock	973.86	23.33
30m rivers + urbanized areas	114.12	2.74
30m rivers + exposed soil	1.54	0.04
30m rivers + water bodies	278.61	6.67
TOTAL	4,173.97	100.00

50M RADIUS AROUND SPRINGS		
TYPE	AREA (ha)	%
50m springs + shrub initial stage	24.37	1.87
50m springs + arboreal initial stage	15.85	1.22
50m springs + middle stage	539.69	41.48
50m springs + advanced stage	53.64	4.12
50m springs + gallery forest	5.48	0.42
50m springs + floodplains	2.59	1.20
50m springs + reforestation	13.17	1.01
50m springs + agriculture and livestock	571.93	43.96
50m springs + urbanized areas	61.78	4.75
50m springs + exposed soil	1.09	0.08
50m springs + water bodies	3.32	0.25
TOTAL	1,292.91	100.00

50M AROUND FLOODPLAINS		
TYPE	AREA (ha)	%
floodplain + shrub initial stage	16.55	3.01
floodplain + arboreal initial stage	10.67	1.94
floodplain + middle stage	126.94	23.13
floodplain + advanced stage	20.89	3.80
floodplain + gallery forest	104.79	19.09
floodplain+ reforestation	4.86	0.88
floodplain + agriculture and livestock	220.47	40.17
floodplain + urbanized areas	15.28	2.78
floodplain + exposed soil	2.10	0.38
floodplain + water bodies	24.05	4.38
TOTAL	5	100.00

SLOPES > 45O		
TYPE	AREA (ha)	%
slope > 45° + shrub initial stage	0.36	1.56
slope > 45° + arboreal initial stage	0.47	2.04
slope > 45° + middle stage	8.37	36.00
slope > 45° + advanced stage	0.80	3.46
slope > 45° + gallery forest	0.16	0.67
slope > 45° + floodplain	0.05	0.24
slope > 45° + reforestation	0.14	0.62
slope > 45° + agriculture and livestock	8.30	35.73
slope > 45° + urbanized areas	4.51	19.42
slope > 45° + exposed soil	0.02	0.09
slope > 45° + water bodies	0.03	0.14
TOTAL	23.21	100.00

50M AROUND LAKES		
TYPE	AREA (HA)	%
lakes + shrub initial stage	9.52	1.52
lakes + arboreal initial stage	2.43	0.39
lakes + middle stage	110.29	17.59
lakes + advanced stage	27.86	4.44
lakes + gallery forest	101.00	16.11
lakes + floodplain	47.49	7.57
lakes + reforestation	7.88	1.26
lakes + agriculture and livestock	288.77	46.05
lakes + urbanized areas	23.76	3.79
lakes + exposed soil	0.05	0.01
lakes + water bodies	2.37	0.34
TOTAL	621.43	100.00

HILLTOPS		
TYPE	AREA (ha)	%
hilltop + shrub initial stage	7.87	2.69
hilltop + arboreal initial stage	2.18	0.75
hilltop + middle stage	82.89	28.34
hilltop + advanced stage	6.95	2.38
hilltop + floodplain	0.21	0.07
hilltop + reforestation	3.16	1.08
hilltop + agriculture and livestock	185.24	63.32
hilltop + urbanized areas	4.02	1.37
TOTAL	292.52	100.00

CRESTS		
TYPE	AREA (ha)	%
crest + shrub initial stage	7.22	8.11
crest + arboreal initial stage	7.42	8.33
crest + middle stage	28.01	31.43
crest + agriculture and livestock	46.48	52.15
TOTAL	89.13	100.00

Areas with constraints on usage in accordance with the Mata Atlântica (Riparian Forest) Law

The natural forms of forest vegetation classified herein as Middle Stage and Advanced Stage of Vegetation Succession (*capoeiras*, *capoeirões* and forests) should be considered areas with constraints on usage, even if not required under the various APPs (Areas of Permanent Preservation, Table 3). If we consider compliance with this legal provision, a total of 3,980.13 ha can be added to the total amount of protected areas, raising the percent of the protected surface area from 30.96% to 47.62%. In Figure 10 the different mapping classifications are superimposed over of the entire area defined as APPs.

4. CONSIDERATIONS OF VEGETATION COVER IN THE MAPPED AREA

If we consider the **Middle** and **Advanced Stages of Vegetation Succession,** the **Gallery Forests** and **Floodplains** as more complex environments with greater biodiversity, these areas are useful in maintaining environmental quality. All these areas combined equal a significant proportion of the mapped surface, 36.78%, that is protected.

Among the different forms of land-use in the Rio Verde Basin, **Agriculture and Livestock**, involving agricultural crops and pastureland, is very significant as it occupies almost half of the surface area (48.56%).

Productive stands or **Reforestations**, using *Pinus* and *Eucalyptus* species, are not very significant in the Rio Verde Basin, occupying only 1.51% of the surface area (360.85ha).

The Urban and/or Built areas total 6.34% of the basin surface and are concentrated mainly in Campo Largo, in the vicinity of highway BR 277, and in the north of the basin within the municipality of Campo Magro.

Although the **Initial Stage of Vegetation Succession** does not have any legal environmental restraints on its conversion in another form of land-use (i.e., agriculture, livestock, urbanization, etc.) it is an initial stage in the process of vegetation re-occupation that, although small in size and limited in biodiversity, is the most immediate form of soil protection as the soil is usually left exposed to weathering and erosion.

As such, the appearance and occupation of shrub and tree species, besides providing the soil with physical protection and nutrients, creates a favorable environment and shelter for birds. Birds, in turn, bring seeds of a greater variety of species which become the foundations of future forests. The major function of this form of land cover is therefore to provide an appropriate environment (microclimate) for the establishment and restoration of the original forest environments naturally and time is the only investment required.

As the succession process continues, the initial plant communities mature and progressively incorporate greater biodiversity and structural complexity. The subsequent **Middle** and **Advanced** stages (*capoeiras*, *capoeirões* and forests) occur, which at this stage fall under strict legal constraints and become increasingly effective in their environmental functions (maintaining soil and water resources and providing flora and fauna resources for humans).

The alluvial plains, here understood as **Floodplains** or swamps on the margins of watercourses, or as Wetlands by current environmental legislation, should be seen as essential components in maintaining drainage channels that regulate water flow during flooding. They are, therefore, inseparable from this function and thus should not be affected by any form of human use.

In these plains, in addition to the prevalent herbaceous formations, tree communities defined as **Gallery Forests** occur. These forests are usually located close to rivers, in better drainage conditions, supported by the "heightening" of river banks (marginal dikes or "shoulders") due to the deposition of sediments of larger granulometry, resulting in soils with greater porosity. For this reason, these environments are more vulnerable to erosion caused by water flow which justifies their classification as areas of permanent preservation by the Brazilian Forest Code since the 1960s.

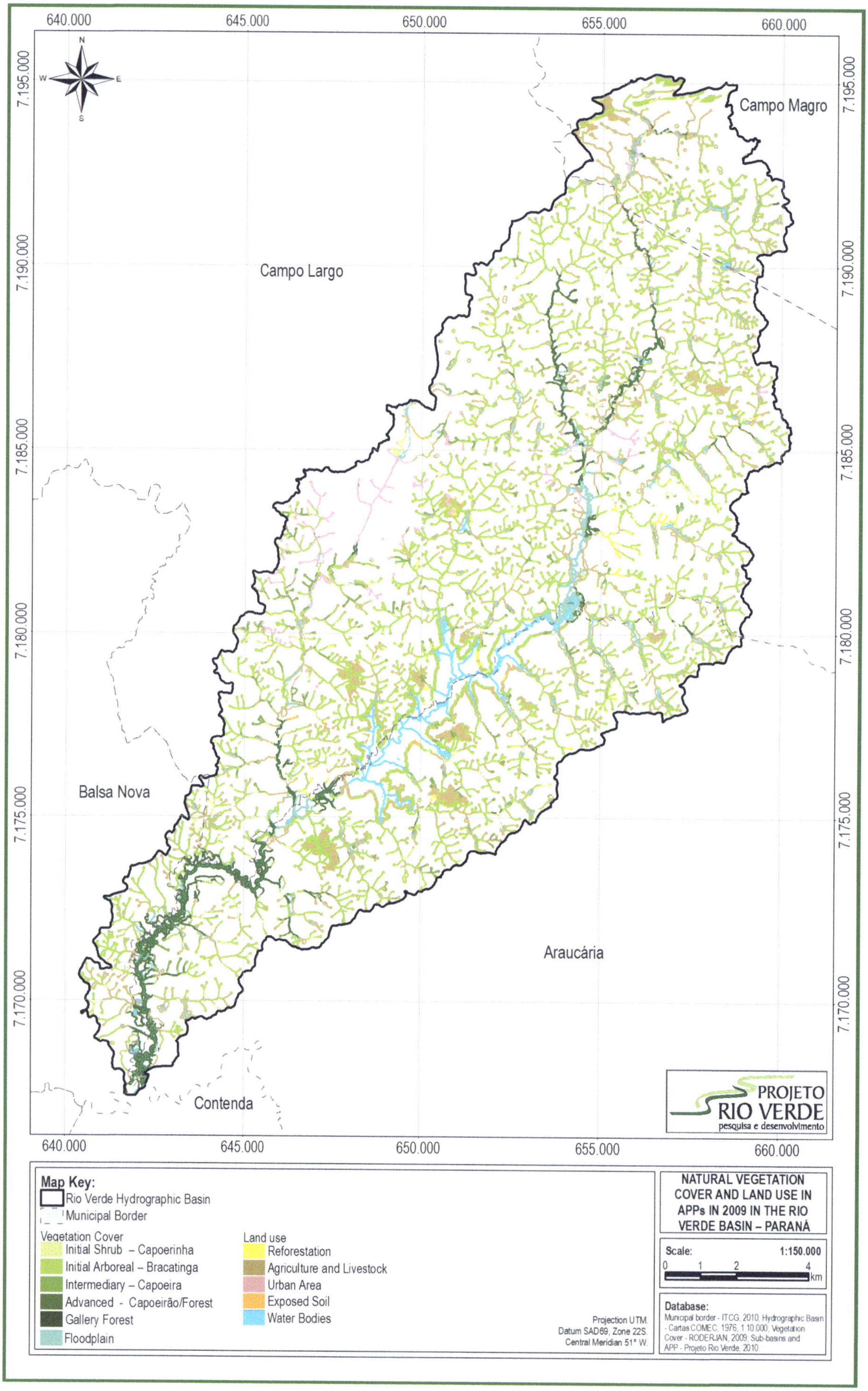

FIGURE 10 – DIFFERENT MAPPING CLASSIFICATIONS OVERLAYING ALL AREAS OF PERMANENT PRESERVATION

5. CONCLUSIONS AND RECOMMENDATIONS

5.1 VEGETATION COVER

The unequivocal functionality and vulnerability of the vegetation cover and the unquestionable richness of the biodiversity that occurs in these areas, urges the unrestricted preservation and conservation of all remaining natural formations. Therefore, we propose:

- compliance with all aspects of the environmental legislation, particularly those related to Areas of Permanent Preservation, with the immediate cessation of any form of use (agriculture, livestock, urbanization and mining), as well as the implementation of programs aimed at the restoration of these areas when degraded. Such a compliance would almost entirely recuperate the connectivity of the existing vegetation communities and therefore the associated fauna, except those already part of the consolidated urban areas, which account for only 3.02% of the entire APPs and are mainly concentrated along highway BR-277;
- for the conversion and restoration of the areas with legal constraints (APPs) that are currently used inappropriately, we suggest the immediate development of a program of seedling production of *bracatinga* (*Mimosa scabrella*), a fast-growing native leguminous tree, which significantly contributes to soil fertilization and "prepares" the land for the establishment and continuity of the succession process. Not considering the portions of the APPs with urban occupation (because they present a complex social problem), the proportion occupied by **Agriculture and Livestock, Reforestation** and **Exposed Soil** constitute 2,481.87 hectares. With conventional tillage spacing of 2 x 2m, conversion of these areas would require an immense production program of 6,204,675 seedlings; at an average cost of R$0.20/seedling, an investment of approximately R$ 1,240,935.00 would be required.
- increase and improve environmental education in primary and secondary levels of public education. The restoration of the APPs will not be effective without local people's awareness and knowledge of the environmental reality in the region. This can only be possible with the development and implementation of a consistent program of environmental education and rural outreach aimed at providing knowledge as well as production and income alternatives for the local population, which in turn will improve their quality of life;
- strengthen the environmental agencies through investment in infrastructure and training of human resources.

5.2 AGRICULTURE AND LIVESTOCK ACTIVITIES AND REFORESTATION

Local demands for produce, cattle, dairy, and wood, must be met without the conversion of the remaining vegetation areas (removal or deforestation) to agriculture and pastureland. Therefore, we suggest:

- maintenance and possible expansion of agriculture and livestock activities to areas already devoid of their original vegetation;
- conversion of these activities to activities with minimal impact on the environment, such as the implementation of infrastructure and training of human resources for the development of leisure and environmental tourism, with investments from the public and private sectors;
- encourage the planting of fast-growing, hardy species (*Pinus* and *Eucalyptus* spp.) to produce raw material for saw mills;
- encourage the use of *bracatinga* (Mimosa scabrella) in restoration programs in degraded areas.

5.3 CONSERVATION UNITS AND URBAN GREEN SPACES

The improvement of the environmental quality, and therefore, the quality of life of the population, requires the protection, rescue, maintenance, research on and appreciation of biodiversity. We suggest strengthening institutions focusing on the creation and maintenance of green spaces and/or municipal conservation units, protecting remnants (of any nature and size) in areas forecast for urban and industrial expansion, providing leisure for the current and future population. Furthermore, this action would strengthen the previous proposal of the restoration of the APPs and their functions.

5.4 TREE PLANTING ON STREETS AND IN PUBLIC PLACES

The revitalization of tree planting along streets, roads and in squares provides countless direct or indirect benefits to residents, tourists and visitors. Therefore, we suggest:

- diagnose and redistribute the established tree populations;
- implement additional and new planting projects;
- provide and train human resources or outsource the service;
- create an infrastructure of seedling production (nurseries and orchards) or outsource the service.

5.5 URBAN AND INDUSTRIAL EXPANSION

The demands of urban and industrial expansion must be met without converting the remaining vegetation areas (removal or deforestation). Therefore, we suggest:

channeling these activities to areas already converted to low productivity agriculture and initial stages of vegetation succession (*capoeirinhas*).

6. REFERENCES

BIGARELLA, J. J., MARQUES FILHO, P. L. & AB'SÁBER. (1961) A. Origem e ambiente de deposição da bacia de Curitiba. *Boletim Paranaense de Geografia*. Curitiba, Nº 4 e 5, p. 71-81.

EMBRAPA. Centro Nacional de Pesquisa de Solos. (1999) *Sistema brasileiro de classificação de solos* – 5ª aproximação. Rio de Janeiro, 412 p.

IBGE – Instituto Brasileiro de Geografia e Estatística. (1992) Manual técnico da vegetação brasileira. *Série Manuais Técnicos em*

Geociências, n. 1, Rio de Janeiro, 92 p.

MAACK, R. (1981) *Geografia física do estado do Paraná*. Curitiba, José Olympio, 450 p.

SALAMUNI, E. (1998) *Tectônica da bacia sedimentar de Curitiba (PR)*. Rio Claro. São Paulo. Doctoral these. Graduation course in Geoscience. Instituto de Geociências e Ciências Exatas. Universidade Estadual Paulista. 211 p.

CHAPTER 7

POTENTIAL AND EMERGING FRAGILITY OF THE SOIL

Everton Passos
Ana Maria Muratori

POTENTIAL AND EMERGING FRAGILITY OF THE SOIL

Everton Passos
Ana Maria Muratori

SUMMARY

The life span of the Rio Verde reservoir is directly related to the environmental dynamics that occur in the basin. Thus, it is essential to comprehend and control the environmental impacts resulting from the characteristics and dynamics of the biophysical environment, as well as anthropogenic interventions and their respective interactions in the transformation of environmental stability. In this context, the purpose of this study was to determine the potential and emerging fragility of the Rio Verde Hydrographic Basin, as the result of geologic-geomorphologic processes, integrating the lithological and tectonic-structural aspects that interact with the past and current climatic characteristics, the relief, the soils and vegetation, in face of human actions within the basin. The methodology used was based on the concept of ecodynamics, which considers the degree of stability of the environment through Empirical Analysis of Natural Environments, advocating for the intersection of automated data with geo-processing techniques and field assessment. After proceeding to the relevant correlations, a Potential Fragility Map and an Emerging Fragility Map of the Rio Verde Hydrographic Basin were obtained to be used in the planning and sustainable management of the use of their natural resources.

KEYWORDS

Declivity, degradation, fragility, geomorphology, environmental impacts.

1. INTRODUCTION

The environment, depending on its genetic characteristics, presents different degrees of fragility which is based on the physical environment. In this context, the development of urban centers in inappropriate areas, such as on steep slopes or in alluvial floodplains, in addition to industry, forestry, mining, agricultural and cattle raising activities are significant. By not considering the capacity and fragility associated with the environment, processes of degradation are accelerated.

This question becomes more worrisome in Brazilian metropolitan regions, such as the Metropolitan Region of Curitiba (RMC), in which increasing population density has occurred randomly in relation to existing natural resources. From this point of view, the impact on water resources is the most worrying.

In the case of the Rio Verde Basin, unplanned human occupation occurs throughout the region but it is particularly concentrated in the municipalities of Campo Largo (Figure 1) and Araucária.

Urban expansion originated largely as a consequence of industrial activities; however, the basin had already been compromised due to extensive primary resource exploitation activities and agricultural practices without proper control (whether through deforestation and occupation of areas that should be preserved, or lack of management practices to ensure soil conservation).

FIGURE 1 – UNPLANNED OCCUPATION OF AREAS OF THE RIO VERDE BASIN

An important indication of the very high emerging fragility in urban areas located upstream of the dam includes urbanization on slopes above 30%, pictured on the left from the municipality of Campo Largo. In the municipality of Campo Magro, there are similar problems in relation to the land-use, as well as the most diverse forms of urban occupation. In the urban area of Campo Magro, located on the headwaters of the Rio Verde, the waste collection system is precarious, as shown above right. Source: PASSOS, 2010.

We can add to these problems the indiscriminate use of pesticides and fertilizers, not to mention mining activities and the transportation of hazardous goods along highways such as BR-277 and PR-423.

When the occupation of an area occurs without proper planning, as is the case in the Rio Verde Basin, it favors the development of high-energy conditions in the environment, through hydrological changes brought about by widespread deforestation, and changes in the characteristics of surface formations, such as the reduction in the permeability of large areas (soil sealing), among other factors.

It is important to note that these modifications occur without significant climate changes to justify them. Anthropogenic activities intensify the morphogenetic processes in which water erosion through diffuse and concentrated runoff predominates. In these conditions, the morphogenesis overcomes the pedogenesis, thus leading to accelerated degradation of the environment.

According to Bigarella & Mazuchowski (1985), these types of environmental degradation occur through laminar or sheet erosion that transports the superficial and fertile horizons of the agricultural soils, truncating them and rendering them infertile. It also occurs through concentrated erosion, creating furrows, ravines, and in extreme cases, gullies, when the terrain is dissected, affecting slopes and drainage basins. The removal of riparian forests from the watercourses increases fluvial erosion, thus modifying and/or destroying natural marginal levees during periods of flooding.

Therefore, according to Justus (1985), widespread anthropogenic activity, without proper planning, is causing many paleo-landforms to regain their functional conditions. This interference is also responsible for new shapes generated through morphogenetic processes related to hydrological conditions similar to the ones that would be caused by climate changes during a dryer period, as shown by the increased concentration of gravel on the surfaces of cultivation areas (Figure 2).

FIGURE 2 – EVIDENCE OF SOIL LOSS

Detail of a cultivation area where the soil was totally eroded and the agriculture is currently undertaken on decomposed rock material which, due to its richness in quartz, creates a detrital cover of gravel composed mainly of this material and sand. Source: PASSOS, 2010

In this sense, studies and investigations that support the evaluation of the natural physical environment according to anthropogenic modifications prove to be extremely important in planning and organizing land-use because they can incorporate social and economic development with the preservation of the environment.

Studies on soil fragility have been conducted mainly to support agriculture and livestock activities. However, based on the approaches used, they can be useful in informing a broader perspective in the planning and management of land-use. In this context, the present study is justified and its goals are presented below.

Considering the information presented above, the general goal of the project was to:

- Analyze the environmental fragility of the Rio Verde Basin.

In order to achieve this, the specific goals were to:

- Conduct a survey of the physical and environmental characteristics and limitations of the Rio Verde Basin;
- Build a physical-environmental database as tool to be used in decision making related to planning and managing the Rio Verde Basin;
- Evaluate physical-environmental characteristics of the Rio Verde Basin in relation to its support capacity and environmental fragility;
- Generate a map of potential fragility and one of emerging fragility of the Rio Verde Basin.

1.1 CHARACTERIZATION OF THE RIO VERDE BASIN

From a geo-environmental point of view, the Rio Verde Basin is located in the Morphostructural Unit of the Orogenic Belt of the Atlantic, in the structural unit defined by Maack (1968) as the First Plateau of Paraná, under the domain of the (Araucária) Mixed Ombrophilous Forest, of which only fragments remain.

The morphological partitioning of the Rio Verde Basin reflects, in part, the partitioning of the Metropolitan Region of Curitiba which is located in the First Plateau of Paraná. This area may be understood from geomorphic zones that can be seen mainly through the relief, sometimes uneven (*Morraria do Açungui*, in the Campo Magro region), sometimes flat (Sedimentary Basin of Curitiba, in the Campo Largo Region), and strongly influenced by the geology (MURATORI, et al.,1987).

Considering the distinct aspects mentioned above and a morphologic point of view (MURATORI, 1966), these subareas of the First Plateau of Paraná are clearly distinct, presenting a rugged, residual, sculpted relief of Crystalophylian rocks from the Açungui Group in the north to northwest portion, that stands out from the pediplain relief of the Curitiba Plateau in the south-central portion, where crystalline basement rocks are present and partially covered by the materials that constitute a small taphrogenic continental basin, known as the Sedimentary Basin of Curitiba (SALAMUNI, 1998; CANALI & MURATORI, 1981; SALAMUNI, EBERT & HASUI, 2004).

The northern portion of the Basin, which encompasses the headwaters of the Rio Verde, is located at the limitrophe of the Mountainous Region of Açungui. The central-southern portion of the basin up to the mouth of the river flowing into the Iguaçu River, is located in the physiographic unit known as the Plateau of Curitiba.

In the northern portion, which corresponds to the Açungui, despite the carving caused by current dissecting processes, features of an ancient morphoclimatogical formation still remain, particularly in areas where quartzite ridges or limestone preserve remnants of former inselbergs, showing evidence of pedimentation. The pediplain of the Campo Largo region (related to the pediplain surface of Curitiba, MURATORI, *op cit.*) represents the western portion of the Curitiba Basin that was the base of material deposition resulting from the pediplanation process and Pleistocene pedimentation.

This surface feature which comprises the largest area of the basin is more dissected and recessed in the eastern and southern portion, constrained by the base level of the Iguaçu River valley with wide plains, which originally featured the Floodplain Mixed Ombrophilous Forest, recognized by Klein (1962) as Riparian Forest and associated with the Edaphic Fields.

The relief of the above mentioned pediplaned surface is locally more preserved than the west-central and the northwest portion of the Basin, in which the composition of the landscape was related to the occurrence of steppe vegetation covering the edaphic fields. These were interspersed with forest fragments of Mixed Ombrophilous Forest with the presence of the Paraná Pine (*Araucaria angustifolia*) emerging from the canopy, in addition to eventual woodlands of *Pinho Bravo* (*Podocarpus lambertii*) which are still found in the area surrounding Campo Largo.

Locally, the pediplain region is made up of an area with hills of soft undulating relief and the occurrence of more developed soils. The interfluves are relatively broad and flattened, with broad concave valleys at their headwaters that are connected to alluvial plains, over which the river course meanders.

From the hydrographic point of view, the Rio Verde Basin is a subsystem of the Iguaçu River Basin and it constitutes a set of sub-basins restricted by divisions of hills and hillocks in general, with hilltops that are remnants of ancient plains and smoothly convex in the less dissected areas. In the areas with more sharp relief, these hilltops are more mamelon in shape, with interfluves that are narrower and more affected by the geology, including structural alignments (layers, flaws or joints) and intrusions that together define the system, generally with northeast to southwest and northwest to southeast directions.

The main drainage of these sub-basins begins in the elevated regions, in small amphitheaters of erosion (grottos, hollows or drainage headwaters) from which channels of first order streams (registered in 1:10,000 topographic maps generated through stereographic interpretation of aerial photographs) converge, forming small streams. In periods of relative drought, the small watercourses of the headwater that have spring sources and in the drainage system constitutes the proto network of drainage, dry out or their volume is lowered, thereby affecting the flow of the main collectors (small streams) which define the micro basin.

The continuity of these small streams is linked to physical factors such as: amounts of rainfall in the region; good annual rain distribution; dense fog that frequently covers the region during the winter and maintains the humidity when the rains are slightly reduced. The main factor in the continuity of these streams is the preservation of the vegetation in the drainage headwaters and along the waterways.

Considering the litho-structural characteristics, the drainage in this region presents various patterns. In the northern part, corresponding to rock formations of the Açungui group, a sub-parallel pattern predominates, where incising tends to form V-shaped valleys, with poorly developed soils (Inceptisols). However, in this same portion, the upper section of the Rio Verde, there is an occasional tendency toward a dendritic pattern especially where the stream gradient is wider with concave valleys. This pattern presents a more significant covering of colluvium over which more developed soils, such as Oxisol and Ultisol soils, are found and generally associated with Inceptisol at points of sharper relief. Finally, in areas with more dissected relief, where slopes are relatively short (less than 200m), declivity is around 20%, and the very weathered crystalline base is exposed, the Inceptisol is frequently associated with the Ultisol in areas where colluvium occurs and is frequently truncated by pebbles. Eventually, Oxisols can occur in residual portions of planation surfaces, related to pediments and colluvial-alluvial ramps.

At the base of the slopes near the valley floors and adjacent to the thalwegs, alluvial or organic materials sometimes occur, in which Gleysols and Histosol are developed, respectively.

The examination of sections in road cuts in the Rio Verde Basin shows the polycyclic nature of the landscape in this section of the Alto Iguaçu River Basin. The relief is preliminarily divided into three units based on the distribution of three main planation surfaces related to the level of erosion or sedimentation (Pd1, P2, P1) as well as the occurrence of terraces, ramps and Holocene alluvial sedimentary plains along the main river and tributaries, forming its flood channel, and, therefore, areas subject to seasonal flooding (PASSOS, 1980-ab).

The interpretation of the various levels found, as well as the sharp changes of slope (knick points) identified and correlated to the mentioned planation surfaces, resulted from exploratory field survey, which should be supplemented with more detailed studies. However, the presence of overlap of colluvium and incipient lines of pebbles in the researched area and its surroundings leaves no doubt. The occurrence of such surfaces and materials allows us to confirm morphoclimatic factors in the evolution of hillsides and thalwegs in question through the identification of correlated deposits mentioned above. This is further supported by the occurrence of alluvial and alluvial-colluvial deposits, as well as their resulting contribution to the current distribution of the soils. In particular, such observations confirm the aspects that have been attributed to the nature and origin of laterite soils (Oxisol) in the Alto Iguaçu and that correlate its lateritic nature to the alternate paleoclimate in favorable and unfavorable stages to its development in the recent past.

According to Bigarella & Mazuchowski (1985), the iron content in Oxisols of the slopes is mainly due to the precipitation of crystalline and amorphous iron compounds in the voids of soils and of colluvial deposits of the slopes. With the dissection of the terrain, the previously deposited iron may be remobilized and form new deposits and soils in the recessed portions of the slope.

Planation surface remnants are recorded in the current morphology of the staggered relief, marked by slope ruptures, identified in features such as shoulders and platforms, cut by erosive trenches and local deposition of colluvium, interspersed with lines of pebbles and thereby producing its topographic sequence[1]. Upon characterization of these surfaces, it is possible to establish a classification of degrees of stability over time when a combined analysis is performed integrating information on the soil characteristics and aspects of land-use.

According to Passos (1987), the geomorphologic units of this region have been, or are subject to, physical or physicochemical stability changes, according to variations in morphogenetic and pedologic processes. These processes remain broadly recorded in the terrain morphology and in the geomorphological structures and allow a temporal-spatial analysis of the Paleosols, while also helping to understand the current topographic sequence.

Besides establishing a clear differentiation of age between several edaphic formations, the topographic sequences provide information about the transformations that have occurred, as well as the sequential deposition of the soils in various landscapes.

Aspects of geomorphology and altitude have a passive influence on the development of the soil profile, while the climate and the vegetation are active factors. Thus, we note that the surface morphology and its modeling agents play an important role in the distribution of the many types of soils.

The influence of the relief in the development and stability of the soil (potential or emerging fragility) is not a simple and direct determining factor of declivity, but rather of its relation with parent material and with drainage, among other environmental factors. Both can be affected by the declivity as well as by the position in the slope.

When the parent material is uniform in the slope, it is possible to find a sequence of similar profiles with deteriorating drainage conditions. However, the parent material does not always present a uniform distribution across the slope. Variations in parent material and topography create a complex interrelation that often makes it difficult to predict the type of originating soil. We must also consider that the soil formation is directly influenced by the characteristics of the parent material or the altered substrate and not specifically by the morphology or the origin of the slope deposit (BIGARELLA, BECKER & PASSOS, 1996).

Considering what occurs in a large part of the Rio Verde Basin, and according to Bigarella & Mazuchowski (1985), human occupation that occurs without adequate planning, favors the occurrence of high-energy conditions, through the hydrological changes provoked by general deforestation, or by changes to the characteristics of surface formations. Such interventions (producing new landscapes) are factors that can cause a potential increase in erosion, among other facts, that result in various environmental impacts. One impact that deserve to be highlighted is the hydric imbalances that affect springs and their respective watercourses, creating numerous disturbances to the population (PASSOS & CANEPARO, Unpublished).

The changes in hydrological conditions result in increased runoff and the alteration of groundwater flow, leading to the local degradation of the groundwater. This degradation can result in the disappearance of springs and, consequently, many first order watercourses become intermittent or disappear altogether.

The increase in runoff implies a higher rate of erosion and the removal of debris from slopes, which are transported to the lowlands, filling them in and silting the beds of rivers and dams. As such, the usual course of a river undergoes significant changes to its channels' patterns, with torrential characteristics.

Thus, anthropogenic activities enhance morphogenetic processes, predominantly water erosion by diffuse runoff (sheet erosion) and concentrated runoff (erosion in furrows or ravines). Under these conditions of morphodynamic instability (TRICART, 1977), morphogenesis overcomes the pedogenesis leading to a process of accelerated environmental degradation (Figure 3).

FIGURE 3 – ACCELERATED EROSION

Crops on slopes without proper soil management, leading to an acceleration of morphogenetic processes (severe loss of soil through erosion), a very common phenomenon in most cultivated areas of the Rio Verde Basin. Source: MURATORI, 2009.

This degradation takes place through sheet erosion, silting the superficial horizons and fertile agricultural soil, truncating and making them infertile. It also occurs

[1] Bigarella and Mazuchowski (1985) highlighted the importance of using the concept of topographic sequence in the environmental interpretation of soils. The integration of the geologic, geomorphologic and pedologic information allows for a better understanding of the soil evolution across space and time, as well as relates pedogenetic processes with environmental variations. According to the authors, the topopraphic sequence not only clarifies the edaphic chronostratigraphy, but also provides information on the transformations of the various soil units.

through concentrated erosion, forming furrows or ravines (Figure 4) that may develop into gullies. In such cases, the terrain is dissected and both slopes and water abstraction basins are affected.

In summary, these geo-environmental degradation processes are closely linked with anthropogenic activity and are significant factors in the degradation of the Basin's water resources. Such processes may create the risk of siltation and eutrophication of the dam.

FIGURE 4 – ACCELERATED EROSION
Exposed soil with cracks due to dryness. The drainage furrow in the middle of the plantation is a technique that enhances erosion in areas of annual cropping (in this case, bean cultivation), as shown in the picture. Source: PASSOS, 2010

2. MATERIALS AND METHODS

2.1 MATERIALS

For this study, we used the resources available from the Rio Verde area, such as academic papers, project reports, various types of maps at different scales, aerial photographs, satellite images, as well as equipment and software available in the Geo-processing Laboratories (LAPIGEO and LAGEAMB) of the Geography Department of the Federal University of Paraná (UFPR), that were acquired through resources provided by *Petrobrás*. We also included information gathered in the field.

2.2 THEORETICAL-METHODOLOGICAL APPROACH

With the objective of the project being the identification of the potential and emerging fragility of the Rio Verde Basin, it is necessary to briefly comment on the methodological approach used to address this goal. Among the various methodological approaches to environmental issues, the systemic approach (BERTALANFFY, 1973; KUMPERA, 1979) was considered the most adequate, given that it allows for environmental changes and organizes interconnected systems based on their differences or their similarities.

This approach allows for different viewpoints. Among the best known and used, the work of the geographer Jean Tricart was chosen, considering that it is the basis of institutional environmental studies, such as the Ecological Economic Zoning (ZEE, for its Portuguese acronym) according to INPE methodology (CREPANI *et al.*, 2000). Tricart (1977; TRICART & KILIAN, 1979) uses as a foundation the interaction of matter and energy flows that shape ecosystems and drive their dynamics and sensitivity to change, considering three major geodynamic processes for classification, as follows:

- Stable Environments – present a slow evolution and are subject to low internal geodynamic activity where the mechanical processes of external activity are also of little importance and pedogenetic processes prevail;
- Intergrade (intermediary) Environment – the dynamic is characterized by interdependency of morphogenesis/pedogenesis. If morphogenesis predominates, the environment becomes unstable, but if processes of pedogenetics predominate, the environment tends to remain stable;
- Unstable Environment – predominance of morphogenesis over pedogenesis. The causes may be related to aggressive bioclimatic conditions that disfavor the presence of vegetation cover and/or a much more intense internal geodynamics, and human action in the distinct environments.

With this approach, the main objective was to use data and scientific knowledge to understand the dynamics of the natural environment and highlight the factors that can define certain land-uses, which are useful in managing water resources. To achieve this goal, the methodological procedures were based on the cartography of natural elements (lithology, relief, vegetation cover, hydrography, human actions) and the identification, localization and analysis of various processes and systems in dynamic interaction.

Such procedures included: collection and analysis of bibliographic and cartographic documents; preparation of field work, with visual interpretation of aerial photographs and images, aiming at constructing a GIS (Geographic Information System) platform; field work to check the background and laboratory data; plotting of the maps from topographic charts and themed maps; analysis of collected data; interpretation; development of the GIS; plotting of maps; and creation of the final report.

Although computer software greatly facilitates diagnosis by using automated procedures, fieldwork is still an essential part of the analysis, as well as following a set of procedures in order to obtain consistent data for the analysis.

Important steps taken during field work were: the observation of the entire landscape and its limits, considering the landscape in the larger context of the relief, hydrography, and vegetation; the observation of natural and anthropogenic elements that compose the landscape; evidence of environmental degradation (erosion, deforestation, pollution, among others); immediate comparison of the data collected through direct observation with aerial photography or images and topographic maps; schematized notes, with succinct descriptions of the observed facts with geographic coordinates or UTM and drawing of profiles with relevant descriptions; quick measurement of soil thickness and others; sample collection.

From this, the methodology applied to the data collection, mapping and analysis of the potential and emerging fragility of the Rio Verde Basin sought to refine the assessment of the information based on the extent and nature of the units studied, as well as the instruments used, considering the preparation of maps of geomorphology, land-use, and vegetation cover, using already existing geological, pedological and climatic information.

This methodology involved three high levels of treatment:

a. Categorization of the relief based on the topographic map, with precise description of the relief forms, according to Ab'Saber (1969);
b. Systematic extraction of information from the surface structure of the landscape, which follows the integrated mapping model presented by Bigarella *et al.* (1979) that, in using digital cartography, considers qualitative and quantitative aspects of the relief, transformed into information layers to construct the relief on a wide scale;
c. Analysis of the physiology of the landscape by means of understanding morphogenic and pedogenic processes, in this case integrated with the concept of ecosystems and adapted to a perspective that defines as resultants of said processes the ecodynamic systematic units suggested by Tricart (1977) and complemented by the mapping-synthesis adapted from Ross (1990), with the indication of soil fragility based on potential and emerging instability.

The classifications of land-use, soils, declivity, as coded in GIS, were given a value on a scale that varies from very low to extreme (see Chart 1). Such valuation is relative and has an exploratory character regarding the relative sensitivity to the emerging fragility of the soils. Although land-use is considered an aggravating factor in the eutrophication of the dam, as it is a factor in the production of sediments and soluble materials favorable to the phenomenon, the soils, combined with declivity and geomorphological aspects, were *a priori* considered with the same weight in the evaluation. Thus, in the creation of the final maps of potential and emerging fragility of the soil, we adopted the value that defines the maximum fragility, for instance: a unit determined by the potential fragility as low or very low under original vegetation cover and thus a stable area, when considered under a modified condition of exposed soils, the unit can present a condition of instability and, consequently, at risk of generating materials that favor eutrophication.

The creation of the geomorphological map relied on images and topographic maps at a scale of 1:10,000 and was supported by observations and measurements taken in the field. The units were classified according to the taxonomic order suggested by Ross (1990), based on

TABLE 1 – SCALE OF VALUES ADOPTED IN THE DETERMINATION OF DEGREE OF FRAGILITY (POTENTIAL OR EMERGENT)

FRAGILITY	VALUE
Insignificant	1
Very Low	2
Low	3
Low to Moderate	4
Moderate	5
Moderate to High	6
High	7
Very High	8

SOURCE: Org.: MURATORI & PASSOS

the taxonomic levels of Demek[2] (1967) and concepts of morphostructure and morphosculpture from Gerasimov[3] (1946) and Gerasimov and Mescherikov[4] (1968), using the following levels: morphostructural units; morphosculptural units; denudational and aggradational forms; units of similar patterns and shapes; mensuration and indication of the forms through morphometric data; linear and punctual forms of the relief.

Firstly, the analysis involved the identification of the relief in different dimensions. By establishing a classification of shapes from greatest to smallest, we reached the level of segments of slopes based on scale, adapting to this level the units suggested by Bigarella *et al.* (1979) and Muratori *et al.* (1987) for the Metropolitan Region of Curitiba.

The following step was the determination of degrees of potential and emerging fragility (with preliminary exploratory tests performed by developing a monographic study in light of the present sub-project, concluded in 2009).

To this end, the relief dissection index, in other words, the intensity of dissection or the intensity of the topographic roughness, which is the first significant indicator of potential fragility that the natural environment presents, was used as a reference. The density of drainage associated with the degree of carving of the combined channels determines the topographic roughness, or the relief dissection index, that enables the definition of the average interfluvial dimension of the homogenous set of shapes or sets of similar shapes.

This way, the work was done with morphometric data of the fluvial dimension versus carving degree of the channels and consequently with classes of average declivities (declivities obtained in MNT generated from topographic map 1:10,000, associated with the pattern of shapes identified in the images), creating a hierarchy of relief dissection indexes. With this hierarchy, an information layer was produced containing the "homogenous units or spots" that properly classified the relief dissection indexes.

---

2 DEMEK, J. Generalization of Geomorphological Maps. In: *Progress Made In Geomorphological Mapping*, Brno, 1967.

3 GERASIMOV, I.P. (1946). Essai d'interprétation geomorphologique du schéma general de La structure geologique de l'URSS. Problèmes de Geographie Physique, Vol. 12, Tzd. Vo AN SSSR, Moscow.

4 GERASIMOV, I.P. & MESCHERIKOV, J.A. Morphostructure. In: *The encyclopedia of geomorphology*. Ed. R.W. Fairbridge, 731-732, New York: Reinhold Book Co., 1968.

The lithopedologic information was ranked from greatest or smallest degree of fragility of the substrate and of the soil considering: the degree of weathering and soil development; its physical and mineral characteristics in relation to the anthropogenic action (land-use); and, above all, implications of pluvial waters. An information layer was also produced with this data, in which "homogenous units or spots" of various types of rock-soils were registered and classified by greatest to smallest degree of fragility in relation to erosion (sheet and in furrows) and to solifluction or mass soil movements.

The data on vegetation cover and land-use that were obtained using satellite images and secondary sources also allowed us to establish a ranking of fragility according to the categories presented in the table below, considering the level of land protection, defining values of fragility according to categories presented in Chart 2.

TABLE 2 – VALUES OF FRAGILITY ATTRIBUTED TO DIFFERENT CATEGORIES OF LAND-USE IN THE RIO VERDE BASIN, 2009

CATEGORY OF LAND USE	CODE	VALUE*
Exposed soil	1	8
Urban	2	8
Agriculture and livestock	3	7
Shrub initial	4	6
Gallery	8	6
Floodplain	10	6
Arboreal-bracatinga initial	5	5
Reforestation	6	4
Intermediary	7	2
Advanced	9	1
Water bodies	11	n/d

FONTE: Org.: MURATORI and PASSOS

The superposition of information layers, elaborated and manipulated in GIS with a few generalizations, allowed for the classification of ecodynamic units or units of morphodynamic behavior, divided basically in two main groups: units of stable ecodynamics or in morphodynamic balance; and unstable units or out of morphodynamic balance. Both unit types (in balance and out of balance) were classified according to all analyzed variables resulting in six degrees of instability from weak to strong (potential to emerging). The forested units, despite being in morphodynamic balance, present degrees of potential instability as a result of geomorphologic and edaphic aspects. The units that are on land without natural forest cover or with various degrees of degradation due to anthropogenic activities, are classified in degrees of emerging instability.

In the present proposition, the classes of declivity used to classify relief units were defined based on the percent limits adapted from Herz and De Biasi (1989) and EMBRAPA (1989):

- <-3%- Class of lands that are not susceptible to erosion and have no limitations, except for those outlined in the legislation (Riparian forest, mangrove, etc.), when in use with simple management practices;
- <-5% – Class that defines the boundaries of urban/industrial use, used internationally as well as in urban planning;
- 5 – 12% – The maximum limit of this class (12%) varies because some sources have adopted a limit of 10% and/or 13%. The difference is minimal because this section defines the maximum limit of the use of mechanization in agriculture (CHIARINI and DONZELLI, 1973);
- 12 – 20% – The class considers the maximum inclination of tolerable risk for plowing the land with a tractor in contour lines;
- 20 – 30% – The limit of 30% is defined by federal legislation (Law 6.766/79) also called Lehmann's Law, that defines the maximum limit for urbanization without constraints, from which all and any form of subdivision will be made based on specific requirements;
- 30 – 45% – The Forestry Code states the limit of 25° (47%) as the limit of clear cut. Exploitation is only allowed if sustained by the cover of forests. Law 4.771/65 of 15/09/65, on the other hand, states that a relief above 45% is characterized as mountainous and is a relief unit that is more restrictive than strongly undulated. It is considered the practical limit for forest exploitation;
- 45 – 75% – Article 10 of the Forestry Code states that in the range between 25° (47%) to 45° (100%), "it is not allowed the cutting of forests, [...] only being tolerated log extraction, when under a sustainable use system, aimed at permanent income." However, it is also justified in being a mountainous area of high natural instability, where even the code itself mentions the need for conservation of hilltops. It is recommended that the escarpment (see classification of relief phases) below them remain equally preserved. This way, it is common sense to indicate this unit as restricted to any anthropogenic action because even with sustainable extraction the natural risks are enhanced during log removal, with the rare exception of when removal is necessary to reduce the risk.
- >75%. Although the Forestry Code has a slope limit of 100% for areas of permanent preservation, as discussed in the previous class, it sensible to classify such a steep relief as restricted from any anthropogenic action because even in sustainable extractive operations, the natural risks are enhanced during log removal, with the rare exception of when it is necessary to reduce the risk.

As with any technique of variable representation that depends on the elevation levels of the relief, it was necessary to construct a base map representing the relief in contour lines. Regarding the scale of the base map and its relation to various applications, the suggestions made by Cunha *et al.* (1989) were considered. They call attention to the fact that "in the representation of the map's energy relief, this observation is directly linked to the equidistance that the different scales provide".

Consequently, the spatial equidistance of the relief only allows contours of average declivity. As such, at smaller scales the result can be very different from reality, particularly in landscapes with low topographic amplitude where important ruptures in the hillside gradient are not registered, giving the false impression of relief homogeneity.

2.2.1 Land elevation model

surveys were based on cartographic sources, scanned and edited for this project, from the COMEC maps (1976), from stereographic interpretation of aerial photographs at a scale of 1:10,000, with 5m contour lines, sheet numbers 233 to 240, 277 to 282, 389 to 396, and 529 to 534 (COMEC numeric codes). These maps were digitalized in Tiff, georeferenced in the Projection System UTM Datum SAD69, and exported to GeoTiff (available through the host institution). Such planialtimetric images were imported to IDRISI (version 15) and vectorized, concatenated into mosaics, and edited in GIS using the software CartaLinx (version 1.2), in the LAPIGEO/LAGEAMB-DEGEO/CT/UFPR. This process generated a georeferenced database called SIG-BRV.

The SIG-BRV, later manipulated in the software IDRISI 16-Taiga, Clark Labs (2009), allowed for the production of the Digital Model of Land-use of the Rio Verde Basin with a spatial resolution (pixel) of 2.5 meters. This model was created using the method of interpolation by triangulation with edge smoothing, based on the mosaic of the map vectors encompassing the basin area and its surroundings.

2.2.2 Hypsometry and declivity

the analysis and the processing of the Rio Verde Basin MNT manipulated in IDRISI enabled the development of the Hypsometry map, which was obtained through resampling of the MNT and the Declivity maps generated with a specific module used for calculating the algorithm proposed by Monmonier (1982) and later reclassified according to Chart 3.

TABLE 3 – SCALE OF VALUES ADOPTED FOR DETERMINING THE RELATIVE DEGREE OF POTENTIAL FRAGILITY ATTRIBUTED TO THE DECLIVITY CLASSES

DECLIVITY CLASSES (%)	FRAGILITY	VALUE*
0-3	Insignificant	1
3-5	Very Low	2
5-12	Low	3
12-30	Low to moderate	4
30-47	Moderate	5
47-75	Moderate to high	6
75-100	High	7
>100	Very High	8

2.2.3 Relief and pedologic cover

The basin soils were subjectively evaluated and classified on a scale adopted for fragility. To develop the scale, the degree of development was considered, i.e., stability from the ecodynamic perspective, according to the suggestions of Tricart (discussed above).

In relation to the main limitations of agriculture in the Basin, as it is the pedologic cover of a spring, there are many constraints to the indiscriminate use of agriculture, although these are not always represented for all mapping units. The units that present more constraints for agriculture are those with more active topography, very strong slopes (undulated and strong undulated relief), and consequently of lower stability and more fragility when used for seasonal cultivation, which generally determines a high emerging fragility.

Among the most common limitations in areas of rugged relief are the soils of greater usage fragility, where direct effects of agriculture mechanization on vulnerability to erosion can be observed and when exposed, were considered units of very high emerging fragility.

TABLE 4 – SCALE OF VALUES ADOPTED IN DETERMINING RELATIVE DEGREE OF POTENTIAL FRAGILITY OF THE RIO VERDE BASIN SOILS

SOIL CATEGORY	CODE	VALUE
Dystrophic Brown Oxisol	1	3
Dystrophic Red Oxisol	2	1
Association of Dystrophic Red Oxisol and Dystrophic Brown Nitisol	4	3
Association of Eutrophic Red Oxisoland Haplic and Eutrophic Inceptisols	5	3
Dystrophic Brown Nitisol	6	4
Association of Dystrophic Brown Nitisol and Dystrophic Red Oxisol	7	3
Association of Dystrophic Red Nitisol and Plintic Dystrophic Red Ultisol	8	6
Abrupt Dystrophic Red Ultisol	9	7
Association of Eutrophic Red Oxisol, Dystrophic Brown Oxisol and Eutrophic Red Nitisol	10	3
Association of Dystrophic Brown Nitisol and Haplic and Dystrophic Inceptisol	11	4
Eutrophic and Haplic Inceptisol	12	5
Dystrophic and Haplic Inceptisol	13	5
Ortic Tiomorphic Gleysol	14	8

3. RESULTS AND DISCUSSION

The identification of environmental fragility of the Rio Verde Basin was established through analysis of the components of the landscape, integrating the biophysical aspects and its transformation due to human activities.

In this context, the potential fragility of the soil was considered a result of geologic and geomorphologic aspects, integrating the lithostructural and tectonic aspects with current and inherited climate characteristics, the soils and the vegetation. From these parameters, classification of emerging fragility resulted from diverse forms of land-use.

3.1 ANALYSIS OF ENVIRONMENTAL FRAGILITY OF THE RIO VERDE BASIN

3.1.1 The Potential Fragility of the Rio Verde Basin

Considering Tricart (1977), the Rio Verde Basin was categorized into geodynamic units of stable, intergrade and unstable environments, which were mapped according to the methodology adapted from Ross (1990) in spatial units, organized by fragility categories, according to Figure 5 and Table 5.

TABLE 5 – POTENTIAL FRAGILITY OF THE RIO VERDE BASIN

FRAGILITY CATEGORIES	POTENTIAL FRAGILITY %
Insignificant	6.2
Low	65.0
Low to moderate	14.5
Moderate	14.3
TOTAL	100.0

Org.: MURATORI and PASSOS
Source: GIS-LAPIGEO (2009) and LAGEAMB (2010)

The categories of potential fragility vary from insignificant, very low, low, low to moderate and moderate. According to the data obtained, we found that from a geoenvironmental perspective, almost the entire basin (86%) is categorized as intergrade or intermediary geodynamic unit.

The following is a description of each category individually:

- Insignificant: does not depend on the lithology, considering that it generally presents a thick pedologic cover (Oxisol) and occurs under the cover of forest (Mixed Ombrophilous Forest), in declivities below 3%, and on elevated surfaces.
- Low: does not depend on the lithology, considering that it generally has a thick pedologic cover of Oxisol and Ultisol, occurring under forest (Mixed Ombrophilous Forest) and steppe (grasslands) cover, in declivities from 3% to 12%, distributed on medium hilltops and along slopes.
- Low to Moderate: lithology from the rocks of the Açungui group with an intermediary pedologic cover of Ultisol and Oxisol, occurring under forest (Mixed Ombrophilous Forest) and associated with patches of steppe (grasslands), in declivities from 12 to 30%, distributed on medium and narrow hilltops and along slopes. It is important to mention that areas of little significance for the study with very steep slopes and declivity above 30% were included in this category.
- Moderate: Lithology related to alluvials under the Alluvial Mixed Ombrophilous Forest, the occurrence of sections of organic deposits under the steppe (grasslands), and hydromorphic soil in declivities of less than 3%.

In summary, it can be concluded that in considering the potential fragility, the natural components of the Rio Verde Basin determine an intermediary stage between morphogenetic and pedogenetic processes, or in other words, interdependence between the stability/instability of the environment.

3.1.2 The Emerging Fragility of the Rio Verde Basin

For the construction of the Emerging Fragility Map of the Rio Verde Basin (Figure 6), the information layers of the geodynamic units established and represented in the Potential Fragility map were integrated with the information layers related to anthropogenic factors, expressed in Table 6.

TABLE 6 – EMERGING FRAGILITY OF THE RIO VERDE BASIN

FRAGILITY CATEGORIES	EMERGING FRAGILITY (%)
Insignificant	0.3
Very Low	21.5
Low	6.6
Low to Moderate	3.0
Moderate	8.0
Moderate to High	3.9
High	48.6
Very High	9.7
TOTAL	100.0

Org.: MURATORI and PASSOS
Source: SIG-LAPIGEO (2009) and LAGEAMB (2010)

The categories of emerging fragility, obtained through the integration of the information layers (PIs) of the natural environment and of anthropogenic factors vary from insignificant, very low, low, low to moderate, moderate, moderate to high, high and very high. According to the data obtained, we found that from a geoenvironmental perspective, almost the entire basin (86%) is categorized as intergrade or intermediary geodynamic unit.

Each category can be described as:

- Insignificant: Preserved areas with forest (Mixed Ombrophilous Forest) cover, that do not depend on lithology, considering that they generally have a thick pedologic cover (Oxisol) in declivities below 3%, located on elevated surfaces (wide hilltops).
- Very Low: preserved areas with secondary forest cover in advanced stage of Mixed Ombrophilous Forest and patches of grassland (steppe) with declivities between 3% and 12%, distributed on medium hilltops and along slopes.
- Low to Moderate: are associated with the lithogy of rocks of the Açungui group under Mixed Ombrophilous Forest and patches of grassland (steppe), in declivities from 12 to 30%, distributed on medium and narrow hilltops and along slopes. Associated with this category are areas of little significance that are not mappable because of scale, with declivities above 30%, and related to categories of Moderate to High with the presence of carbonaceous and phylite rocks.
- Moderate: Lithology of alluvials under the Alluvial Mixed Ombrophilous Forest, the occurrence of organic deposits under grassland cover, and hydromorphic soils in declivities of less than 3%.
- Moderate to High: this category does not depend on lithology and relates to areas modified by anthropogenic activities, where management techniques compromise soil preservation (Figure 7) in areas of slope, in general below 30%, distributed across various types of relief, from low slopes

alongside the river plain to the higher surfaces near the interfluves.

- High: This category does not depend on lithology and relates to areas modified by anthropogenic activities where management techniques can have a more significant impact than the previous category on conservation of the soils in areas of declivity in general above 12%, distributed on narrow hilltops and along slopes.
- Very High: This category does not depend on the lithology or morphology of the land although the situation can be aggravated by a greater degree of susceptibility to destabilization of these components and sensitivity to external agents of morphogenesis. Fundamentally, it relates to areas that have been significantly altered by anthropogenic activities, areas with exposed soils, and water bodies in general (these in particular because of the constant modification of sediment load in suspension and the risk of contamination by soluble materials).

From the data above, Table 3 was prepared to compare the potential and emerging fragility of the Rio Verde Basin and the distribution of stability degrees in relation to the geoenvironmental units.

The data synthetized in Table 7 demonstrates that anthropogenic activities have had a significant impact on the Basin; from a total of 71.2% of the area that originally was included as part of the stable geoenvironmental unit, only 21.8% remains.

With regard to areas of low to moderate fragility, composed primarily of the floodplains, a little more than 55% of the original area remains. This value was reduced from 28.8% of the total basin area to 16% but the intermittent areas of high instability that, according to the results of potential fragility were insignificant to the adopted scale with less than 1% and therefore irrelevant, occupied approximately 62% of the basin based on emerging fragility. This value demonstrates an elevated environmental change due to anthropogenic factors.

The study also identified that the areas classified as moderate to very high emerging fragility at various points in the basin could significantly minimize their degree of instability if the use of conservationist soil management techniques or cultivation were adopted, such as the adoption of contour line systems and tillage. Such procedures would certainly lead to a reduction in the areas categorized as high or very high emerging fragility, reducing them to a category of low emerging fragility. Such actions could contribute significantly to minimizing the risk of eutrophication of the reservoir and conserving the water resources of the Basin.

4. FINAL CONSIDERATIONS

The geoenvironmental characterization of the Rio Verde Basin allowed for the identification of the potential fragility from the point of view of intrinsic characteristics, as well as emerging fragility in the face of destabilization factors represented by anthropogenic activities. While as a whole, the potential fragility was considered as low to moderate, the emerging fragility presented some elevated indexes that are considered critical. This highlights the urgent need to restrict certain uses so as to not exceed the limits of stability and to protect the water of the Basin and as such maintain the stability of the reservoir.

From the ecodynamic perspective of Tricart and from the point of view of environmental preservation of the Rio Verde Basin and the recovery of the areas considered critical, it is absolutely necessary to apply guidelines that structure environmental planning in the Basin. We propose the realization of more detailed studies that facilitate the monitoring of the Rio Verde Basin with the aim of consolidating the institutional mechanisms towards sustainable management.

TABLE 7 – COMPARISON BETWEEN THE POTENTIAL AND EMERGING FRAGILITY OF THE RIO VERDE BASIN

FRAGILITY CATEGORIES	FRAGILITY		GEOENVIRONMENTAL UNIT
	POTENTIAL %	EMERGING %	
Insignificant	6.2	0.3	STABLE
Very Low	65.0	21.5	
Subtotal	71.2	21.8	
Low	14.5	6.6	INTERMEDIARY
Moderate	14.3	9.4	
Subtotal	28.8	16.0	
Moderate to High	0.0	3.9	UNSTABLE
High	0.0	48 6	
Very High	0.0	9.7	
Subtotal	0.0	62.2	
TOTAL	100.0	100.0	

Org.: MURATORI and PASSOS
Source: Maps of Potential Fragility and Emerging Fragility of the Rio Verde Basin

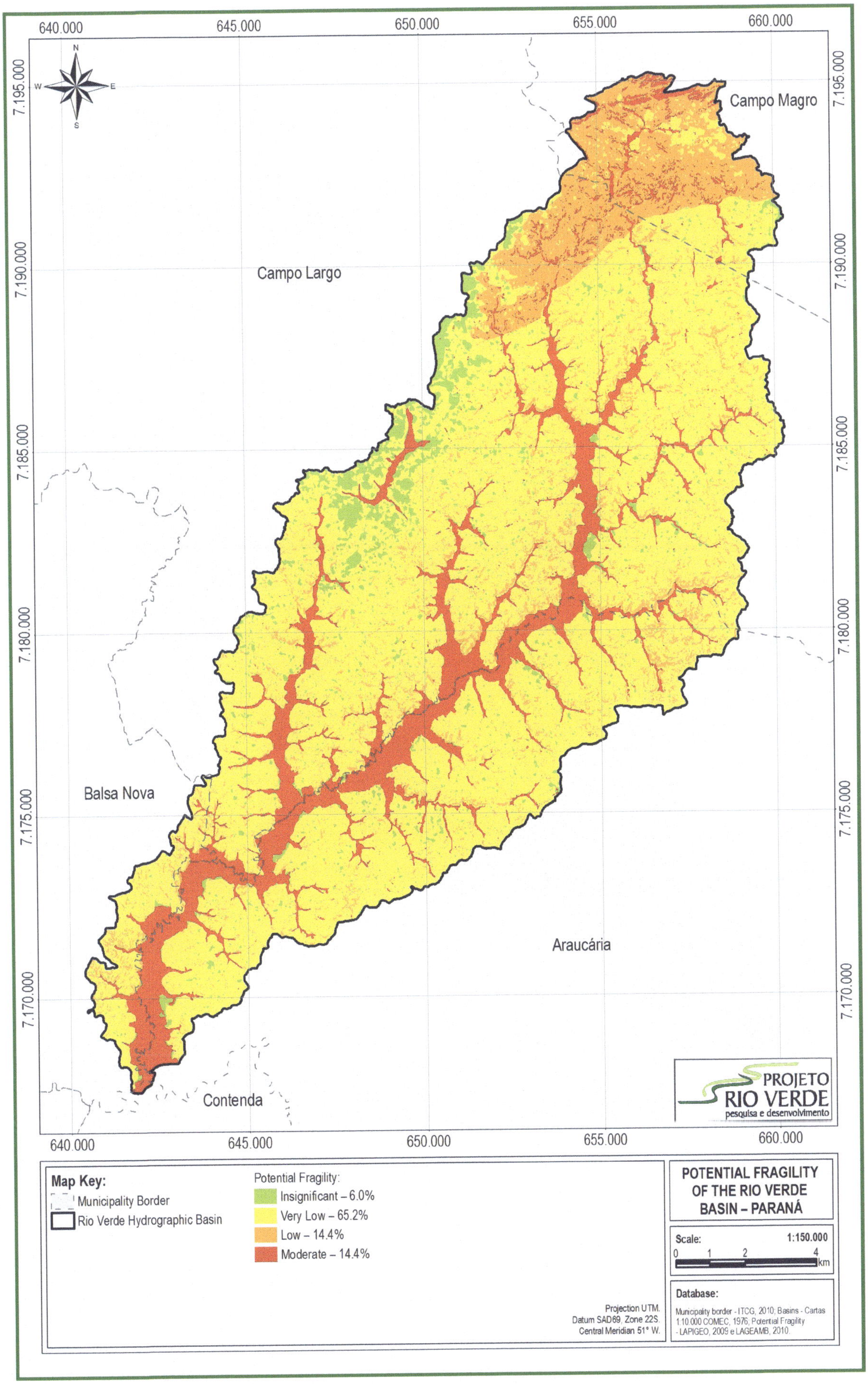

FIGURE 5 – POTENTIAL FRAGILITY MAP OF THE RIO VERDE BASIN
Org.: MURATORI and PASSOS
Source: COMEC (1976); GIS database- LAPIGEO (2009) and LAGEAMB (2010)

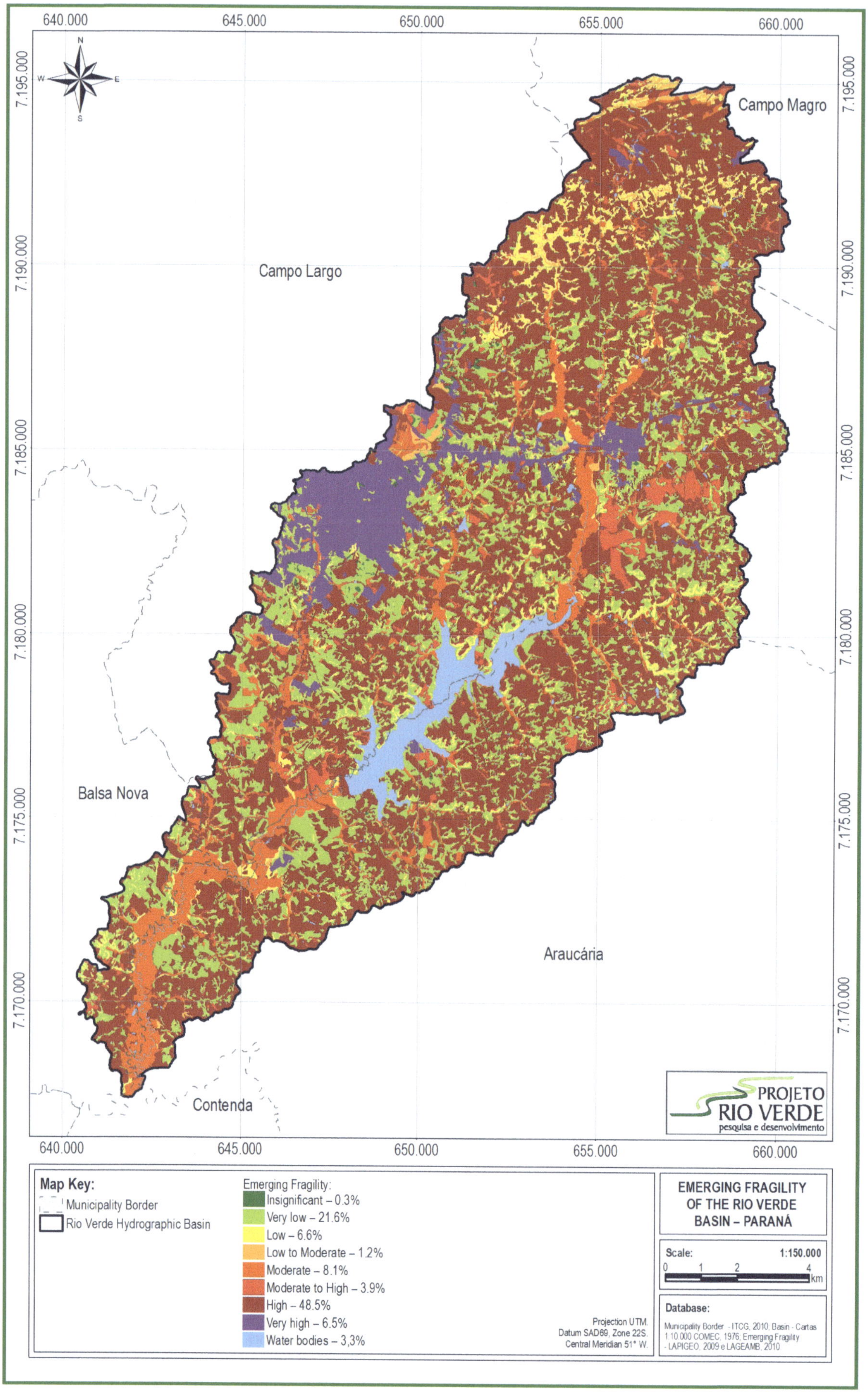

FIGURE 6 – EMERGING FRAGILITY MAP OF THE RIO VERDE BASIN
Org.: MURATORI and PASSOS
Source: COMEC (1976); RODERJAN (2010); SIG-LAPIGEO (2009) and LAGEAMB (2010) database

FIGURE 7 – INADEQUATE MANAGEMENT OF THE SOIL

Illustration of an area in which land-use, although without legal constraints, is improperly managed and with inadequate conservationist techniques. Even with low declivities (less than 12% inclination) soil erosion occurs (observed in field and marked with number 1 in red).

Source: PASSOS, 2010.

5. REFERENCES

AB'SABER, A. N. (1969) Pedimentos e Bacias Detríticas Pleistocênicas em São Paulo. *Geomorfologia*, USP, Instituto de Geografia, (9): 1-12.

BERTALANFFY, L. (1973) von *Teoria Geral dos Sistemas*. Petrópolis: Vozes.

BIGARELLA, J. J. et al. (1979) *Recursos Naturais. Ouro Fino*. Região Metropolitana de Curitiba. Secretaria de Estado do Planejamento, 30 p.

BIGARELLA, J. J. and MAZUCHOWSKI, J. G. (1985) *Visão integrada da problemática da erosão*. Curitiba, ADEA/ABGE, 329 p.

BIGARELLA, J.J; BECKER, R.D. and PASSOS, E. (2003) *Estrutura e origem da paisagem tropical e subtropical*. Florianópolis: Ed. UFSC, v.2., p. 884-939.

CANALI, N. E. and MURATORI A. M. (1981) Síntese da evolução geomorfológica da Bacia Sedimentar de Curitiba. In: *Simp. Regional de Geologia*, 3, Atas, v.2, p. 335-362.

CLARK LABS. (2009) IDRISI 16 Taiga. *Sof ware – Project*. Research center of the Clark University.

CREPANI, E; MEDEIROS, J.S.; AZEVEDO, L.G.; DUARTE, V.; HERNANDES, P.; FLORENZANO, T & BARBOSA, C. (2000) *Sensoriamento Remoto e Geoprocessamento Aplicados ao Zoneamento Ecológico- Econômico e ao Ordenamento Territorial*. São José dos Campos: INPE.

CHIARINI, J.V.; DONZELLI, P.L. (1973) *Levantamento por fotointerpretação das classes de capacidade de uso das terras do Estado de São Paulo*. Campinas: IAC.

EMBRAPA – (1999) Centro Nacional de Pesquisa de Solos, *Sistema Brasileiro de Classificação de Solos*. Brasília: Embrapa Produção de Informação; Rio de Janeiro: Embrapa Solos. 412 p.

HERZ, R. and BIASE, M. (1989) *Critérios e Legendas para Macrozoneamento Costeiro, Comissão Interministeeial para os recursos do Mar- CIRM*, Brasília, p.198.

JUSTUS J. O. (1985) *Subsídios para interpretação morfogenética através da utilização de imagens de radar*. Universidade Federal da Bahia, Salvador. Dissertação de Mestrado, 204 p.

KLEIN, R. M. (1962) Fitofisionomia e notas sobre a vegetação para acompanhar a planta fitogeográfica de partes dos Municípios de Rio Branco do Sul, Bocaiúva do Sul, Almirante Tamandaré e Colombo (PR). *Boletim da Universidade do Paraná*, Curitiba, n. 3, p. 1-33.

KLEIN, R. M.; HATSCHBACH, G. (1962) Fitofisionomia e notas sobre a vegetação para acompanhar a planta fitogeográfica do Município de Curitiba e arredores (Paraná). *Boletim da Universidade do Paraná*, Curitiba, n. 4, p. 1-29.

KUMPERA, V. (1979) *Interpretação sistêmica do planejamento*. São Paulo: Nobel.

MAACK, R. (1968) *Geografia física do Estado do Paraná*. Curitiba: Banco de desenvolvimento do Paraná; Universidade Federal do Paraná; Instituto de Biologia e Pesquisas Tecnológicas, 350 p.

MURATORI, A. (1966) Nota explicativa da folha geológica de Campo Largo. Boletim da Universidade Federal do Paraná. *Geologia* (21): 1-19.

MURATORI, A. M. (Coord.). FERNANDES, A.; PELENS, E. ; SANTOS, E. L.; PASSOS, E. ; ROCHA, H. O.; SCOPEL, I. (1987) *Nota Explicativa da Compartimentação Morfológica da Região Metropolitana de Curitiba*. COMEC.

PASSOS, E. (1987) *Relação entre vertentes e solos em Umbará/ Curitiba-PR*. Master Dissertation. Graduation Program in Soil Science Of the Universidade Federal do Paraná. Advisor Arthur dos Santos Filho. Curitiba.

PASSOS, E. (1980-a) *Mapa de Níveis de Erosão e ou Sedimentação* -Levantamento de recursos naturais na Região Metropolitana de Curitiba. Folha 389 ADEA/COMEC, Curitiba INÈDITO).

PASSOS, E. (1980-b) *Mapa de Níveis de Erosão e ou Sedimentação* -Levantamento de recursos naturais na Região Metropolitana de Curitiba. Folha 390 ADEA/COMEC, Curitiba INÈDITO).

PASSOS, E. Geomorfologia e Solos na Bacia do Rio Verde In: Segundo Seminário PROJETO INTERDISCIPLINAR SOBRE EUTROFIZAÇÃO NO RESERVATÓRIO RIO VERDE, ARAUCÁRIA, PR.

PASSOS, E. and CANEPARO, S. C. *Sistema de Informações Geoambientais na Região Metropolitana de Curitiba* - BACIAS DE MANANCIAIS:BACIA DO RIO MIRINGUAVA - Aptidão dos Solos e Incongruências de Uso da Terra. (Inédito).

PASSOS, E.; MURATORI, A. M. (2008) *Relatório semestral Meio Físico e Modelagem Ambiental Sub-projeto – Fragilidade Potencial e Emergente do Solo*. UFPR/FUNPAR/PETROBRÁS. Curitiba, Outubro.

PETROBRÁS; FUNPAR and UFPR. (2008) Programas do Projeto Interdisciplinar de pesquisa sobre Eutrofização de águas na Bacia do Rio Verde. *Documento Técnico s/n*. Curitiba, março, 130 p.

ROSS, J. L. S. (1990) *Geomorfologia, Ambiente e Planejamento*. Contexto, São Paulo.

SALAMUNI, E. (1998) *Tectônica da Bacia Sedimentar de Curitiba (PR)*. Instituto de Geociências e Ciências Exatas, Universidade Estadual Paulista-Rio Claro. Tese de Doutoramento, 214 p.

SALAMUNI, E; EBERT, H. D and HASUI, Y. (2004) Morfotectônica da Bacia Sedimentar de Curitiba. In: *Revista Brasileira de Geociências*. 34(4):469-478, dezembro.

TRICART, J. (1977) *Ecodinâmica*. Rio de Janeiro, IBGE, Diretoria Técnica, SUPREN.

TRICART and KILIAN, J. (1979) *La Eco-geografía y la ordenación del medio natural*. Elementos críticos. Barcelona: Editorial Anagrama.

SECTION III

HYDROGRAPHIC ASPECTS AND HYDRODYNAMIC

CHAPTER 8

SUPPLY CAPACITY OF THE RIO VERDE RESERVOIR

Mauricio Felga Gobbi
Helder Nocko

SUPPLY CAPACITY OF THE RIO VERDE RESERVOIR

Mauricio Felga Gobbi
Helder Nocko

ABSTRACT

The Rio Verde Reservoir has supplied the President Getúlio Vargas Refinery (REPAR) since the 1970s and it also provides water for urban use through SANEPAR, the local state sanitation company. Little hydrologic (runoff and rainfall) information is available for the Rio Verde Basin which feeds the reservoir. Within the context of current and future shortages of water resources, the expansion of REPAR's activities and a potential increase in demand for public water supply, the intent of this chapter is to complement existing information on the Basin by presenting a study with a different technical perspective to those currently available. As such, this study will provide REPAR and water resource managers with assistance in decision making regarding the management of the reservoir.

KEYWORDS

Hydrologic modeling, rainfall-runoff model, public supply, water availability, markov process.

1. INTRODUCTION

This chapter presents diverse techniques and results aiming at estimating the capacity of the Rio Verde Reservoir to regulate flow rates with varying water consumption scenarios and respective probabilistic rates of error.

The study's main concept is as follows. First, considering there is no historic information on hydrologic inflow rates to the reservoir (runoff), these rates must be generated based on past rainfall data. This is performed by means of a mathematical model which converts rainfall data to runoff. However, data on precipitation is also scarce; no hydrologic station with a long-term record of rainfall exists within the Rio Verde Basin. Therefore, it was necessary to use data from nearby stations. The rainfall-runoff model must also be calibrated; this was accomplished using a comparable neighboring basin (Passaúna) for which runoff data is available, providing a first approximation of the calibration for the Rio Verde Basin. The aim of this study is to estimate the long-term regulating capacity of the reservoir. Thus, a much longer rainfall record is needed than those observed in existing records for the region. The following technique was used to circumvent this limitation: precipitation data were used to generate a calibrated local probability distribution and data for each month was simulated using a Monte-Carlo1 simulation in a Markov process2 with a matrix of transition probabilities. From this, a rainfall record can be generated with the desired number of years and this long-term data series can be used to inform the rainfall-runoff model, which in turn is used to model the temporal evolution of the water volume in the reservoir. Given water abstraction from the reservoir, one can estimate, for instance, the probability of a 1-day failure in water supply, or the estimated number of days of failure in water supply over a one hundred year period.

[1] In a Monte-Carlo simulation, random events (e.g. rainfall) are drawn from a probability distribution deemed appropriate for the specific phenomenon. It is a manner of generating an artificial time series.

[2] Markov Processes are those in which a future event depends only on present conditions and not on past occurrences.

2. REGIONAL HYDROLOGY

Presented within this section are various existing data sources and information on the Rio Verde Basin, as well as sources from neighboring watersheds whose data may contribute to this study. The main information consists of existing studies on the regionalization of flow data for small watersheds in the region (source: IPAGUAS) and rainfall data (source: ANEEL) from hydrologic stations found within the region.

2.1 RIO VERDE BASIN

Rio Verde Basin (indicated in Figure 1 by a purple outline) is found at the approximate geographical coordinates of 25°30′ S (latitude) 49°30′ W (longitude). The basin has a total area of 257 km². In the center of the watershed is the Rio Verde Reservoir, the main focus of the present study. The only hydrologic stations in the Rio Verde Basin are the Formigas station and the Rodeio station, both downstream from the reservoir and strongly influenced by its regulating activity. Adjacent and to the east of the Rio Verde Basin is the Passaúna River Basin.

2.2 STUDIES OF MINIMUM AND MEAN FLOW

Figure 2 shows maps (Source: IPAGUAS) of specific mean and minimum flow values, respectively, within the region of the Rio Verde Basin. These maps were obtained by means of flow regionalization studies for small watersheds in Paraná State (Source: IPAGUAS). One notes that the gradients (spatial variation rates) of these flow values are relatively small in the whole region (between Campo Largo and Curitiba). Mean specific runoff values range between 15 and 17 liters/s/km² (or 1.5×10^{-5} to 1.7×10^{-5} m³/s/km²) and minimum specific runoff values for the region are in the range of 2.5 to 3.5 liters/s/km² (or 2.5×10^{-4} to

3.5×10^{-4} m³/s/km²). For the Rio Verde Basin in particular, studies from CEHPAR show that the estimate of $Q_{7,10}$ flow at the Rodeio station, which has a catchment area of 234 km² (see Figure 1), is 0.59m³/s. For the Rio Verde dam, with a catchment area of 167 m³, $Q_{7,10}$ flow corresponds to 0.42m³/s. The permanence curve for the catchment area of the Rio Verde dam is shown in Figure 3 and provides a Q_{95} flow of 0.66 m³/s.

Studies on flow regulation and regulated volumes for Paraná State using classic techniques of intra-annual regulation (Project HG-77) suggest the volume V necessary to regulate an outflow Q_F can be calculated as: $V = (Q_F - q_T X_E(t))\ t$, where $X_E(t)$ is the minimum (annual) mean flow in given drought period t, and qT is the uniform minimum flow (normalized by the mean) associated with a recurrence time T. These studies show that the Rio Verde Reservoir is able to regulate a little over 70% of the historical mean flow of 25471 L/s, or more specifically, 18751 L/s. This information must be interpreted correctly and used with caution, since: (i) it assumes that the reservoir is completely full at the start of the supposed drought period; (ii) this methodology is based on minimum flow values calculated from probabilistic distributions which are based in turn on relatively short-term data series. In the case of the Rio Verde Basin it consists of data from the nearby Rodeio station, with just 20 years of data, collected after the construction of the Rio Verde dam at a position slightly downstream from it, which greatly reduces the variability of the minimum flow data; (iii) one final important limitation is that this technique does not apply to cases of multi-annual variability. The present study intends to address these limitations by using an alternative approach.

2.3 RAINFALL DATA

Figure 4 shows the location of conventional rainfall stations within the region under study, labeled with their official codes. These stations collect daily rainfall data. The data from all stations had some gaps. When these gaps extended for a period of many consecutive months, the data (or the station itself) was simply discarded. When the number of gaps was relatively small, they were filled in with data from the nearest station. In all cases of gap filling, verifications were carried out to confirm that the data supplied would not cause significant and qualitative changes in the statistics for the respective station.

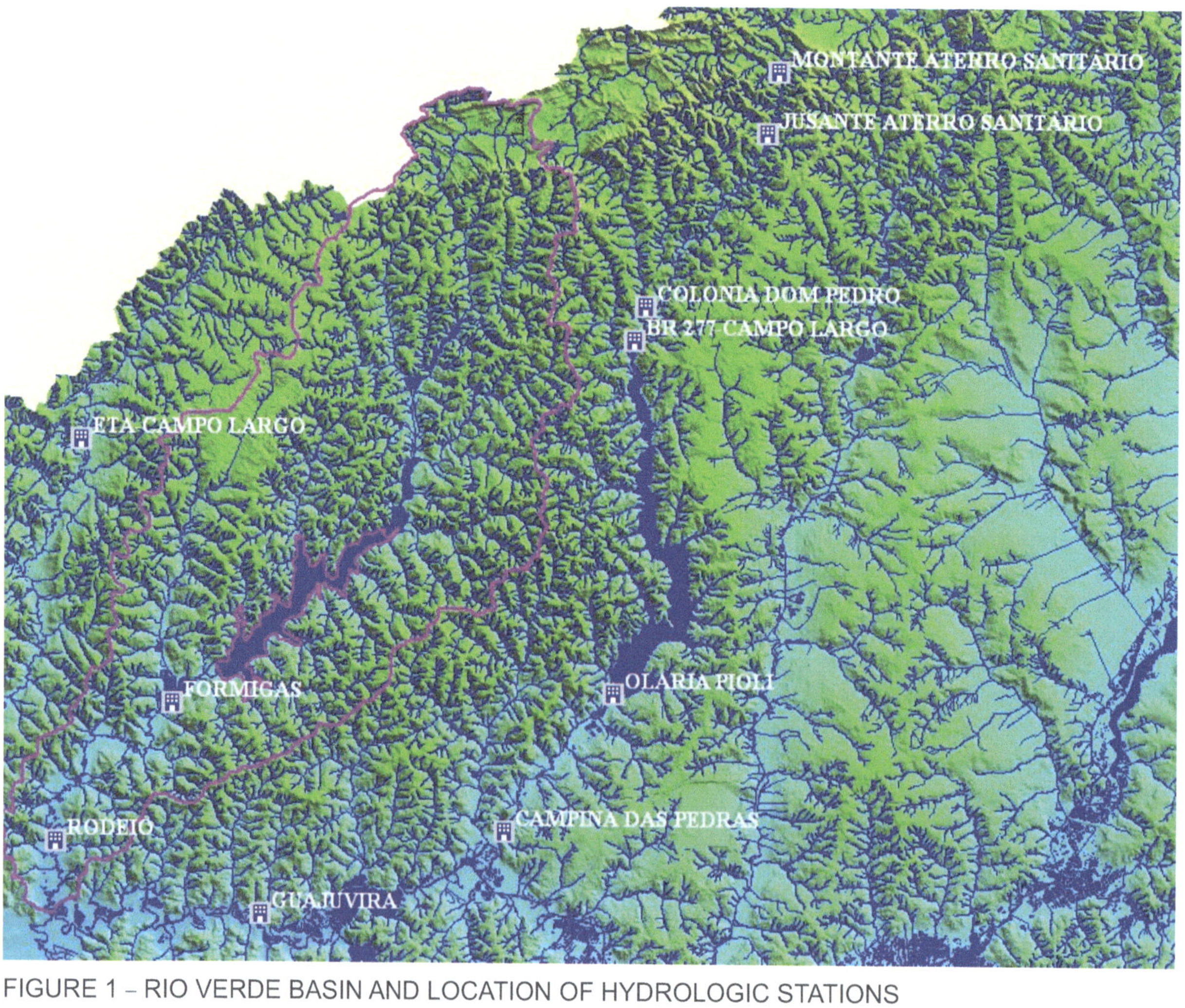

FIGURE 1 – RIO VERDE BASIN AND LOCATION OF HYDROLOGIC STATIONS

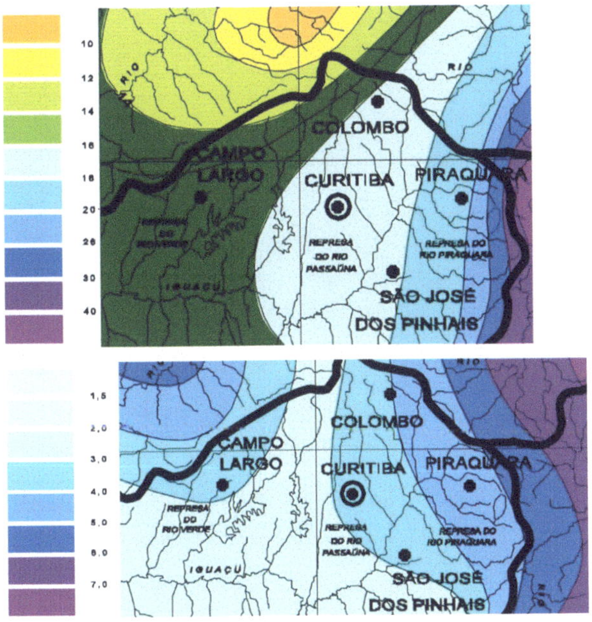

FIGURE 2 – MAP OF SPECIFIC MEAN (ABOVE) AND MINIMUM (BELOW) FLOWS IN liters/km² IN THE RIO VERDE REGION. SOURCE: IPAGUAS

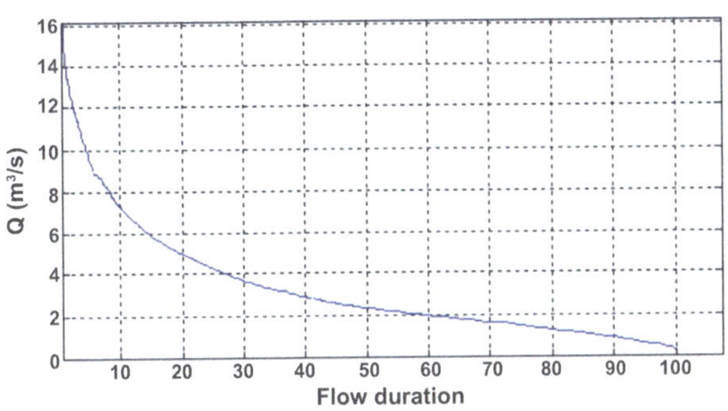

FIGURE 3 – PERMANENCE CURVE FOR THE RIO VERDE DAM (CATCHMENT AREA = 167km²)

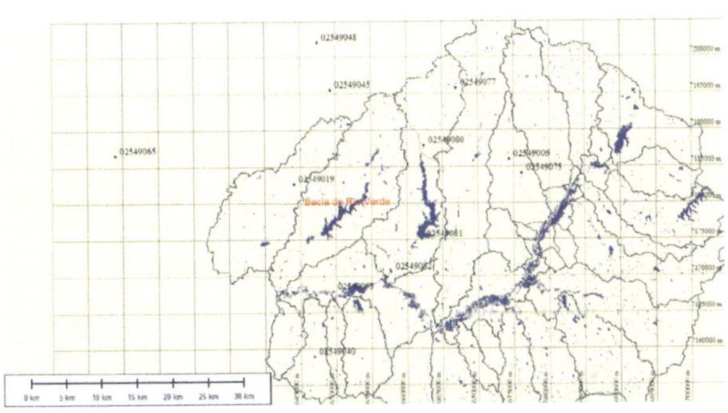

FIGURE 4 – RAINFALL STATIONS WITHIN THE REGION UNDER STUDY

The oldest station in the region is station 02549006, located in Curitiba and it has been collecting data since the late 19th century. If the data from this station could be used as a sufficient representation for the Rio Verde Basin, it would represent an ideal situation since the statistical reliability of this station is unmatched within the region. It is also important that the stations' data be statistically stationary stable, particularly long-term data and drought statistics. To assess the properties of the data from station 02549006 for the present study, this station was compared with other stations located in closer proximity to the watershed of interest (Rio Verde), taking into consideration that these stations do not hold long-term data and only the common period of data collection between the stations may be used in this comparison. The stations used (besides station 02549006) were stations 02549019, 02549040, 02549045 and 02549048 over the period of 01/01/1975 to 12/31/2006 (32 years). The Olaria Pioli telemetry station (see Passaúna Dam in Figure 1) was also used as it possesses rainfall data at 15 minute intervals since 1999. This latter station is not climatologically representative but was used to verify possible problems related to errors in reading conventional rainfall gauges. Such errors are more pronounced in areas of low precipitation, in which relative reading errors are higher.

The first test carried out consisted of calculating empirical cumulative distributions of monthly rainfall (disregarding seasonality) for stations 02549006, 02549019, 02549040, 02549045 and 02549048. Figure 5 plots the comparison of these distributions. Monthly rainfall values may be considered independent events, so it is possible to test whether two cumulative distributions originate from different theoretical distributions. One of the tests most often used for this purpose is Kolmogorov-Smirnoff Test (K-S) (EADIE *et al.*, 1971). The null hypothesis of the test is that the two time series do *not* come from different distributions. If the null hypothesis is rejected, the variables *do* come from different distributions, which casts doubts on the suitability of using data from one station (e.g. station 02549006) as representative of data from another station (e.g. station 02549019). The test provides a *p* value corresponding to the probability (between 0 and 100%, or between 0 and 1) that the rejection of the null hypothesis is a statistical coincidence and not a fact. That is, if the *p* value is low, 5% for instance, one can say that both variables come from different theoretical distributions with 95% certainty. Higher *p* values (i.e., higher than 20%) indicate it is possible or even likely that both variables originate from a single theoretical distribution. Table 1 shows the results of the K-S test for pairs involving station 02549006 and the four other stations used. The results indicate that only station 02549040 seems to present a problem due to high values of cumulative probability in the monthly rainfall range between 100mm and 200mm.

TABLE 1 – KOLMOGOROV-SMIRNOFF TEST. *p* VALUES BELOW 5% CAST DOUBT ON THE USE OF STATION 02549006 FOR THE RIO VERDE BASIN

STATION COMPARED TO 02549006	*p* VALUE
02549019	79%
02549040	10%
02549045	96%
02549048	79%

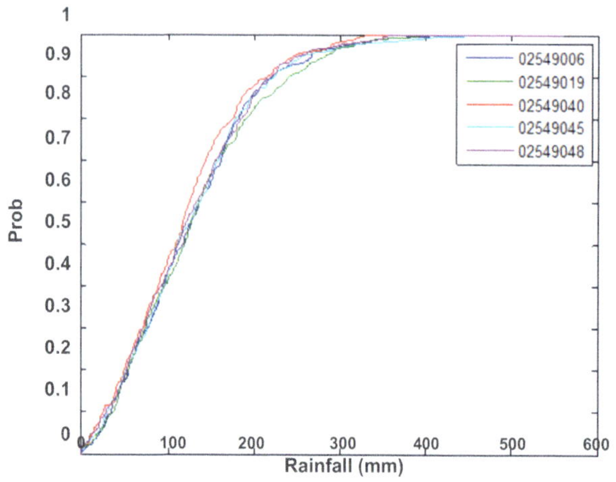

FIGURE 5 – MONTHLY RAINFALL CUMULATIVE PROBABILITY CURVE FOR STATIONS 02549006, 02549019, 02549040, 02549045, AND 02549048

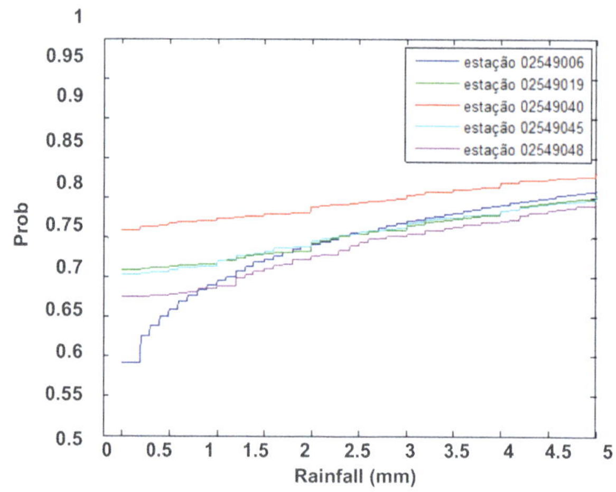

FIGURE 7 – DAILY RAINFALL CUMULATIVE PROBABILITY CURVE FOR STATIONS 02549006, 02549019, 02549040, 02549045 AND 02549048

Figure 6 shows a plot similar to that presented in Figure 5 but without calculating monthly rainfall values, so that the horizontal axis represents daily rainfall and the vertical axis is the cumulative probability of occurrence. Figure 7 shows a close-up the initial values in for small values of precipitation of Figure 6. Daily rainfall cannot be considered a variable for which subsequent events are independent; therefore the K-S test is not recommended in this case. A visual analysis, however, reveals a problem: there is a noticeable difference between stations in the cumulative probability mass at the initial values low precipitation values. In other words, the registered number of days without rain is very different between stations. For rainfall values above zero, the distributions seem to behave similarly. Once again, station 02549040 appears to be the most problematic. The fact that these stations are located in relatively homogenous regions, however, leads us to believe that there may be a real problem with systematic errors in measuring low rainfall values. What is clear is that the number of days without rain recorded at station 02549006 is approximately 59%, whereas at the other stations this number ranges between 67% and 76%.

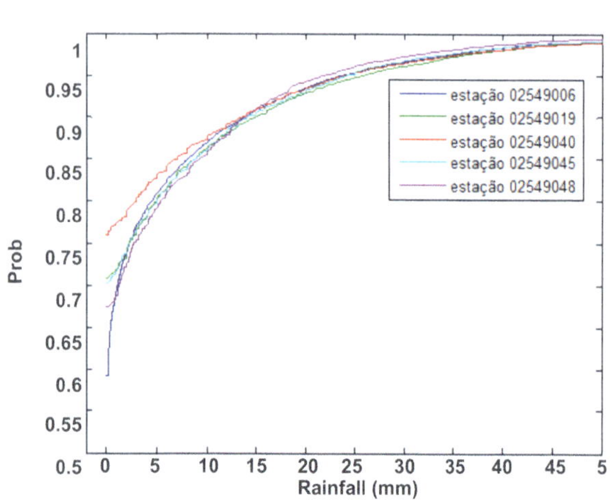

FIGURE 6 – DAILY RAINFALL CUMULATIVE PROBABILITY CURVE FOR STATIONS 02549006, 02549019, 02549040, 02549045 AND 02549048

As noted above, rhere is one automatic telemetry station, Olaria Pioli which is located within the Passaúna River watershed adjacent to the Rio Verde Basin and operated by SIMEPAR, which electronically registers rainfall every 15 minutes. This station started operation in 1999 and has a significant number of gaps from 2006 onwards but the number of gaps between 2000 and 2005 is arerelatively small. As this is a modern station, it is believed that – at least in the first years of operation – it has much higher precision in daily rainfall measurements for two reasons: there are no human errors (operator error) and measurements are taken progressively throughout the day, so it is able to more precisely capture events over the course of a day. Data from this station show a percentage of dry days of 60%. This number agrees with station 02549006 and is much lower than the values indicated by stations 02549019, 02549040, 02549045 and 02549048.

Figure 8 shows the double cumulative rainfall curve between station 02549040 and the mean of the other four stations: 02549006, 02549019, 02549045 and 02549048. One clearly notes a problem with the data from this station. A similar plot is shown in Figure 9; however, it shows the relation between station 02549006 and the mean of stations 02549019, 02549040, 02549045 and 02549048. It can be seen in this plot that the behavior of station 02549006 presents no major anomalies.

The question left concerning the use of the 117 years of data from station 02549006 is whether this series is sufficiently stationarystable. Figure 10 shows the series with 30 day-cumulative rainfall data (monthly rainfall). There is no visual evidence of significant variations in rainfall statistics, so the stability stationarity of the station's rainfall series is accepted for the purposes of this study.

3. RIO VERDE RESERVOIR

Rio Verde Reservoir was built in the mid-1970s by Petrobras to supply the President Getúlio Vargas Refinery (REPAR). The reservoir works run-of-the-river most of the time, with a water level at an elevation of 885.5 m, spilling the excess water that is not pumped to the refinery. Data on water surface area, drainage basin area and bathymetry of the Rio Verde Reservoir can be found in other chapters of this book.

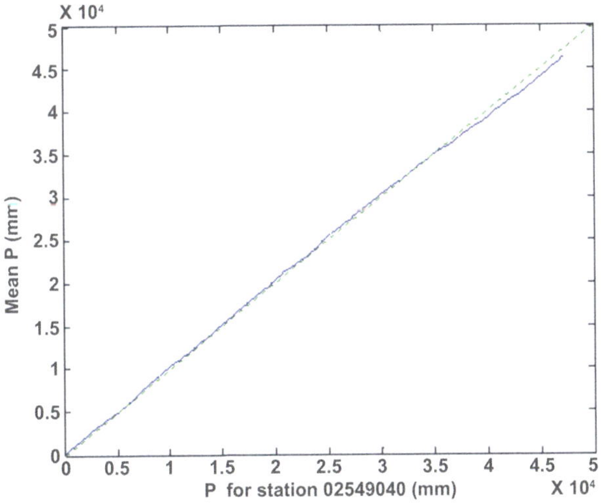

FIGURE 8 – DOUBLE CUMULATIVE RAINFALL CURVE BETWEEN STATION 02549040 AND THE MEAN OF STATIONS 02549006, 02549019, 02549045 AND 02549048

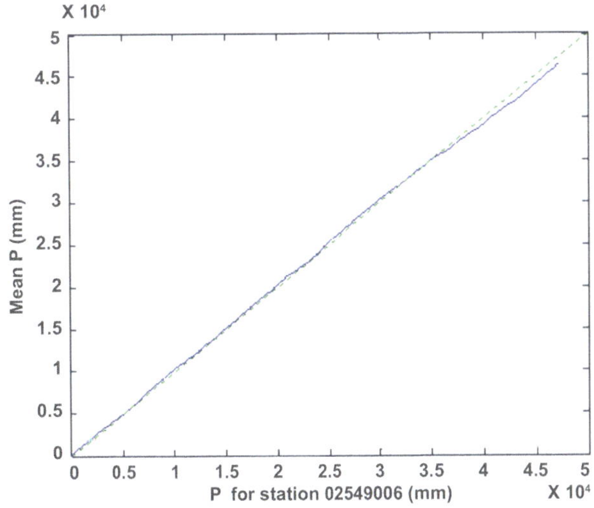

FIGURE 9 – DOUBLE CUMULATIVE RAINFALL CURVE BETWEEN STATION 02549006 AND THE MEAN OF STATIONS 02549019, 02549040, 02549045 AND 02549048

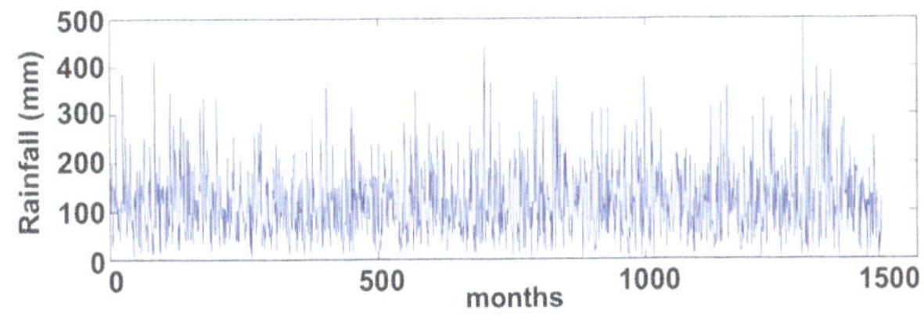

FIGURE 10 – MONTHLY RAINFALL. STATION 02549006

3.1 BATHYMETRY MEASUREMENT

The volume of the Rio Verde Reservoir was determined from two data sources. The main source was the data obtained during the data collection for bathymetry measurements of the reservoir. The second source of information was topographic charts from before the construction of the dam. These maps have been fully digitalized by the team involved in the present project.

The bathymetric survey was carried out between April 7 and April 10, 2008. The water surface of the reservoir during this period was at approximately 885-885.5m. Figures 11 and 12 illustrate the equipment installed on the vessel used during data collection.

A great difficulty faced by the project's bathymetry team was that the reservoir has vast areas with a large amount of vegetation submerged under the water or at the water's surface. No adequate vegetation clearing occurred prior to the filling of the reservoir and for that reason there were great difficulties in accessing various areas of the lake by boat, rendering work almost impossible in these areas (see Figure 13). The results of the bathymetric survey can be seen in the Chapter 12, Hydrodynamics and Transportation in the Rio Verde Reservoir. Figure 14 shows the same result in terms of elevation (meters above sea level) with three cross sections.

FIGURE 11 – COMPUTER SYSTEM AND LEVELING SYSTEM INSTALLED ON THE VESSEL

The elevation-volume curve was calculated up to an elevation of 888m (results presented in Table 2 and Figure 15). It is important to stress that the maximum assumed elevation of 888 relates to the crest of the dam, which cannot be reached during standard operation. Moreover, the volumes considered do not include backwater effects. Taking into consideration backwater effects would result in a slight increase in volume compared to that calculated here.

FIGURE 12 – VESSEL DURING BATHYMETRIC SURVEY NEAR THE DAM

TABLE 2 – ALTITUDE x VOLUME/AREA RELATION FOR THE RIO VERDE RESERVOIR

ALTITUDE (m)	VOLUME (m^3)	AREA (m^2)
870	0	0
871	88	526
872	2068	3959
873	9533	13256
874	61821	177227
875	427932	593799
876	1230442	1003399
877	2468309	1457180
878	4153012	1937132
879	6413160	2532907
880	9217492	3075908
881	12582010	3647825
882	16479408	4157587
883	20927355	4704043
884	25879607	5204554
885	31319126	6127291
885.5	34012645	6493733
886	37806164	6860176
887	45195020	7289502
888	52916670	8234320

FIGURE 13 – ILLUSTRATION OF AREAS WITH EXCESSIVE VEGETATION IN THE RIO VERDE RESERVOIR

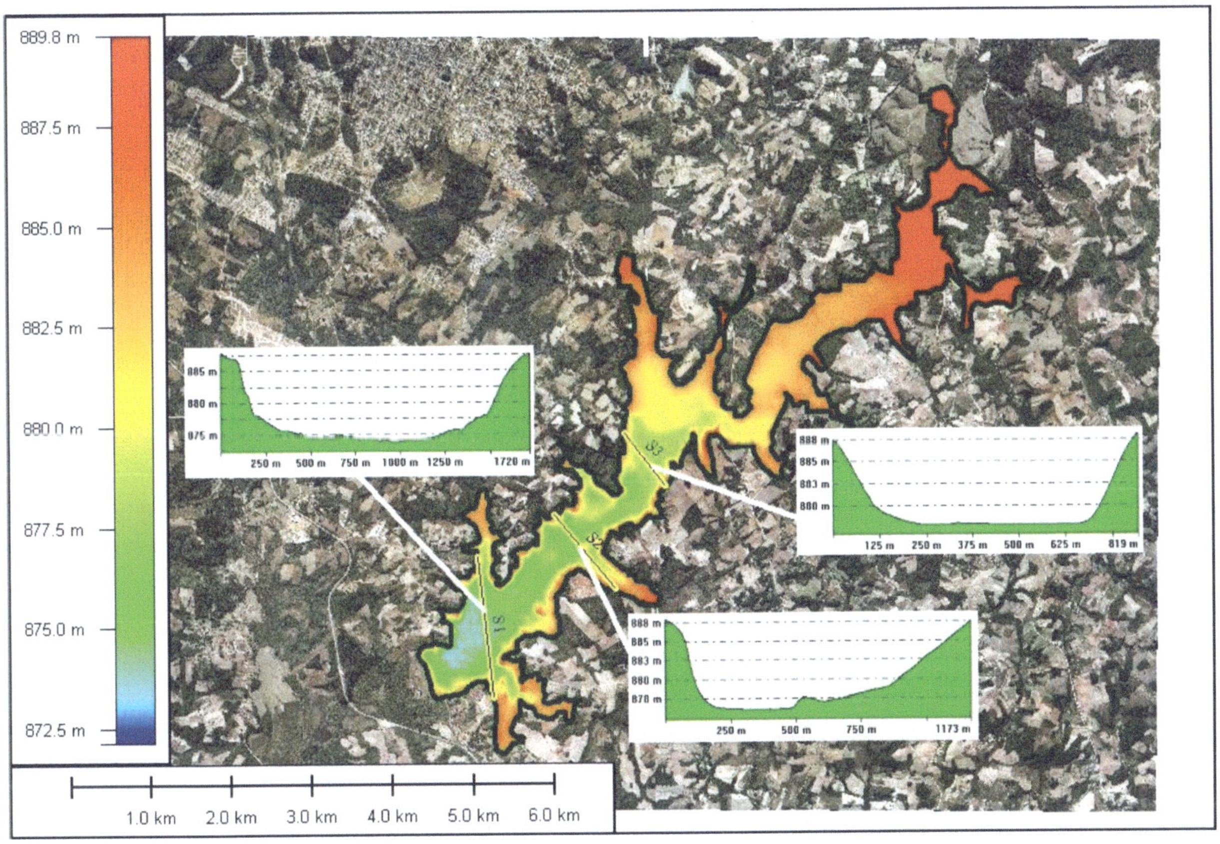

FIGURE 14 – BATHYMETRIC PROFILE OF THE RIO VERDE RESERVOIR IN 3 CROSS SECTIONS

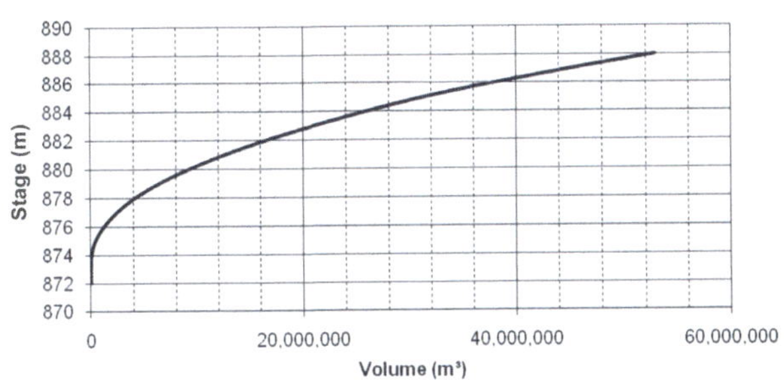

FIGURE 15 – ALTITUDE x VOLUME RELATION FOR THE RIO VERDE RESERVOIR

Besides the spillway, the dam of the Rio Verde Reservoir includes four floodgates located at altitudes of 882.7m, 879.7m, 876.7m and 873.5m. The volumes corresponding to each floodgate may be inferred from Table 2 (intermediate values can easily be interpolated). These floodgates allow for the removal of water from the reservoir at any level, although the quality of the water that is removed likely varies as the water level in the reservoir falls and floodgates at lower levels are used.

4. RAINFALL-RUNOFF MODELING

This section presents the technique used to generate runoff values from rainfall data. As this rainfall-runoff modeling needs calibration and no long-term data on runoff exists for Rio Verde, the solution found was to use the Rio Passaúna basin for calibration, since there is long-term data from a station located upstream from the reservoir.

4.1 TOPMODEL

The model chosen for the purposes of this study work is TOPMODEL (BEVEN & KIRKBY, 1975; BEVEN *et al.*, 1995). This model was chosen for the following reasons:

1. It is a deterministic model, not a stochastic one; that is, the stochastic part of the process is fully constrained by the occurrence of rainfall. If long-term runoff data existed, it would be possible to model runoff through stochastic modeling.
2. It is a conceptual model and not an empirical one. Purely empirical models need long-term high quality data for their calibration. Such data does not exist for the Rio Verde and therefore it is advantageous that this model is based on conservation laws of physics and contains parameters with physical significance.
3. TOPMODEL is one of the few conceptual models which, besides using the law of conservation of mass, also uses information about the slope at each section of the basin. Therefore, the model explicitly contains dynamic components for all the runoff in the basin.
4. It is a simple model. TOPMODEL includes few parameters and therefore is less likely to work just "by luck."

At the watershed mouth, total runoff is given by the sum:

$$q = q_z + q_b, \tag{1}$$

where q_z is surface runoff, q_b is subsurface runoff, and q is total runoff. Surface runoff q_z is given by:

$$q_z = \frac{A_S}{A} p + q_r, \quad (2)$$

where A is the total drainage area up to the section under consideration, A_S is the saturated area, p is un-intercepted rainfall, and q_r is the return flow or water that eventually leaves the ground and emerges as runoff.

Subsurface runoff q_b is made up of contributions q_i from each point i of the watershed, given by:

$$q_i = \frac{K_0}{f} e^{-fz_i} \tan\beta, \quad (3)$$

where z_i is the local groundwater depth, K_0 is the saturated hydraulic conductivity at the surface, and f is the factor of declining hydraulic conductivity with depth. β is an important parameter representing the local slope.

The volume of water needed to saturate the soil is expressed by the so-called storage deficit S. At point i, this deficit (S_i) is related to the depth z_i by:

$$S_i = (\theta_{sat} - \theta_{cc}) z_i, \quad (4)$$

where θ_{sat} is saturation humidity and θ_{cc} is field capacity humidity.

After a rainfall event, the sub-basin transfers the water from the highest elevations to the lowest elevations of the basin. In this manner, subsurface and surface flows accumulate water near valleys and streams, thus increasing the water content in these portions of the basin. These areas then start collecting water during a rainfall event, saturating the soil and creating surface runoff. In order to determine these saturation areas, TOPMODEL uses the concept of a topographic index (*IT*) by means of the following formulation:

$$IT = \ln\left(ac_i / \tan\beta\right). \quad (5)$$

In this formula, ac_i represents the upstream catchment area contributing to each cell's runoff by unit of width (normal to the slope and along a level curve), and tanβ represents a cell's local slope.

4.2 CALIBRATION FOR PASSAÚNA SUB-BASIN

TOPMODEL was applied to the upper portion of the Passaúna River watershed. The mouth defining this sub-basin is located at the Colônia Dom Pedro water level gauging station (see Figure 1). The runoff and rainfall data without any gaps ranged from 02/01/1985 to 07/24/1990. The sub-basin has an area of 92 km². The spatial distribution of the topographic index in the area is shown in Figure 16.

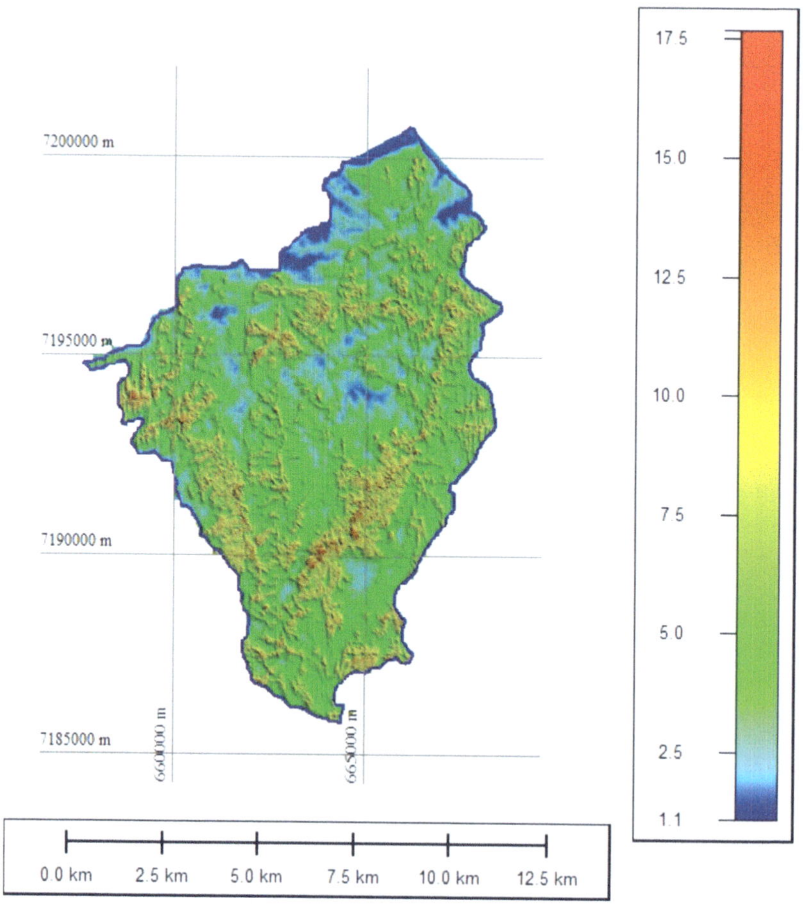

FIGURE 16 – TOPOGRAPHIC INDEX (UNIT ln(m)) OF THE PASSAÚNA SUB-BASIN WHICH FEEDS RUNOFF AT THE COLÔNIA DOM PEDRO STATION. CATCHMENT AREA IS 92 km².

The model's calibration was performed semi-automatically. A rough first adjustment was done automatically by software and followed by manual fine-tuning. Rainfall and runoff data for the chosen period are shown in Figure 17 along with the model's results. Measured runoff shows much larger temporal variation than the simulation for two main reasons: first, the model is a simplified version of reality, using few parameters and mean values of these parameters for the whole basin; secondly, the rainfall values used as the mean for the whole basin come from a single station situated at the basin mouth, which is clearly an approximation since actual rainfall in the basin possesses both spatial and temporal variability (within the time unit of 1 day used in the model).

Despite these limitations, the model works appropriately for long term runoff simulations, since the errors stemming from rainfall variability tend to cancel each other out, and the errors arising from parameter variability tend to disappear when the results are smoothed by calculating temporal means. This is substantiated by comparing mean runoff over nearly six years of simulation. This comparison is shown in Table 3. The relative error between calculated and measured runoff is 0.45%. Both measured and calculated runoff provide values compatible with the results shown in Figure 2. Figure 18 shows the comparison between measured and calculated mean monthly runoff.

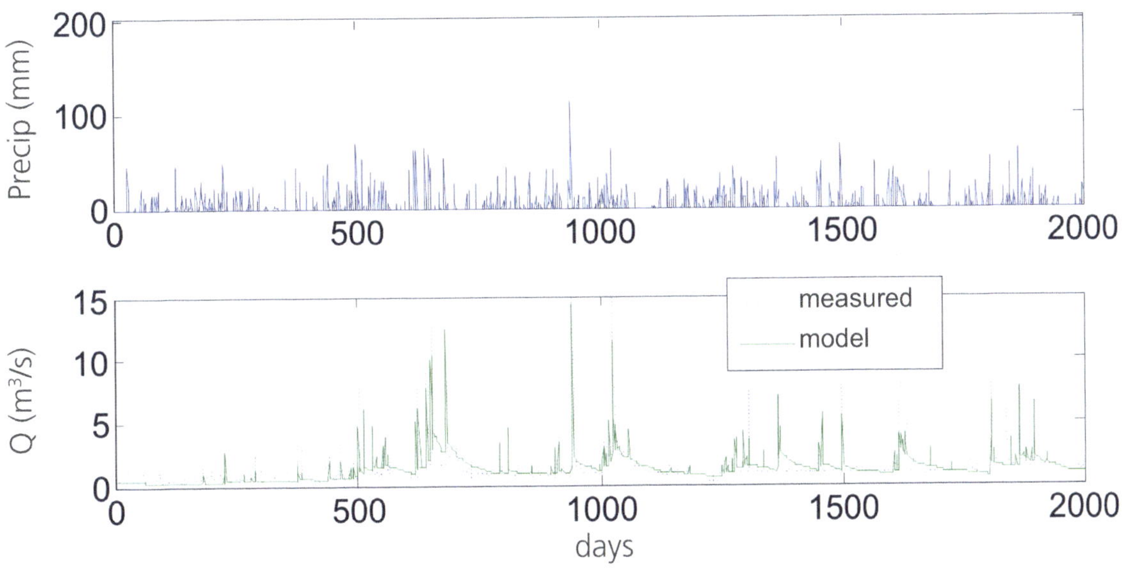

FIGURE 17 – DAILY RAINFALL, MEASURED DAILY RUNOFF, AND CALCULATED DAILY RUNOFF FROM COLÔNIA DOM PEDRO STATION, PASSAÚNA RIVER. PERIOD SHOWN: 02/01/1985 TO 07/24/1990.

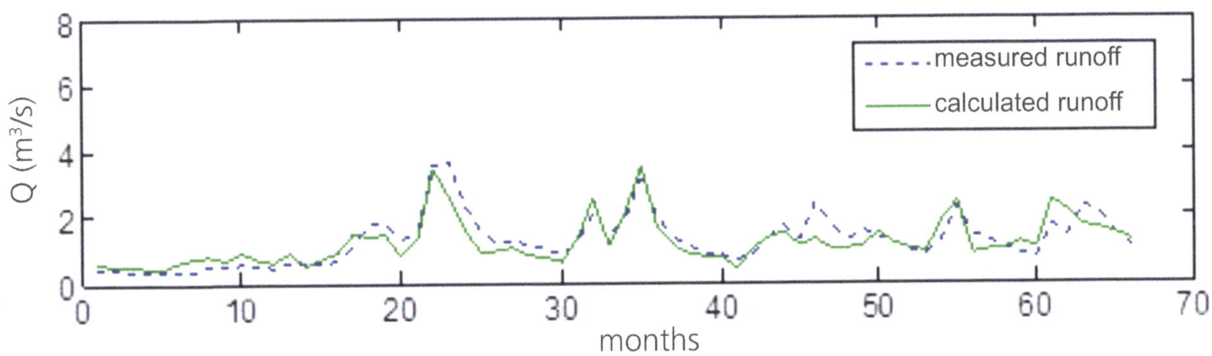

FIGURE 18 – MEASURED MEAN MONTHLY RUNOFF AND CALCULATED MEAN MONTHLY RUNOFF FROM COLÔNIA DOM PEDRO STATION, PASSAÚNA RIVER. (axes: x= months; y= Runoff; in legend: measured runoff (blue line); calculated runoff (green line)

TABLE 3 – COMPARISON BETWEEN MEASURED MEAN RUNOFF AND MODELED MEAN RUNOFF FOR THE SUB-BASIN OF COLÔNIA DOM PEDRO STATION, PASSAÚNA RIVER.

MEASURED MEAN RUNOFF (m^3/s)	MODELED MEAN RUNOFF (m^3/s)	ABSOLUTE ERROR (m^3/s)	RELATIVE ERROR (%)
1.2194	1.2249	0.0055	0.45

4.3 CALIBRATION FOR RIO VERDE SUB-BASIN

No real calibration of the model is possible for the sub-basin feeding the Rio Verde Reservoir due to the lack of local natural runoff data (unregulated by the reservoir). Therefore, the parameters used were the same applied for the Passaúna River, with the exception of the topographic index, which was obtained specifically for the Rio Verde Basin. Preliminarily, the sub-basin up to the Rodeio station was used, which has an area of 236 km² and for which useful long term runoff data (equaling 4.46m³/s) is available. The topographic index map for the Rio Verde sub-basin which drains towards the Rodeio station is shown in Figure 19. The rainfall data used included the entire historical series from station 02549006.

Figure 20 shows part of the runoff simulation for the Rodeio station. The mean long-term value resulting from the simulation was 3.82 m³/s, which compares favorably to the measured long-term mean of 3.92 m³/s (Table 4). It is worth noting that this long-term mean was taken with only 20 years of data and is influenced by the reservoir operation. A simulation that considered this calibration was done for the Reservoir's basin (Figure 21).

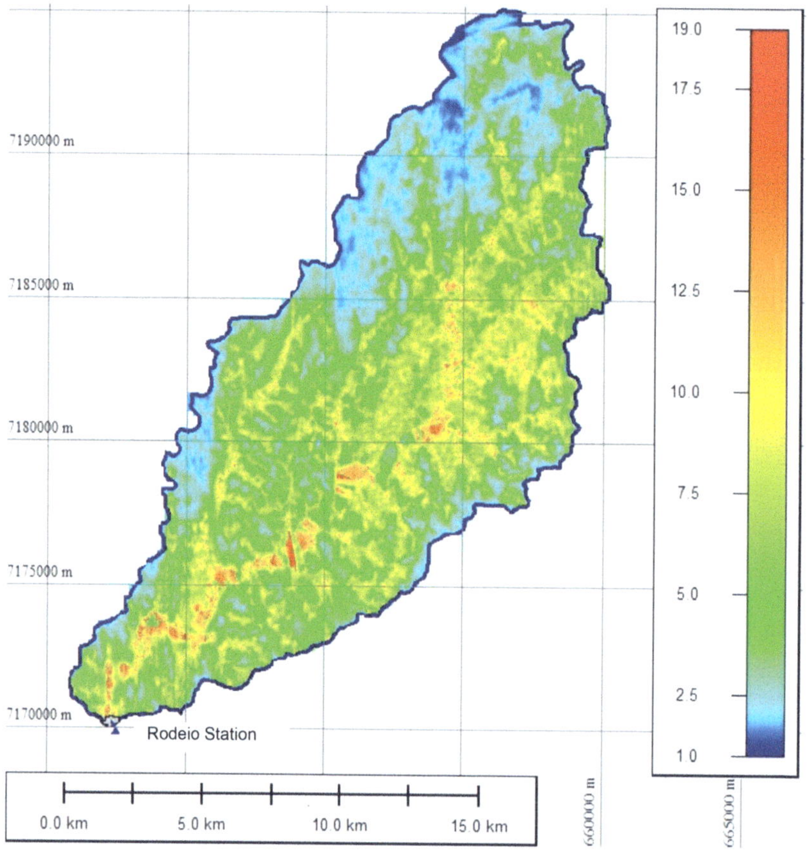

FIGURE 19 – TOPOGRAPHIC INDEX (UNIT ln(m)) FOR THE RIO VERDE SUB-BASIN DRAINING TOWARDS THE RODEIO STATION. CATCHMENT AREA IS 234km².

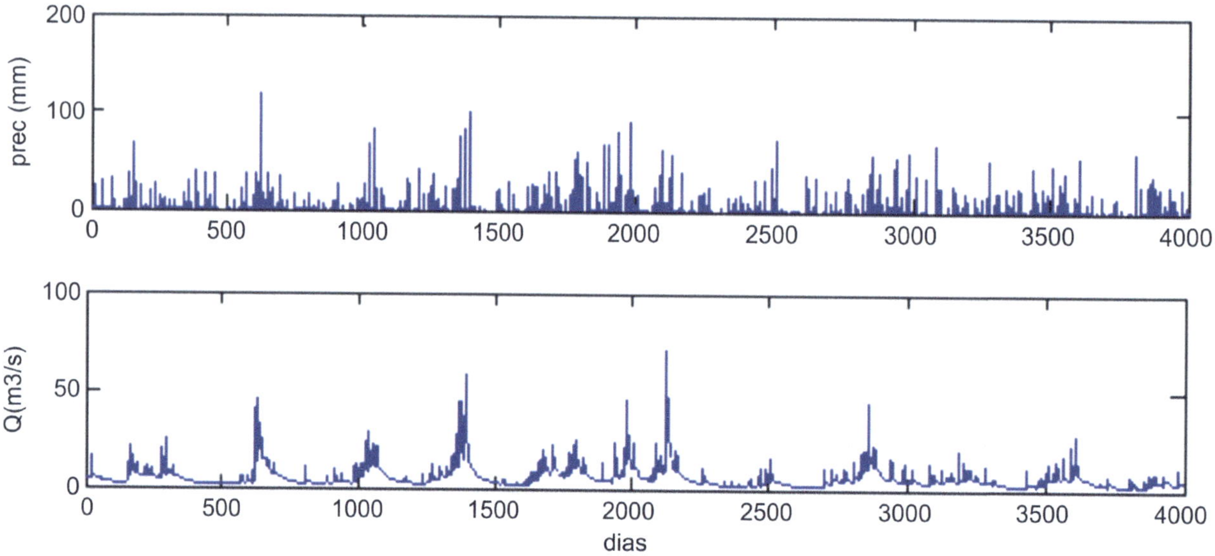

FIGURE 20 – DAILY RAINFALL (STATION 02549006) AND MODELED DAILY RUNOFF AT RODEIO STATION, RIO VERDE. PERIOD SHOWN: 04/24/1993 TO 04/05/2004

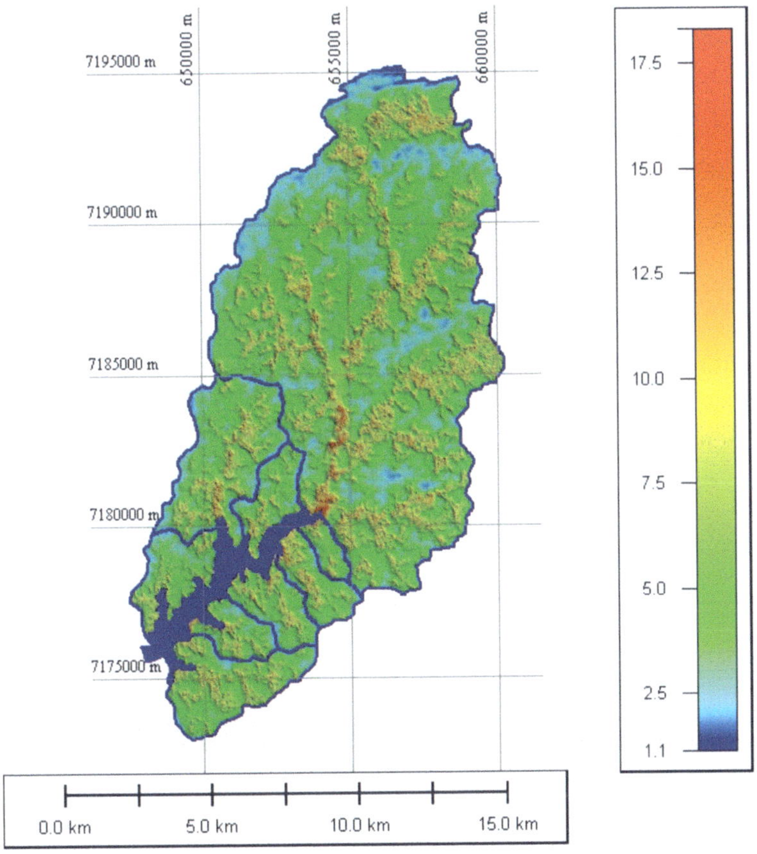

FIGURE 21 – TOPOGRAPHIC INDEX FOR THE RIO VERDE SUB-BASIN DRAINING TOWARDS THE RESERVOIR. CATCHMENT AREA IS 167 km².

TABLE 4 – COMPARISON BETWEEN MEASURED AND MODELED MEAN RUNOFF FOR THE SUB-BASIN OF THE RODEIO STATION IN THE RIO VERDE

MEASURED MEAN RUNOFF (m^3/s)	MODELED MEAN RUNOFF (m^3/s)	ABSOLUTE ERROR (m^3/s)	RELATIVE ERROR (%)
3.82	3.92	0.1	2.55

5. RAINFALL-RUNOFF SIMULATION

This section presents the stochastic model of rainfall generation for an arbitrary time period. In the present study, station 02549006 provided data for the modeling and a series of 1,000 years was generated for use with the model. This series informs the deterministic rainfall-runoff model, calibrated as described in the previous section. The period of 1,000 years was chosen to allow for the estimation of very rare events with reasonable statistical reliability.

5.1 RAINFALL GENERATION METHOD

Rainfall generation was done using a Markov chain multi-state model of daily rainfall and the associated transition probability matrix (TPM) (MEYN, 1993). The criteria for choosing the number of rainfall bands (system states) and the thresholds of rainfall within each system state took into account existing scientific literature (e.g. SRIKANTHAN & McMAHON, 1985) and used trial and error, attempting to preserve the statistical long-term mean. The maximum error accepted was 5% for annual rainfall and 20% for annual rainfall variance. These classes are shown in Table 5.

With the data from station 02549006, a TPM was built which indicates the probability that rainfall on day $n+1$ is state j (j = 1, ...,7), given the state on day n. For each new day of simulation, a random sampling is performed (Monte-Carlo simulation) and provides a class for that day. To determine the value of rainfall within a class, for all states (except for state 1 – no rain) another random sampling is performed and used to calculate rainfall on a given quantile of that state, assuming a probability distribution for each state. In the present study, interconnecting linear distributions were assumed for states 2 to 6 and a Pearson Type III distribution (Ord, 1972) of extreme events was assumed for state 7. The parameters for the latter were calculated by the method of moments using rainfall data above 70 mm/day measured at the station. It is worth noting that 12 matrices (TPM) were produced, one for each month, with their respective distributions. Thus, seasonal dataeffects waswere reasonably reproduced in the simulation.

At the end of the simulation, a proportionality constant c_i, obtained by multiplying the rainfall generated for each year i, may be used to correct the variability of annual mean rainfall (BOUGHTON, 1999):

$$c_i = \frac{M + (T_i - M)F}{T_i} \tag{6}$$

where T_i is the generated annual rainfall, M is the sample mean of annual rainfall, and F is an empirical adjustment factor determined so that simulated total annual rainfall variances reflect those of the measured annual rainfall. Boughton (1999) suggests:

$$F = \frac{\sigma_g}{\sigma_o}, \tag{7}$$

with σ_g and σ_o being the standard deviations of generated and observed annual rainfall, respectively.

With this method, it is then possible to generate rainfall series of arbitrary duration which preserves the long-term rainfall mean and inter-annual long-ter m variances rather well.

TABLE 5 – POSSIBLE STATES OF THE MARKOV-CHAIN ACCORDING TO DAILY RAINFALL VALUES

STATE	RAINFALL (mm)
1	0
2	0.0 - 2.9
3	3.0 - 5.9
4	6.0 - 10.9
5	11.0 - 30.9
6	40.0 - 60.9
7	- +∞

5.2 RUNOFF GENERATION

Rainfall data produced using the techniques presented in Section 5.1 can be used in the previously presented

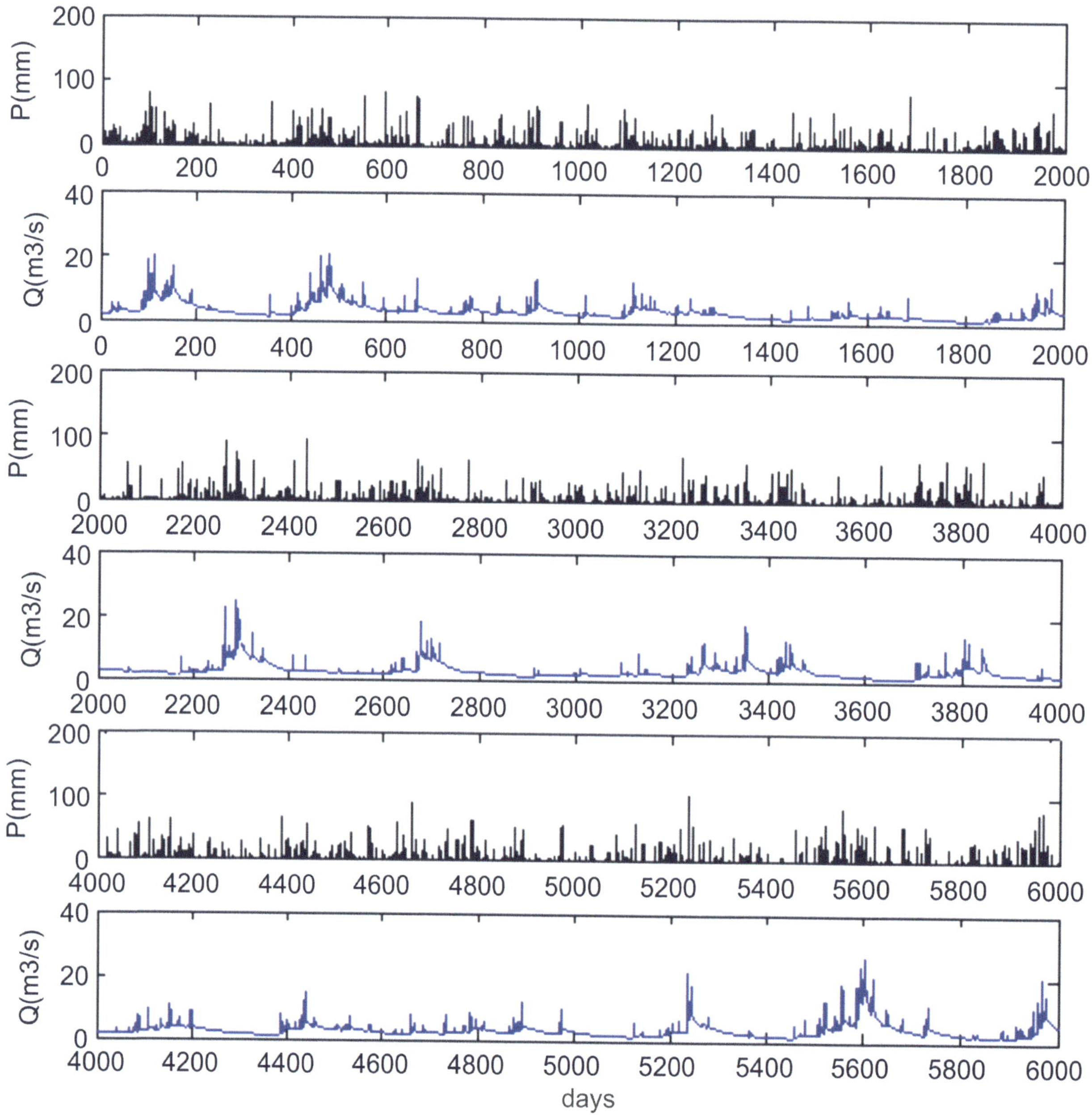

FIGURE 22 – EXAMPLE OF ARTIFICIAL GENERATION AND SIMULATION OF RAINFALL AND RUNOFF IN THE RIO VERDE SUB-BASIN DRAINING TOWARDS THE RESERVOIR. THE PLOTS SHOW A SEQUENCE OF RAINFALL AND RUNOFF SEPARATED IN THREE PANELS

model, calibrated for the area draining the Rio Verde Reservoir, to generate runoff data with an arbitrary duration. Figure 22 illustrates part of the rainfall-runoff data series artificially generated by the techniques described herein.

Such runoff data may be used as natural runoff inflows into the reservoir, enabling the simulation of the capacity of the reservoir to regulate a given outflow. These simulations are covered in the next section.

6. RESERVOIR SIMULATIONS

This section presents a series of simulations for the Rio Verde Reservoir using a long time period (1,000 years) of runoff values generated by the techniques discussed in the previous sections. First shown are simulations of water balance in the reservoir with daily data. Then, cumulative deficits are calculated based on annual runoff values. Such deficit volumes are frequently used as design criteria for reservoir sizing.

Prior to presenting the simulations, some considerations are discussed regarding the water volume of the reservoir (and consequently its water level) and the impact of the reservoir's water level on water quality, using total phosphorus as an indicator.

The Rio Verde Reservoir eutrophication project measured total phosphorus concentrations of about 7.4 mg/m³. Values close to 30 mg/m³ are anticipated for a reservoir to be deemed eutrophic. Consider the following situation: the reservoir is at its maximum volume with a phosphorus concentration of 7.0 mg/m³ and a (typical) phosphorus load of 1,123 kg/year, calculated by Sperling's (1985) formula. If, because of an increased demand, the reservoir level lowers, maintaining the same phosphorus load and the same mean runoff, the question is: at what volume will the reservoir be susceptible to an eutrophic state? The answer to that question also comes from the solution of Sperling's (1985) equation and results in a volume of 8 million m³, which corresponds to the approximate level of the reservoir at an altitude of 880 m. This level is immediately above the second floodgate. Thus, it is considered that if the reservoir level is below the first floodgate, there is a risk of degrading water quality due to an excessive phosphorus load. Therefore, the water level reaching below the first floodgate is used as a critical event for the simulations below and if the level reaches the second floodgate, the reservoir may be compromised with regard to its water quality.

6.1 WATER BALANCE SIMULATIONS FOR THE EXISTING RESERVOIR

In this section we present five examples of simulations for the existing reservoir, with its maximum volume of approximately 34,000,000m³, roughly corresponding to an elevation of 885.5m. The simulations consider a working volume of 30 million m³ (4 million are considered dead volume) and the demands increase progressively from simulations 1 to 3. The simulation scheme is shown in Figure 27. All simulations in this section use 1,000 years and the reservoir is always completely full at the start of each simulation. These simulations are similar to those proposed by Moran (1954), although he used the Probability Transition Matrix and Markov-Chain techniques to advance the reservoir's water volume over time. The results are summarized in Table 4.

The aim of these simulations is to illustrate the type of result that the model may supply, as well as to show how such results must be interpreted. They are therefore didactic simulations.

All water demands and losses (herein referred to simply as *demands*) used in the simulations are shown below:

Demands and Losses

1. Ecological runoff – the ecological runoff used for this study is $½Q_{95}$ = **0.33 m³/s**.
2. Water license to REPAR – license to REPAR in April, 2011 was **0.84 m³/s**.
3. Future expected consumption by REPAR = **0.94 m³/s**.
4. Water license to Fosfertil Company – Fosfertil's water license in April, 2011 was **0.125 m³/s**.
5. license slicense was Water license to SANEPAR (Paraná State Sanitation Company) – SANEPAR's water license in April, 2011 was **0.304 m³/s**.
6. Direct evaporation of the water surface – using Penman's equation for potential evaporation E:

$$E=\frac{\Delta}{\Delta+\gamma}E_r+\frac{\gamma}{\Delta+\gamma}E_a,\ \Delta=\frac{4098e_s}{(T-35{,}7)^2},\ \gamma=\frac{C_p p}{0{,}622 l_v}. \quad (8)$$

   In the equation above, e_s is the saturation vapor pressure, T is air temperature, C_p is the heat capacity of air, p is air pressure, and l_v the latent heat of vaporization. This equation was used with values for a typical year of data (historical mean) of the SIMEPAR weather station, located in Pinhais. Calculation details can be found in Nocko (2004). The value found for direct evaporation was approximately 3.3mm/day, which after multiplication based on the reservoir's area of 6493733 m² at the altitude of 885.5 m (see Table 2) and unit conversion results in approximately 0.25 m³/s. Accordingly, evaporation losses represent approximately **0.25 m³/s**.
7. Infiltration and percolation through the embankment – it is difficult to obtain a precise estimate of these losses without exact knowledge of the distribution of the soil under the embankment and complex three-dimensional computational modeling. A somewhat rough estimate is to use the following equation for the percolated runoff along the approximate length of the embankment B=600 m:

$$Q_{perc}=kH_L\frac{N_f}{N_d}B. \quad (9)$$

   In this equation, k is the soil's hydraulic conductivity, H_L is the reservoir's hydraulic load, and N_f/N_d is the aspect ratio of the current lines between hydraulic equipotential under the embankment. This ratio is unknown for the Rio Verde Reservoir but may be estimated as approximately 0.5. Hydraulic conductivity for a typical silt+sand soil is 3×10^{-6} m/s. The reservoir's load is approximately 13m. The value found for percolation losses is therefore **0.012 m³/s**.

8. Other losses – No detailed studies on illegal abstraction, measurement errors and other losses for the region are available. We thus decided to consider losses of 10% of the total value of the demands outlined above.

The following section describes simulations with arbitrary demand flows intended for use in decision making processes by Rio Verde Reservoir's managers.

- Simulation 1: Current REPAR water license + ecological runoff + current SANEPAR water license + current FOSFERTIL water license

This simulation is carried out with the demands above, totaling 1.599m³/s. First we consider no spillway discharge, that is, the reservoir is kept below its maximum level. For this demand, however, which is slightly higher than in the previous case, the reservoir will be full 80% of the time and there is no spillway discharge 20% of the time. That is, on average for this scenario, for each 100 days, there will be less than 20 days in which the volume of the reservoir will not work to supply this simulation's flow demand.

The expected number of events with no spillway discharge in the 1,000 years of the simulation was 676. The expected time interval between two consecutive events was 434 days. The expected duration of each event (level below maximum) is 107 days.

Figure 24 shows the percentage of decreases in reservoir volume over the 1,000 years of simulation. The greatest decrease is slightly over 60% of the original volume.

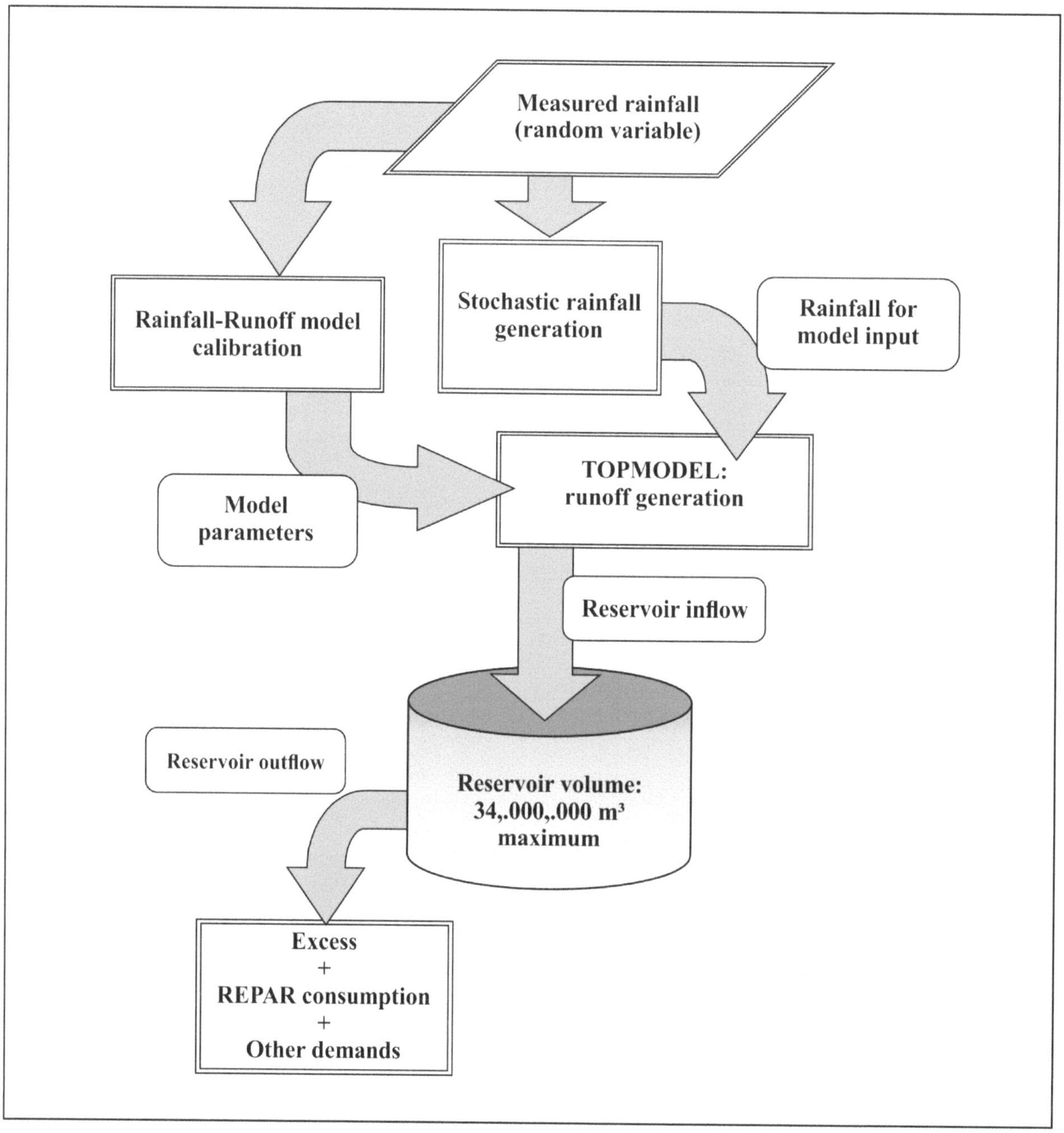

FIGURE 23 – SIMULATION SCHEME

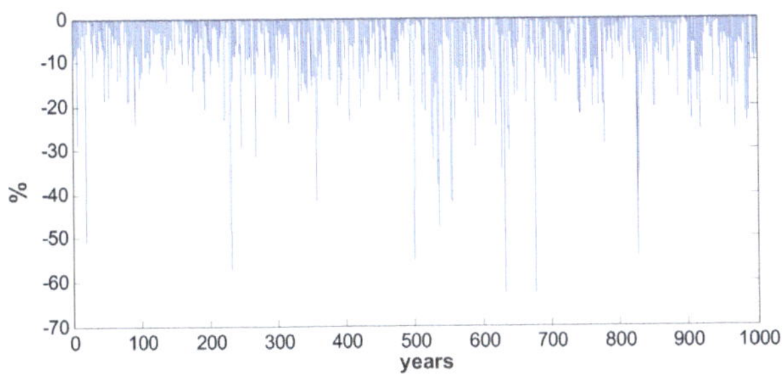

FIGURE 24 – PERCENTAGE DECREASE IN RESERVOIR VOLUME. SIMULATION 1.

For the event in which the water level reaches a *level below the first floodgate* (altitude 882.7 m), this simulation yielded the following results: the expected time percentage in which the event occurred is 0.31%; the number of events expected in 1,000 years is 8; the expected duration of each event is 153 days; and the expected time interval between such events is 155 years.

- Simulation 2: future REPAR consumption + ecological runoff + current SANEPAR water license + current FOSFERTIL water license

The second simulation uses the demands above, totaling 1.7m³/s. As in the previous simulation, the first event considered is that of no spillway discharge (level below maximum). According to the simulation's results, the reservoir was below its maximum level 26% of the time. The expected number of these events (level below maximum) was 774 in 1,000 years, the mean duration of the event was 124 days, and the expected interval between events was 348 days.

Figure 25 shows percentage decreases in reservoir volume over the 1,000 years of simulation (plots for subsequent simulations are similar and shown in Figure 26).

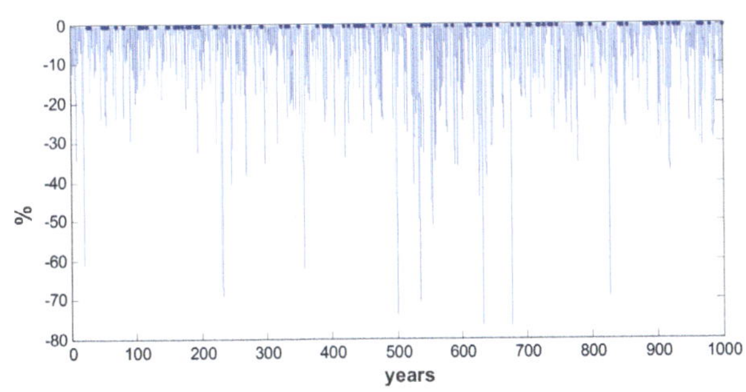

FIGURE 25 – PERCENTAGE DECREASE IN RESERVOIR VOLUME. SIMULATION 2

As in the previous simulation, the event of the reservoir level below the first floodgate was also considered. On average, the reservoir would reach such a state 0.83% of the time. The expected number of such events is 19 (in 1,000 years), each one with an expected duration of 167 days, and with an expected interval between events of approximately 56 years.

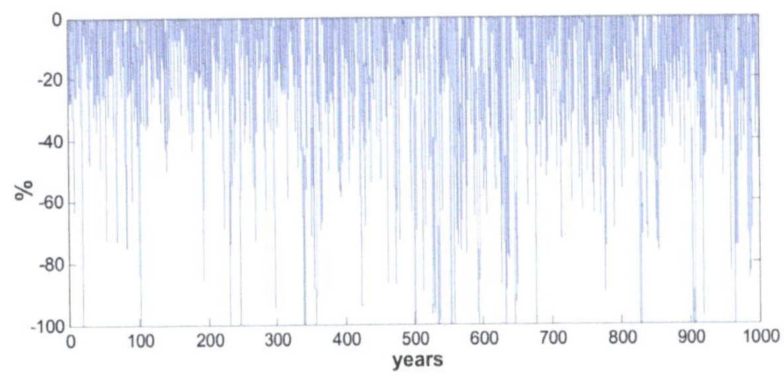

FIGURE 26 – PERCENTAGE DECREASE IN RESERVOIR VOLUME. SIMULATION 3

- Simulation 3: future REPAR consumption + ecological runoff + current SANEPAR water license + current FOSFERTIL water license + Evaporation + percolation + 10% other losses

The third imulation uses the demands above, totaling 2.16 m³/s. The demand for this simulation was increased to include losses due to evaporation, percolation, illegal abstraction, etc.

The following results are found for the event 'level below maximum' (no spillway discharge): the reservoir was below its maximum level 58% of the time; the expected number of such events (level below maximum) was 848 in 1,000 years; the mean duration of the event was 249 days; and the expected interval between events was 182 days.

As in previous simulations, the event of the reservoir reaching a level below the first floodgate was also considered. The reservoir would be in this state 13% of the time. It is expected that this event will occur 160 times (in 1,000 years), each event with an expected length of 290 days, and an expected interval between events of approximately 5.5 years.

In this simulation, the reservoir became completely empty on several occasions (Figure 26).

6.2 SIMULATIONS CONSIDERING FLOODGATE LEVELS AND ARBITRARY DEMANDS

The simulations within this section are probably more useful as a support for decision making processes regarding the increase of water abstraction licenses from the Rio Verde Reservoir.

Any value of outflow from the reservoir can be used in the following graphs, since they have been devised considering a wide range of demand values.

The following simulations were performed for demand values above 1.5 m³/s. The statistics were calculated for the following events:

- Reservoir level below spillway
- Reservoir level below first floodgate (882.7 m)
- Reservoir level below second floodgate (879.7 m)

The statistics resulting from simulations and shown in the figures below for each of the events above are:

- Total percentage of time spent within that event;
- Mean number of events each 100 years;
- Mean length (in years) of each event;
- Mean time interval (in years) between two consecutive events.

Figures 27 through 30 show the variables listed above for the event *Reservoir level below spillway level*.

Figures 31 through 38 show the above mentioned variables for the event *Reservoir level below first floodgate*. In this case there are two plots for each variable, one of them being a close-up of the demand values most likely to occur given current and possible future water licenses, which fall between 1.5 m³/s and 2 m³/s.

Figures 39 through 42 show the variables listed above for the event *Reservoir level below second floodgate*.

The graphs below should be read as follows: the sum of desired water licenses/demands is plotted on the horizontal axis, the vertical axis provides the value of the variable for the event in question (found in the title to each plot).

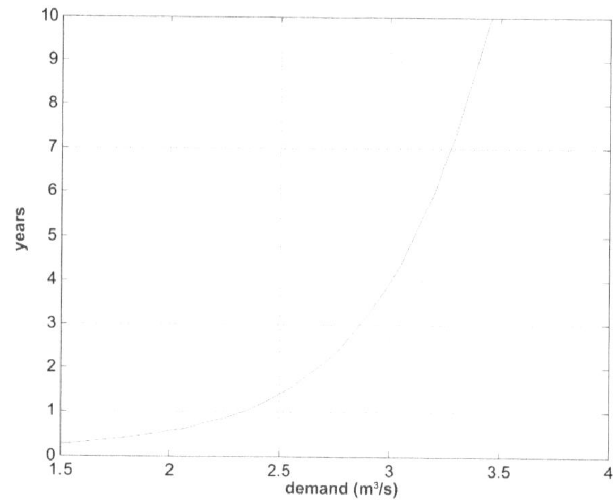

FIGURE 29 – MEAN LENGTH OF EVENTS WITH LEVEL BELOW SPILLWAY

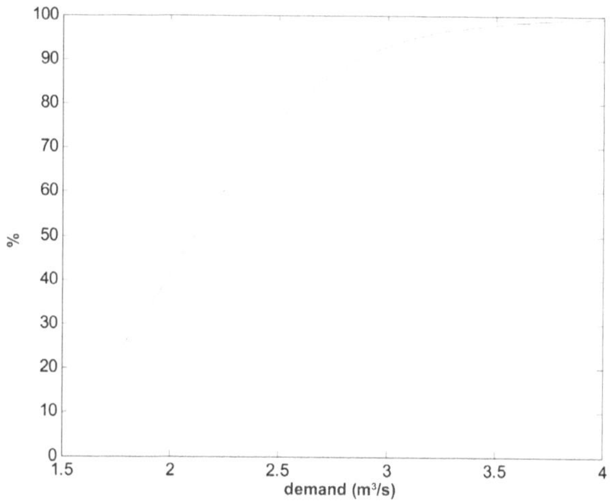

FIGURE 27 – PERCENTAGE OF TIME WITH LEVEL BELOW SPILLWAY AS A FUNCTION OF DEMAND

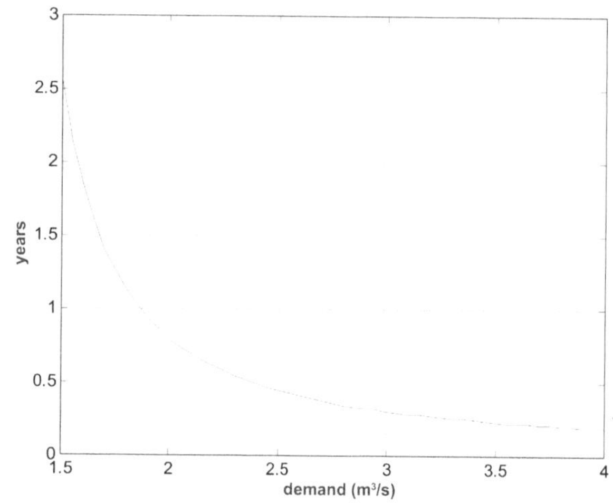

FIGURE 30 – MEAN INTERVAL (YEARS) BETWEEN TWO CONSECUTIVE EVENTS WITH LEVEL BELOW SPILLWAY AS A FUNCTION OF DEMAND.

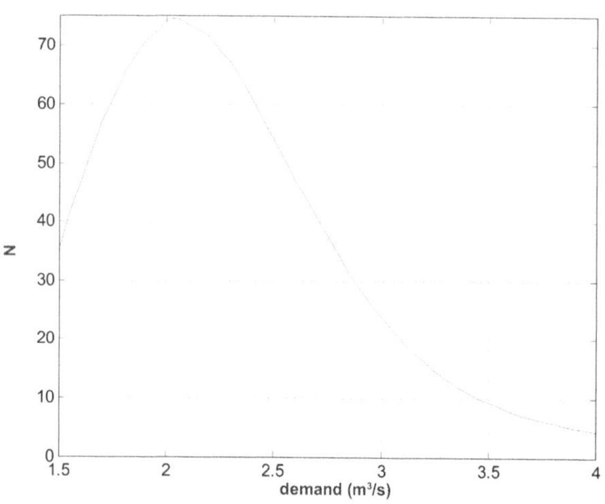

FIGURE 28 – NUMBER OF EVENTS (IN 100 YEARS) WITH LEVEL BELOW SPILLWAY AS A FUNCTION OF DEMAND. NOTE THAT THE NUMBER OF EVENTS STARTS TO FALL (AT APPROX. 2M³/S) WITH THE INCREASE IN DEMAND. THIS IS NOT CONTRADICTORY AS THE EVENTS BECOME MUCH LONGER IN DURATION.

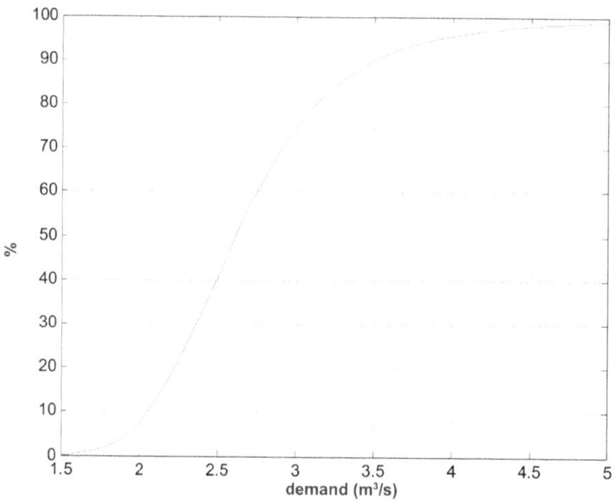

FIGURE 31 – PERCENTAGE OF TIME WITH LEVEL BELOW FIRST FLOODGATE AS A FUNCTION OF DEMAND

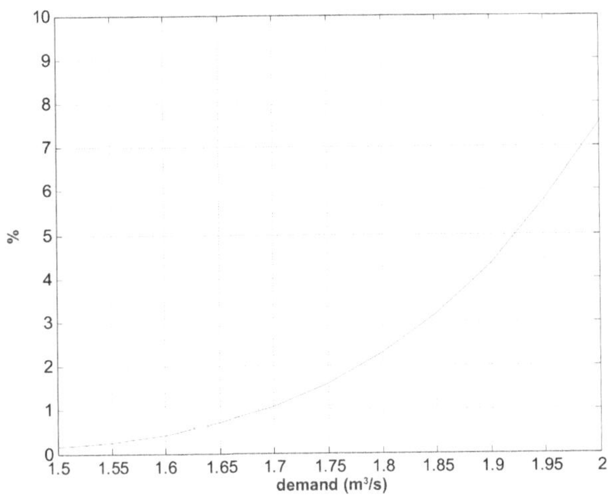

FIGURE 32 – PERCENTAGE TIME WITH LEVEL BELOW FIRST FLOODGATE AS A FUNCTION OF DEMAND (DETAIL)

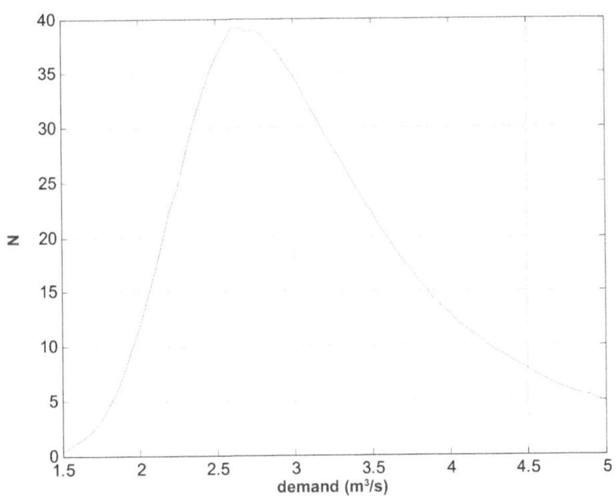

FIGURE 33 – NUMBER OF EVENTS (IN 100 YEARS) WITH LEVEL BELOW FIRST FLOODGATE AS A FUNCTION OF DEMAND. NOTE THAT THE NUMBER OF EVENTS STARTS TO FALL (AT APPROX. $2.7m^3/s$) WITH AN INCREASE IN DEMAND. THIS IS NOT CONTRADICTORY AS THESE EVENTS THEN BECOME MUCH LONGER IN DURATION

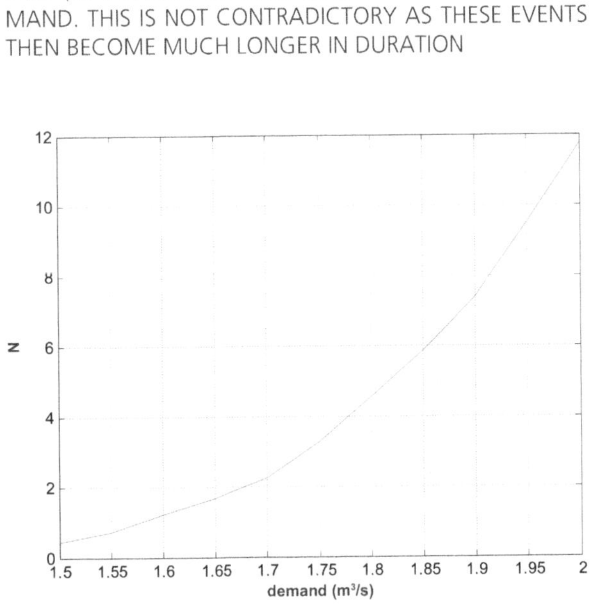

FIGURE 34 – NUMBER OF EVENTS EACH 100 YEARS WITH LEVEL BELOW FIRST FLOODGATE AS A FUNCTION OF DEMAND (DETAIL)

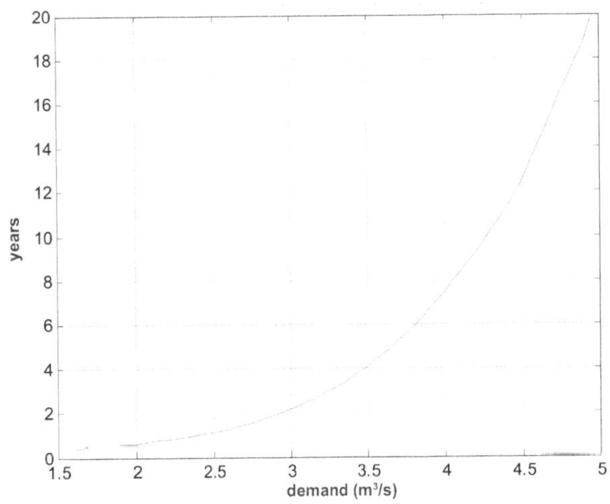

FIGURE 35 – MEAN LENGTH (YEARS) OF EVENTS WITH LEVEL BELOW FIRST FLOODGATE AS A FUNCTION OF DEMAND

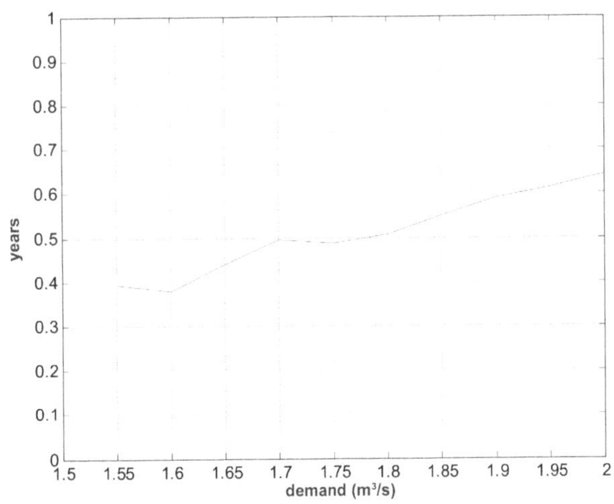

FIGURE 36 – MEAN LENGTH (YEARS) OF EVENTS WITH LEVEL BELOW FIRST FLOODGATE AS A FUNCTION OF DEMAND (DETAIL)

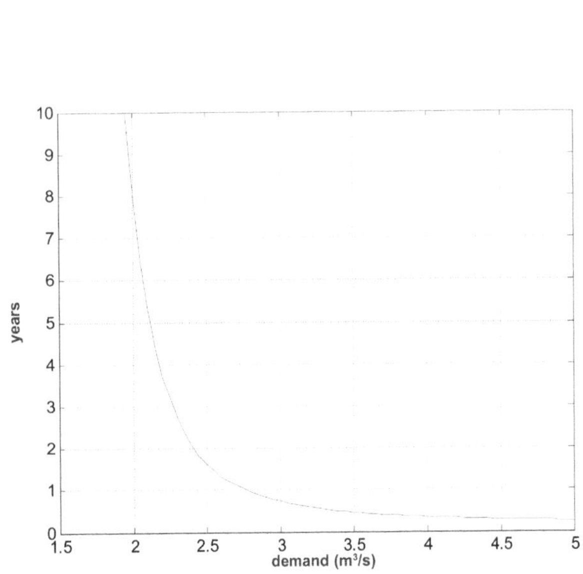

FIGURE 37 – MEAN INTERVAL (YEARS) BETWEEN TWO CONSECUTIVE EVENTS WITH LEVEL BELOW FIRST FLOODGATE AS A FUNCTION OF DEMAND

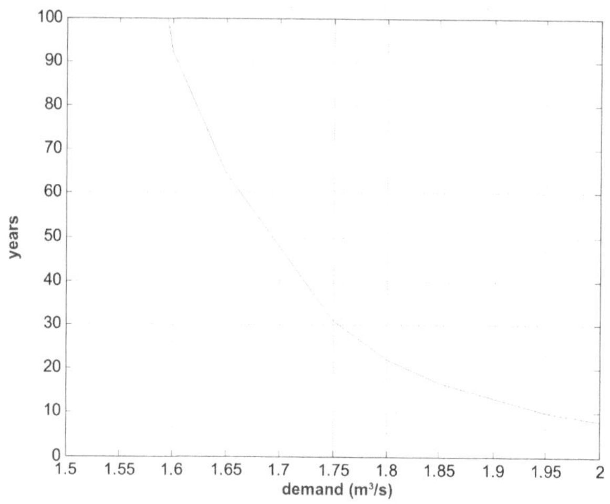

FIGURE 38 – MEAN INTERVAL (YEARS) BETWEEN TWO CONSECUTIVE EVENTS WITH LEVEL BELOW FIRST FLOODGATE AS A FUNCTION OF DEMAND (DETAIL)

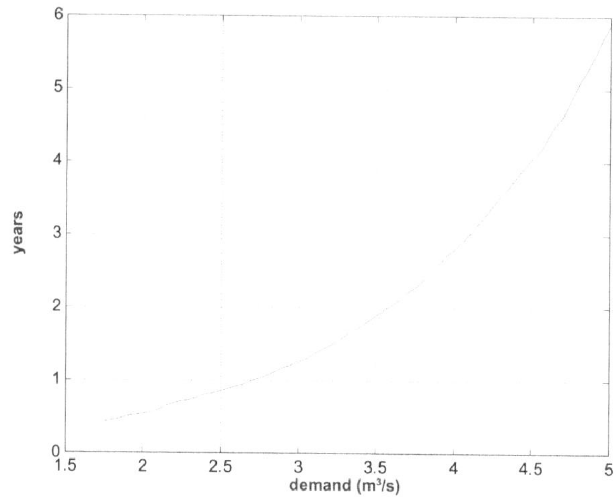

FIGURE 41 – MEAN LENGTH (YEARS) OF EVENTS WITH LEVEL BELOW SECOND FLOODGATE AS A FUNCTION OF DEMAND

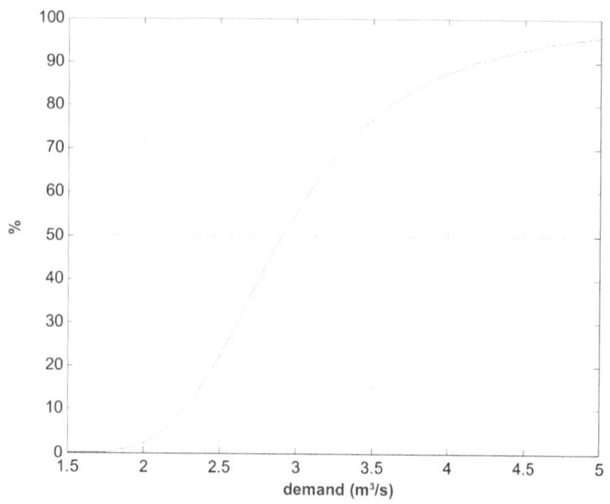

FIGURE 39 – PERCENTAGE OF TIME WITH LEVEL BELOW SECOND FLOODGATE AS A FUNCTION OF DEMAND

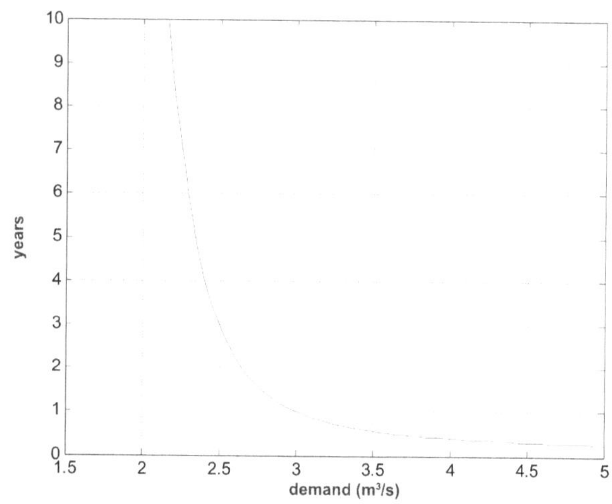

FIGURE 42 – MEAN INTERVAL (YEARS) BETWEEN TWO CONSECUTIVE EVENTS WITH LEVEL BELOW SECOND FLOODGATE AS A FUNCTION OF DEMAND

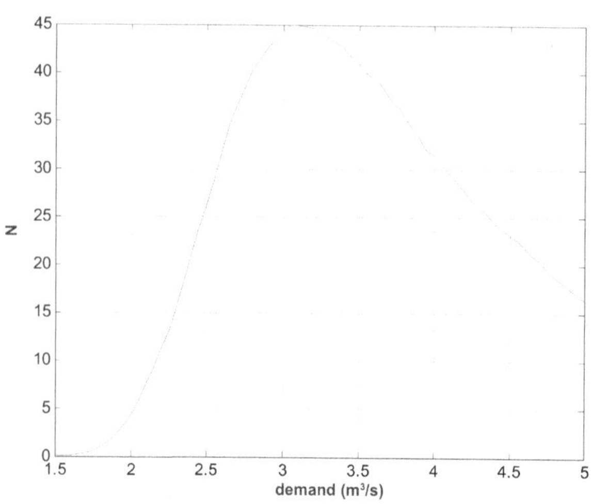

FIGURE 40 – NUMBER OF EVENTS (IN 100 YEARS) WITH LEVEL BELOW SECOND FLOODGATE AS A FUNCTION OF DEMAND. NOTE THAT THE NUMBER OF EVENTS STARTS TO FALL (AT APPROX. 3.1m³/s) WITH THE INCREASE IN DEMAND. THIS IS NOT CONTRADICTORY BECAUSE THESE EVENTS BECOME MUCH LONGER IN DURATION

6.3 SIMULATIONS OF CUMULATIVE STORAGE DEFICIT

Although the previous simulations are sufficient for informing decision making criteria concerning the reservoir, reservoir sizing criteria also exists based on simulations in which inflow and demand are used as input data. One of the variables most often used for determining reservoir volume is the so-called *cumulative storage deficit*.

To render this study more complete, we present in this section simulations which calculate, for arbitrary periods, the expected value of the maximum cumulative storage deficit (GOMIDE, 1975), given inflow and demand values.

With data on inflow and demand from a reservoir, one obtains, for each time period t, a variable X_t (inflow minus outflow). At each time period, one may calculate the partial sum of X_t by:

$$S_t = \sum_{i=1}^{t} X_i \quad , \quad t = 1,2,\ldots,n \tag{9}$$

The maximum cumulative deficit D_n is defined as the highest value for deficit d_j for the period n, by means of S_t, as shown in Figure 43.

The variable D_n is often used in project criteria, such as the volume of the reservoir when the problem at hand is one of multi-annual regulation (in which the X_t data represent annual flow and n would be the number of years). Obviously, as this is a project criterion based on the worst drought event, it cannot be applied when X_t represents daily values, since this would disregard the possibility of two (or more) severe drought events, say, j=5 and j=6, separated by a brief interval with normal flows. In this case we should sum both these d_js instead of using only the highest d_j in the series.

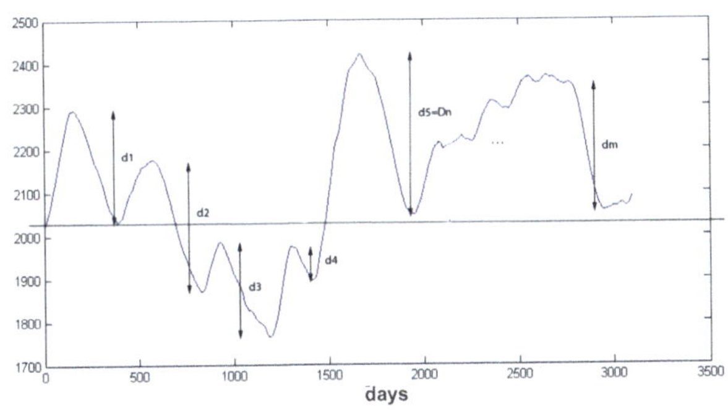

FIGURE 43 – MAXIMUM CUMULATIVE DEFICIT D_n

With this caveat in mind, one may calculate the volume D_n by this criterion and compare it to the maximum volume of the Rio Verde Reservoir using both daily (not recommended, but calculated as a curiosity) and annual data, as a means of verifying the volume of the reservoir based on these criteria.

These simulations were carried out for a hypothetical demand of 1.85 m³/s. Using daily data, calculating the statistics of 20 simulations, each encompassing 1,000 years, the value of the maximum deficit D_n is 20,136,600 m³, with a standard deviation of 3,000,000 m³. This result demonstrates that the calculated volume is relatively small and confirms that the use of D_n is not a good design criterion when using daily flows, which forces one to use longer-term flows.

Using annual flows for the same 20 simulations and considering the lowest D_n for n=1,000 years, the statistics change to a mean volume of 41,111,850 m³, with a standard deviation of 10,096,000 m³. Note that, as expected and according to previous simulations, the volume of 34,000,000 m³ is almost sure to result in failures. However, it is certainly more prudent and realistic to use a shorter time frame n, and consider the maximum deficit in n years with n equaling 10, 20, 40 and 100, for instance. To this end one must simply divide the 20 samples encompassing 1,000 years each into m samples of n years each, so that $n{\times}m$=20,000 years of annual inflows and outflows. The plots for these four values of n are shown in Figures 44 through 47, which present various reservoir demand values (each curve represents a distinct demand value), so one can infer the probability of success (i.e. probability that the highest deficit in n years will be lower than the reservoir volume) for each demand. Note that in all cases, demands higher than 2m³/s rapidly increase the risk of failure in regulating flow.

The interpretation of the results of the D_n success probability plots is as follows: if we take Figures 44 and 46 as examples and we suppose the demand is 2.3m³/s, Figure 44 shows that a volume of 30 million m³ is sufficient to contain a *"10 year maximum deficit"* with a success probability of 70%. That is, for each 100 intervals of 10 years, there will be a failure, on average, in 100-70 = 30 of these periods. Figure 46, similarly shows that the volume of 30 million m³ is sufficient, with a probability of success slightly under 30%, to contain a *"40 year maximum deficit."* That is, a reservoir failure will occur in 70 out of every 100 periods of 40 years.

For an outflow from the reservoir of Q_{tot} =1.85 m³/s, Figure 48 shows success probability curves for the volumes (cumulative maximum deficit) D_{10}, D_{20}, D_{40}, and D_{100}. Note that the success probability for this demand lies over 90% for all four values of n.

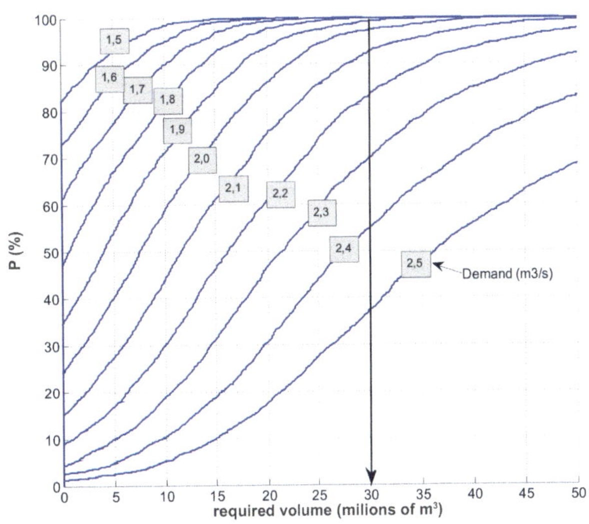

FIGURE 44 – SUCCESS PROBABILITY FOR VOLUME D10 (HIGHEST VOLUME DEFICIT IN 10 YEARS) FOR SEVERAL DEMAND VALUES

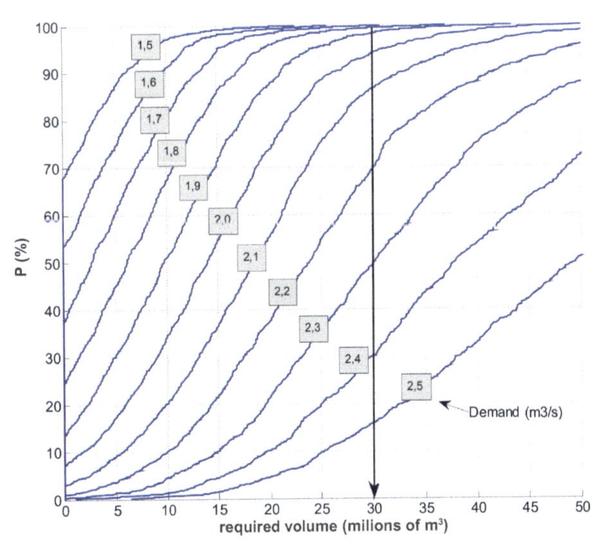

FIGURE 45 – SUCCESS PROBABILITY FOR VOLUME D20 (HIGHEST VOLUME DEFICIT IN 20 YEARS) FOR SEVERAL DEMAND VALUES

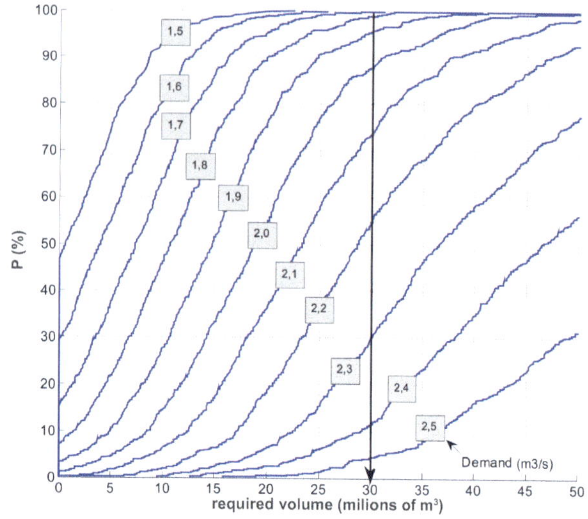

FIGURE 46 – SUCCESS PROBABILITY FOR VOLUME D40 (HIGHEST VOLUME DEFICIT IN 40 YEARS) WITH DIVERSE DEMAND VALUES

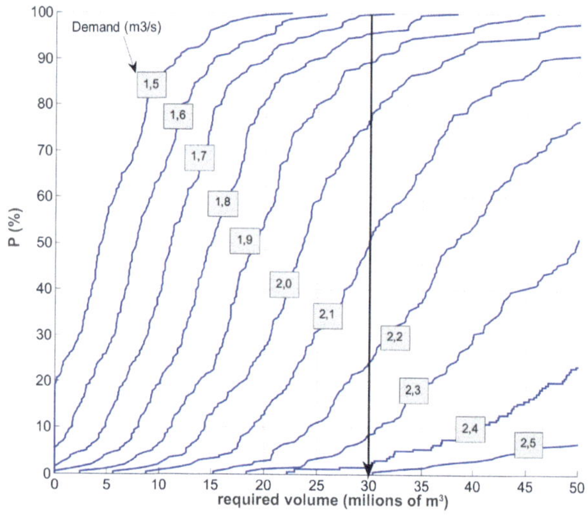

FIGURE 47 – SUCCESS PROBABILITY FOR VOLUME D100 (HIGHEST VOLUME DEFICIT IN 100 YEARS FOR SEVERAL DEMAND VALUES

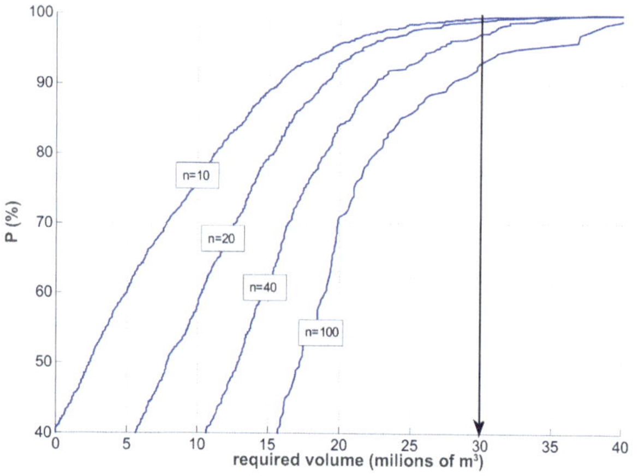

FIGURE 48 – SUCCESS PROBABILITY FOR VOLUME Dn (HIGHEST VOLUME DEFICIT IN n YEARS) FOR n=10, 20, 40 AND 100, WITH A DEMAND Qtot =1.85m³/s

7. CONCLUSIONS

Assuming rainfall data and the regionalization studies used to calibrate the models for this study are correct, rainfall stability in the stations used, direct evaporation from the reservoir to be negligible, and the inexistence of extraordinary demands (leakages, illegal pumping, percolation losses, etc.) beyond those considered in the simulations, one concludes that the Rio Verde Reservoir can regulate current water abstraction licenses (constant flow of 1.689 m³/s) 80% of the time with spillway discharge, with a low probability of working at low volume levels (less than 1% of the time below the first floodgate). It can also be concluded that a small increase in demand to 1.85 m³/s, for instance, will lead the reservoir to operate 46% of the time with its level below the spillway, with a chance of approximately 2.5% of the time operating below half the maximum volume. Further, such a situation suggests the occurrence of approximately one total failure (empty reservoir) every 34 years (mean time) with an expected duration of 15 days.

It can also be concluded that demands that are higher than the current water licenses will place the reservoir in a situation in which the risk of failure increases rapidly with an increase in outflow.

8. REFERENCES

BEVEN, K. J. e KIRKBY, M. J. (1979) *'A physically based variable contributing area model of catchment hydrology'*, Hydrol. Sci. Bull., 24, 43-69.

BEVEN, K., LAMB, R., QUINN, P., ROMANOUWICZ, R., FREER, J. (1995) TOPMODEL. In: SINGH, V.P. (Ed.). *Computer Models of Watershed Hydrology*. Water Resource Publications, Colorado, p.627-668.

BOUGHTON, W.C. (1999) *A daily generating model for water yield and flood studies*. Report 99/9, CRC for Catchment Hydrology, 21p.

EADIE, W.T.; D. DRIJARD, F.E. JAMES, M. ROOS and B. SADOULET. (1971) *Statistical Methods in Experimental Physics*. Amsterdam: North-Holland, p. 269-271.

GOMIDE, F. L. S. (1975) *Range and deficit analysis using Markov chains*, Hydrology Papers, 4, No 79, Colorado State University. Fort Collins. Colorado.

MATHERON, G. (1963) *Principles of geostatistics*. Economic Geology, 58, 1246-1266.

MEYN, S. P. e R. L. TWEEDIE. (1993) *Markov Chains and Stochastic Stability*. London: Springer-Verlag.

MORAN, P. A. P., "A probability theory of dams and storage systems. *Austr. Journal of Appl. Sci.* Vol. 5, 116-124.

NOCKO, H. R., (2004) *Implementação do Modelo Chuva-Vazão TOPMODEL para a Sub-bacia do Rio Pequeno na bacia do Altíssimo Iguaçu/PR*. Projeto Final de Curso de Engenharia Ambiental, UFPR.

ORD J.K.1972. *Families of Frequency Distributions*. Griffin, London.

SPERLING , M. von (1985) *Utilização de gráficos para avaliação preliminary do aporte de fósforo a uma represa*. 13 congresso brasileiro da ABES.

SRIKANTHAN, R. e McMAHON, T.A. (1985) *Stochastic generation of rainfall and evaporation data*. Australian Water Resources Council, Tech. Paper No. 84, AGPS, Canberra, 301pp.

SRIKANTHAN, R. (1995) *Stochastic generation of daily rainfall at a number of sites*. Cooperative research centre for catchment hydrology Australian Water Resources Council, Tech. Report 05/7, 77pp.

CHAPTER 9

CHARACTERIZATION OF THE IONIC AND PARTICULATE SYSTEMS IN THE RESERVOIR

Ricardo Henrique Moreton Godoi
André Virmond Lima Bittencourt
Patricia Yassumoto Hirata
Miguel Jafelicci Junior
José Manoel dos Reis Neto
Rene van Grieken

CHARACTERIZATION OF THE IONIC AND PARTICULATE SYSTEMS IN THE RESERVOIR

Ricardo Henrique Moreton Godoi
André Virmond Lima Bittencourt
Patricia Yassumoto Hirata
Miguel Jafelicci Junior
José Manoel dos Reis Neto
Rene van Grieken

SUMMARY

Suspended particles, as well as dissolved substances, have an important ecological role in water quality by attenuating light, providing reactive surfaces, influencing metabolic activity, and contributing to sediment deposition. The quality of water, particularly in reservoirs, is crucial and is naturally maintained by circulation and sedimentation, which removes phosphorus from the water, for example. In eutrophic reservoirs, these processes are contrasted by the recycling of ions and metals coming from sediments, which can maintain or neutralize the state of eutrophication. The combination of trace elements and microanalysis techniques may be useful in the context of tracking anthropogenic or natural activities close to a water body. These may be used to trace the sources of pollution by means of a chemical fingerprint of such heterogeneous materials, for example, particles in suspension. The magnitude and patterns of ion, metal and particulate systems were evaluated at the Rio Verde Reservoir. Samples were analyzed through energy-dispersive X-ray fluorescence (EDXRF), ion chromatography (IC), inductively coupled plasma - mass spectrometry (ICP-MS), and particle size distribution via Laser and Zeta Potential. Based on the results, the chemical patterns are discussed in relation to heavy metal concentration, suspended matter and ion composition.

KEY WORDS

Reservoir, ICP, zeta potential, x-ray fluorescence, particulate matter.

1. GENERAL CONSIDERATIONS

Any approach regarding the assessment of water in natural environments should consider the environment as a whole, including how different types of reservoirs are interact with the hydrologic cycle. By participating in a cycle, the water can acquire or modify chemical and isotopic features in all aspects of the cycle.

Tundisi *et al.* (1999) point out the existence of few estimates of the contribution of basins to rivers, lakes and reservoirs, as well as of the importance of comparatively evaluating different sub-basins. The authors also emphasize the importance of peaks in the behavior of the reservoir system. These peaks are fundamentally influenced by variable factors, especially climate or varying types of occupation in the basin, but they can be magnified by permanent physical characteristics of the basin, such as geological features. When rocky matter interacts with the superficial environment, its chemical elements are redistributed in accordance with the available phases (TARDY, 1969), which is controlled by the energy content of the system. Each element positions itself in the structure of suitable chemical species so that the system as a whole tends toward lower energy levels, in other words, becomes more stable. This process of seeking stability of the system is not immediate and while some reactions are almost instantaneous, others take a relatively long time. Thus, some minerals weather and their components are very quickly redistributed between the newly formed solid phase and the liquid phase, while others remain metastable for longer periods of time, retaining elements that could be incorporated into the solution until their intercrystalline connections are disrupted. The abundance of a certain chemical element in the water will reflect, in addition to its occurrence, its mobility in that environmental context. More mobile elements are those in which fixing processes in the solid phase are less effective. They have difficulty fixing to existing solid structures, whether as part of a crystal structure or adsorbed on the surface by mineral or organic substances.

The environmental context within a region forms what can be called the regional geochemical landscape. This must be understood in order to evaluate the overall behavior of the chemical elements, which enables a better understanding of the geochemical controlling factors of a reservoir. In part, the set of matter flow systems that access a reservoir can be seen in Figure 1 (BITTENCOURT, 2007).

As the most abundant elements are present in substantial amounts in virtually all natural environments, they represent a valuable indicator of the behavior to be expected from less abundant elements with chemical properties similar to those that are more abundant. For example, it is expected that alkali elements behave similarly to Na and K and alkaline earth elements behave like Ca and Mg, all of them with high mobility and a tendency to easily migrate through water.

The behavior of elements in the superficial cycle establishes a geochemical differentiation process in which

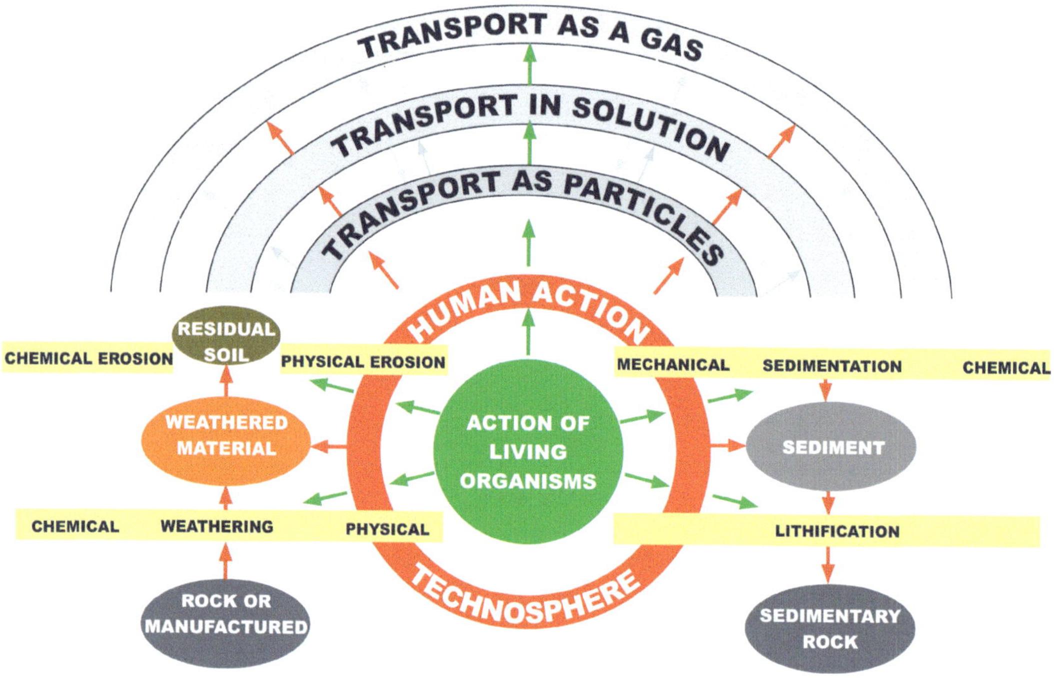

FIGURE 1 – GENERIC FLOW CHART OF THE MOVEMENTS OF MATTER IN THE SURFACE CYCLE. SOURCE BITTENCOURT (2007)

the elements are distributed in chemical species appropriate to the phases present in the environment, ranging from weathering and pedogenesis processes to sedimentation. The chemical elements are differentiated during weathering and pedogenesis. Such differentiation is regulated by factors inherent to the element, such as chemical properties, and factors related to the environment, such as temperature, pH, Eh, the surface potential of a particle, sorption processes and the chemical environment.

Among the chemical characteristics of natural waters, its dissolved metals content is mainly the result of the composition of the rocks within the hydrographic basin and aquifers drained by these waters. The predominant dissolved metal fraction of the vast majority of surface water, besides silicon, which is mainly present as silicic acid (H_4SiO_4), is in solution mainly in the form of simple cations. These cations are: Na^+, K^+, Ca^{2+} and Mg^2. This is directly related to the abundance of elements in the Earth's crust, which is composed in descending order of O, Si, Al, Fe, Ca, Na, K and Mg, while the other elements are found on average in smaller amounts (CLARKE & WASHINGTON, 1924). Since iron and aluminum very easily form insoluble compounds in the physicochemical conditions of the surface environment, they are mainly found in water in the particulate form.

In order to assess the distribution of any element in water, it is essential to know the behavior of the major constituents in solution. They control the presence of minerals in equilibrium with water, which will directly affect the minority element partition between the solution and the particulate phase. One way to assess the distribution of solutes most abundant in water is to systematize a representation through a hydrochemical classification based on the ion charge.

Water quality, particularly in reservoirs, is naturally maintained by the water level, particulate matter sedimentation, and suspension processes in the water column, which play an important role in aquatic ecosystems. Suspended particles attenuate the entry of light into water bodies because they present reactive surfaces with their environment; along with the chemical environment, such reactions can also affect the concentration and stoichiometry of ionic and metallic macroconstituents of the system.

This dynamic of particle suspension and sedimentation can also be associated with the eutrophication process, which is a major concern when the reservoir is used for human water consumption. Eutrophication can be natural or artificial; when natural, it is a slow and continuous process that results in the input of nutrients through rain and surface runoff in the form of particles, or not, which alter the physical and chemical characteristics of the system (see Chapter 10). When eutrophication is anthropogenic, it usually is associated with factors such as the discharge of domestic and industrial effluents, agriculture, or waste disposal on the banks of aquatic ecosystems (RIVERA, 2003). However, when it comes to a water body situated in a region with significant human occupation, such as the Rio Verde Basin, it is difficult to distinguish between independent occupation factors and those arising directly or indirectly from it.

One of the major nutrients, phosphorus (P), can be found in sediments associated with complex salts, such as calcium, iron, aluminum and organic species, or adsorbed on the surface of minerals. Phosphorus is of particular importance in the growth of algae cells and its excess can lead to the eutrophication of a reservoir. Currently, much attention is given to the adsorption/desorption process of P in natural sediments and pure minerals.

Laboratory experiments with Fe (OOH) showed that raising the pH leads to the release of adsorbed P in iron complexes because of the competition with OH-ions (LIJKLEMA, 1977). However, little information on the effects of pH on P sorption processes in natural sediments is available.

The "natural" content of exchangeable phosphorus, as well as its sorption characteristics, was correlated with the composition of sediments such as active iron, aluminum and organic material. In turn, the composition of sediments has been associated with the chemistry of the reservoir, in other words, with hardness, alkalinity and pH, among other parameters.

Importantly, when the "natural" content of phosphorus is significant, the adsorption/desorption processes cannot be ignored. When the sediment containing phosphorus gets in touch with water, P is exchanged with water at the interface until a dynamic balance is reached. The concentrations in the sediment and water are stabilized when the exchange from the sediment to water and exchange from water to sediment become equal.

The purpose of this study is to identify sources and natural processes of geological origin that contribute to water quality in the Basin and, more specifically, in the Rio Verde Reservoir. To this end, we established a network of sample collection of different fractions, including: reservoir sediments, particulate in suspension in water, and free metals in water. Besides the chemical speciation of water and sediment in the reservoir, we tried to understand the influence of pH, zeta potential, and concentration of metals (free and associated with suspended material) on the eutrophic stability of the reservoir. The combination of the techniques of trace elements analysis, microanalysis, and electrophoretic mobility, can be useful in different acute or long-term environmental situations, through a chemical fingerprint of heterogeneous materials, such as suspended particles coming from anthropogenic activities. Specifically in this study, we used the techniques of X-ray fluorescence (XRF), electrophoretic mobility, light scattering for suspended solids and sediment. The identification of the different species of trace metals in the water body of the reservoir was investigated by Atomic Emission Spectrometry through Inductively Coupled Plasma (ICP-MS). The results are interpreted separately and together with the specific goal of identifying compounds that can potentially contribute to the eutrophication of the reservoir.

2. METHODOLOGY

Below we discuss the procedures used for sampling, selection of sampling points, sampling steps, sample preparation and subsequent analysis.

2.1 *SAMPLING POINTS*

Based on available data, such as the map of the Basin and geological units, sampling points were defined across the reservoir in collaboration with other sub-projects. These points were chosen in order to collect material from the reservoir on a spatial scale with both horizontal and vertical distribution.

The hydrogeochemical character of the tributary waterways of the Rio Verde Reservoir was assessed according to the distribution pattern of ionic macroconstituents, total concentration of dissolved solids, and pH.

The hydrochemical characterization of the reservoir's tributary basin was established through the analysis of samples collected at ten points on rivers. Of the tributaries, four are located on the right bank of the basin and four in the left bank, in addition to two collection points on the Rio Verde, one upstream of the reservoir and one downstream of the dam (Figure 2). Two other sample points were added from wells; information for these points, including chemical analyses, were provided by SANEPAR, the Sanitation Company of Paraná. P1 corresponds to a 25 m deep well drilled through limestone into the karst aquifer and is located in the municipality of Campo Magro. P2 is a 35 m deep well, drilled 30 m into weathered migmatite and 5 m into compact migmatite.

The location of sample collection points is shown in Table 1, which also contains data on the two wells located in Campo Magro within the Rio Verde Basin.

Among the samples collected within the reservoir (August and October, 2008), samples were collected at points R1, R3, R4 and Border. However, after experimental analysis and the results of other sub-projects, we concluded that point R3 has characteristics similar to point R2 and, therefore, there was no need to collect from the two points. Thus, point R3 was replaced by point R2 in the sample collection. Another change made to sample collection was the inclusion of a new sample point, R5, the location of which was determined by mathematical models that indicated water stagnation at this site, unlike other points.

Therefore, five sampling points (R1, R2, R4, R5 and Border) were defined and distributed so as to include both the central axis and the border of the reservoir based on two conditions: the relative homogeneity in the spatial distribution and the division between central and lateral regions of the reservoir. This last condition was defined so that any anthropogenic or natural contribution of the reservoir borders could be assessed.

At points R1 and R2, samples were collected at two depths: one on the surface (0 m) and the other based on the variation of temperature in the stratification in the sample point. If the temperature did not significantly vary according to depth, other parameters, such as dissolved oxygen, were taken into consideration to determine sample depth. At point R4, since it is a point with greater depth, samples were taken at four different depths, using the same selection criteria mentioned above. At R5, as well as on the Border, the collection was only taken on the surface (0 m). Due to shallowness of less than 1 meter, the sampling point called Border and the other samples taken from the surface were collected directly into bottles that were acclimatized in the river water. At other depths, water was collected using a 2-liter Van Dorn bottle, also previously acclimatized in the reservoir water. One-liter bottles used for collection were kept in a styrofoam box during sampling.

At each collection station, a sediment sample was taken, using a 30cm "core" collector (an acrylic collector with accessories also in acrylic) without any separation from the vertical profile. The samples were individually packed in polyethylene bags and stored at-5°C.

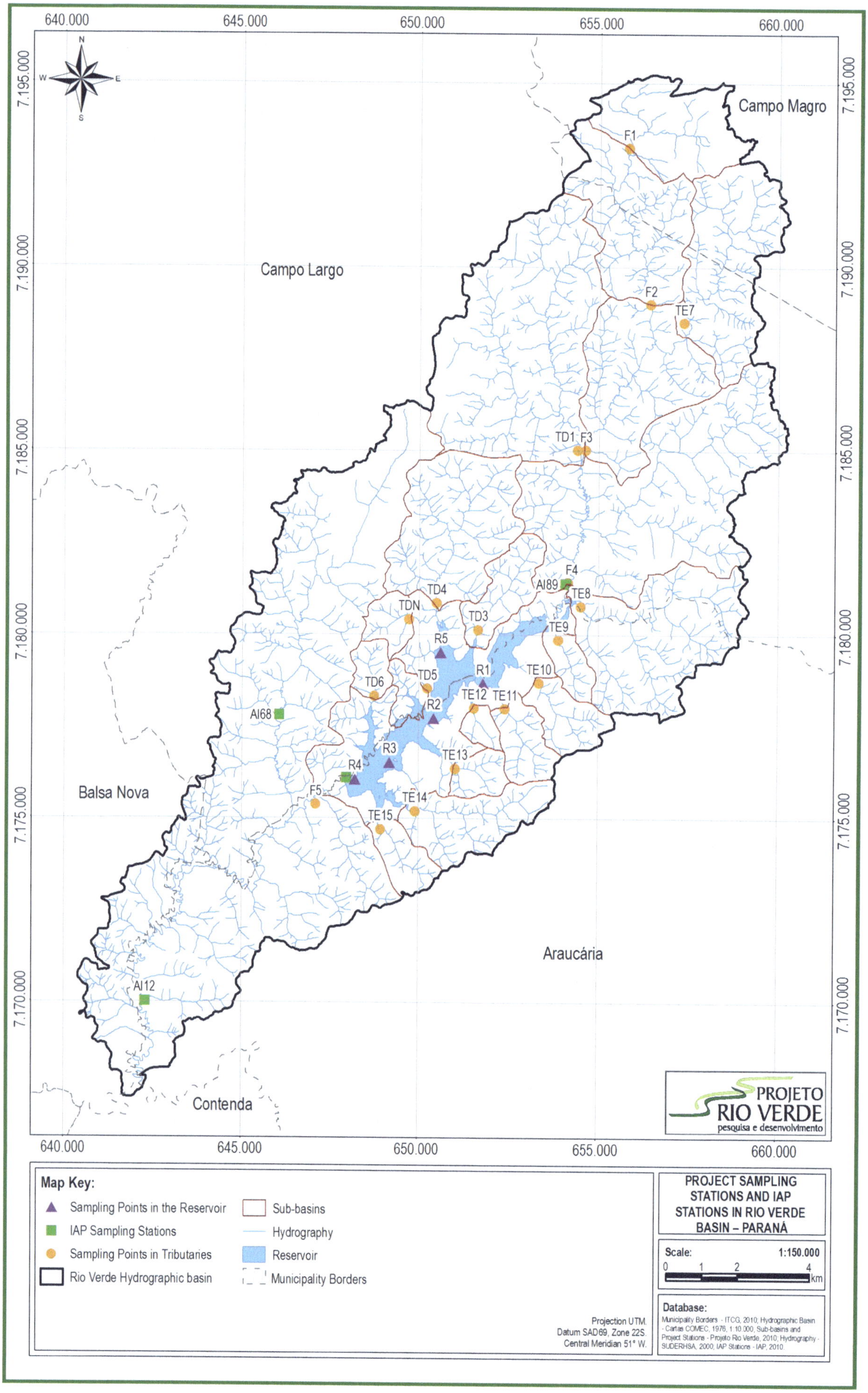

FIGURE 2 – SAMPLING POINTS IN THE RESERVOIR, RIO VERDE AND TRIBUTARIES

TABLE 1 – LOCATION OF SAMPLING POINTS ON THE RESERVOIR TRIBUTARIES

SAMPLING POINT	COORDINATES	
	UTM N	UTM E
F4 – Rio Verde upstream	654,092	7,181,394
F5 – Rio Verde downstream	646,923	7,175,449
D3 – Right bank tributary	651,645	7,180,142
D4 – Right bank tributary	650,470	7,180,837
D5 – Right bank tributary	650,197	7,178,142
D6 – Right bank tributary	648,727	7,178,315
E8 – Left bank tributary	654,464	7,180,706
E10 – Right bank tributary	653,316	7,178,637
E12 – Right bank tributary	651,483	7,177,965
E13 – Right bank tributary	650,605	7,176,454
E14 – Right bank tributary	649,869	7,175,145
P1 – Well in the Atuba Complex in Campo Magro	662,882	7,191,785
P2 – Well in the Karst Aquifer in Campo Magro	656,139	7,193,391

2.2 SAMPLE PREPARATION

Samples of river water were collected on September 10, 2008, in 1000 ml bottles of virgin polyethylene, packed in polystyrene boxes containing ice and sent to the laboratory within 24 hours.

Samples of suspended particulate material for XRF analysis were prepared in the laboratory using vacuum filtration of 500 ml of collected water in polycarbonate membranes with porosity of 0.4 μm (Nucleopore filters). The filtrate was collected in amber bottles and kept refrigerated for later analysis using ICP-MS. 100 mL of water without any previous treatment were stored and kept refrigerated for analysis by electrophoretic mobility and particle size distribution.

For analysis, sediments were thawed at room temperature. About 20 grams of the sample were disaggregated in deionized water for 10 to 20 minutes with addition of sodium pyrophosphate. Suspension was transferred to a cylinder, filled with water and was left to stand for two hours. The supernatant was then collected, dried in an oven at 550°C for two hours, and analyzed according to the X-ray powder diffraction method. The sedimented material was dried at room temperature over an aluminum sample holder and analyzed using the X-ray powder diffraction method. The same dry sediment was also analyzed using X-ray fluorescence.

2.3 ANALYSES

In the Hydrogeological Research Laboratory (LPH) at the Federal University of Paraná (UFPR), the following variables were analyzed according to internationally accepted standards (APHA, 1998): pH, conductivity, total alkalinity, phenolphthalein alkalinity, carbonate, bicarbonate, chloride, sulfate, nitrate, sodium, potassium, calcium, and magnesium.

The value of the concentration of each ion presented in $mg.L^{-1}$ was converted to a per liter equivalent, where cations formed one group and anions another. This ion distribution led to the construction of a composite diagram for a ternary diagram representing both the cationic distribution and the anionic distribution. Orthogonally projecting the points of the triangle of cations and the points of the anion diagram in a square, a Durov diagram was formed (DUROV, 1948) (Figure 3). In view of the presence of sulfate <1 $mg.L^{-1}$ and the constant presence of nitrate, the value of this last anion was added to the axis occupied by sulfate in the diagram.

The modified Durov diagram was supplemented with a xy diagram correlating the variables ionic conductivity and relation in equivalents Na+ $(Ca_2+)^{-1}$.

Thus, it was possible to evaluate the hydrochemical character of water, assess the possibilities of mixtures among different waters, and also comment on the relationship between water and the rocks with which they have been in contact.

Information concerning the elemental mass concentration (bulk) was provided by the dispersive energy of X-ray fluorescence (XRF). Measurements of suspended material filtered in the Nuclepore membranes were performed using an Epsilon-5 (PANalytical, Almelo, The Netherlands). The non-destructive quantitative analysis of the elements was made using the aluminum. The surface potential data were obtained using the zeta potential analysis technique in the Malvern Zetamaster. The measurements were made in the standard module of the equipment using as the conversion factor the zeta potential F (Ka) = 1.5 (Smoluchowski factor for conductive dispersions). Samples were shaken vigorously prior to analysis. The particle size distribution was measured after vigorous agitation by dynamic light scattering with the Malvern Zetamaster.

For ionic speciation of metals Al, B, Fe, Mg, Zn, Ca and Na, an optical emission spectrometer with plasma inductively coupled with radial configuration (Liberty II, Varian Mulgrave, Australia) was used; in the determination of K, an atomic absorption spectrometer (220FS, Varian Mulgrave, Australia) was used. To determine other metals, a mass spectrometer with inductively coupled plasma (810-MS, Varian Mulgrave, Australia) was used.

The X-ray diffractograms of the compounds were obtained on a PHILIPS PW 1830 diffractometer with scanning in the range of 3-70 degrees 2θ, using K_α radiation of copper and graphite monochromator. Based on the x-ray diffraction reflections of the polycrystalline sample collected from the respective diffraction patterns, the d_{hkl} interplanar distances were calculated using Bragg's equation. Comparing the dhkl values obtained with the amounts listed in the ICDD (International Centre for Diffraction Data) records, we can find information about the crystalline phases present in the samples analyzed. Samples were submitted to X-ray diffraction as a total sample and also following treatment with ethylene glycol, preferential orientation and heating at 550°C for two hours.

3. RESULTS AND DISCUSSION

3.1 HYDROGEOCHEMICAL CHARACTERIZATION

The lithological profile on which the Rio Verde Hydrographic Basin is established is roughly the same as the Passaúna River Basin and this similarity refers both to the chemism of the water and the variables discussed herein. Meger (2007) and Meger *et al.* (2007) found a distribution of ion macroconstituents similar to the one identified in this study (Table 2). This fact results from very similar geochemical aspects occurring in both basins. One difference noted between the water of the tributaries of the Passaúna Reservoir and Rio Verde is that all Passaúna tributaries have a mixed-bicarbonated characteristic, while in the case of Rio Verde, water from all sampling points is bicarbonated, ranging from calcic to mixed. Some left bank tributaries are calcic-bicarbonated.

Since the headwaters of the Rio Verde are fed by water with a composition directly influenced by the dissolution of carbonate rocks and as it is the main tributary, the karst section of the basin is the key factor in the reservoir's chemism. A sample taken from a well contains calcium and magnesium, a general characteristic of the waters of the karst aquifer in Paraná (HINDI, 1999), established in dolomitic marble, rocks in which calcium and magnesium predominate over other macroconstituents.

The individual contribution of each lithology that composes the Atuba Complex is not overly evident in Figure 3, since variation in the chemical composition between them is relatively small. These are minerals from the same families with minor variations in the proportion of constituents that have been subjected to the same weathering conditions.

Despite the compositional similarity between the water from the left and right bank tributaries, the Ca^{2+}/Mg^{2+} ratio is evidence of the greater contribution of silicate rocks plus magnesia in the left portion of the basin. This aspect can be viewed in the triangle of cations in Figure 3, built on values listed in Table 3. While the Ca^{2+}/Mg^{2+} ratios of the right bank tributaries vary between 2.3 and 3.1, the left bank tributaries range between 1.6 and 2.1.

In the Guabirotuba Formation, the calcitic levels are common and magnesium is largely fixed to the structure of montmorilonitic clays. This lithology is probably responsible for the more calcic characteristic of water at points D3 and D4.

Figure 3, in the extension of the modified Durov diagram, express the variable dissolved material concentration through conductivity. Here, the higher solute concentration in groundwater from the upper portion of the basin is evident, which would be expected and which affects the concentration of the Rio Verde at its entry into the reservoir at point F4. Waters with less solute loads were found at point D5, followed by E14.

The inverse relationship between calcium and sodium, in this case, does not indicate ion exchange processes since less conductive waters are more sodic and more conductive waters are calcic-magnesic. This is another indication of the lithological control on the water quality of the system. The behavior of the $Na^+ (Ca^{2+})^{-1}$ relation becomes more evident in the supplement to the diagram in Figure 1, where the following distribution is indicated: $Na^+ x(Ca^{2+})^{-1}$= 275.3 conductivity 1.541, with r^2 = 0.9284. This relationship was evaluated ignoring point P2, which has a very different behavior from the others and reflects peculiar local conditions.

The values of nitrate ion found for surface water below 0.9 mg.L^{-1} contrasted with the 4.6 mg.L^{-1} found in well P1, which indicates the vulnerability of the karst aquifer to point source contamination.

TABLE 2 – pH VALUES, ION CONDUCTIVITY AND ION MACROCONSTITUENTS (mg.L^{-1}) OF WATER BODIES IN THE RIO VERDE BASIN

POINTS	pH	COND. µS.cm^{-1}	HCO_3^-	Cl^-	NO_3^-	Na^+	K^+	Ca^{2+}	Mg^{2+} Mg.L^{-1}
F4	7.16	105.2	60.42	6.68	0.54	4.44	1.42	13.77	5.13
F5	7.35	72.1	40.00	4.17	0.20	3.71	1.94	9.39	2.36
D3	6.74	55.5	30.63	1.78	0.33	3.69	1.30	6.80	1.73
D4	7.17	91	47.13	4.40	0.67	5.14	1.83	11.91	2.02
D5	6.93	48.3	29.92	1.56	0.16	4.42	1.45	5.27	1.02
D6	7.21	83.5	50.21	2.23	0.83	4.49	1.31	11.30	2.97
E8	7.16	65.1	37.75	5.68	0.24	4.02	2.30	7.73	2.58
E10	6.96	62.4	37.39	2.12	0.27	3.74	1.16	8.54	2.43
E12	6.57	57.2	34.78	2.39	0.14	3.48	1.15	6.20	2.26
E13	6.76	63.7	36.08	2.9	0.44	3.20	1.30	7.05	3.06
E14	6.87	52.4	28.37	4.07	0.29	4.01	1.92	6.07	1.73
P1	7.71	260	169.1	2.05	4.64	1.50	0.90	32.57	17.96
P2	6.4	158	44.90	4.70	1.04	-	-	7.1	4.2

Values CO_3^{2-} (=0) were not detected; SO42- always <1.0mg.L^{-1}; P1 and P2 data were provided by DHG/SANEPAR.

TABLE 3 - ION MACROCONSTITUENT VALUES IN PERCENTAGE OF EQUIVALENTS PER LITRE FROM WATER BODIES IN THE RIO VERDE BASIN

POINTS	Ca^{2+}	Mg^{2+}	$Na^{+}+K^{+\%}$	HCO^{3+}	$SO_4^{2-}+NO_3^{-}$	$Cl^{-\%}$
F4	50.94	31.77	17.29	80.76	3.99	15.24
F5	53.23	22.40	24.37	82.25	3.11	14.64
D3	49.84	21.23	28.93	85.70	5.80	8.50
D4	57.26	16.26	26.48	80.98	6.11	12.91
D5	45.24	14.66	40.10	88.21	3.94	7.85
D6	53.98	23.75	22.27	86.18	7.28	6.53
E8	45.98	25.69	28.32	76.84	3.42	19.74
E10	51.67	24.62	23.71	87.30	4.25	8.45
E12	45.37	27.69	26.95	86.71	3.11	10.17
E13	44.95	32.66	22.40	82.78	5.86	11.36
E14	44.90	21.42	33.68	76.23	5.10	18.67
P1	50.60	46.70	2.79	95.11	2.92	1.97
P2	36.50	36.23	27.8	77.31	5.80	13.82

Note: In P2 sample, the values for % ($Na^{+}+K^{+}$) were obtained from the ionic balance.

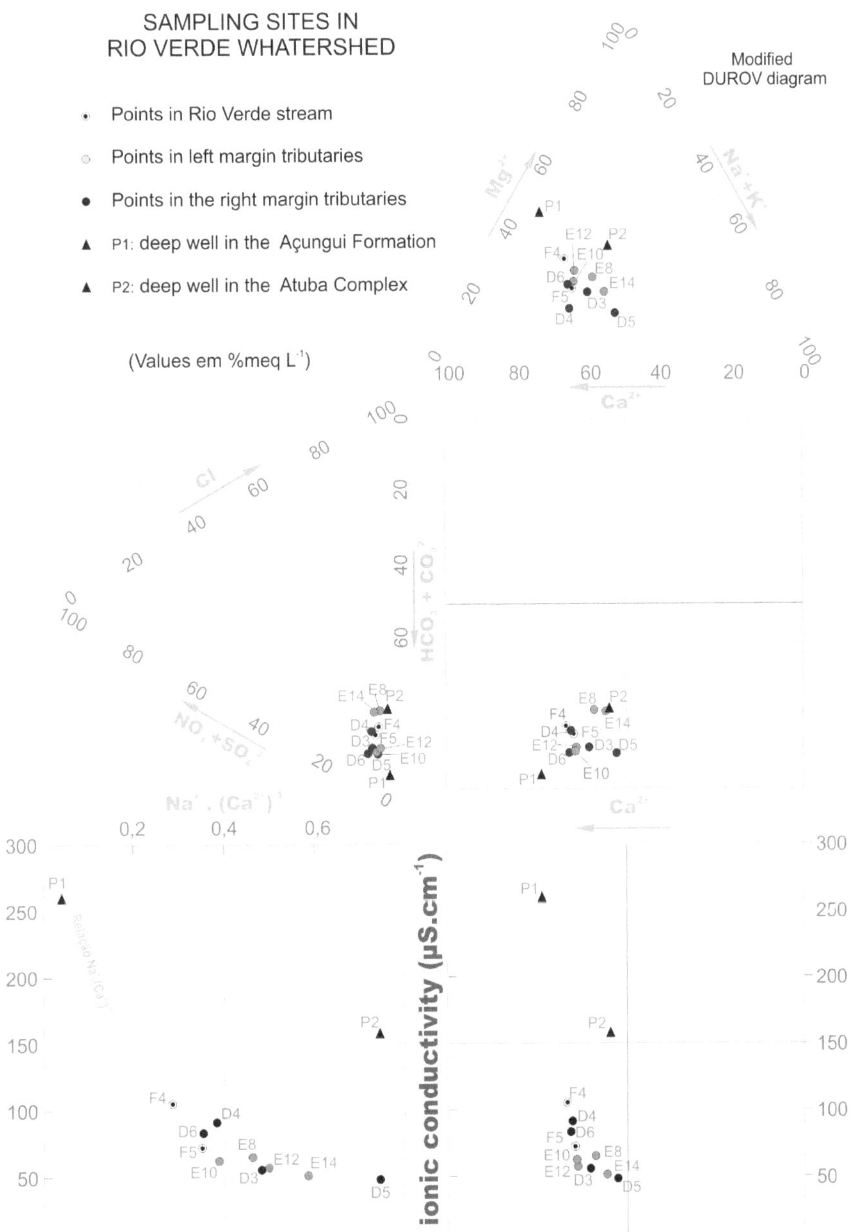

FIGURE 3 – MODIFIED DUROV DIAGRAM, REPRESENTING THE COMPOSITION OF IONIC MACROCONSTITUENTS IN WATER BODIES OF THE RIO VERDE BASIN

TABLE 4 – CONCENTRATION OF ELEMENTS IN FILTERED PARTICULATE MATTER, ANALYZED BY XRF

XRF AUGUST 2008	SAMPLING POINTS											
Elements	Border	R1 (0m)	R1 (4m)	R2 (0m)	R2 (5,5m)	R3 (0m)	R3 (4m)	R4 (0m)	R4 (2.5m)	R4 (5m)	R4 (7.5m)	R5 (0m)
Al	359	432	404			221	220	260	233	226	307	
As	0.3	0.3	0.3			0.2	0.2	0.3	0.3	0.3	0.3	
Ca	25	28	21			16.8	17.0	20.2	18.6	19.1	27	
Cl	7.6	8.9	7.4			8.2	6.9	8.8	8.1	8.6	9.2	
Cr	0.7	0.9	0.7			0.1	0.2	0.4	0.3	0.3	0.5	
Cu	1.3	1696	1.4			1.5	1.3	1.7	1.6	1.6	1.9	
Fe	367	424	378			135	142	160	144	146	196	
K	21	22	21			12.4	12.1	15.9	10541	10.4	15.9	
Mn	61	39	28			47	46	55.8	48	50	64	
Ni	0.2	0.3	0.3			0.1	0.1	0.1	0.1	0.1	0.1	
Pb	1.0	1.6	1.2			1.1	0.8	1.1	0.9	1.1	1.4	
S	7.3	8.1	5.7			7.6	6.5	8.2	7.4	7.1	8.8	
Si	<DL	716	607			626	595	714	652	620	831	
Sr	0	0	0			0	0	0	0	0	0	
Ti	12	13	14			6	6	8	7	7	10	
V	1	1	1			0	0	0	0	0	0	
Zn	1	1	1			1	2	1	1	1	4	

XRF OCTOBER 2008	SAMPLING POINTS							
Elements	Border	R1 (0m)	R1 (4m)	R3 (0m)	R3 (4m)	R4 (0m)	R4 (5m)	R4 (7.5m)
Al	243	479	633	262	208	217	216	215
As	0.2	0.3	0.4	0.3	0.2	0.2	0.2	0.2
Ca	15.5	17.6	23	15.3	12.3	15.3	14.4	13.3
Cl	7.8	7.3	7.3	7.9	6.6	7.4	7.6	7.1
Cr	0.5	1.2	1.6	0.5	0.2	0.4	0.3	0.2
Cu	1.4	1.4	1.3	1.5	1.2	1.4	1.4	1.4
Fe	253	442	601	239	164	187	187	182
K	11.6	14.6	20	11.3	7.5	9.2	9.7	9.2
Mn	25.1	14.8	16.1	20	28	29	32	28
Ni	0.2	0.5	0.6	0.2	0.1	0.1	0.1	0.1
Pb	0.8	0.9	0.9	1.2	1.1	1.0	0.9	1.2
S	7.7	3.9	4.2	4.8	4.5	4.9	5.1	4.8
Si	352	545	714	359	280	316	315	311
Sr	0.2	0.3	0.4	0.2	0.2	0.2	0.2	0.2
Ti	11.3	15.4	20.3	9.1	7.5	9.0	8.6	8.4
V	0.5	0.9	1.2	0.4	0.3	0.3	0.4	0.3
Zn	<DL	<DL	<DL	<DL	<DL	<DL	<DL	<DL

XRF -APRIL 2009	SAMPLING POINTS									
Elements	Border	R1 (0m)	R1 (4m)	R2 (0m)	R2 (5,5m)	R4 (0m)	R4 (4,5m)	R4 (7,5m)	R4 (9,5m)	R5 (0m)
Al	568	159	216	61	100	57	80	57	98	152
As	0.4	0.2	0.2	0.2	0.2	0.2	0.2	0.3	0.1	0.2
Ca	522	12.9	9.1	12.7	14.9	12	20	587	1989	15
Cl	9.1	8.2	8,0	8.2	8.3	7.7	7.9	8.4	28	8.2
Cr	1.9	0.3	0.3	0.1	0.1	0.1	0.1	1.0	5.5	0.3
Cu	0.6	0.3	0.3	0.1	0.4	1.0	0.5	0.7	1.2	0.2
Fe	6686	1907	1593	1378	1806	557	860	1408	51674	1755
K	280	9,0	11.3	4.4	6.7	3.5	5.1	9.0	23	11.6
Mn	1497	300	8.5	509	641	612	1270	6974	33828	362
Ni	0.7	0.1	0.1	0.1	0.1	0.1	0.1	0.1	<DL	0.1
Pb	0.5	0.1	0.1	<DL	0.1	0.2	0.4	0.5	1.2	0.1
S	8.7	4.3	3.8	4.6	5.3	4.7	5.9	7.3	30	5.1
Se	0.1	0.1	0.1	0.1	0.1	0.1	0.1	0.1	0.1	0.1
Si	8182	2327	2639	1294	1855	1045	1500	1185	3395	2738
Sr	0.9	0.2	0.1	0.1	0.2	0.1	0.2	0.6	2.5	0.2
Ti	734	6.6	16.9	3.0	5.4	3.2	5.1	10.7	627	8.0
V	1.4	0.2	0.1	0.1	0.2	0.1	0.2	1.0	6.8	0.2
Zn	1.0	0.4	0.8	5.3	1.3	0.8	2.5	2.6	5.2	0.3

XRF MAY 2009	SAMPLING POINTS									
Elements	Border	R1 (0m)	R1 (3m)	R2 (0m)	R2 (5m)	R4 (0m)	R4 (2m)	R4 (5m)	R4 (7,5m)	R5 (0m)
Al	279	446	700	133	143	114	125	118	129	319
As	0.3	0.3	0.4	0.2	0.2	0.2	0.3	0.3	0.2	0.3
Ca	21.9	256	389	19.2	18.2	16.6	252	17.2	21	259
Cl	8.0	8.5	9.4	8.0	8.1	8.1	8.9	12.9	8.7	8.2
Cr	0.3	1.3	2.5	0.6	0.3	0.6	0.4	0.1	0.1	0.8
Cu	0.3	0.3	0.6	0.3	0.3	0.2	0.5	0.3	0.6	0.2
Fe	2749	4608	7216	2196	2163	2873	3113	3027	3244	3231
K	17.5	24.8	443	9.5	9.7	8.8	10.7	14.3	9.8	21.1
Mn	525	420	607	1567	1364	1789	1795	1798	1915	549
Ni	0.1	0.5	0.8	0.2	0.1	0.1	0.1	0.1	0.1	0.2
Pb	0.1	0.2	0.7	0.1	0.5	0.1	1.1	0.7	0.9	0.2
S	7.0	7.3	8.0	6.3	6.2	6.0	6.9	6.4	6.8	6.7
Se	0.1	0.1	0.1	0.1	0.1	0.1	0.1	0.1	0.1	0.1
Si	4355	5894	9370	2152	2279	1853	2055	1922	2132	4842
Sr	0.3	0.4	0.6	0.2	0.2	0.2	0.3	0.2	0.2	0.3
Ti	11.1	17.2	321	7.6	8.9	8.9	9.5	9.4	10.6	13.3
V	0.4	0.8	1.2	0.3	0.4	0.4	0.5	0.5	0.6	0.6
Zn	0.5	0.6	2.0	0.5	1.2	0.3	2.8	1.3	1.5	0.6

**<DL below the detection limit

Also with respect to the concentrations of anionic species, low sulfate values are caused by the limited presence of sulfides in the rocks, whether they come from the Capiru Formation, Atuba Complex, or mainly the Guabirotuba Formation. In the case of sedimentary rocks, whether metamorphosed or not, it would be reflective of sedimentary environments with oxidant features.

3.2 CHEMICAL COMPOSITION OF PARTICULATE MATTER

The total metal concentration was measured in the water of the Rio Verde Reservoir using the Dispersive Energy X-ray Fluorescence technique. Twenty elements were identified in the samples and the results of this analysis are expressed in $\mu g/L^{-1}$.

In the analysis of the total particulate (bulk), in absolute terms, six elements showed high concentrations that could be presented herein in micrograms per liter, Table 4.

In general, in the first two samples, the results obtained using XRF with samples from the point called Border are lower than those obtained for point R1 at both depths; however, when the concentrations are compared with other points, the concentration profile is the same. However, the sample collection for the month of April 2009 shows that the Border concentrations are higher than at other points, except for point R4 IV, which has a high concentration for most of the studied elements. This is expected since R4 IV is the deepest sample point of the study with a depth of 9.5 m. From this sampling period, we noted a more pronounced variation of concentration at the different sampling points. An example of this is the difference in concentrations of Ca from the sampling points during this collection period. While point R1 presents $9.0 \mu g.L^{-1}$ of Ca, at point R4 IV the concentration reaches $1988.8 \mu g.L^{-1}$. In the month of May, the concentration profile was again similar between the sampling points, with slightly higher concentrations at point R1 (0 m) and R1 (3 m). Comparing the results of samples collected in August and October 2008, we find that concentrations are similar. On the other hand, if compared with the results from April and May 2009, we find a significant increase in the concentration of some elements such as Ca, Fe, Mn, Si and Ti. Fe concentration at the Border sample point in the months of August and October 2008 was $367.3 \mu g.L^{-1}$ and $252.7 \mu g.L^{-1}$, respectively. In the following sample period, this value reached $6685.9 \mu g.L^{-1}$ in April and $2748.9 \mu g.L^{-1}$ in May. A similar study was carried out by Meger *et al.* (2007) in the of Passaúna Reservoir, in the Metropolitan Region of Curitiba. Comparing the results obtained in the first two samples from the Border sample of the Rio Verde Reservoir with the of Passaúna Reservoir, we see that the concentration increased up to ten times for some non-majority elements; however, considering the similar geology of the two reservoirs, these results were expected.

From the August sample at point R1, concentrations are similar across a depth of 0 and 4 meters. This profile is not the same for R1 from the October sample. In this second sample, the elemental values are slightly higher at a depth of 4 m, which is supposedly due to the effect of gravity. Such a concentration profile is repeated for points R3 and R4. The same pattern is found in the sampling of all points in May. On the other hand, in the April collection, it is not possible to define a profile, because the concentrations vary across the elements analyzed and sample points. At point R4, during the 2008 collection, concentrations were similar across all studied depths, 0m, 2.5 m, 5 m, 7.5 m in August, and 0m, 5 m and 7.5m in October. In April 2009, the concentration variation increased with an increase in depth for most of the studied elements. The only exceptions to this behavior were selenium, which presented its highest concentration at 4.5 m, and aluminum and copper, which maintained a similar concentration at different depths. In the samples from May 2009, concentrations of chromium and nickel varied decreasingly with depth. Calcium showed a higher concentration in R4 (2m) with $252.1 \mu g.L^{-1}$, while at other depths, the concentration was about $18.4 \mu g.L^{-1}$. The concentrations of the other elements remained constant. At point R5, results remained almost the same, with a slight increase in the measured concentration in May compared to April. At the Border, the only elements that had a constant concentration were Cl and S, the others varied across the four sampling periods, with the highest concentration of those elements obtained in April 2009.

Based on Table 1, it is evident that the suspended material mainly contains Fe and Al, which are key components of some of the more stable mineral structures formed during the weathering process. As far as mineral components are concerned, the formed organic phases and the complexed soluble fraction must be taken into account. Minerals that contain Fe and Al have been considered the main determinants of the adsorption capacity because of the significant surface area of its compounds, especially oxides, hydroxides and oxyhydroxides. Therefore, further analysis of active Fe and Al should be performed because the proportion of active Al and Fe is probably related to organic matter, trophic state, and water chemistry at the sampling point.

Mn, an element primarily released from ferromagnesian minerals, easily forms in surface oxidizing conditions, hydroxides, oxides and oxyhydroxides, all with great adsorption potential. Contrary to Al and Fe that are disseminated in the soil, Mn is more heterogeneous, which can explain the large variations in concentration in the particulate form, considering different erosion and transport rates at different locations in the basin.

Table 5 compares the measured average elemental concentrations in the Rio Verde Reservoir as well as in the Passaúna Reservoir, which have very similar geological features.

Comparatively, the average concentrations of majority elements in the Rio Verde and Passaúna Reservoirs are of the same order of magnitude, except for Fe, Ca, Mn, Si, which obtained average concentrations considerably higher in the months of April and May 2009. The minority elements, Cl, S, Ti, Zn and P found in Rio Verde, are different from the average values found in the Passaúna Reservoir. The analyses conducted in the Passaúna Reservoir were based on only a few samples; therefore, the average values cannot be deemed definitive.

TABLE 5 - ELEMENTAL CONCENTRATIONS FOUND IN THE RIO VERDE AND PASSAÚNA RESERVOIRS IN μg/L

ELEMENTAL CONCENTRATION OF THE RIO VERDE AND PASSAÚNA RESERVOIRS – XRF					
	RIO VERDE				PASSAÚNA
Elements	August 2008	October 2008	April 2009	May 2009	January 2006
AL	296	309	155	251	240
As	0.3	0.3	0.3	0.3	n.d
Ca	21.5	15.8	319.4	127.3	27.7
Cl	8.2	7.4	10.2	8.9	1.0
Cr	0.5	0.6	1.0	0.7	0.4
Cu	1.5	1.4	0.5	0.4	0.4
Fe	232	282	6962	3442	322
K	15.8	11.7	36.5	56.9	28.7
Mn	49.0	24.4	4600	1233	21.8
Ni	0.1	0.3	0.1	0.2	0.1
P	n.d	n.d	n.d	n.d	6.3
Pb	1.1	1.0	0.3	0.4	1.2
S	7.4	5.0	8.0	6.8	11.3
Si	670	399	2616	3685	323
Sr	0.2	0.2	0.5	0.3	
Ti	9.2	11.2	142.1	41.8	16.1
V	0.4	0.5	1.0	0.6	0.6
Zn	1.4	1.3	2.1	1.2	3.1

* Data presented in μg/l

The average of phosphorus (P) in the Passaúna Reservoir was 6.3 $\mu g.L^{-1}$ and the highest concentration was 27.9 $\mu g.L^{-1}$; however, this element was not identified during the four sampling episodes conducted in the Rio Verde Reservoir. This is a factor in favor of the maintenance of the oligo-mesotrophic conditions of the reservoir. Considering that domestic and industrial sewage are among the main artificial sources of phosphorus, it can be inferred that the contribution of this element from anthropogenic sources is quite low. The average concentration of sulfur (S) of 7.4, 5.0, 8.0 and 6.8 $\mu g.L^{-1}$ found in the Rio Verde Reservoir is up to two times less than the average concentration found in the Passaúna Reservoir (11.3$\mu g.L^{-1}$). The increase in sulfur concentrations in various lakes has been attributed to the transport of gases and particulate matter containing sulfur in the atmosphere and subsequent precipitation with rainfall (ESTEVES, 1998). The concentrations of zinc (Zn) in the suspended material varied; the average from the four sampling episodes were 1.4, 1.3, 2.1 and 1.2 $\mu g.L^{-1}$, which is lower than the average of 3.1 $\mu g.L^{-1}$ found in the Passaúna Reservoir. These values are compatible with various aquatic environments, as noted by Esteves (1998), including the lakes Jacaretinga, Calado, Castanho and Tarumã-Mirim, in the Amazon, with 2.2 $\mu g.L^{-1}$, 3.0 $\mu g.L^{-1}$, 2.9 $\mu g.L^{-1}$, and 4.0 $\mu g.L^{-1}$, respectively. Studies on the Mogi-Guaçu River (MG/SP), on the other hand, show comparatively high values above 30 $\mu g.L^{-1}$ and reaching 200 $\mu g.L^{-1}$ thus demonstrating the anthropogenic effects on its study area. These levels are already having chronic toxic effects on fish. For humans, zinc accumulates in the liver, prostate, pancreas, pancreatic juice and seminal fluid. In human tissues, its concentration ranges from 10 to 200 $mg.L^{-1}$, the highest accumulation found was in the retina and in the prostate, between 500 and 1,000 $mg.L^{-1}$ (ESPINDOLA *et al.* 2003). The significant presence of manganese is common and its sources are ferromagnesian minerals that have been weathered and transformed into oxidized forms mainly present in soils

3.3 ELEMENTAL COMPOSITION OF THE DISSOLVED PHASE

The study of the distribution pattern of trace elements helps us to understand the geochemical processes controlling the hydrochemistry of river systems (ZHANG & SELINUS, 1997). Using the methods of ICP-OES and ICP-MS, this study proposes an estimation of the load of suspended solids and the concentration of trace elements and higher across the Rio Verde reservoir to identify the probable natural and anthropogenic sources of each element. Tables 6 and 7 indicate the concentrations of the main metals found in both surface water and in the water column in terms of the month of the year.

In general, concentrations of each component analyzed were similar at different points and depths in the reservoir for both samples collected in August and October, 2008. It is clear, however, that some elements indicated concentration values higher than others. In the August samples, when Ca, Mg, Na and K were not analyzed, we can see that silicon (Si) showed an average concentration of 3.53 $mg.L^{-1}$, while other elements such as silver (Ag) and copper (Cu) had concentrations ranging from 0.0001 $mg.L^{-1}$ to 0.0016 $mg.L^{-1}$, respectively. In the October samples, calcium (Ca), magnesium (Mg), sodium (Na), potassium (K) and silicon (Si), showed concentrations greater than the evaluated group. Variations found between abundant elements in the particulate matter and dissolved in the water column are directly associated with the crystallinity/solubility of suspended minerals.

Table 8 shows the comparison of the average concentrations found in the Rio Verde Reservoir with other reservoirs from within Brazil and worldwide.

Comparing the results, we note that the average concentrations in the Rio Verde Reservoir are similar to the average concentrations found in the reservoirs of Garonne, Dordogne and Isle in France, as well as The Balaton located in central Europe. The values calculated for the Iraí Reservoir were also similar, with a discrepancy only in the case of manganese (Mn), which showed a concentration of 0.03 mg.L^{-1} at Iraí and 0.0008 mg.L^{-1} in August, and 0.000008 mg.L^{-1} in October at Rio Verde. We can see extremely high concentrations of calcium (Ca) and sodium (Na) in the Wills Creek Reservoir in Ohio, USA. The average concentrations of these elements are 7.98 mg.L^{-1} and 3.86 mg.L^{-1} in Rio Verde and 8.7 and 15.9 mg.L^{-1} in the Bangpra reservoir in Thailand, respectively. In Ohio these concentrations are of 78 mg.L^{-1} and 93 mg.L^{-1}, respectively; this is due to a drainage abatement project in a coal mine located in the vicinity of the reservoir.

The reservoirs of Funil in Itatiaia, Rio de Janeiro, Tucuruí in Paraiba, and Água Fria in Barra da Choça, Bahía, Brazil, showed concentration levels higher than those found in Rio Verde. The Funil Reservoir is the most contaminated, with concentrations of aluminum (Al), arsenic (As), chromium (Cr), iron (Fe), manganese (Mn) and nickel (Ni) above the limits allowed by CONAMA.

The Billings reservoir in Braço do Rio Grande, São Paulo, stands out negatively due to the high concentration of all elements analyzed. This is because it is located in the Alto Tietê Basin. In the 1940s', water from the Tietê River and its tributaries was diverted to increase the flow of the

TABLE 6 – RESULT OF HEAVY METAL ANALYSIS BY ICP-OES AND ICP-MS – AUGUST 2008

ICP - AUGUST 2008									
	SAMPLING POINTS								
Elements	**Border**	**R1(0 m)**	**R1 (4 m)**	**R3 (0 m)**	**R3 (4 m)**	**R4 (0 m)**	**R4 (2,5 m)**	**R4 (5 m)**	**R4 (7,5 m)**
Ag	0.02	0.03	0.01	0.01	0.01	0.02	0.0035	0.01	0.01
As	0.16	0.18	0.19	0.16	0.19	0.16	0.15	0.11	0.18
Ba	22.9	26.1	25.8	25.9	25.7	26.5	26.2	26.5	26.7
Cd	0.01	0.03	0.03	0.03	0.01	0.01	0.09	0.04	0.05
Co	0.05	0.12	0.10	0.07	0.05	0.37	0.17	0.04	0.05
Cr	0.45	0.67	0.44	0.47	0.39	0.38	0.36	0.37	0.46
Cu	0.99	3.69	2.12	1.39	0.78	1.24	1.19	0.83	2.62
Hg	0.48	<DL	<DL	<DL	<DL	<DL	<DL	<DL	<DL
Li	0.23	0.33	0.22	0.24	0.21	0.31	0.19	0.21	0.26
Mn	2.66	1.46	0.73	0.51	0.20	0.31	0.55	0.37	0.30
Mo	0.15	0.30	0.18	0.14	0.33	0.23	0.17	0.13	0.13
Ni	0.29	1.28	0.70	0.53	0.32	0.33	0.37	0.27	1.23
Pb	0.33	0.50	0.36	0.26	0.22	0.21	0.33	0.22	0.30
Sb	0.18	0.10	0.18	0.14	0.19	0.18	0.18	0.19	0.18
Si	3863	3847	4229	3204	3510	3335	3221	3331	3208
Sn	0.27	0.26	0.21	0.22	0.21	0.19	0.22	0.19	0.17
Ti	0.43	0.43	0.43	0.31	0.35	0.27	0.36	0.35	0.38
V	0.17	0.24	0.23	0.22	0.21	0.21	0.19	0.20	0.21

** < DL below the detection limit

TABLE 7 – RESULT OF HEAVY METAL ANALYSIS BY ICP-OES AND ICP-MS – OCTOBER 2008

ICP - AUGUST 2008									
	SAMPLING POINTS								
Elements	**Border**	**R1(0 m)**	**R1 (4 m)**	**R3 (0 m)**	**R3 (4 m)**	**R4 (0 m)**	**R4 (2,5 m)**	**R4 (5 m)**	**R4 (7,5 m)**
Ag	0.02	0.03	0.01	0.01	0.01	0.02	0.0035	0.01	0.01
As	0.16	0.18	0.19	0.16	0.19	0.16	0.15	0.11	0.18
Ba	22.9	26.1	25.8	25.9	25.7	26.5	26.2	26.5	26.7
Cd	0.01	0.03	0.03	0.03	0.01	0.01	0.09	0.04	0.05
Co	0.05	0.12	0.10	0.07	0.05	0.37	0.17	0.04	0.05
Cr	0.45	0.67	0.44	0.47	0.39	0.38	0.36	0.37	0.46
Cu	0.99	3.69	2.12	1.39	0.78	1.24	1.19	0.83	2.62
Hg	0.48	<DL	<DL	<DL	<DL	<DL	<DL	<DL	<DL
Li	0.23	0.33	0.22	0.24	0.21	0.31	0.19	0.21	0.26
Mn	2.66	1.46	0.73	0.51	0.20	0.31	0.55	0.37	0.30
Mo	0.15	0.30	0.18	0.14	0.33	0.23	0.17	0.13	0.13
Ni	0.29	1.28	0.70	0.53	0.32	0.33	0.37	0.27	1.23
Pb	0.33	0.50	0.36	0.26	0.22	0.21	0.33	0.22	0.30
Sb	0.18	0.10	0.18	0.14	0.19	0.18	0.18	0.19	0.18
Si	3863	3847	4229	3204	3510	3335	3221	3331	3208
Sn	0.27	0.26	0.21	0.22	0.21	0.19	0.22	0.19	0.17
Ti	0.43	0.43	0.43	0.31	0.35	0.27	0.36	0.35	0.38
V	0.17	0.24	0.23	0.22	0.21	0.21	0.19	0.20	0.21

** < DL below the detection limit

dam. Needless to say, the river is in critical condition due to the release of residential pollutants and industrial contaminants. Analyzing the quality of water based on the concentrations of dissolved heavy metals measured at several points, we found "subtle" changes in the concentrations of a particular metal. This observation is valid both when comparing surface points and points at different depth.

For purposes of supply, the Ministry of Health (Ordinance no. 518, March 25, 2004) established the responsibilities and procedures for the control and monitoring of water quality for human consumption and its drinkability standards. This rule defines drinking water as water for human consumption of which microbiological, physical, chemical and radioactive parameters meet the drinkability standards and offers no health risks. In totaling the concentrations of the elements analyzed by XRF and ICP, the contribution of solid and dissolved parts, respectively, with the total amount of heavy metals present in the Rio Verde Reservoir, we did not observe levels of heavy metals above those outlined in the Ordinance.

3.4 SURFACE POTENTIAL

When sediment comes in contact with water, the constituent elements of the sediment exchange with the water at the interface until the dynamic balance is reached: processes of adsorption/desorption and precipitation/dissolution. This equilibrium is usually described as the Langmuir adsorption isotherm (Weber *et al.* 1991).

3.4.1 Importance of Zeta Potential for stability formulation

When small particles collide, they can be induced to aggregation. These collisions can be minimized in two ways: steric and electrostatic repulsion. In the absence of the steric effects of colloidal protection, stability of colloidal systems is determined by the balance between attractive and repulsive forces experienced by the particles when they approach each other. If repulsion is mutual, dispersion will resist flocculation. If there is no repulsion, flocculation or coagulation could occur.

pH, electrostatic repulsion and stability formulation is outlined with the examples below. The general rule for electrostatic stability is determined by the value of Zeta potential of + /-30 mV (Figure 4).

A large number of physical and chemical parameters directly influence the zeta potential of the particulate material

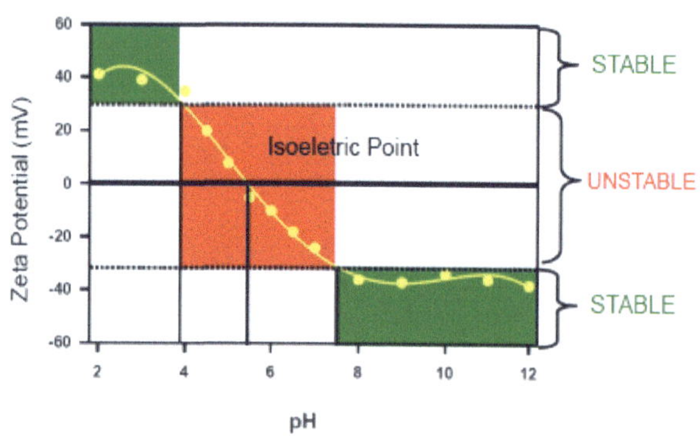

FIGURE 4 – ZETA POTENTIAL VARIATION WITH pH

suspended in water in systems with a homogeneous mineralogy. The chemical parameters that can be modified include the addition of solutes such as: metal salts, salts of electrolytes that alter the ionic strength of the medium (NaCl, NaNO3, KCl), and pH variation. Among the physical parameters, the temperature and viscosity of waters can be included.

3.4.2 The effect of pH and Phosphorus preferential sorption species

The adsorption of phosphorus in adsorbents is determined by the surface load and the state of protonation of phosphorus in the aqueous medium. Studies of soils and sediments in the laboratory show that the absorption of phosphorus varies with pH, but there is little information available on the direct effects of pH on the sorption of P in lake sediments. In the short term, the phosphorus cycle is related to variations of pH in the equilibrium of sorption and kinetics and, in the long term, on the effects of acidification on rates of biological mineralization and sediment composition. In most systems, the sorption of anions, such as phosphate, decreases with increasing pH and the surface load becomes more negative. However, the effect of pH on the adsorption of phosphorus cannot be generalized and depends on each system.

The amount of adsorbed P and the zeta potential of the sediment particles apparently do not show any direct relationship. That is, when a negative zeta potential becomes more negative, the amount of P sorbed remains at the maximum until a certain pH is reached, instead of decreasing rapidly due to electrostatic repulsion as expected.

In fact, the adsorption of phosphorus must be assessed by observing both the chemical nature of the sediment, that is, the adsorption process in relation to chemical affinity, and the electrostatic interactions in the sediment-water interface.

Thus, an evaluation of the electrokinetic profile and pH of the suspended material was made at different points and depths of the reservoir. The results obtained from different sampling points are shown in Table 6 and subsequent figures; they aim to illustrate the zeta potential as a function of the depth profile and pH. The general trend of reduction of the zeta potential with the increase of pH was confirmed (Table 6).

The higher the zeta potential, the greater the repulsive forces and the more stable the colloidal systems, Zeta potential of +/-30 mV. The results of this research indicate that the zeta potential varies from -15 to -40 mV in the reservoir in the samples collected in 2008. Thus, only two points, R3 (0 m) and R4 (2.5 m), taken in August, and point R3 (0 m), in October, can be considered as regions with colloidal stability (Figure 3). However, the fact that the pH in the same sampled points differs when the sampling periods are compared must be taken into consideration. Other relevant information is that the average particle diameter also differs when comparing the sampling places and period. Such changes between the August and October may be related to the fact that weather conditions in the two samples were different. While, in August, we had a lot of rain during the week prior to the sample collection, in October, it was a sunny day.

TABLE 8 – Comparison of selective trace element and heavy metal concentrations in the Green river reservoir in Brazil with other studies (μg L-1)

	AVERAGE CONCENTRATION OF TRACE AND HEAVY METALS USING ICP (μg L^{-1})												
	Green Reservoir		Rio Grande/SP	Itatiaia/RJ	Tucurui/PA	Barra da Choça/BA	Curitiba/PR	Ohio/USA	Thailand	France			Central Europe
Elements	ago/08	out/08	Billings reservoir	Funil reservoir	reservoir	Água Fria reservoir	Iraí reservoir	Wills Creek Reservoir	Bangpra	Garonne	Dordogne	Isle	Lake Balaton
Ag	0,13	0,02		4,04	0,46								
Al		13,5	3,14	355	115		0,15	1,50	<DL				
As	0,16	0,09		0,03					0,00				
Ba	25,9	29	87	69	41		0,01	0,02	0,01				
Be		0,06			2,71			<DL					
Ca		7983						78	8,70				
Cd	0,03	0,01	11,4	0,62	0,04	11,5		<DL	<DL	0,05	0,02	0,03	0,00
Co	0,11	0,06		10,3	0,07								0,07
Cr	0,45	1,64	48	127	0,42			0,00	<DL				
Cu	1,65	1,47	1518		0,61	0,31		<DL	0,00	1.23	0.89	1,30	0,48
Fe		14,3	6,0	797	359		0,31	0,25	0,09				
Hg	0,49	1,44											
K		1676							6,70				
Li	0,25	0,44											
Mg		3909						7,60	2,90				
Mn	0,79	0,01	316	113	20		0,03	0,08	0,15				
Mo	0,20	0,15			0,06								
Ni	0,60	0,47	19,0	65	1,54			0,00	0,00				0,60
Pb	0,31	<DL	63	26	0,49	4,46		0,01	<0,002	0,14	0,11	0,15	0,09
Si	353	1760						2,86	1,30				
Sn	0,22	0,04											
Sr		43			27		0,02						
Zn	<DL	23,3	186			5,27		<DL	0,10	3.07	2.49	5,00	0,85
Zr		0,32											

* < DL below the detection limit

TABLE 9 – ZETA POTENTIAL AND PH OF SUSPENDED MATTER SAMPLED IN DIFFERENT STATIONS OVER 2008-2009

	SAMPLING POINTS	Margin	R1 (0 m)	R1 (4 m)	R2 (0 m)	R2 (4 m)	R3 (0 m)	R3 (4 m)	R4 (0 m)	R4 (2,5 m)	R4 (5m)	R4 (7,5 m)	R5 (0 m)
Aug 2008	Zeta Potential (mV)			-17,6			-24,8	-21,9	-19,4	-29,3	-19,1	-17,5	
	pH		7,1	7,1			7,0	7,2	7,2	7,0	6,8	6,7	
October 2008	Zeta Potential (mV)		-16,4	-19,2			-39,3				-13,6		
	pH		7,4	7,1			7,5	7,4	7,2	7,4	7,1	6,4	
April 2009	Zeta Potential (mV)		-9,2	-14,6	-7,0	-10,2			-13,1	-14,3	-9,9	-11,6	-14,0
	pH		7,4	7,3	7,4	7,2			7,3	7,3	7,2	6,5	7,4
May 2009	Zeta Potential (mV)	-14,0	-11,3	-11,7	-13,4	-10,9			-8,6	-9,9	-9,6		-10,1
	pH		7,4	7,2	7,2	7,2			7,0	7,0	7,0	7,1	7,4
June 2009	Zeta Potential (mV)		-14,4	-15,7	-16,4	-14,0			-16,2	-15,6	-15,2	-14,8	-12,0
	pH		7,3	7,4	7,4	7,4			7,1	7,2	7,1	7,1	7,3
July 2009	Zeta Potential (mV)	-14,9	-15,0	-14,6	-15,6	-15,9			-12,0	-13,6	-14,1	-13,8	
	pH		7,4	7,3	7,4	7,4			7,0	7,1	7,1	7,0	

The results of the zeta potential of the 2009 samples range from -7 to -16.4mV and they are more unstable, as there is no point from any of the four samples which is in the stability range. We must consider, however, that point R3 (0m), in which colloidal stability was verified in the 2008 samplings, was not sampled in the subsequent year. In August and October 2008, we observed lower values of pH at greater depths: 6.8, 6.7 and 6.4, from R4 III (5m) and R4 IV (7.5 m) in August, and R4 IV (7.5 m) in October, respectively. This pattern was only confirmed in the sample from April 2009, in which the pH was 6.5 for R4 IV (9.5 m). In the other samples, the pH values showed little variation ranging from 7.0 to 7.4.

The results of average diameters (nm) showed significant fluctuations. They varied among the different sampling points, depths, and dates of collections, and we did not see any relationship between these changes and the pH, zeta potential, or climatic conditions. In general terms, the negative zeta potential can be associated with the adsorption of organic acids. This information is relevant because the suspension of particulate material in the reservoir, as well as subsequent treatments and filtration, will be impacted by tendencies of surface potential due to natural and anthropogenic events occurring in the reservoir and/or tributaries. The zeta potential measurements in this study were similar to the variations expected in surface water. The zeta potential in the waters of the Rio Verde Reservoir did not show a definite trend with respect to pH.

3.5 SEDIMENT COMPOSITION

Sediment samples collected at the four sampling points within the reservoir were subjected to X-ray diffraction for mineralogical determination and X-ray fluorescence to assess its chemical composition.

In all samples, kaolinite and quartz were identified as major mineral components (Figure 5). Gibbsite aluminum hydroxide and goethite iron oxyhydroxide were found in all samples. Other phyllosilicates such as chlorite and muscovite or illite were detected in the samples subjected to treatment. In samples R1, R2 and R3, the presence of a clay mineral with ratio Si:Al of 2:1 was found, probably a montmorillonite. At sample point R3, the results indicate the presence of chlorite.

The entire mineralogical framework indicates the prevalence of detritus material or material generated during the weathering of the basin rocks, or composed of resistate minerals, the latter present in soils as components of the original rocks. Quartz, muscovite/illite, and probably chlorite would be present in the original rocks, while kaolinite, goethite, and gibbsite were generated mainly in soils and weathered mantle of rocks. 2:1 minerals, in turn, originate most likely in sedimentary rocks of the Guabirotuba Formation and to a lesser extent may also come from weathered biotite bands of gneissic rocks.

As for chemism (Table 10), the gradual increase of silica in the sediments, from upstream to downstream, is noteworthy (Figura 6). This behavior can be linked to the deposition of diatoms formed in the lake and progressively precipitated downstream, in the lake itself. Iron content has a contrasting behavior, as it is higher in R1 than in R4. In this case, higher levels upstream can be understood as a reflection of the higher density of ferruginous compounds, products of basin erosion, which are transported by the Rio Verde and tend to precipitate with the loss of transport capacity within the lake.

TABLE 10 – CHEMICAL COMPOSITION OF SEDIMENTS

COMPOSITION %	SAMPLING POINTS			
	R1	R2	R3	R4
SiO	36.1	40.6	44	50.5
Al_2O_3	25	24.8	25.9	21.4
Volatile (loss on ignition)	21	18.6	18.8	14.8
Fe_2O_3	15.3	13.4	8.3	9.5
K_2O	0.5	0.3	0.4	1.3
TiO_2	0.8	1.1	1.2	0.9
MgO	0.4	0.3	0.5	0.5
SO_3	0.2	0.2	0.1	0.3
P_2O_5	0.3	0.2	0.3	0.2
CaO	0.2	0.1	0.2	0.1
BaO	0.1	0.1	0.1	0.1
MnO	0.1	0.1	0.2	0.1
CeO_2	0.1			

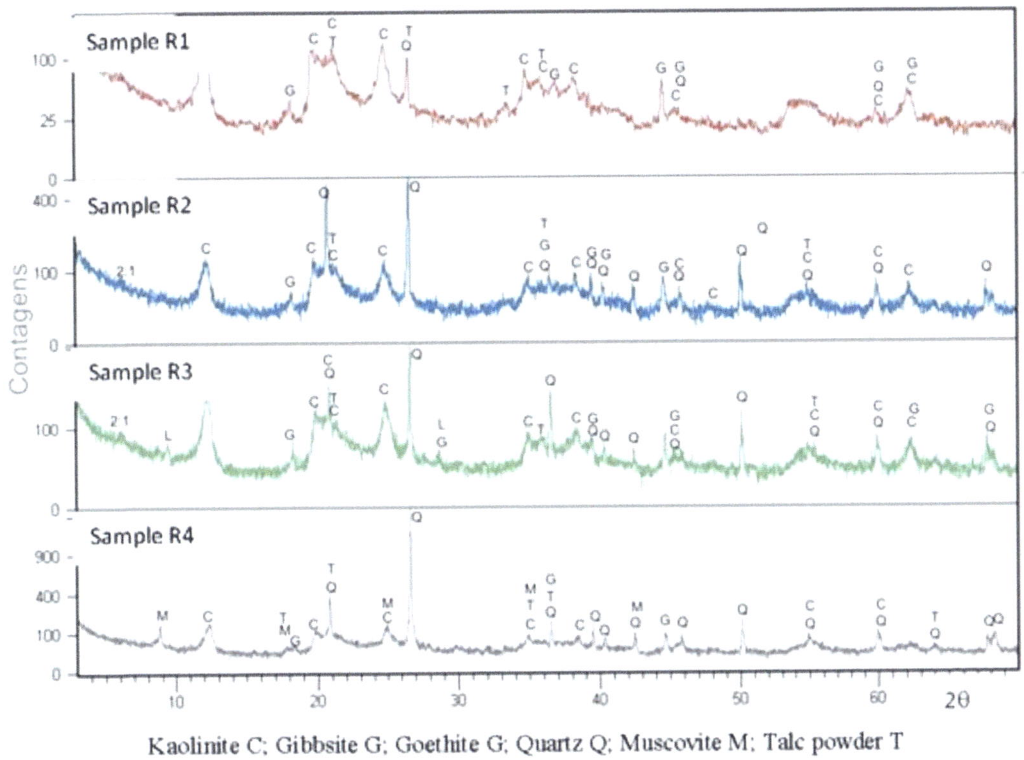

FIGURE 5 – X-RAY DIFFRACTION (CUKA) OBTAINED FROM NATURAL SEDIMENT SAMPLES COLLECTED IN THE RIO VERDE RESERVOIR

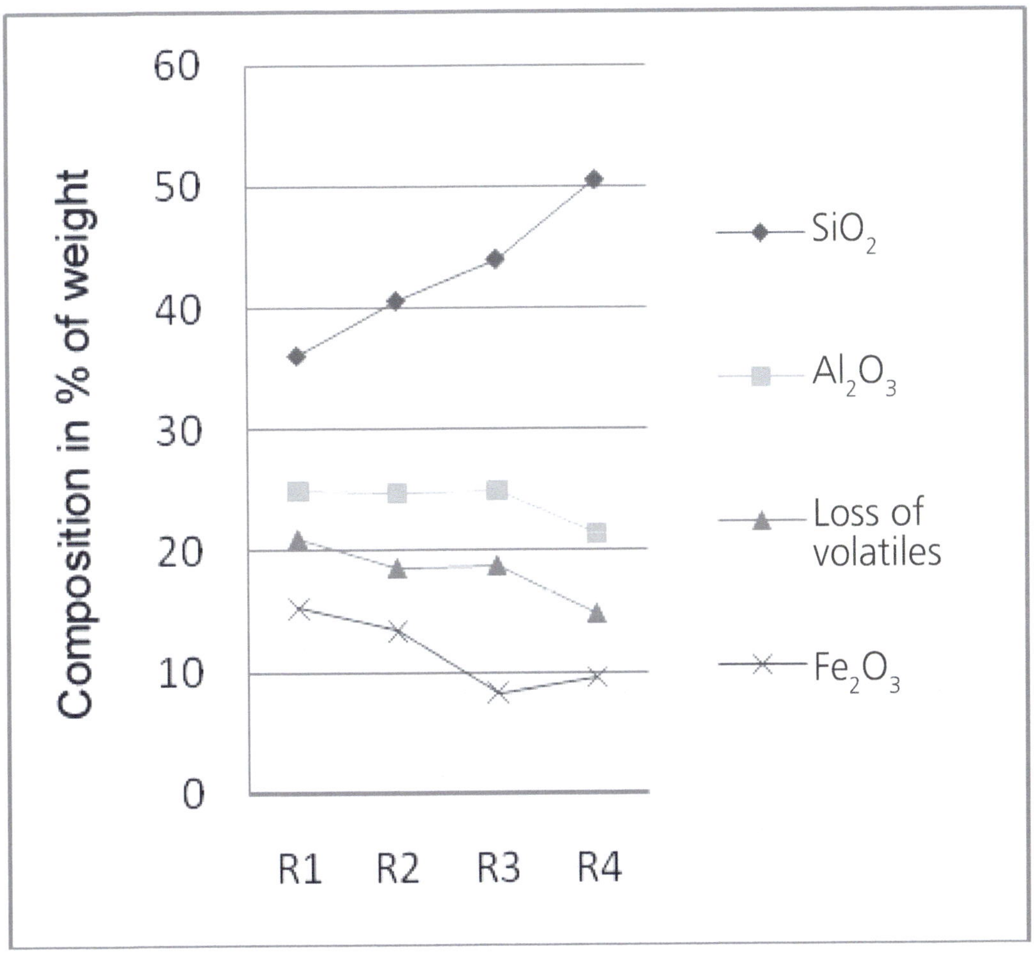

FIGURE 6 – VARIATION IN THE CHEMICAL COMPOSITION OF CHEMICAL MACROCONSTITUENTS IN THE SEDIMENTS OF THE RIO VERDE RESERVOIR

The decrease in the amount of volatiles is interpreted as a consequence of the more rapid deposition upstream of larger vegetal debris carried from the hydrographic basin to the lake by the main tributary.

Less abundant elements, such as phosphorus, do not show regular variation, which is also the case for Ti, Mg, S, Ca, Ba, and Mn. The presence of cerium, in amounts detectable by the methodology employed, is not peculiar considering the significant presence of rare-earths in the Guabirotuba Formation first mentioned by Coutinho (1955).

4. CONCLUSIONS

Information generated in this study on ionic macroconstituents, suspended particulate matter, and sediments in the Rio Verde Reservoir, allowed a spatial characterization of the elements coming from the hydrographic basin and their concentrations at the time of study. This information allowed us to evaluate different parameters that may be indicative of the quality of the water in the reservoir.

The composition, especially of metals in solution in the waters of the tributaries of Rio Verde Reservoir, is the result of the lithology of the basin, with a strong influence from the karst section. The hydrochemical characteristic of the waters is calcic-bicarbonated to mixed-bicarbonated, with low contents of sulfate.

In light of the findings on the availability of trace metals in the water of the reservoir, including the contributions in the solid phase (particulate matter) and the dissolved phase, we may conclude that the reservoir water is not jeopardized by the presence of toxic metals.

The minimal presence of phosphorus, both as suspended particulate matter and as water soluble matter, indicates that pH and zeta potential are in a range that promotes the adsorption of phosphorus in the sediment and/or that there are no significant anthropogenic sources of phosphorus. Phosphorus sorption processes have complex mechanisms and are influenced by various factors and as such further studies should be performed.

The lithological characteristics and the chemism of sediments studied represent the geology of the reservoir's drainage basin. Based on the expertise employed in this study, there were no significant chemical changes that indicate recent anthropogenic activities in the reservoir or coming from tributaries. Finally, we may conclude that this study can serve as the foundation for future environmental and geochemical investigations in the region, once concentration ranges and reference relationships for monitoring have been established.

5. REFERENCES

ANDREOLI, C.V.; CARNEIRO, C. (2005) Gestão integrada de mananciais de abastecimento eutrofizados. Sanepar; FINEP. Curitiba, 500 p.

ALEXANDRINO, D. M.; SANTOS, J. S.; SANTOS, M. L. P.; LACERDA, J. J. J. Avaliação da contaminação do reservatório Água Fria, Barra do Choça BA;por Cd,Pb,Cu e Zn provenientes de atividades agrícolas. Departamento de Ciências Naturais – Universidade Estadual do Sudoeste da Bahia. Vitória da Conquista, Bahia.

APHA (1998) Standard Methods for the Examination of Water and Wastewater 20th ed. American Public Health Association, Washington D.C.

BITTENCOURT, A.V.L.; TUBBS, D. (2007) Hidrogeoquímica e aplicações em prospecção geoquímica. In: Otávio Augusto Boni Licht, Carlos Siqueira Bandeira de Mello e Cássio Roberto da Silva (Org.). Prospecção Geoquímica. Sociedade Brasileira de Geoquímica.

BOSTELMANN, E. Avaliação da concentração de metais em amostras de sedimentos do Reservatório Billings, Braço Rio Grande, São Paulo, Brasil. Dissertação de Mestrado. Instituto de Pesquisas Energéticas e Nucleares (IPEN).

CLARKE, F.W.;WASHINGTON, H.S. (1924) The composition of the Earth Crust, U.S. Geological Survey Profess. Paper 127, 177 p.

COUTINHO, J.M.V. (1955) Lantanita de Curitiba, Paraná. Boletim Fac. Fil. Ciênc. Letr. USP 186 (Mineralogia 13), p. 119-126.

DUROV, S.A. (1948) Natural Waters and graphic representation of their composition. Dokl. Acad. Nauk. SSSR v.59, p. 87-90.

ESPÍNDOLA, E.L.G.; BRIGANTE, J. (2003) Avaliação das modificações na qualidade da água do rio Mogi-guaçu: uma análise temporal. In: Limnologia fluvial: um estudo do rio Mogi-guaçu. São Carlos: RIMA.

ESTEVES, F.A. (1998) Fundamentos da limnologia. 2. ed. Rio de janeiro: Interciência.

FILHO, A.R.R. (2006) Avaliação da influência do reservatório do Funil na qualidade da água do rio Paraíba do Sul. Dissertação de Pós-graduação.Universidade de São Paulo Instituto de Química, São Paulo, 46 p.

HINDI, E.C. (1999) Caracterização hidroquímica e hidrogeológica das fontes cársticas das bacias dos rios Tumiri, Água Comprida, Fervida e das Onças – Colombo, PR. Dissertação – Setor de Ciências da Terra, Universidade Federal do Paraná. Curitiba, 127 f.

KRUAWAL, K.; SACHER, F.; WERNER, A.; MULLER, J.; KNEPPER, T.P. (2004) Chemical water quality in Thailand and its impacts on the drinking water production in Thailand.

LIJKLEMA, L. (1977) The role ofiron in the exchange of phosphate between water and sediments. In: GOLTERMAN, H.L. (Ed.). Interaction between Sediments and Freshwater. W. Junk. Publ., The Hague, p. 313-317.

MASSON, M.; BLANC, G. ; SCHÄFER, J. (2006) Geochemical signals and source contributions to heavy metal (Cd, Zn, Pb, Cu) fluxes into the Gironde estuary via its major tributaries.

MEGER, D. G. (2007) Material particulado suspenso e macroconstituintes iônicos em um reservatório de abastecimento: o caso do rio Passaúna, Curitiba, Paraná, Brasil. Master Dissertation. Universidade Positivo, Curitiba, Paraná, 143 p.

MEGER, D. G.; BITTENCOURT, A.V.L. ; GODOI, A.F.L. ; VERMAAK, S.P. ; CARVALHO FILHO, M. A. S. ; LAGOS, P. E. D. ; CARNEIRO, C. M. S. ; MAKAROVSKA, Y. ; GRIEKEN, R. V. ; GODOI, R. H. M. (2007) Assessment of suspended particles in selected reservoir of Curitiba City, Southern Brazil. In: 10th European Workshop on Modern developments and applications in Microbeam Analysis, 2007, Belgium. Book of Abstracts of EMAS 2007 – European Microbeam Analysis Society. Antwerp, 280 p.

NGUYEN, H.L.; LEERMAKERS, M. ; OSÁN,J.; TÖRÖK, S. ; BAEYENS,W. (2004) Heavy metals in Lake Balaton: water column suspended matter, sediment and biota.

OLIVEIRA, G.R. F.;PEREIRA,S. F. P. P.; SARAIVA, A.C.F;LIMA, M.A; OLIVEIRA, J.S.; SILVA, J.S.;MORAIS,S.H.S, MIRANDA,R.G;ALVES,D.T.V. Avaliação da Presença de Metais Pesados em Água do reservatório de Tucuruí-PA.

RIVERA, E. A. C. (2003) Modelo sistêmico para compreender o processo de eutrofização em um reservatório de água, Campinas, São Paulo.Unicamp.

TARDY, Y. (1969) Géochimie des eaux d'altérations. Etude dês arènes et dês eaux de quelques massifs critalins d'Europe et d'Afrique. Mém, Serv. Carte Géologique Als. Lorr, Strasbourg, 31, 199 p.

TUNDISI, J.G.; TUNDISI, T.M.; ROCHA, O. Ecossistemas de Águas

Interiores, In: REBOUÇAS, A.C.; BRAGA, B. e TUNDISI J, G. (Org.). Águas Doces do Brasil: Escrituras. São Paulo, 717p.

WALKER, H.; TAERAKUL,P.; LAMMINEN, M.; HE, Y; TRAINA, S.; WHITLATCH, E. Short- and Long-Term Behavior of Fixated FGD Material Grout at the Roberts-Dawson Mine. (http://www.flyash.info/2005/214wal.pdf).

WEBER JR., W.J., MCGINLEY, P.M., KATZ, L.E. (1991) Sorption phenomenon in subsurgace systems: concepts, models and effects on contaminant fate and transport. Wat. Res. 25 (5), p. 499-528.

ZHANG, C. & SELINUS, O. (1997) Spatial analyses for copper lead and zinc contents in sediments of the Yantze River Basin. The Science of the Total Environment, p. 251-262.

CHAPTER 10

ASSESSING THE POLLUTION POTENTIAL OF LAKE BASINS: CONCENTRATIONS AND LOADS

Charles Carneiro
Cleverson V. Andreoli
Kelly C. Gutseit
Jonas H. Kondageski
Patricia E. D. Lagos
Hugo H. Okumoto

ASSESSING THE POLLUTION POTENTIAL OF LAKE BASINS: CONCENTRATIONS AND LOADS

Charles Carneiro
Cleverson V. Andreoli
Kelly C. Gutseit
Jonas H. Kondageski
Patricia E. D. Lagos
Hugo H. Okumoto

SUMMARY

Reservoirs experience increases in nutrient levels due to both human actions and natural processes, which play a dominant role in the eutrophication process. The goal of this study was to qualify and quantify the loads that contribute to the Rio Verde Reservoir Araucária, Paraná State, Brazil, and to propose criteria for classifying the potential pollution impact of such loads in tributaries, reservoirs and hydrographic sub-basins. Water samples for physicochemical analysis were collected from 13 tributaries in different sub-basins between 2008 and 2010. The results were used to build a classification matrix (2PAM - Pollution Potential Assessment Matrix) based on the pollution potential. This matrix was based on three different perspectives related to pollution: i) stream water quality; ii) reservoir ecosystem equilibrium; and iii) sub-basin management. Four variables were assessed (total phosphorus, total nitrogen, BOD and COD) and weighted based on concentration, total load and unit-area load. Each parameter was weighted based on the analyzed scenarios and priority levels for management were set. The 2PAM matrix provided a different way to analyze concentrations and loads, enabling managers to prioritize action plans according to the desired use of the basin and this analysis can be replicated in other reservoir basins. From the perspective of river basin management, sub-basins F4 and TE-10 of the Rio Verde Reservoir were classified as having the highest pollution potential and should therefore be prioritized. From the perspective of stream water quality, as well as from the perspective of reservoir ecosystem equilibrium, the tributaries TE10, F1 and Mouth F4 were identified as having the highest pollution potential.

KEYWORDS

Nitrogen, phosphorus, pollution potential, load.

1. INTRODUCTION

All actions and activities performed in a drainage area will directly or indirectly affect water quality: untreated sewage increases the risk of waterborne diseases; industrial and organic effluents reduce dissolved oxygen and generate chemical contamination; erosion increases the turbidity of the water, carries nutrients, and causes siltation; pesticides used in agriculture can be washed down-stream causing problems for aquatic biota and for humans who consume the water; and sewage, even if treated, along with misguided agricultural practices increase trophic levels of water, among others.

World-wide, eutrophication is regarded as one of the five major water-supply related problems. Considering that the pollutants noted above significantly accelerate and enhance eutrophication in reservoirs, it is essential to identify the factors influencing eutrophication in order to support prevention management and decision-making.

The reservoir eutrophication process is a consequence of the relationship among several climatological, hydrological, morphological, physicochemical, and biological factors that occur not only in the drainage basin but also in the reservoir itself. Human activities contribute the most to the acceleration of this process.

Eutrophic environments are characterized as having water with limited transparency, shallow depth and high nutrient levels. There are several factors involved in the process of algal blooms or the proliferation of macrophytes; however, most researchers have focused on assessing the impact of organic matter, silica, nitrogen, and phosphorus on microalgae growth, with phosphorus playing a key role in algal blooms in tropical and subtropical water bodies. The high concentration of nutrients (especially nitrogen and phosphorus) can result in significant primary production of macrophytes and microalgae. In the case of an overpopulation of cyanobacteria, water quality may be seriously affected, especially in water sources used for public water supply.

In Brazil, several reservoirs have experienced serious problems with algal blooms. In the state of Paraná, the cities of Curitiba, Londrina, Pinhão, Foz do Iguaçu, and Ponta Grossa have been dealing with similar problems and recently the Rio Verde Reservoir in Araucária has begun to draw the attention of local managers. The Rio Verde Reservoir showed the first evidence of elevated trophic levels in 2005, when the presence of the potentially toxic cyanobacteria *Cylindrospermopsis raciborskii* was confirmed in densities greater than 90,000 cells/mL. This reservoir is affected by both natural and anthropogenic processes, particularly in relation to rural and urban development. The problems arising from the density of populations around the reservoir include soil sealing, irregular releases of domestic and industrial effluents, and pollution from cemeteries and gas stations. In rural areas, the problems include soil over-use, the intensive use of fertilizers, riparian deforestation, and improper land-use management.

2. RELIABILITY OF WATER QUALITY ASSESSMENTS

This discussion focuses on the importance of reliable data in assessing water quality. Here, the intention is to discuss the reliability of water quality assessments that are performed routinely, from two particular perspectives: i) reliability of sampling and analysis; and ii) the use and interpretation of data.

When evaluating the reliability of water sampling and analysis, it is not uncommon to observe problems in the data. The main question is whether there is a particular problem during the process of data collection or if the result obtained is different from that expected and the explanation is not immediately obvious.

In most cases, operational problems are to blame, such as data collection, analytical procedures, or even data processing. As for the second situation, researchers should always be diligent in the analysis so that new information does not go unnoticed and they should feel free to leave the obvious aside so that important and relevant information may become clear.

Too often we find that there is a problem with the data and then questions arise relating to the initial stages of research, such as data collection. At this stage many questions can be asked, including: In what climatic conditions were the samples collected? Was there adequate preservation of the sample? Was there adequate cleaning of the sampling materials? Was there any interference or contamination? Between the sampling and the analysis, several other steps and activities take place and at this point it is very difficult to determine where the error occurred.

There are two primary types of errors: determined and undetermined. Certain determined errors are caused by failures in the analytical procedure or in the equipment used and can therefore be avoided. When these errors are constant, they may go undetected, sometimes either underestimating or over-estimating the true values (JEFFERY *et al.*, 1992). Therefore, even with a high correlation across several repetitions, the values might not be the real ones. Undetermined errors cause variations above and below the actual values, they are inconsistent and often small and thus are difficult to detect. They may be caused, for example, by problems in the calibration process (ROBINSON, 1990), or systematically during sampling. Thus, we could say that errors accumulate at every step of the process and control over these errors is quite difficult. Statistics can be used to try to identify and minimize uncertainties; however, the best alternative is to manage every step of the process, from beginning to end, with the utmost care and attention. As such, standard procedures and training are crucial.

Another serious problem is the ability of samples to represent the ecosystem being studied. It is also possible that samples do not actually portray the conditions of the area. Sampling for water quality monitoring should also seek to verify the influence of all climatic variables, especially rainfall. However, much of the systematic collections carried out by the government agencies that are required to do so are only performed under stable weather conditions, i.e. with no rain. Studies have shown that water samples collected in the presence and absence of rain are extremely different from each other, varying by more than 1,000% (CARNEIRO, 2008).

Another common problem is the format and reliability of historical data. With regards to the formats used in historical data, information is often presented in different units and scales. Phosphorus, for example, is found in reports as phosphorus, total phosphorus, reactive phosphorus, orthophosphate, total phosphate, or reactive phosphates, etc. There is no problem in making the appropriate conversion; however, the problem lies in determining the reliability that the unit being described is the really one being used. Furthermore, in using data from many years ago information is often systematically presented in one way and in subsequent measurements it is changed. Therefore, the question remains as to whether it is possible to trust and use the necessary conversions or if that there was misunderstanding in the data units employed. This problem happens frequently in studies using long-term data in their analysis. It is therefore necessary to promote not only the standardization of sampling, but also the recovery of missing historical data. This problem is not new, but it persists.

Reliability is a particular problem in long-term databases. Considering all the above-mentioned factors related to sampling procedures and laboratory analysis, the reliability of these data is frequently questioned. Once the technical problems related to water quality have been assessed, it is important to address the interpretation of generated data.

In assessing water quality, there are some standard assumptions (most of which are relevant and true), such as: the presence of ammonium compounds in water indicates the occurrence of a recent pollution episode; high levels of nitric compounds suggests a non-recent pollution episode; the presence of *Escherichia coli* is associated with the presence of sewage; high levels of phosphorus is mainly due to sewage and (or) agricultural and industrial activities; high COD/BOD ratio indicates pollution that is predominantly chemical, etc. However, it is important to note that when it comes to environmental issues, several interrelated and dynamic aspects are inherent and therefore all aspects must be considered in order to minimize misinterpretations. For example, the presence of nitrogen is commonly associated with organic discharges and nitrogen fertilizers. However, the rain water itself may bring small amounts of dissolved ammonia or nitric acid from lightening or even from industrial gas emissions that release nitrogen gases in the atmosphere.

It is important to note that the assessment of water quality requires not only technical knowledge, but also scientific rigor and focused, relevant, and objective research.

3. CONSIDERATIONS RELATED TO LOAD ASSESSMENTS

3.1 POINT SOURCE AND DIFFUSE POLLUTION

Loads that cause surface water contamination can come from various sources. According to Bollmann, Carneiro & Pegorini (2005), the main pollutants that deteriorate water quality include:

- Suspended and dissolved solids: transported to water bodies by urban and rural runoff, increasing water turbidity;
- Organic material: from rural, industrial and urban areas, point source or diffuse loads, reducing the availability of dissolved oxygen for aquatic biota;
- Nutrients: mainly nitrogen and phosphorus, providing a rapid and intense increase in primary production of macrophytes and other associated problems;
- Heavy metals, organic pollutants and other hazardous substances: accumulate in sediments and in the biota;
- Bacteria and viruses: potentially disease causing.

Pollutant loads move into watercourses in namely two ways: point source and nonpoint source (diffuse), contributions. Point source contributions correspond to localized natural or anthropogenic releases. Diffuse contributions are dispersed discharges into the environment that are directly related to meteorological factors, mainly rainfall, which is responsible for much of the solubilization and input of nutrients into watercourses. As such, diffuse contributions are difficult to quantify and control.

In general, industrial releases and discharges, regular or not, are the main cause of point source pollution; on the other hand, non-point source discharges are mainly the result of agricultural activities. Sewage, drainage and runoff of urbanized areas can be characterized in as both, sometimes concentrating in storm drains (point source), or draining into lower areas through runoff (diffuse).

A striking feature of diffuse loads is the fact that the discharge of this load is intermittent and is related to rainfall events. Thus, it is assumed that the diffuse load enters into a water body when the runoff caused by rainfall carries such a load.

Considering that rainwater is the main vehicle of diffuse pollution, it is essential to include weather conditions, especially rainfall, in water assessment. Rainwater promotes the entrainment of particulate and dissolved particles. The intensity of the rainfall, in association with the chemical and physical properties of the soil, moisture, and type of land-use/land-cover, determines the strength and magnitude of the runoff and interflow. The influence of each of these variables is dependent on the climate and the activities in the environment, whether rural or urban.

Due to intense urbanization over recent decades, there has been an increase in soil sealing, leading to significant increases in runoff (Figure 1). Runoff from urban areas carries most of the solid and liquid waste deposited on roofs, streets, sidewalks, promoting a type of "cleansing" of the city, with the runoff eventually reaching water bodies.

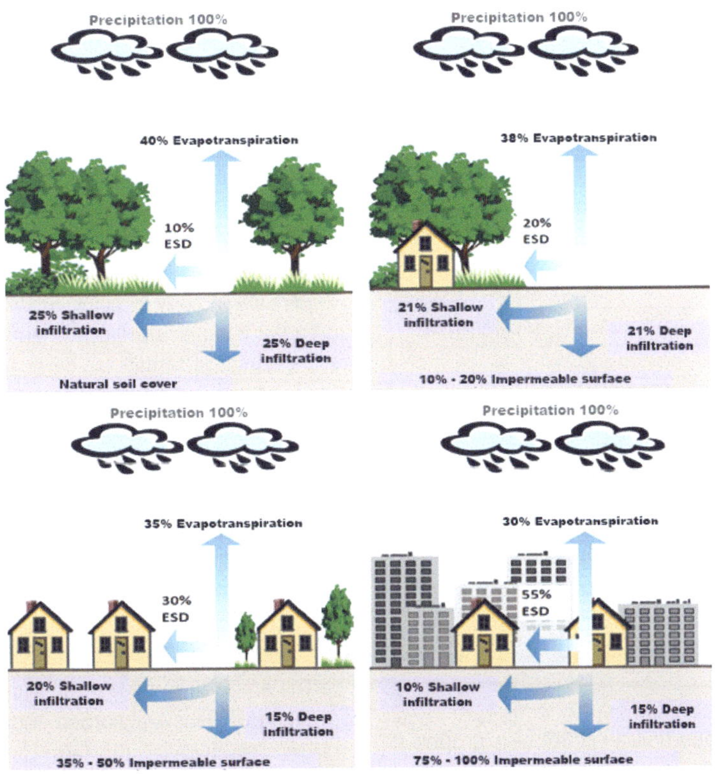

FIGURE 1 - DIAGRAM OF THE EFFECT OF URBAN DEVELOPMENT ON HYDROLOGICAL BEHAVIOR
Note: ESD: direct surface outflow
Source: Adapted from Fisrwg (1998); Zahed Filho & Porto (2010)

Tomaz (2006) points out that at least an average of 25% of the water pollution comes from diffuse sources; however, this number can be significantly higher, even in urban environments where point source loads usually prevail. In domestic wastewater, the concentrations of total phosphorus are usually around 6.5 to 9.0 mg/L and 2.0 to 7.0 mg / L for orthophosphate (METCALF & EDDY, 2003; SPERLING, 1996).

Generally, water flow can be divided into: 1) *runoff*, which represents the flow over the surface of the soil and its multiple channels; 2) *interflow*, which represents the flow that occurs in the sub-surface, i.e., in the soil profile; 3) *groundwater flow*, which is the flow of water into the aquifer. In general, runoff and groundwater flow predominate.

Agricultural areas also produce a significant diffuse nutrient contribution, especially given the large amount of fertilizers used for some crops. In studying water bodies, Cordeiro Netto & Dutra Filho (1981) observed that agricultural activities and dumping of illegal sewage were the major anthropogenic sources of phosphorus. The problem intensifies when there is a lack of management practices and conservation measures. These conditions enhance the effects of water quality degradation. According to studies conducted on north-American catchments, the type of crop may influence the amount of soluble phosphorus that reaches water bodies. By way of comparison, a study by Sharpley & Halvorson (1994) found that in areas with 90% forest cover, the concentration of soluble phosphorus in the surface runoff reached 0.009 mg/L, while in areas with 90% agricultural crops, this concentration reached 0.071 mg/L. Further, Jorgensen (1989) compiled studies on the estimated export of pollution from different geological substrates, considering different land uses (Table 1). In Brazilian soils in general, the natural levels of pollution are considered low. The content of total phosphorus in the soil, including organic and inorganic, ranges from about 0.01 to 0.30%.

TABLE 1 – EXPORTATION COEFICIENTS FOR TOTAL P (mg/m²/year) TAKING INTO ACCOUNT BASIN GEOLOGY

	TYPE OF ROCK	
	IGNEOUS	**SEDIMENTARY**
Forest		
Interval	0.7 - 9	7 - 18
Average	4.7	11.7
Forest + Pasture		
Interval	6 - 12	11 - 37
Average	10.2	23.3
Agriculture		
Citrus	18	
Pasture	15 – 75	
Cultivated areas	22 - 100	

SOURCE: JORGENSEN (1989)

Salas and Martino *apud* Silva (2006) also proposed some Total Phosphorus export coefficients based on land use, as shown in Table 2.

TABLE 2 – EXPORTATION COEFICIENTS OF TOTAL P (mg/m².year) RELATED TO LAND USE

LAND USE	TOTAL P
Urban	100
Agriculture	50
Forest	10

SOURCE: Salas and Martino – apud Silva (2006).

According to Eiger *et al.* (1999), the sources of diffuse loads in rural areas include: agricultural and livestock activities, mining, leisure farms and natural areas. For agricultural activities, the loads depend on the type of crop and the phase of the production cycle (which includes soil preparation, sowing, growth, and harvest), while for production of livestock the loads depend on the type of activity and techniques being used. Mining has a high pollution potential depending on the type of ore and the adoption of appropriate extraction techniques. Rural recreational and leisure properties dispose of sewage and domestic waste generated from household activities, while natural areas can produce diffuse loads, for example, through decomposing organic plant matter carried to waterways through runoff.

Furthermore, Eiger *et al.* (1999) note additional sources of diffuse loads in rural areas, including: domestic sewage dumped directly into rivers; solid waste dumped directly into rivers or the drainage system; waste arising from the decay of pavements; residues from tires, oils and lubricants, etc. Moreover, atmospheric deposition may also be considered as a source of diffuse loads, although usually on a minimal scale.

3.2 NUTRIENTS: QUANTITY X AVAILABILITY

The availability of a given chemical element can be understood as the fraction of the element actually capable of being absorbed or used. The availability of an element depends on many correlated and uncorrelated variables and it is highly complex, especially for elements that occur at trace levels and unstable conditions. Such variables include physical, chemical and biological processes. Water molecules are very effective in solubilizing other molecules; it can stimulate numerous chemical reactions, resulting in compounds of higher or lower toxicity. Among the intervening factors, the following are noteworthy: temperature, types of associated organisms, pH of the water, and oxygen content.

Once in the aquatic ecosystem, chemical elements can be found in different aspects of the ecosystem (water column, sediment, interstitial water) and in different states (organic/inorganic, suspended/settled/dissolved, complexed/isolated, etc.).

Bioavailability is a pharmacokinetic term which refers to the amount and speed at which a chemical element or active ingredient is absorbed and becomes available at the target site; however, this term can be used in the eco-physiology of aquatic environments as the fraction that can be effectively utilized by the biota. Often, a significant amount of a given compound exists in an environment but in a condition of non-biological availability, i.e. not available for direct absorption.

The biological availability of phosphorus in aquatic ecosystems is limited in environments with little human interference with values that are usually below 30µg/L. Although soil phosphorus levels are higher compared to water levels, the majority of the phosphorus is in a condition of low availability.

More than 90% of soil P is in a non-labile form fixed to the clay and is thus not readily available. Research by Pierzinzky *et al.* (2000) showed that 50 to 70% of P exists in an inorganic form in soil but in relatively stable bonds. Of these, an average of 1mg/dm^3 is labile and only 0.03 mg/dm^3 exists as inorganic phosphate in the soil solution, a very minimal amount critical to plant development. These amounts can vary widely depending on a several factors. However, a concentration of 0.03 mg/dm^3 also hinders cyanobacterial blooms in aquatic ecosystems. On the other hand, in addition to soil dissolved phosphorus, adsorbed and absorbed P, as well as aluminosilicates (clays) should be considered, which in most basins are transported to water bodies through erosion, thus representing the major source of phosphorus.

Although the values of orthophosphates are minimal, they can be quite significant considering that these forms are readily absorbed by microalgae. The soil, rocks and mineral loads are diffuse, usually minimal, and difficult to control; nevertheless, they can have a significant impact on the trophic states of aquatic ecosystems.

3.3 FLOW EVALUATION

In order to estimate the contribution of the organic and inorganic load, it is necessary to know the hydrological regime of the tributaries, as well as variations in flow. In general, the relationship between the concentration of dissolved materials and the flow is inverse, except in the case of the so-called "*valetão* effect" (Portuguese term) which refers to the wetting front of a rainfall event, carrying solid and liquid materials accumulated in ducts, channels or even in the soil, normally with a high pollutant concentration. This effect is more pronounced when there is significant soil sealing.

The rate of river flow is the volume of water that runs through a cross section of the river over a period of time (SANTOS *et al.*, 2001). There are different ways to measure flow: 1) the conventional method of measuring and integrating the distribution of velocity; 2) the acoustic method; 3) the volumetric method; 4) the chemical method; 5) the use of regular geometric devices; and 6) measurement using floats.

The conventional method uses measurements of the distribution of velocity in a cross-section and requires the calculation of the area of the section and the average velocity of the flow passing through the section. The cross-sectional area is calculated by measuring the width and depth of the river at a number of points (verticals) along the section. In order to obtain the total flow, the midsection method must be used, which involves multiplying the mean velocity in each vertical section by the area. The area is calculated as the depth of the vertical section by the sum of the midsections to the adjacent verticals. The sum of these partial flows provides the full flow of the section. Through a series of measurements of flow, an elevation-discharge ratio is defined covering a range of water elevations observed at the station, i.e. the key-curve or discharge curve (SANTOS *et al.* 2001; UFRS, 2007)

3.4 CONCENTRATIONS X LOADS

Considering the spatial area, the chemical composition of river water varies widely depending on lithology, vegetation, land-use, and the presence of point source or diffuse loads (SANTOS *et al.*, 2001). To a significant extent, the amount of suspended and dissolved particles depends on the relative contribution of runoff to the river flow, which fluctuates across space and time.

Several physical, chemical, biological and hydrological variables influence the dynamics and nutrient concentrations in aquatic ecosystems. With some difficulty it is possible to quantify the influence of most of these variables through a mass balance. This enables better control of water quality and assists in decision-making. Commonly, dissolved oxygen, salinity, hydrogen ion potential, alkalinity, carbon, nutrient content (especially compounds of nitrogen and phosphorus), heavy metals and solids are the parameters used for monitoring lakes and reservoirs (BOLLMANN, CARNEIRO & PEGORINI, 2005).

In analyzing water quality, especially in lentic environments, it is essential to consider two aspects: concentration and load. Other variables are relevant when examining loads, such as: flow rate, total load, and specific area load. Conducting the analysis under different weather conditions will better reflect actual conditions of land-use in the basin and thus provide more reliable technical information for the development of preventive and mitigation strategies.

Although Brazilian legal regulations (CONAMA 357/2005) identifies concentration as the measurement to be used, load levels may have a greater impact on trophic levels, especially if there is a dam or long residence time. Therefore, concentration levels tend to have a greater significance in intermittent evaluations in lotic environments, while loads tend to be more significant in lentic environments. However, both measurements will be considered.

3.5 FLOW HYDROGRAPH

River discharge is the result of the interaction of precipitation in the basin with the characteristics of the basin itself, which in turn influences infiltration, storage and evapotranspiration.

Typically, the flow can be divided into two parts: surface flow and groundwater flow, as shown in Figure 2.

When rainfall occurs in the basin, the initial rainfall that reaches the soil infiltrates completely, thus feeding the underground reservoir. If rainfall is prolonged to the point of soil saturation, the water no longer infiltrates but begins to flow over the ground to lower elevations; this process is known as runoff. This water flow is an important event within a basin, especially when considering the contribution of diffuse loads to the water body as runoff is responsible for transporting solid and dissolved material to water bodies.

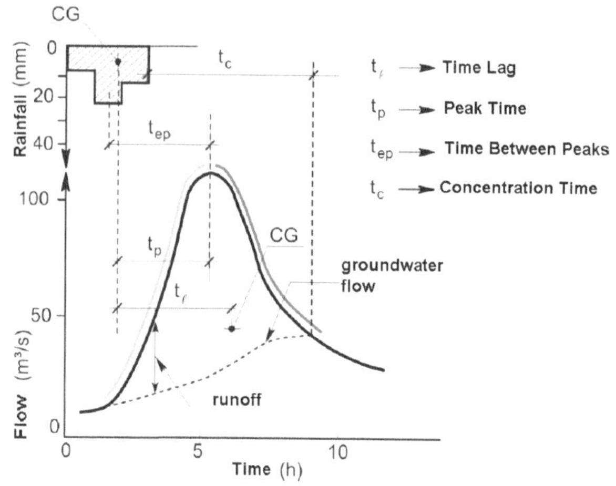

FIGURE 2 – HYDROGRAM AND RESPONSE TO RAINFALL
SOURCE: COLLISCHONN & TASSI, 2010

On the other hand, after a long period without rain, rivers are 'loaded' slowly and gradually by water stored in the underground reservoir. This slow flow, which occurs during drought, can be called groundwater flow, or base flow.

When assessing the contribution of pollutant load to a water body, it is important to distinguish if the river flow is only due to base flow or if there is pollutant contribution through runoff. This information allows for important conclusions to be made about the water body. For example, if any pollutant load is detected in a water body during a period which mostly consists of base flow, one may conclude that there is a high probability that the load cannot be attributed to diffuse sources, since during flow recession, no runoff caries any load to the water body. On the other hand, the intensity of the input of loads can be related directly to the magnitude of the runoff.

Thus, by understanding the hydrograph of the water body to be studied, it is possible to separate runoff periods, when potential diffuse loads to the river occur. During periods of mostly groundwater flow, the pollutant load, if existent, should be point source. After separating periods of surface and groundwater flow, which may correspond to drought and floods, the loads monitored in each are distinguished. The loads are divided into two groups: base load, which is the load monitored in recession flow; and diffuse load, understood as the load associated with rainfall.

4. EVALUATING WATER QUALITY VARIABLES

All activities that interfere with hydrologic processes in the catchment will be reflected in water quality. The socioeconomic situation of communities within the basin, the geographical position of the water body, geomorphology of the basin, anthropogenic activities, and land-use, among other factors, must be considered simultaneously in evaluating water quality. The interaction of both internal and external factors has an impact on water quality, thus the monitoring plan and evaluation of water quality should identify all of the activities that interfere with the hydrologic process.

While several parameters have been used as indicators of water quality, some specific parameters can enable a better diagnosis and these often depend on type and purpose of the evaluation. In some cases, more specific and specialized analyses might also be necessary. Nevertheless, commonly used basic physical and chemical parameters provide relevant and consistent information on water quality assessment and these parameters are often used to guide basin management and water monitoring plans.

The assessment of the Rio Verde reservoir basin is presented below as a case study which aims to provide greater insight into this analytical approach. The parameters evaluated in this study were determined based on current Brazilian legislation, CONAMA Resolution n° 357/2005 for fresh water. The parameters selected for the analysis include those that provide the best description of water quality, especially those used to determine trophic states, as well as parameters that reflect ecosystem dynamics and occasional contaminants. This study aimed to evaluate the water quality of the tributaries and the influence of land-use within the sub-basins on the reservoir.

4.1 METHODS AND PROCEDURES

4.1.1 Selection of Sampling Stations

The definition of sampling stations is crucial in evaluating water quality. The criteria for defining sampling station must consider the location of buildings and urban areas, different uses of sub-basins and their potential impact, optimization of sampling logistics, and budget.

In the Rio Verde Reservoir Basin, 13 sampling stations were established in the direct tributaries: five areas located in tributaries on the right bank of the reservoir, eight areas located in tributaries on the left bank. Two sampling areas were placed in indirect tributaries and five sampling areas were located on the Rio Verde River, the main tributary of the reservoir, four upstream and one downstream of the reservoir (after the spillway), totaling 20 sampling areas.

4.1.2 Monitoring Period

The time frame for evaluating various parameters should vary according to the program objectives, logistics and characteristics of the study area. Considering that it is necessary to identify and characterize the various point source and diffuse loads, sampling should be carried out and organized so that it includes different weather conditions, from periods of drought to intense rainfall.

Sampling occurred between July 2008 and February 2009 and between March and December 2010. Samples were taken during critical flow periods, as well as periods of intense and prolonged rainfall. Samples taken during rainfall allowed for the assessment of diffuse loads, although none of the analyses specifically evaluated the wetting front (an important consideration for future research studies).

Monthly sampling was taken following the ABNT/NBR 9897 recommendations (ABNT, 1987) and in order to address the needs of hydrodynamic modeling projects, sampling was extended for a few more months.

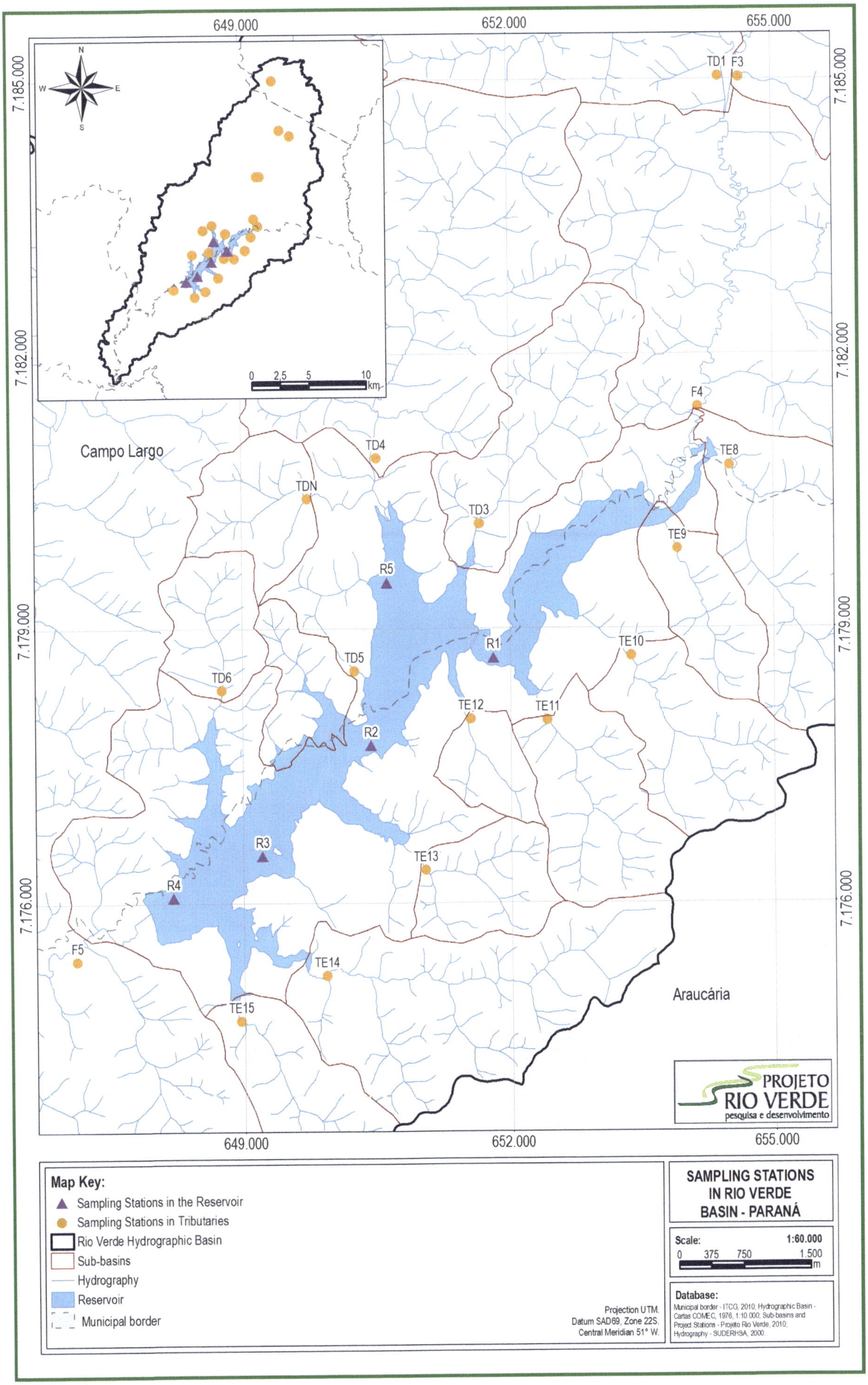

FIGURE 3 – LOCATION OF SAMPLING STATIONS IN THE RIO VERDE BASIN, ARAUCÁRIA, PR

4.1.3 Selection of variables and analytical control

There are a large number of parameters related to water quality that can be used in monitoring programs. As noted above, the selection of the variables should depend directly on the objective of the study; however, other factors usually influence sampling area selection, such as laboratory availability, funding, the physical features of the basin, accessibility, etc. At the very least, the selected variables must show the potential interferences in the aquatic ecosystem in terms of basin use and possible sources of pollution.

Data was collected for the following parameters during the monitoring conducted at the Rio Verde Reservoir: water temperature, electrical conductivity, COD, BOD, total-P, reactive-P (orthophosphate), nitrite, nitrate, pH, total solids, fixed solids, volatile solids, total coliform, and *Escherichia coli* density. Conductivity and water temperature were measured *in situ* with the aid of a Lutron CD-4303® conductivity meter. The flow rate was measured at all points simultaneously during sample collection with the aid of a gaging station (Newton) coupled to a counter (Hengstler, model Tico 731) (Figure 4). Precipitation and air temperature were obtained from meteorological station installed in the Rio Verde Reservoir and from the Bom Jesus meteorological station (Institute of Water - Instituto das Águas).

The methods used for sampling, preserving and transporting samples, as well as laboratory procedures, followed the recommendations of the Standard Methods for the Examination of Water and Wastewater (APHA, 2005). Samples were collected with the aid of a bucket or the immersion of a vial to assess the concentration of nutrients in the water at depths of 0-20cm. They were then placed in coolers, protected from light and heat and sent to the laboratory for analysis. The maximum time between sampling and transportation to the laboratory was about eight hours. Samples were collected from all areas on the same day.

The analysis of the BOD, COD and Total Solids in this case study only considers the samples taken during the period from 2008 to 2009.

4.2 DISCUSSION OF THE MONITORING RESULTS

4.2.1 Rainfall

There are numerous climatological variables that can interfere with the nutrient budget (inputs and outputs): rainfall affects turbidity and trophic levels as the result of the solubilization and transportation of substances; temperature influences evaporation rates, dynamics of plankton communities, and water profiles; whereas, wind affects the stratification of the water column, re-suspension of nutrients, and reproductive processes, particularly in reservoirs. However, we emphasize that it is impossible to analyze concentrations and loads without considering the effect of rainfall. Rainfall is the main vehicle driving diffuse contributions and in most cases it is responsible for the largest pollutant load, especially from urbanized areas and poorly managed agricultural activities.

Some basic information regarding rainfall quantity and intensity is important taking into account water quality assessments, including: data covering as long a time range as possible; reliable data (the meteorological station must be located in close proximity to the study area); seasonal information (rainy/dry seasons); and when and for how long it rained before sampling (daily data for a minimum of 10 days prior to sampling). Another very important aspect is to understand the intensity of the rainfall (mm/h), which represents the quantity of rain over a period time: 20 mm of rainfall in one hour is quite different from 20mm in five hours. Rainfall intensity, therefore, has a significant impact on the form and conditions of entrainment and solubilization.

Figure 5 shows the rates of rainfall during the study period. These values were applied to the Rio Verde Reservoir water quality analysis; especially those related to total-P, reactive-P, ammonia-N and nitrate-N assessments collected during 2010.

As shown in Figure 5, and as suggested above, it is important to check the rainfall conditions before analyzing water quality. To avoid misinterpretations, each observed value should be related to weather conditions, especially

FIGURE 4 – MEASUREMENT OF FLOW WITH AND WITHOUT USE OF BOAT

when considering the overestimation and underestimation of data. Over a sequence of 10 sampling events, rainfall occurred on the days when samples 1, 4 and 8 were taken, with rainfall between 25 and 35mm; it rained 62 mm one day before sample 9 was taken. However, the wetting front was not sampled. Sampling occurred during rainfall events.

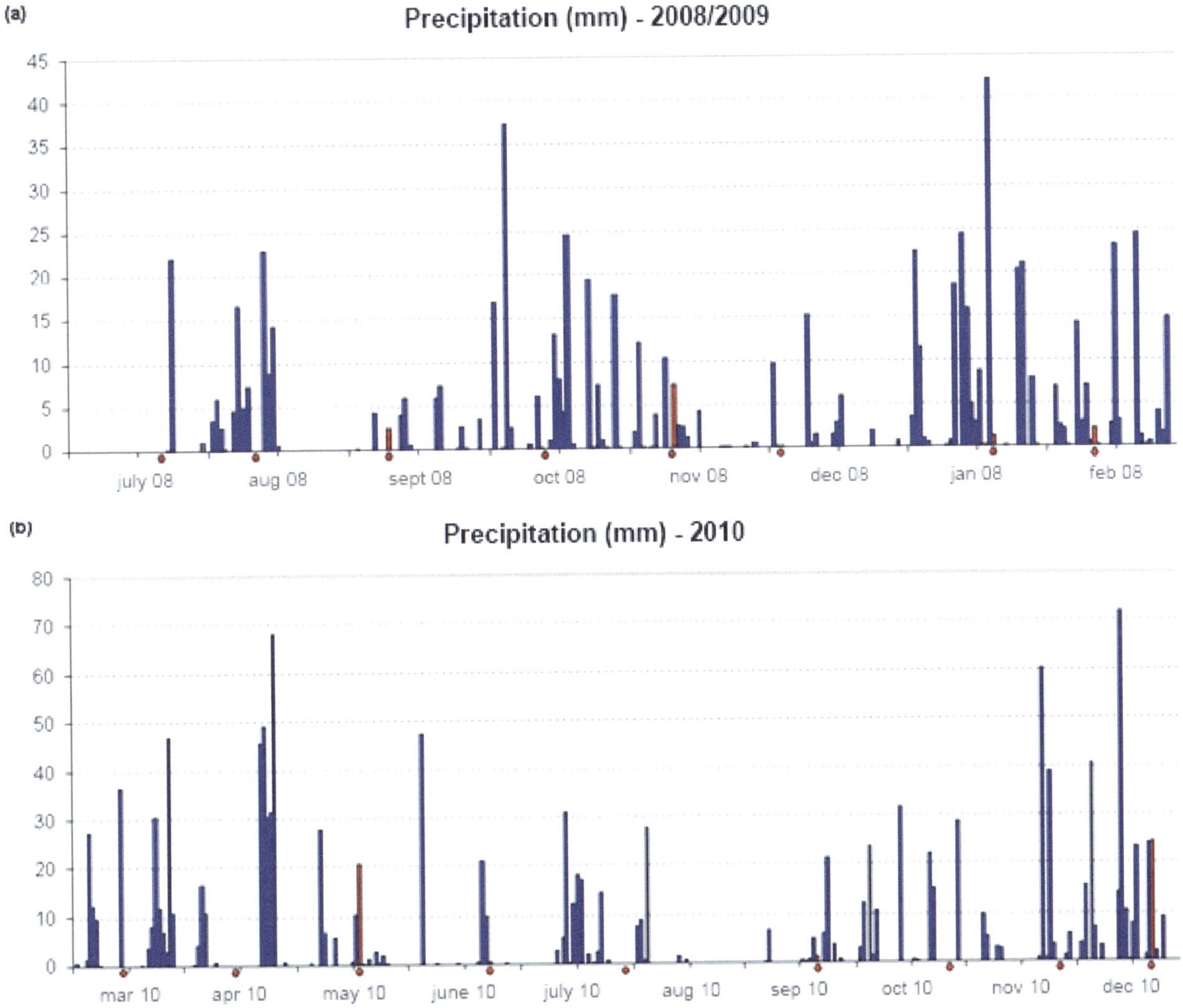

FIGURE 5 – RAINFALL DURING (a) 2008/2009 AND (b) 2010, RIO VERDE RESERVOIR BASIN
NOTE: Red dots (●) represent sampling days.

4.2.2 Nutrients

4.2.2.1 Phosphorus - concentration and load

Figure 6 shows values for total-P and reactive-P in four main sampling areas: F4, F5, TD4 and TE10; F4 and F5 are the input and output of the reservoir, respectively; TD4 is an important tributary on the right bank; and TE10 another important tributary on the left bank. The sequence represents 300 days of monitoring over the period from March to December 2010. Total-P box plots, with percentiles of 25, 50, 75 and 100%, are shown in Figure 7 for the same period.

As discussed above in "Concentrations and Loads", both aspects should be taken into consideration. An integrated analysis is performed herein to better evaluate the potential impact of each sub-basin. Initially we discuss the concentration.

The highest total-P content was lower than 50μg/L, the value recommended by Brazilian legislation CONAMA 357/2005, considering a direct tributary under in a lentic environment and a class II river (see Figure 6). Considering the median (50% percentile), which means the mid-point of the data with 50% of measurements above and 50% below, all sample areas showed median values below the maximum allowed (VMP) of 50μg/L. Except for the results from TE10 sub-basin, even the 75% quartiles presented values lower than the maximum permitted (see Figure7). In the majority of cases, high total-P values were associated with high levels of reactive-P or orthophosphate (correlation r = 0.55), which is the form of phosphate readily available to phytoplankton (Figure 6).

Among the values of total-P that exceeded 50μg/ L, TE10 presented three instances out of 10 samples, while F5 was the only area that did not present any value exceeding this limit. These results suggest that the reservoir is retaining phosphorus; even without the occurrence of large contributions over time, these loads can be significant. Obviously, the values above the recommended limit can result in algae blooms. However, these amounts indicate that the reservoir is intermittently receiving a P load higher than it should. Studies carried out by Vollenweider in the 1960s indicated the occurrence of cyanobacteria even in extremely low concentrations of phosphorus at 18µg/L (VOLLENWEIDER, 1968). It is possible that the natural geopedology-cal load from the catchment area is enough to promote or maintain these phosphorus levels in the water. According to the discussion presented in Chapter 5, the labile-P val-

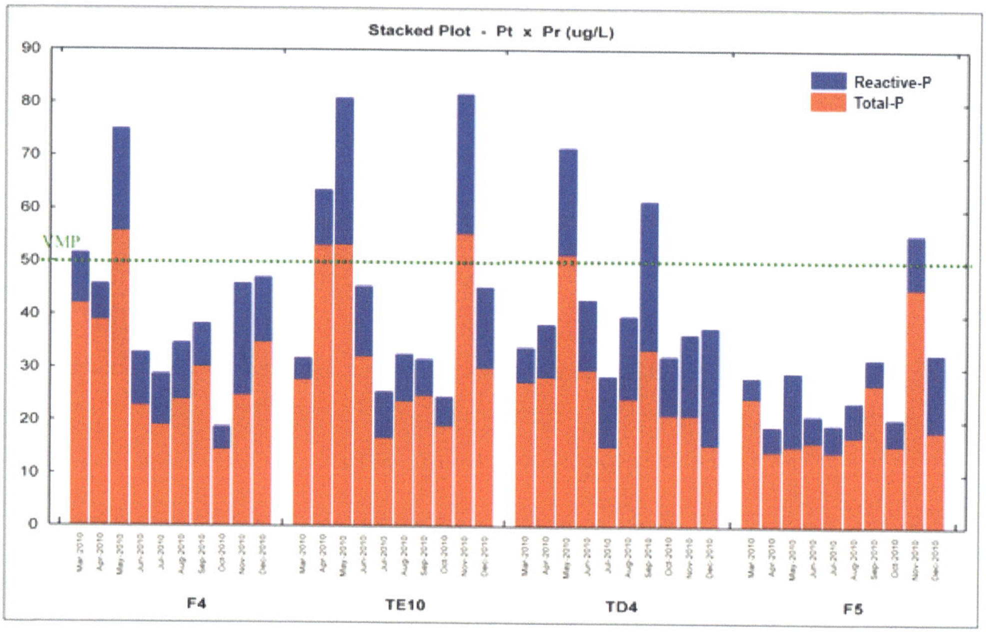

FIGURE 6 – CONCENTRATIONS OF TOTAL-P AND REACTIVE-P IN 4 SAMPLING AREAS AS A FUNCTION OF TIME (MAR. – DEC. 2010), RIO VERDE RESERVOIR

NOTE: 1 - VMP - maximum allowable value according to CONAMA Resolution 357/2005 - 50µg/L for direct tributary in lentic environment and class II rivers.

Values for reactive-P (in blue) are not cumulative on the values of total-P (in red)

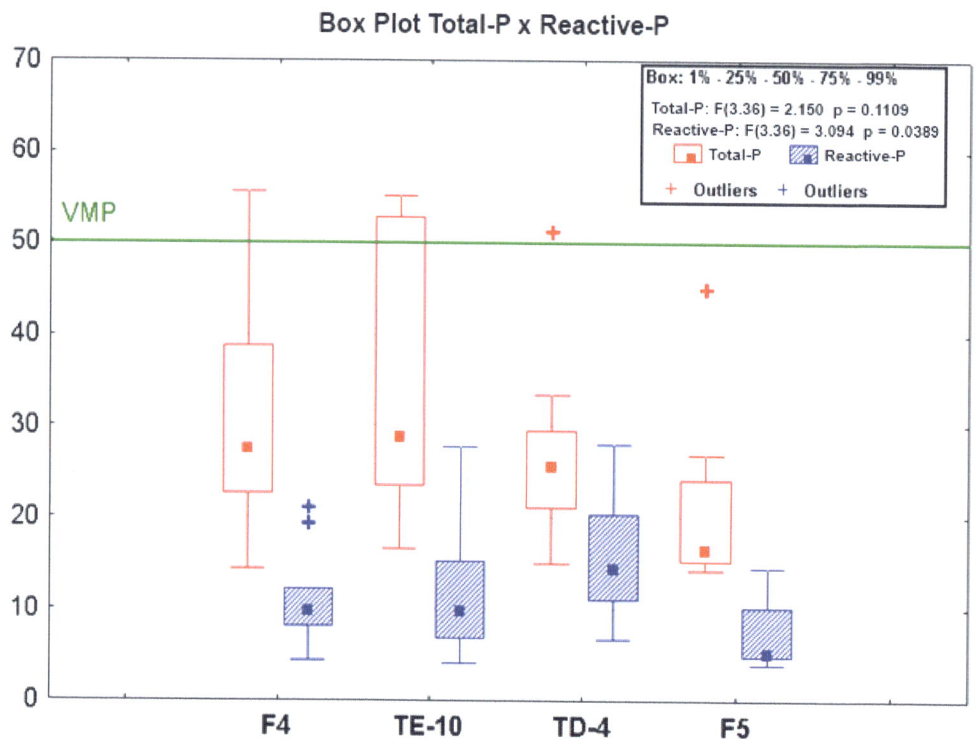

FIGURE 7 – PERCENTILES (25, 50, 75, 100%) FOR TOTAL-P IN 4 SAMPLING AREAS AS A FUNCTION OF TIME (10 SAMPLES: MAR. – DEC. 2010), RIO VERDE RESERVOIR

NOTE: VMP - maximum allowable value according to Resolution 357/2005 - 50µg/L in a direct tributary in a lentic environment, and a class II river.

ues in the soil are quite high which is likely due to the use of phosphorus-based fertilizers.

The historically reported phosphorus levels for the reservoir have been relatively low. According to the Environmental Institute (IAP), between the years of 1999 and 2008 (IAP 2004 and IAP 2009), the mean value of total-P was 25.2 µg/L, ranging between 4 to 110 µg/L and more than 75% of values (third quartile) were below average. The low levels recorded by the IAP are consistent with the low levels presented herein. However, in more recent samples (2008/2009) the mean values for total-P have substantially increased: the mean value from the surface water was 46 µg/L, ranging from 10 to 118µg/L. In the deepest waters the mean was 48 µg/L varying from 10 to 120 µg/L. Indeed, further investigation is required to better understand these differences.

In the present condition of the Rio Verde reservoir, cyanobacteria have not been found in significant numbers. It appears that another factor related to P input is limiting the cyanobacteria growth, such as the ratio of P:N. Other neighboring reservoirs under similar weather conditions, such as Lake Alagados in the city of Ponta Grossa, frequently have had cyanobacteria blooms even under mesotrophic-eutrophic conditions (mean range 30 - 70µg/L, SANEPAR unpublished data).

The interquartile range, or the variability between the values for the interval between the lower quartile (25%) and the upper quartile (75%), varied greatly between the sampling areas and was more pronounced at point TE10. As for the other sampling areas, they showed high content "stability" or less variability (see Fig. 7). The results taken during sample 3, in May 2010, were the main results responsible for the increase in the interquartile range in the sampling areas, with the exception of point F5, which is located downstream and thus being settled by the dam. Only a few other samples showed high values. Another important aspect of the data collection is that samples taken to assess phosphorus concentrations occurred during rain, including samples 1, 4 and 7 (taken in March, June and September 2010, respectively), as well as sample 9 (November 2010) which was preceded the day before by 62 mm rain; however, the samples were not significantly influenced by precipitation. This suggests that either these sub-basins have excellent conditions of preservation and protection, or that the runoff occurred exclusively within the wetting front, an aspect that was not addressed in this study. Point TE10 was the only area in which the 62 mm rainfall had any impact on concentration levels, despite the proximity of the sampling areas and the intensity of the rain. However, this fluctuation occurred only for P and it was not perceived for the studied forms of N. On the other hand, sample 3 (May 2010) presented high levels of phosphorus even in dry conditions (with no rain occurring in the five days prior to sampling); therefore, the contribution of phosphorus through diffuse loads can be excluded. The same reasoning applies for point source contributions, as the tributaries of different sub-basins showed similar increases in levels of total-P. Thus, further investigation should be conducted in order to identify the cause of this fluctuation. As for the increases in phosphorus levels without the occurrence of rainfall, they could be explained by point source contributions, for example, through discharges from fishing or manure.

It is important to carefully evaluate the land-use occurring in sub-basins in order to identify the source of these consistently higher values. In basins occupied by humans and with poor agricultural management, it is common to find that the wetting front carries high amounts of nutrients, varying according to rural or urban land-use. In a study carried out by Carneiro (2008), who evaluated the influence of rainfall on river water quality, the author found that even with a minimal amount of rainfall (4mm), well managed agricultural and farming areas had no significant impact on water quality, while urban areas increased total P up to 35%. With a medium rainfall (16 mm), the total-P increase was about 1,000%; the diffuse load represented almost 95% of pollution in that water body. A peak of 3,000 µg/L was observed soon after the increased flow of the river, demonstrating the so-called "valetão effect" (described above), which is typical in urban areas, resulting in P-load increases of over 2,000%. Therefore, studies must take into account the effect of rainfall on the sub-basins in order to establish a suitable nutrient budget. In the water quality assessment in the Rio Verde basin, the influence of rainfall on P levels was observed. This becomes even more important when the load is also considered;the load considers the aspect of flow, which is strongly influenced by rainfall. It should be emphasized that, at first, rainfall events vary according to quantity and intensity and tend to enhance nutrient concentrations due to runoff. Subsequently, it is possible that dilution occurs due to the increased flow and the entrainment of the main load.

For the four tributaries shown in Figure 6, there was no significant variation between the sub-basins. All sampling areas presented significant variation at specific moments. However, there are several sub-basins with low total-P values, such as the sub-basins TD3, TD6, TE9 and TE15, suggesting that they are the most preserved areas.

Considering the equilibrium of a given aquatic ecosystem, it is insufficient to only analyze nutrient concentrations. It is also necessary to technically assess the load (concentration x flow) and consider total load (mass/time) and relative load that relates to discharge across the catchment area (mass/time/area).

Figure 8 shows total-P and reactive-P values for four sampling stations: the Rio Verde River (F4), the main tributary of the Rio Verde Reservoir, was obviously the main source of total-P load to the reservoir (85% of the total load) due to high flow levels (an average of 5.1 kg/day in the study period and an amplitude between 1.8 kg/day and 10.3 kg/day). Similar to the results found for the concentration, the highest load value (10.3 observed in the sample taken March 2010) was not caused by precipitation.

F4 represents the last point downstream in the main Rio Verde channel and F3, F2 and F1, are all located upstream. By analyzing these points, one can observe that the section F4 "Mouth" presented the highest total load values (68%, 5,1 kg/day) for total-P. The section "Mouth F4" represents the section between points F3 and F4, downstream of the highway BR 277. The F3 section was also somewhat significant with load values of 15%. Besides F4, the load of the TE10 tributary was also significant, accounting for approximately 6.4% of the total load input into the reservoir. The other tributaries, under conditions of normal rainfall have low flow levels, resulting in low values of total load (Figure 9-I). Currently, the total P discharge of Rio Verde sub-basins is around 6 kg/day.

Considering the specific load (load/day/km²), the sub-basin with the greatest pollution potential was TE10 (43.7%) and then F4 (23.5%), with the F4 and F3 "Imouth" as the largest contributor. The water quality was considered good in five out of the 13 evaluated samples: TE7 (section F4), TD5, TD3, TD6, TE15 and TE13 presented very low load values <2.4% (Figure 9-II). Therefore, considering both total and relative load, TE10 and the F4 "Imouth" require further attention, where basin recovery actions should be a priority.

Figure 9-III shows a simple balance of inputs and outputs of total-P in the reservoir allowing us to verify that the amount that enters the reservoir is much higher (89%) than the amount that exits the reservoir. This provides real

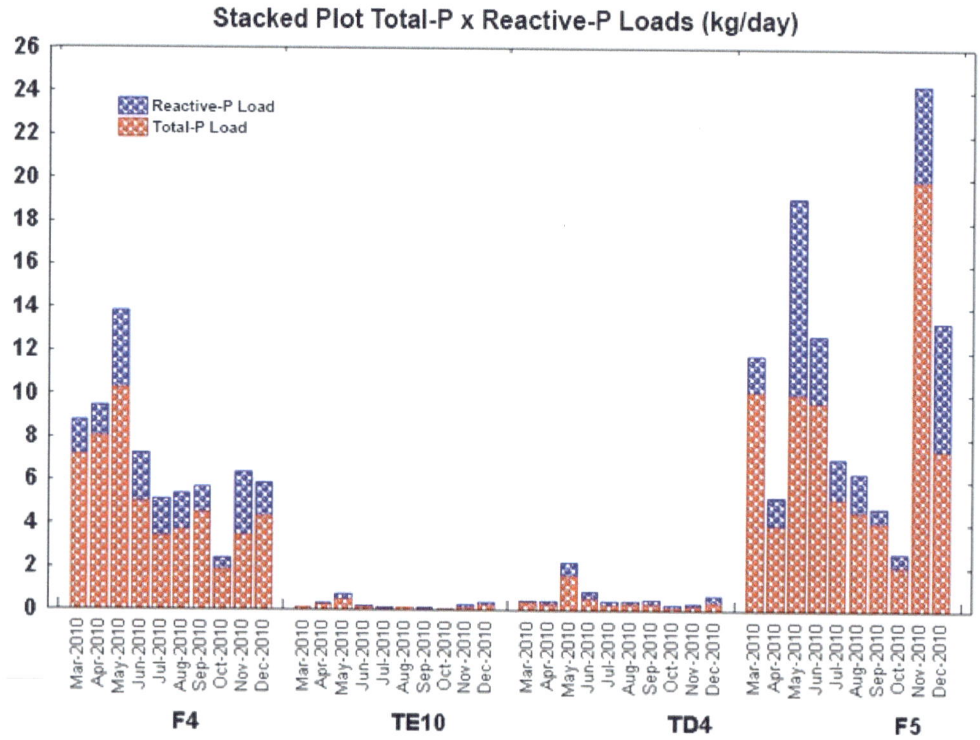

FIGURE 8 – TOTAL-P AND REACTIVE P- LOADS FOR FOUR SAMPLING STATIONS (MAR. – DEC. 2010), RIO VERDE RESERVOIR

evidence that the reservoir functions properly as a physical barrier to the output of phosphorus.

4.2.2.2 Nitrogen: concentration and load

The main nitrogen forms commonly assessed in relation to water quality are: Total-N, N-NH_4^+,-NO_2^-N, N-NO_3^-, and organic-N. Many other forms of nitrogen can be found from both natural and non-natural sources in aquatic environments; however, organic forms of nitrogen usually predominate and occur as numerous different structures.

Water commonly presents a pH between 6 and 8. Under more acidic conditions, the amino, amide and ammoniac forms of nitrogen are prevalent. In more alkaline conditions, there is a rise in the nitrification process. On the other hand, ammoniac forms are the first substances produced during the decomposition of organic materials and nitrification is the subsequent process. In some aquatic environments, nitric forms of nitrogen can be generated by deammonification (ESTEVES, 1998). NO_2^- compounds are very unstable; therefore, they generally occur in minute quantities. On the other hand, the gaseous forms of NO, N_2O, N_2 and particularly NH_3, can eventually accumulate in significant concentrations as a result of these processes. The denitrification process that occurs under anaerobic conditions in some organisms (particularly *Pseudomonas*, *Bacillus*, and *Paracoccus*) has the ability to obtain O_2 from NO_2^- and NO_3^-, and generates most of these nitrogen gases. The pH also has a direct influence on the formation of NH_3.

Ammonium forms are often used as indicators of recent pollution events in aquatic ecosystems and are largely influenced by temperature, pH and oxygen levels in the environment. As for nitric forms, which leach easily into soils, they can accumulate in water bodies altering normal concentration levels and can endanger human health. Therefore, nitric forms do not always represent earlier pollution episodes as they often erode from soils and are leached into water bodies from surrounding areas, particularly in areas of intensive agriculture.

More information about the dynamics of nitrogen compounds in water can be found in Mccarthy (1975),

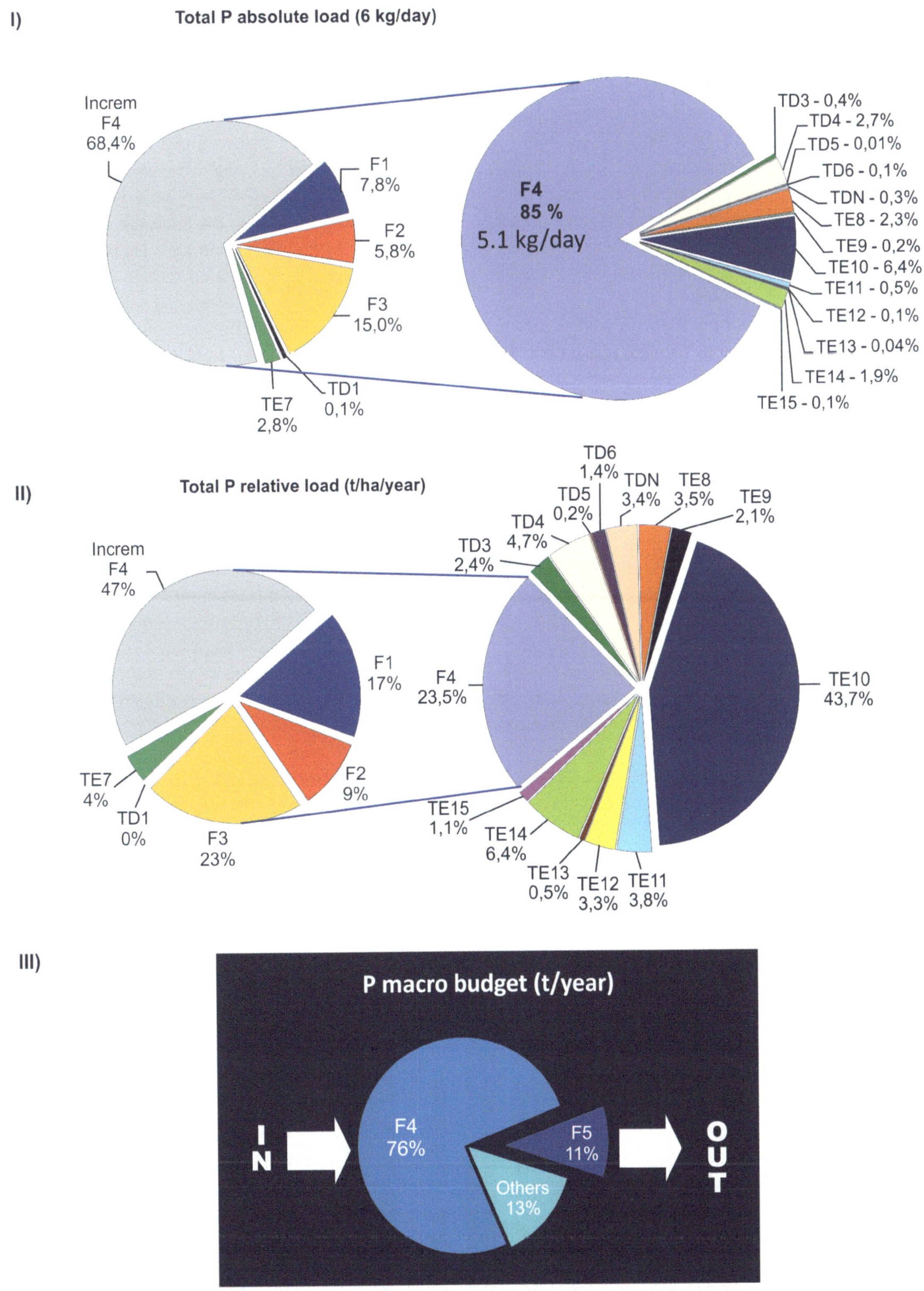

FIGURA 9 – RELATION BETWEEN TRIBUTARIES AND TOTAL-P LOADS IN THE RIO VERDE RESERVOIR: I) TOTAL LOADS OF TOTAL-P, II) RELATIVE LOADS OF TOTAL-P, AND III) TOTAL-P MACRO-BUDGET

Snoeyink & Jenkins (1980), Esteves (1998), Pankow (1997).

Figure 10 shows the observed values of ammonium-N, nitrate-N and nitrite-N at four monitoring areas (the same as those used in the phosphorus analysis): F4, TD4, TE10 and F5; F4 and F5 are located at the entry and exit of the reservoir, respectively; stations TD4 (right bank) and TE10 (left bank) are located in two of the basin's direct tributaries. Ten samples were taken at each sampling point over 300 days between the months of March and December 2010. As mentioned above, the samples were taken in a variety of climatic conditions (Figure 5). Figure 11 shows the box plot dispersion of ammonium-N, nitrate-N and nitrite-N from March to December, 2010, with percentiles of 25, 50, 75 and 100%.

It is worth highlighting once again the importance of the integrated analysis of both "Concentration and Loads", even if these aspects are treated separately.

For ammonium-N, all values, even the outliers, were well below the maximum levels recommended by CONAMA 357/05 of 3.7 mg/L, for class II rivers and pH lower than 7.5. The ammonium-N values were between 0.05 and 0.5 mg/L (with an average of 0.29 mg/L, about 10 times less than the recommended limit), while pH values were between 6.1 and 7.5 (average of 7.0), suggesting that these environments are well preserved in relation to nitrogen compounds, especially from recent organic loads (Figures 10 and 11).

The observed values for nitrate-N, as well as for ammonium-N, were also well below the limits recommended by CONAMA 357/05 of 10mg/L. The values ranged between 0.05 and 0.45 mg/L, with an average of 0.28 mg/L (about 35 times less than the recommended limit). The nitrate-N concentrations were also very low and similar to those observed for ammonium-N. These levels suggest that there has been no significant recent organic contribution to the reservoir. There appears to be low-level, regular inputs of nitrogen compounds in which the nitrification rate is similar to the new organic nitrogen input. It is also possible that the source of both variables are minerals, since the catchment areas of these basins are heavily influenced by agricultural fertilizers, especially urea. Nitrification is a process that occurs naturally in the soil under aerobic conditions and both nitrate-N and ammonium-N are easily leachable, especially nitrate. Therefore, these nitrogen compounds can easily reach water bodies by adsorption to mineral clays, solubilization through soils, or percolation by rainwater. The use of nitrogen based fertilizers around the reservoir can have this impact even without industrial or sewage inputs. Even with proper management and conservation practices, there is a detectable presence of these ions in adjacent waters.

As with the other forms of nitrogen, the values for nitrite-N were well below those required by CONAMA 357/05 with a limit of 1 mg/L. The values for nitrite-N were between 0 and 0.02 mg/L, with an average of 0.01 mg/L, about 100 times less than the limits allowed by law.

Considering the median, or 50% percentile (the central positioning of the data), as well as the upper quartiles and outliers, all sampling areas had values below the MAV of 50 µg/L. The variability between quartiles for nitrite-N was negligible. As for ammonium-N, the results showed a wide range across all sampling areas, even at point F5, the exit point from the reservoir. Point F5 was the only sample area that showed variability for nitrate-N but the variability was significant (F test of $p = 0.230$). Ammonium-N, with an F test result of $p = 0.948$, had significant variability (Figure 7).

Similar to phosphorus, there was no direct influence of rainfall on the concentrations of the nitrogen compounds. None of the samples produced values of total ammonium-N, nitrate-N and nitrite-N greater than 1 mg/L. But if we consider the organic portion of nitrogen, total-N values would certainly be between 1 and 1.5 mg/L, which are still considered average values. Although the values for ammonium-N and nitrate-N were much lower than the legally recommended levels, it is important to analyze the results over time in order to better understand the inputs and outputs of the basin and provide supplementary information to the phosphorus analysis.

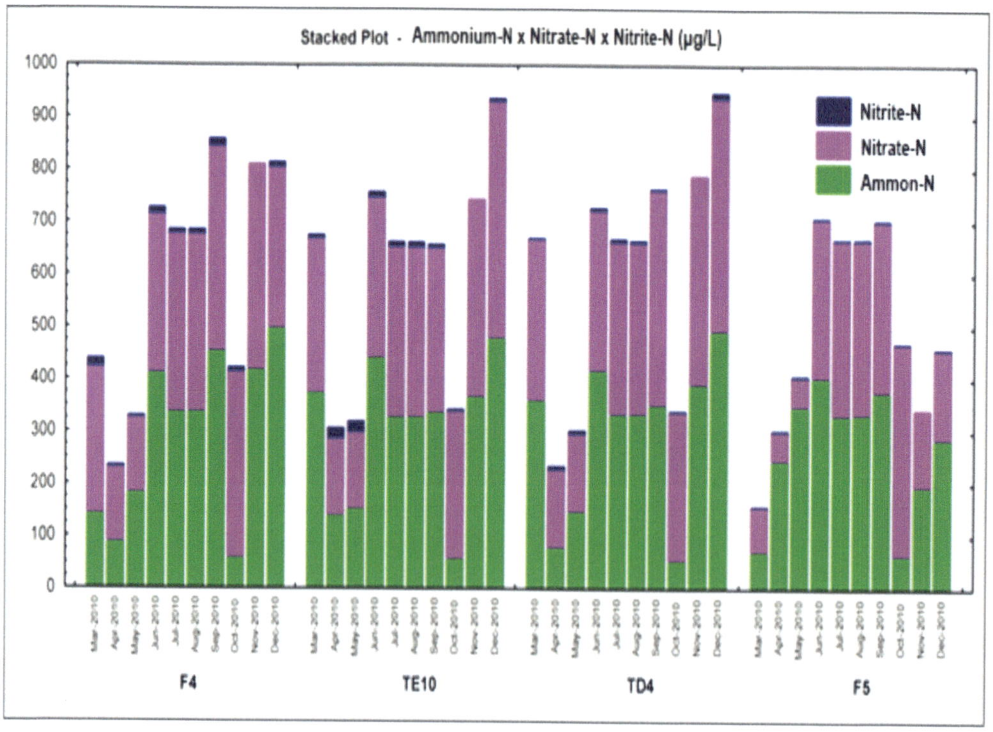

FIGURE 10 – CONCENTRATION FOR AMMONIUM-N, NITRATE-N AND NITRITE-N (IN µg/L) FROM 4 SAMPLING AREAS (MAR. – DEC. 2010), RIO VERDE RESERVOIR

NOTE: Limits CONAMA 357/2005: 3700µg/L for ammonium-N (pH 7.5); 10,000 µg/L for nitrate-N; 1000 µg/L for nitrite-N.

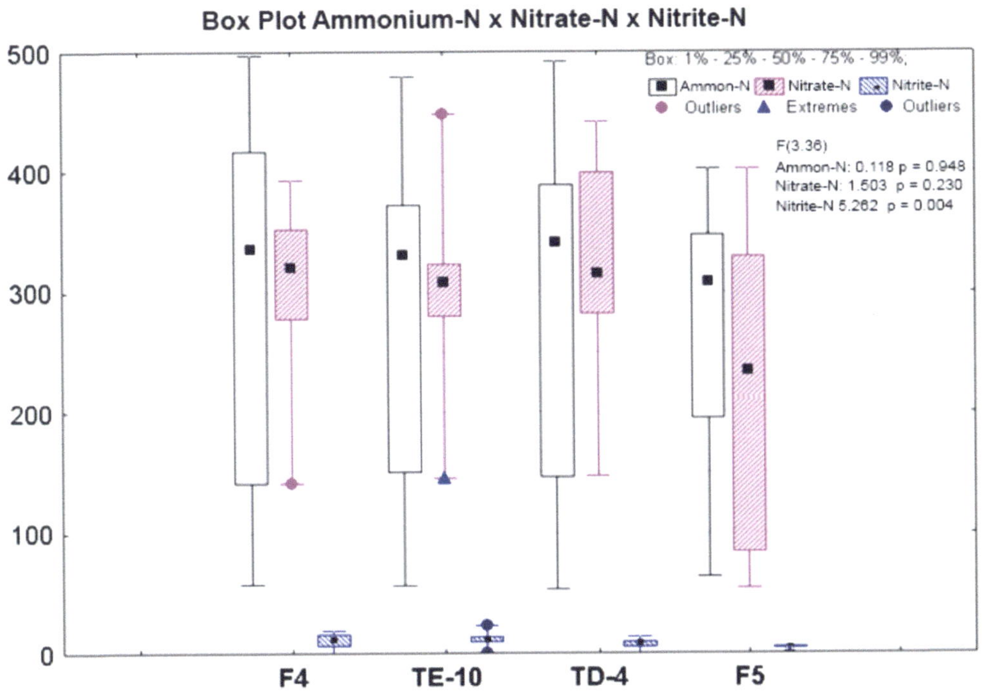

FIGURE 11 – PERCENTILES (25, 50, 75, 100%) FOR AMMONIUM-N, NITRATE-N AND NITRITE-N (µg/L) FROM 4 SAMPLING AREAS (MAR – DEC 2010), RIO VERDE RESERVOIR

Considering the nitrogen load values in the four assessed areas, the total loads of the two tributaries (TE10 and TD4) were not significant (Figure 12). At point F4, the results showed values for ammonium-N of 47.4 kg/day and 47.8 kg/day for nitrate-N, similar values with a relatively high correlation (r = 0.67). The values collected at point F4 were largely responsible for the supply of nitrogen loads, accounting for 77% of the total (Figure 13). In looking at the F4 sub-basin through sections, section F3 produced higher load values (50.5%) even in comparison to "Mouth" F4, which is downstream of F3, includes urban areas, and was the most compromised section in the phosphorus analysis (Figure 12). Furthermore, "Mouth" F4 and F1, both sections of F4-Rio Verde, also had significant contributions. Currently, the total N discharge of Rio Verde sub-basins is around 126 kg/day.

At certain times we observed discharge values from the reservoir (at point F5) higher than the input values; however, and as with the results for phosphorus, the sum of the loads from all tributaries suggests that the reservoir retains significant nitrogen load when compared to the losses via the channel's spillway (point F5; Figures 12 and 13).

Considering the relativity of loads (kg/area - Figure 13-ii) there was no tributary that presented values significantly different than the rest; however, points TE10, TE11, TDN, TE14 and F4 (F3 section) each had values of approximiately 10%, representing sections with higher contributions of total-N in relation to the area of the sub-basin.

4.2.2.3 BOD, COD and Total Solids: concentration and load

The Biochemical Oxygen Demand (BOD) is used as an index of water quality and consists of determining the oxygen required for the oxidation of organic materials. This process is mainly carried out by heterotrophic aerobic bacteria and protozoa, which convert oxygen into organic carbon (CO_2), hydrogen (H_2O), oxygen (H_2O) and nitrogen (NO_2^- or NO_3^-). The Chemical Oxygen Demand (COD) relates to the oxidation of organic materials, biodegradable or not, using an oxidizing agent. This measure allows for the verification of the presence of substances in water which are resistant to biological degradation.

The BOD values were considered very low for all samples analyzed (average = 1.2 mg/L - Figure 14). CONAMA Resolution 357 recommends maximum BOD values of up to 5.0 mg/L of O_2 for class II rivers. According to IAP (unpublished data), considering long-range data from 2006 to 2010, the BOD values ranged between 2-4 mg/L and are similar to those observed in this study.

The values of COD were in many cases higher than expected, especially given the low values of BOD (Figure 14). Several COD values were greater than 10mg/L, suggesting that non-biologically degradable substances contribute to the reservoir, or that oxidation throughout analysis, reaching inorganic levels. In general, we observed that during the months of August, 2008, and January, 2009, the results showed higher values of COD than during other periods for the vast majority of sampling areas. In January there was significant rainfall on the sampling day, which may have been the cause of these higher values. Unlike the 2010 samples (analysis of N and P concentrations discussed above), the 2008 samples were influenced by rainfall and possibly considering the wetting front. IAP data (unpublished data) for 2006-2010 also detected similar variations between 3 and 35 mg/L.

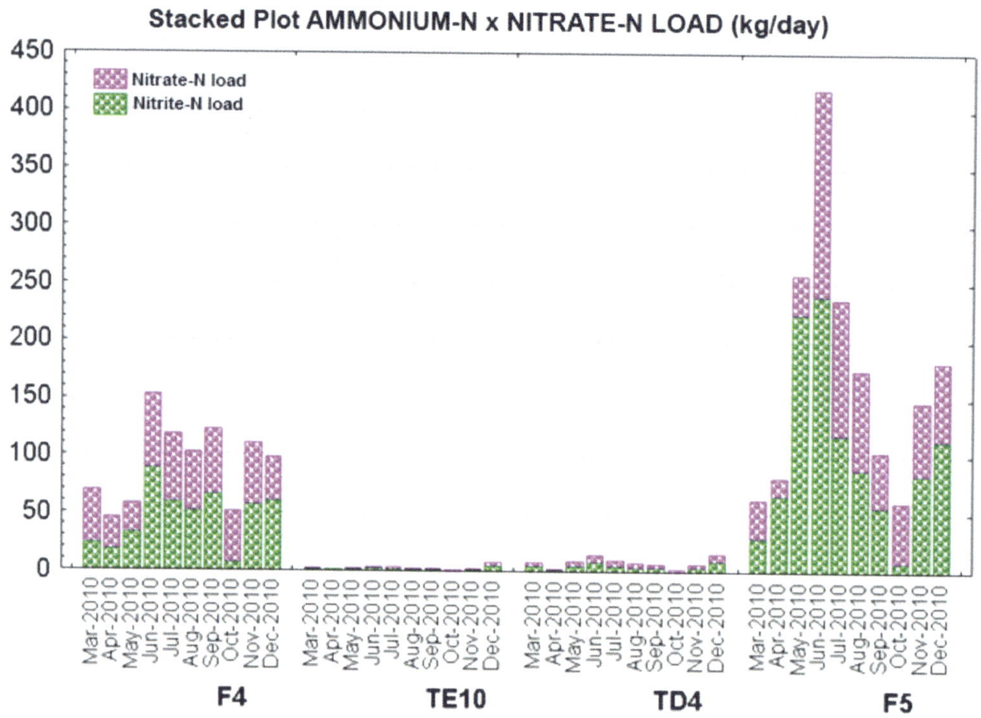

FIGURE 12 – AMMONIUM-N AND NITRATE-N LOADS (kg/day) FROM 4 SAMPLING STATIONS (MAR. – DEC. 2010), RIO VERDE RESERVOIR

The BOD/COD ratio ranged from 0.13 to 0.41, with an average of 0.25. Therefore, 25% of the COD is biologically degradable and the remaining 75% depends on chemical oxidation or a greater length of time for effective biodegradation. A low BOD/COD ratio means that much of the sampled fraction is not biologically degradable and depends on other forms of oxidation. On the other hand, a high BOD/COD ratio means that the majority of the fraction is biologically degradable, if there are no agents in the environment that may interfere with the microbiological activity.

Considering all collected samples, no correlation or trend between BOD and COD (r = 0.049) was observed; however, after further analysis of each sub-basin, some correlations were observed in F1 (r = 0.48), TDN (r = 0.41), F2 (r = 0.73), TE7 (r = 0.71), TD1 (r = 0.52) and TD6 (r = 0.61).

The parameter Total Solids represents the existing solid load in the environment and it is usually related to the presence of silt and clay (plus other materials) as well as organic matter. From this parameter we can infer the amount of particulate matter that is contributing to the water body. There are different ways of classifying solids based on the objectives of the analysis, such as fixed and volatile, suspended, particulate, and dissolved, allowing for the collection of targeted information that assess the dynamics of solids in the water body.

Average values of total solids ranged between 80 and 100 mg/L, with some exceptions; the seventh sampling event, which occurred in January, was conducted during rainfall and produced higher values. We also observed higher values at sampling areas F1 and F2, near the urban center of Campo Magro, and their location in urban areas likely explains this situation (Figure 15). Data obtained between 2006 and 2010 by IAP at point F4 (SANEPAR, unpublished data) ranged from 28 to 369 mg/L, with no identifiable pattern.

Regarding the total loads of BOD, COD and Total Solids, and again due to the flow of the tributary, point F4 in the Rio Verde River represents the largest contribution to the three variables: 78% or 401.5 kg/day of BOD; 79.3% or 4819.6 kg/day of COD; and 77.4% or 51.1 tons/day of solids. As mentioned in the section "Selection of Sampling Areas", the catchment area of the F4 sampling area, was subdivided into sections because of its size (F3, F2, F1, TD1, TE7). Our results showed that the greatest contribution of BOD and COD load occurs after point F3, the F4 "Mouth" section (Figures 16 and 17), demonstrating the influence of urbanization and industry in the region near highway BR277. As for total solids, both the F3 sub-basin as well as the F4 "Mouth" section, recorded significant total load (Figure 18).

In the direct tributaries of the reservoir, points TE8 and TD4 are noteworthy as together they account for about 11 and 12% of the total load of BOD and COD, respectively. Therefore, these sub-basins should receive greater attention (Figures 16 and 17). Both are adjacent to the "mouth F4" section; TD4 is located in close proximity to a large urban area in an adjacent sub-basin (urban center of Campo Largo), and TE8 is in an area of intense agricultural activity. Observing the solid load, only TD4 presented a significant value. On the other hand, the concentration values observed for these variables are not different from most other sub-basins and flow is the factor that increased the total load values in the TD4 and TE8 sub-basins. Nevertheless, as the load is considered high, it is important to closely monitor these areas.

Considering load relativity, namely the relationship between load and area, TD4 again appears as a priority

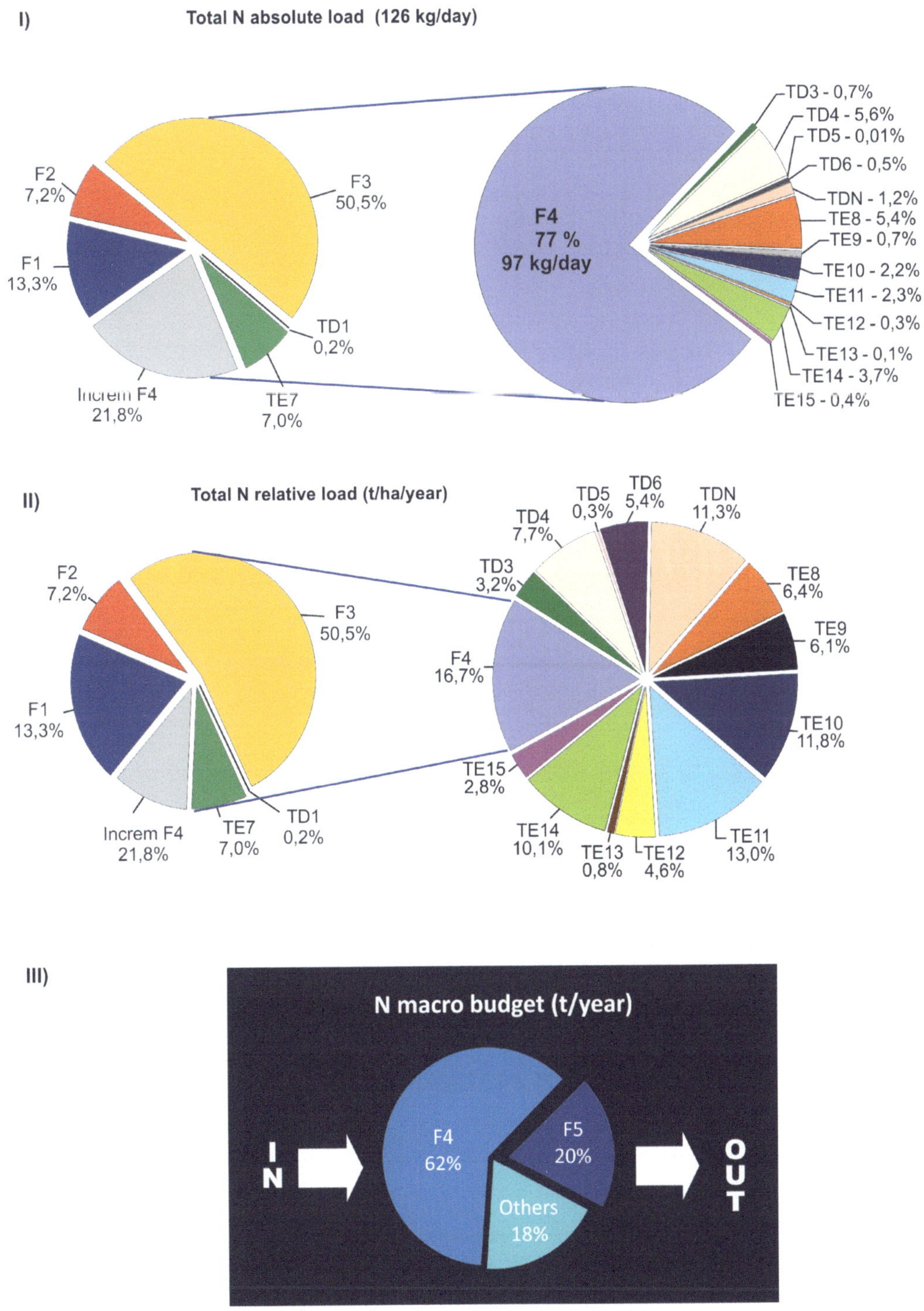

FIGURE 13 – RELATION BETWEEN TRIBUTARIES AND TOTAL-N LOADS IN THE RIO VERDE RESERVOIR: I) TOTAL LOADS OF TOTAL-N; II) RELATIVE LOADS OF TOTAL-N; AND III) TOTAL-N MACRO-BUDGET

NOTE: F4 = F1 + F2 + F3 + TE7 + TD1 + "Mouth" F4

sub-basin for intervention; The "mouth F4" section, which is one of the sections of F4 between BR277 and the reservoir, also showed a significant relative load. Furthermore, high values of relative load were also found for: TDN, which is located next to TD4 and therefore adjacent to a large urban area (part of the urban area of Campo Largo); and TE11 and TE10, located near TE8, which are areas of intense agricultural activity. However, as discussed above, total load was low. TE8 presented low impact, considering the relative load. The values for BOD, COD and Total Solids point to the same sub-basins as priorities for intervention, thus providing further supporting information for intervention activities that can be developed in these locations in the future (Figures 19, 20 and 21).

Because these are tributaries of lentic environments with the potential for eutrophication and the proliferation of algae and macrophytes, it is recommended that preventive and mitigating activities are given priority in the sub-basins where greater total loads were detected, followed by activities in the sub-basins that present higher relative loads and concentration. The guiding principle is that the greater the amount of nutrient input, the greater the potential for an increase in trophic levels and consequently the greater risk of algae blooms. However, it is necessary to emphasize that both parameters (concentrations and loads) are important and can guide preventative activities.

Considering the sequence of eight sampling events between 2008 and 2009 and the average loads of BOD, COD and Total Solids, a simplified balance of loads was established, taking into account the difference of discharge values due to the total contribution entering the reservoir via tributaries and the total amount exiting the reservoir via the spillway. In other words, we evaluated the amount of load that enters and exits the reservoir in relation to the tributaries over time (Figure 22). Our analysis showed that for COD and total solids, the load input to the reservoir is higher than the output; therefore, the environment is gradually retaining sediments and compounds, especially those of inorganic degradation, since there seems to be a balance between input and output for BOD.

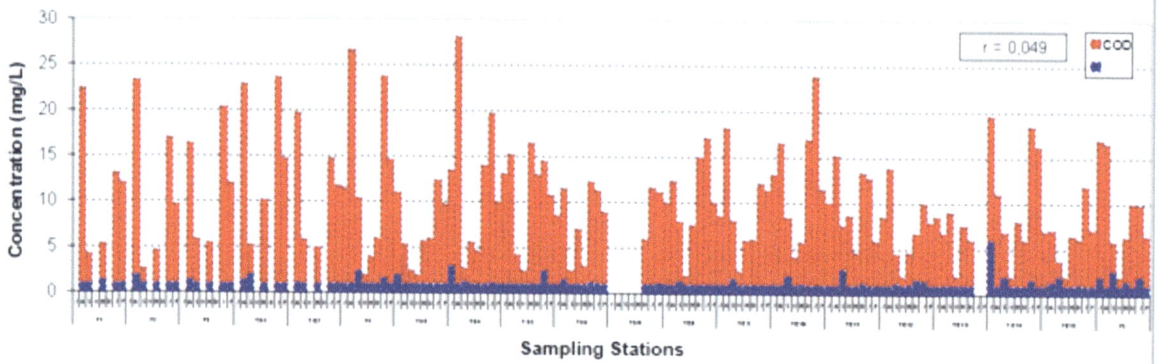

FIGURE 14 – BOD AND COD VALUES FROM 20 SAMPLING STATIONS (JUL – FEB 2008/2009), RIO VERDE RESERVOIR

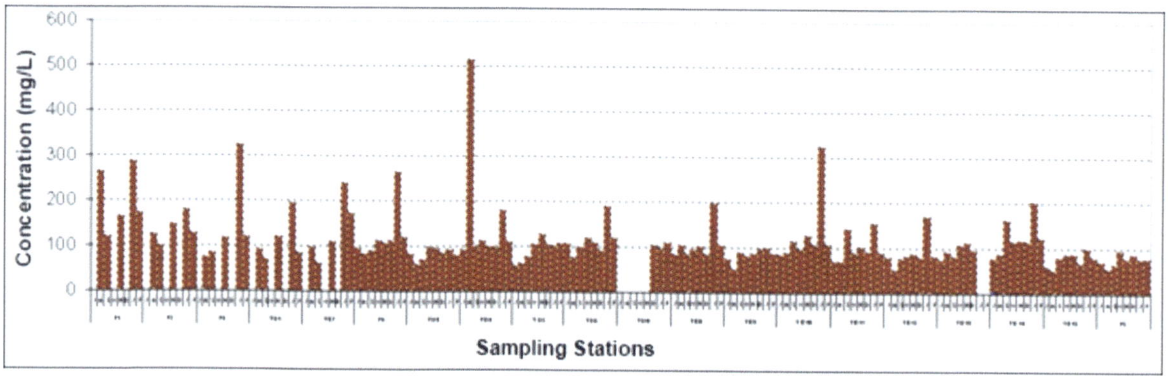

FIGURE 15 – TOTAL SOLIDS FOR 20 SAMPLING STATIONS (JUL – FEB 2008/2009), RIO VERDE RESERVOIR

NOTE: Blank spaces mean no analysis performed.

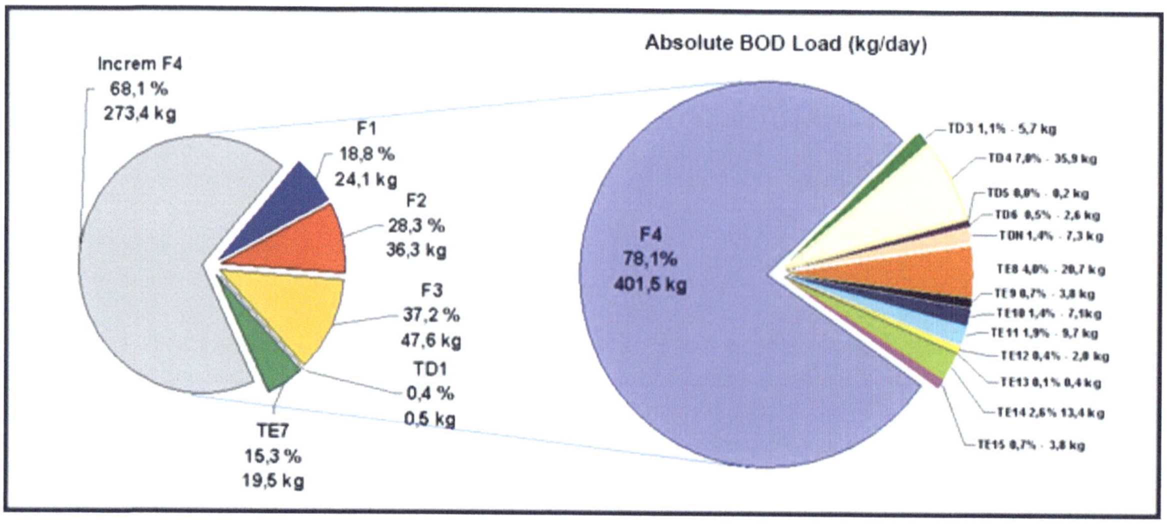

FIGURE 16 – MEAN TOTAL BOD LOADS IN 20 SAMPLING STATIONS (JUL. – FEB.), RIO VERDE RESERVOIR

NOTE: F4 = F1 + F2 + F3 + TE7 + TD1 + "Mouth" F4

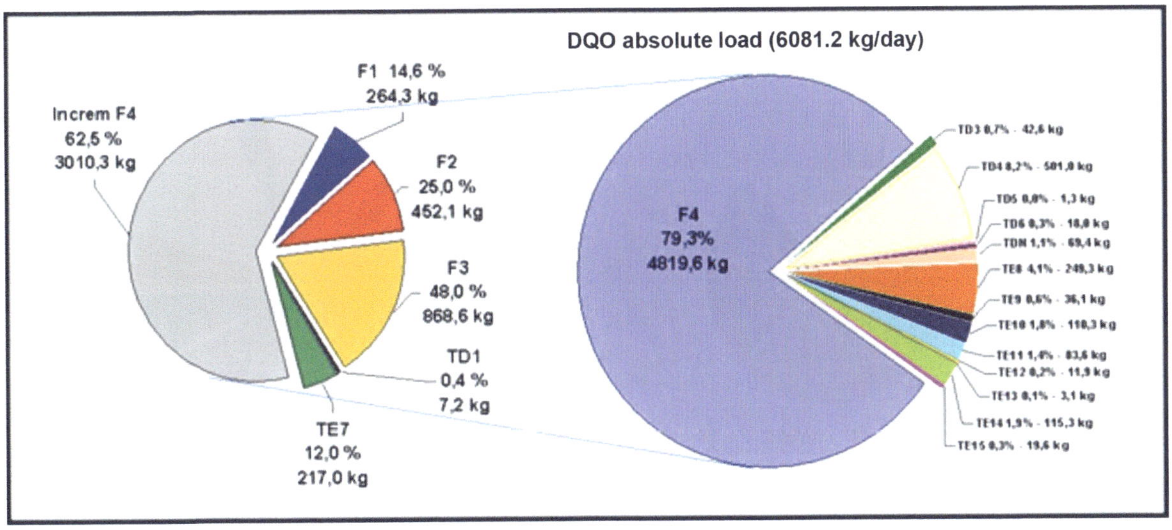

FIGURE 17 – MEAN TOTAL COD LOADS AT 20 SAMPLING STATIONS (JUL – FEB 2008/2009), RIO VERDE RESERVOIR

NOTE: F4 = F1 + F2 + F3 + TE7 + TD1 + "Mouth" F4

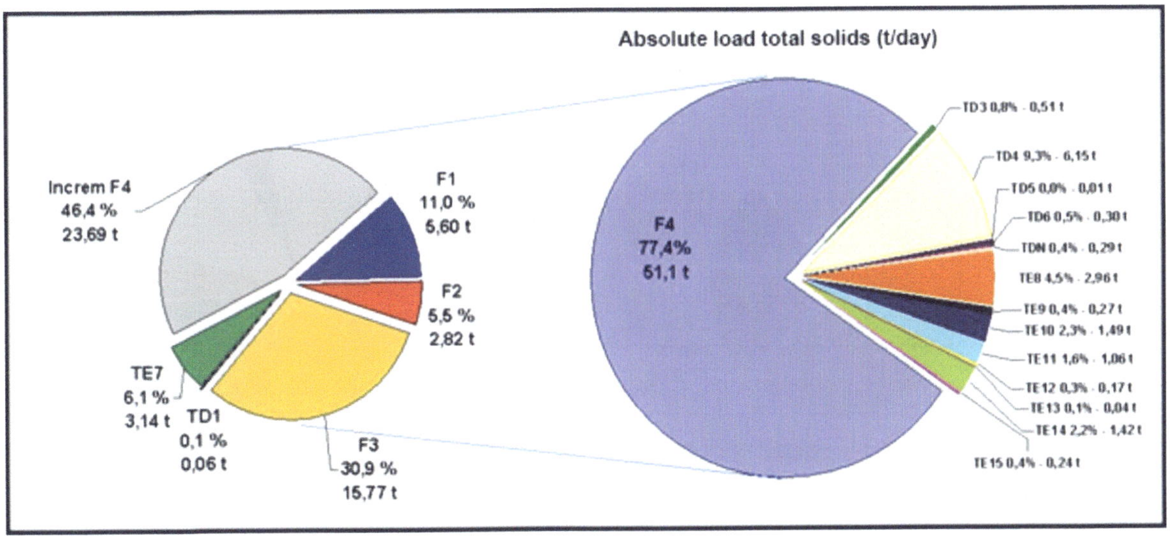

FIGURE 18 – MEAN TOTAL LOADS FOR TOTAL SOLIDS AT 20 SAMPLING STATIONS (JUL – FEB 2008/2009), RIO VERDE RESERVOIR

NOTE: F4 = F1 + F2 + F3 + TE7 + TD1 +" Mouth" F4

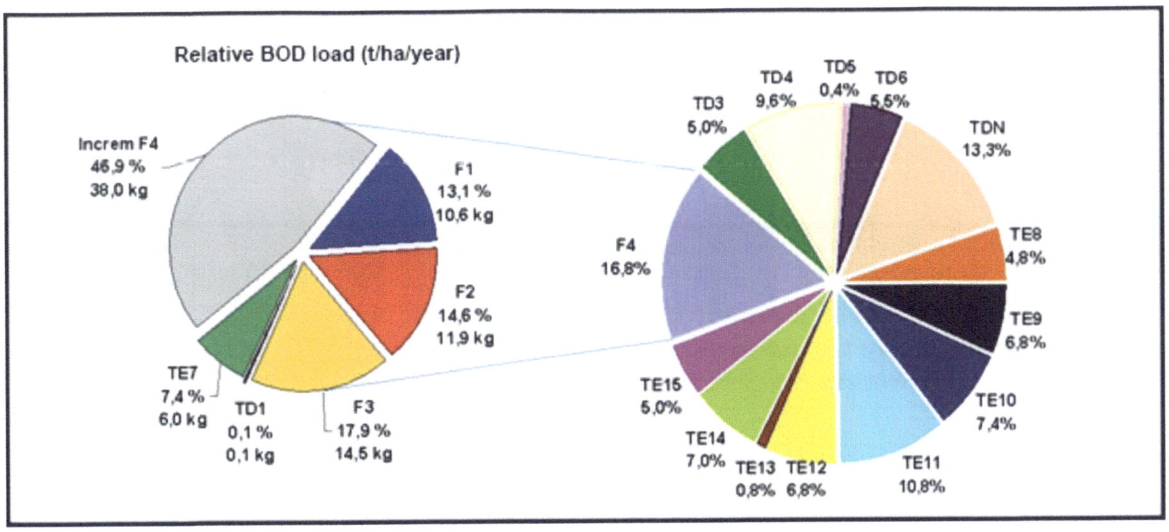

FIGURE 19 – MEAN RELATIVE BOD LOADS AT 20 SAMPLING STATIONS (JUL – FEB 2008/2009), RIO VERDE RESERVOIR

NOTE: F4 = F1 + F2 + F3 + TE7 + TD1 + "Mouth" F4

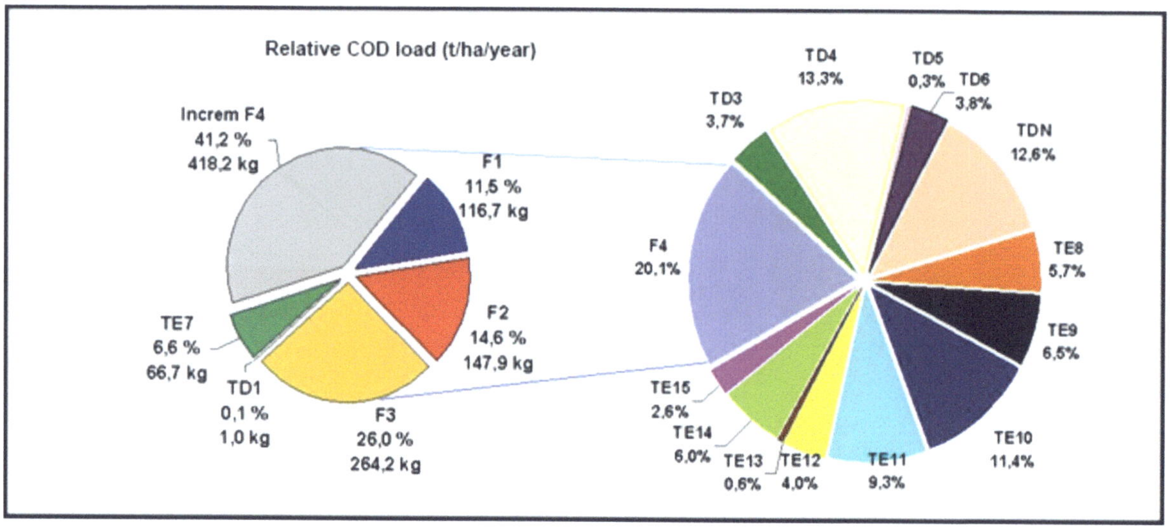

FIGURE 20 – MEAN RELATIVE COD LOADS AT 20 SAMPLING STATIONS (JUL. – FEB. 2008/2009), RIO VERDE RESERVOIR
NOTE: F4 = F1 + F2 + F3 + TE7 + TD1 + "Mouth" F4

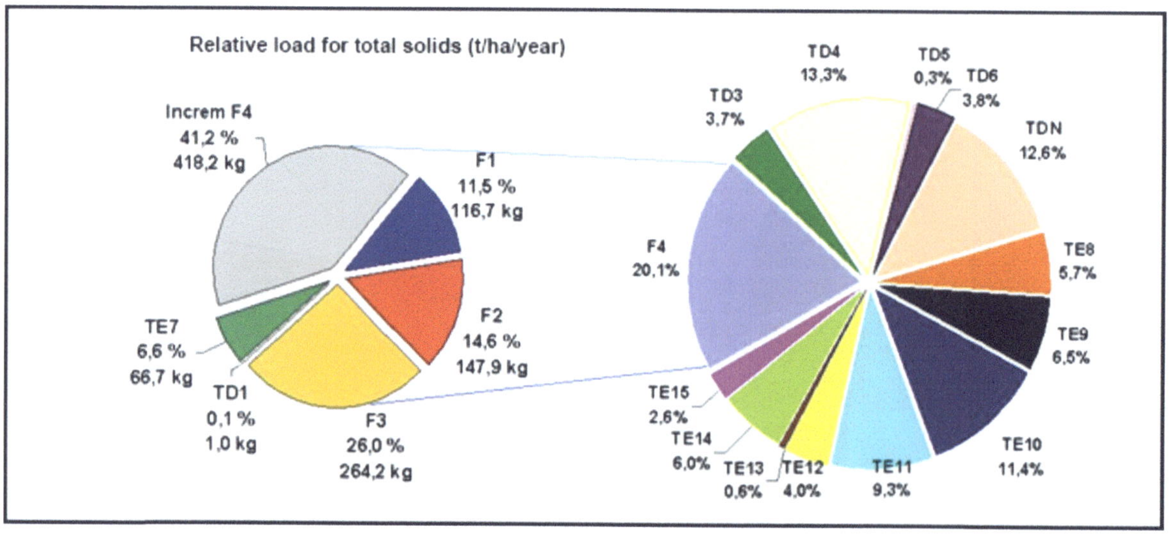

FIGURE 21 – MEAN RELATIVE LOADS FOR TOTAL SOLIDS AT 20 SAMPLING STATIONS (JUL – FEB 2008/2009), RIO VERDE RESERVOIR
NOTE: F4 = F1 + F2 + F3 + TE7 + TD1 + "Mouth" F4

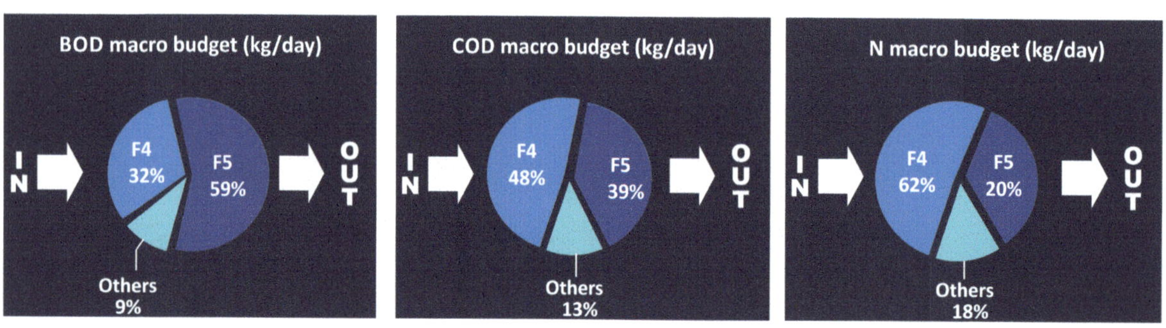

FIGURE 22 – SIMPLIFIED BUDGET IN THE RIO VERDE RESERVOIR FOR TOTAL BOD LOAD, TOTAL COD LOAD AND TOTAL SOLIDS LOAD. (MEAN VALUES, JUL – FEB 2008/2009)
NOTE: OTHERS → TD3, TD4, TD5, TD6, TDN, TE8, TE9, TE10, TE11, TE12, TE13, TE14 and TE15

4.2.3 Land Use in the Rio Verde Basin

The Rio Verde Basin presents different classes of land-use, including urban areas, agriculture, reforestation, and natural vegetation. In general, agricultural and livestock activities are predominant in this basin as well as areas of native vegetation (Figure 23).

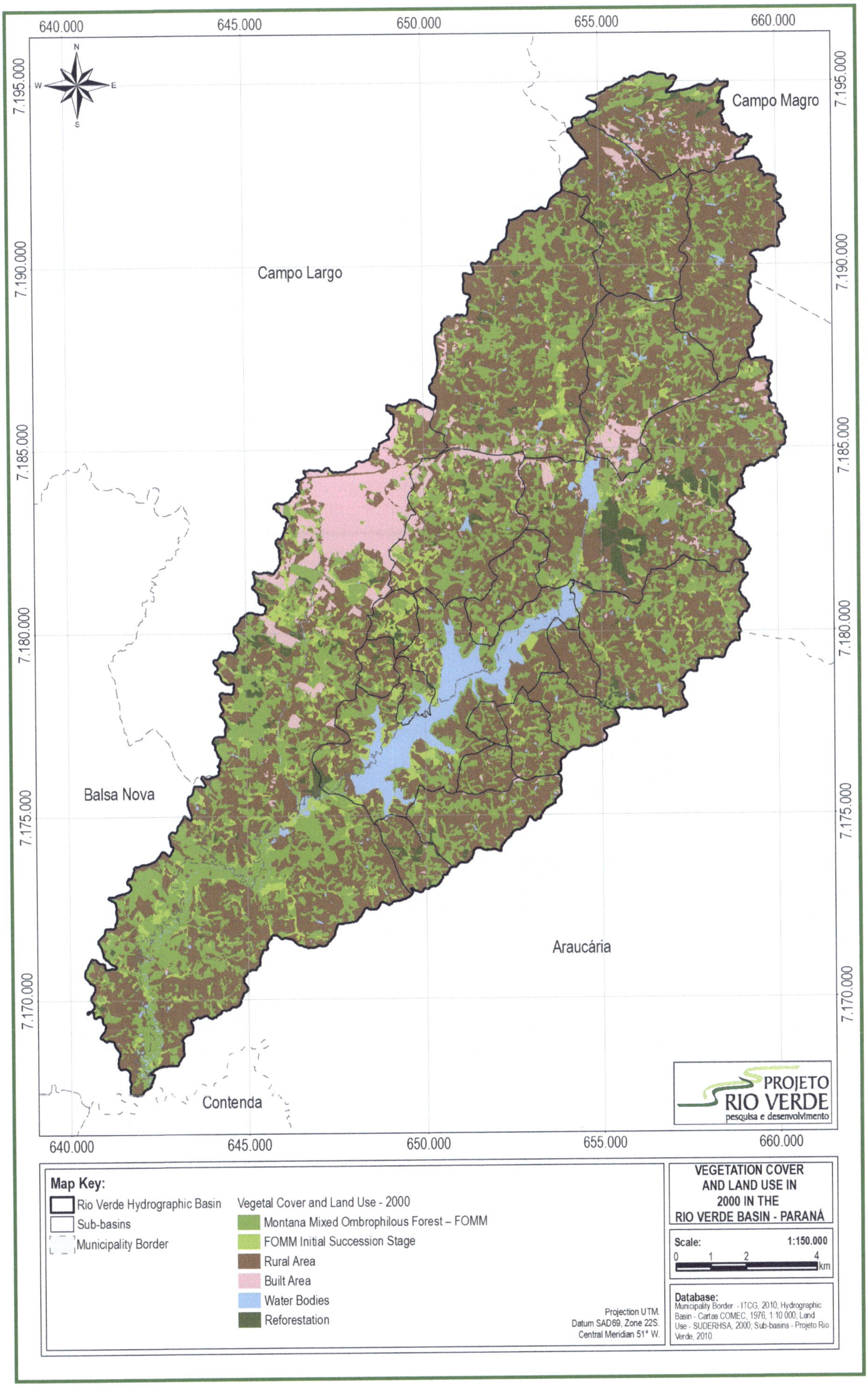

FIGURE 23 – LAND-USE MAP IN THE RIO VERDE RESERVOIR BASIN

There is a small urban area at the basin head, in the municipality of Campo Magro, and a more significant urban area in Campo Largo, at the mid-point of the basin. However, this significant urban area is outside the drainage area of the Rio Verde reservoir, as it is located in a sub-basin that flows into the Rio Verde River, downstream of the dam. In the sub-basin upstream of point F4, there is a small urban area within the municipalities of Campo Magro (the head) and Campo Largo. There is also a small urban area at the head of the TD4 sub-basin. For the other sub-basins that drain into the reservoir, both on the right and left bank, there is almost no urbanization, only agricultural activities and natural forest. The municipalities of Araucária and Balsa Nova, although located partially within the basin, have little urban development within the basin.

The evaluation of land-use in the Rio Verde Basin is important as the input of pollutants and nutrient loads to the rivers, and consequently the reservoir, is closely linked to the way the land is used. Thus, agricultural activities, which occur in much of the basin, have the potential to significantly contribute nutrients derived from agricultural fertilizers, notably phosphate and nitrogen. In this case, the input of these nutrients is mainly diffuse through runoff along the soil surface during rainfall. Besides causing damage to the environment due to the input of nutrients to the rivers, surface runoff results in losses for farmers, since the artificially applied nutrients are lost, affecting crop growth, or additional fertilization is required, which increases production costs.

In turn, in urban areas, the potential pollution through point source loads and others contaminants is also very significant, mainly due to illegal dumping of sewage into the rainwater drainage system and intermittent releases of domestic sewage into the rivers.

Table 3 shows the areas of land-use classes for each of the sub-basins considered in this study which will be compared to total and relative loads measured during monitoring.

From a macro analysis, some important aspects of the land-use related to the measured loads must be highlighted. Figure 23 and Table 3 show that there is a predominance of areas occupied by agricultural activities, approximately 56% of the total basin area. Another important land-cover is the interim forest, *capoeira*, which occupies 28% of the total area. All other LULC are less significant in the basin.

According to Figures 9 and 13, the F4 sub-basin is responsible for most of the input nutrient load to the reservoir (85% of total phosphorus and 77% of total nitrogen). Given that the area of this sub-basin is considerably greater than the others, the total load also tends to be higher. In terms of relative load, or the total load divided by area, F4 contributed approximately 23.5% of total phosphorus and 16.7% of total nitrogen (Figures 9 and 13). The F4 sub-basin presents heterogeneous land-use activities, the majority of which are urban and agricultural areas, which may be contributing significantly to the load input to the reservoir (Table 3).

When analyzing only the relative load of phosphorus, the TE-10 sub-basin is problematic as it represents 43.7% of the total. Land-use in this sub-basin only includes rural areas and vegetation, with no urban area. Therefore, the high phosphorus load is likely due to the interference of agricultural activities, possibly with the heavy use of phosphate fertilizers.

Sub-basins TD-3, TD-N, TD-5, TD-6, TE-9, TE11, TE12, TE-13, TE-15, which have land-use patterns of both preserved vegetation and agriculture, are the most preserved areas with regards to LULC and present lower phosphorus loads into reservoir. However, sub-basins TD-N, TE-11, TE-14, and F-3, present nitrogen loads that are relatively high and should be closely monitored.

The measured load values were also compared with theoretical estimates of load export, based on average rates of loss from soils, considering different land uses. While we did not expect a high correlation between these estimates, the values should be relatively consistent. Thus, we adopted a simplified methodology to assess the potential load of phosphorus in the Rio Verde Basin, based on the drainage area of each sub-basin and the export coefficients relating to land use (proposed by Salas & Martino *apud* Silva (2006) and Jorgensen (1989) and presented in section 3.1). The coefficients used were: 50 mg/m^2.year for agriculture and livestock; 100 mg/m^2.year for urban use and exposed soil; 10 mg/m^2.year for areas with vegetation, native or not. For purposes of simplification, we used the Grove export coefficient for areas with any type of vegetation.

The equation used to estimate total phosphorus in the sub-basin is represented by:

$$C_j = \sum_{i=1}^{n} E_i A_{ij}$$

Where *Cj* is the load in sub-basin "j"; E_i is the export coefficient for land-use "i"; A_{ij} is the area of basin "j" in which occurs the usage type "i".

Given that the land-use within the Rio Verde River sub-basins is known, we estimated the phosphorus loads in these sub-basins using the export coefficient and the results of load estimates are shown in Table 4.

While the loads estimated by the export coefficient showed values different from those measured in the sub-basins, the relative scale was consistent. For example, for sub-basins F4, TD4 and TE10, our results showed average loads of approximately 1900 kg P/year, 165 kg P/year and 60 kg P/year, respectively, while estimates were calculated as 3444 kg P/year, 444 kg P/year and 95 kg P/year, respectively.

We observed that agriculture represents the largest source of phosphorus load in the basin, followed by urban areas. This demonstrates the need for control of agricultural areas in order to reduce nutrient loads to the Rio Verde Reservoir. The significant impact of agriculture in relation to other uses is the result of the predominance of this land-use activity in the basin. Urban land-use, although it is minimal in the basin and occupies approximately 3.6% of the area, contributes approximately 10% of total phosphorus input into the reservoir, thus demonstrating the potential impact of this type of land-use. These results reinforce the need to carefully monitor and control urban sprawl in the Rio Verde Basin, otherwise the impact on water quality of the Basin and the Rio Verde Reservoir will be significant.

TABLE 3 – LAND-USE AND VEGETATION COVER AREAS IN THE RIO VERDE RESERVOIR BASIN

Percentage of Land Use Classes and Vegetation Cover											
Sub-Basin	Agriculture Livestock	Water bodies	Urban Area	Advanced Forest "Capoeirão"	Riparian forest	Initial forest "Bracatinga"	Initial forest "Capoeirinha"	Interim forest "Capoeira"	Reforestation	Exposed Soil	Floodplain
F-4	57.2%	0.7%	4.0%	2.8%	2.4%	0.9%	2.2%	25.1%	2.3%	0.0%	2.4%
TD-3	54.3%	0.9%	-	7.9%	-	0.2%	2.4%	32.0%	1.5%	-	0.7%
TD-4	40.2%	0.9%	12.8%	3.4%	-	1.6%	2.8%	37.3%	0.9%	0.0%	0.2%
TD-N	52.1%	0.1%	1.8%	9.9%	-	5.7%	2.6%	27.7%	-	-	-
TD-5	35.0%	16.7%	-	1.6%	-	0.6%	5.8%	39.4%	0.9%	-	-
TD-6	64.1%	-		0.7%	-	3.0%	1.8%	29.4%	0.3%	-	0.6%
TE-8	54.4%	0.7%	-	2.5%	0.9%	0.1%	1.2%	34.4%	2.0%	-	3.8%
TE-9	70.3%	2.8%	-	6.3%	-	-	1.1%	12.6%	0.2%	-	6.6%
TE-10	53.5%	0.5%	-	2.3%	-	-	0.1%	35.8%	7.2%	-	0.7%
TE-11	61.9%	0.5%	-	3.5%	-	0.1%	1.2%	30.3%	1.0%	-	1.5%
TE-12	60.0%	-	-	12.1%	-	-	-	26.5%	-	-	1.4%
TE-13	68.4%	0.1%	-	5.4%	0.6%	0.4%	0.8%	24.3%	-	-	-
TE-14	62.2%	0.6%	-	3.2%	0.6%	3.3%	1.6%	26.5%	0.5%	-	1.4%
TE-15	55.9%	0.4%	-	4.6%	-	8.4%	0.4%	29.7%	-	-	0.5%
Total	55.8%	0.8%	3.6%	3.2%	1.6%	1.1%	2.0%	27.6%	2.0%	0.0%	2.1%

TABLE 4 – ESTIMATES FOR TOTAL P LOADS IN THE RIO VERDE BASIN TAKING INTO ACCOUNT DIFFERENT LAND-USE (kg/year)

Land Use and Total P Load (kg/year)										
Sub-Basin	Agriculture Livestock	Urban Areas	Advanced Forest "Capoeirão"	Riparian forest	Initial forest "Bracatinga"	Initial forest "Capoeirinha"	Interm forest "Capoeira"	Reforestation	Exposed Soil	P Load (kg/year)
F-4	2,722.7	377.6	27.1	22.4	9.0	20.7	239.2	219	4.2	3,444.8
TD-3	98.8	0.0	2.9	0.0	0.1	0.9	11.6	0.6	0.0	114.8
TD-4	237.9	151.3	4.0	0.0	1.9	3.3	44.2	1.1	0.3	443.9
TD-N	45.2	3.2	1.7	0.0	1.0	0.4	4.8	0.0	0.0	56.4
TD-5	21.9	0.0	0.2	0.0	0.1	0.7	4.9	0.1	0.0	28.0
TD-6	48.6	0.0	0.1	0.0	0.5	0.3	4.5	0.1	0.0	53.9
TE-8	372.1	0.0	3.5	1.2	0.2	1.6	47.1	2.8	0.0	428.4
TE-9	61.4	0.0	1.1	0.0	0.0	0.2	2.2	0.0	0.0	64.9
TE-10	81.4	0.0	0.7	0.0	0.0	0.0	10.9	2.2	0.0	95.2
TE-11	88.1	0.0	1.0	0.0	0.0	0.3	8.6	0.3	0.0	98.3
TE-12	28.2	0.0	1.1	0.0	0.0	0.0	2.5	0.0	0.0	31.9
TE-13	58.5	0.0	0.9	0.1	0.1	0.1	4.2	0.0	0.0	63.9
TE-14	187.2	0.0	1.9	0.4	2.0	0.9	16.0	0.3	0.0	208.7
TE-15	66.3	0.0	1.1	0.0	2.0	0.1	7.1	0.0	0.0	76.6
t-Load	4,118.3	532.1	47.4	24.0	16.7	29.6	407.7	29.3	4.5	5,209.6

5. FINAL CONSIDERATIONS

As mentioned above, the identification of the pollution potential of hydrographic basins should be based on the analysis of both **concentration** and **load**, as well as consider **total loads** and **relative loads,** maximum **climate variability** (especially rainfall), and **different land-uses and management**. Furthermore, **the data quality** obtained by monitoring is important in the assessment of water quality. These aspects, when integrated, can provide reliable results. Without considering all of these factors, the results can lead to a misinterpretation of the areas that need to be prioritized in monitoring and mitigation activities.

Thus, considering all these variables that play a role in the phenomenon of eutrophication, we have used the data generated for the Rio Verde Reservoir as the focus of this case study. From these analytical guidelines, some important considerations could be observed. In order to systematize this information we have created a Pollution Potential Assessment Matrix (2PAM).

The purpose of this matrix was categorizes the environments based on collected data and the weighting system employed. The matrix was built based on different perspectives, parameters and factors. However, questions needed to be addressed such as: How can we establish this pollution potential? What criteria are needed to assess pollution potential? Which variables should be used? From which perspective should the pollution potential be addressed: environmental, social, and/or legal?

In this proposed matrix for the Rio Verde basin, we analyzed the basin from three perspectives using four chemical parameters and three factors, as shown in the Table 5.

Obviously, there are several points of view that can be analyzed and often one point of view, while just as important, may not necessarily have a direct relationship with another. For example, we can consider two perspectives: water quality of the river and ecosystem balance of the reservoir. From the legal perspective, the guiding parameter considered is the concentration in both the river and the reservoir. Therefore, in considering the quality of river water, this parameter seems to be more relevant, even from the environmental point of view. However, if we consider a lentic environment (reservoir/lake), the load appears to be more relevant, since a given tributary with a flow rate of 100 L/s and 50 ppb total-P contributes much more phosphorus compared to a tributary of 500 ppb but with a flow rate of only 5 L/s. The first tributary contributes twice the amount of phosphorus to the lentic body, which can be more problematic considering the eutrophication potential of the environment. Therefore, it is essential to define the perspective of analysis and then determine the pollution potential as assessing each situation is quite different.

In order to ensure that each situation (perspective) is analyzed consistently, and the factors of analysis are interpreted correctly, it is necessary to establish a weighting system for each factor (concentration, relative load and total load) depending on the perspective to be examined. For the Rio Verde Reservoir, several indices have been proposed, which are shown in Table 6.

From the weights assigned to each parameter and interpretations of the graphics depicting relative load and total load (Figures 6-21), the data and information taken from the analysis were plotted on a matrix of simple analysis (Table 7).

In order to better focus future activities and to determine which activities should take place in the basin, reservoir or rivers, priority levels were established (priority 1, priority 2, and priority 3) according to the degree of concentration and loads of total phosphorus, total nitrogen, BOD and COD. As we decreased the level of priority, we decided to reduce one tenth the weight value in order to facilitate the differentiation and identification of priority levels. The different weights were given between 0.8 and 2 to each factor. The weights were arbitrarily assigned and ranked to each factor based on the following: i) importance of perspective in the ecosystem, e.g.: total load for lentic systems, concentration for lotic systems, and relative load (load/m2) for land use management; ii) preliminary verification of the weighting system in its possibility to distinguish different environments; iii) local legislation, which mandates concentration as the reference parameter for river water quality; and iv) reference literature and research studies in other local basins.

TABLE 5 – PERSPECTIVES CONSIDERED IN THE PREPARATION OF THE POLLUTION POTENTIAL ASSESSMENT MATRIX FOR THE RIO VERDE BASIN

PERSPECTIVES	PARAMETERS	FACTORS
Stream water quality Reservoir ecosystem equilibrium Basin management	Total phosphorus Total nitrogen BOD and COD	Concentration Total load Relative load

TABLE 6 – WEIGHT-VALUES FOR 2PAM FACTORS, VARYING BY PERSPECTIVE

PERSPECTIVE	CONCENTRATION FACTOR	TOTAL LOAD FACTOR	RELATIVE LOAD FACTOR
Considering stream water quality	(2.0)	(1.0)	(1.5)
Considering reservoir ecosystem equilibrium	(1.0)	(2.0)	(1.5)
Considering basin management	(1.0)	(1.5)	(2.0)

Then, depending on the degree of deterioration, a classification system was established to differentiate the sub-basins with the most extreme water quality degradation in order to establish priorities for action, ranging from very low to very high priority.

In applying the weighted values to the analyzed sub-basins (tributaries), we obtained interpretive matrices for each proposed perspective: river water quality, reservoir ecosystem equilibrium, and sub-basin management; shown in Tables 8, 9 and 10, respectively.

As shown in the matrices and graphs of concentrations and loads, the Rio Verde Basin can still be considered a basin that is relatively well preserved. Considering river water quality, as well as the reservoir ecosystem equilibrium, streams TE-10, F1 and Mouth-F4 presented the highest pollution potential based on our analysis using 2PAM. From a sub-basin management perspective, the sub-basins Mouth-F4 and TE-10 were defined as having the highest pollution potential according to the 2PAM classification matrix; therefore, managers must focus their attention on these sub-basins.

The 2PAM - **Pollution Potential Assessment Matrix** – is a simple and straightforward way to analyze water quality that allows us to assess the data from varying perspectives. It is possible to adapt the 2PAM, as well as to establish different levels and weights for the variables; however, it is important to highlight that there are many factors that influence assessments of water quality and the goals of the water quality assessment can vary. It is essential that these factors and considerations are assessed in the early stages of strategic planning and logistical monitoring.

The final interpretive matrices demonstrate a novel way to analyze concentrations and loads in reservoir basins in order to determine the sub-basins that should be carefully managed. Thus, we try to facilitate the prioritization of activities in order to avoid wasteful efforts in ecosystems that have high levels of preservation. As such, it is worth mentioning again that it is essential to define the sub-basin that should be given the highest priority for intervention. Although reservoir-river-basins are systemically interconnected, some priorities must be defined in order to maximize intervention efforts.

Finally, the 2PAM can use existent data and can be easily replicated or adapted to other lake basins, by modifying the assessed parameters, e.g. inclusion of dissolved oxygen.

TABLE 7 – ANALYSIS OF SUB-BASINS THROUGH THE 2PAM FOR TRIBUTARIES IN THE RIO VERDE RESERVOIR BASIN

PARAMETER: TOTAL-P			
Deterioration Level	Pollution Potential of Rivers		
	Factor Concentration	**Factor Total Load**	**Factor Relative Load**
↑	TE10	Mouth F4	TE10
	TE14	F3	Mouth F4
	F1	TE10	F3
	TD1, TE7, TD5, TD3, TD6, TE15 and TE13	TD1, TE7, TD5, TD3, TD6, TE9, TE11, TE12, TE15 and TE13	TD1, TE7, TD5, TD3, TD6, TE15 and TE13
PARAMETER: TOTAL-N			
Deterioration Level	Pollution Potential of Rivers		
	Factor Concentration	**Factor Total Load**	**Factor Relative Load**
↑	TE14	F3	TE11
	F1	Mouth F4	TE10
	TE10	F1	TDN
	TD1, TE7, TD5, TD3, TD6, TE15 and TE13	TD1, TD5, TD3, TD6, TE15, TE13 TE9	TD1, TE7, F2, TD5, TE13 TE15
PARAMETER: BOD AND COD			
Deterioration Level	Pollution Potential of Rivers		
	Factor Concentration	**Factor Total Load**	**Factor Relative Load**
↑	TD1	Mouth F4	TD4
	TE10	F3	TDN
	Mouth F4	F1 and F2	TE10 and Mouth F4
	TD5, TD6, TDN, TE9, TE12,	TD1, TD3, TD5, TD6, TE9, TE12, TE13, TE15	TD1, TE13, TD5, TD6, TE15

TABLE 8 – SYNTHESIS MATRIX CONSIDERING " RIVER WATER QUALITY "

PRIORITY FOR ACTION	**SUB-BASIN/TRIBUTARY**		
	P	**N**	**BOD and COD**
VERY HIGH	TE10	F1*	Mouth F4 *
HIGH	Mouth F4 *	TE10	TE10
MEDIUM	F3 *	TE14	TD1
NORMAL	F1	F3* Mouth F4 * TE10, TE11	F1* F2* F3* TD4
LOW	TD1, TE7, TD3, TD5, TD6, TE15, TE13	TE15, TE13	TD5, TD6
VERY LOW	TE11 and TE12	F2 * TE7, TD3 TD6, TE9	TE12, TD3, TE9, TD1, TE13, TE15, TDN

NOTE: * Sub-sections of F4, i.e. F1, F2, F3, and Mouth F4, comprise the F4 sub-basin.

TABLE 9 – SYNTHESIS MATRIX CONSIDERING "RESERVOIR ECOSYSTEM EQUILIBRIUM"

PRIORITY FOR ACTION	SUB-BASIN/TRIBUTARY		
	P	N	BOD and COD
VERY HIGH	TE10	F1 *	Mouth F4 *
HIGH	Mouth F4 *	TE10	TE10
MEDIUM	F3 *	F3 *	F3 *
NORMAL	TE14, TE12, F1 *	TE14, TE11, TDN, Mouth F4 *	TDN, F1*, F2 * TD4, TD1
LOW	TE7, TD3, TD6, TD1, TD5, TE15, TE13	TD1, TD5, TE15, TE13	TD6, TD5
VERY LOW	TE11 TE9, TE12	TE7, F2 * TD3, TD6	TE9, TE12, TD3, TD1, TE13, TE15, TDN

NOTE: * Sub-sections of F4, i.e. F1, F2, F3, and Mouth F4, comprise the F4 sub-basin.

TABLE 10 – SYNTHESIS MATRIX CONSIDERING "BASIN MANAGEMENT"

PRIORITY FOR ACTION	SUB-BASIN/TRIBUTARY		
	P	N	BOD and COD
VERY HIGH	TE10	TE10	Mouth F4 *
HIGH	Mouth F4 *	TE11	TE10
MEDIUM	F3 *	F1 *)	TD4
NORMAL	TE10, TD4, TE11 e TE12, F1*, F2*, F3*, Mouth F4*	F3*, Mouth F4* TDN, TE14	F1*, F2*, F3* TDN, TD1
LOW	TD3, TE7, TD1, TD5, TD6, TE13, TE15,	TD1, TD5, TE13, TE15,	TD5, TD6
VERY LOW	TE11, TE12, TE9	TE9, TE7, TD3, TD6, F2*	TE12, TE9, TD1, TE13, TE15, TDN

NOTE: * Sub-sections of F4, i.e. F1, F2, F3, and Mouth F4, comprise the F4 sub-basin.

6. REFERENCES

ABNT – Associação Brasileira de Normas Técnicas. (1987) *Planejamento de amostragem de efluentes liquidos e corpos receptores – Procedimento*. ABNT/NBR 9897.

APHA – AMERICAN PUBLIC HEALTH ASSOCIATION. (2005) *Standard methods for the examination of water and wastewater.* 21. ed. Washington, DC, 1600 p.

BOLLMANN, H.A.; CARNEIRO,C.; PEGORINI, E.S.; Qualidade da água e dinâmica de nutrientes. In: ANDREOLI, C.V.; CARNEIRO, C. (2005) (Eds.). *Gestão Integrada de Mananciais de Abastecimento Eutrofizados.* Curitiba: Sanepar Finep, 500 p.

CARNEIRO, C.A. (2008) *Movimentação por veiculação hídrica do fósforo e outros macroconstituintes químicos sob a influência de fatores naturais e tecnogênicos na bacia do rio Timbú, tributário do reservatório Iraí – Região Metropolitana de Curitiba.* Thesis (Doctorate) – Universidade Federal do Paraná, Curitiba. 222 p.

COLLISCHONN, W.; TASSI, R. (2010) *Introduzindo Hidrologia.* Porto Alegre: IPH/UFRGS, 151 p.

CONAMA – Conselho Nacional de Meio Ambiente – *Resolution 357*, from March 17, 2005. Dispõe sobre a classificação dos corpos de água e diretrizes ambientais para o seu enquadramento, bem como estabelece as condições e padrões de lançamento de efluentes, e dá outras providências. Diário Oficial da União, Brasília, 2005. (Provides for the classification of water bodies and environmental guidelines for their classification as well as establishes the conditions and standards for effluent discharge and other measures.(Provides for the classification of water bodies and environmental guidelines for its classification and establishes the conditions and patterns for effluent discharge, and other measures. Official Gazette,Brasilia,2005)

CORDEIRO NETTO, O.M.; DUTRA FILHO, D.O. (1981) Aporte de fósforo no Lago Paranoá – Brasilia. In: *Anais do XI Congresso Brasileiro de Engenharia Sanitária e Ambiental.* Fortaleza – CE, September 20 to 25.

EIGER, S.; ARANHA, C.H.; GOMES, M.C.A.; PEREZ, L.; BUENO NETO, A.S. (1999) Modelagem Matemática de Cargas Difusas na Bacia do Guarapiranga, São Paulo. In: *XIII Simpósio Brasileiro de Recursos Hídricos.* Belo Horizonte.

ESTEVES, F. A. (1998) *Fundamentos de Limnologia.* 2. ed. Rio de Janeiro: Interciência, 602 p.

FISRWG – Federal Interagency Stream Restoration Working Group. (1998) *Stream Corridor Restoration: Principles, Processes and Practices.* N.° 0120-A; Docs n.° A 57.6/2:EN 3/PT 653.

IAP – Instituto Ambiental do Paraná. (2009) *Relatório da qualidade das águas dos reservatórios do estado do Paraná, no período de 2005 a 2008.* Curitiba, 122 p.

IAP – Instituto Ambiental do Paraná. (2004) *Monitoramento da qualidade das águas dos reservatórios do estado do Paraná, no período de 1999 a 2004.* Curitiba, 77 p.

JEFFERY, G.H.; BASSET, J.; MENDHAM, J.; DENNEY, R.C. (1992) *Vogel – Análise química quantitativa.* 5. ed. Rio de Janeiro, 712 p.

JORGENSEN, S. E. (1989) Use of models. In: JORGENSEN, S. E.; VOLLENWEIDER, R. A. *Guidelines of lake management – v.1. Principles of lake management.* Otsu: ILEC – UNEP, p. 71-98.

MCCARTHY, J.J.; TAYLOR, W.R.; TAFT, J.L.; (1975) The Dynamics of Nitrogen and Phosphorus Cycling in the Open Waters of the Chesapeake Bay. *Marine Chemistry in the Coastal Environment,* June 1, p. 664-681.

METCALF; EDDY. (1997) *Wastewater engineering: treatment, disposal, reuse.* 4. ed. New York: McGraw Hill Book, 2003. 1815p.

PANKOW, J.F. *Aquatic Chemical Concepts.* Chelsea: Lewis Publishers, 673 p.

PIERZYNSKI, G.M.; SIMS, J.T.; VANCE, G.F. (2000) *Soil phosphorus and environmental quality. In: Soils and Environmental Quality.* Boca Raton: Florida, p.155-207.

ROBINSON, J.W. (1990) *Atomic Spectroscopy*. New York: Marcel Dekker, Inc., 300 p.

SANTOS, I.; FILL, H.D.; SUGAI, M.V.B.; BUBA, H.; KISHI, R.T.; MARONE, E.; LAUTERT, L.F.C. (2001) *Hidrometria Aplicada*. Curitiba: Instituto de Tecnologia para o Desenvolvimento.

SHARPLEY, A.N.; HALVORSON, A.D. (1994) The management of soil phosphorus availability and its impact on surface water quality. In: LAL, R.; STEWART, B.A. (Eds.). *Soil process and water quality: advances in soil science*. Boca Raton: Lewis publishers, 398 p.

SILVA, H.L.G. (2006) *Modelagem bidimensional do fósforo com vistas a gestão de bacias hidrográficas – estudo de caso: reservatório de Fiú, Paraná*. Master dissertation, 113 p.

SNOEYINK, V.L.; JENKINS, D. *Water Chemistry*. (1980) USA: John Wiley and Sons, Inc., 463 p.

SPERLING, M.V. (1996) *Princípios básicos do tratamento de esgotos*. Vol. 2. Belo Horizonte: Departamento de Engenharia Sanitária e Ambiental – UFMG, 211 p.

TOMAZ, P. (2006) *Poluição difusa*. São Paulo: Navegar, 446 p.

UFRS – Universidade Federal do Rio Grande do Sul. (2007) *Alguns Fundamentos em Hidrologia*. Capítulo 5 – Medição de Vazão. UFRS.

VOLLENWEIDER, R.A. (1968) *Scientific Fundamentals of the Eutrophication of Lakes and Flowing Waters, with Particular Reference to Nitrogen and Phosphorus as Factors in Eutrophication*. OECD, Paris, 220 p.

ZAHED FILHO, K; PORTO, R.L. (2010) *Inundações Urbanas*. Escola Politécnica da Universidade de São Paulo, São Paulo, 44 p.

CHAPTER 11

SEDIMENT

Charles Carneiro
Kelly C. Gutseit
Cleverson V. Andreoli
Rubia L. D. Lima

SEDIMENT

Charles Carneiro
Kelly C. Gutseit
Cleverson V. Andreoli
Rubia L. D. Lima

SUMMARY

The present study evaluated the sediment of the Rio Verde Reservoir, located in the Curitiba metropolitan region. To this end, samples of the sediment profile were collected before and after the planting season (November, 2009 and June, 2010), by scuba diving and collecting samples using a 50cm PVC core. In 2009, the sediment was sampled at five points in the reservoir and in 2010 samples were taken at two points. Furthermore, interstitial water, water-sediment interface and water column were also analysed. The parameters assessed from the sediment samples included: grain size (granulometry); carbon and organic matter content; and the concentration of phosphorus, nitrogen, heavy metals and pesticides. Water samples were also collected for nutrient and pesticide assessment in the post-planting period. The reservoir sediment is an inorganic silt (organic matter content < 10%), with a neutral pH and very negative redox potential (an average of -160mV). The concentration of Dissolved Oxygen in the bottom water of the reservoir was higher than 6.5mg/L at both points sampled in 2010. Total phosphorus concentrations in the sediment were low (33.2µg/g maximum). In the interstitial water, maximum values were approximately 350µg/L. The maximum concentration of total nitrogen in the sediment was 4.391µg/g. Both, the concentration of P and N were below the levels recommended in CONAMA legislation 344/2004. Pesticides were not detected in the sediment. Chromium was the only heavy metal detected with concentrations greater than the maximum values outlined by current legislation. However, due to the reducing state of the sediment, we believe that chromium does not have an immediate negative effect on the biota. However, more detailed and longer-term studies are needed to evaluate chromium toxicity on the biota at the concentration found in the Rio Verde reservoir. Considering the physicochemical states of the sediment and the underlying water, we also speculate that a release of P into the water column, although minimal, is possible in anoxic conditions. However, the impact of this release on the phytoplankton population is unknown and should be further investigated.

KEYWORDS

Sediment, reservoir, phosphorus.

1. SEDIMENTS

Sediment is the outcome of the interaction of all the processes that occur in an aquatic ecosystem and sediment composition reflects the evolution and the intensity of the processes to which these ecosystems have been subjected (ESTEVES, 1998). Deforestation and agricultural practices in the region surrounding dams, lakes and rivers contribute to the high rate of sediment transport from terrestrial to the aquatic systems, particularly in periods of intense rainfall (TUNDISI, 2008).

The composition and distribution of the sediment depend on the geology of the region, the characteristics of the terrestrial vegetation, topography, meteorological characteristics, and chemical, biological, limnological and/or hydrological factors, including the movement of water (WARD, 1992).

The sediment of lakes and reservoirs is composed of complex mineral aggregates, organic and inorganic compounds, and water (CHRISTOPHORIDIS & FYTIANOS, 2006). In most ecosystems it is possible to distinguish two layers in the sediment: a recent or biological layer that presents an elevated concentration of organic matter and is in direct contact with the water column; and a permanent layer characterized by the low content of organic matter and commonly by anaerobiosis (ESTEVES, 1998). The permanent layer is composed of two different phases: particulate matter (solid phase), which is the energy source for benthic organisms; and the liquid phase which is comprised of interstitial water, with high concentrations of nutrients that may be diffused into the water column and have an important role in the primary productivity of aquatic ecosystems (ESTEVES, 1998).

Depending on various biological and physicochemical factors, sediments act as either a source or a sink of a large number of organic and inorganic compounds, particularly phosphorus (BOERS *et al.*, 1998; SØNDERGAARD *et al.*, 2003). Through the release of phosphorus, sediments can increase the concentration of this element in the water column, a process known as "Internal Loading" (GONSIORCZYK *et al.*, 1998; KIM *et al.*, 2003; SØNDERGAARD *et al.*, 2003; CHRISTOPHORIDIS & FYTIANOS, 2006), which can lead to the eutrophication of the water body.

2. SEDIMENT DYNAMICS

Sediment dynamics involves the formation of precipitate from chemical, physical and biological processes resulting from interactions of allochthonous and autochthonous particles. The presence of autochthonous particulate material is directly related to the system's productivity, while the presence of allochthonous particulate material depends on the land-use and occupation of the drainage basin.

As it is formed by organic and mineral particles, sediment has a high potential for retaining and releasing both polluting and non-polluting compounds. Several factors influence sorption processes (absorption, adsorption and desorption); however, the composition and structure of the organic and mineral material is particularly noteworthy. The sorption mechanisms of sediment elements are related to their chemical characteristics: cation sorption, anion sorption, metal complexation in the organic matter, precipitation of inorganic ions, and ternary complexes.

The concentration of chemical elements in the sediment is significantly higher than in the water column. In addition to deposition and accumulation of substances in the sediment, several biological, physical and chemical processes occur which are essential to the maintenance of the metabolism of the entire system, such as: i) flow of energy and matter: carbon deposition and recycling; ii) nutrient cycling: reactions of fixation and release to the water column; and iii) environment contamination: accumulation and release of pollutants (BOLLMANN et al., 2005). Most of these processes are interrelated and occur in both the solid and liquid phases of the sediment.

Through a physical, chemical and biological assessment of the stratigraphy of the sediment, it is possible to estimate previous contaminations, potential influences on and evolution of the aquatic ecosystem, as well as obtain important information for the management of the water system.

2.1 SOLID PHASE

The particulate material is comprised of: cyclic and acyclic organic compounds in several stages which are generally adsorbed on aluminosilicates, diverse mineral particles, rock fragments, carbonates and precipitated compounds of iron, manganese and aluminum, and other associated elements.

The ratio of mineral to organic fractions usually determines the type of sediment: organic sediments present an organic content higher than 10% DM (dry matter). Depending on whether the organic portion is autochthonous or allochthonous, these sediments may be classified as "gyttja" or "dy", respectively (ESTEVES, 1998; SCHÄFER, 1984). Inorganic sediments are characterized by an organic material content below 10% DM.

In areas of good basin preservation, mainly in forested regions, the organic fraction is high and it is the main source of energy for organism populations, including benthic organisms. However, other allochthonous sources of organic material such as agricultural and livestock activities may also be significant.

Organic structures in the sediment vary in size and particles can reach the size of colloidal structures comparable to clays. Such structures are electrically charged, predominantly negative, and therefore have the capacity to retain cationic compounds, although they do not have the same stability as aluminosilicates (BRADY, 1989; MELLO et al., 1984). Organic structures can, however, form stable complexes with metals, thus increasing the capacity of displacing metal pollutants (DICK & MARTINAZZO, 2006). Organic structures can also generate positive charges and adsorb anions; however, this occurs much less frequently than negative charges. The charge originates in the dissociation of carboxylic (-COOH), enolic (-OH) and phenolic (benzene-OH) groups and are pH dependent (BRADY, 1989). Thus, in a medium that is acidic, the capacity of adsorption through the organic compound is reduced because the deprotonation of H^+ is more difficult.

Understanding the organic matter content in sediments is important to evaluate the sediment's potential in the processes of sorption of chemical elements or compounds, including nitrogen and phosphorus.

Mineral portions are predominantly allochthonous, transported to water sources through erosion, runoff and rural and urban drainage, as well as through point source pollution from poorly managed activities in the basin.

The granulometric composition of sediment is mainly characterized by fine fractions (sand, silt and clay) and it rarely includes a significant amount of coarser fractions. This physical characteristic provides reactivity to the sediment because thin particles, mainly clays (aluminosilicates) and colloids, react with the compounds in the environment.

Isomorphic substitution and the breakdown of crystals promote an unbalance in the ionic equilibrium in clays, generating charge. When the balance is positive, there is a predisposition for anion adsorption; when the balance is negative, adsorption of cations occurs (BRADY, 1989; MELLO *et al.*, 1984)

2.2 LIQUID PHASE

Interstitial water is very important in the exchange of elements from the sediment to the water column because it can have high concentrations of some phosphorus and nitrogen compounds, such as phosphates, ammonia and nitrates. Nutrient availability for plants and aquatic microorganisms is related to the spectrum of nutrient resources and regeneration processes determined by chemical processes, mainly the redox potential (TUNDISI, 2008).

Desorption and dissolution processes and organic substance decomposition are related to the mobilization of elements at the water-sediment interface. On the other hand, diffusion, gas emissions, biologic disturbances and turbulence produced by wind are related to the transfer of elements to the hypolimnion (WETZEL, 2001; TUNDISI, 2008).

Those processes will be further discussed in item 3, "Sediment and phosphorus".

2.3 SEDIMENT INPUTS

Diffuse pollution is the major factor in the transportation of sediments and other materials to water bodies. Put simply, diffuse pollution can be considered the pollution load that essentially enters water courses during rainfall.

The expansion of urbanization in recent decades increased soil sealing in cities, leading to an increase in runoff. On the other hand, agricultural activities also produce a significant input of nutrients, particularly given the large amounts of fertilizers used on some crops. Problems related to runoff and the input of nutrients is more serious when there is a lack of adequate management and conservation practices. These conditions create diffuse pollution loads that negatively affect water quality through the accumulation of elements and chemical substances, as well as organic and mineral materials, in the bottom sediments of water bodies.

These elements and materials can be transported to water bodies through runoff, which is the flow over the soil surface and its multiple channels and the main phenomenon responsible for erosion; to a lesser extent transportation occurs through subsurface flow or interflow, which is the flow beneath the surface, through the soil profile, and also through groundwater flow.

Due to the fact that rainfall is spatially and temporally random, it is extremely difficult to pinpoint the input of sediments and pollutants into watercourses. Further, it is difficult to forecast and quantify diffuse pollution which is often the main contributing factor to eutrophication. Thus, the evaluation of the bottom material of these environments allows us to determine, with some accuracy, the contributions to and the usage of the basin. The issue of load contribution is discussed more extensively in Chapter 10 – Evaluation of Concentration and Loads.

3. SEDIMENT AND PHOSPHORUS

Phosphorus is an essential element for all organisms because it is used in fundamental processes such as the storage and transfer of genetic information (DNA and RNA), cell metabolism (several enzymes) and in the energy system of cells (adenosine triphosphato – ATP) (BRÖNMARK & HANSSON, 2005).

Even when stationary or fixed to the sediment, phosphorus may be carried to water bodies. Due to the high affinity of phosphorus to soil materials, particularly aluminosilicates, this element has low mobility and, consequently, is infrequently lost through lixiviation. Therefore, runoff and erosion are the predominant ways by which phosphorus in transported to water bodies. Table 1 presents the main forms of phosphorus in aquatic ecosystems. The processes of traction, suspension and solubilization can transport practically all forms of phosphate, each creating distinct deposition characteristics in the system.

According to Filippelli & Delaney (1996) and Boers *et al.* (1998), phosphorus retention in aquatic systems begins with the deposition of particulate compounds in the sediment and the effect of hydrochemical processes on the suspended material or material undergoing sedimentation. Furthermore, the processes that occur at the water-sediment interface are also very significant in phosphorus retention (ESTEVES, 1998; TUNDISI, 2000; BAUMGARTEN *et al.*, 2001).

Recycling of phosphorus in the sediment to the water column can occur through at least five different mechanisms: diffusion of dissolved phosphorus through the water-sediment interface; sediment resuspension caused by wind or propelling currents; bioturbation by benthic invertebrates and fish; absorption by plants with subsequent excretion into the water column; and the release of phosphorus during plant decomposition (REDDY *et al.*, 1999). The importance of each mechanism in the phosphorus cycle varies greatly among lakes and reservoirs because it depends on the chemical, physical and biological state of each ecosystem. Changes within a lake or reservoir or changes over time must also be taken into account. Considering these mechanisms and their variations, the sediment can either act as a source or a sink of P.

Phosphorus resuspension from the sediment to the water column may be substantial, particularly when external inputs of phosphorus are not significant. Stumm & Morgan (1996) and Esteves (1998) considered that the water-sediment interface can act as permanent or temporary reservoir of phosphorus and its release usually occurs in anoxic environments.

Several factors influence the rate of phosphate exchange between sediment and hypolimnion. According to Carvalho (1995), water temperature, pH and redox potential are directly or indirectly implicated in these processes. The type and degree of phosphate complexation with materials from the sediment are also important.

Temperature, besides catalyzing chemical and biological reactions, affects oxygen solubility in water. When temperatures increase, the solubility of the gas decreases. Temperature also directly affects water density, causing changes in thermal stratification; colder water is heavier and denser, thus causing its displacement to lower depths (Figure 1). The specific thermal stabilization of each stratum may result in chemical stratification, a common process, particularly in still, deep waters.

The pH controls the solubility of chemical species and compounds; in acidic conditions, soluble forms are favored, while in conditions tending toward the alkaline, complexation and chelation reactions predominate. The pH also determines the surface charge density of some minerals which influences cation retention and releasing capacity.

The redox potential of the environment affects the degree of oxidation of chemical species, thus influencing

TABLE 1 – MAJOR FORMS OF P IN AQUATIC ECOSYSTEMS

PHOSPHATE	SOLUBLE FORM	INSOLUBLE FORMS
Inorganic	H2PO4-, HPO42-, PO43- (Ortophosphates)	Clay-phosphate complexes
	FeHPO4+ (ferric phosphate) CaH2PO4+ (calcium phosphate)	Metal-mineral hydroxides complexes (Ex: apatite – Ca10(OH)2(PO4)6
Organic	Dissolved organic compounds: phosphatases, phospholipids, inositol, phosphoproteins, etc.	Phosphorus complexed to organic matter

Source: adapted from Stum & Morgan (1981).

their solubility and chemical reactivity. Usually, a layer of ferric phosphate (oxidized layer) can form which creates an interaction barrier between the sediment and the underlying water. The intensity of phosphate transportation through this layer depends on the degree of anoxia or re-oxygenation of the system that alters the state of oxidation or reduction of iron. In a state of oxidation with a high redox potential (400 or 500 mg), iron and manganese (Fe^{3+} and Mn^{3+}) are insoluble, forming precipitates or complexes in the form of hydroxides. In reduced states (Fe^{2+} and Mn^{2+}) iron and manganese are soluble and free of complexation (CHRISTOPHORIDIS & FYTIANOS, 2006; TUNDISI, 2008), i.e., without phosphorus precipitation.

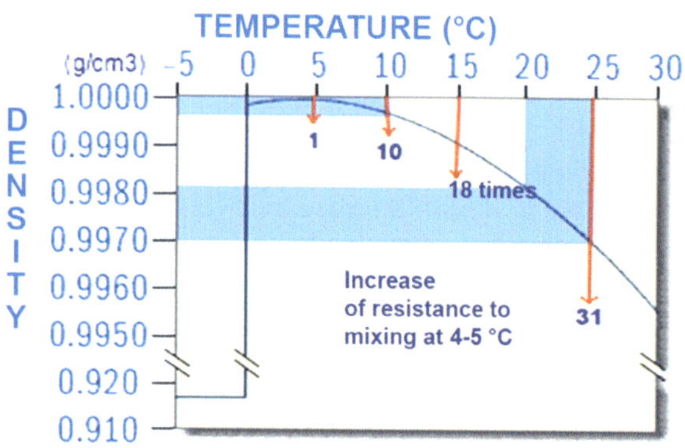

FIGURE 1 – RELATION OF TEMPERATURE X WATER DENSITY
Source: Adapted from http://www.ufrrj.br/institutos/it/de/acidentes/tem.htm.

According to Esteves (1998), when the concentration of Fe^{3+} is higher than that of PO_4^{3+}, practically all orthophosphates are precipitated; on the other hand, when the concentration of Fe^{3+} is the same or lower than that of PO_4^{3+}, orthophosphates are not completely precipitated, even in a state of oxygen availability in the environment. However, as the medium is reduced, the reduction of Fe^{3+} to Fe^{2+} occurs and the reduced ion solubilizes in the medium, thus releasing phosphates that are bound to their functional groups (SCHENATO *et al.*, 2008). In situations of redox potential lower than + 200mV, the release of part of the phosphorus sorbed on the surface or inside these iron compounds occurs, likely at sites with lower binding energy (more information in PERSSON & JANSSON, 1988; ESTEVES, 1998, NOVAIS & SMYTH, 1998; BOERS *et al.*, 1993; DILLON & RIGLER, 1974; BOLLMANN *et al.*, 2005).

An important aspect to be considered in this process is the possibility of the water-sediment interface maintaining a high degree of oxidation. Thus, if this oxidation layer between sediment and water is maintained through turbulence, profile management or artificial aeration, compounds in the sediment and interstitial water are retained.

Phosphorus adsorption capacity is an important indicator of the phosphorus reserve in sediments; however, it seems to have very little impact on the rate and amount of P released (desorbed) to the water column. The release of phosphorus is related more directly to changes in binding energy, which is mainly influenced by chemical parameters (PELLEGRINI, 2005), and also by the oxidation level of the water-sediment interface if there are enough ions and favorable conditions for the precipitation of the available phosphorus.

Phosphorus adsorption on clay minerals may lead to a situation where the release of P is more difficult. In such cases, adsorption is the initial phase of the process, with a subsequent formation of immobilized phosphorus, called non-labile P. According to Barrow (1983 and 1985) *apud* Novais & Smyth (1999), initially only simple coordinate bonds are formed on the surface of oxy-hydroxides and aluminosilicates; subsequently and more gradually, phosphate penetration occurs through imperfections or between crystals.

Although adsorption and exchange of certain anions occurs by simple reactions, sometimes the reactions are more complex, such as for phosphates or sulfates, due to specific reactions between these anions and certain sediment components; thus, $H_2PO_4^-$ ion can react with a protonated hydroxyl group leaving the solution and becoming a more complex molecule in the solid phase instead of remaining as an easily adsorbed anionic form.

Iron oxides and hydroxides adsorb phosphates through specific high energy bonds. Figure 2 shows P adsorption on a Fe oxy-hydroxide. Adsorption of complex phosphorus anions is performed by negative charges (OH^- groups) of organic matter, coloids and clay minerals associated to Fe and Al oxy-hydroxides, and some 1:1 clays and allophane clays. However, we must highlight that some studies (BAHIA FILHO, 1982; SMITH, 1996 *apud* NOVAIS & SMYTH, 1999; FERNANDEZ, 1995) have shown that oxides, hydroxides and oxy-hydroxides of Fe and Al, especially goethite, have a significant influence on phosphorus adsorption and immobilization.

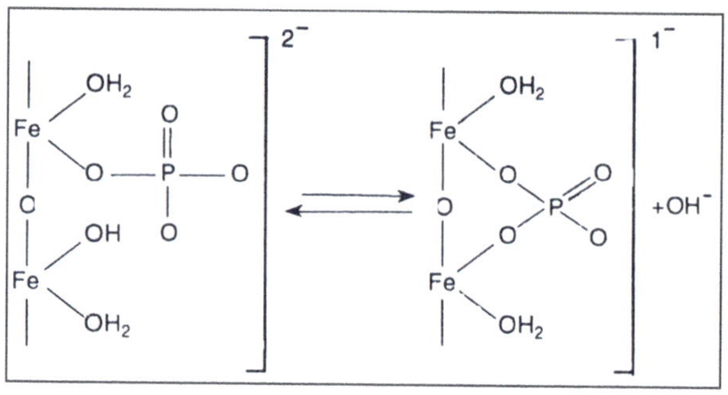

FIGURE 2 – REPRESENTATION OF PHOSPHATE ADSORPTION ON MONO AND BI-DENTATE BINDINGS OF Fe OXY-HYDROXIDE LEADING TO THE FORMATION OF NON-LABILE P
SOURCE: Novais & Smyth (1999).

4. SEDIMENT AND NITROGEN

Similar to phosphorus, nitrogen can limit the growth of populations of aquatic organisms because it is used in the synthesis of proteins and amino acids (STERNER & ELSER, 2002). Nitrogen may enter aquatic ecosystems through precipitation, fixation of atmospheric nitrogen (N_2) or input through surface or underground drainage areas (BRÖN-

MARK & HANSSON, 2005). The main sources of nitrogen for aquatic plants and microorganisms are nitrate, nitrite and ammonia, as well as some dissolved organic compounds such as amino acids and urea.

In freshwater aquatic ecosystems, only some cyanobacteria, such as Anabaena and Cylindrospermopsis, can fixate the available atmospheric nitrogen and as such phosphorus is usually the limiting element. In polluted lakes with high P concentrations, nitrogen can assume this function. Nitrogen concentrations in lakes can vary greatly from about 100μg/L to 6000μg/L (BRÖNMARK & HANSSON, 2005).

The decomposition of organic matter in the sediment releases NH_4^+, which can be: i) adsorbed on sediment particles; ii) assimilated and directly metabolized by the sediment's biota; or iii) transformed into nitrate (nitrification) by bacteria. Nitrate (NO_3^-) present in bottom water can be diffused into the sediment. Denitrification in aquatic sediments is an important way of removing nitrogen from aquatic environments because it converts NO_3^- into N_2 in conditions of low levels of dissolved oxygen. A large proportion of the NO_3^- produced in the sediment is denitrified and a large part of the organic-N (75 – >95%) mineralized in the sediment is lost through both nitrification and denitrification (WETZEL, 2001). More information on nitrogen dynamics in water can be obtained in McCarthy (1975), Snoeyink & Jenkins (1980), Esteves (1998), and Pankow (1997).

5. SEDIMENT AND PESTICIDES

According to Federal Law 98.816, Article 3, Paragraph I, the term pesticide is defined as the set of products and agents of physical, chemical and biological processes that are intended for use in the sectors of production, storage, and processing of agricultural products and pastures, and the protection of native and planted forests, and other ecosystems (ANDREI, 2005).

5.1 USE OF PESTICIDES AND THEIR ACTIVE INGREDIENTS

Brazil is currently the second largest market for phytosanitary products, trailing only the United States. Virtually all agriculture in Brazil that is economically significant is dependent on the intensive use of pesticides. Organic farming, which reduces the use of agrochemicals, is still an insignificant portion of the market. Brazil is responsible for the consumption of about 50% of the pesticides used in Latin America, a trade that is worth about US$ 2.56 billion (SINDAG, 2006).

The active ingredient of a pesticide is directly related to its chemical composition. It is therefore possible to classify pesticides into three categories: a) inorganics: made up of fluorides, arsenic, copper and mercury compounds; b) natural organics: including rotenone, pyrethrum, nicotine; and c) chlorinated hydrocarbons (organochlorines), organic phosphates (organophosphates), thiocarbamates (MATOS, 2001).

Considering the mode or mechanism of action, pesticides can be: acetolactato-synthase inhibitors; acetyil-coenzyme A-carboxylase inhibitors; inhibitors of photosynthesis; synthetic auxin and enolpiruvil-shikimate 3-phosphate synthase inhibitors. The mode of action relates to the first enzyme, protein or biochemical step that affects the object.

5.2 ACCUMULATION, LIXIVIATION AND PERCOLATION POTENTIAL IN THE SOIL

Sediment contamination is caused by organic or inorganic toxic substances. However, the presence of contaminants in the sediment does not necessarily affect water quality or local biota. It is possible that these compounds and elements are in forms that are unavailable or difficult to dissolve and therefore are unlikely to cause damage to the ecosystem.

The sediment's capacity to accumulate stable compounds and foster the formation of certain compounds, makes it one of the most important aspects in assessing the level of contamination of aquatic ecosystems (ESTEVES, 1998).

The movement of pesticides in the soil occurs mainly via water that flows on the surface of the soil (torrent) or water that percolates through the soil profile (lixiviation). These two transport mechanisms can result in the contamination of water resources. According to Piasarolo *et al.* (2008), the contamination of water courses with pesticide residues (such as atrazine, aldicarb, organochlorines, etc.) in agricultural areas occurs mainly through runoff.

According to Guimarães (1987), the interdependent factors that determine the destination of pesticides in the environment are: decomposition (chemical, photochemical or biologic); volatilization; lixiviation; lateral drag; solubility and partition coefficient; adsorption/desorption processes; and removal by plants and microorganisms. These processes result in the persistence, degradation, mobility and bioaccumulation of the chemical compound.

In response to the problems of groundwater contamination, numerous studies have discussed solute transport in porous media. Generally the transportation of contaminants in soils is related to advection, dispersion and loss or gain of solute mass processes, as a result of reactions or decay (either chemical, biochemical or radioactive) (FREEZE & CHERRY, 1979).

Moncada (2004) and Nascentes (2006) presented several chemical and biological processes and the factors that contribute to the reaction of soil-solute interaction. Those processes are listed in Table 2.

All processes presented in Table 2 are linked to the retention/delay of the contaminant in the medium. Solutes may be sorbed on the surface of mineral particles by organic carbon or suffer chemical precipitation. They may be subjected to degradation, suffer redox reactions, and radioactive compounds may suffer decay.

The degradation process consists of changes in the chemical structure of the pesticide, caused by chemical reactions with soil organisms. The degradation of pesticides in the environment occurs through chemical, physical and biological processes. The speed of degradation depends on factors such as the chemical structure, temperature, humidity, soil composition, water acidity and salinity, microbial populations, and plant species, among others.

TABLE 2 – CHEMICAL AND BIOLOGICAL PROCESSES AND THE FACTORS THAT CONTRIBUTE TO THE REACTION OF SOIL--SOLUTE INTERACTION CEC: CATION EXCHANGE CAPACITY; OM: ORGANIC MATTER

REACTIONS	CHARACTERISTICS		
	SOLUTION	SOIL	ENVIRONMENT
Adsorption	Concentration	Granulometry	Climatic conditions
Ion exchange	pH	Mineralogy	Hydrogeology
Precipitation	Density	CEC	Atmospheric Pressure
Oxi-Reduction	BOD, QOD	Content and type of OM	Redox Potential
Complexation	Polarity	Distribution of voids	Microorganisms
Biodegradation	Solubility	Saturation degree	Temperature
Decay	Vapor Pressure	Type of Cations	Aerobic/anaerobic conditions

SOURCE: adapted from Moncada (2004) and Nascentes (2006).

Organochlorines are more stable than organophosphates and carbamates when subjected to the same conditions (BARBOSA, 2004).

Biodegradation occurs when the enzymes excreted from the metabolic process of microorganisms come in contact with pesticide molecules, inside or outside microbial cells, resulting in a series of reactions, such as: oxidation, reduction, hydrolysis, etc. (LAVORENTI, 1996).

The transformation of the molecule through degradation may be complete, creating CO_2, H_2O and mineral salts, or incomplete, creating metabolites. Usually the metabolites formed in the process are less toxic than their original form, although occasionally the resulting product is more toxic than the original molecule (COX, 1997).

According to Felsot & Dzantor (1990), pesticide degradation may be encouraged when organic molecules are added to the soil, because they supply energy and nutrients to microorganisms which are able to promote the degradation of the molecule.

Depending on the physicochemical characteristics of the pesticide residue, once in the water the residue can either: bind to suspended particulate material, be deposited in the sediment at the bottom, or it can be absorbed by microorganisms and then detoxified or accumulated. Some pesticides and metabolites may also return to the atmosphere through volatilization. Thus, there is clearly a continuous interaction between pesticides, the sediment and water, which is influenced by water movement, turbulence and temperature (WHITE & RASMUSSEN, 1985).

The persistence of pesticides in the soil is quite variable, as shown in Table 3. Some pesticides remain in the soil for several days, such as Malathion and Parathion (organophosphates), while others persist for several years, such as Chlordane and DDT (organochlorines). This information is important in evaluating the pollution potential of pesticides.

6. SEDIMENT AND METALS + ARSENIC

Heavy metals are a group of trace elements with atomic densities greater than 6g cm^{-3}. They are highly reactive and in normal conditions do not degrade. The major sources of heavy metals in aquatic ecosystems are through the weathering of rocks and past volcanic activities, erosion and runoff, industrial activities, domestic sewage, fertilizers and agrochemicals.

The heavy metals associated with sediments are usually classified as residual and non-residual. Residual elements stem from the crystalline structure of minerals (silicate matrix) of the sediment. Non-residual heavy metals are incorporated into the sediment by physical or chemical processes such as adsorption, precipitation and complexation with organic and inorganic substances. In general, Cr, Ni, Cu, Zn, Hg, Pb, Cd are the result of anthropogenic activities and thus they are non-residual.

Some metals, such as Hg, As and Cr, can accumulate in the aquatic biota and can occasionally cause health disorders, including cancer.

TABLE 3 – GENERAL PERSISTENCE IN SOIL OF SOME PESTICIDES

PESTICIDE	PERSISTENCE IN SOIL
Chlorinated hydrocarbons insecticides insecticidesinseticideslorados	> 18 months
Ureas, triazines and picloran herbicides	18 months
Acid benzoic and amida herbicides	12 months
Phenoxys, toluidines and nitrite herbicides herbicides	6 months
Carbamate and aliphatic herbicides	3 months
Phosphate insecticides	3 months
Chlordane	5 years
DDT	4 years
BHC, Dieldrin	3 years
Heptaclhor, aldrin	2 years
Diazinon	12 weeks
Disulfon	4 weeks
Malathion, parathion	1week
Propazine, pichloron	18 months
Simazine	12 months
Atrazine, monuron	10 months
CDAA, dicamba	2 months
2,4 – D	1 month
TCA	12 months

Source: Macedo, 2002

7. SEDIMENT AND POLYCHLORINATED BIPHENYLS: PCBS

PCBs are organochlorine compounds resulting from the reaction of the biphenyl group with anhydrous chlorine in the presence of a catalyst. The main factors that affect the availability of PCBs in the sediment are: organic matter content, surface contact and the octanol-water partition coefficient.

PCBs are resistant to oxidation and reduction and they are thermally stable and non-flammable. As such, they are used widely in industry but are also persist in the environment and are an environmental contaminant.

Currently in Brazil, records indicate that PCBs are not produced in the country. Restrictions on their use were implemented by Inter-ministerial Ordinance 19 in 1981, which prohibits the manufacture, marketing and use of PCBs nationwide. However, the use of equipment that had been used for the application of PCBs was allowed until they were replaced.

A study by Penteado & Vaz (2001) outlined the main sources of PCBs in the environment: i) industrial and urban effluents with PCBs; ii) accidents or spills during the handling of PCBs and (or) fluids containing PCBs; iii) vaporization of compounds with PCBs; iv) leaks in transformers, capacitors or heat exchangers; v) decomposition/removal of paints and preservatives; vi) leaks of hydraulic fluids containing PCBs; vii) irregular storage of residues containing PCBs or contaminated residue; and viii) smoke from the incineration of products containing PCBs.

8. SEDIMENT AND POLYCYCLIC AROMATIC HYDROCARBONS – PAHS

Polycyclic aromatic hydrocarbons (PAHs) are hydrophobic crystalline compounds with a high melting point under normal temperature and pressure conditions. They have from 2 to 7 condensed aromatic rings, with properties similar to benzene. They are occur widely and are found in complex mixtures in all aspects of the environment, especially in the sediment (FRONZA, 2006).

PAHs are formed during the incomplete combustion of organic matter (XAVIER, 2005), originating from both human and natural activities (BETTIN & FRANCO, 2005; FRONZA, 2006). The main sources of PAHs in the environment are: microbial activity, disposal of residues from waste incineration, production of asphalt and creosote oil, burning of fossil fuels, domestic sewage and industrial effluents (CETESB, 2001). Some important scientific studies have established an association between the presence of PAHs with the original contaminants, as well as the relationship between their different compounds.

Due to their low solubility in water, they present high persistence levels in the environment. PAHs are easily adsorbed on suspended particulate matter, thus they tend to accumulate in bottom material. In addition, PAHs do not have a significant capacity to re-solubilize in the water column as compared to metals and, consequently, they accumulate in the sediment.

9. SEDIMENT ANALYSIS

9.1 SAMPLING STRATEGIES AND PROCEDURES

Much of the disturbances occurring in the drainage area are directly or indirectly detected in not only the water, but also in the sediment and sediment analysis can elucidate the activities that occur in the drainage area. Thus, a thorough water quality assessment and monitoring plan should include systematic evaluations of the sediment, allowing the diagnosis of impacts that otherwise could go unnoticed. Integrating assessments of the physical and chemical states of the water and sediment can help identify possible impacts generated by activities occurring in the basin.

Furthermore, similar to water quality assessments, several environmental variables allow for the identification of the physical, chemical and biological states of the sediment and therefore the purpose of the assessment must be defined. Certain variables may specify and clarify a given diagnosis; however, it is necessary to clearly understand the purpose of the investigation, as well as which variables are more adequate for identifying and quantifying specific impacts. In some situations more detailed analyses are required. Nevertheless, basic biological, physical and chemical parameters have been widely used in sediment studies that have provided consistent data. They are thus frequently used in environmental management and monitoring plans.

The evaluation of the quality of several variables in the sediment in the Rio Verde Reservoir, Araucária, Paraná, is presented below as a case study which seeks to facilitate the evaluation of the proposed research. This study represents the first physicochemical evaluation of the sediment of this reservoir since it was filled more than 40 years ago. The definition of the parameters of this study was supported by CONAMA Resolution 344/2004, DD 195-2005- E/ CETESB, along with the technical knowledge of the researchers. The purpose was to evaluate the physical and chemical quality of the reservoir sediment as a function of its usage, as well as to assess the influence of the sub-basins on the drainage area. As such, this assessment can be replicated in other areas; however, in all cases, the selection of points, parameters, laboratories, logistics, and available resources must be carefully evaluated and adapted to the situation. We recommend thorough consideration of item 2 – RELIABILITY OF WATER QUALITY ASSESSMENTS presented in Chapter 10, because many of these considerations can and must be applied to sediment analyses.

9.1.1 Fundamentals of sediment evaluation

CONAMA Resolution 344/2004 is the only national legislation on sediment evaluation; however, this legislation focuses on the dredging of marine and inland waters and sediment reuse in other activities. We therefore chose to analyze major contaminants and elements that are implicated in the eutrophication process and are present in the legislation, as well as the guiding principles of the Environmental Sanitation Technology Company (Companhia de Tecnologia de Saneamento Ambiental – CETESB), entitled "Establishment of guiding values for soils and groundwater in the state of São Paulo" ("Estabelecimento de Valores Orientadores para Solos e Águas Subterrâneas no Estado de São Paulo"). The guiding principles were first published in the Official Gazzette; Business; São Paulo, 111 (203), October 26, 2001, and were later repealed by Board Decision 195-2005- E, in November 23, 2005, and replaced by the "New Guiding Values for Soils and Groundwater in the state of São Paulo" ("Novos Valores Orientadores para Solos e Águas Subterrâneas no Estado de São Paulo"). This Board Decision 195/2005 – CETESB implements different assessment levels and consequently updates content levels for elements and compounds, as well as quality, prevention and intervention levels.

9.1.2 Analytical parameters

The following parameters were analyzed: phosphorus, nitrogen, pesticides, metals, arsenic, pH, ORP, granulometry and organic matter.

The analyses of pesticides in the sediment included the following compounds recommended in CONAMA Resolution 344/2004 and CETESB (n.°195-2005-E): Cyanide Benzo (b) fluoranthene, benzo (g, h, i) perylene, benzo (k) fluoranthene, indeno (1,2,3, cd) pyrene, 1,2,3,4-tetrachlorobenzene, 1,2, 3.5-Tetaclorobenzeno, 1,2,4,5-Tetaclorobenzeno, Chlordane (cis and trans), pentachlorophenol, Percent Solids, BHC Alpha, Beta BHC, BHC Delta, Gamma BHC, cis Chlorane , trans Chlorane, DDD (isomers), DDE (isomers), DDT (isomers), Dieldrin, Endrin, PCBs (sum-list Dutch), Benzo (a) anthracene, benzo (a) pyrene, Chrysene, Dibenzo (a, h) anthracene, Acenaftaleno, Acenaphthylene, anthracene, Phenanthrene, fluoranthene, fluorene, 2-methylnaphthalene, Naphthalene, pyrene, and sum of PAHs.

Metal analyses included the following elements: Aluminum, Antimony, Arsenic, Barium, Beryllium, Bismuth, Cadmium, Calcium, Lead, Cobalt, Copper, Chromium, Tin, Strontium, Iron, Phosphorus, Lithium, Magnesium, Manganese, Mercury, molybdenum, Nickel, Potassium, Silver, Selenium, Sodium, Vanadium, Zinc.

9.1.3 Sampling Areas

Sampling areas were established for sediment analysis based on the representativeness of each site, the geographical position in the reservoir, entry point of tributaries, depth and hydrodynamics of the reservoir. As such, five sample points were established in the Rio Verde Reservoir: R1, R2, R3, R4 and R5 (Figure 3, Table 4).

Two sampling periods were realized, one between October and November, 2009, and another in June, 2010. The scheduling of sample collection was implemented in order to coincide with agricultural activities; the first sampling period was conducted at the beginning of the planting season during soil preparation (October/November 2009) and the second was performed after the harvest (June, 2010), when pesticides and fertilizers are no longer applied.

9.1.4 Sample collection, preservation, packaging and analytical procedures

Sediment samples were obtained by scuba diving and collecting with the use of a PVC core of 50cm in length and 10cm internal diameter. The samples taken at points R1 and R4 are 50cm in length, but for the remaining points samples are only 30cm in length due to the difficulty of inserting the core into the sediment. During the October/November 2009 collection period, samples were collected at all five points in the reservoir. In June, 2010, for logistical reasons, only points R1 and R4 were sampled. The samples were sectioned into predetermined layers immediately after removal from the water. The layers were defined as: surface sediment (0 – 10cm), intermediate sediment (10 – 30cm), and deep sediment (30 – 50cm) (Table 5). Only surface and intermediate layers were obtained from the 30cm samples.

All samples collected were stored in clean containers appropriate for each type of analysis and kept refrigerated and protected from the light until they reached the laboratory. Sample collection and preservation followed the criteria outlined by the Environmental Protection Agency (EPA, 2006) for pesticides and the 'Standard methods for the examination of water and wastewater' (APHA, 2005) for the other parameters.

In addition to the samples mentioned above, samples from the interstitial water and the water-sediment interface were collected for phosphorus analysis in June, 2010. Nitrogen concentration was also evaluated in the water-sediment interface and in the sediment. The samples of interstitial water were collected with a 10 cm core that was transported to the lab in a closed cooler with ice. The core was only opened at the point of removal of a sub-sample. The sub-sample was processed by centrifuging at 3000 rpm, at 4°C, for 5 minutes. The supernatant (about 50 mL) was collected and frozen. In order to collect the water-sediment interface sample, a Van Dorn bottle was used and we assumed this sample as representative of the bottom water. To this end, the bottle was slowly submerged in a location where no sampling had taken place that day and the sample was taken only when the equipment touched the bottom of the reservoir.

During the June, 2010, campaign, a sediment sample of 50cm was obtained at points R1 and R4 to measure redox potential (ORP) and pH. For these measurements, a PVC core, identical to that used to obtain sediment samples, was perforated at 5cm intervals (Figure 4). The holes were individually sealed with duct tape, which were removed at the time of measurement and immediately replaced. The measurements were performed with the aid of a Digimed DM-2P portable pH meter equipped with a Model DME-CV2 pH Electrode and a combined platinum redox annular diffusion electrode (DMR-CP2).

The physicochemical parameters of the water column, such as temperature, pH, dissolved oxygen and conductivity, were measured in the field every 0.5m with the aid of a Horiba model U22XD® multiparameter sensor during both the 2009 and 2010 collection periods. In 2010, in addition to the sediment samples, water samples were also collected for the analysis of phosphorus, nitrogen and pesticides (Table 4). The water samples were obtained at two or three depths according to the euphotic zone, estimated by the Secchi depth and with the aid of a Van Dorn bottle. These samples were mixed to create a composite sample of the euphotic zone. The water samples were protected from light and heat during transportation to the laboratory. Sub-samples were then pre-filtered through a glass fiber filter (GF-3 Macherey-Nagel®, 0.6 μm pore diameter) for the analysis of dissolved N and P. The samples were stored in a freezer until they were sent for analysis.

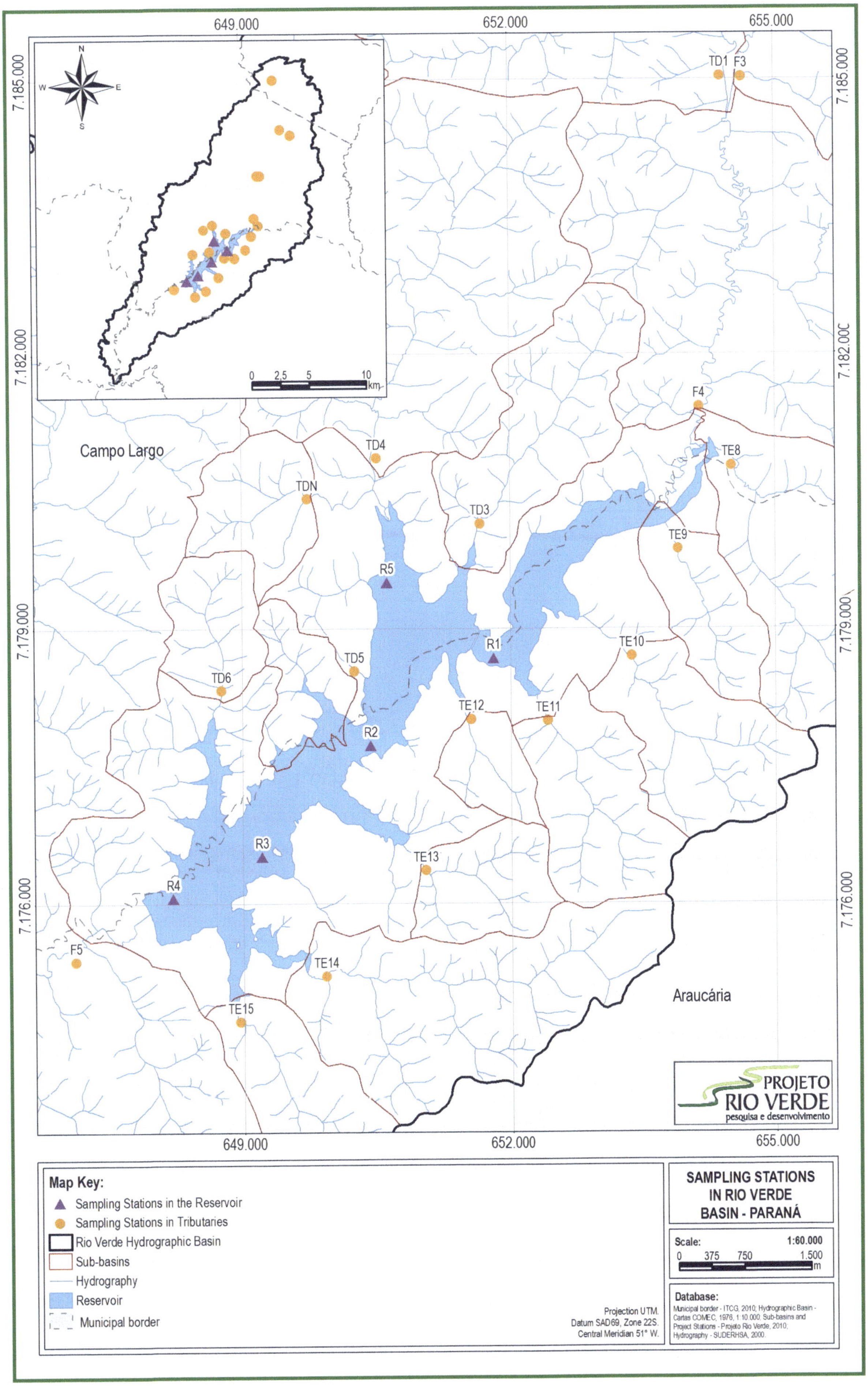

FIGURE 3 – REDOX POTENTIAL AND pH MEASUREMENT IN THE RIO VERDE RESERVOIR SEDIMENT

TABLE 4 – PARAMETERS AND RESPECTIVE SAMPLING POINTS FOR SEDIMENT EVALUATION IN RIO VERDE RESERVOIR OVER TWO PERIODS: 2009 AND 2010

PARAMETER	OCT./NOV. 2009	JUL. 2010
Pesticides (sediment)	All points *	R4 and R1 *
Pesticides (water)		R4 and R1
P and N (sediment)	All points *	R4 and R1 **
P and N (water column)		R4 and R1
P (interstitial water)		R4 and R1
P and N (interface water-sediment)		R4 and R1
pH and ORP (sediment)		R4 and R1 **
Granulometry in Org Matter (sediment)	All points **	R4 and R1 *
Metals (sediment)		R4 and R1 ***

NOTE: * Only surface sediment (0 – 10cm); ** 50cm core *** 30cm core.

TABLE 5 – LOCATION OF SAMPLING POINTS IN THE RIO VERDE RESERVOIR AND DEPTH OF THE SEDIMENT LAYERS SAMPLED

POINT	COORDINATES UTM (SAD69)	PROFILE (cm)	PARTICULARITIES
R1	x:0651812 y:7178676	0-10 10-30 30-50	Region closest to the head of the reservoir
R2	x:0650436 y:7177720	0-10 10-30	Intermediate region
R3	X: 0649212 y: 7176510	0-10 10-30	Intermediate region
R4	x:0648214 y:7176058	0–10 10–30 30-50	Deepest region, next to the dam
R5	x: 0650625 y: 7179505	0-10 10-30	Region of low water speed

FIGURE 4 – MAP OF SAMPLING POINTS

9.2 ANALYTICAL RESULTS AND DISCUSSION

9.2.1 pH, DO and temperature

As mentioned above, some auxiliary parameters were concurrently evaluated for the water with the goal of providing further support for the sediment quality analysis.

Figures 5 and 6 present the results of pH, dissolved oxygen (DO) and temperature of the water column for points R1 and R4.

The temperature data showed a more pronounced vertical gradient in 2009 than in 2010 at both sample points. R1 presented higher temperatures in both sampling periods than the deeper point R4.

The pH was relatively homogenous in the water column over both sampling points and periods (means of 6.8 to 7.1 for R1 and R4, respectively) and within normal values.

The DO concentrations in the epilimnion were consistently above 5 mg/L for all samples. On the other hand, DO concentrations in the hypolimnion were lower in 2009 (2.5mg/L in R1 and 3.4mg/L in R4) which occurred during a cold period. The electrical conductivity was also measured with an average value of 0.091 (± 0.004 standard deviation) at R1 and 0.077 (± 0.001 standard deviation) at R4. As with the pH values, these values are very similar.

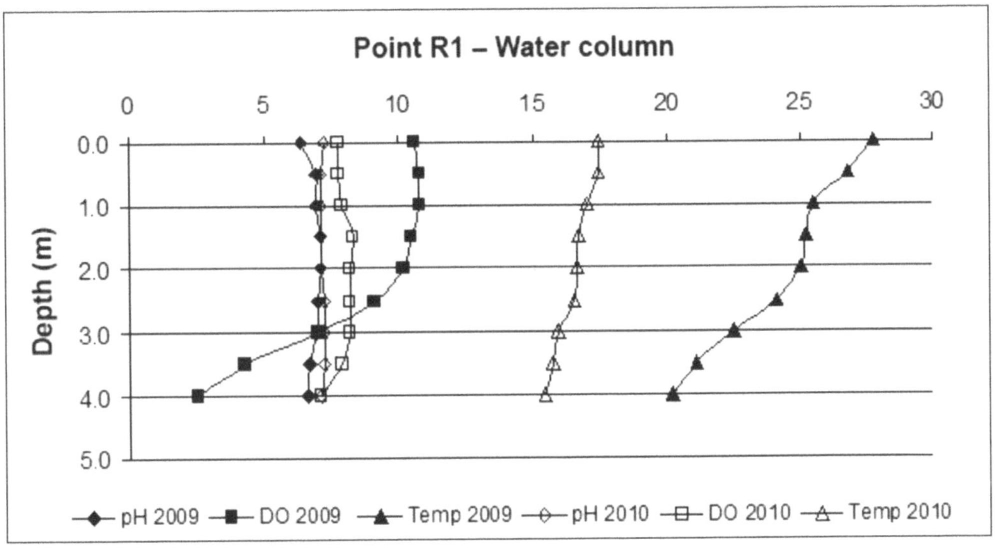

FIGURE 5 – pH, DISSOLVED OXYGEN (DO) CONCENTRATION AND TEMPERATURE (TEMP) IN THE WATER COLUMN AT POINT R1 OF THE RIO VERDE RESERVOIR IN 2009 AND 2010.

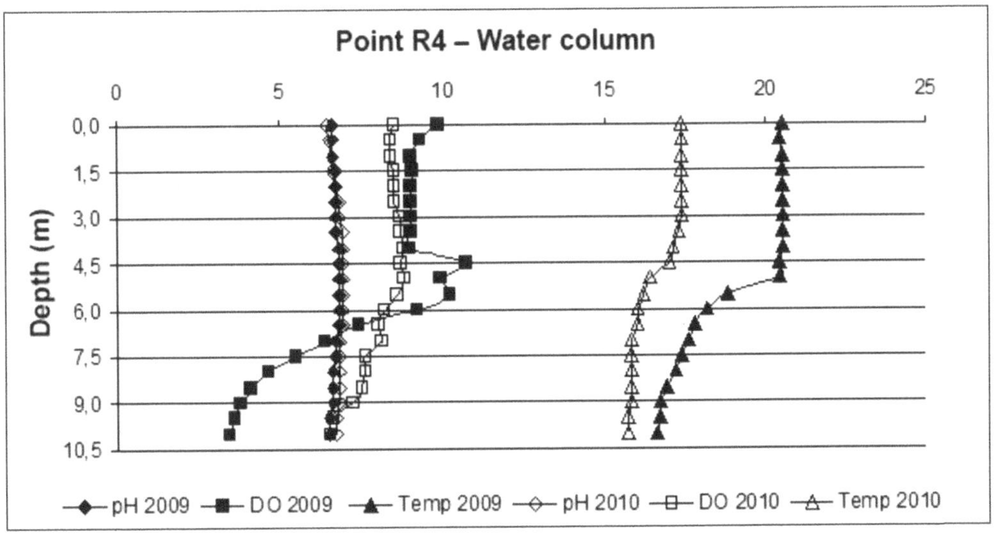

FIGURE 6 – pH, DISSOLVED OXYGEN (DO) CONCENTRATION AND TEMPERATURE (TEMP) IN THE WATER COLUMN AT POINT R4 OF THE RIO VERDE RESERVOIR IN 2009 AND 2010

9.2.2 Granulometry

Table 6 shows the results of the granulometric analyses of the sediment from the Rio Verde Reservoir.

In the Rio Verde Reservoir there is a prevalence of soils that are rich in fine sediments. The clay fraction is particularly prevalent with sizes smaller than 0.002mm in most of the sampled points. Only at point R5 did the results show lower values for the clay fraction. However, the proportion of silt, which is also thin (between 0.002 and 0.06mm), was high (ABNT NBR 6502/95).

The granulometric composition of the fractions at R5 was different than the other sample points. While all other points showed variations in clay and silt content, they presented a more homogenous pattern with respect to percentages of sand which reached a maximum of 1% in the surface layer. For the sand fraction at point R5, values of 22% were observed in the surface layer and 12% in the deepest layer (10 to 30cm), suggesting that the sub-basins

TABLE 6 – GRANULOMETRIC ANALYSIS IN 3 LAYERS (I = 0 – 10cm; II = 10 – 30cm; III = 30 – 50cm) OF THE SEDIMENT PROFILE FROM THE RIO VERDE RESERVOIR, 2009/2010

SAMPLE	YEAR	% GRAVEL	% SAND	% SILT	% CLAY
R1Sed I	2009	0	0.09	29.34	70.57
	2010	0	0.44	20.02	79.55
R1Sed II	2009	0	0.10	12.49	87.41
R1Sed III	2009	0	0.07	31.80	68.14
R2Sed I	2009	0	0.37	26.06	73.56
R2Sed II	2009	0	0.55	27.73	71.72
R3Sed I	2009	0	1.02	28.01	70.97
R3Sed II	2009	0	2.05	31.87	66.08
R4Sed I	2009	0	0.11	30.98	68.91
	2010	0.32	4.93	21.47	73.28
R4Sed II	2009	0	0.30	24.69	75.01
R4Sed III	2009	0	0.87	33.24	65.89
R5Sed I	2009	0.09	22.13	72.12	5.66
R5Sed II	2009	0.18	12.30	83.07	4.45

that contribute to this part of the reservoir have more sandy soil in comparison to the other areas in the basin.

The abundance of fine sediments at the sample locations across the reservoir was expected because of the lotic environment, the dam and the characteristics of the basin soils (CALLISTO & ESTEVES, 1996).

9.2.3 pH and ORP

pH is an important parameter in controlling metallic ion precipitation, mobility, and bioavailability in the sediment (ESTEVES, 1998). The redox potential (ORP) is significant in the dynamics related to the ions contained in the sediment and water column, particularly nutrients such as nitrogen and phosphorus. This parameter is important for understanding and assessing the sorption processes (adsorption, absorption and desorption) of nutrients in the sediment.

The sediment pH was assessed in the 2010 sampling period. The values were relatively homogenous across the whole profile (Figure 7), with the mean at point R4 (7.1) slightly higher than that at R1 (6.7) and similar to those observed in the water column.

On the other hand, the ORP evaluated in 2010 presented a greater variation at both points (R1 and R4) (Figure 8), with a mean value at R4 (-172mV) lower than at R1 (-151mV). These results suggest that the sediment has a lower reduction potential but it is not completely anoxic, which would present electronegativity values of about -300mV (SCHENATO *et al.*, 2008).

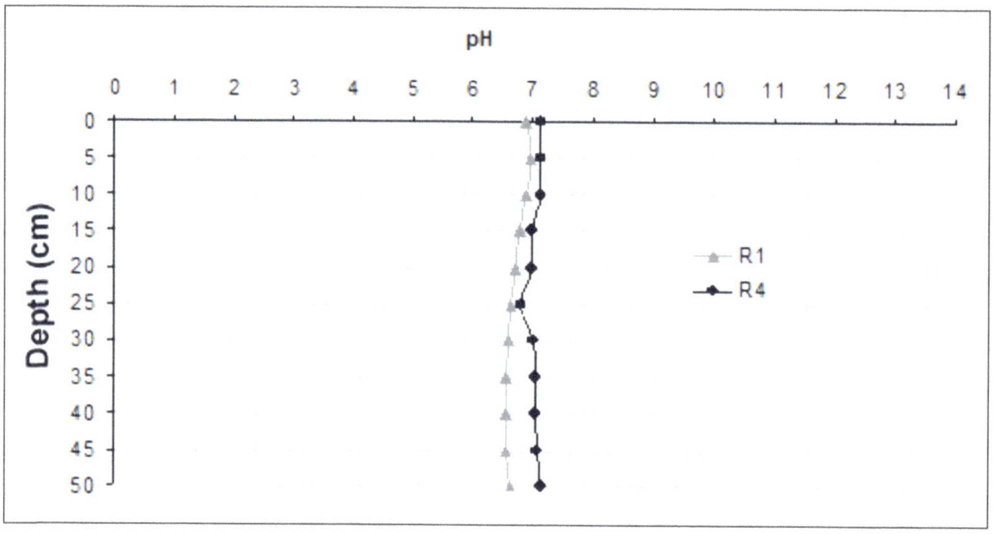

FIGURE 7 – pH VALUES IN THE SEDIMENT PROFILE (0-50cm) AT R1 AND R4 IN 2010 FROM THE RIO VERDE RESERVOIR

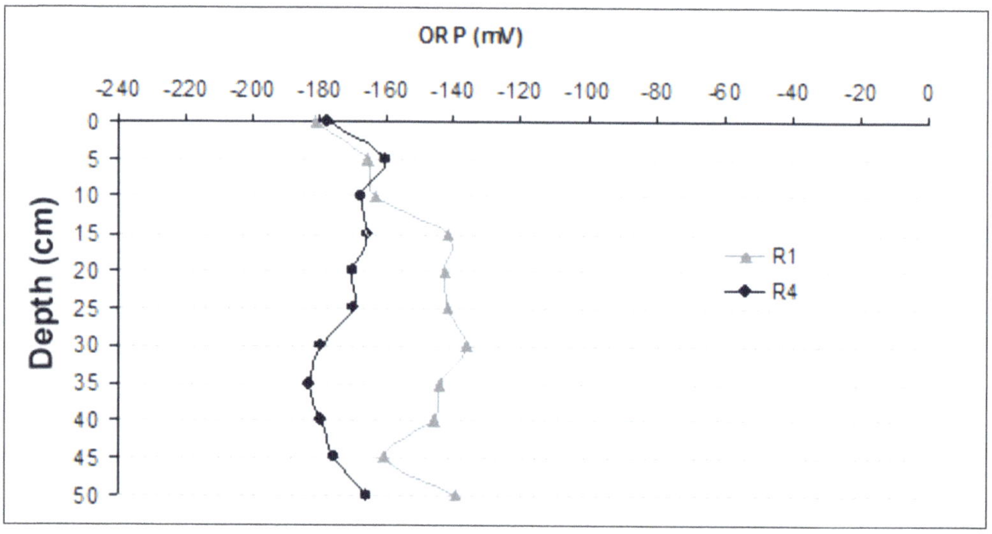

FIGURE 8 – REDOX POTENTIAL VALUES (mV) OF THE SEDIMENT PROFILE (0-50cm) AT R1 AND R4 IN 2010 FROM THE RIO VERDE RESERVOIR

9.2.4 Organic matter content

The organic matter content varied between 3 and 12% and the most significant variation was not related to differences at sampling points but to the different sampling periods. In 2010, the highest organic matter content values were detected (R1 – 12.4 % and R4 – 10.6%), which were practically double those observed in 2009 (Table 7).

Studies conducted in the Iraí reservoir (CARNEIRO *et al.*, 2004), which is located in the same Alto Iguaçu hydrographic basin, showed lower values of organic matter than those seen in this study, averaging about 1-2%, in accumulation areas close to the dam. Therefore, despite the predominance of inorganic material in the sediment, with levels of organic matter lower than 10%, these results suggest intense cycling in the reservoir and gradual deposition of organic content beyond that of the benthic community, which according to Heggie *et al.* (2002), can contribute significantly to the organic biomass to the sediment. It is important to note, however, that the reservoir is more than 40 years old and total vegetation removal was not performed before the reservoir was filled.

A study on the Australian coast by Heggie *et al.* (2002) reported that in order to avoid eutrophication, the carbon contribution to water bodies should be lower than 40mmol m^{-2} day^{-1}, i.e. lower than 2,000t ha^{-1} $year^{-1}$. The Rio Verde Reservoir presented mean supplied BOD values at 500kg day^{-1}.

9.2.5 Phosphorus

9.2.5.1 P concentration

The analysis of the samples from 2009 showed that the concentration of total P in the surface sediment (0 – 10cm) was higher at points R3, R4 and R5, with the first two located in accumulation areas closer to the dam, and R5 a sampling area with low water circulation speed (Figure 9). On the other hand, in 2010, the total P concentration was relatively higher at point R1 (closer to the head of the reservoir) than at R4, indicating an increase in recent contributions.

These values (between 12 and 33mg g^{-1}) are relatively low if compared with other Brazilian reservoirs of the same age (FRANZEN, 2009; SILVA *et al.*, 2001; SANTOS, 2003). Franzen (2009), in a study conducted in three reservoirs of the Salto System in Rio Grande do Sul State, found total phosphorus values at about 1200 mg g^{-1} in the surface sediment (0 – 20cm). According to the author, these values may be considered intermediate if compared with the guideline values presented in the "Sediment Quality Guideline of Ontario" (CANADIAN, 1993), or even high when compared to total phosphorus levels found in other lentic systems such as Wind Lake (644mg kg^{-1}) in Canada. This variation demonstrates that there are different ways of classifying phosphorus levels and it is important to consider values from similar systems when conducting comparisons, particularly in relation to climate and morphological conditions.

The values of total P are about 60 times below the alert levels as outlined by CONAMA Resolution 344/2004.

TABLE 7 – ORGANIC MATTER CONTENT (%) IN DIFFERENT SEDIMENT LAYERS OF THE RIO VERDE RESERVOIR, 2009 AND 2010

SAMPLING POINTS		R1		R2	R3	R4		R5
Year		2009	2010	2009	2009	2009	2010	2009
% Organic Matter in the sediment layers	0-10cm	4.8	12.4	4.7	3.3	7.2	10.6	6.9
	10-30 cm	4.9	-	7.4	6.9	3.2	-	5.4
	30-50 cm	4.5	-	-	-	5.6	-	-

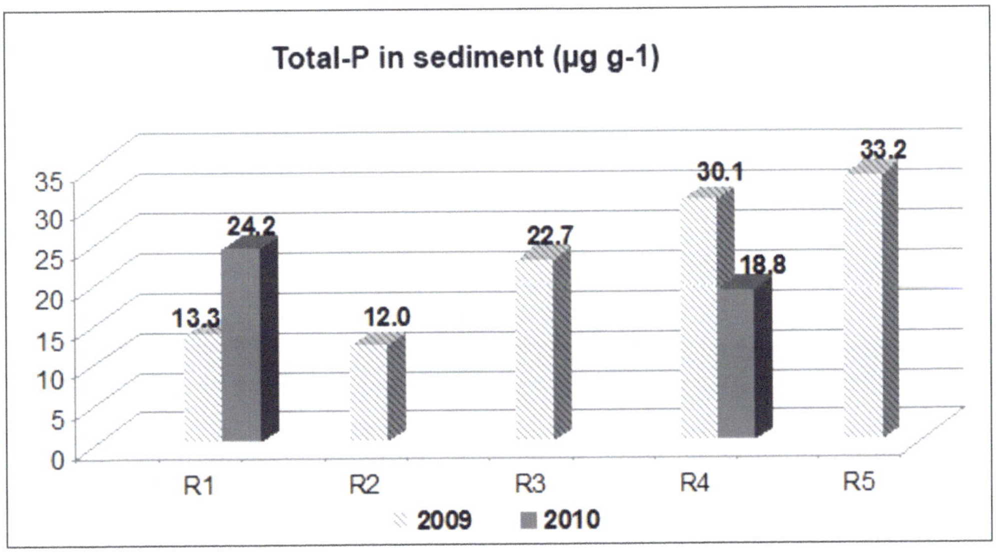

FIGURE 9 – TOTAL PHOSPHORUS CONCENTRATION (mg g-1) IN THE SURFACE SEDIMENT (0 – 10 cm) AT ALL SAMPLING POINTS IN RIO VERDE RESERVOIR, 2009 AND 2010

The observed pattern of organic P in the sediment was similar to the pattern observed for total P, i.e., the values were higher at points R3, R4 and R5, with R3 and R4 considered accumulation areas, and R5 an area with low water circulation (Figure 10). Again, in the following year the highest values were observed at point R1, the closest to the mouth of the main tributary, suggesting that in 2010 there was an increase of input loads.

Organic P represents about 50% to 77% of total P content. Schenato (2009), in a study of phosphorus dynamics in Rio Grande do Sul State, observed that the phosphorus present in the sediment from the sub-basin with anthropogenic inputs was predominantly geochemical, while in areas less affected by anthropogenic disturbances, the phosphorus from biological sources prevailed. This can explain the predominance of the organic fraction in the sediment of Rio Verde Reservoir because its basin experiences little anthropogenic disturbance, if compared to other basins. In relation to agriculture, there are several areas with this type of activity in the basin; however, according to the analysis of "concentrations and loads" (Chapter 10), most of the sub-basins have good soil management and preservation, since high phosphorus levels in the soil have not yet been detected.

The inorganic P concentrations in the surface sediment (0 – 10cm) were also slightly higher at points R3, R4 and R5 than at points R1 and R2 in 2009. In 2010, the concentrations at both studied points were similar (Figure 11). Unlike in 2010, the results from 2009 did not show high levels at the points of entry into the reservoir thus indicating that in 2010 the contributions increased.

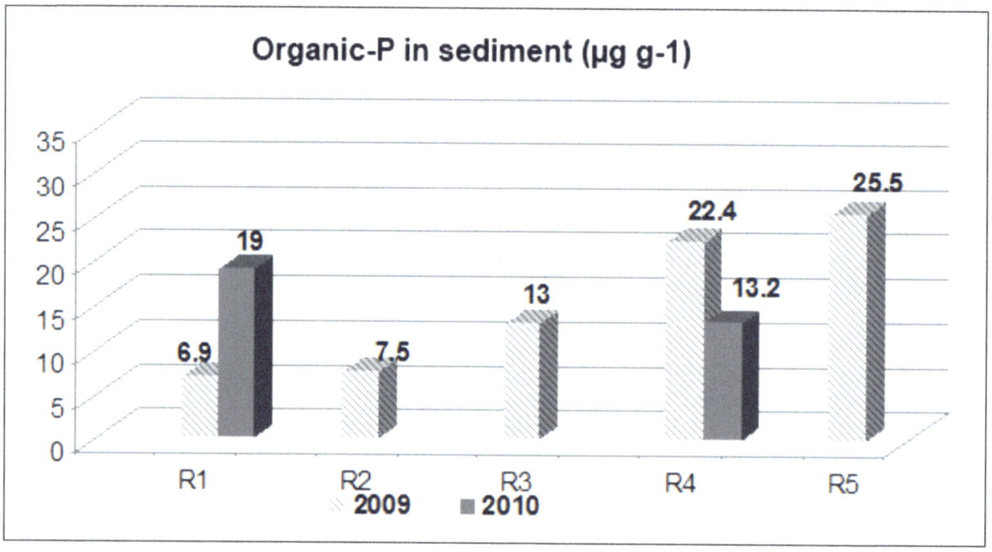

FIGURE 10 – ORGANIC PHOSPHORUS CONCENTRATION (mg g-1) IN SURFACE SEDIMENT (0 – 10 cm) FROM ALL SAMPLING POINTS OF THE RIO VERDE RESERVOIR, 2009 AND 2010

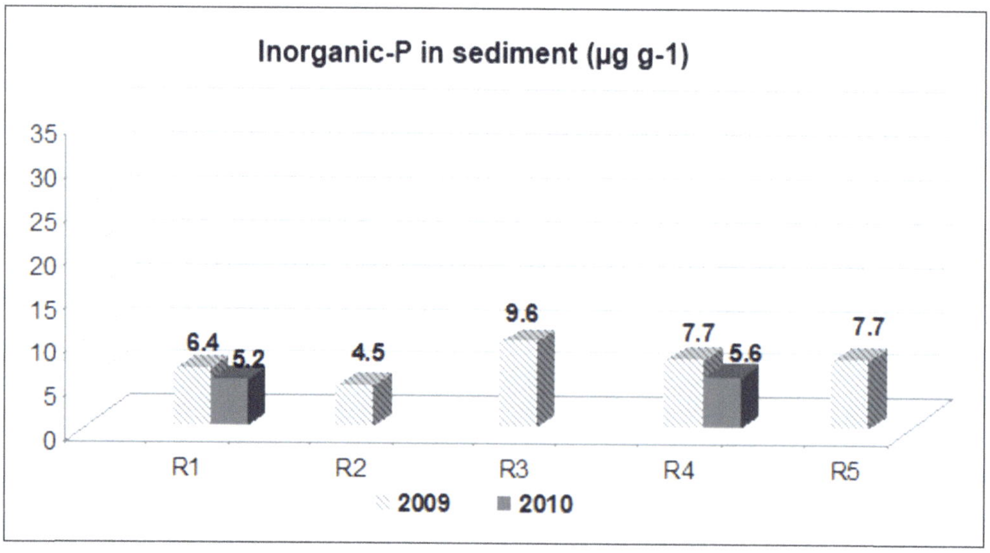

FIGURE 11 – INORGANIC PHOSPHORUS CONCENTRATION ($\mu g\ g^{-1}$) IN SURFACE SEDIMENT (0 – 10cm) AT ALL SAMPLING POINTS OF THE RIO VERDE RESERVOIR, 2009 AND 2010

The concentration of labile P (the fraction of the solid portion more easily available) in the surface sediment (0 – 10 cm) was very similar at points R1, R2 and R3, in 2009. Point R4 presented the highest labile P value (8.1µg/g) in that year and point R5 the lowest value (3.0µg/g).

In general, the values of labile P (Figure 12) were similar to those of inorganic P (Figure 11). A very significant correlation (r = 0.98) was observed between the inorganic and labile P fraction values registered in 2010 (points R1 and R4), while in 2009 there was no correlation (r = 0.02).

At some sample points, the labile P values were higher than inorganic P, suggesting the labilization of some organic forms during the analytical procedure. Due to the high clay contents in the sediment (Table 6), it was expected that the ratio of P_{inorg} / P_{labile} would be significantly lower.

Table 8 shows values for total P, organic P and inorganic P in the sediment profile in the samples collected in 2010. In addition to the surface layer, two other layers at different depths were analyzed at points R1 (beginning of the reservoir and higher water circulation speed) and R4 (closest point to the dam and reduced water circulation speed).

At point R1, the concentration of the different forms of phosphorus was quite similar across the three layers analyzed (0 – 10; 10 – 30cm and 30 – 50cm). On the other hand, at point R4 which is closest to the dam, phosphorus concentrations were higher in the most superficial layer (0 – 10cm) of the sediment whereas the values of phosphorus in deeper layers were very low. This suggests a recent increase in the input load into the reservoir. Considering that at R1 the concentration of phosphorus was similar in both the deep and surface layers, this suggests a very significant physical displacement of sediment or an intense displacement of phosphorus in recent years in the area near R4. The low values of total, inorganic and organic phosphorus in the deeper layers at R4 were also accompanied by low labile P values at the same depth (Table 8).

In comparing these two sampling areas, higher values of total P were expected at R4 because of the proximity to the dam, a known area of accumulation and low water circulation speed, and furthermore, the values recorded in 2009 had been higher at R4 when compared to R1. However, the results did not meet our expectations and more detailed studies must be conducted in order to better understand this phenomenon. Situations like this are not uncommon and for this reason it is very important to ensure the quality of sampling and laboratory procedures. If doubts exist, they should be further examined to better understand the dynamism and complexity of the system (see Chapter 10, Section 2 – Reliability of Water Quality Assessments).

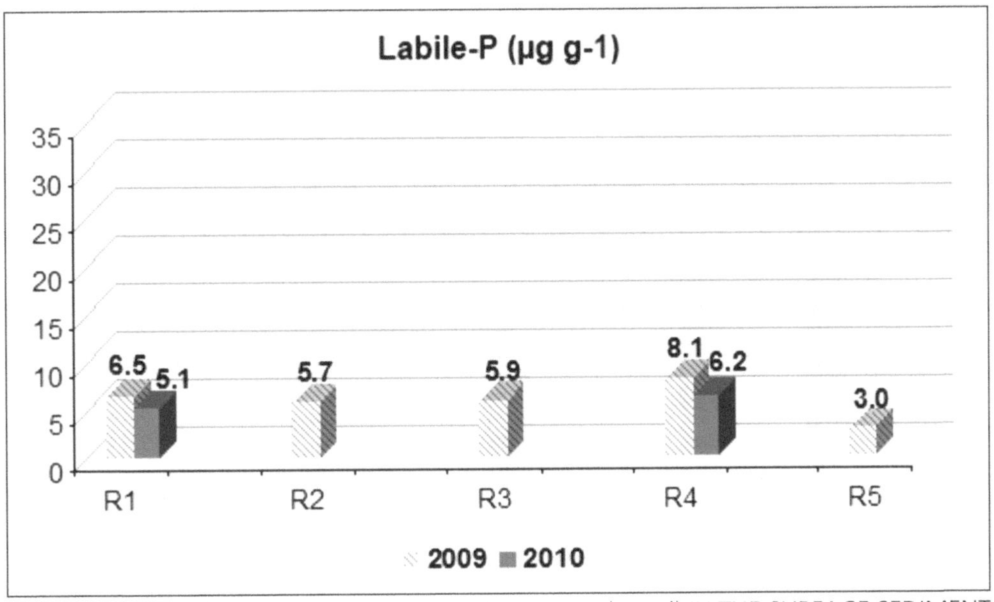

FIGURE 12 – LABILE PHOSPHORUS CONCENTRATION (µg g^{-1}) IN THE SURFACE SEDIMENT (0 – 10cm) AT ALL SAMPLING POINTS OF THE RIO VERDE RESERVOIR, 2009 AND 2010

TABLE 8 – TOTAL PHOSPHORUS (Total-P), ORGANIC PHOSPHORUS (Organic-P), INORGANIC PHOSPHORUS (Inorganic-P) AND LABILE PHOSPHORUS (Labile-P) CONCENTRATION VALUES (µg g-1) IN 3 LAYERS OF THE SEDIMENT PROFILE : I = 0 – 10cm; II = 10 – 30cm ; III = 30 – 50cm – RIO VERDE RESERVOIR, 2010

	TOTAL-P	ORG.-P	INORG.-P	LAB-P
R1 I	24.2	19.0	5.2	5.1
R1 II	25.0	19.4	5.5	5.1
R1 III	26.9	18.4	8.5	7.0
R4 I	18.8	13.2	5.6	6.2
R4 II	2.1	1.1	0.9	1.2
R4 III	0.6	0.2	0.4	1.0

9.2.5.2 Concentration of phosphorus in interstitial water, water-sediment interface and water column

The sampling points R1 and R4 were analyzed to assess and compare P concentrations in interstitial water, water-sediment interface and the water column. The results of the analyses showed that the concentrations of P_{total} and P-PO_4^- at both sampling points were significantly higher in interstitial water than in the water-sediment interface or the water column. The concentration of P-PO_4 observed in the water column was only 1% to 2% of the concentration observed in the interstitial water, indicating the potential of this component for retaining and (or) releasing ortophosphates. With a large stock of phosphorus in the sediment that can be easily released to the water column, and considering the redox potential of the sediment (-151 to -171mV – Figure 8), it is important that the hypolimnion and particularly the water-sediment interface is kept aerated and oxidized, minimizing phosphorus re-suspension. Several studies have shown that the interstitial water is a component rich in nutrients that systematically disperses to higher waters (BEZERRA, 1987; ISHII, 1987; OHLE, 1976). Many authors prefer to associate interstitial water with trophic levels in lakes rather than with the particulate material of the sediment (OHLE, 1976, ESTEVES 1998) because of a greater correlation of phosphorus values in these components. It is important to note, that when comparing concentrations of P-PO_4^- between the interface and the water column, very similar values are obtained. However, if we consider only P_{total}, there is a disparity between microenvironments, with higher values presented at the water-sediment interface, especially at point R1(Figure 13) and possibly in the form of organic phosphate.

The concentrations of P_{total} and PO_4^- in the interstitial water and in the water-sediment interface were higher at point R4, an area of slower circulation than R1, which is located at the beginning of the reservoir and thus with greater water speed. However, these concentration levels in the interstitial water may be considered low when compared to other lakes. A study conducted by Ohle (1976) showed nutrient values for some temperate lakes were related to trophic states, where oligotrophic environments had values around 20ug L^{-1} of P_{total}, mesotrophic environments 140 – 170ug L^{-1}, and eutrophic environments values between 1,000 and 2,200 20ug L^{-1}.

In the water column, P_{total} concentrations were higher in R1; again, suggesting a recent increase of phosphate loads into the reservoir, although these values are very low and characteristic of an oligotrophic environment.

Considering the discussion above, we can say that the water column and the particulate material of the sediment of the Rio Verde Reservoir present some oligotrophic characteristics, while the interstitial water of the same sediment presents meso-eutrophic characteristics. These results suggest that the reservoir is able to keep the hypolimnion aerated most of the time, with little thermal-chemical stratifications, otherwise the values in the water profile would be significantly higher: a situation for which there is some potential.

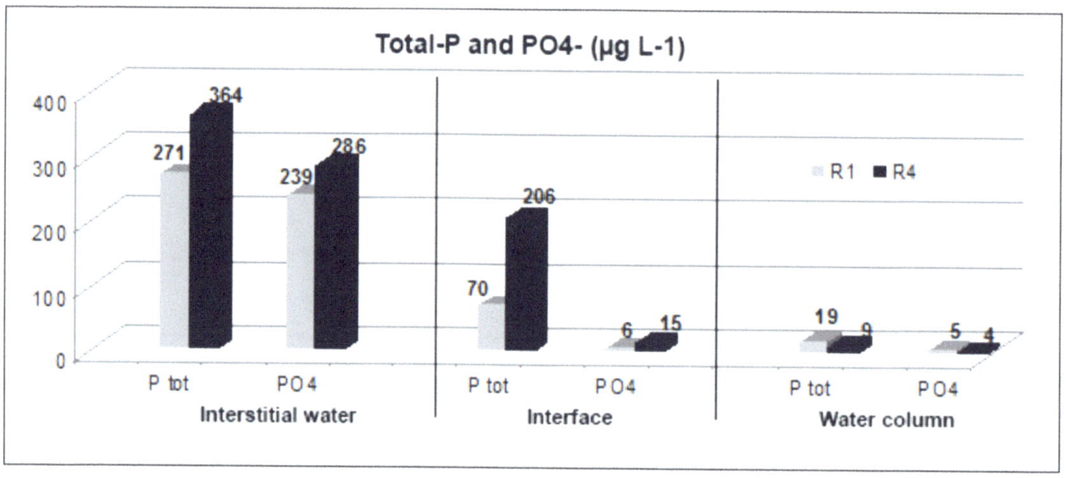

FIGURE 13 – CONCENTRATION OF TOTAL PHOSPHORUS AND PHOSPHATE (mg L-1) IN THE INTERSTITIAL WATER, WATER-SEDIMENT INTERFACE AT POINTS R1 AND R4 IN THE RIO VERDE RESERVOIR, 2010

NOTE: Ptotal in the sediment in µg g-1 is shown in Figure 9, with R1: 13.3 (2009) and 24.4 (2010) and R4: 30.1(2009) and 18.8 (2010)

9.2.5.3 Evaluation of phosphorus release

Temperature can play an active part in phosphorus dynamics in the sediment, reducing the adsorption and indirectly increasing biological activity. Considering that organic phosphorus is most prevalent in the sediment, it is important to explore microbial activity as a part of the phosphorus release process. During mineralization of organic matter, phosphorus can be indirectly mobilized, either by consumption of nitrate and sulfate or by the formation of methane (CHRISTOPHORIDIS & FYTIANOS, 2006). However, the phosphorus released by this biological process may become unavailable when there is oxidized Fe and when environmental conditions are favorable to adsorption/precipitation.

At high densities, benthic invertebrates and benthophagic fish may considerably disturb the sediment, thus transferring by diffusion phosphorus from the sediment to the water column (WETZEL, 1999). In the Rio Verde

Reservoir, consistently high catch levels in the samples of *Geophagus brasiliensis* (*acará*), a species of benthophagic fish, was observed which may be a factor in phosphorus diffusion. Another process that may contribute to phosphorus diffusion from the sediment to the water column is gas production. Gases produced by biological activity, such as CO_2, CH_4 and N_2, form bubbles which contribute to the movement of phosphorus as the gases rise to the surface. In the Rio Verde Reservoir, the release of bubbles from the sediment was observed on several occasions when collecting samples for the project, along with the capture of surface methane caused by the high degree of decomposition of the remaining plant debris from the filling of the reservoir 40 years ago.

Complexation reactions between PO_4^- and Fe or Al is favored in conditions of neutral to slightly acidic pH (FRANZEN, 2009) which are consistent with the conditions found in the Rio Verde Reservoir. However, the complexation P-Al is probably a secondary reaction in the reservoir, considering that at pH values greater than 5.6 to 5.8, Al^{3+} ions precipitate as $Al(OH)_3$. This suggests a greater importance of the P – Fe bond. In the field, we observed a reddish color to the sediment which is caused by the presence of significant amounts of iron. This indicates that, in the inorganic component, Fe may be one of the major regulators of P dynamics in the Rio Verde Reservoir.

The low phosphorus concentrations found in the water column and the water-sediment interface are not consistent with the high values observed in the interstitial water. In addition, the redox potential of the sediment was consistently electronegative (below +200mV), i.e., favorable for phosphorus desorption. This suggests that the water-sediment interface is frequently in an aerobiotic state, with the establishment of an oxidation microzone. Other factors which also affect phosphorus release from the sediment, such as pH, temperature and bioturbation, had little effect on element resuspension to the water column. These results can be explained by the levels of dissolved oxygen (DO) observed at the bottom of the reservoir. Even under the stratification conditions in 2009, the values of DO in the bottom layer were higher than 5 mg/L (Figures 5 and 6), forming an oxidation layer that minimizes the diffusion of interstitial desorbed phosphorus to the waters in higher layers. In 2010, DO concentrations in the bottom waters of the reservoir were higher than 6.5mg/L.

Generally, anoxic states favor the mechanisms of P release from the sediment. It is estimated that in an anaerobic state the levels of phosphorus resuspension from the sediment to the water column may reach 1,000 times the values observed in aerobic states (WETZEL, 2001). Despite the fact that our analysis produced satisfactory DO values that are consistent with the phosphorus content in the water column, states of anoxia and hypoxia are quite common, particularly between November and April. This can lead to a gradual release to the water column from the bottom (internal loading). Consequently, there would be a risk of promoting the proliferation of cyanobacteria microalgae (see Chapter 13).

As previously discussed, it is important to note that the levels of total P in the sediment are considered low, varying from 13.3 to 33.2mg g^{-1}, well below the alert levels outlined in the legislation (CONAMA 344/2004). However, the values encountered in the interstitial water are not as low as that found in the sediment, suggesting that attention must be paid to reservoir management, including restricting new contributions and constantly monitoring phosphorus levels in the water column. We emphasize that continuous monitoring is necessary in order to better understand whether phosphorus concentration in sediment components (including solid phase, liquid phase and interface) will significantly affect the equilibrium of the phytoplankton community of the reservoir in the future.

The assessment of phosphorus release from the sediment does not depend solely on the analysis of phosphorus fractions in the sediment and water. The physicochemical characteristics of the sediment and of the environment where it occurs, as well as biological activities of the ecosystem, should be taken into consideration. All these processes may be significant in phosphorus dynamics in the sediment of the Rio Verde Reservoir. However, the quantification of phosphorus release from the sediment to the water column, especially the dynamics of the interstitial water and the importance of the different processes that contribute to this release, should be better studied through controlled experiments in the laboratory. The present study presents an initial assessment of the phosphorus concentration in the sediment that can be used as a reference for further studies on the dynamics and importance of this element in the reservoir.

9.2.6 Nitrogen

9.2.6.1 N concentration

The N_{total} values in the surface sediment of the Rio Verde Reservoir varied from 2,400 to 4,400µg g^{-1}, considering the samples taken in both 2009 and 2010. These values are below the alert levels indicated by CONAMA 344/2004 (Figure 14).

The total nitrogen (N_{total}) values in 2010 were 15 to 30% lower than the values obtained in 2009. Of the five points evaluated in 2009, point R5 had the highest total N value in the surface sediment (0 – 10cm) and R4 the lowest: both points are in areas of low water speed (Figure 14).

A study developed by Prada & Oliveira (2006) in a reservoir under good conditions of preservation (essentially forested) in its drainage basin, identified values of about 3,200 to 6,600µg g^{-1}, demonstrating that basins with little agricultural or anthropogenic disturbance can also have high concentrations of organic nitrogen and total N can reach values similar to those reported for anthropogenically disturbed environments.

In lotic environments, sediment stabilization and accumulation is unusual and significantly lower total N values are expected, even when compared to more disturbed environments. The surface sediment of the Itajaí-Açu River, Santa Catarina State, was assessed in a study conducted by Susin *et al.* (2007), where values between 350 and 2,200µg g^{-1} were observed. Another study in the Barigui River, which is also a lotic environment, located in the same hydrographic unit as the current study and is known to receive a large sewage load, detected values in the order of

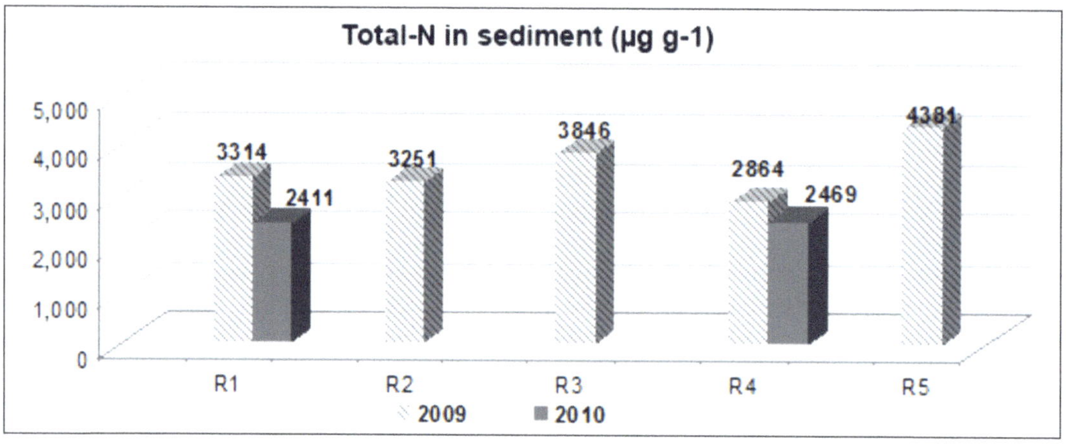

FIGURE 14 – TOTAL NITROGEN CONCENTRATION (mg g-1) IN THE SURFACE SEDIMENT (0 – 10cm) AT ALL SAMPLED POINTS OF THE RIO VERDE RESERVOIR, 2009 AND 2010

362 to 1,778µg g^{-1} (FROEHNER & MARTINS 2008). These values are lower than those observed in the relatively well preserved environment of the Rio Verde Reservoir. It is important to study another reservoir in this hydrographic basin with similar preservation conditions to better understand these results. One environment that would be worth analyzing is the Piraquara I Reservoir, located in the Metropolitan Region of Curitiba.

In examining the profiles sampled in 2010 (Table 8), the values of N_{total} were similar at all points sampled, about 2,400µg g^{-1} (mean 2,422 µg/g ± 52 SD). These values are significantly lower than those observed in 2009 (surface layer) (Figure 14).

Unlike the results observed for phosphorus, total nitrogen concentrations in the water-sediment interface were lower than the values observed in the water column, suggesting recent inputs. On the other hand, both the values observed for the interface and the water column were similar across the sampled points, indicating a certain degree of homogeneity in the reservoir.

9.2.7 Metals and Arsenic

Table 9 shows the values for the metals and arsenic present in the sediment of the Rio Verde Reservoir at two sampling points (R1 and R4) and three depths in the sediment profile (I = 0 – 10cm; II = 10 – 30cm ; III = 30 – 50cm).

The elements arsenic and cadmium were not detected in the analyses (detection limits of 10 and 1mg/kg, respectively). However, contamination level 1 established by CONAMA 344/04 is 5.9mg/kg for arsenic and 0.6mg/kg for cadmium, values below the detection limits used in this study. The detection limits used are nevertheless below contamination level 2.

The heavy metal chromium presented the highest concentration in the sediment, varying from 100 to 139mg/kg. These results are above the values recommended by CONAMA resolution 344/04. Bonai (2007) found similar levels of chromium in the sediment of the Itá reservoir in Santa Catarina varying from 74.3 to 119.5mg/kg at five sample points. Silvano (2003), studying a pond downstream of mining activities also in Santa Catarina, found higher values, between 215 and 325mg/kg. If we consider the base levels established by Bowen (1979) of approximately 72mg/kg, the levels suggested by Fôrstner & Wittmann (1983) of 77 mg/kg, and contamination level 2 of the CONAMA resolution 344/04 at 90mg/kg, the sediment of the reservoir can be characterized as having excessive chromium. The concentration of chromium in water should be lower than 1µg/L.

Chromium can have different oxidation levels from Cr^{2+} to Cr^{6+}. The trivalent form is the most stable and common in nature and it is essential for the functioning of living organisms. The hexavalent form is highly toxic and carcinogenic. Long-term exposure to chromium can result in the accumulation of the metal in certain tissues, mainly the liver and spleen (FREITAS, 2006), and can cause cutaneous ulcers, dermatitis, nasal inflammation, lung cancer, among others. The presence of chromium

TABLE 8 – TOTAL N CONCENTRATION VALUES (µg g-1) IN THREE LAYERS OF THE SEDIMENT PROFILE: I = 0 – 10cm; II = 10 – 30cm ; III = 30 – 50cm; TOTAL N IN THE WATER-SEDIMENT INTERFACE (µg L-1); TOTAL N (µg g-1) IN THE WATER COLUMN (µg L-1), RIO VERDE RESERVOIR, 2010

PROFILE	TOTAL-N SEDIMENT	TOTAL-N WATER-SEDIMENT INTERFACE	TOTAL-N WATER COLUMN
R1 I R1 II R1 III	2.411 2.372 2.505	177	283
R4 I R4 II R4 III	2.469 2.367 2.408	179	293

ions in water can indicate industrial pollution, such as residues and effluents from the production of anodized aluminum, stainless steel, paints, pigments, explosives, paper and photographs. Clearly, it is important to better understand the industrial areas and activities in the basin, upstream of the reservoir.

Aquatic organisms exhibit significant variation in sensitivity to this element (BRANCO, 1972; ABNT, 1997). The impact of chromium on the biota of the Rio Verde Reservoir, therefore, depends on its oxidation state which is directly affected by pH and redox potential (Eh). In environments with low Eh and tendencies toward acidic pH, the species of Cr^{3+} are most prevalent as cations Cr^{3+}, $CrOH^{2+}$, $Cr(OH)_2^{+}$, $Cr(OH)_3^{0}$ and anions $Cr(OH)_4^{-}$. Under oxidizing states and a more alkaline pH, the species of Cr^{6+} predominate as anions $HCrO_4^{-}$ and CrO_4^{2-} (BERTOLO *et al.*, 2009). Given that the pH of the Rio Verde Reservoir sediment is neutral on average, the redox potential may be a determining factor in the chromium oxidation state. Since Eh values are electronegative in the reservoir, it is likely that chromium exists in the most reduced state. Despite being found naturally, the long-term effects of Cr^{3+} on the biota, at the concentrations found in the Rio Verde Reservoir, should be further investigated.

9.2.8 Contaminants in the Sediment and Water: Pesticides, PCBs and PAHs

The analyses of pesticides in the sediment samples revealed that in both study years (2009 and 2010) all the compounds chosen for analysis and included in the current legislation (CONAMA/344 and CETESB/2005), were below the detection limits of the analytic procedures employed. They are thus within the permitted levels.

In the water samples, the values found for all pesticides included in CONAMA resolution 357/2005 were below the maximum authorized levels.

Through a survey of the points of sale for pesticides in the Rio Verde Basin, we developed a list of the 22 most commonly used pesticide compounds. Of this list, nine were herbicides, eight insecticides and five fungicides (Table 10).

CONAMA resolution 344/2004 includes substances with a greater persistence potential in the environment, such as organochlorines and PAHs. Due to their chemical structure, these compounds are most likely to be found in the sediment. As can be seen in Table 10, the chemical groups of the pesticides used in the Basin, especially the insecticides (carbamates, pyrethroids and organophosphates), are less persistent in the environment as compared to organochlorines.

TABLE 9 – METAL CONTENTS IN THE SEDIMENT AT POINTS R1 AND R4, RIO VERDE RESERVOIR, 2010

METALS	R1 – I	R1 – II	R1 – III	R4 – I	R4 – II	R4 – III	CONAMA 344		UNIT
							LEVEL 1	LEVEL 2	
Aluminum	2.97	3.09	3.01	3.86	4.43	4.35			g/100g
Antimony	ND 10	ND 10	ND 10	ND 10	ND 10	ND 10			mg/kg
Arsenic	ND 10	ND 10	ND 10	ND 10	ND 10	ND 10	5.9	17	mg/kg
Barium	138	144	138	136	61	35			mg/kg
Beryllium	1.4	1.4	1.4	0.8	0.4	0.2			mg/kg
Bismuth	ND 10	ND 10	ND 10	ND 10	ND 10	ND 10			mg/kg
Cadmium	ND 1	ND 1	ND 1	ND 1	ND 1	ND 1	0.6	3.5	mg/kg
Calcium	932	1007	900	636	418	224			mg/kg
Lead	19	26	19	20	16	17	35	91.3	mg/kg
Cobalt	18	18	19	14	6	4			mg/kg
Copper	43	46	46	25	18	15	35.7	197	mg/kg
Chromium	120	129	126	102	139	100	37.3	90	mg/kg
Tin	ND 10	ND 10	ND 10	ND 10	ND 10	ND 10			mg/kg
Strontium	9	10	10	7	4	4			mg/kg
Iron	7.37	7.67	7.27	6.86	5.58	4.5			mg/kg
Phosphorus	1008	1077	1068	721	312	225			mg/kg
Lithium	8	8	8	7	8	7			mg/kg
Magnesium	1176	1223	1091	819	290	172			mg/kg
Manganese	499	542	473	813	381	204			mg/kg
Mercury	0.13	0.11	0.14	0.1	0.08	0.08	0.17	0.488	mg/kg
Molybdenum	ND 10	ND 10	ND 10	ND 10	ND 10	ND 10			mg/kg
Nickel	26	28	26	14	13	7	18	35.9	mg/kg
Potassium	943	1008	903	820	326	200			mg/kg
Silver	ND 4	ND 4	ND 4	ND 4	ND 4	ND 4			mg/kg
Selenium	ND 10	ND 10	ND 10	ND 10	ND 10	ND 10			mg/kg
Sodium	99	87	10	79	55	48			mg/kg
Vanadium	11	117	118	101	109	88			mg/kg
Zinc	55	60	58	38	20	16	123	315	mg/kg

NOTE: Data highlighted in light gray in the table refer to inorganic elements, potentially pollutants, discussed in CONAMA resolution 344/04
ND: Not-detected.
Level 1: threshold below which there is a probable impact on the biota
Level 2: threshold above which there is a probable adverse impact on the biota

TABLE 10 – MAIN PESTICIDES USED IN THE RIO VERDE BASIN

CHEMICAL GROUP	COMPOUND	CLASS	CULTURE
Organophosphates	Chlorpyrifos	I	Potatoes, corn, beans, tobacco, wheat, sorghum and vegetable crops
	Parathion	I	Potatoes, beans, corn and wheat
	Methamidophos	I	Beans, potatoes, tomato and wheat
Carbamates	Carbaryl	I	Beans, tobacco, corn and vegetable crops
	methomyl	I	Potatoes, corn, wheat and vegetable crops
	Mancozeb	F	Potatoes, grapes, tobacco, wheat, beans and vegetable crops
Pyrethroid	Lambda Cyhalothrin	I	Beans, tobacco, corn, vegetable crops and wheat
	Deltamethrin	I	Corn, sorghum, wheat, beans, potatoes and tomato
Acetanilide	Metolachlor	H	Corn and beans
Propionate Aryloxy phenoxy	Buti Fluazifop-P + Fomesafen	H	Potatoes, vegetable crops, beans and tobacco
Gycine derivative	Glyphosato	H	Pasture, corn, grapes, wheat, conifer and eucalyptus
Triazine	Atrazine	H	Corn, conifers and sorghum
bipyridyl	Paraquat	H	Wheat, corn, grapes and beans
Strobilurin	Azoxystrobin	F	Potatoes, beans and vegetable crops
Phthalonitrile	Chlorothalonil	F	Potatoes, beans, grasslands, corn, vegetable crops, wheat and grapes
Acetamide	Cymoxanil	F	Potatoes, grapes, tomato and onion
Triketone	Mesotrione	H	Corn
Benzothiazine	Bentazone	H	Beans, corn and wheat
Benzonitrile	Ioxnil	H	Onion
Organic tin	Fentin Hydroxide	F	Potatoes, beans and vegetable crops
Sulfonylurea	Nicosulfuron	H	Corn

NOTE: I = insecticide; H = herbicide; F = fungicide.

Organophosphate pesticides are non-persistent and thus they represent an improvement on organochlorines. However, they usually have a severe toxic effect on humans and other mammals as compared to organochlorines. Organophosphates decompose in days or weeks and, therefore, they are rarely found in food chains and sediment (BAIRD, 2002). An example is Parathion, an organophosphate of the phosphorothioate group, which is very toxic but persists in the soil on average one week.

As with organophosphates, carbamates also have a very short life-cycle in the environment: they react with water, decomposing into simple, non-toxic products. Pyrethroids, as well as some herbicides, are xenobiotic and are analogous to compounds found in nature. Considering that pesticide degradation occurs predominantly through biological processes, due to the enzymes present in animals, plants and microorganisms, the persistence of these compounds in the environment are also considered low.

Based on a general analysis comparing the persistence of the chemical groups of pesticides, we note that there is a low probability of accumulation, and therefore low probability of detection, in the sediment of the most widely used compounds in the Rio Verde Basin. On the other hand, our analysis is based on the substances outlined in CONAMA resolution 344/2004, and thus our results can be explained in part by the fact that organochlorines are not used in the basin.

In a more detailed examination of the persistence and contamination potential of pesticides, the Goss, EPA and GUS indices were evaluated. The assessment of surface water pollution (Goss Index) and groundwater (EPA and GUS criteria) resulted in the identification of the following priority compounds to be analyzed in the basin: for surface waters – Atrazine, glyphosato, mancozeb, metolachloro, chlorpyrifos and metonil; and for sediment – mancozeb, chlorpyrifos, endosulfan and lambda-cyhalothrin. Although the models used do not indicate a high impact potential for parathion and carbaryl, these molecules were considered as priorities for investigation: parathion in the sediment and carbaryil in the water.

It is worth noting, that among the 22 most used pesticides in the Rio Verde Basin (Table 10), only three (atrazine, carbaryl and glyphosate) are included in CONAMA resolution 357/2005. These three pesticides are indicated as potential water pollutants in the basin, considering the GUS and EPA criteria and the Goss index, as described above.

On the other hand, all the compounds analyzed in the water were below the detection limit. This result may be associated with the collection time, which occurred in June, a period in which agricultural activity in the region is greatly diminished because most of the crops in the basin are summer crops.

10. FINAL CONSIDERATIONS

The amount of phosphorus in the water sediment and water-sediment interface may be considered low; however, the phosphorus found in the interstitial water is relatively high, suggesting that the ecosystem requires constant monitoring.

This phosphorus may be released from the sediment component if the oxygenation states of the water bottom are changed. As phosphorus concentrations in the sediment are low, temporary periods of anoxia in the hypolimnion likely will not significantly influence the phytoplankton community. Nevertheless, we suggest monitoring both water and sediment. We also recommend that specific experiments are conducted that assess the potential release of phosphorus from the sediment and the interstitial water, in order to better understand the intervening dynamics in the reservoir.

Considering the results of the analysis of pesticides and nutrients in both water and sediment, we conclude that the Rio Verde Hydrographic Basin does not contribute significant pollutant loads to the reservoir.

The contamination of the sediment with chromium was detected at levels above those allowed by the current legislation. Due to the current physical and chemical state of the reservoir, this excess does not seem to pose a threat to the biota. However, further investigation of the possible causes of chromium accumulation must be conducted as it may become a more serious problem in the future.

11. REFERENCES

ABNT – Associação Brasileira de Normas Técnicas. (1984) *Grãos de solo que passam na peneira de 4,8mm – Determinação da massa específica.* NBR 6508.

ABNT – Associação Brasileira de Normas Técnicas. (1997) *Água – Determinação de cromo – Método da espectrometria de absorção atômica por chama.* NBR 13814.

ANDREI, E. (2005) *Compêndio de defensivos agrícolas.* 7. ed. São Paulo: Andrei Editora Ltda. 1141p.

APHA – AMERICAN PUBLIC HEALTH ASSOCIATION. (2005) *Standard methods for the examination of water and wastewater.* 21. ed. Washington, DC, 1600p.

BAHIA FILHO, A.F.C. (1982) *Índices de disponibilidade de fósforo em latossolos do planalto central com diferentes características texturais e mineralógicas.* Tese de Doutorado, Viçosa, 179p.

BAIRD, C. (2002) *Química Ambiental.* 2. ed., Porto Alegre: Bookman, 368p.

BARBOSA, L.C.A. (2004) *Os pesticidas, o homem e o meio ambiente.* Viçosa: UFV, 215p.

BAUMGARTEN, M.D.Z.;NIENCHESKI, L.F.H.; VEECK, L. (2001) Nutrientes na coluna da água e na água intersticial de sedimentos de uma enseada rasa estuariana com aportes de origem antrópica (RS – Brasil). *Atlântica*, v.23, p. 101-116.

BERTOLO, R.A.; MARCOLAN, L.N.O.; BOUROTTE, C.L.M. (2009) Relações Água-Rocha e a Hidrogeoquímica do Cromo na Água Subterrânea de Poços de Monitoramento Multiníveis de Urânia, SP, Brasil. Geol. USP, *Sér. cient.*, São Paulo, v. 9, n. 2, Jun.

BETTIN, S.M.; FRANCO, D.W. (2005) Hidrocarbonetos policíclicos aromáticos (HPAs) em aguardentes. *Ciênc. Tecnol. Aliment.*, v. 25, n. 2, Abr./Jun., p. 234-238.

BEZERRA, M.A.O. (1987) *Contribuição ao estudo limnológico da represa de Três Marias (MG) com ênfase no ciclo do nitrogênio.* Master dissertation. São Carlos – UFSCar, 127p.

BOERS, P.C.M., CAPPENBERG, Th.E.; VAN RAAPHORST, W. (1993) (Eds) Proceeding of the Third International Workshop on Phosphorus in Sediments. *Hydrobiologia*, 253, 376p.

BOERS, P.C.M.; VAN RAAPHORST, W.; VAN DER MOLEN, D.T. (1998) Phosphorus retention in sediments. *Wat. Sci. Tech.*, v. 37, n. 3, p. 31-39.

BOLLMANN, H.A., CARNEIRO, C., PEGORINI, E.S. (2005) Qualidade da Água e Dinâmica de Nutrientes In: ANDREOLI, C.V. (Ed.). *Gestão Integrada de Mananciais de Abastecimento Eutrofizados.* Curitiba: Graf. Capital Ltda., p. 215-270.

BONAI, N.C. (2007) *Distribuição de metais pesados nos sedimentos do reservatório da Usina Hidrelétrica de Itá e avaliação de eficiência de Eichhornia crassipes na remediação de águas contaminadas pela atividade de criação de suínos.* Dissertation (Master's degree) – Universidade Comunitária Regional de Chapecó, 66p.

BOWEN, H. (1979)*Environmental Chemistry of the Elements.* Academic Press, London: Academic Press, 348p.

BRADY, N.)C. (1989) *Natureza e propriedade dos solos.* 7. ed., Rio de Janeiro: Livr. Freitas Bastos,898p.

BRANCO, S.M.(1972) *Poluição: a morte de nossos rios.* São Paulo: Cetesb, 157p.

BRÖNMARK, C.; HANSSON, L-A. (2005) *The biology of lakes and ponds.* 2. ed. New York: Oxford University Press, 285p.

CALLISTO, M.; ESTEVES, F.A.(1996) Composição granulométrica do sedimento de um lago amazônico impactado por rejeito de bauxita e um lago natural (Pará, Brasil). *Acta Limnologica Brasiliensia*, v. 8, p. 115-126.

CANADIAN – (1993) *Sediment quality guidelines for the protection of aquatic life*, Hull: National Guidelines and Standards Office.

CARNEIRO, C.A. (2008) *Movimentação por veiculação hídrica do fósforo e outros macroconstituintes químicos sob a influência de fatores naturais e tecnogênicos na bacia do rio Timbú, tributário do reservatório Iraí – Região Metropolitana de Curitiba.* Thesis (Doctorate) – Universidade Federal do Paraná, Curitiba, 222 p.

CARNEIRO, C.A., PEGORINI, E.S., ANDREOLI, C.V., SILVA, N.G.M. (2004) *Avaliação da Distribuição Temporal do Fósforo e dos Elementos K, Ca, Mg e Na em Água e Sedimento no Reservatório Iraí, Pinhais-PR*, Brasil. Sanare – Revista Técnica da Sanepar, v 21, n 21, p. 33-41.

CARVALHO, I.G. (1995) *Fundamentos da geoquímica dos processos exógenos.* Salvador: Bureau, 211p.

CETESB – Companhia de Tecnologia de Saneamento Ambiental. (2001) *Relatório de estabelecimento de Valores Orientadores para Solos e Águas Subterrâneas no Estado de São Paulo.* São Paulo, 245p.

CETESB – *Companhia de Tecnologia de Saneamento Ambiental do Estado de São Paulo* – Decisão de Diretoria n.° 195-2005-E, de 23 de novembro de 2005. Dispõe sobre a aprovação dos valores orientadores para solos e águas subterrâneas no estado de São Paulo – 2005, em substituição aos valores orientadores de 2001, e dá outras providências. Diário Oficial do Estado de São Paulo, Poder Executivo, São Paulo, 2005. (Board Decision 195-2005-E, from November 23, 2005. Provides for the adoption of guiding values for soil and groundwater in the state of São Paulo – 2005, replacing the guiding values of 2001 and other provisions. Official Gazette of the state of São Paulo, the Executive Power, São Paulo-2005)

CHRISTOPHORIDIS, C.; FYTIANOS, K. (2006) Conditions Affecting the Release of Phosphorus from Surface Lake Sediments. *J. Environ. Qual.*, v. 35, p. 1181-1192.

CONAMA – (2004) *Conselho Nacional de Meio Ambiente* – Resolução n.° 344, de 25 de Março de 2004. Estabelece diretrizes gerais e os procedimentos mínimos para a avaliação do material a ser dragado em águas jurisdicionais brasileiras, e dá outras providências. Diário Oficial da União, Brasília, 2004. (Establishes general guidelines and procedures for the evaluation of the minimum material to be dredged in Brazilian waters, and other measures, Official Gazette, Brasilia.

CONAMA – (2005) *Conselho Nacional de Meio Ambiente* – Resolução n.° 357, de 17 de Março de 2005. Dispõe sobre a classificação dos corpos de água e diretrizes ambientais para o seu enquadramento, bem como estabelece as condições e padrões de lançamento de efluentes, e dá outras providências. Diário Oficial da União, Brasília. (provides for the classification of water bodies and environmental guidelines for their adequacy and establishes the conditions and standards for effluent discharge and other measures. Official Gazette, Brasilia, 2005).

COX, C. (1997) Chlorothalonyl. *Journal Pesticide Reform*, v. 17, p.14-20.

DICK, D.P.; MARTINAZZO, R. (2006) *Matéria orgânica em ambientes terrestres e aquáticos: compartimentos, composição e reações* In: POLETO, C., MERTEN, G. H. (Eds). *Qualidade dos Sedimentos.*

Associação Brasileira de Recursos Hídricos - ABRH , Porto Alegre, p. 141-179.

DILLON, P.J.; RIGLER, F.H. (1974) The phosphoruschlorophyll relationship in lakes. *Limnology and Oceanography*, 19, p. 767-773.

EPA – (1974) *Environmental Protection Agency*. Waste oil study. Report to Congress. Authorized by Section 104 (m), P.L. 92-500, U.S. Environmental Protection Agency, Washington, D.C. (PB-257-693) April.

ESTEVES, F.A. (1998) *Fundamentos de Limnologia*. Rio de Janeiro: Interciência, 2. ed., 602p.

FELSOT, A.S.; DZANTOR, E.K. (1990) Enhancing biodegradation for detoxification of herbicide waste in soil. In: RACKE, K.D.; COATS J.R., (Eds.). *Enhanced biodegradation of pesticides in the environment*. Washington, DC: ACS, p. 249-268.

FERNANDEZ, I.E.J. (1995) *Reversibilidade de fósforo não-lábil em diferentes solos em condições naturais e quando submetidos à redução microbiológica ou química*. Tese de Doutorado. Viçosa, 94p.

FILIPPELLI, G.M. & DELANEY, M.L. (1996) *Phosphorus geochemistry of equatorial Pacific sediments*. Geochimica et Cosmochimica Acta, v. 60, n. 9, p. 1479-1495.

FORSTNER, U.; WITTMANN, G.T.W. (1983) *Metal Pollution in the Aquatic Environment*. Berlin: Springer-Verlag, 486p.

FRANZEN, M. (2009) *Dinâmica do fósforo na interface água sedimento em reservatórios*. Dissertação (Doutorado) – Universidade Federal do Rio Grande do Sul, Porto Alegre, 225p.

FREEZE, R.A.; CHERRY, J.A. (1979) *Groundwater*. Englewood Cliffs, NJ: Prentice-Hall, Inc., 604p.

FREITAS, T.C.M. (2006) *O cromo na indústria de curtumes de Mato Grosso do Sul, Brasil: Aspectos ecológicos*. Thesis (Doctorate) – Convênio Universidade de Brasília, Programa Multiinstitucional de Pós-Graduação em Ciências da Saúde, Campo Grande, 120p.

FRONZA, L. (2006) *Capacidade de Liberação de Hidrocarbonetos dos Sedimentos de Áreas Contaminadas do Estuário da Lagoa dos Patos – RS*. Dissertation (Master's degree) – Fundação Universidade Federal de Rio Grande, Programa de Pós-Graduação em Oceanografia Física, Química e Geológica, Rio Grande, 116p.

GONSIORCZYK, T.; CASPER, P.; KOSCHEL. (1998) Phosphorus--binding forms in the sediment of na oligotrophic and na eutrophic hardwater lake of the Baltic Lake District (Germany). *Wat. Sci. Tech.*, v. 37, n. 3, p. 51-58.

GUIMARÃES, G.L. (1987) Impactos ecológicos do uso de herbicidas ao meio ambiente. *Séria Técnica IPEF*, v. 4. n. 12, Piracicaba, p. 159-180.

GUNKEL, G.; RUETER, K.; CASALLAS, J.; SOBRAL, M.C. (2003) Estudos da limnologia do reservatório de Tapacurá em Pernambuco: problemas da gestão de reservatórios do semiárido brasileiro. *XV Simpósio Brasileiro de Recursos Hídricos*, Resumo, Curitiba.

HEGGIE, D., FREDERICKS, D., SMITH, C., RADKE, L., MURRAY, E., ATKINSON, I. (2002) *Sediment Nutrient Fluxes and Eutrophication*, Geoscience Australia, Canberra, p.165-168.

ISHII, I.H. *Contribuição ao estudo do ciclo do carbono na represa de Três Marias, MG*. Master dissertation. São Carlos – UFSCar, 1987. 159p.

KIM, L.H.; CHOI, E.; STENSTROM, M.K. (2003) Sediment characteristics, phosphorus types and phosphorus release rates between river and lake sediments. *Chemosphere*, Los Angeles, v. 50, n. 1, Jan. p. 53-61.

LAVORENTI, A. (1996) Comportamento dos herbicidas no meio ambiente. WORKSHOP SOBRE BIODEGRADAÇÃO, *Anais*, Jaguariúna: Empresa Brasileira de Pesquisa Agropecuária, Campinas, p. 81-115.

MATOS, A.T. (2001) *Uso racional dos recursos naturais e seus reflexos no meio ambiente* – Poluição ambiental e seus efeitos. Módulo G. Brasília: Abeas – Associação Brasileira de Educação Agrícola Superior, 121p.

McCARTHY, J.J.; TAYLOR, W.R.; TAFT, J.L. (1975) The Dynamics of Nitrogen and Phosphorus Cycling in the Open Waters of the Chesapeake Bay. *Marine Chemistry in the Coastal Environment*. June 1, p. 664-681.

MELLO, F.A.F.; BRASIL SOBRINHO, M.O.C.; ARZOLLA, S.; SILVEIRA, R.I.; NETTO, A.C.; KIEHL, J.C. (1984) *Fertilidade do solo*. 2. ed., Piracicaba, 400p.

MONCADA, M.P.H. (2004) *Estudos em Laboratório de Características de Colapso e Transporte de Solutos Associados à Infiltração de Licor Cáustico em um Solo Laterítico*. Dissertação (Mestrado) – Pontifícia Universidade Católica do Rio de Janeiro, Civil Engineering Graduation Program , Rio de Janeiro, 219p.

NASCENTES, R. (2006) *Estudo da mobilidade de metais pesados em um solo residual compactado*. Thesis (Doctorate) – Universidade Federal de Viçosa, Civil Engineering Graduation Program , Viçosa, MG, 153p.

NOVAIS, R.F.; SMYTH, T.J. (1999) *Fósforo em solo e planta em condições tropicais*. Viçosa, MG, 399p.

OHLE, W. (1976) General consideration on environmental problems of lakes. In: *International Congress human environment*. Kyoto, p. 383-390.

PANKOW, J.F. (1997) *Aquatic Chemical Concepts*, Lewis Publishers, Chelsea, 673p.

FROEHNER, S; MARTINS, R.F. (2008) *Avaliação da composição química de sedimentos do Rio Barigui na Região Metropolitana de Curitiba*, Quím. Nova, v. 31, n. 8, p. 2020-2026.

PELLEGRINI, J.R. (2005) *Fósforo na água e no sedimento na microbacia hidrográfica do Arroio Lino – Agudo – RS*. Dissertation (Master) – Universidade Federal de Santa Maria, Graduation Program in Soil Science, Field of Concentration in Chemical Processes, Santa Maria, 85p.

PENTEADO, J.C.P.; VAZ, J.M. (2001) O legado das bifenilas policloradas (PCBs). *Quim. Nova*, v. 24, n. 3, p. 390-398.

PERSSON, G.; JANSSON, M. (1988) Phosphorus in freshwater ecosystems. *Hydrobiologia*, 170, 340p.

PIASAROLO, L.; RIGITANO, R.L.O.R.; GUERREIRO, M.C. (2008) Influência da polaridade de pesticidas não-iônicos sobre sua sorção em um latossolo. *Ciênc. agrotec.*, Lavras, v. 32, n. 6, Nov/Dez, p. 1802-1809.

PRADA, S.M.; OLIVEIRA, E. (2006) *Distribuição de nutrientes (C, N e P) em testemunhos de sedimentos do reservatório das Graças, Cotia – SP*, São Paulo, 29a Reunião Anual da Sociedade Brasileira de Química.

REDDY, K.R.; O'CONNOR, G.A.; SCHELSKE, C.L. (1999) *Phosphorus biogeochemistry in subtropical ecosystems*. Boca Raton, Florida: CRC Press LLC, 707p.

SANTOS, I.R. (2003) *Geoquímica dos elementos metálicos e matéria orgânica em sedimentos superficiais da Lagoa Mirim, RS, Brasil*. Monografia. Rio Grande-RS, 53p.

SCHÄFER, A. (1984) *Fundamentos de ecologia e biogeografia das águas continentais*. Porto Alegre: Ed. da Universidade UFRGS, 532p.

SCHENATO, R.B. (2009) *Dinâmica do fósforo de sedimentos de uma pequena bacia hidrográfica de cabeceira*. Dissertation (Master's degree) – Universidade Federal de Santa Maria, Centro de Ciências Rurais, Graduation Program in Soil Science, Santa Maria, 105p.

SCHENATO, R.B.; RHEINHEIMER, D.S.; SANTANNA, M.A.; TOLEDO, J.A.; TIECHER, T. (2008) Dessorção de fósforo de sedimentos erodidos de solos de encostas sob diferentes condições de oxi--redução. *FertBio*.

SILVA, C.S.; LAYBAUER, L.; DILLENBURG, S.R. (2001) Incremento de nutrientes e metais pesados no registro sedimentar da Laguna de Tramandaí, Litoral Norte do Rio Grande do Sul, Brasil. VIII Congresso ABEQUA. *Anais...*, Imbé-RS.

SILVANO, J. (2003) *Avaliação de metais na água, no sedimento e nos peixes da Lagoa Azul formada por lavra de mineração de carvão a céu aberto, Siderópolis – SC*. Dissertation (Master's degree) – Universidade Federal do Rio Grande do Sul, Graduation Program in Ecology , Porto Alegre, 75p.

SINDAG – Sindicato Nacional da Indústria de Produtos para Defesa Agrícola. (2006) *Agrotóxicos ilegais*. Available in: <http://www.sindag.com.br>

SNOEYINK, V. L., JENKINS, D. (1980) *Water Chemistry*. John Wiley and Sons, Inc., USA, 463 p.

SONDERGAARD, M.J.; JENS, P.; JEPPESEN, E. (2003) Role of sediment and internal loading of phosphorus in shallow lakes. *Hydrobiologia*, Amsterdam, v. 506, n. 2, Nov., p. 135-145.

STERNER, R.W.; ELSER, J.J. (2002) *Ecological Stoichiometry*. Princeton, NJ: Princeton University Press, 584p.

STUMM, W.; MORGAN, J.J. (1981) *Aquatic chemistry: An introduction emphasizing chemical equilibria in natural waters*. 2. ed. New York: John Wiley & Sons, 780p.

STUMM, W.; MORGAN, J.J. (1996) *Aquatic chemistry: Chemical equilibria and rates in natural waters*. 3. ed. New York: John Wiley & Sons, 1022p.

SUSIN, J., GONÇALVES JR, A.C., SILVA, M.R. (2007) *Distribuição de nitrogênio e fósforo em sedimentos de superfície do rio Itajaí-Açu*. 30ª Reunião Anual da Sociedade Brasileira de Química, São Paulo.

TUNDISI, J.G. (2000) Limnologia e gerenciamento integrado de recursos hídricos: avanços conceituais e metodológicos. *Ciência & Ambiente*, v. 1, n. 21, p. 9-20.

TUNDISI, J.G.; TUNDISI, T.M. (2008) *Limnologia*. São Paulo: Oficina de Textos, 631p.

WARD, J.V. (1992) *Aquatic insect ecology: biology and habitat*. New York: John Wiley & Sons, Inc., 438p.

WETZEL, R.G. (1999) Organic phosphorus mineralization in soils and sediments. In: REDDY, K.R.; O'CONNOR, G.A.; SCHELSKE, C.L.. (Eds.). *Phosphorus Biogeochemistry in Subtropical Ecosystems*. Boca Raton, Florida: CRC Press LLC, p. 225-245.

WETZEL, R.G. (2001) *Limnology: Lake and River Ecosystems*. 3. ed. San Diego: Academic Press, 1006p.

XAVIER, L.F.W.; MOREIRA, I.M.N.S.; HIGARASHI, M.M.; MOREIRA, J.C.; FERREIRA, L.F.V.; OLIVEIRA, A.S. (2005) Fotodegradação de Hidrocarbonetos Policíclicos Aromáticos em Placas de Sílica Impregnadas com Dióxido de Titânio. *Quím. Nova*, v. 28, n. 3, Out., p. 409-413.

CHAPTER 12

HYDRODYNAMICS AND TRANSPORTATION IN THE RIO VERDE RESERVOIR

Cynara da Nóbrega Cunha
Guilherme Franz
Ana Paula Kelm Soares
Jose Eduardo Gonçalves
Paulo Cesar Colonna Rosman

HYDRODYNAMICS AND TRANSPORTATION IN THE RIO VERDE RESERVOIR

Cynara da Nóbrega Cunha
Guilherme Franz
Ana Paula Kelm Soares
Jose Eduardo Gonçalves
Paulo Cesar Colonna Rosman

ABSTRACT

The characteristics of hydrodynamic circulation and the transport of contaminants is discussed in this work in order to add to our understanding of the phenomenon of eutrophication in the Rio Verde Reservoir. Models used in the analysis are part of SisBaHiA® (in Portuguese, Sistema Base de Hidrodinamica Ambiental). SisBaHiA® is a set of models that is able simulate coastal and inland water bodies. In this study, we used the horizontal two-dimensional hydrodynamic circulation model (2DH) and the vertically integrated model of water quality and eutrophication for non-conservative and passive scalars. Results from environmental modeling show that the chosen models are suitable and can be used quantitatively in characterizing the excessive algae growth in reservoirs, thus providing the foundation for new applications and developments. Despite the errors associated with boundary conditions and the simplifications considered herein, the model has demonstrated its potential to adequately simulate the data set collected from the reservoir. Based on the results of the hydrodynamic model, we noted the significant influence of wind in determining hydrodynamic circulation in the Rio Verde Reservoir. Inflows/outflows only generate impacts in circulation in their immediate surrounding regions and do not contribute to the circulation of the reservoir as a whole.

KEY WORDS

Water quality models, hydrodynamic circulation, reservoir.

1. INTRODUCTION

The characteristics of reservoirs place them in an intermediate position between rivers and lakes and they can be divided into three zones based on different physical, chemical and biological properties. In relation to surface area, reservoir drainage basins of are typically larger than natural lake basins, as well as narrower and more elongated. Thus, reservoirs receive only a small fraction of their inflow directly from runoff in the drainage basin and most of the water, nutrients and sediments are usually supplied by one or two major tributaries.

Water quality in reservoirs is closely linked to average residence time. In large, deep reservoirs, residence time can be years, while in narrow and shallow reservoirs, residence time can be only a few weeks. If the residence time is short, the quality of the water in the reservoir can be mainly determined by the quality of tributaries. However, if the residence time is long, one can expect a significant effect from sources coming from the surface or bottom of the reservoir as well as biological activity (FISCHER *et al.* 1979).

The various phenomena that occur within reservoirs hinder the understanding of system behavior and its analysis in different scenarios. The mathematical representation of a lacustrine system for the study of water quality, considering all its physical, chemical and biological processes, can be quite complex. Water quality models are able to assess the impacts generated by the release of pollutants and it is possible to understand phenomena such as excessive growth of algae due to eutrophication.

Quantitative ecological modeling of reservoirs is a major challenge, especially considering that transport mechanisms and physical, chemical and biological processes active in the water body are vast and require an interdisciplinary approach. However, the development of mathematical models that seek to quantify these processes has been the object of study in recent decades. The complexity of the system must be reflected in the modeling process, ensuring that the model has the capaity to understand the behavior of the environmental system and, thus, evaluate the effects that various actions can have on its behavior.

Data needed for studies and projects focusing on water resource management are generally scarce and obtained from only a few points across the area of interest. Water quality models are indispensable tools to integrate this spatially dispersed information, allowing its interpolation and extrapolation across the entire area of interest. In addition, new scenarios may be simulated to analyze changes in the medium. The spatial dynamics of the movement of pollutants can also be better understood with the use of these models (ROSMAN, 2000).

In lakes and reservoirs, the spatial scale of phenomena related to water quality occurs preferentially in the vertical direction, whereas the dominant transport processes in lakes and reservoirs occur along the depth, with few transverse and longitudinal variations. Another important characteristic of reservoirs, related to depth and low longitudinal velocity, is thermal and chemical stratification. Solar radiation, wind, and inlet and outlet flow, related to residence time, influence the processes of stratification.

The vertical dynamic observed in reservoirs is related to the presence or absence of stratification which, in the case of continental reservoirs, relates to variations in temperature. The penetration of light into the water column defines the region where photosynthesis can occur and thus determines the migration of species to the surface. In shallow lakes, wind has a more significant impact on the vertical mixing processes, in addition to increasing turbidity and decreasing light penetration. Another factor to be considered is the distribution of nutrients throughout the water column. As distribution is not uniform, it is possible that species adapted to regions with higher concentrations of nutrients can have an advantage over other species, especially in periods of decreased concentration.

The distribution of a substance within a body of water is performed by advective and diffusive transport mechanisms. Advection can be defined as the movement of the substance associated with the resolvable velocity field in the scale of interest. The velocity field in bodies of water with open surfaces, such as reservoirs, is determined through the use of hydrodynamic models, which represent a set of basic input data for water quality models.

Diffusion can be defined as any movement in non-resolvable scale, which must be modeled in terms of resolvable magnitudes. In large reservoirs or lakes, wind is the primary agent in transmitting random movement to water (CHAPRA, 1997).

Water quality models are based on the principle of mass conservation, representing advection-diffusion and changes in concentrations of substances due to chemical, physical and biological processes. When the substance is passive, that is, the velocity field is independent from its concentration, the study of substance transport is a problem that is disengaged from hydrodynamic modeling.

Water quality models have various dimensions according to the body of water of interest. Water quality models with zero dimensions can be used to estimate the spatial average of pollutant concentration. In this case, it is assumed that the pollutant is thoroughly mixed within the water body and hydrodynamic circulation data is not required. Models in one dimension are usually used to simulate rivers; the geometry of the system is represented by a linear network of segments. Variations in water quality parameters occur along the longitude as water is transported from one segment to another. These models are also used in deep lakes and those with a small surface area to simulate the vertical variation of temperature and other quality parameters.

Water quality models applied to reservoirs often have two or three dimensions to represent spatial heterogeneities. When two-dimensional models are used, a new term is added to the transport equation to represent the loss of differentiated advection along the vertical and/or horizontal direction. This process is called dispersion (CHAPRA, 1997).

Physical, chemical and biological processes can increase or decrease the concentration of substances within the system. The chemical and biochemical reactions can be modeled by defining their kinetics, which is related to the velocity or rate at which these reactions occur. Reactions can be classified according to their order, which expresses how the concentration of a particular reagent varies. Although the order of reactions needs not necessarily be a whole number, most significant reactions used in water quality modeling have an order equal to zero, one or two.

Zero-order reactions are independent of the concentration of reactants. Examples include the oxidation of ammonia nitrogen to nitrite nitrogen and glucose oxidation by aerobic bacteria. However, these biochemical reactions slow down when substrate concentrations approach zero. In the case of first-order reactions, the rate of change is directly proportional to the reagent concentration. Examples of these reactions include dissolution and removal of gases in water and the decomposition of organic matter by bacteria in tests to define the biochemical oxygen demand (BOD) in a water sample. Second-order reactions are those in which the reaction rate is proportional to the square of the concentration of one reactant or to the product of concentration of two different reagents. An important use of second-order reactions is the representation of biological transformations.

Knowledge of the thermal regime of reservoirs and lakes is particularly important for three reasons: effluent discharges at different temperatures can cause negative effects on the aquatic ecosystem; temperature influences chemical and biological reactions; and temperature variations affect the density of water and, consequently, changes transport processes, forming thermal stratification. The water temperature is the main variable responsible for maintaining the life of aquatic systems, affecting the growth and development of living beings (ANGELOCCI & VILLA NOVA, 1995).

The solar radiation that reaches the surface of the lake is transformed into heat, especially near the surface. This transfer process occurs slowly, producing layers of different temperatures and hence different densities. Wind is the principle force able to mix these layers. If the energy generated by the wind is not sufficient to promote mixing, the water column is stratified for long periods of time. Thus, stratification can be quite resistant to mixing, forming three distinct layers: epilimnion, hypolimnion and metalimnion. The mixed layer can be restricted to the epilimnion or move to hypolimnion, causing vertical movement throughout the water column, mainly resulting from the wind.

Thus, temperature variations can occur on daily and seasonal scales, generating stratification and de-stratification events within these scales. This vertical circulation influences the distribution of nutrients in the water column: if stratification occurs, nutrients are quickly consumed by phytoplankton; when de-stratification takes place, there is a homogeneous distribution of nutrients throughout the water column, which does not cause any shortage of nutrients in the epilimnion.

Temperature variations noted throughout the water column in lakes and reservoirs depend on wind and other energy flows to which these water bodies are subjected. Several studies have shown the influence of solar radiation on the formation of temperature profiles. Sperling *et al.* (2004) show the thermal structure of a developing lake, highlighting the temporal variation of thermal stability of the water column over two years. Lopes & Bicudo (2001) review the existing literature on lake systems in the Amazon and Southeast regions of Brazil and present an investigation of diurnal and daily variations for different physical properties in a subtropical, shallow and oligotrophic reservoir for two distinct periods, dry and rainy. Lawson & Anderson (2007) discuss a de-stratification system installed in the Elsion lake which is able to increase the levels of dissolved oxygen.

In the literature there are a vast array of ecological models. Jorgensen *et al.* (1996) show that approximately 1000 models have been described and referenced in the current literature. These models demonstrate an evolution that began in the 1920s with the work of Streeter and Phelps on rivers, up to current models, which involve the simulation of toxic substances, organic pollutants, metals and sediment-water interactions (FRAGOSO JR., 2009). Within this context, SisBaHiA® (in Portuguese, Sistema Base de Hidrodinamica Ambiental) has been developed since 1987 by the Coastal and Oceanographic Engineering Department, Oceanic Engineering Program, Federal University of Rio de Janeiro (COPPE/UFRJ). SisBaHiA® is able to model coastal and inland water bodies and it consists of: a three-dimensional hydrodynamic circulation model (3D) or a two-dimensional horizontal hydrodynamic circulation model (2DH); an Eulerian advective-diffusive transport model with reactions; a water quality and eutrophication model; a Lagrangian deterministic transport model; a probabilistic model; as well as a wave generation model. The Water Quality And Eutrophication Model of SisBaHiA® is an Eulerian advective-diffusive transport model vertically integrated for non-conservative scales, or in other words, the modification of concentration is the result of physical, chemical and biological processes. In this case, the velocity field used is an average in the vertical direction and no description of the velocity profile is allowed. Further details on SisBaHiA® can be obtained through www.sisbahia.coppe.ufrj.br.

This study presents the development of environmental modeling for assessing the hydrodynamics and transport of contaminants in the Rio Verde Reservoir. The model allows us to simulate, predict and assess eutrophication relating to biological data and water quality. As such, it provides support for decision making by defining the effectiveness of mitigation measures and assessing reservoir zones where higher algae concentration may potentially occur.

2. DESCRIPTION OF THE RIO VERDE RESERVOIR

The Rio Verde Reservoir is located in the Metropolitan Region of Curitiba (RMC), at a longitude of 490 31'W and latitude of 250 31'S, having a surface area of approximately 5,971,731.0 m^2, an average volume of 25,643,732.0 m^3, and an average depth of 5,642 m. The reservoir is part of the Rio Verde Hydrographic Basin. Further details on land-use, soil type, morphology and geology of the basin that forms the tributaries of the Rio Verde Reservoir can be found in Chapters 1 through 5. The bathymetry of the reservoir, shown in Figure 1, was obtained from bathymetric data collection carried out from April 7 to 10, 2008. Due to the presence of submerged trunks and excessive vegetation in some parts of the reservoir, it was not possible to perform bathymetry throughout the entire reservoir. Therefore, it was necessary to use topographic maps to determine depth where data collection was not possible. The elevation volume ratio of the reservoir can also be seen in Chapter 9. The Rio Verde Reservoir is oriented in a Northeast (NE)-Southwest (SW) direction, with a length of approximately 7500 m and a maximum width of 1300 m. The reservoir has two distinct regions: one near the dam, with greater depths, and another near the Rio Verde, with shallower depths. Another feature of the reservoir is the dendritic system formed by branches where small rivers drain into the reservoir and these regions are characterized by high residence time.

The main tributary of the reservoir is the Rio Verde, accounting for 70% of the total tributary flow; other smaller tributaries also contribute with low flow levels. Figure 2 shows the flow rates obtained by sampling or using the SWAT model (Soil and Water Assessment Tool). Further details on the determination of flow rates can be found in Chapter 8. In relation to outflows, two exit points were set, one in the catchment area, close to the dam and with a permanent value of 0.83 m3/s, and another at the spillway, such that the sum of inflows is equal to the sum of outflows.

The climate of the region, according to the Koeppen classification, is Cfb - temperate, with cool summers, frequent frosts, and no defined dry season. Meteorological data obtained from a station installed near the dam of the Rio Verde Reservoir (on a slab located at water intake, Latitude 25°31'36,83"S and Longitude 49°31'39,07") confirmed the classification of the region's climate. This station provides wind direction and speed, air temperature, solar radiation, and precipitation. Temperatures are taken every 15 minutes. Sensors are installed approximately 10 meters from the open water surface.

Figure 3, 4, 5 and 6 show the values of solar radiation, air temperature, humidity, and precipitation, collected at the meteorological station between August 2008 and August 2010 and used for the temperature model. The maximum values of solar radiation measured at the station are dependent on cloud cover, showing a variation between winter and summer. The maximum solar radiation recorded is 1442.0 W/m^2. However, taking into account the moving average over a period of 30 days, this variation is no longer visible.

Air temperatures vary with the seasons, with values close to zero in the month of May. The average temperature during the period is 17.12°C, with a maximum value of 33.79 °C and minimum of 0.04°C. Humidity values show little annual variation, with average values of 80.0%, a maximum value of 100.0%, and a minimum value of 17.26%. Data on rainfall that reaches the reservoir can be seen in Figure 6. There is no clearly defined period of drought or flooding; during summer, rains occur with greater volume and during the winter it is generally drier.

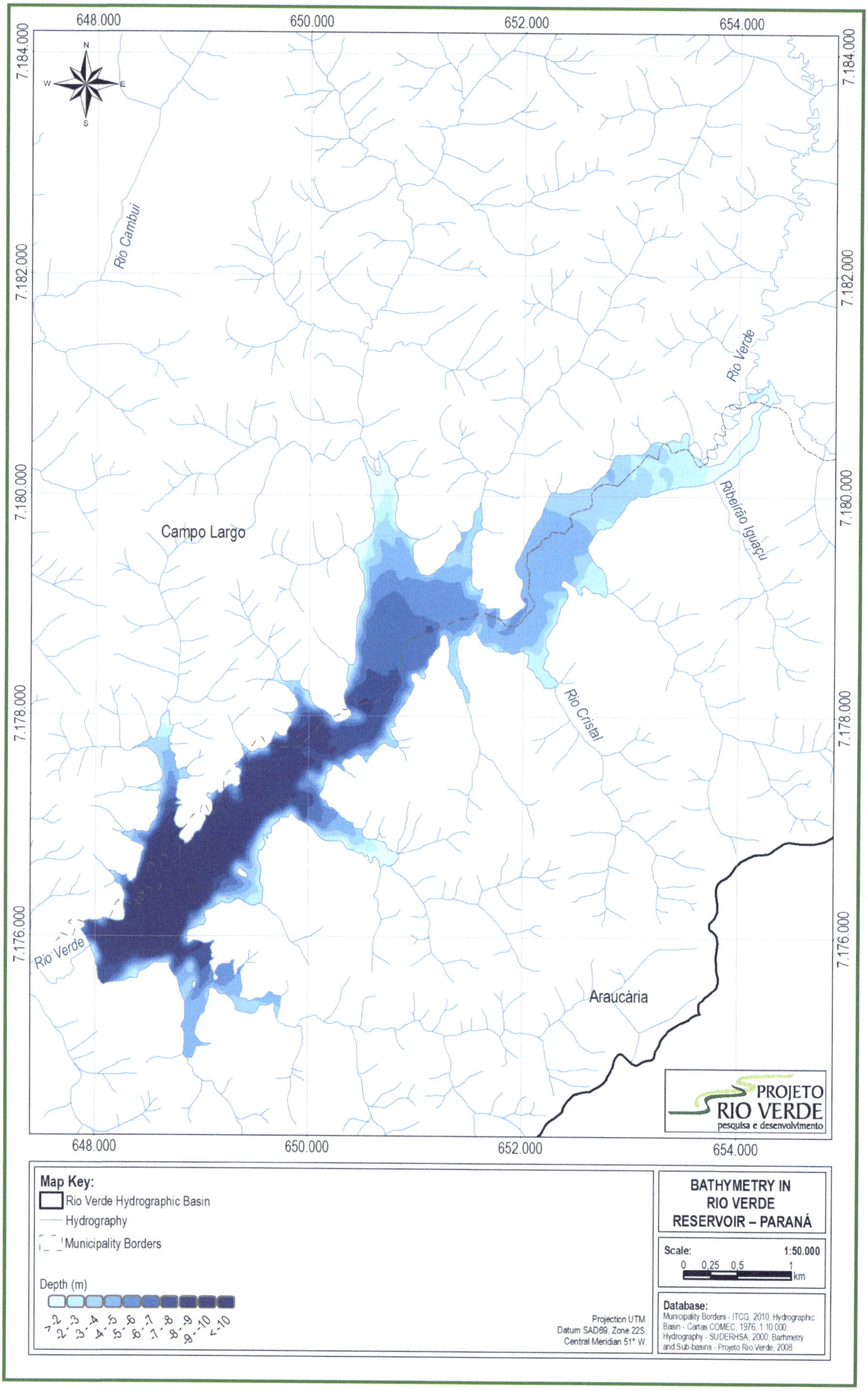

FIGURE 1 – TOPOGRAPHY OF THE RESERVOIR BOTTOM IN THE MODELING DOMAIN AS DEFINED BY THE DISCRETIZATION MESH WITH THE LOCATION OF FLOW MEASUREMENT STATIONS

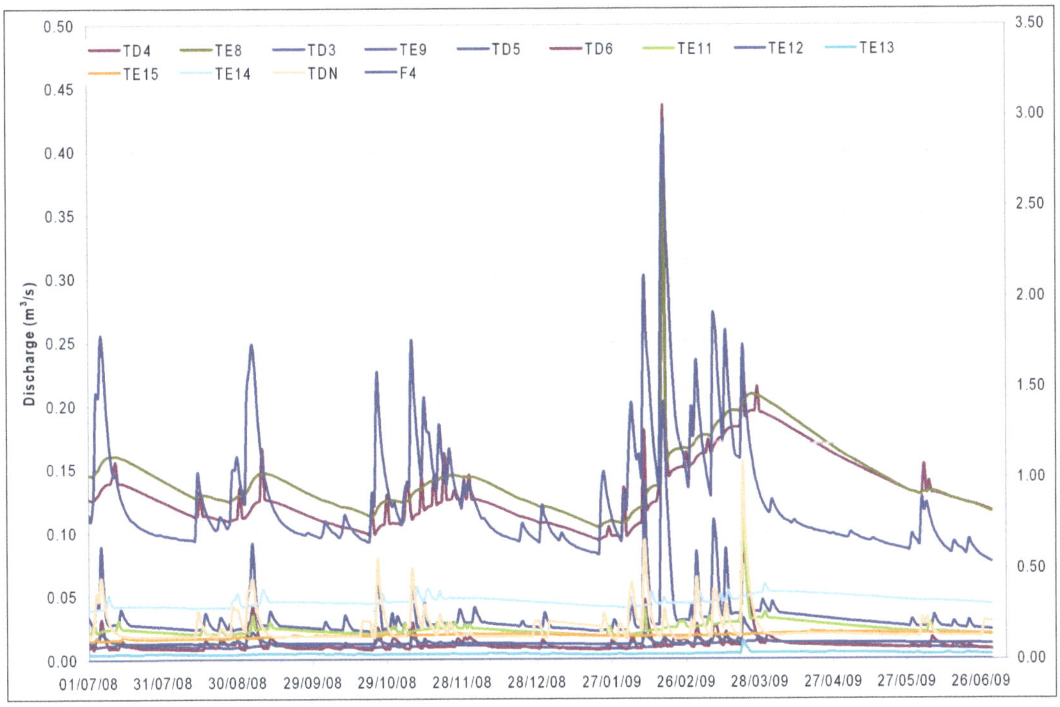

FIGURE 2 – DISCHARGE (MEASURED AND/OR FROM SWAT) IN THE TRIBUTARIES OF THE RIO VERDE RESERVOIR

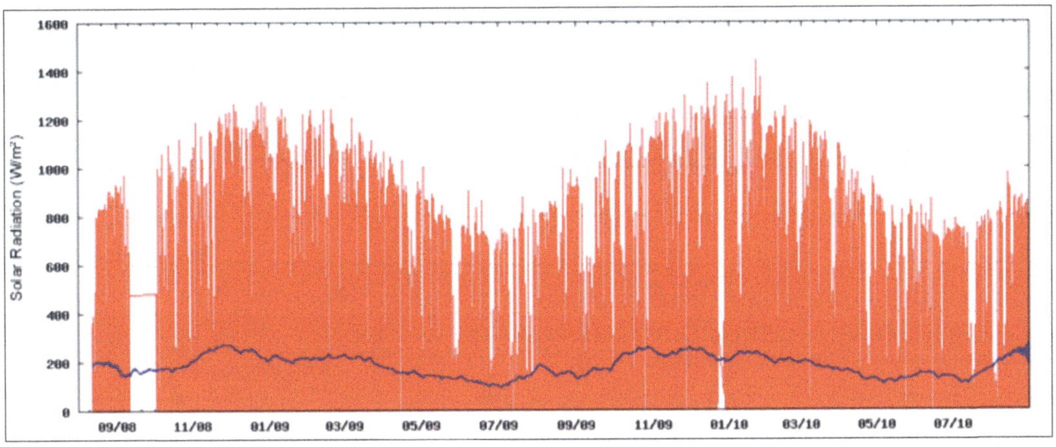

FIGURE 3 – DATA ON SOLAR RADIATION IN W/m² MEASURED AT THE WEATHER STATION IN THE RIO VERDE RESERVOIR. THE CONTINUOUS LINE REPRESENTS THE AVERAGE FOR A 30 DAY PERIOD

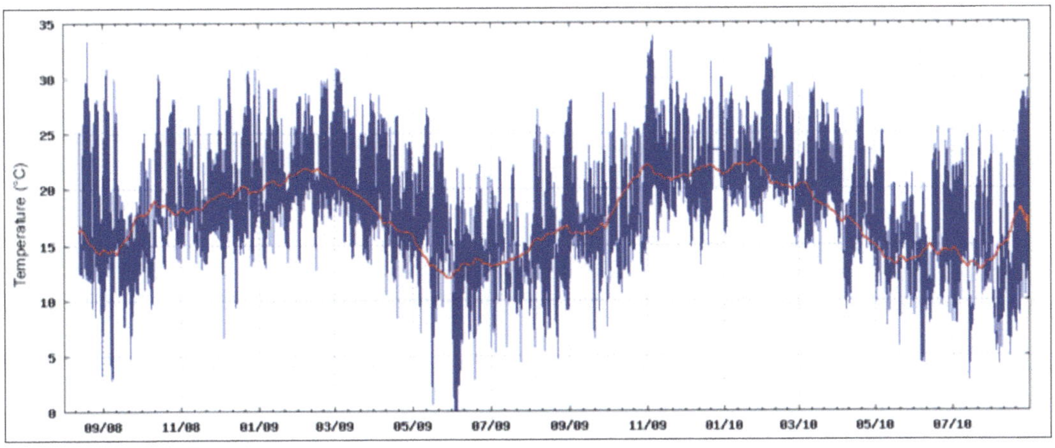

FIGURE 4 – DATA ON AIR TEMPERATURE (°C) MEASURED AT THE WEATHER STATION OF THE RIO VERDE RESERVOIR. THE CONTINUOUS LINE REPRESENTS THE AVERAGE FOR A 30 DAY PERIOD

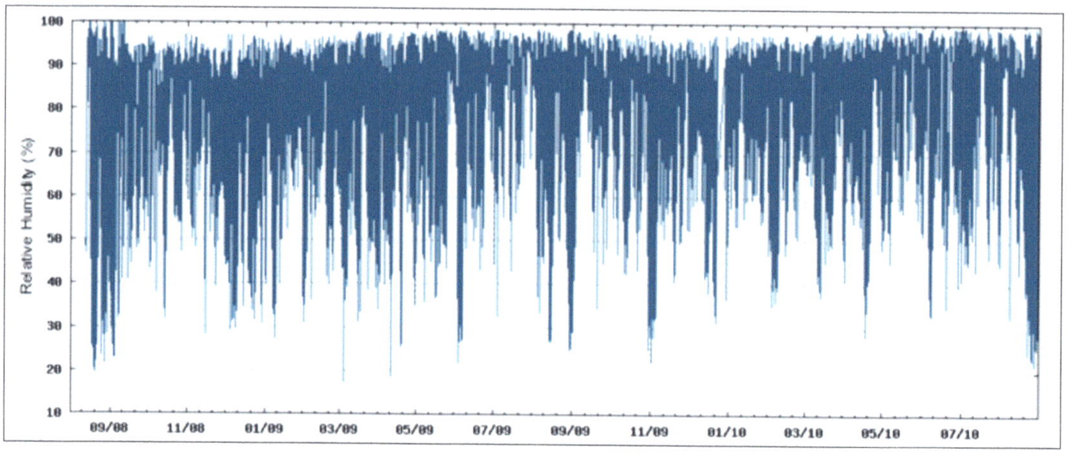

FIGURE 5 – DATA ON RELATIVE HUMIDITY (%) MEASURED AT THE WEATHER STATION OF THE RIO VERDE RESERVOIR

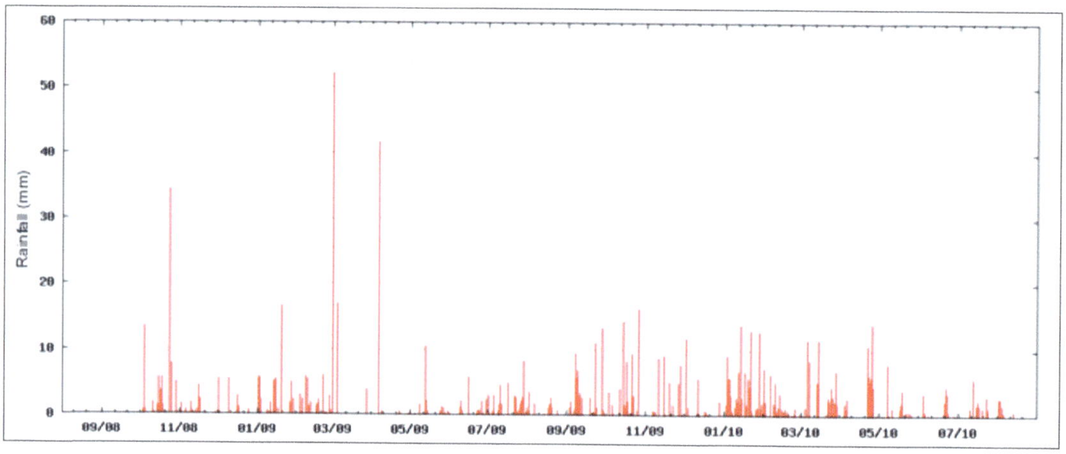

FIGURE 6 – DATA ON RAINFALL (mm) MEASURED AT THE WEATHER STATION OF THE RIO VERDE RESERVOIR

The concentration values of various substances in the tributaries, which are used as environmental conditions in the model of water quality, are shown in Figure 7. BOD concentrations in the rivers are consistently low, less than 3.0 mg/L, except at point TE14, where BOD is equal to 6.0 mg/L in July 2008. Most of the concentrations in the tributaries of the Rio Verde Reservoir are 1.0 mg/L, which is the limit of detection, and from this we can conclude that the load of organic matter contributing to the reservoir is minimal.

The Rio Verde Reservoir presents a dynamic system of stratification, characterized by a reasonable stratification in summer and a mixed water column during the winter. Figure 8 and 9 show the temperature profiles measured at stations R2 and R4, respectively, between July 2008 and July 2009. We can see that the two stations have similar behavior in relation to stratification: between April and September, the water column shows no stratification and can be considered well mixed; from October, with the arrival of summer, the water column begins to stratify, with the largest gradients found in late summer (March), a behavior typical of subtropical lakes. One can also see that January 2009 was an unusual month with velocity values less than in December, which demonstrates a month with mild temperatures, as can be seen in air temperatures shown in Figure 4.

The specific mass profile can be defined based on the temperature data, taking into account that, in the reservoir, there is no salinity gradient. Considering a binary system, the density in g/cm³, can be written as:

$$\rho(T) = \frac{1+A}{B+0.698A} \;;\; A = 5890 + 38T - 0.375T^2 \text{ e} \qquad (1)$$

$$B = 1779.5 + 11.25T - 0.0745T^2$$

where T is the water temperature in degrees. The profiles of the specific mass at R4 are shown in Figure 10. It is clearly observed that the density gradients generated are small, indicating a limited capacity for vertical mixing in the water column due to buoyancy. The largest gradient occurred in February 2009 ($15.61.10^{-5}$g.cm^{-3}.m^{-1}) and the lowest occurred in May 2009 ($0.17.10^{-5}$g.cm^{-3}.m^{-1}), showing that the water column stratifies in summer and de-stratifies in winter.

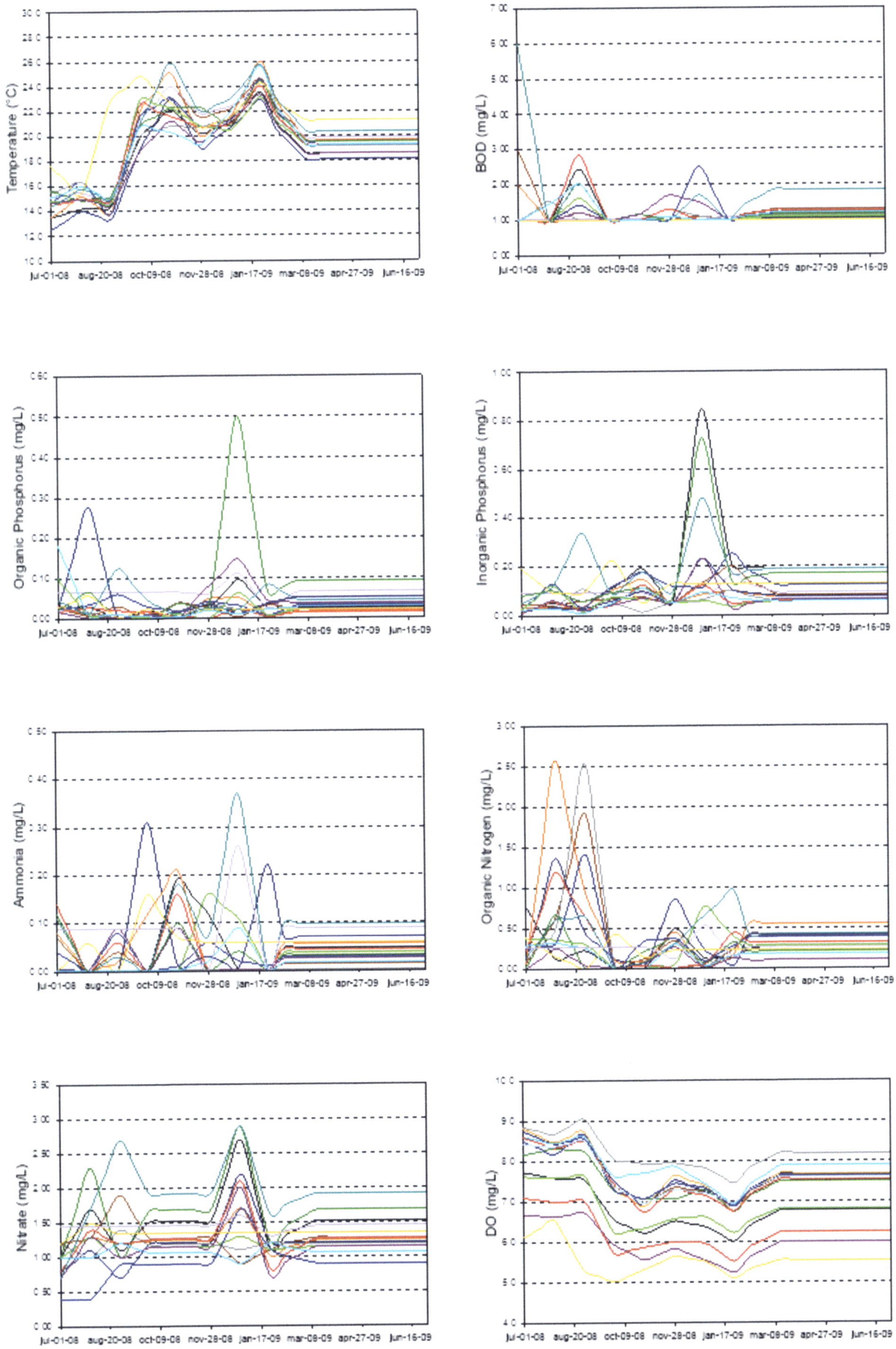

FIGURE 7 – CONCENTRATIONS OF VARIOUS SUBSTANCES IN THE TRIBUTARIES OF THE RIO VERDE RESERVOIR

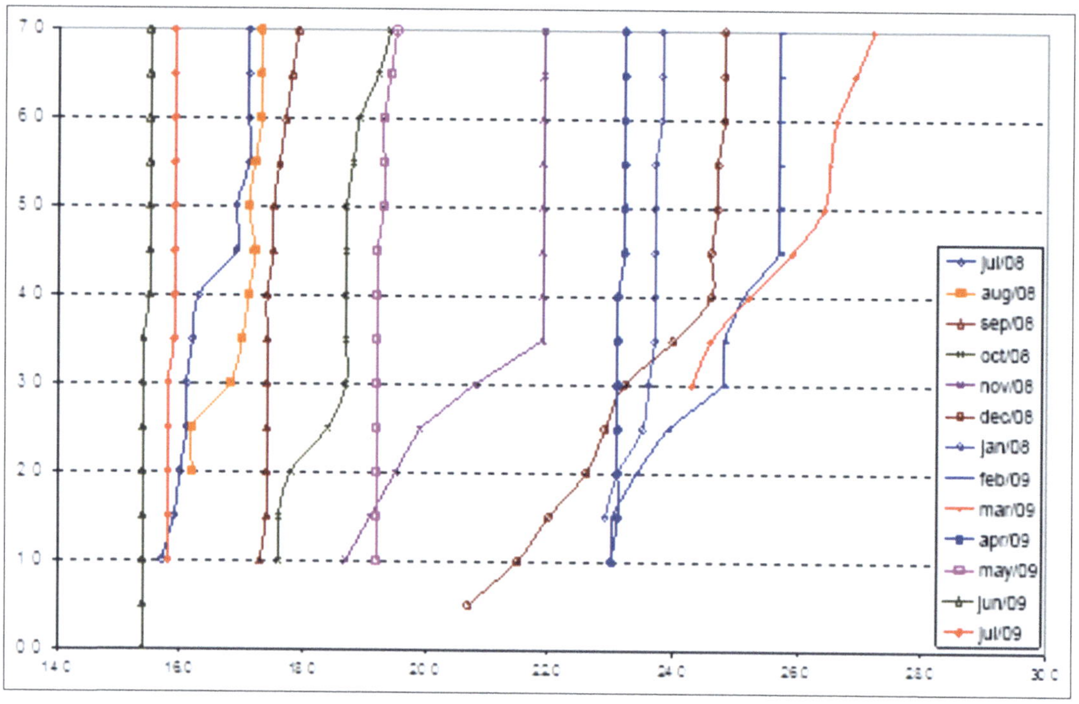

FIGURE 8 – TEMPERATURE PROFILES AT STATION R2 BETWEEN JULY 2008 AND JULY 2009

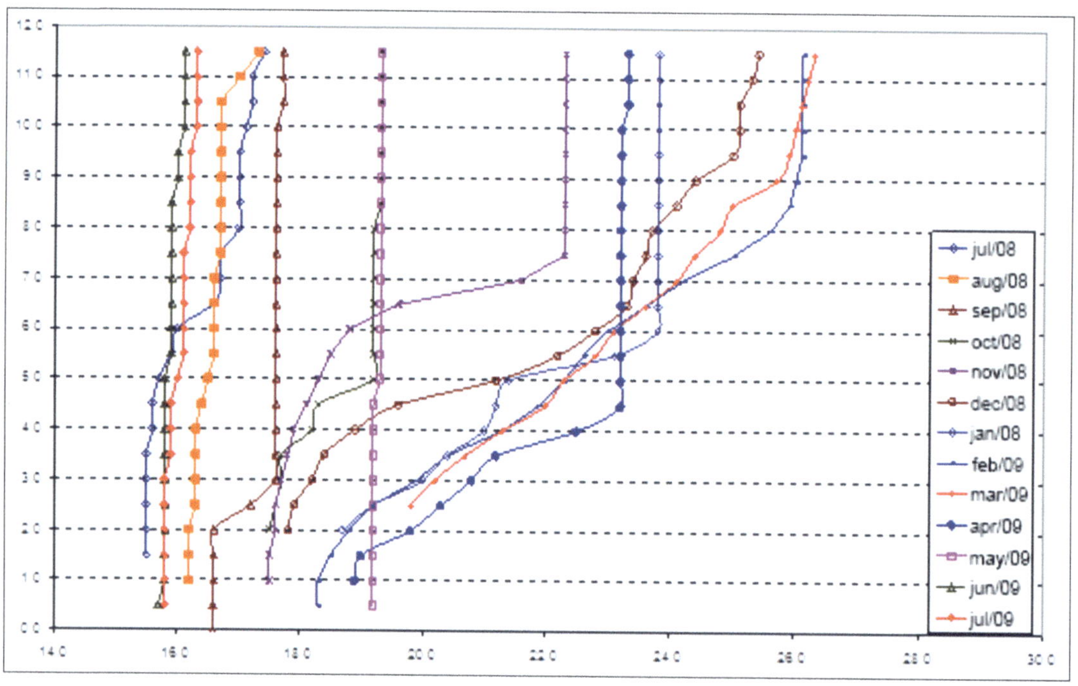

FIGURE 9 – TEMPERATURE PROFILES AT STATION R4 BETWEEN JULY 2008 AND JULY 2009

2.1 HYDRODYNAMICS AND TRANSPORTATION IN THE RIO VERDE RESERVOIR

The ratio between depth and width of the reservoir is about 1:230 and between depth and length is of 1:1330. This suggests a predominance of horizontal velocities and that velocities in the vertical direction are sufficiently low so that the hydrostatic pressure distribution can be assumed. Besides, shallow reservoirs should present a smooth thermal stratification, as can be seen in the data obtained from the Rio Verde Reservoir.

A criterion for evaluating the importance of the stratification process and input and output circulation data of the reservoir, and consequently the distribution of mass in the reservoir, is the densimetric Froude number, defined as:

$$F_D = \left(\frac{L}{h_m}\right)\left(\frac{Q}{V_T}\right)\sqrt{\frac{\rho h_m}{\Delta\rho g}} \qquad (2)$$

where L is the length of the reservoir (7500 m), h_m is the

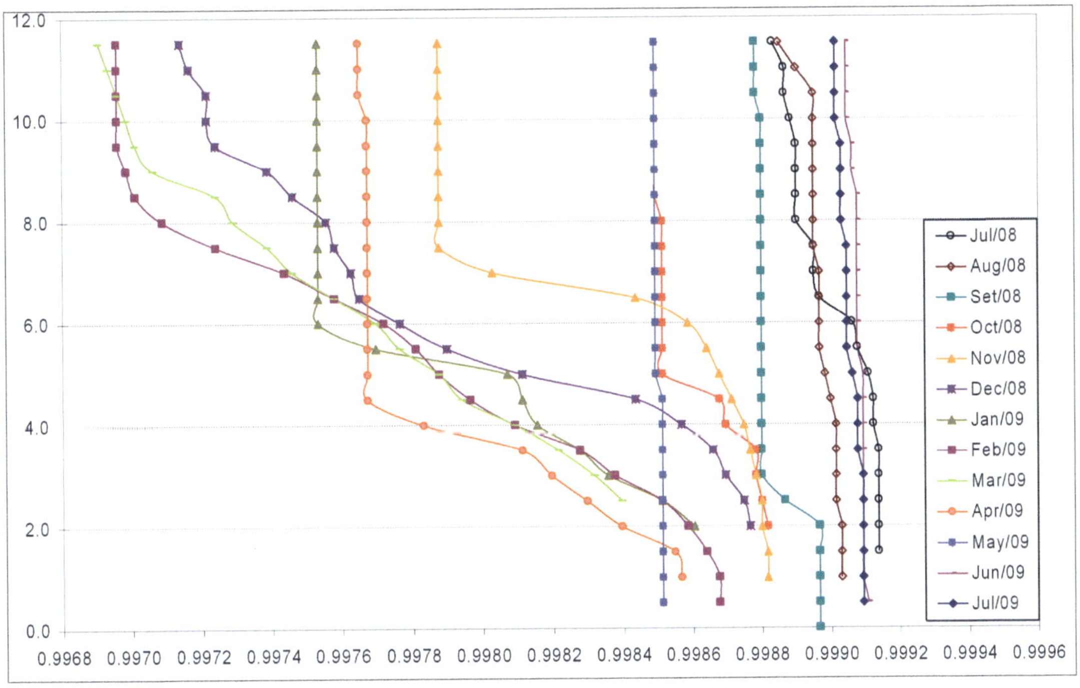

FIGURE 10 – SPECIFIC MASS PROFILES (g/cm^3) AT STATION R4 BETWEEN JULY 2008 AND JULY 2009

average depth (5.6642 m), Q is the average of inflows (m^3/s), Vt is the average volume (25,643,732 m^3), g is the acceleration of gravity (9.81 m/s^2), ρ is the specific mass of water (1.000 g/cm^3), and $\Delta\rho$ is the average specific mass gradient between the bottom and the surface. In average summer conditions, and considering the month of highest gradient (February 2009), the result is $F_D = 0.001$. In average winter conditions, and considering the month of lowest gradient (May 2009), $F_D = 0.01$.

The wind patterns noted in the Rio Verde Reservoir demonstrates that wind mostly blows Northeast (NE), which coincides with the alignment of the reservoir axis thus generating a good trajectory for wind in this direction, with approximately 4850 m. Wind is expected to be the major force in hydrodynamic circulation and as such the major force in mass transport in this reservoir. In two regions, one near the dam where the main water intake and the spillway are located, and the other near the Rio Verde, the main tributary of the reservoir, the hydrodynamic flow and transport are dominated by inflows and outflows, forming a "near field." Outside this region, circulation and transport are determined by the wind.

3. CIRCULATION IN THE RIO VERDE RESERVOIR

The hydrodynamic circulation pattern is a set of phenomena essential for determining the transport of simulated substances. For this, we used the SisBsHiA® two-dimensional hydrodynamic model, considering the inflow from tributaries, the intake flow, and the wind measured at the weather station installed at the dam as driving forces. The simulation was performed for a period of 364 days, between July 01, 2008 and June 30, 2009, the same period of calibration used for the water quality model.

The SisBaHiA® spatial discretization scheme allows an optimal representation of uneven edges and complex bathymetries like those found in reservoirs, preferably made via biquadratic quadrilateral finite elements. However, spatial discretization can also be made using quadratic triangular finite elements, or a combination of both types of elements. Such a method of space discretization is potentially of the fourth order. Time discretization is based on implicit finite difference method with truncation error of the second order.

For the development of the modeling process, it is necessary to build a finite element mesh in the modeling domain. Figure 11 shows the developed mesh and the established contours. We established two types of contours: open contour, with three nodes defined close to the spillway, and land boundaries. Except for the nodes corresponding to the rivers, the spillway and the intake, all nodes were considered impermeable, with normal speed equal to zero. The hydrodynamic model was tested with the values of flow and wind from July 01, 2008. The wind was considered variable in time and spatially homogeneous. The amplitude of the equivalent background roughness (e), needed for the definition of friction tensions, was established by regions: in regions with large amount of submerged trees, bottom roughness amplitude values were increased (with values around 0.130 m). In other regions of the reservoir, we considered the predominance of silty-clay sediments, with values around 0.015 m. Figure 12 shows the spatial distribution of this parameter.

The parameters used in the numerical simulation of hydrodynamic circulation set for the viscous stress terms are shown in Table 1. SisBaHiA® uses few parameters related to the calibration of these terms, maintaining the accuracy of the results. Details on the hydrodynamic circulation model and parameters involved can be found in Rosman (2000).

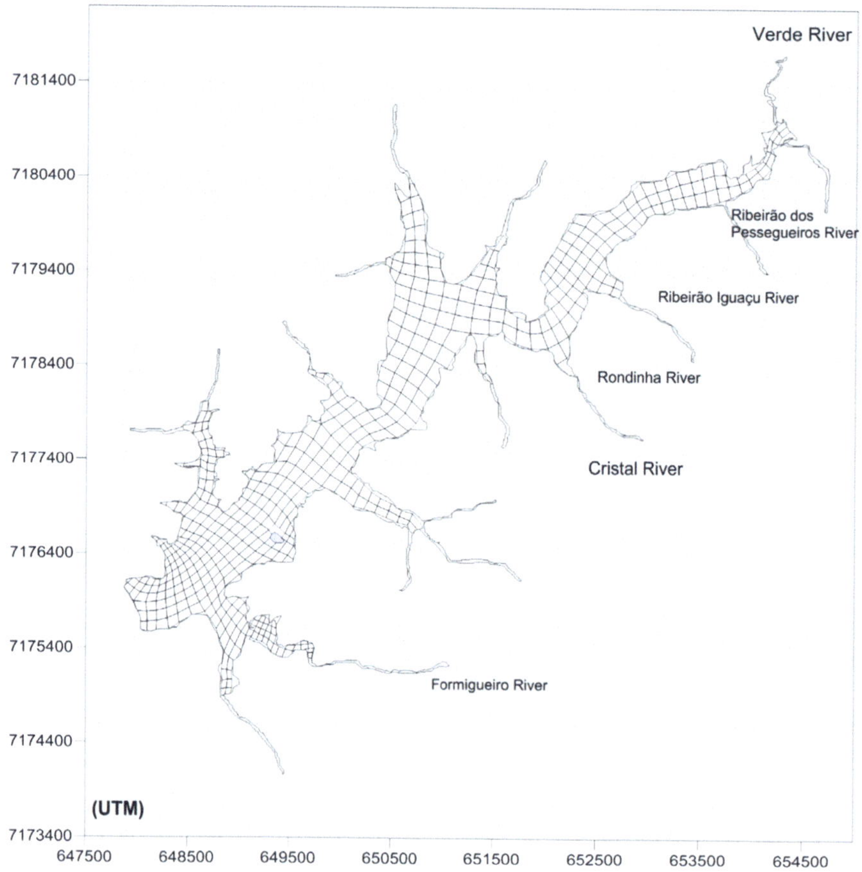

FIGURE 11 – MODELING DOMAIN FOR THE RIO VERDE RESERVOIR SYSTEM SHOWING THE MESH WITH 507 FINITE ELEMENTS AND 2402 NODES

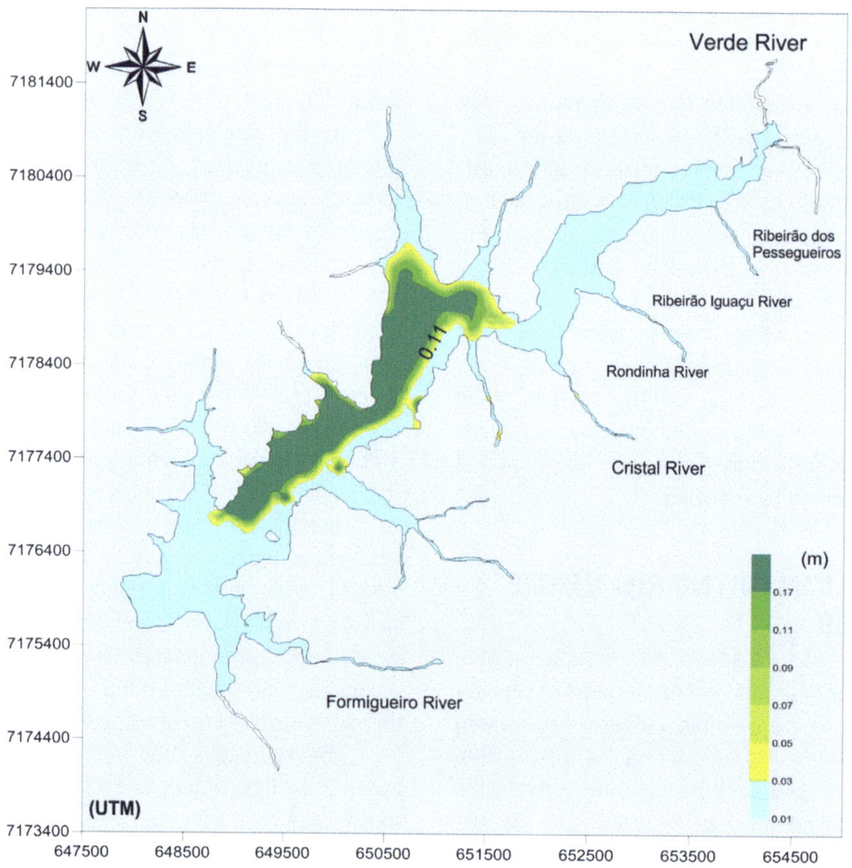

FIGURE 12 – VISUALIZATION OF BOTTOM ROUGHNESS AMPLITUDE IN THE MODELING DOMAIN AS SEEN BY THE DISCRETIZATION MESH PRESENTED IN FIGURE 11

TABLE 1 – PARAMETERS USED IN HYDRODYNAMIC CIRCULATION

PARAMETERS	VALUES
αx (scale parameter in the x dimension)	1.0
αy (scale parameter in the y dimension)	1.0
Δt (s)	100.0
Von Karman Constant	0.405
Maximum Courant Number	55.7
Average Courant Number	22.7

The results of the two-dimensional hydrodynamic circulation model of the Rio Verde Reservoir, including the field of vertically averaged currents, shall be considered qualitatively. Figures 13 and 15 show the spatial pattern of currents in two different situations: in the dry season (September 29, 2008) and the rainy season (March 28, 2009). The velocity fields show a strong relationship with the local wind pattern (Figures 14 and 16) through the presence of a clockwise vortex formed in the region near the dam (lentic region), which intensifies depending on the wind pattern.

During the dry season, represented by the day September 29, 2008, there is a northeasterly wind with a speed of approximately 6.0 m/s during the day. At night, the wind speed is reduced. The currents follow the wind direction, forming two vortices: a large vortex, clockwise in the central region of the reservoir, and a smaller, counterclockwise one. In the regions near the spillway and the Rio Verde, there is an intensification of the currents due to local effects of inflows/outflows.

In the rainy season, represented by day March 28, 2009, the wind shows a very weak pattern, only intensifying later in the day. Under these conditions, only the clockwise vortex in the central portion of the reservoir is formed. There is also an intensification of the currents in the region near the Rio Verde, caused by floodsing observed during this period. Taking into account these patterns, it can be stated that the hydrodynamic circulation, and consequently the horizontal transport, is strongly dependent on the wind and that inflows/outflows generate a localized circulation.

3.1 RESIDENCE TIME

Once the hydrodynamic circulation pattern is determined, it is possible to assess the residence time in the reservoir. The residence time is a key concept to estimate the scale of transport that occurs in a reservoir, allowing us to determine the capacity to eliminate contaminants released into a water body and also define areas of a greater or lesser stagnation.

SisBaHiA® allows the calculation of the residence times based on the use of the Lagrangian transport model (ROSMAN, 2000). Put simply, SisBaHiA® evenly distributes an amount of particles within the domain, with a constant spacing. At the end of the simulation, the time it takes for each particle to leave the domain is calculated. This determines the residence time for the region "marked" by the particle. In the Rio Verde Reservoir, particles were released with spacing of 20.0 x 20.0 meters.

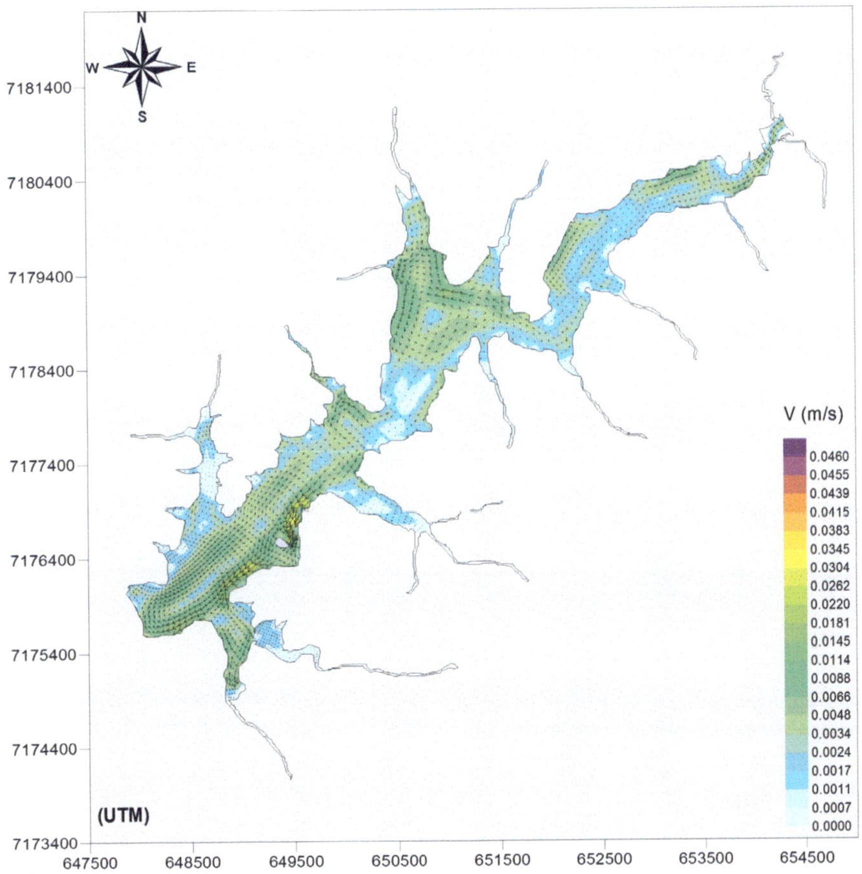

FIGURE 13 – CIRCULATION PATTERN OBTAINED WITH SISBAHIA® FOR SEPTEMBER 29, 2008

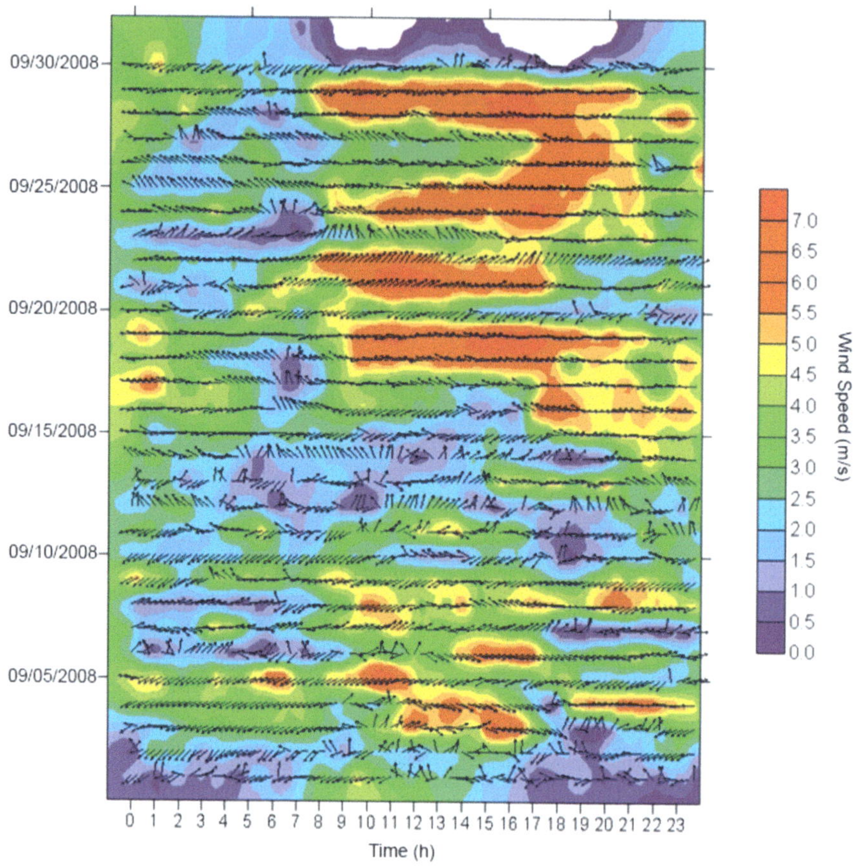

FIGURE 14 – WIND PATTERN ON SEPTEMBER 2008 AT THE RIO VERDE RESERVOIR

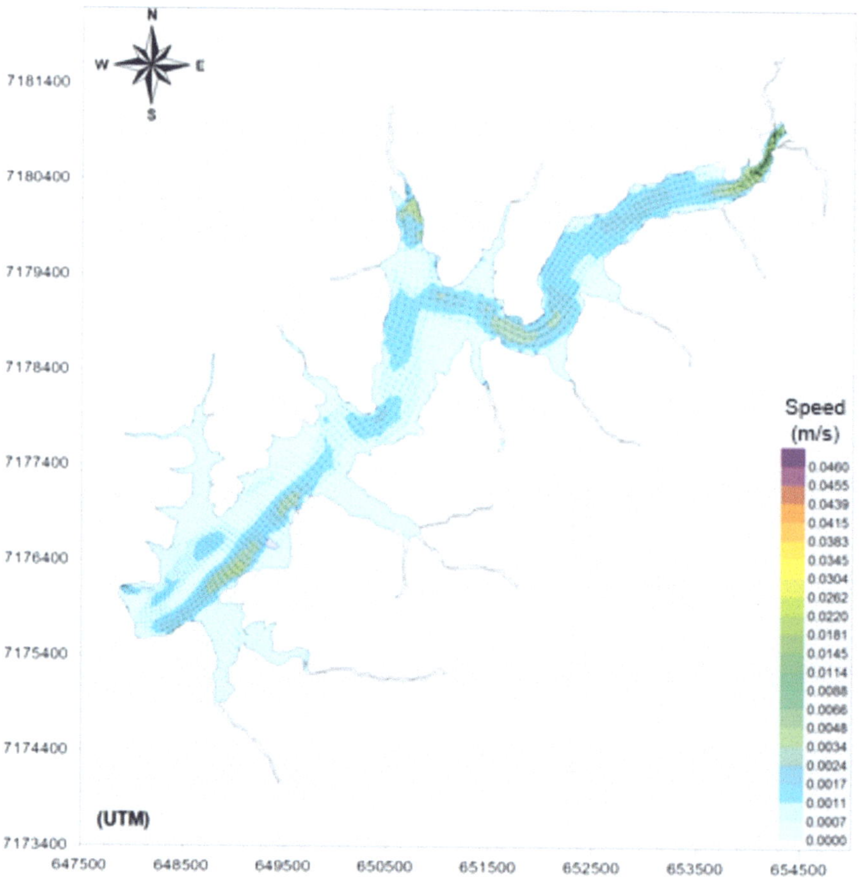

FIGURE 15 – CIRCULATION PATTERN OBTAINED WITH SISBAHIA® FROM MARCH 24, 2009

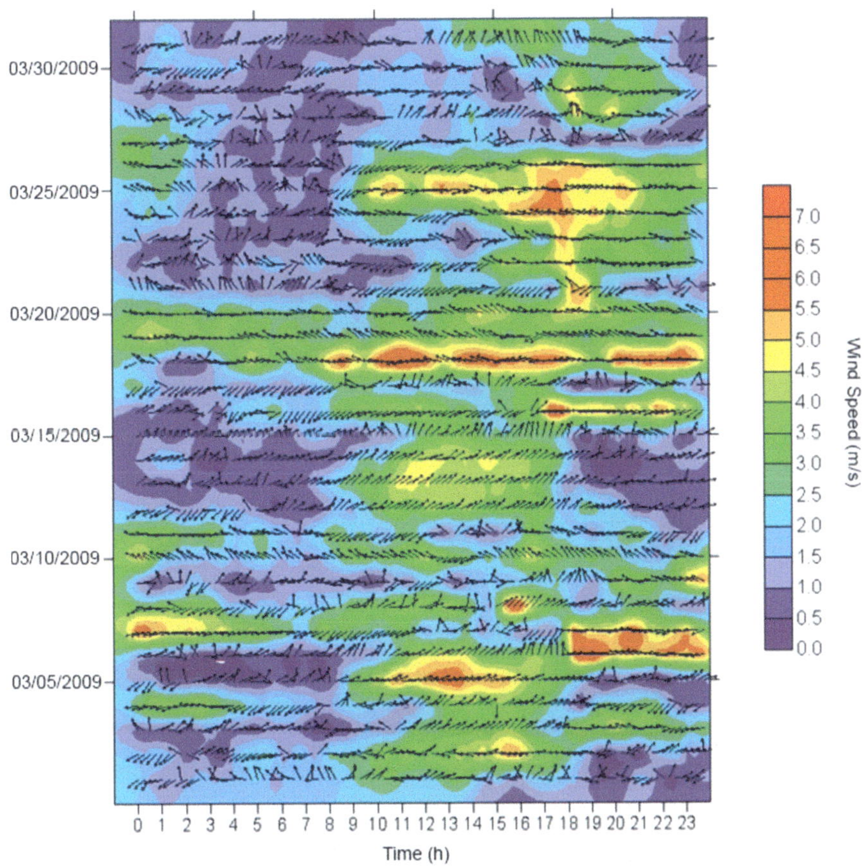

FIGURE 16 – WIND PATTERN ON MARCH 2009 AT THE RIO VERDE RESERVOIR

From the hydrodynamic circulation obtained for 364 days, it is possible to set cycles, or repeat the circulation for up to 1092 days, or three cycles of 364 days. Thus, a simulation time sufficient to characterize the residence times in the reservoir was obtained.

Figure 17 shows the spatial distribution of residence times obtained with SisBaHiA ®, with an average of 218 days. Taking into account the spatial distribution of residence time, we can define three distinct regions: the first is formed by the arms of the reservoir where the residence time is high, exceeding one year; the second is the region close to the Rio Verde, which also has a high residence time of about 450 days; and the third is the central region of the reservoir, with a residence time of less than 150 days. From the spatial distribution of residence times, it is also possible to define the regions most prone to stagnation; regions with high residence times are places where eutrophication processes may occur more frequently.

4. HORIZONTAL MASS TRANSPORTATION IN THE RIO VERDE RESERVOIR

4.1 ADDUCTIVE-DIFFUSIVE TRANSPORT EQUATION

The problem of modeling the parameters of water quality involves two distinct aspects: the space and time dimension of transport, and understanding the chemical processes related to nutrients and their interactions with biological agents. In this sense, considering that the stratification period at Rio Verde Reservoir is limited and that the greatest interest lays in understanding the horizontal distribution of nutrients, it is possible to determine that a two-dimensional model for transport can be used, responding satisfactorily to the observed variations. The vertical integration suggests that this model can be applied to shallow water bodies that have horizontal dimensions prevalent over the vertical dimension, with a well-mixed water column, where the vertical stratification is of little relevance. Time scales are dependent on measurements performed, with significant daily variations. The chemical processes of nutrients comprise the biochemical cycles for each nutrient separately and their interactions (FRAGOSO Jr. *et al.*, 2009). Based on the requirements listed above, and supported by the circulation as defined in item 3, the model used in this study to assess water quality was SisBaHiA®. In SisBaHiA®, the velocity field is known, that is, the scalar transport does not change the hydrodynamics of the receiving water body. As a consequence, the modeling of the hydrodynamic pattern of the water body and the modeling of the scalar transport are unrelated issues. However, for a realistic simulation of the scalar transport to occur, an adequate simulation of the hydrodynamic field is necessary. In relation to the transport mechanisms of a given substance in a water body, the advective process is usually dominant thus suggesting a large dependency between the hydrodynamic simulation and the transport process. However, diffusive processes are also very important for

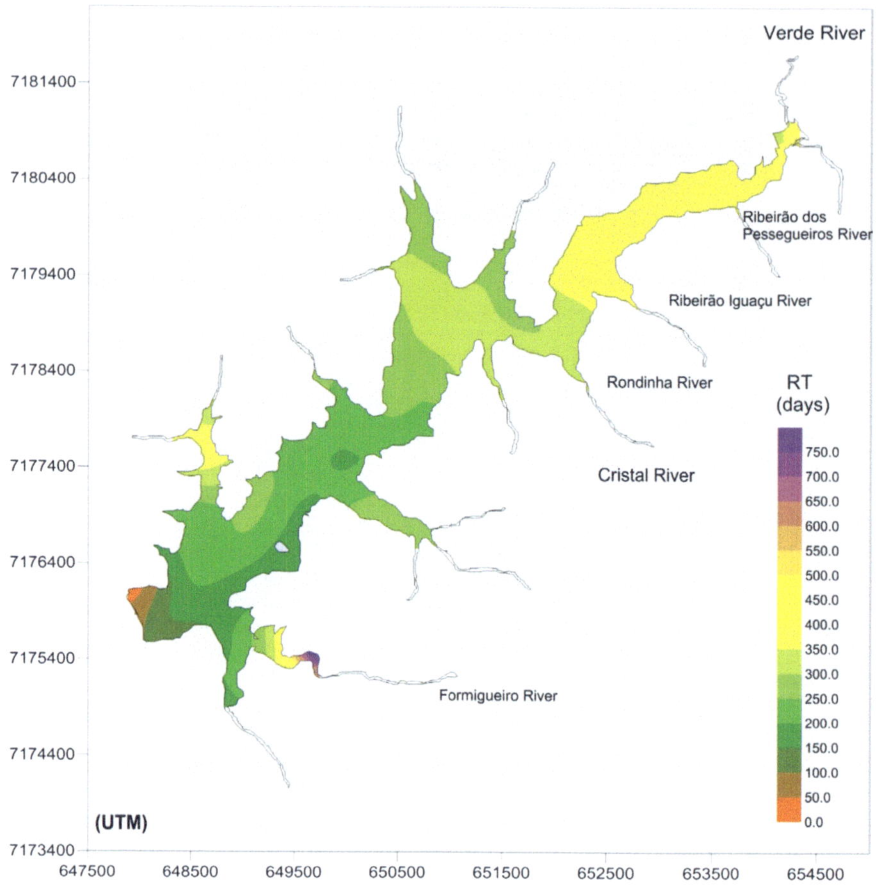

FIGURE 17 – RESIDENCE TIME IN THE RIO VERDE RESERVOIR FOR A SIMULATION OF 1092 DAYS OBTAINED USING SISBAHIA®

the definition of transport. In the SisBaHiA® MQA, diffusion is modeled according to filtering techniques. The vertically integrated equation, describing the transport of scalar to large-scale variables and using the filtering technique for modeling turbulent stresses, is given by (ROSMAN, 2000):

$$\frac{\partial C_j}{\partial t}+U_i\frac{\partial C_j}{\partial x_i}=\frac{1}{H}\frac{\partial}{\partial x_i}\left(H\left[D_{ij}d_{jk}+\frac{\Lambda_k^2}{12}\left|\frac{\partial U_i}{\partial x_k}\right|\right]\frac{\partial C_j}{\partial x_k}\right)\pm\sum R_j \quad (2)$$

where Cj is the concentration of the variable of interest j, Ui are the velocity components in the x_i direction vertically integrated, ΣR_j represents the processes of transformation of the variable j, H is the height of the water column, D_{ij} is the tensor representing the turbulent mass diffusion coefficient, δ_{jk} represents the Kronecker delta, and Λ_k (equal to $\alpha_k\Delta_k$) is the filter width in the dimension x_k, with α_k being a parameter of scale. Typical α_k values calibrated for diffusion/advection are of 0.25 to 2.0, with the usual value of 1.0. Further details on the transport model can be found in Cunha *et al.* (2002). Variables considered in the SisBaHiA®, which describe the nutrients, are shown in Table 2.

The plankton community consists of bacteria, uni- and multi-cellular algae (phytoplankton), and invertebrates (zooplankton), characterized by the ability to float in water. In continental water, representatives of virtually all groups of algae can be found and all these groups have chlorophyll-a in their cells. Therefore, the analysis of chlorophyll-a is an important indicator of the phytoplankton community's biomass and is used in the SisBaHIA® as the variable that identifies phytoplankton biomass.

TABLE 2 – VARIABLES CONSIDERED IN THE SISBAHIA®

SYMBOL	VARIABLE	UNIT
C9	Chlorophyll-a	µgChla/L
C4	Herbivore Zooplankton	mgC/L
C7	Organic Nitrogen	mgNO/L
C1	Ammoniac Nitrogen	mgNA/L
C2	Nitrate Nitrogen	mgNI/L
C8	Organic Phosphorus	mgPO/L
C3	Inorganic Phosphorus	mgP/L
C5	Biochemical Demand for Oxygen	mgO/L
C6	Dissolved Oxygen	mgO/L
T	Temperature	°C
S	Salinity	mg/L

4.2 PHYSICAL, CHEMICAL AND BIOLOGICAL PROCESSES

There are several interactions between nutrients generated through the processes modeled in the SisBaHIA®. Biogeochemical cycles for each nutrient and their interactions with phytoplankton dynamics are shown in the following figures: Figure 18, which represents the processes related to phytoplankton dynamics; Figure 19, which represents the processes related to the nitrogen cycle; and

Figure 20, which represents processes related to the phosphorus cycle. The dynamics of Dissolved Oxygen and Biochemical Oxygen Demand in water are modeled according to the processes outlined in Figure 21.

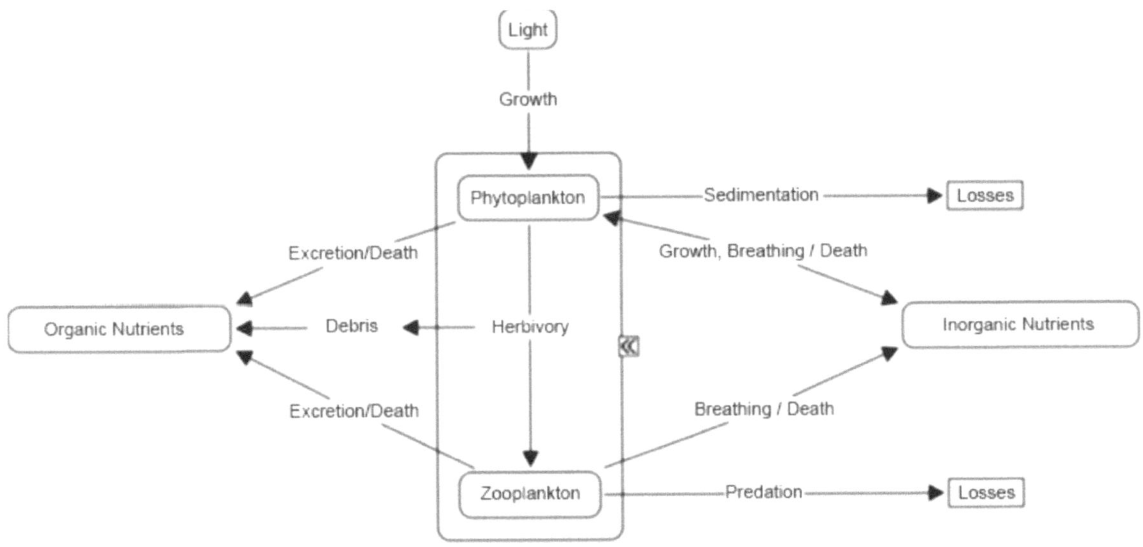

FIGURE 18 – PROCESSES REPRESENTED IN THE PHYTOPLANKTON DYNAMICS

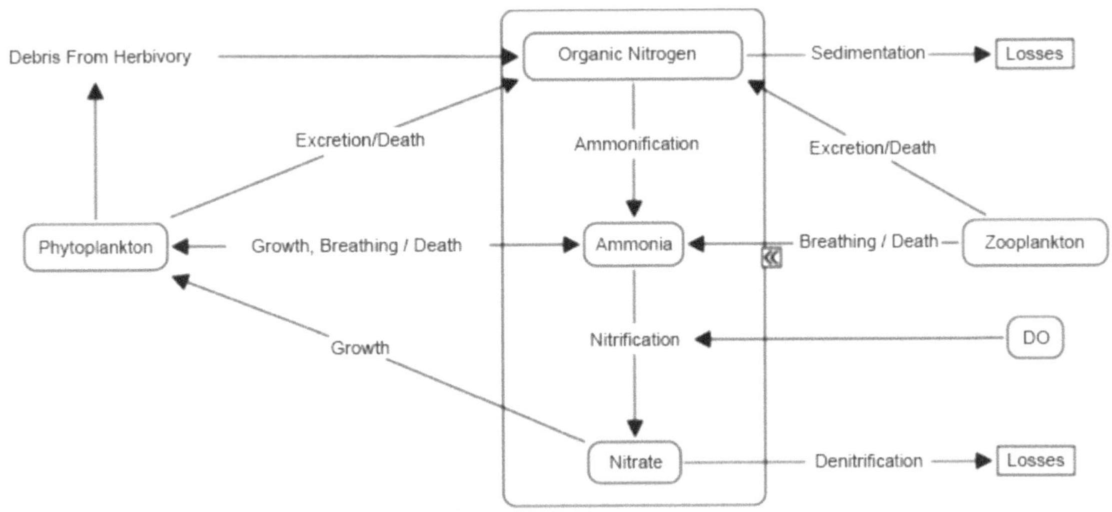

FIGURE 19 – PROCESSES REPRESENTED IN THE NITROGEN CYCLE

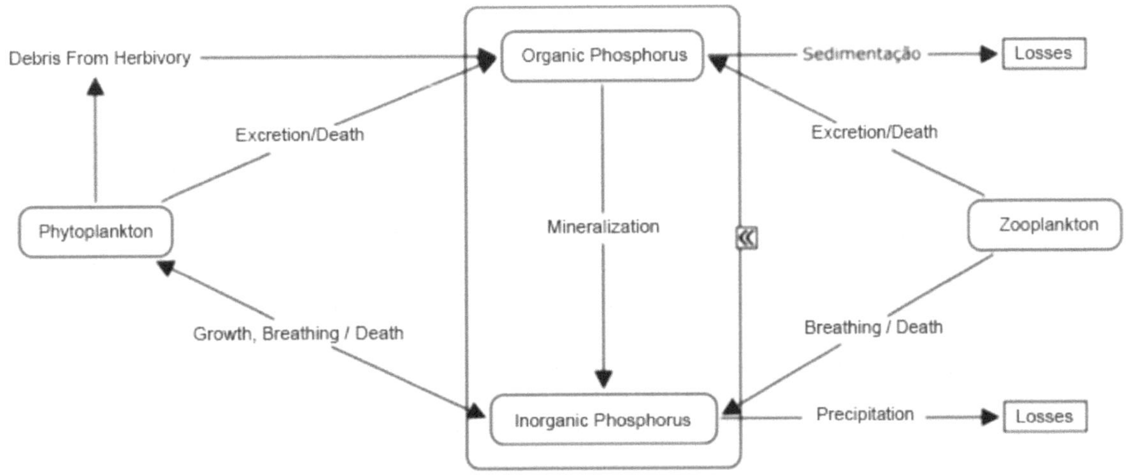

FIGURE 20 – PROCESSES REPRESENTED IN THE PHOSPHORUS CYCLE

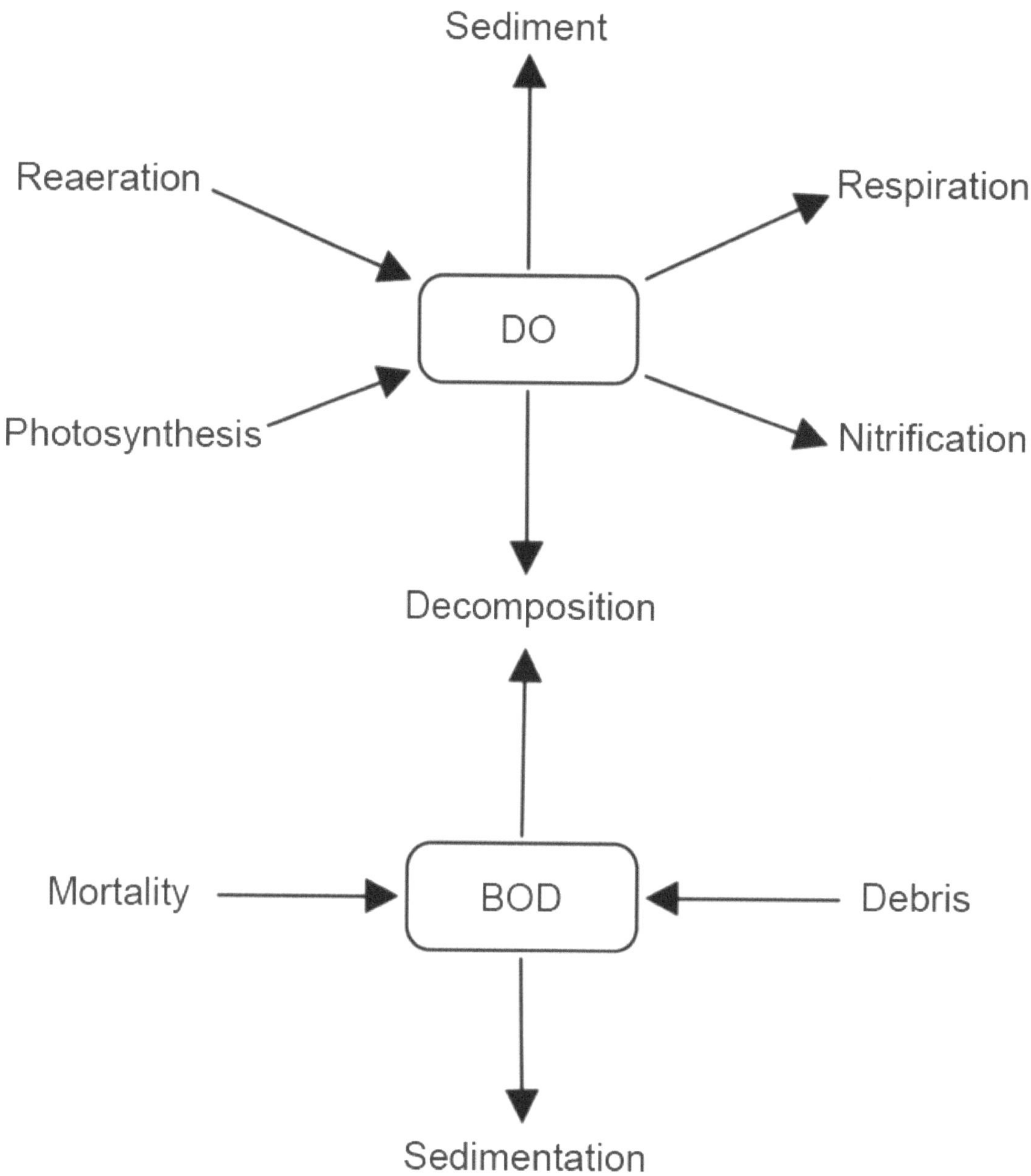

FIGURE 21 – PROCESSES REPRESENTED IN DO-BDO DYNAMICS

Different forms of nitrogen and phosphorus in water can be modeled from the transformation processes described above. The distribution of these different forms of nutrients is complex, mainly due to the large number of species and transformation processes involved. The distribution of species is defined to identify problems related to water quality which are associated with the species and the influence on the limnological conditions of the reservoir. Considering the problem of simple mathematical formulations represening complex processes, and the lack of information about the spatial and temporal variability of the coefficients related to the transformation processes, simplifications can lead to a reduction in the accuracy of the results obtained by the model. However, the accuracy of the results can meet the project objectives. Mathematical formulations of the transformation processes used by the SisBaHiA® are shown below. In Table 3, constants used in the SisBaHiA® (Chapra, 1997) are presented.

$$\Sigma R_5 = k_g C_9 - k_{ra} C_9 - k_{ea} C_9 - k_{gz} C_9 - \frac{V_s}{H} C_9 \tag{3}$$

$$\Sigma R_4 = r_{\mathrm{ca}} E_{\mathrm{z}} k_{\mathrm{gz}} C_9 C_4 - k_{\mathrm{rz}} C_4 - k_{ez} C_4 - k_{gzc} C_4 \tag{4}$$

$$\Sigma R_7 = -k_{71} C_7 - \frac{V_{s3}\left(1 - f_7\right)}{H} C_7 + r_{na}\left(1 - E_z\right) k_{gz} C_9 C_4 + r_{na} k_{ra} f_{on} C_9 + \left(r_{na}/r_{ca}\right) k_{rz} f_{on} C_4 + r_{na} k_{ea} f_{on} C_9 + \left(r_{na}/r_{ca}\right) k_{ez} f_{on} C_4 \tag{5}$$

$$\Sigma R_1 = r_{na}k_{ra}(1-f_{on})C_9 + \frac{r_{na}}{r_{ca}}k_{rz}(1-f_{on})C_4 + r_{na}k_{ea}(1-f_{on})C_9 + (r_{na}/r_{ca})k_{ez}(1-f_{on})C_4 - \frac{C_1}{(k_{am}+C_1)}r_{na}f_{uN}k_gC_9 + k_{71}C_7 - k_{12}C_1 \quad (6)$$

$$\Sigma R_2 = k_{12}C_1 - k_{2D}C_2 - \left(1-\frac{C_1}{(k_{am}+C_1)}\right)r_{na}f_{uN}k_gC_9 \quad (7)$$

$$\Sigma R_8 = r_{pa}K_{ra}f_{op}C_9 + (r_{pa}/r_{ca})K_{rz}f_{op}C_4 + r_{pa}K_{ea}f_{op}C_9 + (r_{pa}/r_{ca})K_{ez}f_{op}C_4 + r_{pa}(1-E_z)K_{gz}C_9C_4 - K_{83}C_8 - \frac{V_{s3}(1-f_{D8})}{H}C_8 \quad (8)$$

$$\Sigma R_3 = r_{pa}k_{ra}(1-f_{op})C_9 + (r_{pa}/r_{ca})\,k_{rz}(1-f_{op})C_4 + r_{pa}k_{ea}(1-f_{op})C_9 + (r_{pa}/r_{ca})\,k_{ez}(1-f_{op})C_4 - r_{pa}f_{uP}k_gC_9 + k_{83}C_8 - \frac{v_{fr}}{H}C \quad (9)$$

$$\Sigma R_6 = k_a(O_s - C_6) - k_DC_5 - r_{on}k_{12}C_1 + r_{oc}r_{ca}k_gC_9 - r_{oc}r_{ca}k_{ra}C_9 - r_{oc}k_{rz}C_4 - \frac{SOD}{H} \quad (10)$$

$$\Sigma R_5 = -k_DC_5 - \frac{v_{s3}(1-f_{D5})}{H}C_5 + r_{oc}r_{ca}(1-E_z)k_{gz}C_9 + r_{oc}r_{ca}k_{ea}C_9 + r_{oc}k_{ez}C_4 \quad (11)$$

The growth rate of phytoplankton k_g varies with the intensity of light, temperature, and the availability of nutrients. The adjustment of the rate of growth with temperature is accomplished by means of an exponential formula, the Arrhenius equation. This formula is also used to adjust the remaining rates considered in the model that vary with temperature. To determine the limitation of growth due to nutrients, the Michaelis-Menten kinetics is used and the effects of each nutrient are combined according to the minimum limiting factor. Only phosphorus and nitrogen are represented as limiting nutrients. The effect of light intensity on algae growth was represented by the Steele equation integrated along the depth, which considers the effects of photoinhibition. As we used average data for solar radiation at 15 minute intervals, the photoperiod (expressed as the fraction of the day with sunlight) was omitted from the Steele equation (CHAPRA, 1997). The formulation of the rate of growth of phytoplankton can be described as:

$$k_g = k_g(20)\theta^{T-20}\frac{2.718}{k_eH}\left(e^{-\frac{I_0}{I_s}e^{-k_eH}} - e^{-\frac{I_0}{I_s}}\right)\min\left(\frac{N}{k_{sN}+N},\frac{P}{k_{sP}+P}\right) \quad (12)$$

where k_g (20) is the maximum growth rate at 20°C under optimal conditions of light and excessive nutrients, θ is the correction factor for temperature *T*, *H* is the total depth of the water column, I_0 is the intensity of light on the surface, I_S is the optimal intensity of light, k_{sN} and k_{sP} are half-saturation constants for phosphorus and nitrogen, *P* is the concentration of reactive phosphate, and *N* the sum of ammonia nitrogen and nitrate nitrogen concentrations.

The attenuation of light through the water column is defined by the Beer-Lambert law. The light extinction coefficient k_e used was:

$$k_e = k_e' + 0.0088a + 0.054a^{2/3} \quad (13)$$

where k_e'' is the light extinction coefficient due to the absorption by suspended particles, with the exception of algae, and *a* is the concentration of algae (μgChla/L). A constant value of 0.3m^{-1} to k'_e was adopted as it is a typical value for reservoirs (CHAPRA, 1997).

4.3 TEMPERATURE MODEL

The water body temperature depends on the heat exchange between the water surface and the atmosphere and, hence, the distribution of energy throughout the water column. The balance of the heat flow may be written as (THOMANN AND MÜLLER, 1987):

$$Hn = Hs + Ha - Hbr - He - Hc \quad (14)$$

where, *Hn* is the total surface heat flow at the air-water interface (W/m^2), *Hs* is the net solar shortwave radiation (W/m^2), *Ha* is the net atmospheric longwave radiation (W/m^2), *Hbr* is the longwave back radiation from the water (W/m^2), *He* is evaporation (W/m^2), and *Hc* is conduction (W/m^2). The temperature variation due to external sources and sinks can be obtained based on the following relationship:

$$\sum T = \frac{Hn}{\rho cH} \quad (15)$$

where, *c* is the specific heart, Cal/kg°C, ρ is the specific mass of water and *Hn* is the total heat flow per unit area. The short-wave atmospheric radiation was measured directly at the weather station; the long wave atmospheric radiation is estimated as a function of air temperature and relative humidity, values obtained at the weather station; heat flow related to long wave radiation in water toward atmosphere is determined as a function of temperature at the water surface; heat flow through conduction and evaporation are defined according to the wind intensity in the reservoir (also obtained at the weather station) and according to the gradient between air and water temperature.

TABLE 3 – DESCRIPTION OF CONSTANTS CONSIDERED IN THE SISBAHIA®, RANGE OF PERMITTED VARIATION (VARIATION) AND VALUE ADOPTED USED IN THE MODELING OF THE RIO VERDE RESERVOIR (VALUE)

SYMBOL	PARAMETERS	VARIATION	VALUE ADOPTED
$k_g(20)$	Phytoplankton growth rate of at 20°C (d^{-1})	0.2-8.0	2.0
θ_g	Temperature coefficient for phytoplankton growth	1.01-1.2	1.066
k_{sN}	Half-Saturation Constant for Nitrogen (μgN L^{-1})	1.4 – 400.0	25.0-200.0
k_{sP}	Half-Saturation Constant for Phosphorus (μgP L^{-1})	0.5 – 80.0	20.0
I_s	Optimal light level (ly d^{-1})	200.0-350.0	250.0
k_{sa}	Half saturation constant for predation of zooplankton on algae (μgChla L^{-1})	2.0 – 25.0	15.0
θ_{gz}	Temperature correction factor for phytoplankton predation by zooplankton		1.08
$k_{gz}(20)$	Grazing Rate 20°C (m^3 gC^{-1} d^{-1})	0.5 – 5.0	1.0
$k_{ra}(20)$	Phytoplankton losses due to respiration and excretion at 20°C (d^{-1})	0.005 – 0.8	0.05
θ_{ra}	Temperature correction for respiration and excretion		1.08
$k_{rz}(20)$	Zooplankton losses due to respiration and excretion to 20°C (d^{-1})	0.001 – 0.36	0.01
θ_{rz}	Temperature correction for respiration and excretion		1.08
E_z	Efficiency predation of zooplankton on algae	0.4 – 0.8	0.6
θ_{gzc}	Temperature correction for predation losses		1.08
$k_{gzc}(20)$	Predation losses rate at 20°C (d^{-1})	0.001 – 0.1	0.01
f_{on}	Fraction of death and respiration of phytoplankton recycled to organic nitrogen		0.5
k_{am}	Half saturation constant for preference of Ammonia Nitrogen (μgN L^{-1})		50.0
r_{oc}	Quantity of oxygen consumed in the decomposition of one gram of organic carbon (gO gC^{-1})		2.67
r_{pa}	Ratio of Phosphorus to chlorophyll in the phytoplankton (gP $gChla^{-1}$)	0.5-1.0	1.0
r_{na}	Ratio of Nitrogen to chlorophyll in the phytoplankton (gN $gChla^{-1}$)	2.7-29.0	7.2
r_{ca}	Ratio of carbon to chlorophyll in the phytoplankton (gC $gChla^{-1}$)	10.0-100.0	50.0
θ_{12}	Temperature coefficient for nitrification	1.02-1.08	1.08
θ_{2D}	Temperature coefficient for denitrification	1.02-1.09	1.045
θ_{71}	Temperature coefficient for mineralization	1.02-1.09	1.08
θ_{83}	Temperature coefficient for mineralization	1.02-1.09	1.08
θ_a	Temperature coefficient for reaeration	1.008-1.047	1.024
θ_D	Temperature coefficient for deoxygenation	1.02-1.15	1.047
θ_s	Temperature coefficient for sediment oxygen demand		1.08
f_{D5}	Fraction of dissolved DBO in the water column	0.1-0.9	0.5
f_{D7}	Fraction of dissolved organic nitrogen in the water column	0.1-1.0	1.0
f_{D8}	Fraction of dissolved organic phosphorus in the water column	0.1-1.0	0.85
f_{op}	Fraction of dead and respired of phytoplankton in phosphorus cycle	0.1-0.9	0.5
$k_{12}(20)$	Nitrification coefficient at 20°C (d^{-1})	0.03-0.9	0.1
$K_{2D}(20)$	Denitrification coefficient at 20°C (d^{-1})	0-1.0	0.1
$k_{71}(20)$	Organic nitrogen mineralization coefficient at 20°C (d^{-1})	0.001-0.2	0.03
$k_{83}(20)$	Organic phosphorus mineralization rate at 200C (d^{-1})	0.001-0.8	0.03
$k_a(20)$	Reaeration coefficient at 20°C (d^{-1})	0.1-5.0	1.38
$k_D(20)$	Deoxygenation coefficient at 20°C (d^{-1})	0.01-1.5	0.2
k_{DBO}	Half saturation constant for oxidation of BOD (mgO_2 L^{-1})		0.5
k_{NIT}	Half-saturation constant for DO limitation in the nitrification process (mgO_2 L^{-1})	0.5-2.0	0.5
k_{NO3}	Half-saturation constant for DO limitation in the denitrification process (mgO_2 L^{-1})		0.1
k_{ea}	Phytoplankton mortality rate (d^{-1})	0.003-0.17	0.01-0.1
k_{ez}	Zooplankton mortality rate (d^{-1})	0.001- 0.125	0.005
$S_{OD}(20)$	Sediment oxygen demand at 20°C (gO_2 m^{-2} d^{-1})	0.2 – 4.0	1.0
v_{s3}	Organic matter settling velocity (m d^{-1})	0.2 – 2.3	1.0
v_{s4}	Phytoplankton settling velocity (m d^{-1})	0 – 30.0	0.05
v_{fr}	Inorganic sediment settling velocity (m d^{-1})	-	0.02

4.4 WATER QUALITY MODEL IN THE RIO VERDE RESERVOIR

The results obtained with the water quality model demonstrate the potential of this tool to support the management decisions related to the reservoir. With continuous development and research, this tool can support a greater understanding of the mechanisms related to eutrophication. The water quality model developed for the Rio Verde Reservoir allows us to define and understand the mechanisms related to nutrient dynamics and their implications for phytoplankton biomass. The model presented herein is intended to unite research and reservoir management in order to identify the best ways to manage these systems.

4.4.1 Parameters used for the water quality model

The general parameters used in the numerical simulation of advection and diffusion can be seen in Table 4. The turbulent diffusion coefficient represents the mixture due to turbulence generated mainly by the bottom and can be divided in the longitudinal and transverse directions. Fisher (1979) parameterized the diffusion coefficient as a function of the longitudinal and transversal dispersion ranges, used to facilitate calibration, and the characteristic friction velocity. Constant values provided by the research can be added to this model. Details on the transport model and parameters considered can be found in Cunha *et al.* (2006). The parameters related to the transformation processes used from the water quality model are shown in Table 3.

TABLE 4 – PARAMETERS USED IN THE NUMERIC SIMULATION OF THE ADVECTIVE-DIFFUSIVE TRANSPORT MODELING

PARAMETERS	VALUE
D_{xx} (m^2/s)	2.0
D_{xy} (m^2/s)	0.5
D_{yy} (m^2/s)	1.0
Δt (s)	300.0
Maximum Peclèt Number	10.0
α (longitudinal dispersion scale)	1.0
β (transversal dispersion scale)	1.0
α_x (scale parameter in the *x* dimension)	1.0
α_y (scale parameter in the y dimension)	1.0
A_t (scale parameter in the *t* dimension)	1.0

The entire reservoir was defined as having these initial concentrations: Temperature - C_T $(x, y, 0)$ C = 16.34°C; Organic Nitrogen - C_7 $(x, y, 0)$ = 0.19 mg N/L; Ammoniac Nitrogen - C_1 $(_{x, y,}$ 0) = 0.09 mg N/L; Nitrate Nitrogen - C_2 $(x, y, 0)$ = 0.23 mg N/L Biochemical Oxygen Demand - C_5 $(x, y, 0)$ = 1.31 mg O_2/L,; Dissolved Oxygen - C_6 $(x, y, 0)$ = 5.84 mg O_2/L; Chlorophyll-a - C_9 $(x, y, 0)$ = 5.4646 μg/L; Zooplankton - C_Z $(x, y, 0)$ = 0.0 mg/L; Organic Phosphorus - C_4 $(x, y, 0)$ = 0.02 mg P/L; Inorganic Phosphorus - C_P $(x, y, 0)$ = 0.011 mg P/L. Such parameters represent the averages of collected data at sampling points in the reservoir on July 16, 2008. For non-zero discharge nodes, or the sections corresponding to rivers, concentration values were defined for the variables of the water quality model. These values were calculated from collected sample data during the project (Figure 7). No concentration value was defined for the salinity variable and for the herbivore zooplankton variable. The zooplankton variable was not simulated as no analysis of these organisms was available in mgC/L (the unit used in this model) at the Rio Verde Reservoir. As the analysis of concentrations of water quality parameters in the tributaries of the reservoir were only conducted between July 2008 and February 2009, the time series of concentrations for the simulation of 364 days was filled with average values from February 2009 onward. Chlorophyll-a concentrations were not measured in the tributaries and permanent values of 1.0 μg/L were assumed. In relation to the dissolved oxygen, only one measurement was taken in the tributaries in December 2008. Thus, the values of the boundary conditions of DO were defined by calculating the oxygen saturation, as a function of water temperature, and subtracting the oxygen deficit in each tributary determined by data from the only collected measurement.

4.4.2 Results - Temperature

The results of temperature variation modeling in the Rio Verde Reservoir, presented in this section, must be considered quantitatively, since it is possible to calibrate the model within the simulated time range. The temperature is used as the calibration parameter for the variables related to advective and diffusive transport. Measurements for comparison are available between 2008 and 2009. The long-term temperature data simulated with SisBaHiA® were compared with two measured results: the average values of the vertical velocity profiles at stations R1, R2, R3 and R4 between July 2008 and June 2009, and time series measured at the intake station, obtained in April 2009.

The results represent 364 days of simulation, between July 1, 2008 and June 30, 2009. Taking into consideration point R1, located in the innermost part of the reservoir, where depths are shallower and there is less influence of the open water, the temperatures measured demonstrate the process of heating that occurs during summer and cooling during winter. However, from the spatial distribution of temperature obtained with SisBaHiA® (Figures 23 to 27), we can see that station R1 is strongly influenced by the boundary conditions imposed by the Rio Verde. In this sense, the differences between the measured values and those obtained with SisBaHiA® can be explained by the inappropriate assumption of boundary conditions.

At R2, also located in the innermost part of the reservoir, where the influence of the boundary conditions is less signficant, the values obtained with SisBaHiA® are close to the measured values. The same pattern is repeated at station R3. That is, in these stations, temperature variations occur mainly through the variation of heat flow which occurs on the water-air interface, which is accounted for reasonably in the model. Looking at the results, the model was able to characterize, in a quantitative manner, the temperature variation in the Rio Verde Reservoir.

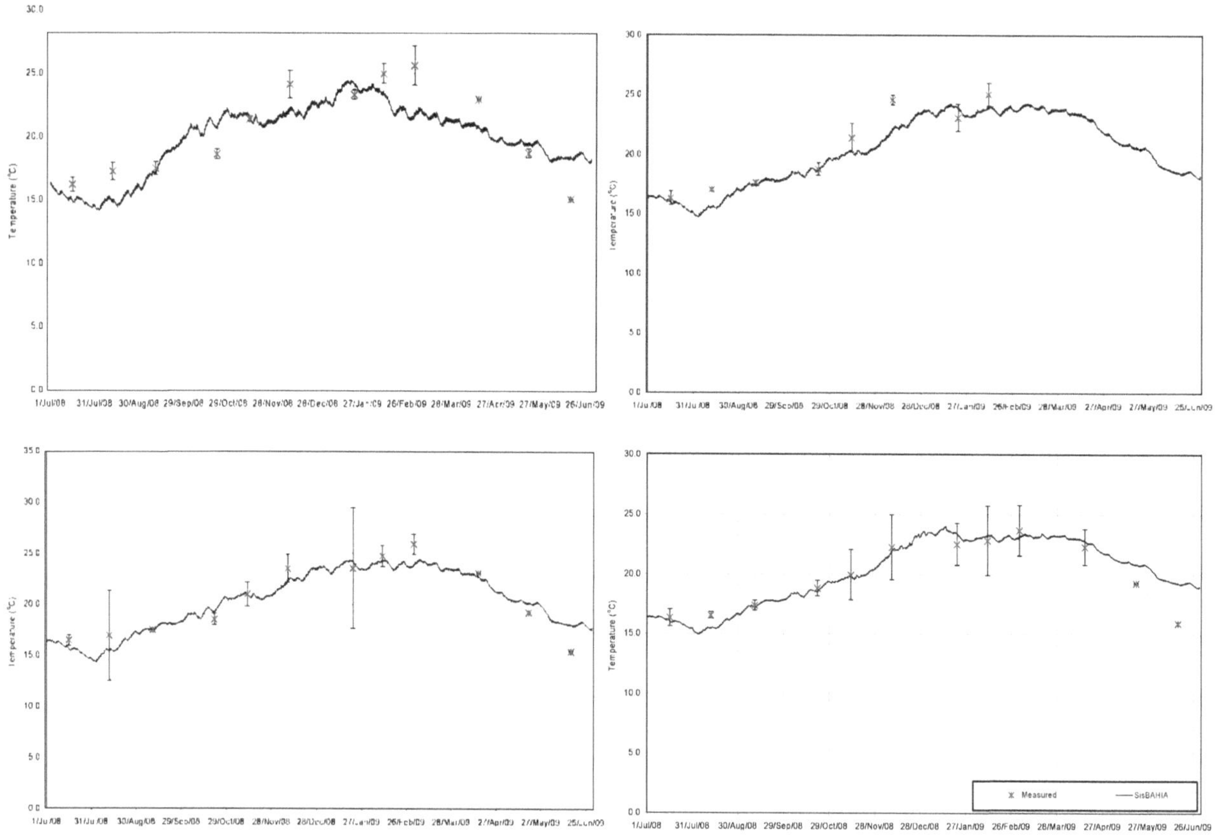

FIGURE 22 – TEMPERATURES MEASURED AT STATION R1, R2, R3 AND R4 (COUNTERCLOCKWISE) AND MODELED BY SISBAHIA®

Figures 23 to 27 show the spatial distribution of temperature in the Rio Verde Reservoir simulated by SisBaHiA® at different points in time. Note that the reservoir is almost homogeneous with respect to the horizontal gradients, since the horizontal temperature variations are concentrated in regions close to rivers. We can also see that there is no major variation in temperature distribution according to season, namely in winter (Figure 23) and summer (Figure 25); gradients remain similar. However, the reservoir undergoes a warming (summer) and cooling (winter) process as a whole. These variations seem consistent enough, considering that the model calculates the average vertical temperature.

It can also be concluded that the Rio Verde Reservoir is a water body which presents temperature between 14.0 and 26.0°C. It is noteworthy that with this application, it is possible to illustrate the possibilities of SisBaHiA® in environmental modeling, as it is quantitatively conclusive with regard to the distribution of temperature in the Rio Verde Reservoir.

4.4.3 Results – Water quality parameters

This section shows the results of environmental modeling of water quality parameters using SisBaHiA®. Whereas the temperature was used as the calibration parameter for the variables related to the advective and diffusive transport, calibration for modeling water quality only refers to the transformation processes. In order to quantify the performance of SisBaHiA®, we calculated the average relative errors between the results obtained with the model and the observations at the monitoring stations. Table 5 shows the results obtained for the calibration. We can see that the average relative error for temperature is minimal at all monitoring stations; at station R1, where the influence of the boundary conditions is significant, errors are significant, demonstrating the difficulty of obtaining results in a distributed model in a time scale of hours, with boundary conditions observed on a time scale of days. In this regard, it is noted that at station R4, where the influence of boundary conditions is not significant, the results obtained by the model best fit the collected data.

TABLE 5 – AVERAGE RELATIVE ERRORS FOUND IN THE MODEL CALIBRATION

VARIABLE POINT	T	NA	NI	a	NO	P	PO	OD	L
R1	0.08	0.62	1.11	0.31	4.08	0.94	2.00	0.15	0.26
R2	0.05	0.54	0.30	0.49	0.99	0.46	0.58	0.11	0.28
R3	0.04	1.56	0.33	0.59	1.02	0.53	0.70	0.14	0.19
R4	0.04	0.84	0.31	0.42	0.60	0.56	1.18	0.30	0.26

One of the results obtained with the model are the concentrations of modeled parameters. Figures 28 to 35 show the concentration found in two different moments: in the dry season (Sepember 29, 2008, 00:00) and rainy season (March 28, 2009, 00:00). Some substances, such as nitrate nitrogen and inorganic phosphorus, have little variation between the two points in time, but others, such as organic nitrogen, ammoniac nitrogen and organic phosphorus, show significant variation between both time points. This is due to differences between advective and diffusive transports observed during these moments, as stated in the previous item. The BOD concentration changes little and this is clearly the result of the fact that observed values revolve around the detection limits. DO concentration reflects the observed processes, with a greater consumption of oxygen due to the presence of more nutrients. Chlorophyll-a concentrations are higher in March than in September, probably due to the greater availability of solar energy, favoring primary production.

The main environmental problem related to water systems is usually due to excessive domestic or industrial pollutants released into the aquatic environment, as well as an increased demand for water, which reduces the flows and promotes contamination. Water quality models are technological tools capable of assessing the impacts generated by the release of pollutants in a given water body, which may be more or less accurate, depending on the assumptions made in the formulation of the mathematical model. A necessary, but not sufficient, condition for water quality models to have predictive power is that they are able to simulate conditions observed *a priori* for a certain time interval. Within this context, it is possible to study the effects of variations in input loads on the Rio Verde Reservoir.

In the management of water bodies, it is important to determine the nutrient that is primarily limiting phytoplankton growth. This is normally done using nitrogen and phosphorus. The ratio between nitrogen and phosphorus in the biomass is approximately 7.2. Thus, a ratio of nitrogen to phosphorus that is less than 7.2 suggests that nitrogen is the limiting nutrient. On the other hand, if the ratio is greater, it implies that phosphorus is the limiting nutrient. However, this simple analysis is subject to variation, for example, due to the variability of the stoichiometry of plants (THOMANN and MUELLER, 1987; CHAPRA, 1997). According to Wetzel (1981), continental water ecosystems are more sensitive to inputs of phosphorus while nitrogen often limits primary production in estuarine and marine ecosystems. According to Salas and Martino (1991), in the majority of tropical lakes in Latin America phosphorus is the limiting factor in the growth of algae.

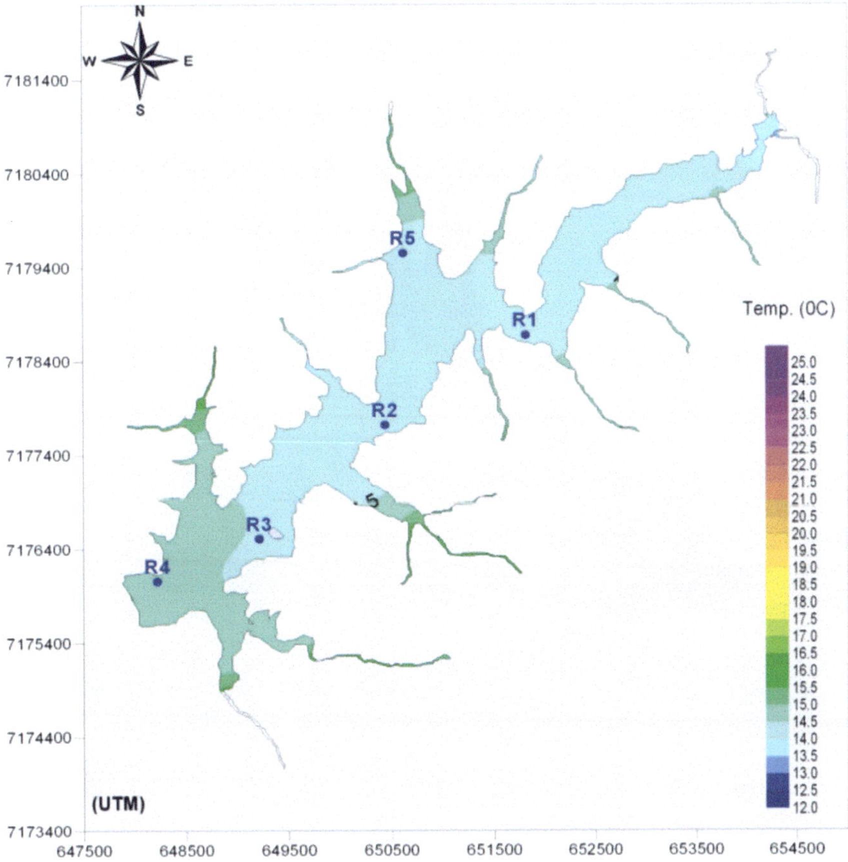

FIGURE 23 – TEMPERATURE DISTRIBUTION SIMULATED WITH SISBAHIA® ON JULY 31, 2008

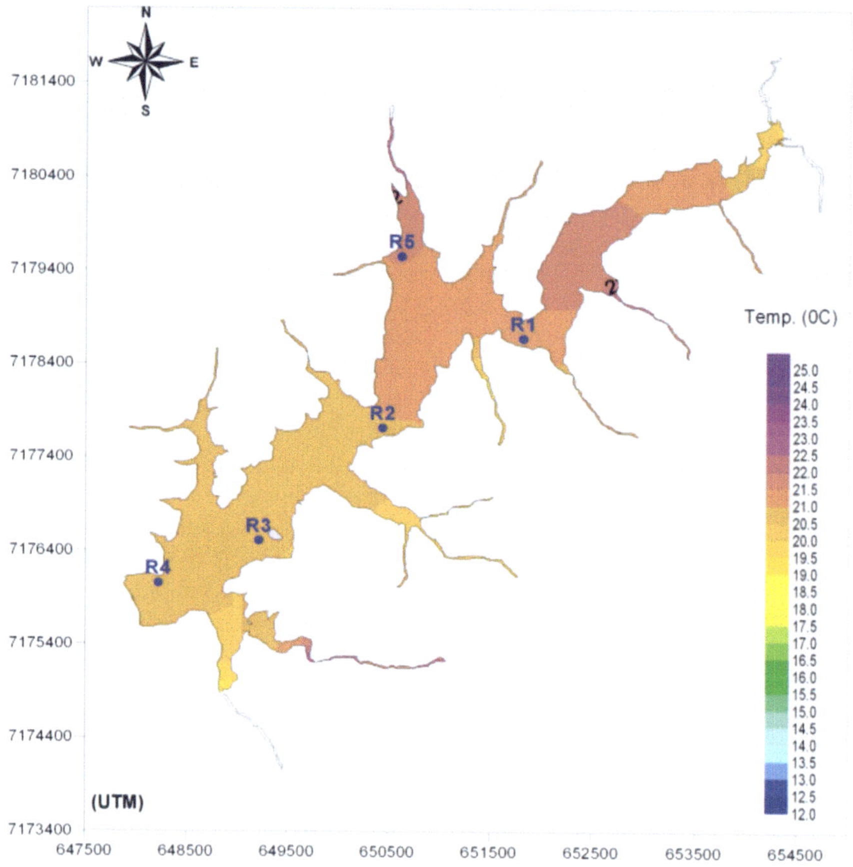

FIGURE 24 – TEMPERATURE DISTRIBUTION SIMULATED WITH SISBAHIA® ON NOVEMBER 17, 2008

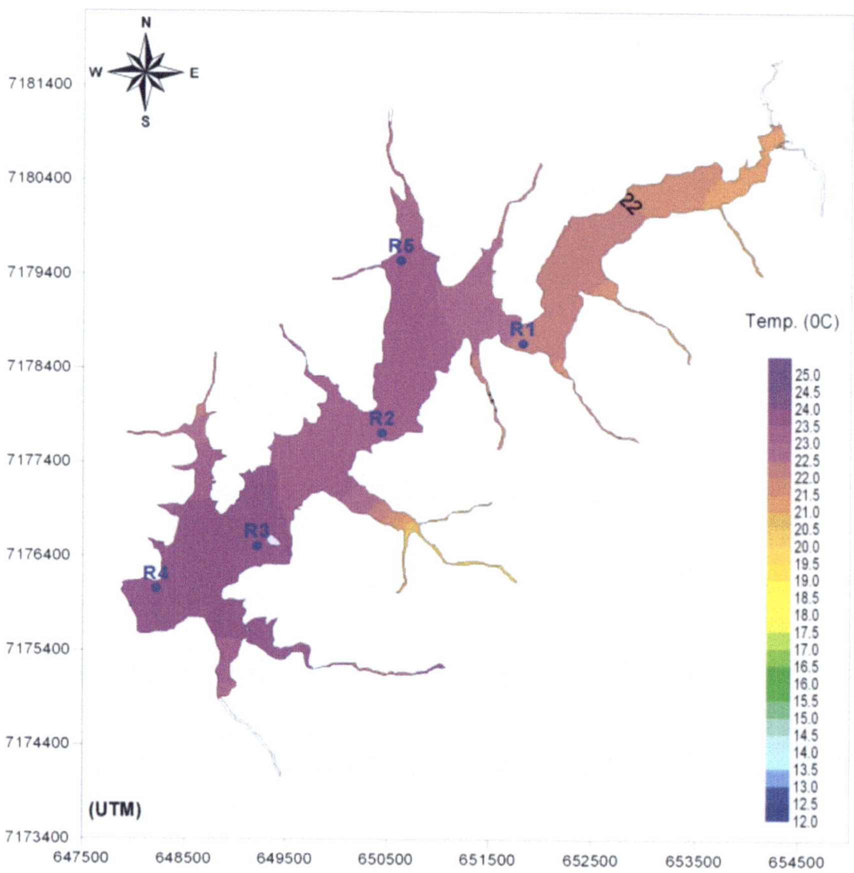

FIGURE 25 – TEMPERATURE DISTRIBUTION SIMULATED WITH SISBAHIA® ON DECEMBER 31,2008

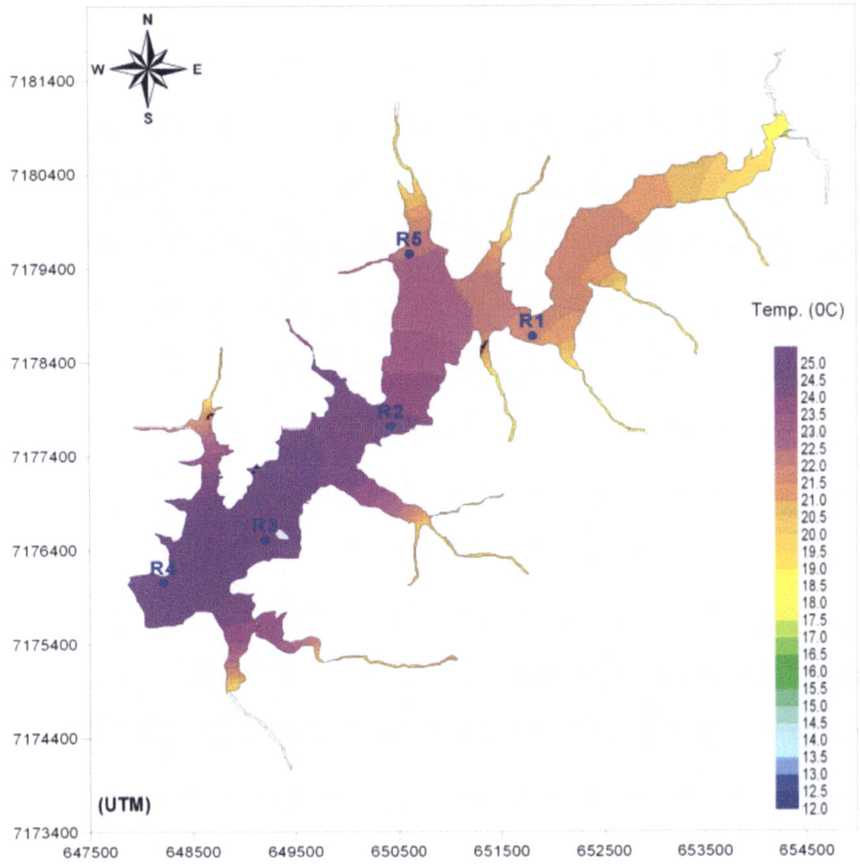

FIGURE 26 – TEMPERATURE DISTRIBUTION SIMULATED WITH SISBAHIA® ON MARCH 7, 2009

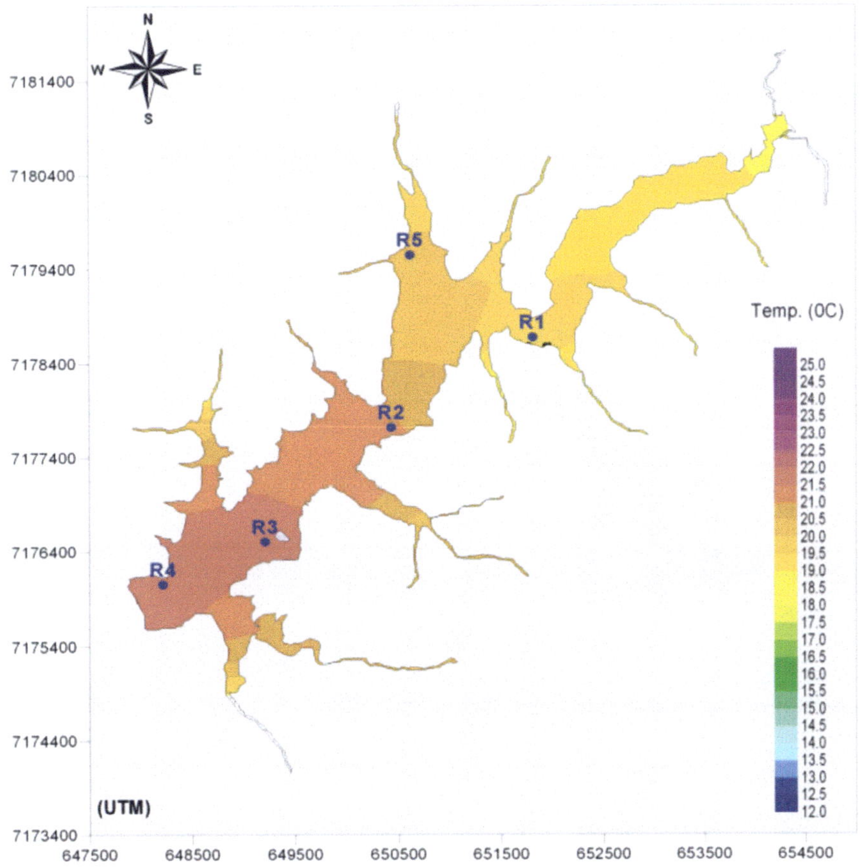

FIGURE 27 – TEMPERATURE DISTRIBUTION SIMULATED WITH SISBAHIA® ON MAY 5, 2009

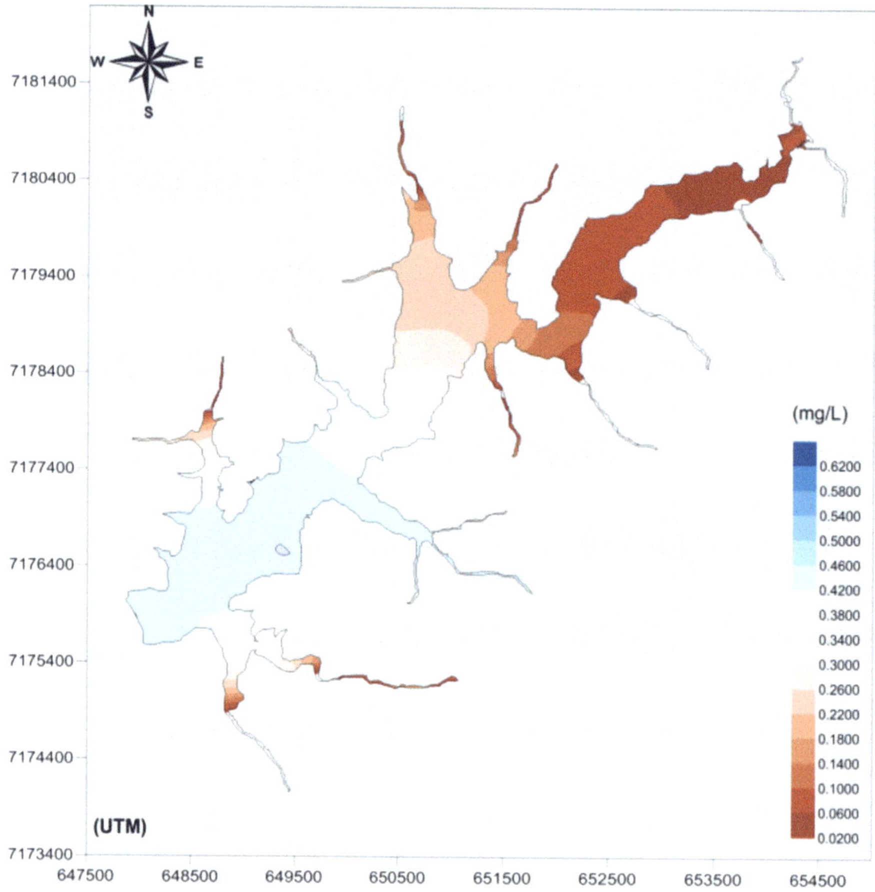

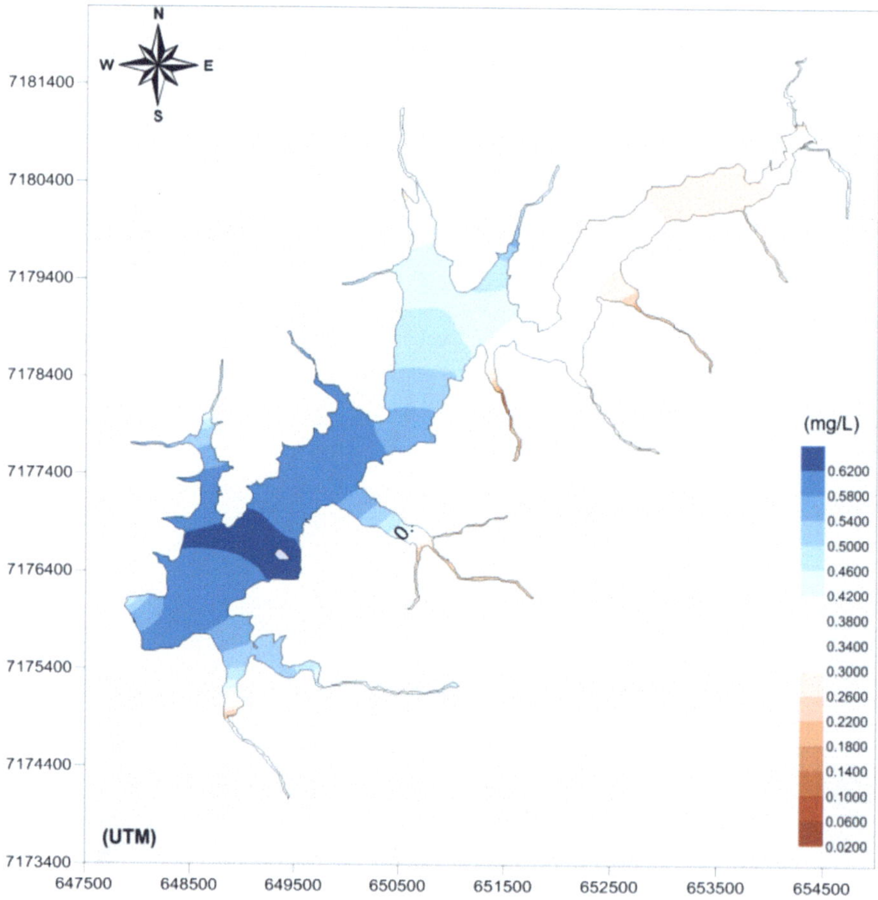

FIGURE 28 – CONCENTRATIONS OF SIMULATED ORGANIC NITROGEN FOR SEPTEMBER 29, 2008 00:00 AND MARCH 24, 2009 00:00

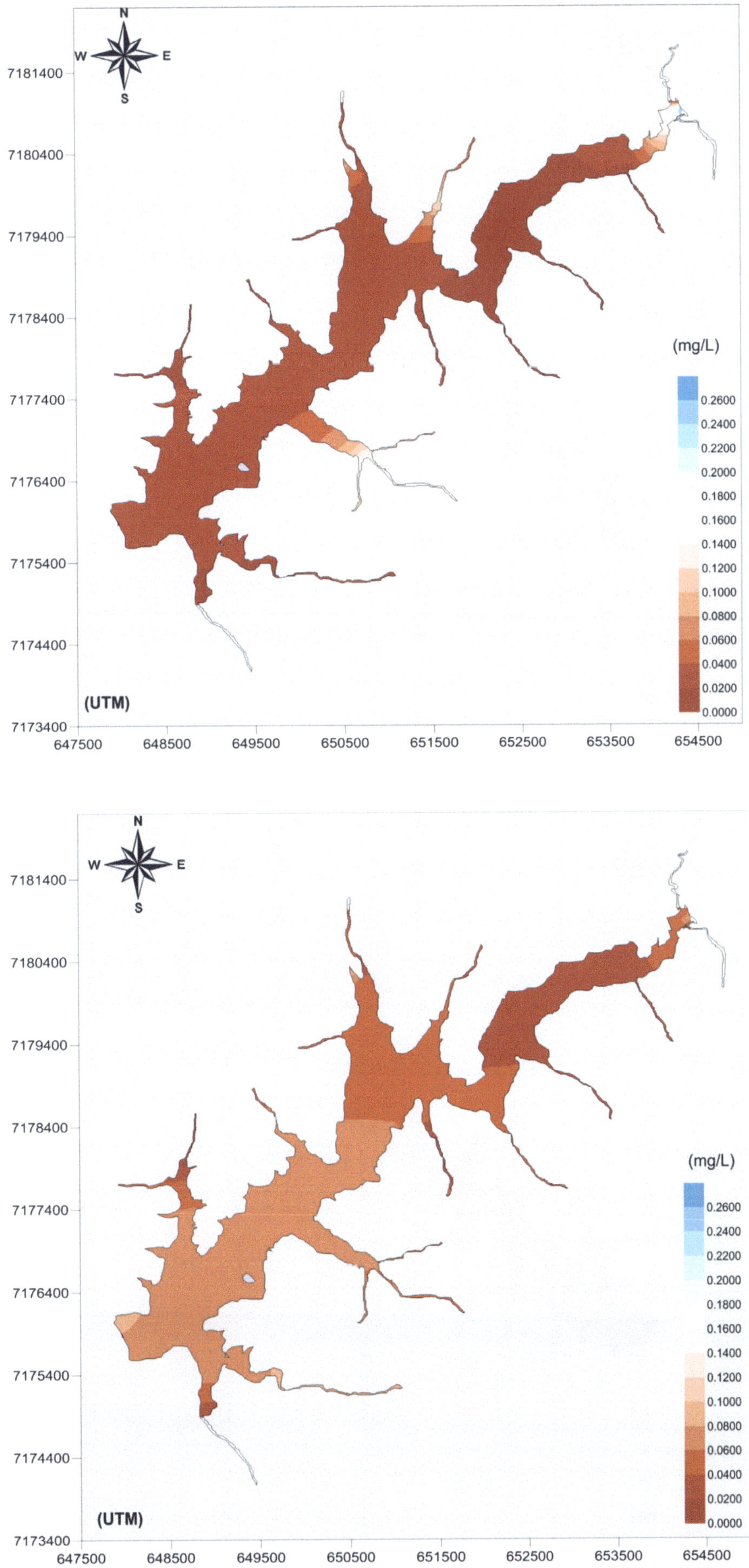

FIGURE 29 – CONCENTRATIONS OF SIMULATED AMMONIA NITROGEN FOR SEPTEMBER 29, 2008 00:00 AND MARCH 24, 2009 00:00

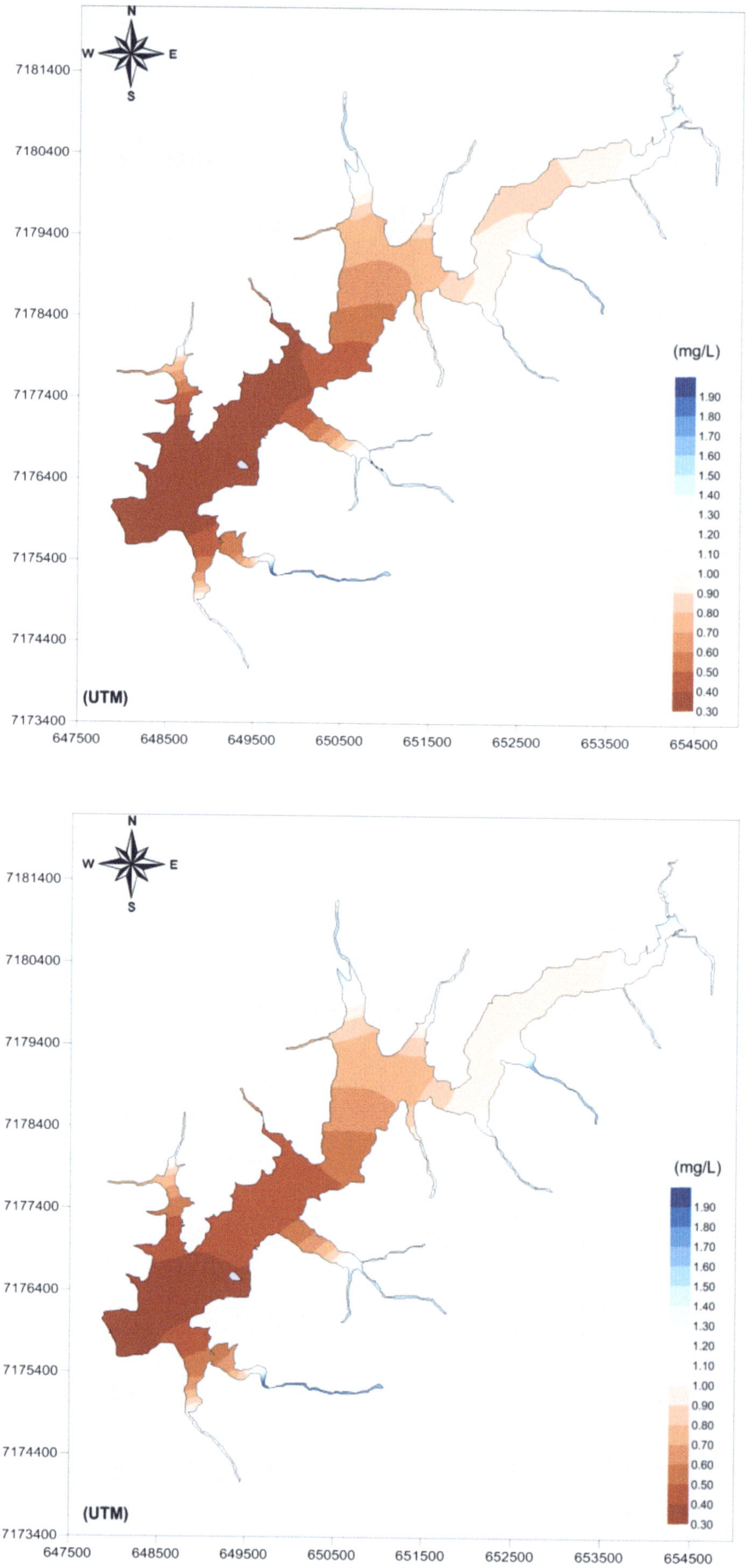

FIGURE 30 – CONCENTRATION OF NITROGEN NITRATE SIMULATED FOR SEPTEMBER 29, 2008 00:00 AND MARCH 24, 2009 00:00

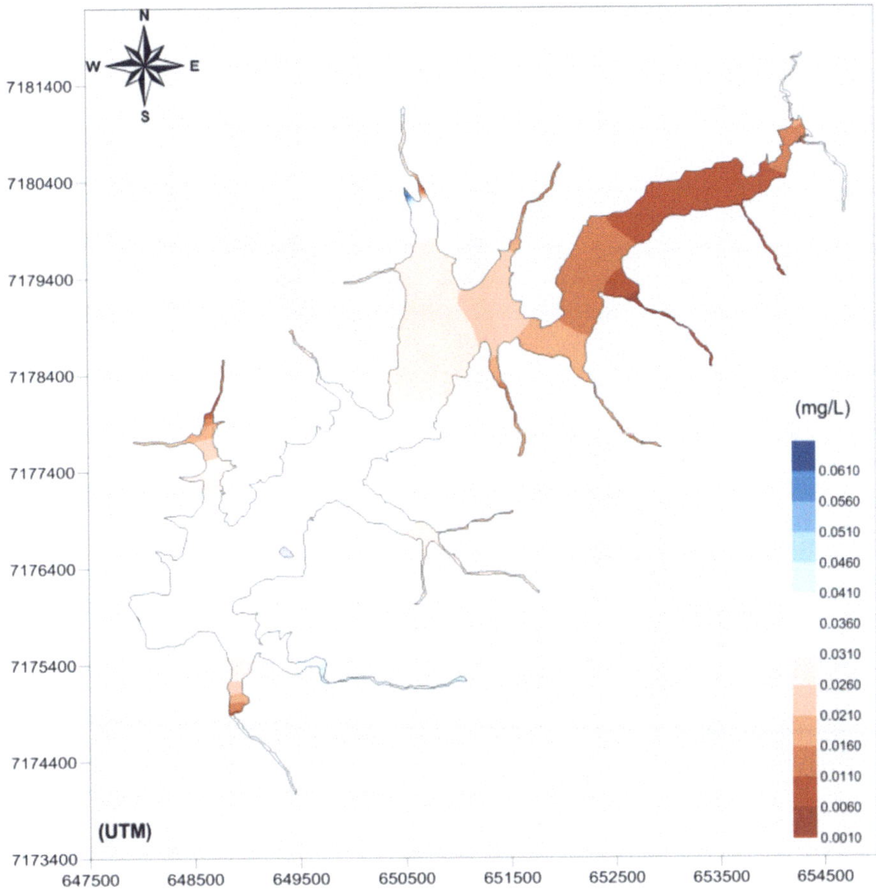

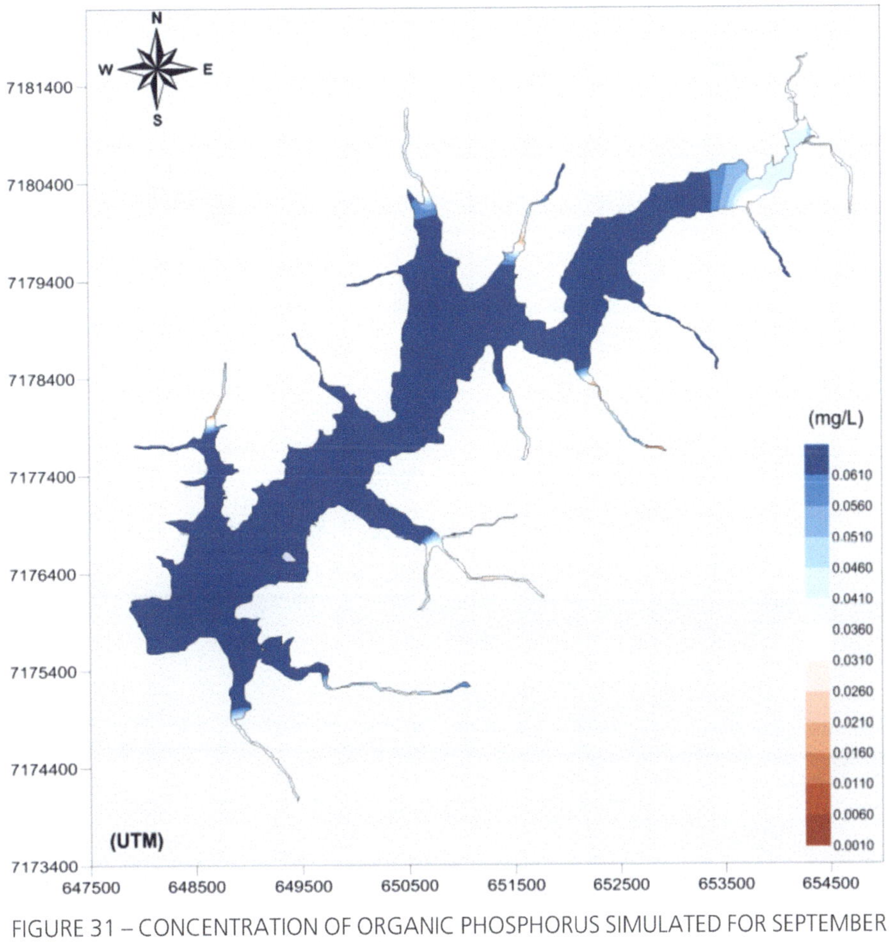

FIGURE 31 – CONCENTRATION OF ORGANIC PHOSPHORUS SIMULATED FOR SEPTEMBER 29, 2008 00:00 AND MARCH 24, 2009 00:00

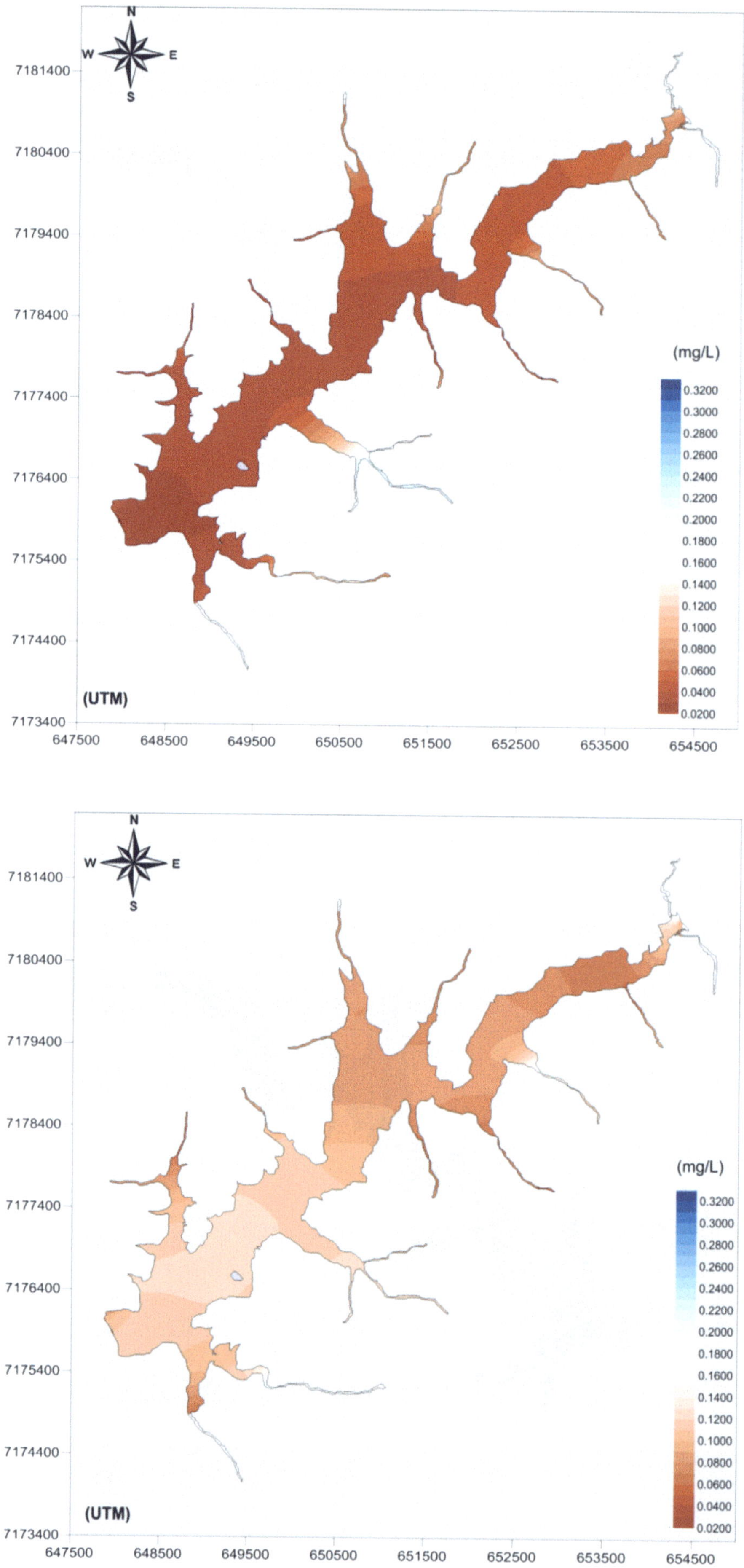

FIGURE 32 – CONCENTRATIONS OF INORGANIC PHOSPHORUS SIMULATED FOR SEPTEMBER 29, 2008 00:00 AND MARCH 24, 2009 00:00

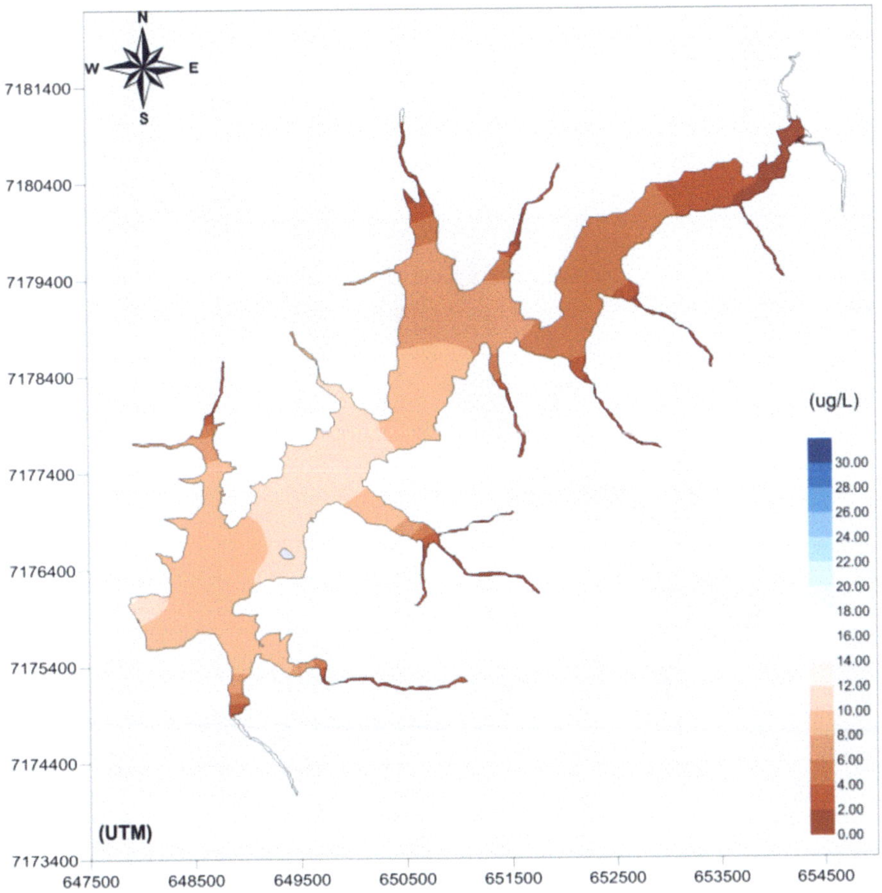

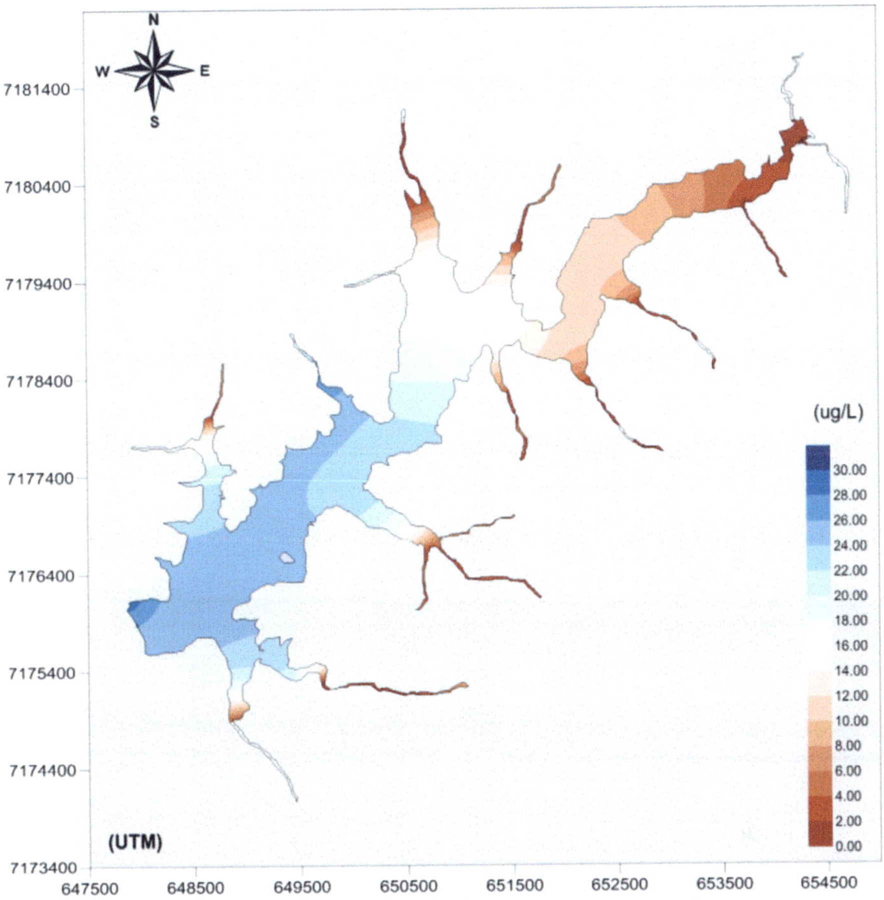

FIGURE 33 – CONCENTRATIONS OF CHLOROPHYLL-A SIMULATED FOR SEPTEMBER 29, 2008 00:00 AND MARCH 24, 2009 00:00

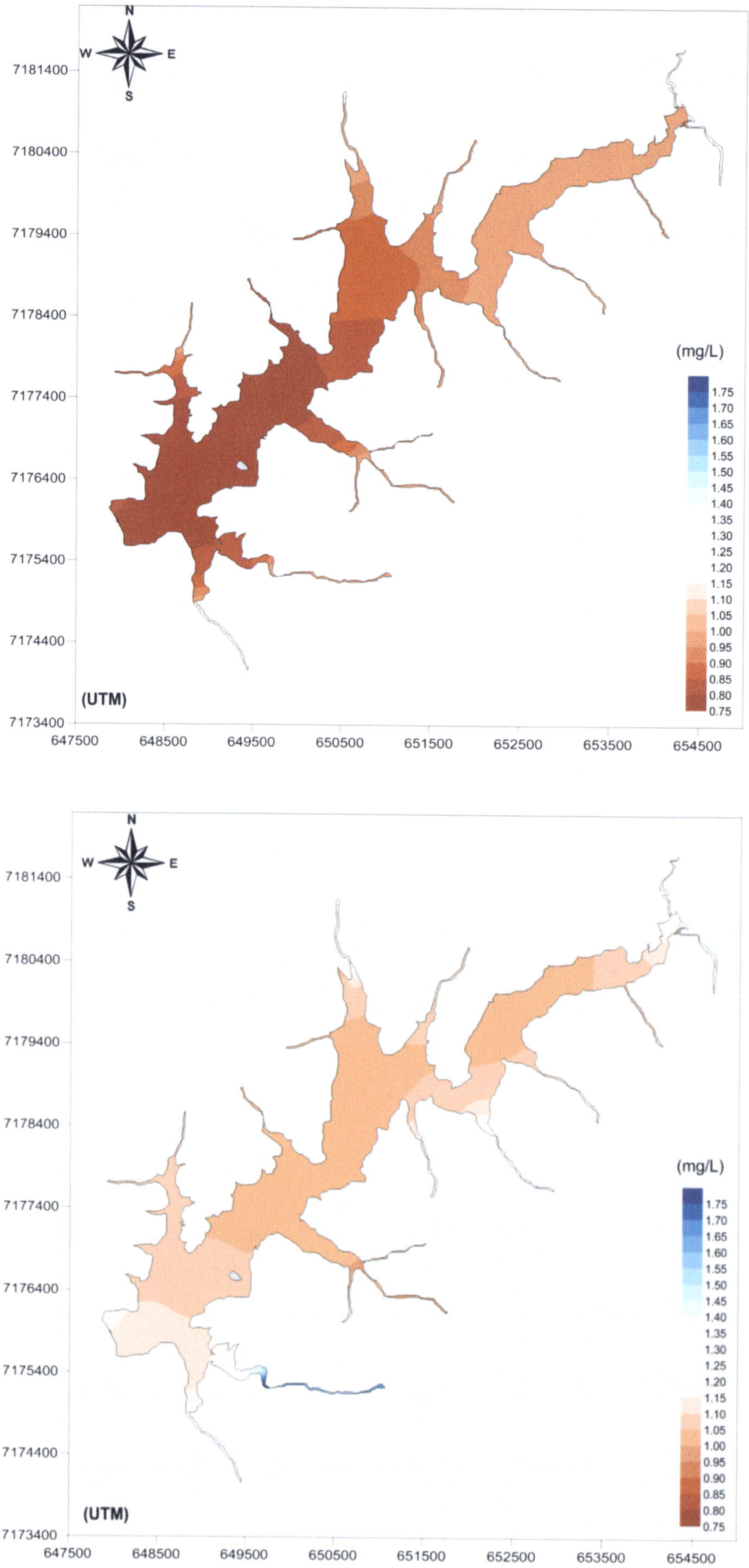

FIGURE 34 – CONCENTRATIONS OF THE BIOCHEMICAL OXYGEN DEMAND SIMULATED FOR SEPTEMBER 29, 2008 00:00 AND MARCH 24, 2009 00:00

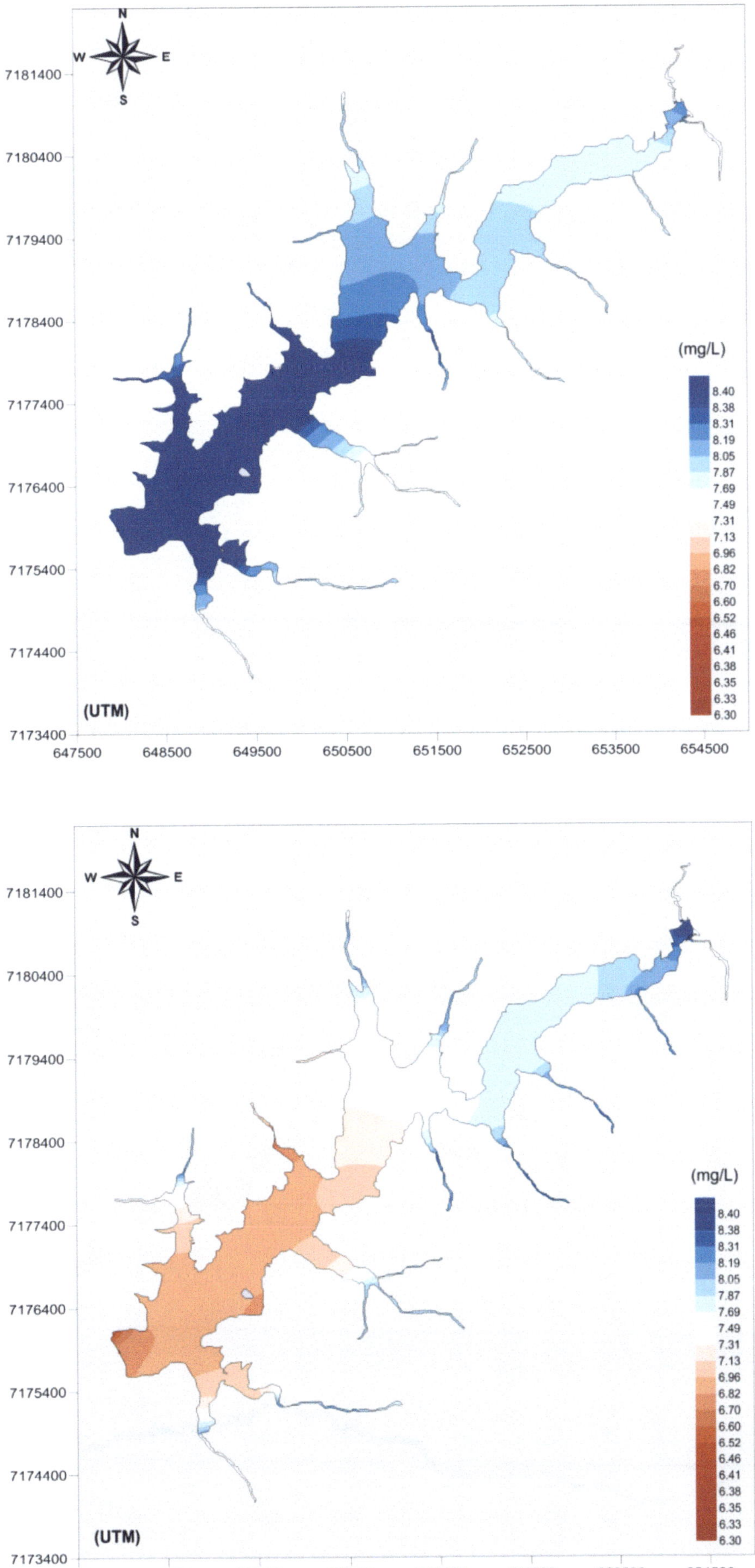

FIGURE 35 – CONCENTRATIONS OF DISSOLVED OXYGEN SIMULATED FOR SEPTEMBER 29, 2008 00:00 AND MARCH 24, 2009 00:00

The limiting nutrient in the Rio Verde Reservoir was defined by determining the ratio between the average concentration of inorganic nitrogen (ammonia nitrogen and nitrate nitrogen) and inorganic phosphate (reactive), obtained at the sampling points located inside the reservoir. The values found were:

R1: N/P=8.60;

R2: N/P=5.38;

R3: N/P=9.23;

R4: N/P=8.24.

Thus, changes in phosphorus concentration should have a significant impact on phytoplankton dynamics. Simulations were performed with SiSBaHiA® considering scenarios with changes in phosphorus concentrations in the tributaries of the Rio Verde Reservoir. In scenario 1, the concentrations of inorganic phosphorus and organic phosphorus were maintained and in scenario 2, the concentrations in the tributaries were multiplied by ten. In both scenarios, the same initial conditions, boundary conditions of other variables and the model parameters were maintained.

Some conclusions can be obtained from considering an increase of the phosphorus load: the response in the reservoir is not homogeneous, as there is a greater increase in peak concentrations. If we establish point R4 as the control point for this experiment, we can consider that if the phosphorus load increases 10 times, Chlorophyll-a concentration at station R4 should increase significantly.

The model's response shows that increased concentrations of phosphorus in effluents discharged by rivers into the Rio Verde Reservoir quickly deteriorates water quality. However, Chlorophyll-a is not a good parameter to measure variations in pollutants, since it responds differently to load variation. These conclusions are valid for a hypothetical case in which the increase of the load was carried out randomly and the biological and chemical conditions of the reservoir do not change. However, they show the potential of the SisBaHiA® water quality model in solving problems related to the integrated management of water resources, with a focus on quality and quantity of water.

Water bodies can be classified according to their trophic state, from lowest to the highest productivity as: ultraoligotrophic, oligotrophic, mesotrophic, eutrophic, and hypereutrophic. The quantification of trophic state may be performed in terms of total phosphorus or concentrations of Chlorophyll-a, and from water transparency obtained with the Secchi disk. Table 6 shows the percentage of the simulated time period in which the reservoir remained in each trophic state, considering the first scenario and the second scenario using the chlorophyll results obtained through SisBaHiA® at point R4. We can see that there are significant differences between the two scenarios, where the majority of the simulated time period showed trophic states between mesotrophic and supereutrophic and neither of the two scenarios reached the hypereutrophic level.

TABLE 6 – TROPHIC LEVELS FOUND AT POINT R4 CONSIDERING TWO DIFFERENT SCENARIOS

TROPHIC LEVEL	SCENARIO 1 (%)	SCENARIO 2-10XP (%)
Ultraoligotrophic	0.0	0.0
Oligotrophic	0.9	0.0
Mesotrophic	87.4	11.4
Eutrophic	11.7	37.7
Supereutrophic	0.0	50.9
Hypereutrophic	0.0	0.0

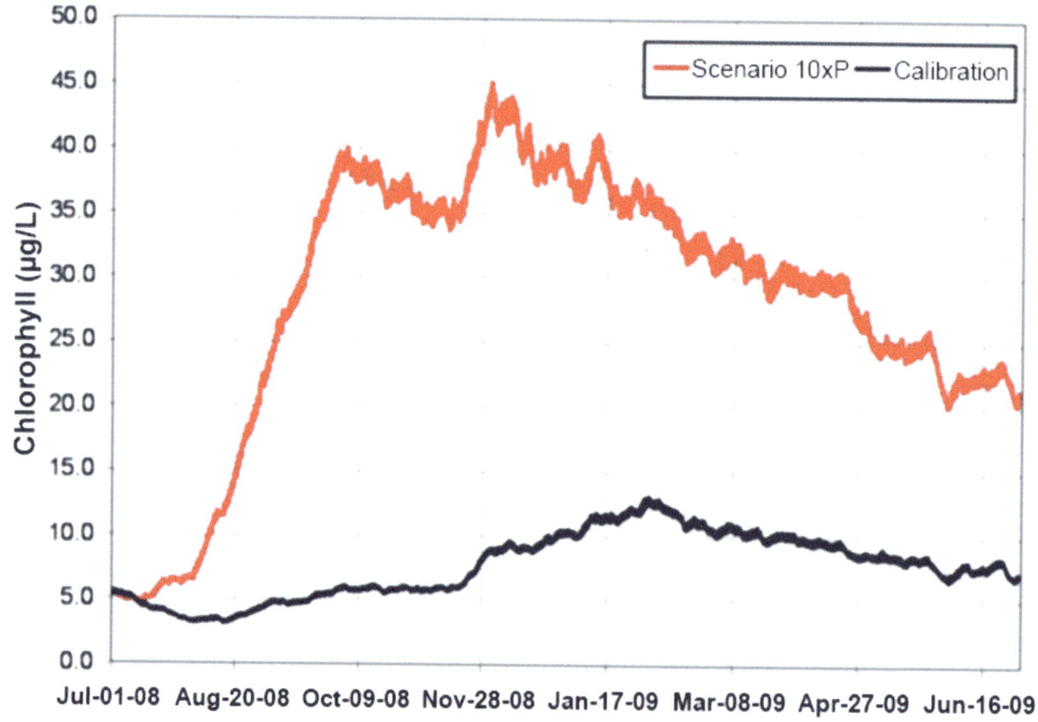

FIGURE 36 – COMPARISON BETWEEN THE CONCENTRATIONS OF CHLOROPHYLL-A FOR TWO SCENARIOS: CURRENT AND WITH INCREASED PHOSPHORUS LOAD AT STATION R4

5. CONCLUSIONS

The results presented in this paper show that environmental modeling of the Rio Verde Reservoir is appropriate and can be used quantitatively to study the phenomenon of eutrophication and excessive growth of algae in reservoirs. As such, this study provides the foundation for new applications and developments. Despite the errors associated with boundary conditions and simplification, the model has demonstrated its potential to adequately represent the data obtained from the reservoir.

One can consider the modeling of temperature variation based on the simulation of other parameters with similar behavior. Whereas the advective and diffusive transport is independent from the parameter modeled, with the correct definition of the chemical, physical and biological processes involved in each water quality parameter, it is possible to characterize the distribution of these parameters in the Rio Verde Reservoir.

The frequency of water quality data measurement in the tributaries of the Rio Verde Reservoir indicates the problems that can be associated with the definition of the concentration boundary conditions of variables considered in SisBaHiA®, which may cause significant errors in estimating the loads of these variables. The use of hydrological models can help define flows from rainfall data as they are able to generate loads of nutrients and organic matter. They are therefore useful tools in the modeling water bodies and can be used in further studies.

Based on the results of the hydrodynamic model, we noted the significant influence of wind on hydrodynamic circulation in the Rio Verde Reservoir. In addition to intensifying the levels, wind causes changes in the direction of the velocity vectors, contributing to the mixture of substances discharged into the reservoir. Thus, the importance of the proper use of wind data in modeling studies in reservoirs is clear. Inflows/outflows generate impacts in circulation only in their immediate surroundings and do not contribute to the circulation of the reservoir as a whole.

The simulated scenarios with increased load of phosphorus released into the reservoir showed consistent results. However, the results of the simulations are illustrative only, since SisBaHiA® has not been validated for this new data set. In addition, there was no data on zooplankton that could be used to verify the results of SisBaHiA®. The average relative errors calculated in the model calibration can serve as a comparison for future studies.

SisBaHiA® is a relatively simple model that does not consider the latest developments in ecological modeling and water quality. Thus, further studies should be conducted in order to develop the model and integrate, for example, different groups of phytoplankton and zooplankton, and the variation in the stoichiometric ratios of these organisms. However, the increased complexity leads to significantly more data necessary in order to calibrate and validate the model, making its implementation difficult. Despite the many possibilities for future development, SisBAHIA® is vertically integrated and its applicability is therefore limited to shallow water bodies. Thus, further studies should be conducted to consider the process of thermal stratification, with the separation of the water column in different layers, thus increasing the aplicability of the model. In addition, a new section may be created to represent sediments at the bottom, which would better simulate interactions at the sediment-water interface.

6. REFERENCES

ANGELOCCI, L. R. & VILLA NOVA, N. A.(1995) Variações da temperatura da água de um pequeno lago artificial ao longo de um ano em Piracicaba – SP, *Sci. Agric.*, 52: 431-438.

CHAPRA, S. C. (1997) *Surface Water-Quality Modeling*. USA: McGraw-Hill Companies.

CUNHA, C. L. N., MONTEIRO, T., ROSMAN, P. C. C. (2002) Modelagem Bidimensional de Transporte de Escalares Não-conservativos em corpos d'água Rasos. *Revista Brasileira de Recursos Hídricos*, Brasil, 7, no2.

CUNHA, C. L. N.; ROSMAN, P. C. C.; FERREIRA, A. P.; MONTEIRO, T. C. N. (2006) Hydrodynamics and water quality models applied to Sepetiba Bay. *Continental Shelf Research, 26:* 1940-1953.

FISCHER, H. B. (Ed.) (1979), *Mixing in Inland and Coastal Waters*. Academic Press Inc., New York.

FRAGOSO, Jr., FERREIRA, T. F., MARQUES, D. M.(2009) *Modelagem Ecológica em Ecossistemas aquáticos*. Oficina de textos.

JORGENSEN, S. E., HALLING-SORENSEN, B., NEILSEN, S. N. (1996) *Handbook of environmental and ecological modeling*. Lewis Publishers, New York.

LAWSOM, R. & ANDERSON, M. A. (2007) Stratification and mixing in lake elsinore, california: An assessment of axial flow pumps for improving water quality in a shallow eutrophic lake. *Water Research, 41:* 4457-4467.

LOPES, M. R. M.; BICUDO, C. E. M. (2001) Short term variation of physical characteristics of a shallow subtropical oligotrophic reservoir, southeast Brazil. *Acta Limnologica Brasiliensia, 13*: 87-98.

ROSMAN, P. C. C. (2000) *Referência Técnica do* SISBAHIA – SISTEMA BASE DE HIDRODINÂMICA AMBIENTAL, Programa COPPE: Engenharia Oceânica, Área de Engenharia Costeira e Oceanográfica, Rio de Janeiro, Brasil.

SALAS, H. J. & MARTINO, P. (1991). A simplified phosphorus trophic state model for warm-water tropical lakes. *Water Research, 25*: 341-350.

SPERLING, E. V., JARDIM, F. A., GRANDCHAMP, C. A. P. (2004) Qualidade da água durante a formação de lagos profundos em cavas de mineração: Estudo de caso do lago de águas claras - MG, *Engenharia sanitária e ambiental, 9:* 250-259.

THOMANN, R. V. and MULLER, J. A. (1987) *Principle of Surface Water Quality Modeling and Control,* Harper and Row, New York.

WETZEL, R. G. (1981). *Limologia*. Barcelona: Eiciones Omega.

SECTION IV

AQUATIC COMMUNITIES

CHAPTER 13

PHYTOPLANKTON ECOLOGY IN THE RIO VERDE RESERVOIR

Luciano Felício Fernandes
Kelly Cristiany Gutseit
Juliana Wojciechowski
Patricia Esther Duarte Lagos
Christine da Fonseca Xavier

PHYTOPLANKTON ECOLOGY IN THE RIO VERDE RESERVOIR

Luciano Felício Fernandes
Kelly Cristiany Gutseit
Juliana Wojciechowski
Patricia Esther Duarte Lagos
Christine da Fonseca Xavier

ABSTRACT

In this chapter, we present the results from the analysis of phytoplankton ecology in the Rio Verde Reservoir in order to describe the community and its most abundant species and to assess the causes of the observed variations in the annual cycle between 2008 and 2009. Some recommendations on how to deal with potential cyanobacteria blooms are also presented and briefly discussed. Data were colected monthly between July 2008 and July 2009 at five stations established along the reservoir, and samples were colected from variable depths. The reservoir is warm monomictic in the lentic region, and tends to be mixed throughout the year in the lotic region. As a general pattern, the concentrations of nutrients in the Rio Verde Reservoir were low, and this feature reflected on the phytoplankton densities and its species composition. The phytoplankton was dominated by colonial mucilaginous Chlorophyceae (functional groups **X1** and **J**), especially during spring and summer, adapted to the periods of thermal stratification and higher light intensity. During the circulation of the water column in autumn and winter, the diatoms of functional groups **C** and **B**, which are more tolerant to low temperatures and reduced luminosity, increased their relative contribution to the phytoplankton community. Coincidentally, silicate concentrations increased in these periods (>0.8 mg/L), thus supporting the growth of several species, especially *Aulacoseira ambigua*, *Asterionella formosa* and *Fragilaria crotonensis*. Although detected in the reservoir, Cyanobacteria occurred in reduced abundances throughout the study. Toxic species recorded included *Dolichospermum planctonicum*, *Cylindrospermopsis raciborskii*, *Microcystis wesenbergii* and *Microcystis aeruginosa*. A brief discussion on previous blooms of *Cylindrospermospis raciborskii* is also provided.

KEYWORDS

Harmful algae, functional groups, subtropical reservoir, phytoplankton - Paraná.

1. INTRODUCTION

The continental aquatic ecosystems can be compartmentalized according to three basic processes that regulate the carbon flow in the trophic web: primary production by microalgae (periphyton and phytoplankton) and macrophytes, consumption by mixotrophs and heterotrophs (protists to vertebrate metazoans), and decomposition and re-mineraliziation of the organic matter and ions by bacteria and fungi.

Phytoplankton plays an important role in lentic water bodies (or "semi enclosed" systems, like lakes and reservoirs), because it represents the base of the plankton trophic web, and transfers the inorganic carbon to higher trophic levels. Phytoplankton can be conceptualized as the "community of photoautotrophic prokaryote and eukaryote organisms, suspended in the water column and with locomotion not enough to overcome the movement of water" (FERNANDES *et al.*, 2005).

In lakes and reservoirs, periods of greater phytoplankton development are common when there are significant alterations in the factors that regulate the cells' growth, thus intensifying the carbon dynamics in the trophic web, usually benefitting the aquatic ecosystem as a whole. The most important regulating factors in algae growth are light in the visible spectrum (400-700mm); temperature; micronutrients and macronutrients; concentrations of oxygen and carbonic gas, physical stability of the water column; competition and herbivory by zooplankton (FERNANDES *et al.*, 2005; REYNOLDS, 2006).

In Brazil, the periods of phytoplankton growth vary significantly from region to region but they can be roughly categorized based on latitudinal gradients. In the north, mid-west and northeastern regions, there is a greater dependence on the dry and wet seasons to regulate the supply of nutrients, considering that the water temperature and light intensity are relatively constant and non-limiting across the year (ESTEVES, 1998). In the southern and southeastern regions, areas with greater contrasts in climate, more intensive growths or blooms have been recorded during two seasons. In autumn, temperature reduction and increase of winds promote the circulation of the water column, leading to the re-suspension of nutrients from the hypolimnion. During this season, with lower water temperature (<17°C), less intense luminosity and a shorter photoperiod, it can normally be observed the predominance of diatoms either in large reservoirs (e.g. Itaipu) (DOMINGUES *et al.*, 2001) or in the small ones (e.g. Piraquara and Passaúna) (COQUEMALA *et al.*, 2006). During spring and part of the summer, the rainy period increases the input of nutrients associated with rains, and the water heating lead to stratification of lakes and reservoirs. During this season, chlorophytes and nanoplanktonic phytoflagellates tend to dominate the phytoplankton community (FERNANDES *et al.*, 2005). This pattern is common in oligotrophic to mesotrophic reservoirs; however, in eutrophic environments, dominance of cyanobacteria is observed in the spring and summer, and a larger contribution of diatoms in the autumn and winter, but still a predominance of cyanobacteria (ESTEVES, 1998; FERNANDES *et al.*, 2005a).

Brazilian reservoirs have suffered a series of anthropogenic impacts partially compromising the physical-chemical quality of the water, increasing the costs of treatment and water supply for the urban populations. The two traditional problems that deteriorate the water quality are: (**I**) the accelerated growth of cities and waste disposal (sewage, served water) into rivers; and (**II**) the progressive sealing of the soils, simultaneous to the deforestation of the (water retentive) forests, resulting in reduced infiltration into the soil and a reduction in available groundwater. In the last two decades, a third factor has become important in the process of water resources degradation in Brazil: (**III**) the blooms of harmful microalgae, as a consequence of artificial eutrophication. The term eutrophication is conceived here as the excessive input of organic material and nutrients, exaggeratedly pumping the aquatic ecosystem and causing a large increase in productivity by the aquatic organisms, affecting other biogeochemical cycles (ESTEVES, 1998; REBOUÇAS *et al.*, 2002; ANDREOLI & CARNEIRO, 2005).

Similar to what has been happening in the majority of Brazilian water bodies, the reservoirs of Paraná State have been suffering the impacts of anthropogenic interferfence, usually in the form of excessive nutrient input (especially nitrogen and phosphorus available to phytoplankton), as well as organic matter. As a consequence, the cell growth is stimulated, generating high densities and even frequent blooms of some species (IAP, 2009; COQUEMALA *et al.*, 2006; FERNANDES *et al.*, 2005). Unfortunately, in these situations the cyanobacteria tend to be dominant, specially the toxic species, requiring not only the removal of these cells during water treatment, but also the removal or inactivation of their toxins.

In the Metropolitan Region of Curitiba, there are four small reservoirs that supply the city of Curitiba and smaller municipalities: Iraí, Piraquara I and II and Passaúna. Other reservoirs are becoming operational like the Rio Verde, and the Miringuava. The reservoirs of Iraí and Passaúna suffer from frequent blooms of toxic cyanobacteria, usually *Microcystis aeruginosa, M. wesenbergii, Dolichospermum planctonicum* and *Cylindrospermopsis raciborskii* (FERNANDES et al., 2005; COQUEMALA et al., 2006). Moreover, due their low water quality, they are classified from critically degraded to polluted, (IAP, 2009). The causes of these problems are an excess of nutrients through point and diffuse sources, like domestic sewage, industrial waste-water, and excessive agricultural fertilizers, in addition to the almost complete deforestation of the river basins in which the reservoirs are located (ANDREOLI & CARNEIRO, 2005).

On the other hand, the Rio Verde Reservoir, focus of the present study, presents a good water quality that is reflected in relatively low concentrations of nutrients, turbidity and chlorophyll-a, not to mention satisfactory levels of chemical indicators (IAP, 2009). In April 2005, during the environmental monitoring program of reservoirs administered by the Environmental Institute of Paraná (*Instituto Ambiental do Paraná*, IAP) of the State Environment Secretariat (SEMA), a bloom of *C. raciborskii* was detected in the Rio Verde Reservoir, reaching approximately 100,000 cells/mL (IAP, 2009) a toxin level potentially harmful for human consumption, according to Ordinance 518/2004 of the Ministry of Health. The Brazilian oil company Petrobrás, concerned with the possibility of an emerging anthropogenic impact, supported a broader environemental analysis of the Rio Verde Basin. Finally, as Sanepar will shortly begin using the water from the Rio Verde Reservoir for public water supply, the first research on the phytoplankton community of this reservoir was conducted.

In this chapter, we present the results of the investigation on the phytoplankton ecology in the Rio Verde Reservoir with the goal of characterizing the community and its most abundant species, and determining the causes of observed variations in the annual cycle between 2008 and 2009. Some recommendations to handle potential future cyanobacteria blooms are also provided and briefly discussed.

2. MATERIALS AND METHODS

The Rio Verde Reservoir (25°31`30"S, 49°31`30"W), completed in 1976, is located in the municipality of Araucária, Metropolitan Region of Curitiba, an is stuated in the Alto Iguaçu Basin. The reservoir is located in the climatic zone classified as Cfb type according to Köeppen (1936); that is, subtropical with humid and fresh summers, mild winters, and common occurrence of frosts. The reservoir area is 5.9 km^2, with an average volume of 25.6 million of m^3 and residence time of approximately 218 days. The maximum and medium depths are 11 and 5.6 m respectively.

Field samplings were carried out monthly between July 2008 and July 2009, at five stations established in the reservoir (Figure 1). The sampling stations were chosen based on hydrological characteristics of the reservoir, in such a way that station R1 represented the lotic region, stations R2 and R3 the intermediary region and R4 the lacustrine region. Since stations R2 and R3 were fairly similar to stations R2 and R3, and in order to facilitate the interpretation of data, only the results for stations R1 and R4 are presented in this study.

The meteorological data (solar radiation, rainfall, and intensity and direction of wind) were obtained from a meteorological station located at the dam of the reservoir (see chapter 12 of this book).

Water temperature, pH, dissolved oxygen and conductivity were measured in the field every 0.5 meters using the Horiba U22XD Multi Parameter Probe. (Table 1). Water samples were collected from two depths in the station R1 and four depths in the station R4 for nutrient analysis (total phosphorus, reactive soluble phosphorus – RSP, total Kjedahl nitrogen, ammoniac nitrogen, organic nitrogen, nitrate, nitrite, silicate), total solids, chlorophyll-a concentration and enumeration of phytoplankton cells. The sampling depths were determined based on the extension of the photic zone, estimated from readings of water transparency using a Secchi disk, according to Esteves (1998). These samples were obtained using a Van Dorn bottle and kept under refrigeration and protected from light during transportation to the lab. Analyses of cyanotoxins were conducted for all collections, from a sub-surface sample at each sampling station.

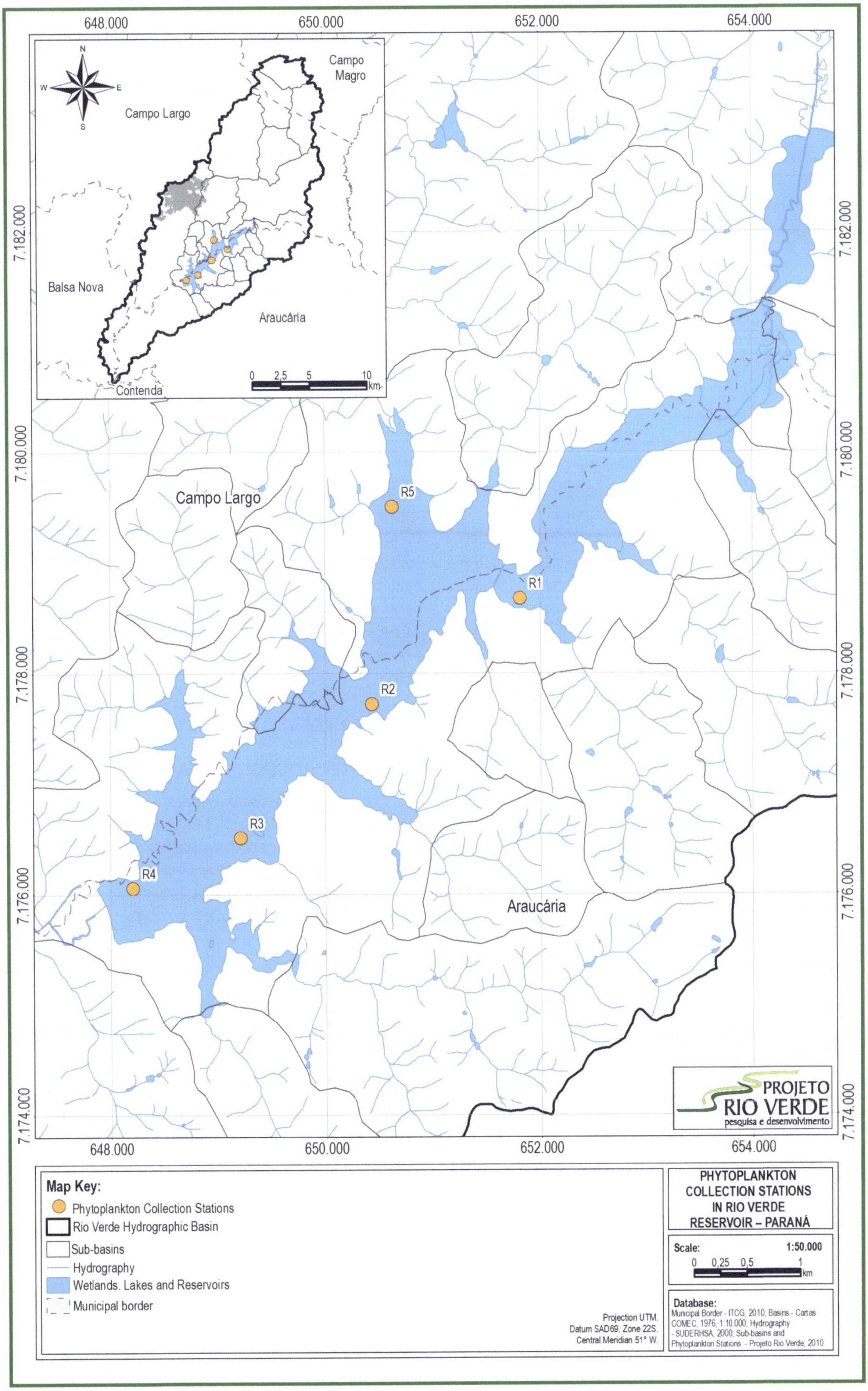

FIGURE 1 – MAP OF THE RIO VERDE RESERVOIR WITH THE LOCATION OF SAMPLING STATIONS

TABLE 1 – PARAMETERS COLLECTED DURING THE SAMPLING PERIOD OF IN THE RIO VERDE RESERVOIR

PARAMETERS		
IN FIELD	**LABORATORY**	**METEOROLOGICAL STATION**
Water temperature	Total Phosphorus	Air temperature
pH	Reactive Phosphorus (RSP)	Solar radiation
Dissolved oxygen	Total K Nitrogen	Rainfall
Conductivity	Ammonia nitrogen	Wind speed
Transparency	Organic nitrogen	Wind direction
	Nitrate	
	Nitrite	
	Silicate	
	Total solids	
	Phytoplankton	
	Chlorophyll-a	
	Cyanotoxins	

Nutrient analyses (nitrogenized compounds, phophorus and silicate), as well as total solids were analyzed according to the "Standard Methods for the Examination of Water and Wastewater" (APHA, 1995). Chlorophyll-a was cold extracted according to the Nusch method (1980).

Samples for quantitative analysis of the phytoplankton community were collected from the chosen depths using the Van Dorn bottle, added to amber flasks and preserved with acetic lugol. For qualitative analysis of the phytoplankton, samples were collected using a plankton net (0.6 x 1.3 m) with mesh opening of 20 µm, placed in polyethylene flasks and observed, still alive, in light microscopic in the same day of the collection. Sub-samples were preserved in Transeau solution for identification, adding equal parts of fixative to the sample. The density of organisms (cells/mL) of the most abundant groups and species of phytoplankton was measured according to the Utermöhl technique (1958), using an inverted Olympus IX70 microscope. A minimum number of 70 fields or at least 100 cells of each one of the two most abundant species within the sample was counted.

The classification of Reynolds *et al.* (2002) was used to describe the phytoplankton functional groups of the reservoir. Only the species with relative densities higher than 3% in relation to the total were chosen.

A Principal Components Analysis (PCA) was used to explore the physicochemical data using the software STATISTICA 6.0. Variables included in the analysis were: pH, conductivity, dissolved oxygen, temperature, total dissolved solids, silicate, total phosphorus, phosphate and total nitrogen. The data were previously transformed (log10 (x+1)) to meet the assumption of normality and homogeneity of analysis variances (DYTHAM, 2003). Only components with *eigenvalues* greater than 1 were retained, and variables with correlations greater than 0.5 in relation to the components were considered important.

3. RESULTS

3.1 METEOROLOGICAL DATA

The study area can be separated in two metereological periods, according to the air temperature data (Figure 2), solar radiation (Figure 3) and rainfall (Figure 4). The period of spring-summer, from September to March, is marked by warmer temperatures and elevated levels of average solar radiation and rainfall. In the period of autumn-winter, from April to August, the temperatures are colder, with less rainfall and shorter days.

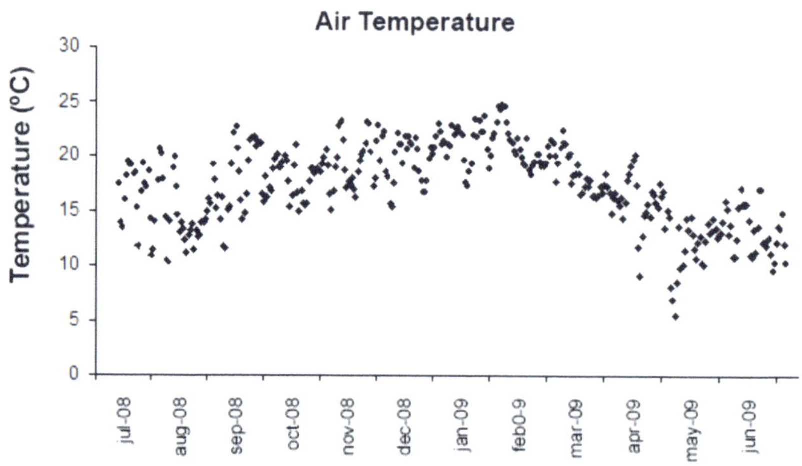

FIGURE 2 – AVERAGE DAILY AIR TEMPERATURE (°C) DATA RECORDED DURING THE STUDY PERIOD FROM THE METEOROLOGICAL STATION OF RIO VERDE RESERVOIR

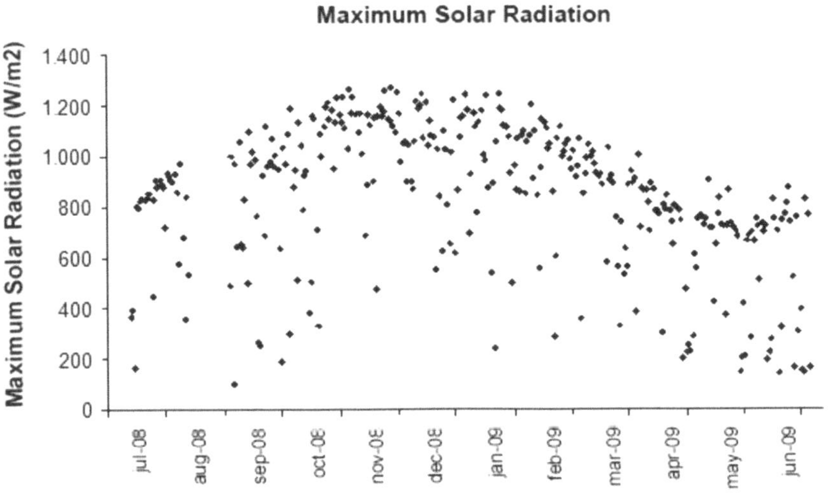

FIGURE 3 – MAXIMUM SOLAR RADIATION (W/m²) DURING THE SAMPLING PERIOD FROM THE METEOROLOGICAL STATION OF RIO VERDE RESERVOIR

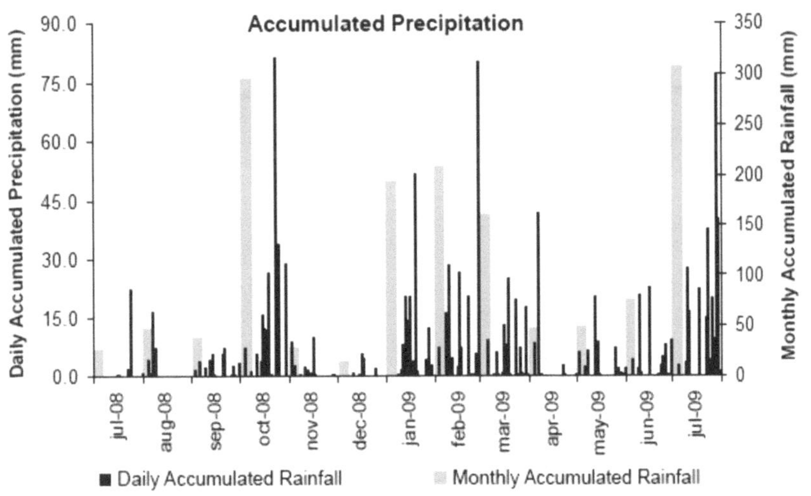

FIGURE 4 – MONTHLY ACCUMULATED RAINFALL BETWEEN OCTOBER 2008 AND JULY 2009. NOTICE THAT THE AXES OF RAINFALL ARE NOT TO SCALE

The average daily air temperature varied from 5.5°C to 24.8°C (Figure 2). The highest temperatures were registered in the period of spring-summer, when values oscillated between 10.3°C and 24.8°C. In the autumn-winter, the temperatures varied from 5.5°C to 22.5°C. The solar radiation varied from 100 to 1273 W/m², with the higher values occurring during the period of spring-summer, and the lowest during autumn-winter (Figure 3).

Total annual rainfall was 1205 mm. The months of October 2008, January to March, and July 2009 were the rainiest of the study period (Figure 4).

3.2 PHYSICOCHEMICAL PARAMETERS IN THE RIO VERDE RESERVOIR

Station R4 (lentic environment, near the dam of the reservoir)

In Table 2, physicochemical data is presented for the sampling period in the Rio Verde Reservoir. Water temperature in the station R4 varied between 15.5°C (July 2008) and 26.3°C (March 2009), with an average of 19.4°C. The highest temperatures occurred in spring-summer (from September to March) and the lowest in autumn-winter (from April to August). Between November and April, thermal stratification became well established (Figure 5). On average, the epilimnion was 3.3°C warmest than the hypolimnion, with the maximum difference recorded in February 2009, when the difference reached to 7.8° C. From July to October 2008 and from May to February 2009, water column was homogenous, with a vertical gradient below 2°C. The highest concentration of dissolved oxygen (DO) detected in the station R4 was 9.5 mg/L in July 2009 in surface layers of the water column. During warm months, when the reservoir was stratified, there was a strong DO concentration gradient, with values decreasing towards the bottom. Between September and December of 2008 and between January and April of 2009, the hypolimnion reached the anoxic levels (Figure 5). The pH values were more elevated during summer months, oscillating between 6 (July 2008) and 8 (February 2009). During the warmest months, a sharp gradient of pH was noticed, with alkaline surface layers and slightly acidic hypolimnion. During winter, pH values were more homogenous, and consistently around 7. Conductivity reached maximum value (237 μS/cm) in April 2009 in the hypolimnion of the station R4 and minimum (83 μS/cm) in October and November 2008 in the same region. A gradient with greater values was noted near the bottom, especially in the warmest months (Figure 5). The depth of the Secchi disk oscillated from 1.3 to 3.5 m throughout the sampling period.

During the sampling period, total phosphorus varied from approximately 0.01 to 0.14 mg/L in the station R4. The lowest concentrations were measured in May 2009, and the highest in October and November 2008 (Table 2). The reactive soluble phosphorus (RSP) varied from 0.002 to 0.12 mg/L. Similar to the results for total phosphorus, the highest concentrations of phosphate occurred in October and November 2008. The lowest value was observed in July 2008 (Table 2). Total nitrogen presented minimum of 0.02 and maximum of 3.06 mg/L. The highest values were recorded near the bottom during the stratification period and the lowest measured in the surface layers in April 2009. Silicate values varied from 0.1 mg/L (July 2009) to 3.4 mg/L (January 2009); lower concentrations were recorded in October and November 2008, and May and June 2009. During summer higher values were detected over the water column; a maximum peak occurring in January 2009 (Figure 6).

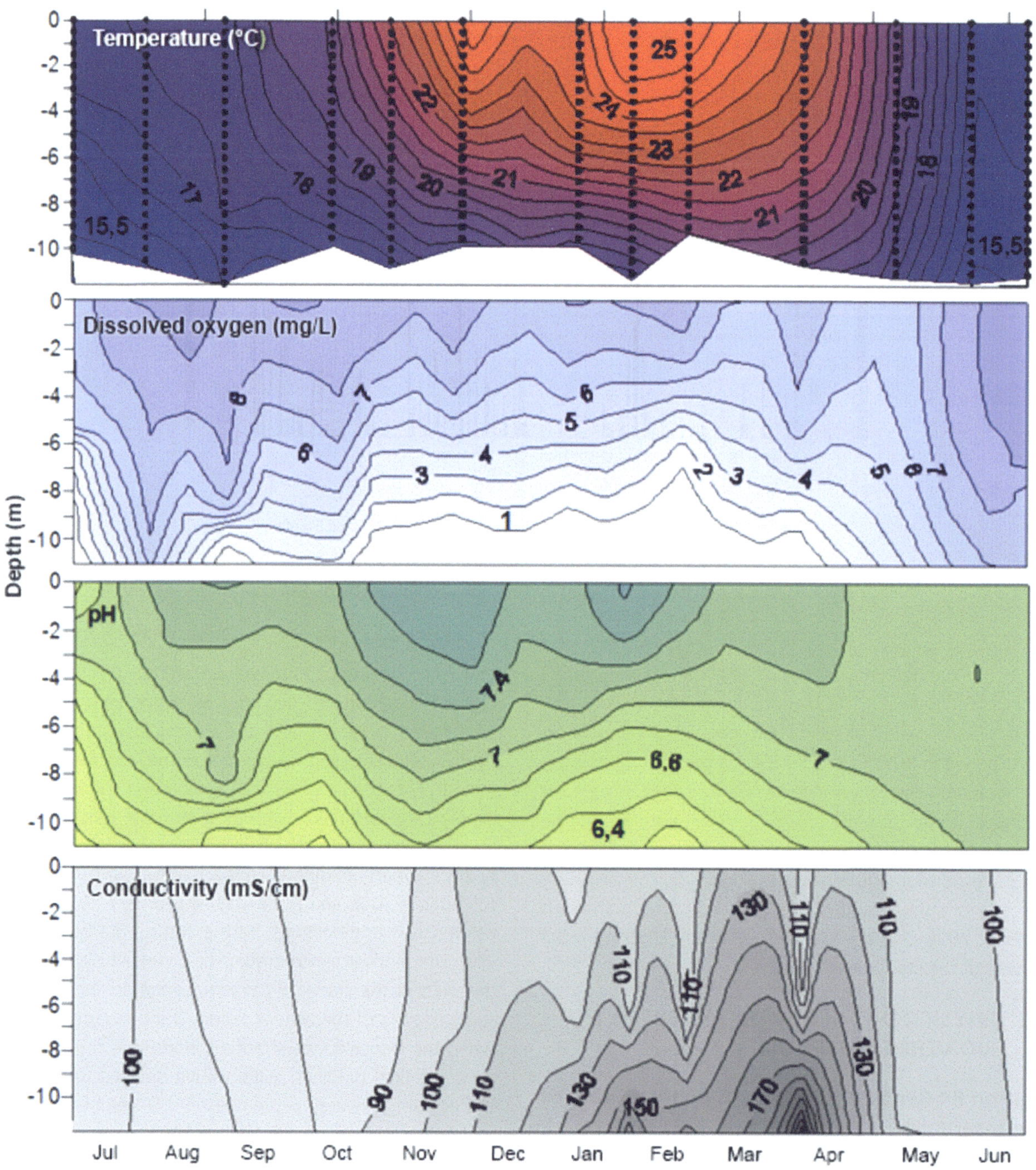

FIGURE 5 – PROFILES OF TEMPERATURE (°C), DISSOLVED OXYGEN (μg/L), pH AND CONDUCTIVITY (μS/cm) IN THE STATION R4 OF THE RIO VERDE RESERVOIR BETWEEN JULY 2008 AND JULY 2009

TABLE 2 – PHYSICOCHEMICAL DATA (MEAN, MINIMUM AND MAXIMUM BETWEEN BRACKETS) MEASURED IN THE RIO VERDE RESERVOIR FROM JULY 2008 TO JULY 2009

Station	Date	Temperature °C	Secchi m	pH	DO mg/L	Conductivity µS/cm	Total solids mg/L	Total phosphorus mg/L	Reactive soluble phosphorus mg/L	Total nitrogen mg/L	Silicate mg/L	Chlorophyll-a µg/L
R1	jul/08	16.2 (15.3 - 17.0)	1.2	6.4 (6.2 - 6.5)	6.0 (5.7 - 6.2)	108 (105 - 112)	75 (73 - 76)	0.04 (0.03 - 0.04)	0.02 (0.01 - 0.04)	0.002 (0.002 - 0.002)	1.2 (0.9 - 1.5)	3.9 (3.1 - 4.6)
	aug/08	17.3 (16.0 - 17.9)	0.8	7.2 (7.1 - 7.2)	9.6 (8.7 - 10.3)	119 (117 - 121)	100 (95 - 105)	0.08 (0.04 - 0.11)	0.06 (0.03 - 0.08)	0.26 (0.03 - 0.49)	1.4 (1.3 - 1.5)	4.8 (4.6 - 5.0)
	sep/08	17.5 (17.1 - 18.9)	0.8	7.1 (7.0 - 7.2)	8.2 (7.5 - 9.5)	104 (102 - 104)	65 (62 - 67)	0.02 (0.02 - 0.02)	0.01 (0.01 - 0.01)	0.07 (0.06 - 0.07)	1.4 (1.3 - 1.5)	7.2 (7.0 - 7.3)
	oct/08	18.6 (18.1 - 19.3)	0.8	7.2 (7.1 - 7.4)	6.9 (6.4 - 7.9)	91 (90 - 93)	83 (81 - 84)	0.11 (0.08 - 0.14)	0.11 (0.08 - 0.14)	0.22 (0.12 - 0.32)	1.6 (1.2 - 2.1)	4.8 (4.7 - 4.9)
	nov/08	21.4 (21.0 - 21.5)	1.4	7.3 (7.0 - 7.5)	7.1 (3.5 - 7.7)	91 (90 - 92)	74 (72 - 75)	0.10 (0.06 - 0.13)	0.09 (0.06 - 0.13)	0.09 (0.002 - 0.17)	1.5 (1.2 - 1.8)	5.8 (5.0 - 6.5)
	dec/08	24.1 (22.5 - 25.3)	2.2	7.7 (7.4 - 7.8)	8.7 (7.2 - 9.1)	109 (102 - 112)	68 (66 - 70)	0.05 (0.05 - 0.05)	0.05 (0.05 - 0.05)	0.12 (0.09 - 0.15)	2.7 (2.4 - 2.9)	8.1 (5.2 - 11.1)
	jan/09	23.3 (22.7 - 23.7)	1.1	7.3 (6.6 - 9.7)	5.1 (1.0 - 6.7)	104 (97 - 107)	103 (96 - 109)	0.05 (0.04 - 0.06)	0.03 (0.02 - 0.05)	0.002 (0.002 - 0.002)	3.0 (3.0 - 3.1)	8.3 (5.6 - 11.0)
	feb/09	25.0 (23.4 - 25.6)	2.1	7.5 (6.8 - 7.7)	6.4 (0.0 - 8.1)	107 (105 - 118)	72 (57 - 86)	0.05 (0.04 - 0.06)	0.04 (0.03 - 0.05)	0.18 (0.15 - 0.21)	3.7 (3.0 - 4.3)	12.4 (5.9 - 18.9)
	mar/09	25.6 (23.4 - 27.8)	2.0	7.6 (6.8 - 7.8)	6.6 (0.6 - 8.8)	112 (109 - 116)	N.R.	0.04 (0.03 - 0.04)	0.04 (0.03 - 0.04)	0.17 (0.11 - 0.22)	4.6 (4.1 - 5.1)	12.3 (11.8 - 12.7)
	apr/09	23.0 (22.9 - 23.0)	1.4	7.3 (7.3 - 7.4)	7.1 (6.9 - 7.2)	118 (118 - 119)	67 (62 - 71)	0.19 (0.17 - 0.20)	0.01 (0.01 - 0.01)	0.36 (0.23 - 0.48)	4.7 (3.8 - 5.5)	9.9 (9.6 - 10.2)
	may/09	18.7 (18.4 - 19.4)	1.0	7.3 (7.2 - 7.4)	7.4 (6.7 - 7.8)	116 (115 - 117)	99 (96 - 101)	0.05 (0.01 - 0.08)	0.01 (0.01 - 0.01)	0.61 (0.52 - 0.69)	2.4 (1.9 - 2.9)	7.0 (6.7 - 7.2)
	jun/09	15.1 (15.1 - 15.1)	1.1	7.4 (7.3 - 7.4)	8.6 (7.8 - 8.8)	107 (107 - 108)	55 (30 - 79)	0.04 (0.04 - 0.05)	0.04 (0.04 - 0.05)	0.002 (0.002 - 0.002)	2.2 (1.8 - 2.6)	8.3 (7.7 - 8.9)
	jul/09	16.1 (15.7 - 16.4)	1.3	7.3 (7.3 - 7.4)	9.4 (8.2 - 10.7)	111 (110 - 112)	54 (51 - 56)	0.04 (0.04 - 0.05)	0.03 (0.03 - 0.04)	0.41 (0.33 - 0.48)	1.0 (0.9 - 1.1)	3.9 (2.8 - 5.0)
R4	jul/08	16.3 (15.5 - 17.4)	2.3	6.4 (6.0 - 6.9)	4.5 (0.6 - 7.0)	91 (89 - 93)	69 (63 - 81)	0.03 (0.02 - 0.04)	0.01 (0.01 - 0.02)	0.31 (0.21 - 0.38)	1.2 (0.7 - 1.4)	5.2 (2.8 - 7.4)
	aug/08	16.6 (16.2 - 17.3)	1.5	6.8 (6.6 - 7.2)	7.9 (7.3 - 9.2)	101 (101 - 102)	87 (50 - 121)	0.05 (0.05 - 0.05)	0.03 (0.02 - 0.03)	0.38 (0.14 - 0.64)	1.2 (1.0 - 1.2)	5.2 (4.4 - 5.8)
	sep/08	17.4 (16.6 - 17.7)	1.3	7.0 (6.1 - 7.5)	6.6 (0.0 - 8.5)	98 (96 - 102)	58 (49 - 66)	0.02 (0.01 - 0.03)	0.01 (0.01 - 0.03)	0.15 (0.04 - 0.23)	1.2 (1.0 - 1.4)	10.0 (7.6 - 11.2)
	oct/08	18.8 (17.5 - 19.3)	2.9	7.0 (6.1 - 7.4)	7.3 (3.5 - 8.6)	85 (83 - 87)	69 (63 - 78)	0.02 (0.01 - 0.04)	0.02 (0.01 - 0.03)	0.26 (0.17 - 0.31)	0.2 (0.1 - 0.5)	6.5 (4.6 - 7.6)
	nov/08	20.0 (17.5 - 22.3)	2.0	7.3 (6.7 - 8.0)	4.4 (0.0 - 9.0)	86 (83 - 93)	142 (90 - 297)	0.06 (0.01 - 0.12)	0.07 (0.03 - 0.11)	0.23 (0.02 - 0.61)	0.5 (0.3 - 0.6)	6.4 (5.0 - 7.4)
	dec/08	22.3 (17.8 - 25.4)	3.5	7.4 (6.6 - 7.9)	5.6 (0.0 - 8.8)	100 (95 - 106)	69 (66 - 71)	0.05 (0.05 - 0.05)	0.05 (0.05 - 0.05)	0.11 (0.002 - 0.34)	0.6 (0.4 - 0.7)	4.7 (2.5 - 8.4)
	jan/09	22.5 (18.7 - 23.8)	1.9	7.1 (6.4 - 7.6)	4.9 (0.0 - 8.0)	104 (99 - 137)	150 (88 - 316)	0.02 (0.01 - 0.02)	0.01 (0.01 - 0.02)	0.46 (0.07 - 1.00)	2.8 (1.9 - 3.4)	6.1 (5.3 - 6.7)
	feb/09	22.8 (18.3 - 26.1)	3.0	7.0 (6.2 - 8.0)	3.7 (0.0 - 8.7)	121 (94 - 176)	184 (49 - 553)	0.07 (0.05 - 0.09)	0.05 (0.04 - 0.08)	0.30 (0.10 - 0.56)	0.6 (0.4 - 0.7)	14.4 (5.8 - 30.7)
	mar/09	23.7 (19.8 - 26.3)	2.3	7.0 (6.2 - 7.8)	4.1 (0.0 - 8.7)	113 (95 - 156)	N.R.	0.04 (0.03 - 0.07)	0.03 (0.02 - 0.07)	1.22 (0.34 - 2.01)	0.9 (0.8 - 0.1)	9.6 (6.8 - 13.0)
	apr/09	22.3 (18.9 - 23.3)	1.8	7.1 (6.5 - 7.3)	5.2 (0.0 - 7.4)	133 (109 - 237)	76 (62 - 87)	0.10 (0.07 - 0.12)	0.01 (0.01 - 0.01)	0.35 (0.002 - 1.41)	0.9 (0.8 - 0.1)	9.4 (6.5 - 11.7)
	may/09	19.3 (19.2 - 19.3)	1.5	7.0 (7.0 - 7.1)	5.2 (4.5 - 5.7)	109 (109 - 109)	77 (71 - 82)	0.01 (0.01 - 0.01)	0.01 (0.01 - 0.01)	0.94 (0.43 - 1.24)	0.4 (0.3 - 0.5)	6.2 (5.7 - 6.5)
	jun/09	15.9 (15.7 - 16.1)	1.6	7.1 (7.1 - 7.2)	8.2 (7.7 - 8.7)	104 (104 - 104)	60 (53 - 66)	0.03 (0.03 - 0.04)	0.02 (0.02 - 0.02)	0.24 (0.002 - 0.81)	0.3 (0.2 - 0.4)	6.8 (3.6 - 8.9)
	jul/09	16.0 (15.8 - 16.3)	2.4	7.1 (7.0 - 7.1)	8.8 (7.2 - 9.5)	96 (96 - 97)	156 (141 - 165)	0.03 (0.03 - 0.05)	0.02 (0.01 - 0.02)	0.17 (0.002 - 0.57)	0.3 (0.1 - 0.3)	5.8 (3.4 - 12.9)

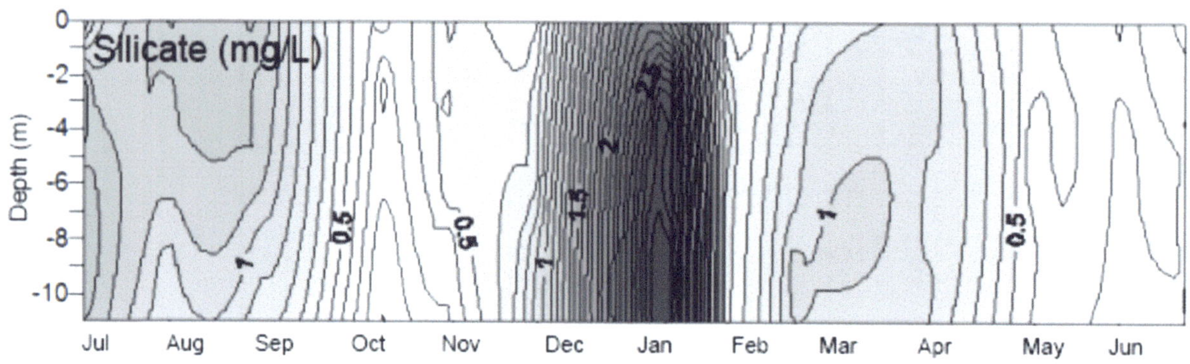

FIGURE 6 – SILICATE PROFILE (µg/L) AT THE STATION R4 IN THE RIO VERDE RESERVOIR FROM JULY 2008 TO JULY 2009

Station R1 (lotic environment, near the headwaters of the reservoir)

At the station R1, where the average depth is four meters, there was no stratification of the water column for the majoruty of the period, except briefly from January to March 2009. Differences in temperature between the epilimnion and hypolimnion were usually low, averaging 1.5°C. The region was fully mixed in June 2009. In March 2009, the gradient of temperature reached a maximum of 4.4°C. Concentrations of dissolved oxygen were similar to and followed the same variation of the station R4, with a gradient more evident during the warmest months. In February 2009, the hypolimnion reached anoxic levels. Maximum value of DO was 10.7 mg/L in July 2009 (Table 2). Regarding pH, maximum value was 7.8 during warmer months and minimum was 6.2 in the deepest layer in July 2008. Conductivity had a minimum of 90 µS/cm in the deepest layers in October and November 2008 and maximum of 121 µS/cm in August 2008 (Table 2). The depth of the Secchi disk varied from 0.8 to 2.2 m.

Total phosphorus varied from 0.01 (May 2009) to 0.20 mg/L (April 2009), and the RSP varied from 0.01 to 0.14 mg/L (average 0.042 mg/L) (Table 2). In relation to total nitrogen, the minimum was 0.025 mg/L in June 2009 and the maximum 1.03 mg/L in May 2009. Silicate varied from 0.87 mg/L to 5.46 mg/L;, the lowest concentrations recorded in July 2009 and the highest in March and April 2009 (Table 2).

3.3 PHYSICOCHEMICAL CHARACTERIZATION OF THE RESERVOIR FROM THE PRINCIPAL COMPONENTS ANALYSIS (PCA)

PCA clearly distinguished the stations R1 and R4; R1 presented lotic characteristics in relation to the R4, which behaved as a typically lacustrine area. Axes 1 and 2 of the ordination explained 24% and 20% of the data variation, respectively (Table 3). Three groupings were discriminated (represented by circles in Figure 7), of which Groups Ia and Ib delimitd the period of water column stratification (November to April), and Group II delimits the circulation period (May to October). Group Ia clustered more superficial samples of staion R4, that is, in the photic zone and during the stratification, associated specifically with higher values of temperature and depleted nitrogen in the water column. A small subgroup of samples from station R1 was also included in Groups Ia and Ib, revealing a short period of physical or chemical stratification in the lotic zone from January to March. This stratification probably extended across the reservoir, based on an analysis of valid raw data from the stations R2 and R3 (data not presented). Group Ib comprised samples from or near the hypolimnion that are more affected by high conductivity, high concentrations of nitrogen (especially ammonia) and low values of oxygen or even anoxia. Group II includes samples from station R1 throughout the year and from station R4 during circulation. This grouping indicates that the two sampling areas presented similar physical and chemical characteristics due to the homogenization of the water column. Group IIwas likely more affected by low temperatures and lower nutrient concentrations.

3.4 PHYTOPLANKTON: TAXONOMIC COMPOSITION

The analysis of the phytoplankton community resulted in the identification of about 130 infrageneric taxa (Table 4) distributed among the groups Chlorophyceae (45.1%), Cyanobacteria (8.8%), Bacillariophyceae (13.9%), Cryptophyceae (3.3%), Chrysophyceae (2.1%), Zygnemaphyceae (1.2%), Euglenophyceae (1.0%), Dinophyceae (0.1%) and others (5.3%).

Chlorophyceans dominated the phytoplankton in terms of species richeness, with 43 infrageneric taxa. The most representative genera were *Closteriopsis* (Figure 8G), *Dictyosphaerium* (Figure 8H), *Elakatothrix* (Figure 8J), *Monoraphidium* (Figure 8D) and *Scenedesmus* (Figure 8E, 8F). Among the diatoms (Bacillariophyceae) 25 genera were registered, mostly planktonic, such as *Asterionella* (Figure 9A), *Aulacoseira* (Figure 9E), *Fragilaria* (Figure 9C, 9D) and *Urosolenia* (Figure 9F, 9G, 9H). Fifteen taxa of Cyanobacteria were identified, including the toxigenic *Dolichospermum* (Figure 10G), *Aphanocapsa* (Figure 10A, 10B) and *Microcystis* (Figure 10H). Representative taxa from this group were *Aphanocapsa* (Figure 10A, 10B), *Merismopedia* (Figure 10C), *Microcystis* (Figure 10H), *Woronichinia* and a non-identified filamentous species (Figure 10E, 10F),. In Cryptophyceae, the most common genus was *Cryptomonas* (Figure 11A, 11B, 11C). Only six taxa of Chrysophyceae were recorded, *Dynobrion* being the most representative. In Euglenophyceae, seven genera were found, among them *Trachelomonas*, *Phacus* and *Euglena*. Genera of Dinophyceae and Xanthophyceae were rare, with *Gymnodinium* and *Peridinium* identified from the first family and *Centritractus* and *Isthmochloron* from the second.

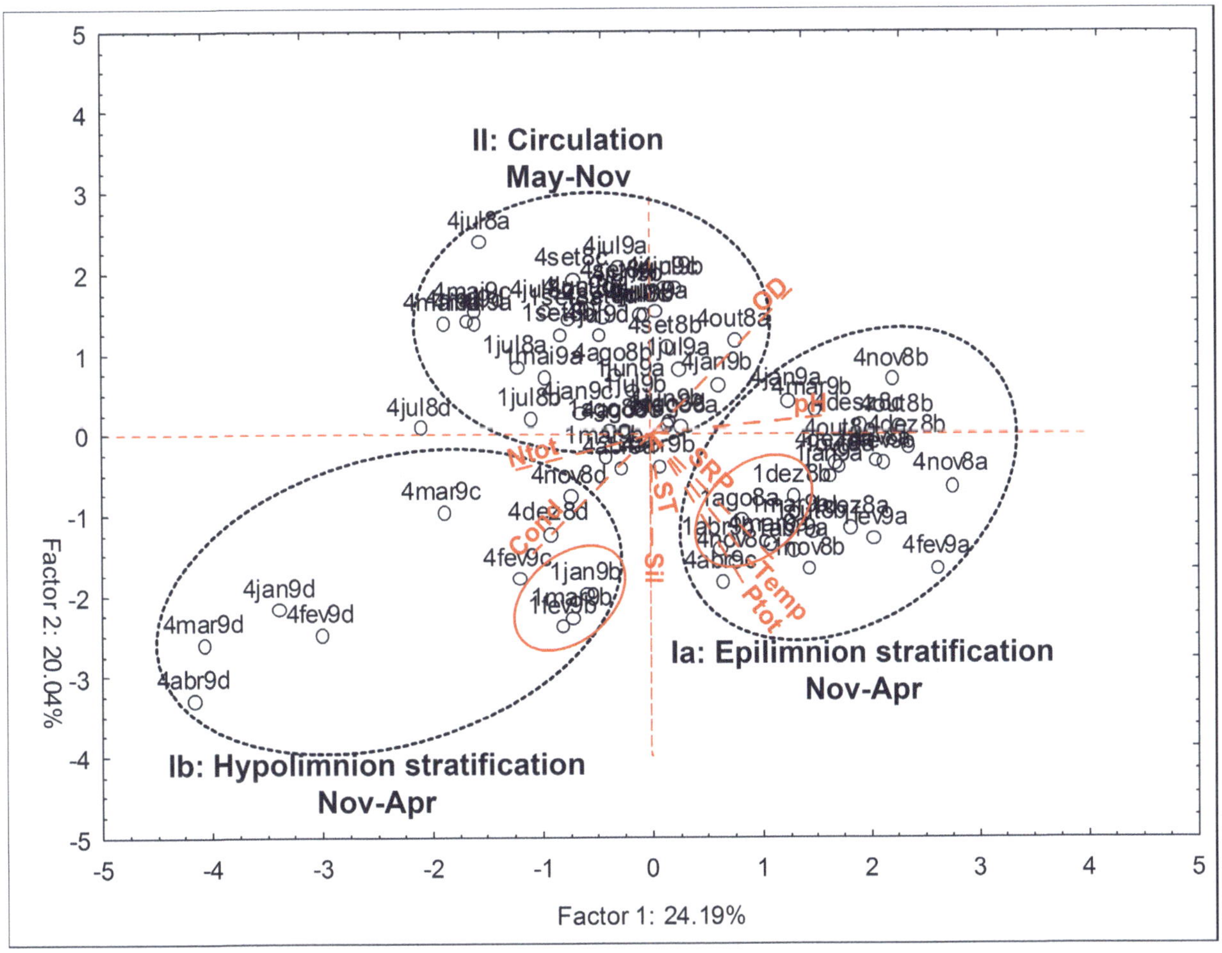

FIGURE 7 – PRINCIPAL COMPONENTS ANALYSIS OF THE PHYSICOCHEMICAL PARAMETERS, CHARACTERIZING THE SAMPLING STATIONS R1 (HEADWATER OF THE RESERVOIR) AND R4 (NEXT TO THE DAM) IN THE RIO VERDE RESERVOIR ACCORDING TO THE SPATIAL AND TEMPORAL DISTRIBUTION OF THE SAMPLES. TWO GREAT GROUPS WERE DISCRIMINATED, SEPARATING THE PERIOD OF STRATIFICATION (GROUP I) AND OF CIRCULATION (GROUP II) OF THE WATER COLUMN. THE **GROUP I** WAS SUBDIVIDED BY THE SAMPLES OF EPILIMNION (**Ia**) AND HYPOLIMNION (**Ib**), RESPECTIVELY, SPECIALLY THE ONES FROM THE R4 STATION. FOR CLARITY, RED CIRCLES HIGHLIGHT THE SAMPLES OF THE STATION R1. THEREFORE, THE R1, THOUGH PRESENTING A MORE LOTIC CHARACTERISTIC, ALSO APPEARED RELATIVELY STRATIFIED BETWEEN DECEMBER AND APRIL. THE **GROUP II** CLUSTERED SAMPLES FROM THE TWO STATIONS DURING RESERVOIR CIRCULATION. DETAILS IN THE TEXT. **CAPTION**:THE NUMBERS 1 AND 4 REFER TO STATIONS IN THE RESERVOIR (R1 AND R4) FOLLOWED BY MONTH, YEAR AND DEPTH OF THE SAMPLE (a= surface, b and c=intermediary; d= near the bottom).

TABLE 3 – RESULT OF THE PRINCIPAL COMPONENTS ANALYSIS OF THE PHYSICOCHEMICAL PARAMETERS IN THE RIO VERDE RESERVOIR

	PCA1	PCA2	PCA3
Eigenvalue	2,2	1,8	1,5
%Variância total	24,2	20,0	16,2
pH	0,78	0,08	0,41
Condutividade	-0,50	-0,48	0,38
OD	0,60	0,61	0,21
Temperatura	0,42	0,53	0,47
N total	-0,58	-0,14	0,6
P total	0,40	-0,62	-0,41
SRP	0,52	-0,53	-0,49
Silicato	0,01	-0,44	0,62
Sólidos totais	0,02	-0,18	-0,27

TABLE 4 – LIST OF INFRAGENERIC TAXA FOUND IN THE RIO VERDE RESERVOIR.

	CHLOROPHYCEAE	
Ankistrodesmus sp1	*Desmodesmus* sp2	*Lagerheimia* sp.
Ankistrodesmus sp2	*Diacanthos* sp.	*Micractinuim* sp.
Ankistrodesmus sp3	*Dictiosphaerium* sp.	*Monoraphidium minutum*
Bothrycoccus sp.	*Dydimocistis* sp.	*Oocystis* sp.
Closteriopsis longissima	*Elakatothrix* sp.	*Pediastrum duplex*
Closteriopsis sp.	*Eutetramorus* sp.	*Pediastrum simplex*
Coelastrum cambricum	*Eutetramorus* sp2	*Pediastrum tetras*
Coelastrum reticulatum	*Golenkinia* spp.	*Scenedesmus disciformis*
Crucigenia fenestrata	*Hyaloraphidium* sp.	*Scenedesmus ecornis*
Crucigenia tetrapedia	*Keratococcus* braunii	*Schroederia* sp.
Crucigeniela sp.	*Kirchneriella* sp1	*Tetraedron minimum*
Desmodesmus denticulatus	*Kirchneriella* sp2	*Treubaria setigera*
Desmodesmus sp1	*Koliella* sp.	
	CYANOPHYCEAE	
Aphanocapsa spp.	Cyanophyceae sp3	*Pseudoanabaena* sp.
Anabaena sp.	Filamentosa NI 1	*Spirulina* sp1
Chroococcus sp.	Filamentosa NI 2	*Spirulina* sp2
Cyanophyceae sp1	*Merismopedia* spp.	*Wronichinia* sp.
Cyanophyceae sp2	*Microcystis* spp	
	BACILLARIOPHYCEAE	
Asterionela formosa	*Aulacoseira* sp1	*Penada* NI
Achnanthidium sp.	*Cyclotela* sp.	*Tabellaria* sp.
Aulacoseira alpigena	*Discostella stilligera*	*Thalassiosira* sp.
Aulacoseira ambigua	*Fragilaria crotonensis*	*Urosolenia longiseta*
Aulacoseira angustissima	*Fragilaria* sp1	*Urosolenia* sp1
Aulacoseira granulata	*Fragilaria* sp2	*Urosolenia* sp2
	CRYPTOPHYCEAE	
Cryptomonas sp.		Cryptophyceae NI
	CHRYSOPHYCEAE	
Chrysophyceae NI1	Dynobrion spp.	Synura sp.
Chrysophyceae NI2	Malomonas spp.	
	ZYGNEMAPHYCEAE	
Closterium sp.	*Staurastrum arctiscon*	*Staurastrum* sp1
Cosmarium spp.	*Staurastrum aversum*	*Staurastrum* sp2
Euastrum sp.	*Staurastrum cristatum*	*Teilingia* sp.
	EUGLENOPHYCEAE	
Euglena sp.	*Trachelomonas* sp2	*Trachelomonas* sp4
Phacus suecicus	*Trachelomonas* sp3	*Trachelomonas volvocina*
Trachelomonas sp1		
	DINOPHYCEAE	
Gymnodinium sp.		Peridinium sp.
	XANTOPHYCEAE	
Centritractus sp.		Istmocloron lobulatum
	FITOFLAGELADOS	
Chlamydomonas sp.	*Desmarella moniliformes*	Nanoflagelado NI2
Chrysococcus sp.	Nanoflagelado NI1	

Fitoflagelados = Phytoflagellates
Nanoflagellate NI1
Nanoflagellate NI2

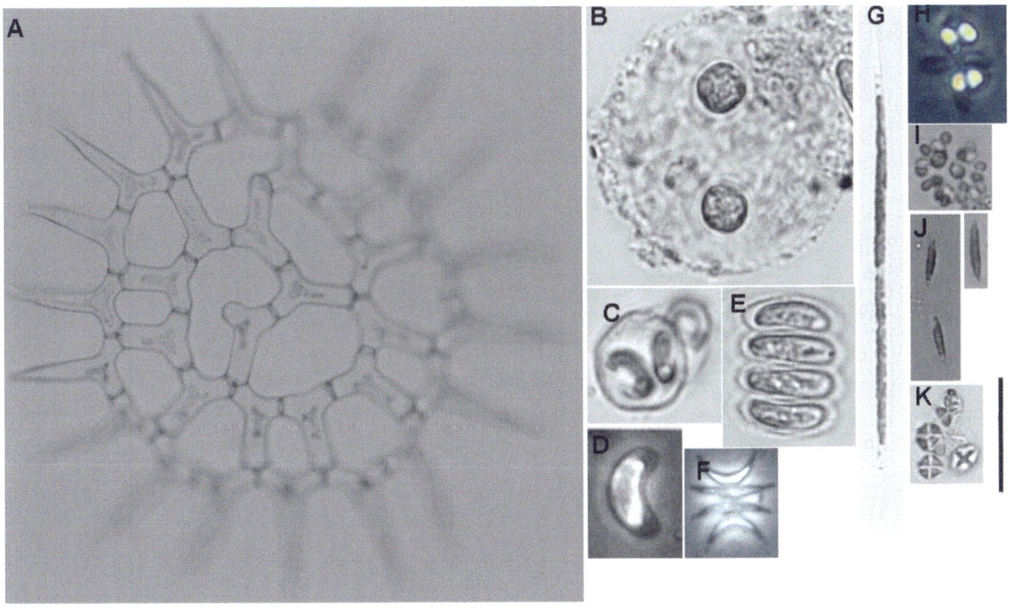

FIGURE 8 – SOME COMMONLY CHLOROPHYCEANS FOUND IN THE RIO VERDE RESERVOIR. A) *Pediastrum simplex*; B) *Eutetramorus fottii*; C) *Nephrochlamys* sp.; D) *Monoraphidium minutum*; E) *Scenedesmus* sp.; F) *Scenedesmus acuminatus*; G) *Closteriopsis longissima*; H-I) *Dyctiosphaerium* spp.; J) *Elakatothrix* sp.; K) *Tetrastrum triangulare*. Scale = 10 µm

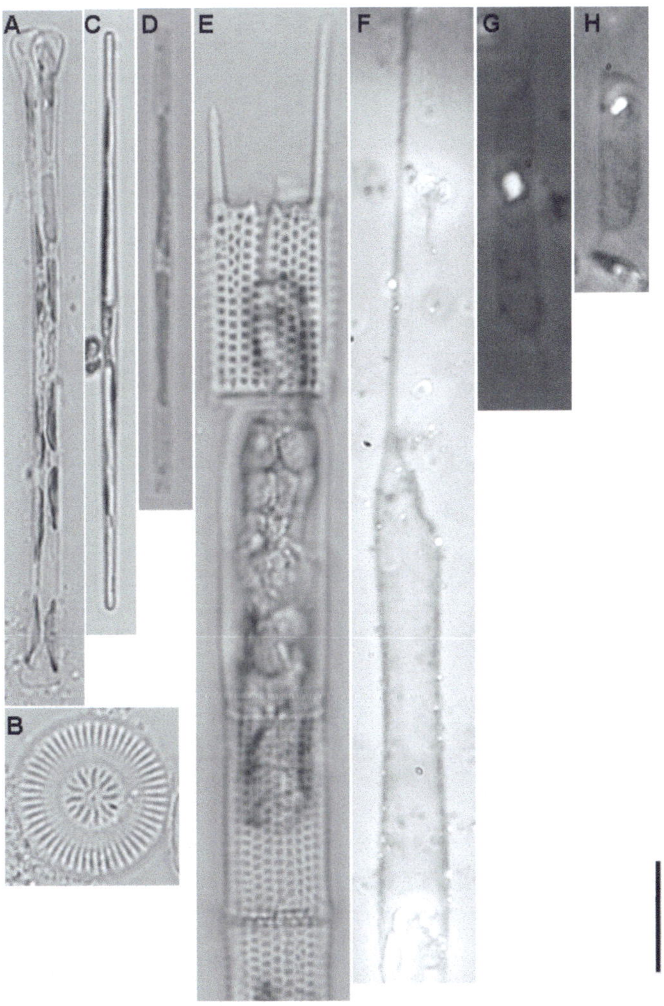

FIGURE 9 – SOME DIATOMS OBSERVED IN THE RIO VERDE RESERVOIR. A) *Asterionella formosa*; B) *Discostella stelligera*; C-D) *Fragilaria* spp.; E) *Aulacoseira granulata*; F) Urosolenia *longiseta*; G-H) *Urosolenia* spp. scale bar corresponds to 10 µm

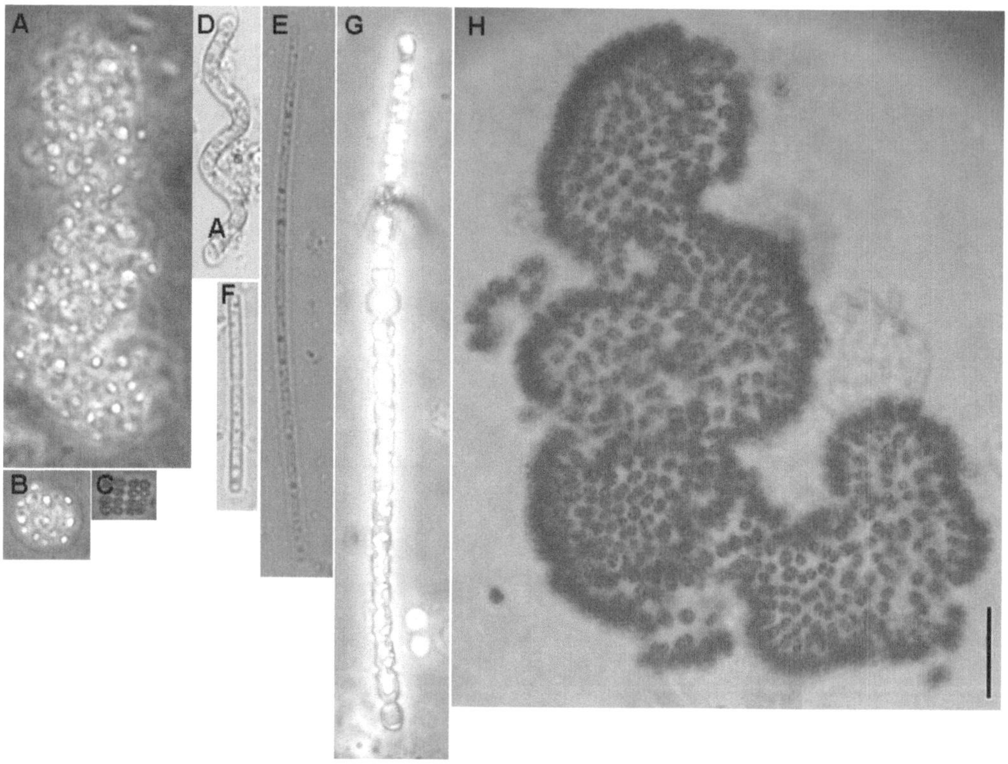

FIGURE 10 – SOME CYANOBACTERIA SPECIES OBSERVED IN THE RIO VERDE RESERVOIR. A-B) *Aphanocapsa* spp.; C) *Merismopaedia* sp.; D) *Arthrospira* sp.; E-F) FILAMENTOUS SPECIES NOT IDENTIFIED; G) *Dolichospermum planctonicum*; H) *Microcystis* sp..Scale = 10 μm

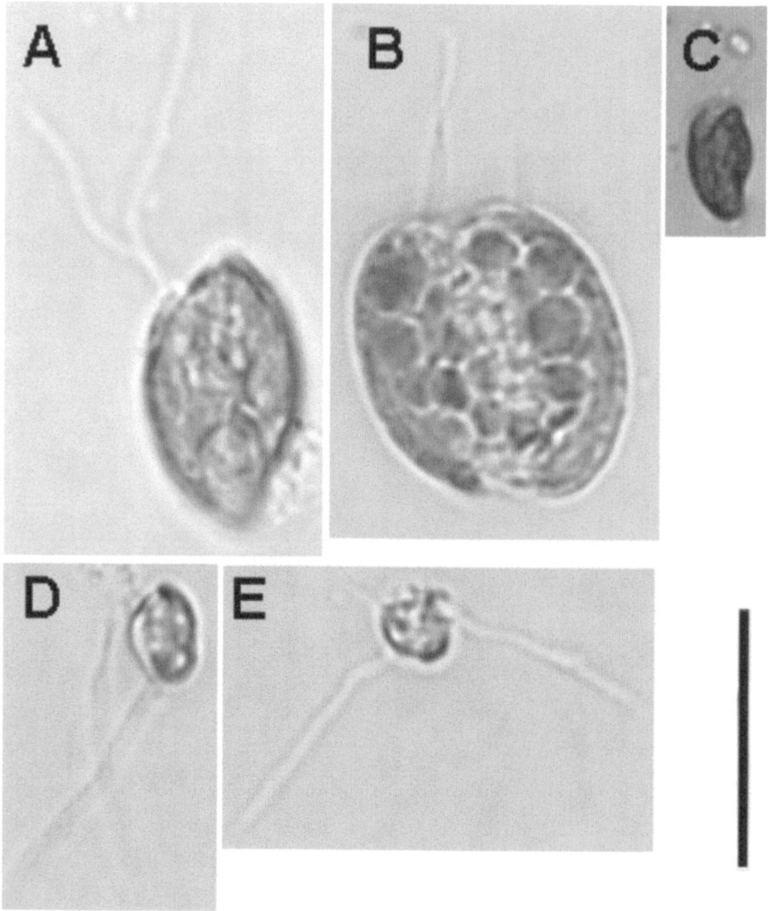

FIGURE 11 – FLAGELLATES RECORDED IN THE RIO VERDE RESERVOIR. A-C) SPECIES OF *Cryptomonas*; D-E) NANOFLAGELLATE.Scale = 10 μm

3.5 CELL DENSITIES OF THE PHYTOPLANKTON AND CHLOROPHYLL-A

Station R4 (dam of the reservoir)

Chlorophytes and nanoflagellates are the most commonn groups regarding relative contribution for the total phytoplankton density (Figure 12) throughout the sampling period.

In Figure 13, it can be observed the profile of the total phytoplankton density throughout the study period, varying from 1.6 to 9.2x10^4 cells/mL. The highest values occurred near the surface from February to the end of April 2009, with predominance of colonial Chlorococcales. At the end of April 2009, strong wind gusts prompted the circulation of the water column. Diatoms were abundant at the end of August 2008 and in January 2009. Colonial mucilagenous chlorophytes dominated the samples especially from February to April 2009.

The annual variation and vertical distribution of chlorophyll-a (Figure 14) were consistent with the recorded abundances of phytoplankton. Cell densities oscillated from 2.5 to 30.7 µg/L, and the highest values were concentrated in January to April 2009. In general, the concentrations of chlorophyll-a were lower than 8.0 µg/L.

Numerically important species during the study period were nanoflagellates, *Dictyosphaerium* spp., *Monoraphidium minutum*, *Cyclotella* spp., *Eutetramorus fottii*, *Crucigenia tetrapedia*, *Elakatothrix gelatinosa* and *Scenedesmus* spp. (Figure 15). In July and August of 2008 and 2009 and at the end of January 2009, diatoms became important, oscillating between 1,000 and 3,000 cells/mL; the majority of the species were *Aulacoseira ambigua*, *A. granulata*, *Asterionella formosa*, *Fragilaria crotonensis* and *Cyclotella* spp. (Figure 15). Cryptophyceans occurred in abundance usually greater than 500 cells/mL during the study period, reaching up to 1200 cells/mL in February 2009 (Figure 15).

The environmental descriptor species (according to REYNOLDS *et al.*, 2002) found in the Rio Verde Reservoir fell in six associations: **A**, **C**, **J**, **P**, **X1** and **Y**. They are detailed in Table 5.

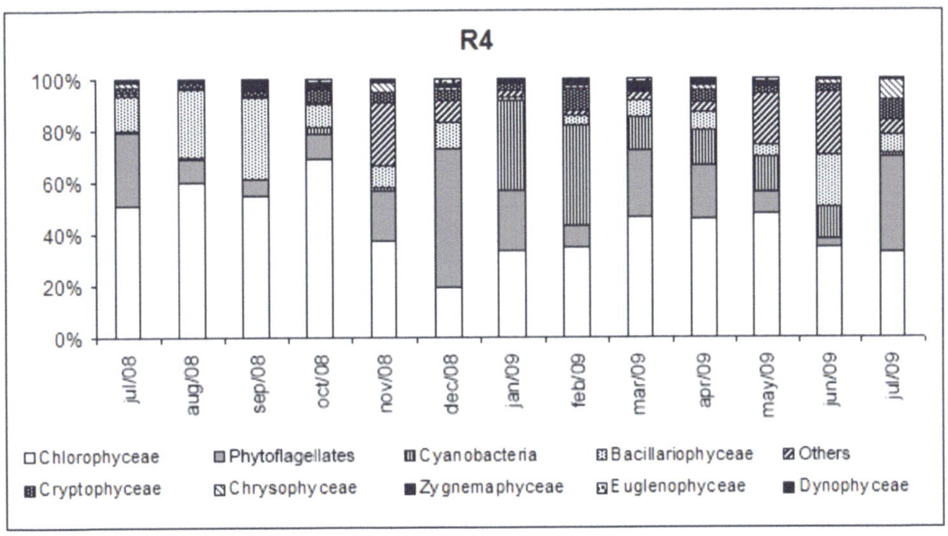

FIGURE 12 – CONTRIBUTION (%) OF THE DIFFERENT CLASSES OF PHYTOPLANKTON TO THE TOTAL DENSITY AT STATION R4 IN THE RIO VERDE RESERVOIR , JULY 2008 TO JUNE 2009

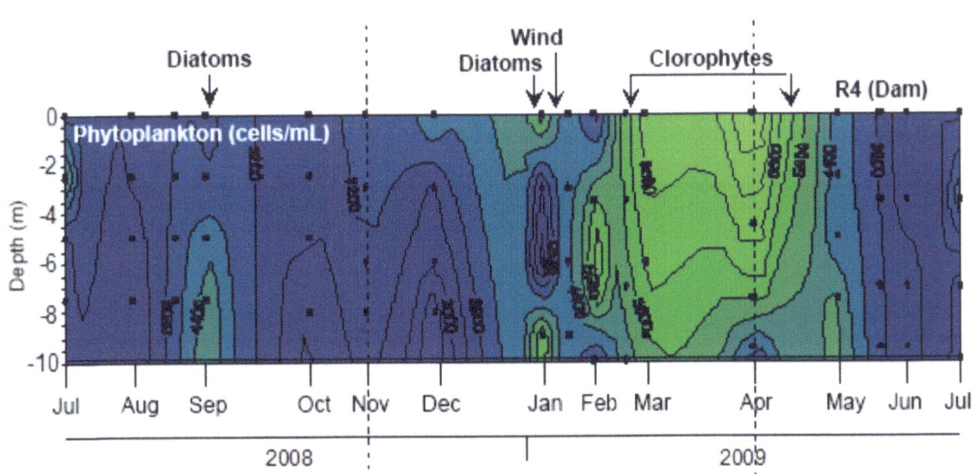

FIGURE 13 – SEASONAL VARIATION OF PHYTOPLANKTON (CELLS/ML) IN THE RIO VERDE RESERVOIR 2008/2009 AT STATION R4

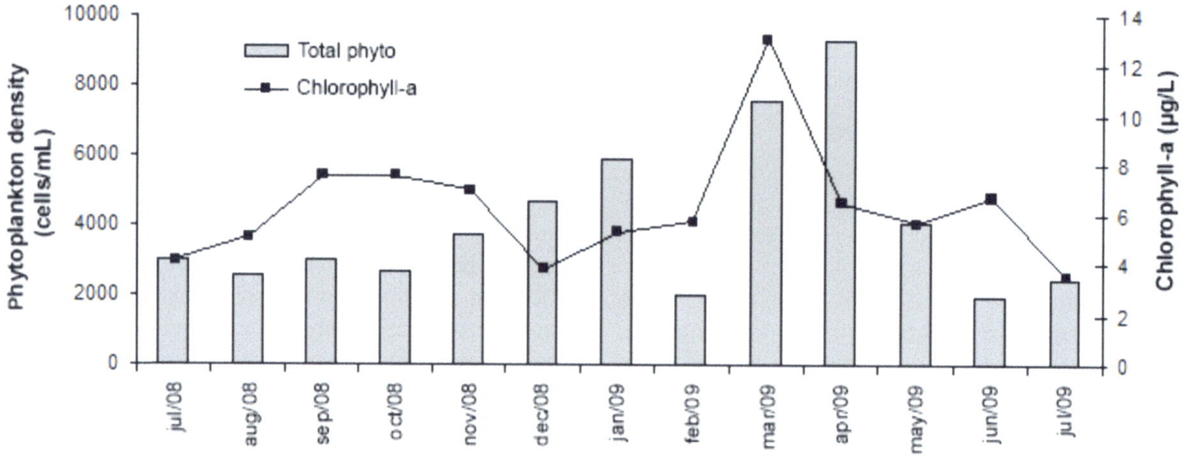

FIGURE 14 – ANNUAL VARIATION OF THE TOTAL DENSITY OF PHYTOPLANKTON (CELL/mL) AND CONCENTRATION OF CHLOROPHYLL-A (µg/L) IN THE RIO VERDE RESERVOIR FROM JULY 2008 TO JUNE 2009 AT STATION R4. ONLY SURFACE SAMPLES WERE CONSIDERED

TABLE 5 – PHYTOPLANKTON SPECIES DEFINED AS ENVIRONMENTAL DESCRIPTORS OF THE RIO VERDE RESERVOIR FOR EACH STATION OF THE YEAR AND THEIR RESPECTIVE FUNCTIONAL GROUPS

SPECIES	SEASON	FUNCTIONAL GROUP
Nanoflagelados	ALL	**X2**
Dictyosphaerium spp.	ALL	**F**
Monoraphidium minutum	ALL	**X1**
Cryptomonas sp.	Spring/Summer	**Y**
Scenedesmus ecornis	Spring/Aut	**J**
Cyclotella spp.	Spring/Summer/Winter	**A**
Asterionella formosa	Autumm/Winter	**C**
Dynobrion spp.	Summer	**E**
Fragilaria spp.	Summer	**P**
Aphanocapsa sp.	Autumm	**K**

Description of functional groups (from Reynolds *et al.*, 2002):
X2: Species inhabiting the mixed layer of shallow mesotrophic lakes, tolerant to stratification and sensitive to depth mix and herbivore
F: Colonial chlorophytes favored by clear epilimnion, sensitive to turbulence and CO_2 deficiency , tolerant to nutrient depletion.
X1: Algae of shallow, homogenous and enriched waters,. tolerant to stratification; suffer from nutrient depletion and herbivory
Y: Species present in small lakes with enriched water; tolerant to low luminosity intensity; limited by herbivory.
J: Organisms of shallow and enriched waters that require high luminosity.
A: Microalgae of lake with clear and homogeneous water; tolerate nutrient deficiency and are sensitive to high pH.
C: Algae of small to medium reservoirs with homogeneous and eutrophic water. Resist high luminosity and carbon deprivation; sensitive to the lack of silica and to stratification.
E: Species of small oligotrophic lakes, tolerant to nutrient shortage (with possibility of mixotrophy) and sensitive to CO_2 deficiency.
P: Species of eutrophic epilimnion. Tolerant to low light and carbon deficiency; sensitive to stratification and silica exhaustion.
K: Organisms adapted to nutrient deprivation, sensitive to deep mixture.

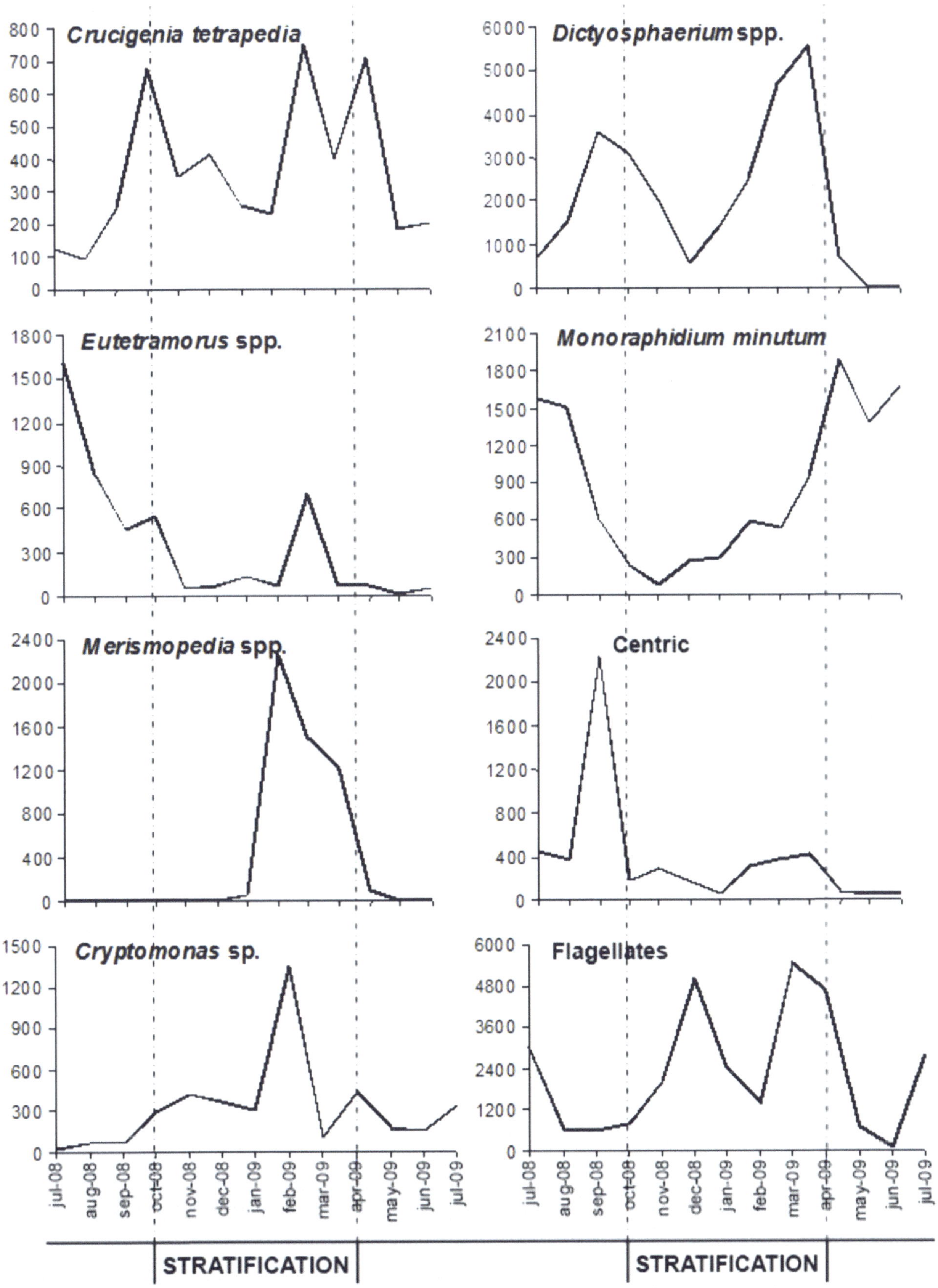

FIGURE 15 – DENSITY (CELLS/mL) AT STATION R4 OF SPECIES DEFINED AS ENVIRONMENTAL DESCRIPTORS OF THE RIO VERDE RESERVOIR, JULY 2008 TO JUNE 2009

Station R1 (lotic environment, next to the headwaters of the reservoir)

Since this station was relatively shallow, due to practical reasons only the data collected from surface waters were considered in the majority of the results.

In general, the annual variation of the phytoplankton classes and total densities of the phytoplankton in the station R1 were similar to those observed at station R4 (Figures 16 and 17). Chlorophytes and nanoflagellates were important throughout the sampling period, and chryptofyceans increased in number in September 2008 and June and July 2009 (Figure 16). The greatest densities of phytoplankton were recorded from December 2008 to March 2009, reaching 6,000 to 10,000 cells/mL, alternating with months of much lower density (2,000 to 6,000 cells/mL) in the other months (Figure 17). Chlorophyll-a varied from 2.8 to 12.7 µg/L, with high values being detected during January to April 2009 (Figure 17). Except for the more significant contribution of species like *Asterionella formosa* and *Dinobryon* sp., the descriptor species were identical to those found in station R4 (Figure 18).

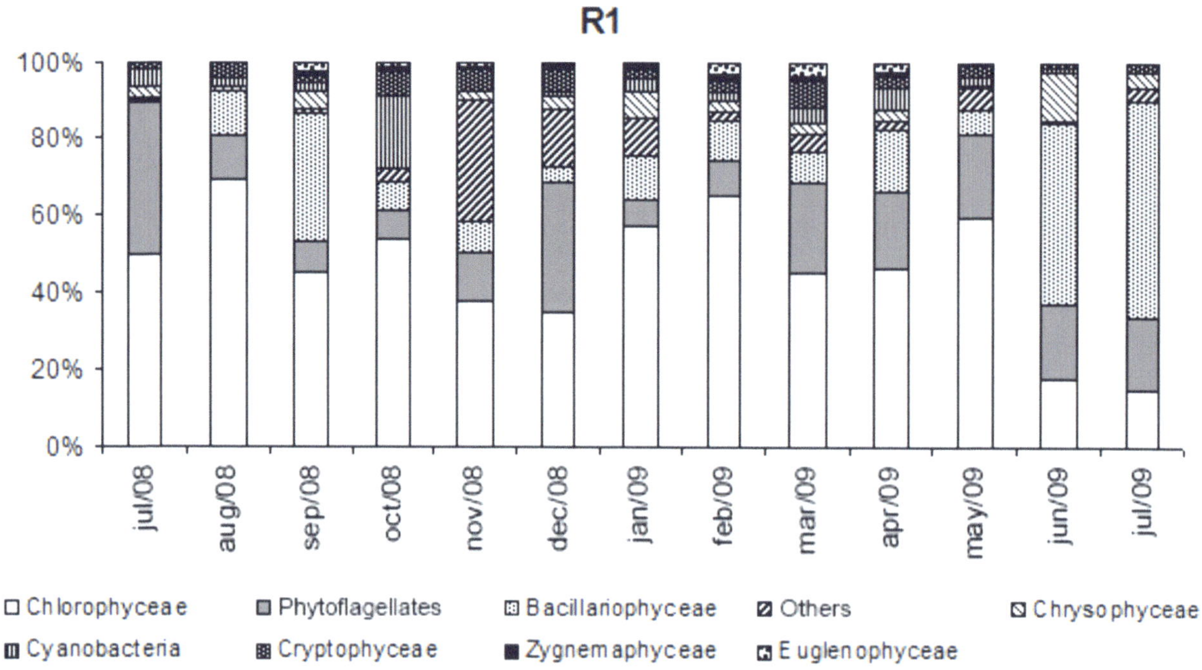

FIGURE 16 – CONTRIBUTION (%) OF THE DIFFERENT CLASSES OF PHYTOPLANKTON TO THE TOTAL DENSITY AT STATION R1 IN THE RIO VERDE RESERVOIR, JULY 2008 TO JULY 2009.

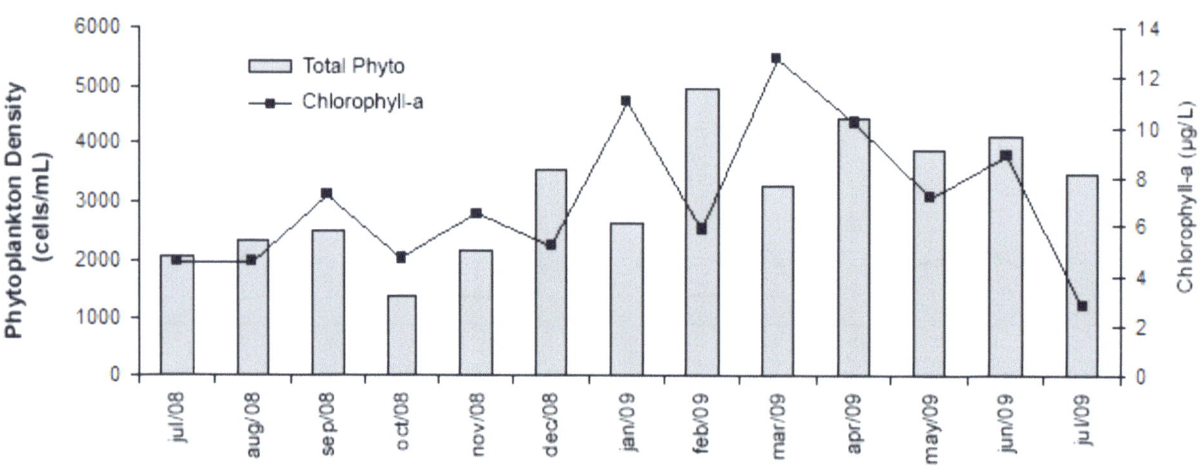

FIGURE 17 – ANNUAL VARIATION OF THE TOTAL DENSITY OF PHYTOPLANKTON (CELLS/mL) AND CHLOROPHYLL-A CONCENTRATION (µg/L) AT STATION R1 IN THE RIO VERDE RESERVOIR, JULY 2008 TO JULY 2009. ONLY SURFACE SAMPLES WERE CONSIDERED

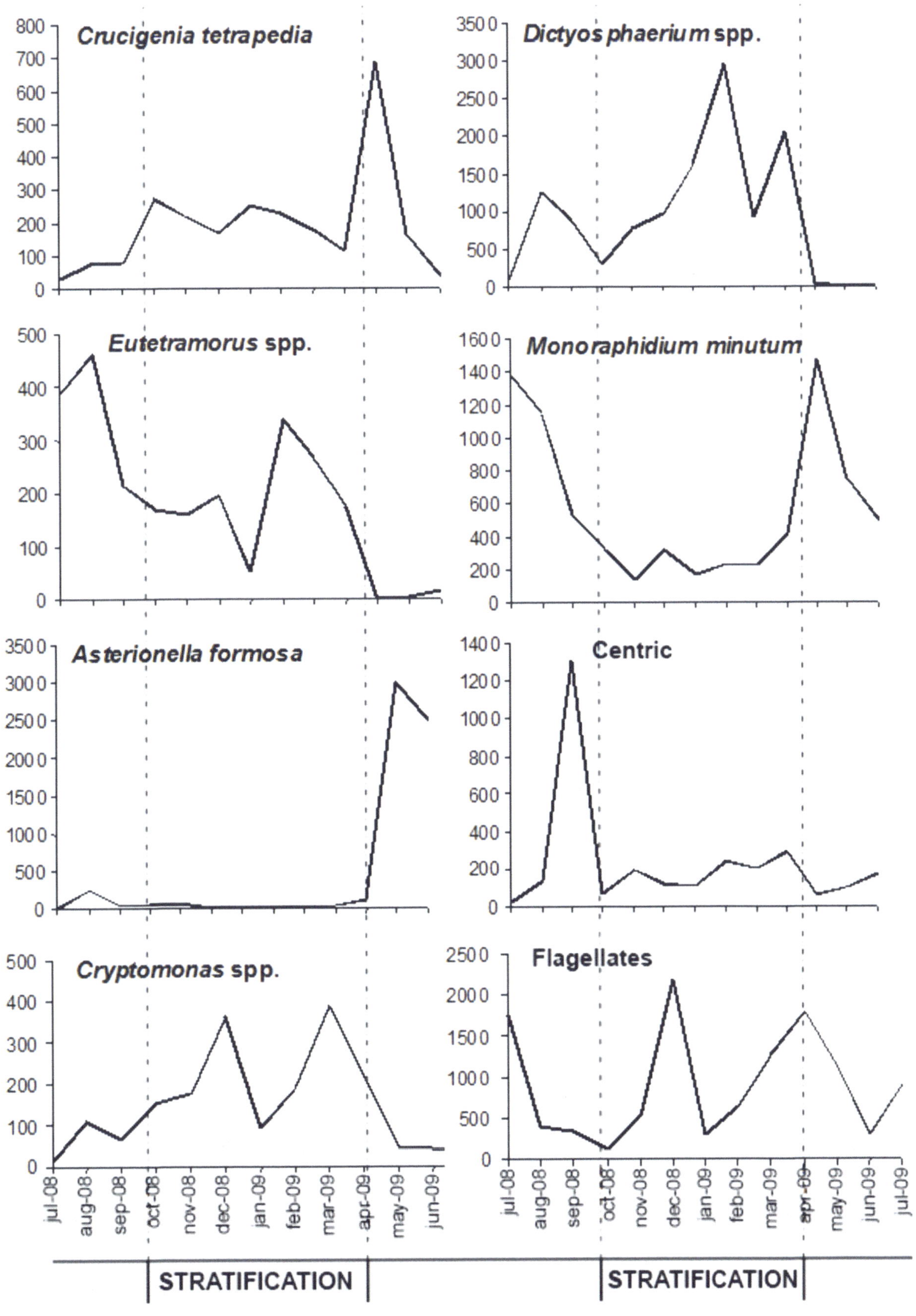

FIGURE 18 – CELL DENSITIES (CELLS/mL) OF INDICATOR SPECIES AT STATION R1 IN THE RIO VERDE RESERVOIR, JULY 2008 TO JULY 2009

4. DISCUSSION

Although the Rio Verde Reservoir is relatively shallow, with a maximum depth of 11 meters, in its lacustrine region presented environmental characteristics typical of a warm monomictic water body. This is likely because it is located between mountainranges, which reduces wind action and due to the lower influence of the tributary river flows. The establishment of the epilimnion began in November, lasting until the end of April. The hypolimnion presented a rather shallow depth, when compared to the volume occupied by the epilimnion. Even so, a persistent anoxic layer with elevated concentrations of ammonia (unpublished data) was detected throughout the period of thermal stratification in the summer. Therefore, Rio Verde Reservoir shares several physical characteristics with other small reservoirs of the Metropolitan Region of Curitiba, like the Piraquara, Passaúna and Parigot de Souza (ZEHNDER-ALVES, 2003; COQUEMALA, 2003; BORGES, 2006), and is thus subjected to the same environmental forces that regulate thermal stability. Wind regime and heat exchange with the atmosphere promote the circulation observed in the autumn and winter, and the subsequent heat convection generates a homogenous water column around 15 – 16°C. From the spring on, the stratification process of the water column begins, favored by the increases of the day length and the light intensity. These phenomena will bring important impacts on the phytoplankton, discussed further below. Obviously, this classification does not apply to the lotic and transitional zones of the Rio Verde Reservoir, where the water column is mostly homogenous and other factors are responsible for its physical and chemical dynamics, such as the entrance flow from the main river and its tributaries, re-suspension of bottom nutrients, etc. (KIMMEL & GROEGER, 1984).

Overall, nutrient concentrations in the Rio Verde Reservoir were low, which were reflected in the densities of phytoplankton and the species composition. The highest values of cell density were found during the summer, reaching 9,625 cells/mL, at least one order of magnitude lower than other eutrophic reservoirs in Brazil (CALIJURI *et al.*, 2002; GENTIL *et al.*, 2008). In comparison with other reservoirs in Paraná and other regions of Brazil, the nutrient concentrations in the Rio Verde Reservoir indicate the environment as mesotrophic, with limited input of nutrients, generating a phytoplanktonic biomass that is regulated by phosphorus and (or) nitrogen. In mesotrophic or oligotrophic tropical reservoirs located at similar latitudes, such as Piraquara I the same functional groups are found and are equally limited by nutrients, mainly phosphorus (ZEHNDER-ALVES, 2003).

Historically, the water parameters collected from the Rio Verde Reservoir every six month by IAP (since 1998) indicate a *moderately degraded* environment (IAP, 2004; IAP, 2009). According to this data, dissolved inorganic nitrogen varied from 0.04 to 0.54 mg/L and total inorganic phosphorus from 0.004 to 0.06 mg/L. The N:P ratio calculated from this database is 31 on average, indicating phosphorus as an important regulator of phytoplankton growth in the reservoir, according to the 106C:16N:1P Redfield ratio (1958). The data obtained from the monitoring conducted from March 2010 to September 2010, after the current study was conducted, showed nitrate and ammonia varying from 0.007 to 0.37 mg/L, and 0.07 to 0.4 mg/L, respectively, and reactive phosphorus from 0.003 to 0.01 mg/L (unpublished data). These levels are consistent with the historical data gathered by IAP.

During the study period, phytoplankton was dominated by colonial mucilaginous Chlorococcales, especially during spring and summer. In the classification of the functional groups of Reynolds *et al.* (2002), the chlorophytes of the Rio Verde can be categorized into **F, J** and **X1** groups. They are species adapted to periods of stratification with high light intensity; thus the production of mucilage reduces the rate of sedimentation, which is an important adaptation in thermally stratified waters. Various species of Chlorococcales, including those found in this work, have the ability to assimilate phosphorus and nitrogen from dissolved organic sources (BERMAN, 1990; BERMAN & BRONK, 2003). This can explain the growth of these species even in periods of lower nutrient levels. In Rio Verde, *Pediastrum* was abundant during the summer, when the environmental conditions favored its growth in relation to the other microalgae groups. Species of this genus grow better in environments with higher nitrogen concentration , or with a high N:P ratio, high temperatures and saturating solar radiation (BERMAN *et al.*, 1997).

Some diatoms grow better during the water circulation because the cells are re-suspended to the photic zone, simultaneously to the release of nutrients that were generated in the hypolimnion during summer and the remineralization of silicate. In this period of destratification and circulation in the autumn and winter, diatoms of groups **C** and **B**, more tolerant to the low temperature and reduced light radiation, increased their relative contribution to the phytoplankton community. Coincidentally, silicate concentrations also augmented during these periods, reaching values (>0.8mg/L) enough for supporting cell growth, though insufficient to promote a bloom. Current literature suggests a silicate value of 1 mg/L to support the intensive reproduction of diatoms like *Aulacoseira ambigua* and *Asterionella formosa* (REYNOLDS, 1997).

The dominant diatoms in the Rio Verde Reservoir were *Aulacoseira ambigua*, *Asterionella formosa* and *Fragilaria crotonensis*, presenting a higher specific growth rate than the other algae groups (SANDGREN, 1988). Additionally, even after temperature reduction from May to August 2009, the diatoms remained at a higher density than other phytoplankton species. Takano & Hino (1996), by comparing diatoms, chlorophytes and cryptophyceans under culture conditions, concluded that diatoms tend to keep their photosynthetic efficiency, saturating the photosynthesis even under low light and lower temperature, provided that they have optimal nutrient conditions, especially silicate and phosphate. *Asterionella formosa*, abundant in Rio Verde in August 2008 and in May and June 2009, also presents the capacity of storing excess phosphorus, usable during the periods of nutrient depletion (SANDGREN, 1988). Such eco physiological adaptations should also be important in the Rio Verde Reservoir, where temperature is lower in winter (15.5 – 17.7°C) and light intensity is approximately half (800 W/m2) of values recorded during summer.

Therefore, the chemical properties and the monomictic pattern of the Rio Verde reservoir established an environment more favorable to the dominance of colonial or solitary chlorophytes (functional groups **F, J, X1**), periodically altered by the re suspension of nutrients and higher development of diatoms (functional group **C**) during the autumn and winter circulation.

Possible causes of the *Cylindrospermopsis raciborskii* bloom in 2005

During this project, cyanobacteria were found only in very low concentrations during the counting in the inverted microscope. However, in April 2005, IAP technicians recorded a bloom of *Cylindrospemopsis raciborskii* in some calmer areas of the reservoir, which lasted for 8 to 10 days. Chlorophyll-a concentration was 21.7 µg/L and densities of these toxic cyanobacterium reached 96,000 cells/mL, a value much higher than the values recommended in regulation MS 2914/2011. Nutrients levels measured only once during the bloom were low, possibly due to cell uptake, suggesting the bloom was already in its decaying phase. Total phosphorus and dissolved nitrogen concentrations in the photic zone were 0.007 mg/L and 0.041 mg/L, respectively. Near the base of the epilimnion, ammonia values were one order of magnitude greater, reaching 0.37 mg/L. This was the first and only recorded cyanobacteria bloom in the reservoir since the beginning of the environmental monitoring program by IAP.

Cylindrospermopsis is a toxic cyanobacterium, that produces saxitoxin and cylindrospermopsin, and It is very common in eutrophic tropical water bodies. Recently, blooms of this cyanobacterium have been attributed to the global climatic changes, such as increasing temperatures, changes in the hydrodynamics of lentic environments, etc. Another alarming observation is the finding that the species is expanding to temperate regions, modifying the composition and abundance of the phytoplankton communities (PADISÁK, 1997; VIDAL & KRUK, 2008). Usually, strategies of diazotrophs for optimizing cell growth involve the production of nitrogen fixing heterocytes, accumulation of phosphorus in excess , active migration in the water column (through gas vesicles) along with tolerance to low light intensity, and affinity for ammonia and its filamentous thallus which reduces the herbivory by the zooplankton (PADISÁK, 1997; FERNANDES *et al*., 2005a). *C. raciborskii* grows slower than colonial cyanobacteria like *Microcystis* spp., which are generally abundant or dominant in most of the reservoirs in the world. Due to the adaptive strategies mentioned above, massive growths of the species have been frequently observed right after the bloom of other phytoplankton species, especially those with rapid intrinsic growth rate, such as *Microcystis* spp, or diatoms. For instance, Fernandes *et al.* (2005), studying the blooms of *Microcystis aeruginosa* in the Iraí Reservoir, recorded that *C. raciborskii* replaced *Microcystis* in May, when *Mycrocystis* growth decayed. The authors registered low concentrations of nutrients during the bloom, associated with consumption of nutrients by phytoplankton and the dry season. Finally, they concluded that *C. raciborskii* benefited from: the fixation of gaseous nitrogen (since the majority of the thricomes presented heterocytes); the utilization of accumulated phosphorus; elevated temperatures; and competitive exclusion of other species of the pelagic zone which were incapable of efficiently attaining growth during the period of nutrient depletion. On the other hand, as the temperatures lowered and the Iraí Reservoir began to experience events of whole water circulation, *C. raciborskii* was replaced by diatoms and *Microcystis* spp.. Tucci & Sant'Anna (2003) also detected high concentrations of *C. raciborskii* in the Lake of Garças, São Paulo, and associated its summer dominance to the thermal stability of the water column, the utilization of accumulated phosphorus, and adaptation to low light intensity. In their study, aquatic macrophytes were responsible by the depletion of phosphorus, while nitrogen was abundant during the whole period; therefore heterocytes were not observed.

Finally, due to the lack of field data , it becomes extremely difficult to assess the causes of the bloom of *C. raciborskii* in 2005. On the other hand, the review of ecological characteristics of *C. raciborskii*, as discussed above, can help to understanding the mechanisms underlying the intensive growth of this species in the Rio Verde Reservoir and in other water bodies in Paraná. In future, this information can help unraveling the environmental triggers that promote the blooms, in addition to providing useful tools for more effective monitoring measures of the Rio Verde Reservoir.

5. REFERENCES

ANDREOLI, C.; CARNEIRO, C. (2005) Gestão Integrada de Mananciais de Abastecimento Eutrofizados. Companhia de Saneamento do Paraná – SANEPAR e FINEP, 494 pp.

APHA – AMERICAN PUBLIC HEALTH ASSOCIATION. (1995) Standard methods for the examination of water and wastewater. Washington, DC.

BERMAN, T.; BRONK, D. A. (2003) Dissolved organic nitrogen: a dynamic participant in aquatic ecosystems. Aquat. Microb. Ecol., 31: p. 279-305.

BERMAN, T.; HADAS, O.; KAPLAN, B. (1977) Uptake and respiration of organic compounds and heterotrophic growth in *Pediastrum duplex* (Meyen). Freshwater Biology, 7:(5), October, p. 495-502.

BERMAN, T.; CHAVA, S. (1999) Algal growth on organic compounds. J. Plankt. Res., 21:(8), p. 1423-1437.

CALIJURI, M. C.; DOS SANTOS, A. C. A.; JATI, S. (2002) Temporal changes in the phytoplankton community structure in a tropical and eutrophic reservoir (Barra Bonita – S.P. – Brazil). J. Plankt. Res., 24:(7), p. 617-634.

COQUEMALA, V.; FERNANDES, L.F.; BRANDINI, N. (2006) Variação sazonal do fitoplâncton no reservatório Passaúna, Araucária, Paraná. Resumos da X reunião brasileira da SBFic, Espírito Santo.

DOMINGUES, L. L.; BRUNKOW, R. F.; WOSIACK, A. C. (2001) Composição dos grupos de fitoplâncton no Reservatório de Itaipú. Livro de Resumos do "VIII Congresso Brasileiro de Limnologia". João Pessoa, Paraíba, 2 a 6 de setembro.

DYTHAM, C. Choosing and Using Statistics: A Biologist's Guide. 2. ed., Blackwell Science Ltd.: Malden, MA, USA, 2003, 264 pp.

ESTEVES, F. A. (1998) Fundamentos de Limnologia. 2. ed., Interciência: São Paulo, 602 pp.

FERNANDES, L. F.; LAGOS, P. D.; WOSIACK, A. C.; PACHECO, C.

V.; DOMINGUES, L.; ZENHDER-ALVES, L.; COQUEMALA, V. (2005) Comunidades Fitoplanctônicas em Ambientes Lênticos. In: Gestão Integrada de Mananciais de Abastecimento Eutrofizados. ANDREOLI, C. V.; CARNEIRO, C. Curitiba: Sanepar Finep, p. 303-366.

FERNANDES, L. F.; WOSIACK, A. C.; PACHECO, C. V.; DOMINGUES, L.; LAGOS, P. E. D. (2005a) Cianobactéria e cianotoxinas. In: Gestão Integrada de Mananciais de Abastecimento Eutrofizados. ANDREOLI, C. V.; CARNEIRO, C.. Curitiba: Sanepar Finep, p. 367-388.

GENTIL, R. C.; TUCCI, A.; SANT'ANNA, C. L. (2008) Dinâmica da comunidade fitoplanctônica e aspectos sanitários de um lago urbano eutrófico em São Paulo, SP. Hoehnea, 35:(2), p. 265-280.

IAP – Instituto Ambiental do Paraná. (2004) Monitoramento da qualidade das águas dos reservatórios do estado do Paraná, no período de 1999 a 2004. Curitiba, 13 p.

IAP – Instituto Ambiental do Paraná. (2009) Monitoramento da qualidade das águas dos reservatórios do estado do Paraná, no período de 2005 a 2008. Curitiba, 120 p.

KIMMEL, B. L.; GROEGER, A. W. (1984) Factors controlling primary production in lakes and reservoirs. In: Lake and reservoir management. Proc. North American Lake Management Society Symposium. EPA 440/5/84-001. U.S.E.P.A., Washington D.C., 604 pp.

KÖEPPEN, W. (1936) Das Geographische System der Klimatologie. 44 pp.

KOMÁRKOVÁ, J.; LAUDARES-SILVA, R.; SENNA, P. A. C. (1999) Extreme morphology of Cylindrospermopsis raciborskii (Nostocales, Cyanobacteria) in the Lagoa do Peri, a freshwater coastal lagoon, Santa Catarina, Brazil. Algological Studies, 94: p. 207-222.

NUSCH, E. A. (1980) Comparation of diferent methods for chlorophyll and phaeopigment determination. Arch. Hydrobiol. Beih. Sttutgart, 14: p. 14-36.

PADISÁK, J. (1997) Cylindrospermopsis raciborskii (Woloszynnska) Seenayya et Subba Raju, an expanding, highly adaptive cyanobacterium: worldwide distribution and review of its ecology. Archiv fur Hydrobiology, 107: p. 563-593.

REBOUÇAS, A. C., BRAGA, B.; TUNDISI, J. G. (2002) Águas doces no Brasil: capital ecológico, uso e conservação. 2 ed., São Paulo: Escrituras Editora, 703 p.

REYNOLDS, C. S. (1997) Vegetation process in the pelagic: a model for ecosystem theory. In: KINNE, O. Excellency in ecology. Germany, Ecology Institute, 371 p.

REYNOLDS, C. S. (2006) Ecology of phytoplankton. Cambridge University Press, Cambridge, 535 p.

REYNOLDS, C. S.; HUSZAR, V.; KRUK, C.; NASELLI-FLORES, L.; MELO, S. (2002) Towards a functional classification of the freshwater phytoplankton. J. Plankton Res., 24, p. 417-428.

SANDGREN, C. D. (1988) Growth and reproductive strategies of freshwater phytoplankton. Cambridge University Press, Oxford, 442p.

TAKANO, K.; HINO, S. (1996) The effect of silicon concentration on replacement of dominant diatom species in a silicon-rich lake. Jpn. J. Limnol., 57, p. 153-62.

TUCCI, A.; SANT'ANNA, C. L. (2003) Cylindrospermopsis raciborskii (Woloszynska) Seenayya & Subba Raju (Cyanobacteria): variação semanal e relações com fatores ambientais em um reservatório eutrófico, São Paulo, SP, Brasil. Rev. bras. Bot., 26:(1), Mar.

UTERMÖHL, H. (1958) Zur Vervollkommnung der quantitativen Phytoplankton-Methodik. Komitee fur Limnologische Methoden, (9).

VENRICK, E. L. (1978) How many cells to count? In: SOURNIA, A. (Ed.) Phytoplankton manual. Paris: Unesco, p. 167-180.

VIDAL, L.; KRUK, C. (2008) Cylindrospermopsis raciborskii (Cyanobacteria) extends its distribution to Latitude 34°53'S: taxonomical and ecological features in Uruguayan eutrophic lakes. Pan-American Journal of Aquatic Sciences, 3:(2), p. 142-151.

ZEHNDER-ALVES, L. (2003) Fitoplâncton do reservatório Piraquara I, Piraquara, Paraná: Fatores reguladores das dinâmicas temporal e vertical. Dissertação (Mestrado em Botânica) – Setor de Ciências Biológicas, Universidade Federal do Paraná, Curitiba.

CHAPTER 14

ZOOPLANKTON

Moacyr Serafim Júnior
Gilmar Perbiche-Neves
Cláudia Fileto
Danilo Augusto de Oliveira Naliato

ZOOPLANKTON

Moacyr Serafim Júnior
Gilmar Perbiche-Neves
Cláudia Fileto
Danilo Augusto de Oliveira Naliato

SUMMARY

The spatial and temporal variation of the zooplankton was studied over an annual cycle at three collection points along the reservoir in the fluvial, transition and lacustrine regions. Altogether, 63 zooplankton species were identified, with 40 Rotifera, 15 Cladocera, and eight Copepoda. Among the Rotifera, 23 species were considered to be rare (<25%), 11 were considered as frequent (25-75%), and seven as constant (>75%). Regarding the Cladocera, seven species were considered rare, seven frequent, and just one was considered as constant. One Copepoda species was considered rare, four were frequent, and three constant. An increase in diversity of the zooplankton groups was found during the spring and summer months. The most abundant Rotifera species were: *Conochilus unicornis*, *Kellicotia bostoniensis,* and *Polyarthra vulgaris*, all above 18,000ind.m^{-3}. Among the Cladocera, *Bosmina hagmanni* was the most abundant species, with an average value of 20,313ind.m^{-3}, followed by *Ceriodaphnia cornuta* var. *rigaudi* with 3,912ind.m^{-3}. Among the Copepoda, nauplii of Cyclopoid prevailed, with an average value of 47,756ind.m^{-3}. For adults, higher abundances (above 1,300ind.m^{-3}) were observed for *Thermocyclops minutus*, *Thermocyclops decipiens,* and *Microcyclops anceps*; for Calanoida, *Notodiaptomus spiniger* was the most abundant. The richness of species of zooplankton found in the present study was consistent with that expected for oligo-mesotrophic reservoirs.

KEYWORDS

Rotifera, cladocera, copepoda, reservoir, trophic state.

1. INTRODUCTION

Zooplankton are a phylogenetically diverse community, encompassing groups of protozoans and metazoans with body sizes ranging from a few micrometers to several millimeters (BOZELLI & HUSZAR, 2003). In continental aquatic environments, zooplankton are mainly represented by Protozoa, Rotifera, Cladocera and Copepoda. These groups are generally listed as zooplankton species and they prevail in artificial reservoirs. Occasionally, some other groups are found, such as Diptera (Chaoboridae) and Turbellaria (ROCHA *et al.*, 1999).

These organisms exhibit significant variety in dietary habits, actively participating in nutrient cycling and the maintenance of trophic chains in aquatic environments. Some are generalist filters that may have high filtration rates, while others are able to actively select their prey. Most often, phytoplankton are the main food source of zooplankton but zooplankton may be carnivores and dentritivorous. They also serve as the food source for other invertebrates and fish of different classes and mainly small sizes (MARGALEF, 1983; TUNDISI & MATSUMURA-TUNDISI, 2008).

Figure 1 presents a simplified diagram of the relationship of the main zooplankton groups with other levels in the aquatic trophic chain. Zooplankton are the main factor responsible for the flow of energy and matter between primary producers and higher order consumers. Also outlined in Figure 1 are trophic chain regulation mechanisms, such as "top down and bottom up," and human interventions, such as the construction of reservoirs that directly affect the dynamics of the physical, chemical, and biological variables of water bodies.

The spatial structure and temporal dynamics of zooplankton in artificial reservoirs is related to longitudinal and lateral gradients, where the highest density and diversity are expected in transition zones between the main river and the reservoir and between tributaries and the reservoir (backwater zones) (MARZOLF, 1990; VAN DEN BRINK *et al.*, 1994; SERAFIM-JUNIOR, 2002). Other driving forces, such as climate and residence time in the water, can interfere with the structure and dynamics of this community in reservoirs (TUNDISI *et al.*, 1991).

Over the last several decades, studies were conducted on the zooplankton in Brazilian reservoirs. Most analyses concentrate on the South and Southeast regions, in the hydrographic basins of the Paraná, Paranapanema and Iguaçu Rivers.

Lansac-Tôha *et al.* (2005) studied the richness and abundance of the zooplankton in 31 reservoirs in the state of Paraná, particularly the Vossoroca, Passaúna, Parigot de Souza, Guaricana and Iraí, located in the Metropolitan Region of Curitiba. Other studies on the longitudinal distribution of zooplankton biomass were carried out by: Bonecker *et al.* (2006); Velho *et al.* (2006); Bettler & Bonecker (2007); Bini *et al.* (2007); and Bonecker *et al.* (2007).

studies of note include thoseRiver on A series of environmental and social studies were carried out at the Iraí Reservoir, due to its importance in Curitiba's water supply. Among them, we would like to highlight the study of the zooplankton community performed by Serafim-Júnior *et al.* (2005, 2010), Perbiche-Neves *et al.* (2007), and Ghidini *et al.* (2009). The results obtained from these studies showed that assemblages of rotifers, cladocerans and copepods present an irregular variation in relation to ecological cha-

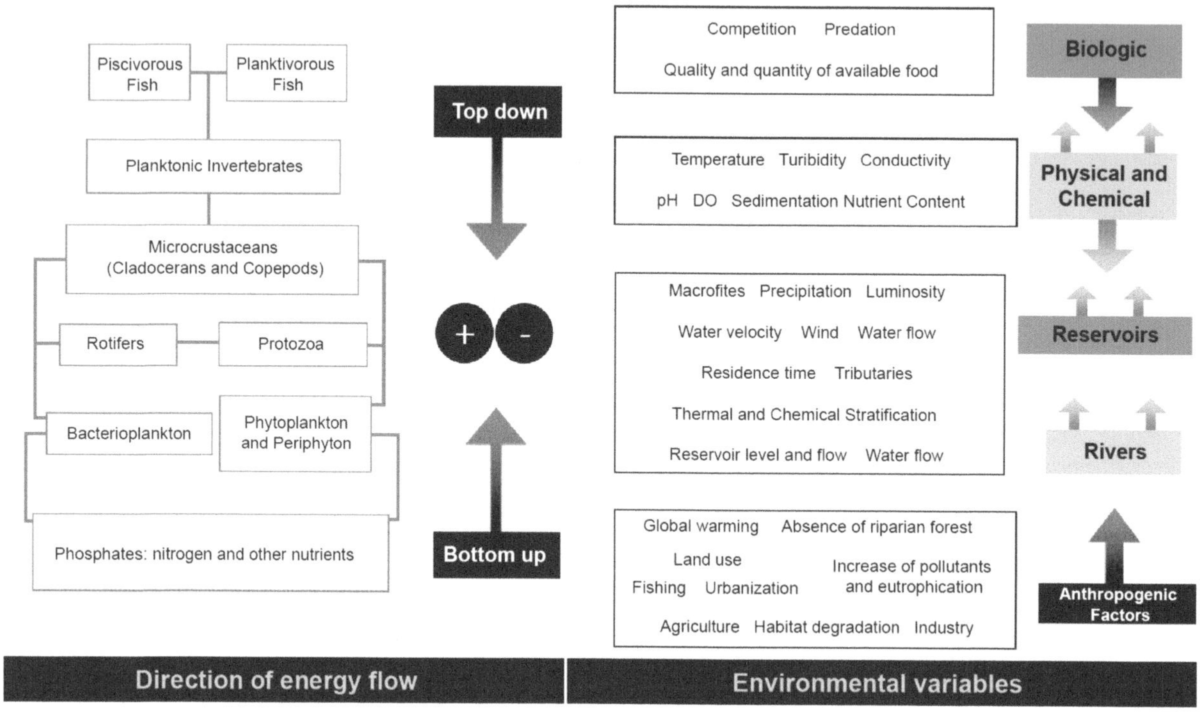

FIGURE 1 – SIMPLIFIED DIAGRAM OF THE AQUATIC TROPHIC CHAIN HIGHLIGHTING THE DIRECTION OF ENERGY FLOW AND THE MAIN ENVIRONMENTAL VARIABLES

racteristics. Their analyses associated this irregularity with the eutrophic state and recent formation of the reservoir.

In the Southeast, studies of note include those developed in cascade reservoirs in the Paranapanema River focusing on the composition and diversity of the zooplankton (SAMPAIO *et al.*, 2001; NOGUEIRA 2001; PANARELLI *et al.*, 2003, NOGUEIRA *et al.*, 2006; SARTORI *et al.*, 2009; PERBICHE-NEVES & NOGUEIRA, 2010).

Among the reservoirs with high anthropogenic influence, one of the most studied is the reservoir in Barra Bonita, São Paulo, in the middle section of the Tietê River. In this reservoir, Matsumura-Tundisi & Tundisi (2005) noted a high diversity index of the zooplankton giving rise to the perception that moderate eutrophication can lead to greater richness, while in extreme cases of eutrophication it becomes irregular and poor. The zooplankton biomass was assessed in different trophic states by Sendacz *et al.* (2006) in supply reservoirs in the upper Tietê basin, similar to the one assessed herein (comparable in size, shape, depth, among other characteristics). The effects of eutrophication on the seasonal distribution, size and biomass of zooplankton were analyzed in the Pampulha Reservoir, Minas Gerais, by Pinto-Coelho *et al.* (1998, 2005).

However, an ecological study of the zooplankton community of the Rio Verde reservoir has not yet been developed; our study represents the first analysis of zooplankton in this reservoir. As such, understanding the structure and the dynamics of this community in different sections of the reservoir is an important aid in water monitoring and management.

2. MATERIALS AND METHODS

Qualitative and quantitative samples were collected along the central axis of the reservoir between July 2008 and June 2009, at sampling points R01, R02 and R04, which correspond to lotic, intermediate and lentic areas, respectively (Figure 2). Thirty-six samples were collected with a 64µm mesh conical net in vertical hauls in the water column (one meter from the bottom to the surface). The volume filtered in each haul was calculated using the formula for the volume of a cylinder, considering the haul distance (m) as the height. The material retained in the net was fixed with a 4% formaldehyde solution.

In the laboratory, samples were studied using optical and stereoscopic microscopes, through 1 mL sub-samples in acrylic cuvettes and a Sedgewick-Rafter camera, until a minimum of 200 adult individuals of each group (Rotifera, Cladocera and Copepoda) and 100 individuals in the nauplius form for copepods (Cyclopoida and Calanoida) was reached. The abundance values were converted to individuals per cubic meter. A specialized bibliography for taxa identification was used (Rotifera: KOSTE, 1978; Cladocera: SMIRNOV, 1986; KORÍNEK, 1987; ELMOOR-LOUREIRO, 1997; HOLLWEDEL *et al.*, 2003 and ELMOOR-LOUREIRO *et al.*, 2004; ELMOOR-LOUREIRO, 2007; and Copepoda: SENDACZ & KUBO, 1982; REID, 1985; MATSUMURA-TUNDISI, 1986; DUSSART & DEFAYE, 1995; ROCHA, 1998; SANTOS-SILVA, 2000; PAGGI, 2001).

Some basic ecological attributes of the zooplankton community were assessed, including: composition, frequency (later the Dajoz index was used), abundance, richness, diversity (alpha, with Shannon Wiener index), and the correlation between abundance and environmental variables. Higher taxonomic levels, such as families, orders, or even, genera not identified to species, were not included because they are treated separately with specific diversity indexes.

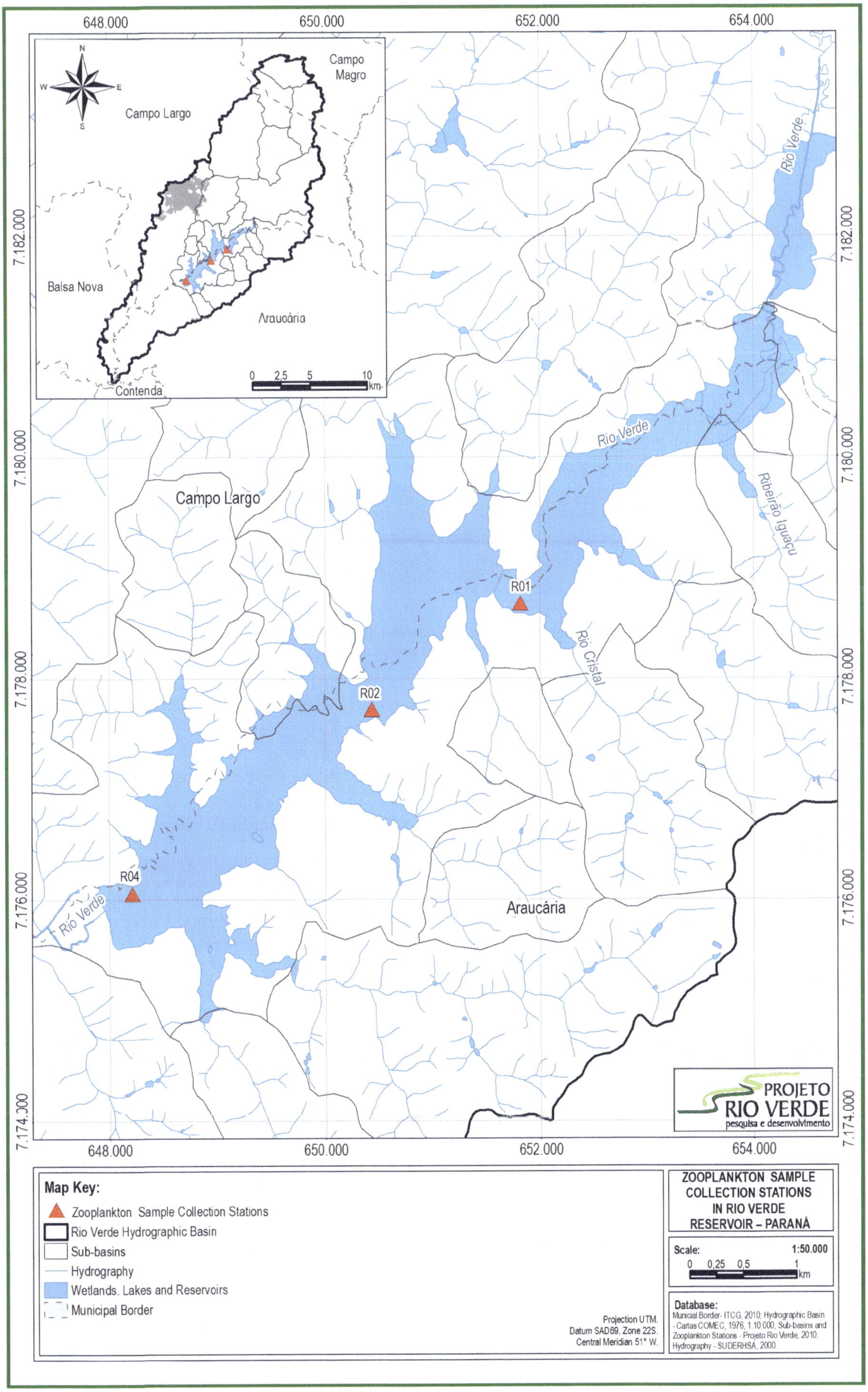

FIGURE 2 – ZOOPLANKTON SAMPLE COLLECTION POINTS AT THE RIO VERDE RESERVOIR. IN THE TEXT, THE POINTS ARE CALLED R1, R2 AND R4

Abundance data, after transformation in log (X+1), presented normal distribution and were processed using the bifactorial analysis of variance (ANOVA two-way, with factors: sample points and months). Several taxa that were not identified in some months throughout the study period were excluded from the analysis as they did not meet the test conditions.

The abundance was also correlated with environmental variables (morphometric, physical, chemical, and biological), after being logarithmized, by canonical correspondence analysis (CCA) using 1,000 permutations. Two-dimensional graphs were generated; one for each canonic variable (CV), showing significant correlations, with a maximum limit of significance level at 0.17.

The following environmental variables were used in the CCA: depth, water temperature, pH, dissolved oxygen, conductivity, transparency, total nitrogen, total phosphorous, content of total suspended solids, chlorophyll-a and total phytoplankton (data of the two last variables were taken from Chapter 13, Ecology of the phytoplankton in the Rio Verde Reservoir).

Finally, the abundance of the zooplankton groups, the richness and diversity were correlated with rainfall through Pearson correlation, for parametric data. Rainfall data were taken from four locations within the Rio Verde Hydrographic Basin close to the dam (municipalities of Campo Largo, Araucária and Almirante Tamandaré) and the information was provided by the *Instituto de Águas do Paraná* (Water Institute of Paraná; formerly SUDERHSA). ANOVA, CCA and Pearson correlation were processed using the freeware R Cran Project (2010).

3. RESULTS

3.1 ECOLOGICAL ATTRIBUTES

Altogether, 63 zooplankton species were recognized, with 40 rotifers, 15 cladocerans and eight copepods. The identified taxa are listed in Table 1, with respective mean abundances, standard deviation and frequency of occurrence.

Regarding the frequency of occurrence in the samples, of the 40 rotifer species, 23 were considered rare (<25%), 11 frequent (25-75%), and seven constant (>75%). Among the cladocerans, seven species were rare, seven were frequent, and only one was constant. Among the copepods, one species was considered rare, four frequent, and three constant.

The rotifer species that had a higher mean abundance were *Conochilus unicornis, Kellicotia bostoniensis* and *Polyarthra vulgaris*, all above 18,000ind.m^{-3}. *Keratella cochlearis, Conochilus coenobasis* and *Collotheca* sp. were also abundant with means higher than 6,400ind.m^{-3}. Among the cladocerans, *Bosmina hagmanni* was the most abundant species, with a mean value of 20,313ind.m^{-3}, followed by *Cerioaphnia cornuta* var. *rigaudi* with 3,912ind.m^{-3}. Among the copepods, the larvae, young, and adult forms of Cyclopoida were more abundant than Calanoida. A high mean value of 47,756ind.m^{-3} was observed for nauplii of Cyclopoida. Among adult individuals, the higher abundances (above 1,300ind.m^{-3}) were observed for *Thermocyclops minutus, Thermocyclops decipiens* and *Microcyclops anceps.* Of the two species of Calanoida, *Notodiaptomus spiniger* was the most abundant.

Figure 3 shows images of some microcrustaceans species found during the study (Cladocera and Copepoda).

Table 2 presents mean values of richness, diversity, equitability, and total abundance. Rotifera presented the highest mean richness. The values for Cladocera and Copepoda were similar, although Cladocera showed higher richness levels. Diversity values were higher for rotifers and copepods and the copepods were more homogenous in relation to equitability (Table 2). Finally, rotifers were also the most abundant, followed by cladocerans and copepods; we note that the immature forms of copepods were not included in this calculation although they were the prevailing individuals, as shown in Table 1.

The zooplankton groups presented significant variability of richness, diversity and equitability across the points and months sampled. In general, there was a greater number of species during the spring and summer months and this abundance was more distinct for cladocerans and copepods. The same trend was verified for diversity, with low values found during the winter of the first year sampled. An opposite trend was verified for equitability (Figure 4) and rotifers were the least homogenous compared to all microcrustaceans.

The analysis of variance used to identify spatial differences in relation to abundance of the zooplankton groups indicated low heterogeneity. A significant difference was verified for: two rotifer species; the total abundance of rotifers; and nauplii, copepodites and adult individuals of Calanoida (Table 3). All taxa mentioned above showed a decreasing trend toward the lacustrine zone (Figure 5).

Mainly for copepods, significant differences were found among the study months. The rotifers, *Conochilus coenobasis*, *Collotheca* sp.1, and *Keratella lenzi,* showed a strong decreases during winter and spring months (Figure 6a). *Ceriodaphnia cornuta* var. *cornuta* was less abundant in winter and remained abundant between November 2008, and April 2009 (Figure 6b). *Microcyclops anceps* and *Thermocyclops decipiens* showed similar variations up to December 2008, when their densities began to alternate (Figure 6b). Nauplii and copepodites of Cyclopoida presented higher values in the summer months, while the adult individuals decreased in July, February, and May (Figure 6c).

3.2 INFLUENCE OF ENVIRONMENTAL VARIABLES ON THE ZOOPLANKTON

Significant correlations were not found between rainfall and ecological attributes of the zooplankton groups (Figure 7), although the abundance and equitability of cladocerans and phytoplankton showed high level of correlation, near the minimum significance level. Rainfall negatively affected the abundance of cladocerans and phytoplankton.

TABLE 1 – LIST OF TAXA, TAXA ABBREVIATION (ABBR.), FREQUENCY (FR), MEAN ABUNDANCE (MEAN), AND STANDARD DEVIATION (SD) FOR THE STUDIED MONTHS. HIGHEST VALUES ARE IN BOLD

TAXA	ABBR.	FR(%)	MEAN	SD
Rotifera				
Anaureopsis navicula Rousselet, 1911	Anav	50	367	727
Ascomorpha ecaudis Pety, 1850	Aeca	11	197	758
Ascomorpha saltans Bartsch, 1870	Asal	72	2132	3025
Ascomorpha ovalis Carlin, 1943	Aova	8	31	126
Asplanchnella sieboldi (de Beauchamp, 1951)	Asie	3	2	14
Brachionus angularis Gosse, 1851	Bang	3	18	106
Brachionus falcatus Zacharias, 1898	Bfal	42	1532	3278
Collotheca sp1	Colsp1	86	**6434**	7655
Collotheca sp2	Colsp2	3	56	338
Conochilus coenobasis Skorikov, 1914	Ccoe	**100**	**8391**	7800
Conochilus unicornis Rousselet, 1892	Cuni	**97**	**26175**	25906
Epiphanes clavatula (Ehrenberg, 1832)	Epcla	3	8	48
Euchlanis dilatata Ehrenberg, 1832	Edil	3	4	23
Filinia longiseta (Ehrenberg, 1834)	Flon	3	12	69
Filinia opoliensis (Zacharias, 1898)	Fopo	22	325	939
Gastropus hyptopus (Ehrenberg, 1838)	Ghyp	50	827	1418
Hexarthra intermedia brasiliensis (Hauer, 1953)	Hintbra	39	3175	7455
Kellicottia bostoniensis (Rousselet, 1908)	Kbos	**100**	**20675**	25816
Keratella americana Carlin, 1943	Kame	**94**	3297	4105
Keratella cochlearis (Gosse, 1851)	Kcoc	**100**	**9963**	11040
Keratella lenzi (Hauer, 1953)	Klen	75	3222	5520
Keratella tropica (Apstein, 1907)	Ktro	39	342	725
Lecane bulla (Gosse, 1851)	Lbul	3	5	28
Lecane cornuta (O. F. Muller, 1786)	Lcor	3	2	14
Lecane luna (Muller, 1776)	Llun	3	19	117
Lecane lunaris (Ehrenberg, 1832)	Lris	3	2	12
Lecane stichaea Harring, 1913	Lsti	3	7	44
Notommata sp.	Nosp	3	5	28
Plationus patulus patulus (O. F. Muller, 1786)	Ppa	31	571	1408
Plationus patulus macracanthus (Daday, 1905)	Ppama	3	2	14
Polyarthra vulgaris Carlin, 1943	Pvul	**94**	**18663**	32192
Synchaeta cf. *stylata* Wierzejski, 1893	Spect	67	3755	18683
Synchaeta sp.	Sysp	11	90	338
Trichocerca bicristata (Gosse, 1887)	Tbic	6	15	66
Trichocerca capucina (Wierzejski e Zacharias, 1893)	Tcap	17	52	121
Trichocerca cylindrica chattoni (Beauchamp, 1907)	Tcych	28	245	610
Trichocerca elongata (Gosse, 1886)	Telo	3	12	61
Trichocerca pusilla (Lauterborn, 1898)	Tpus	11	33	96
Trichocerca similis grandis (Hauer, 1965).	Tsigra	6	40	199
Trichocerca rattus (O. F. Muller, 1776)	Trat	3	5	30
Bdeloidea	Bde	33	123	215
Cladocera				
Alona guttata Sars, 1862	Agut	3	7	41
Alona monacantha Sars, 1901	Amon	3	1	6
Bosmina hagmanni Stingelin, 1904	Bhag	**94**	**20313**	25488
Bosminopsis deitersi Richard, 1895	Bdei	3	1	6
Daphnia gessneri Herbst, 1967	Dges	72	1106	1563
Daphnia parvula Fordyce, 1901	Dpar	53	702	1299
Ceriodaphnia cornuta Sars, 1885	Ccorc	61	1532	3809

Ceriodaphnia cornuta var. *rigaudi*	Ccorri	64	**3912**	9947
Ceriodaphnia reticulata (Jurine, 1820)	Cret	47	363	505
Ceriodaphnia silvestrii Daday, 1902	Csil	11	25	74
Diaphanosoma birgei Korinek, 1981	Dbir	22	193	439
Diaphanosoma brevireme Sars, 1901	Dbre	33	645	1807
Diaphanossoma sp.	Disp1	6	19	81
Moina minuta Hansen, 1899	Mmin	64	535	765
Simocephalus latirostris Stingelin, 1906	Slat	3	8	48
Copepoda				
Acanthocyclops robustus Marsh (1983).	Arob	33	428	888
Notodiaptomus spiniger (Brian, 1925)	Nspin	75	983	1995
Notodiaptomus incompositus (Brian, 1925).	Ninc	61	549	908
Microcyclops anceps (Richard, 1897)	Manc	78	1369	1794
Thermocyclops decipiens (Kiefer, 1929)	Tdec	81	1951	2111
Thermocyclops inversus (Kiefer, 1936)	Tinv	67	853	1046
Thermocyclops minutus (Lowndes, 1934)	Tmin	81	**2737**	4459
Tropocyclops prasinus (Fischer, 1860)	Tpras	3	13	77
Calanoida Nauplii	NCa	**100**	**13855**	14552
Cyclopoida Nauplii	NCy	**100**	**47756**	46906
Calanoida Copepodite	CCa	**97**	4181	5956
Cyclopoida Copepodite	CCy	**100**	**18595**	19107
Harpacticoida	Har	8	10	43
Ergasilidae	Erga	3	5	27

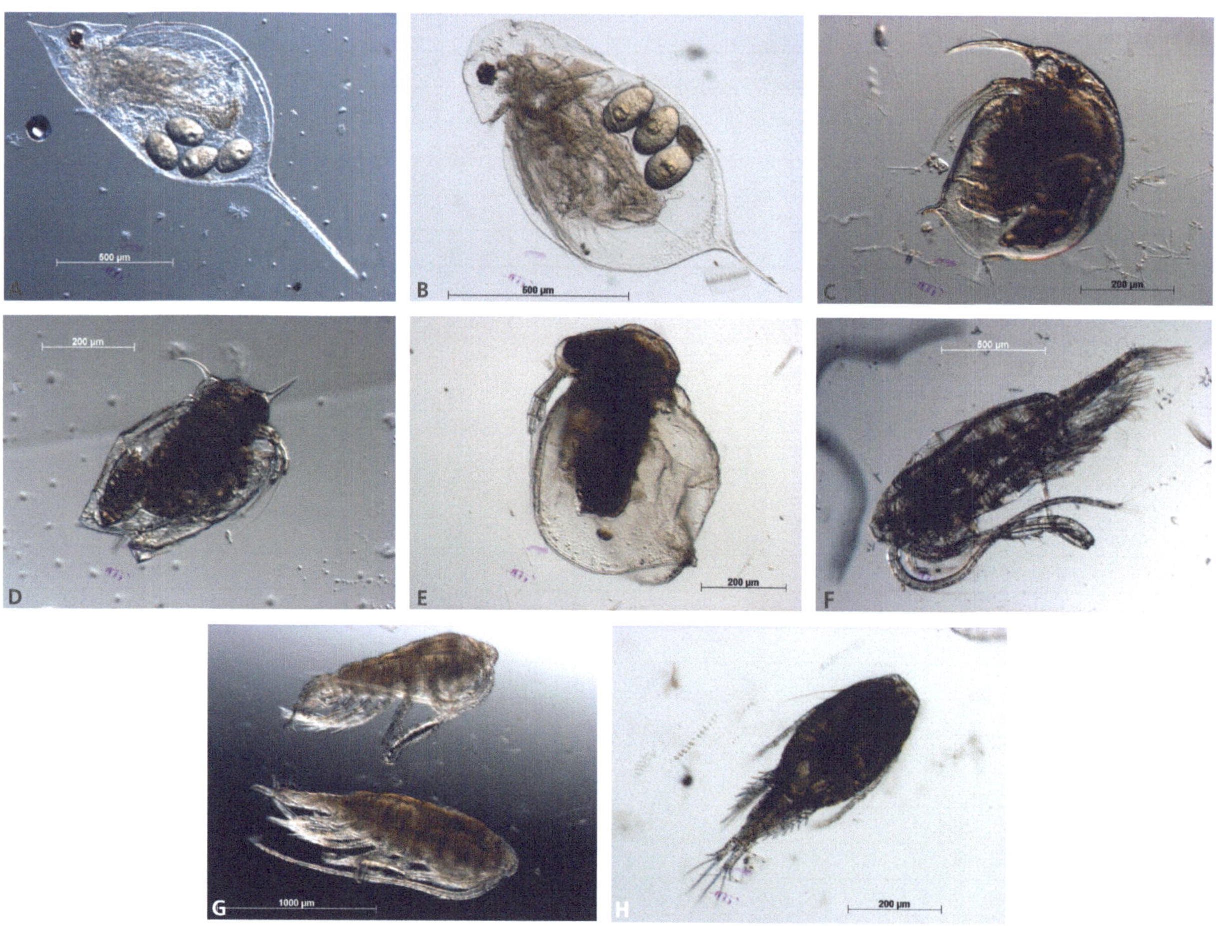

FIGURE 3 – IMAGES OF SOME OF THE IDENTIFIED MICROCRUSTACEANS SPECIES: A. *DAPHNIA GESSNERI* (♀), B. *DAPHNIA PARVULA* (♀), C. *BOSMINA HAGMANNI* (♀), D. *CERIODAPHNIA CORNUTA* (♀), E. *CERIODAPHNIA SILVESTRII* (♀), F. *NOTODIAPTOMUS INCOMPOSITUS* (♀), G. *NOTODIAPTOMUS SPINIGER* (♀ E ♀), H. *THERMOCYCLOPS INVERSUS* (♀)

TABLE 2 – MEAN VALUES (± STANDARD DEVIATION) OF THE ECOLOGIC ATTRIBUTES ANALYZED (RICHNESS-S, DIVERSITY-H, EQUITABILITY-E AND TOTAL ABUNDANCE (ind.m^{-3})) OF THE ZOOPLANKTON GROUPS

GROUPS/ATTRIBUTES	S	H	E	ABUNDANCE
Rotifera	13.02±2.57	1.78±0.31	0.48±0.08	110,710±48,163
Cladocera	5.39±1.97	0.92±0.46	0.53±0.14	29,361±24,781
Copepoda	4.78±1.27	1.22±0.35	0.79±0.08	8,883±5,891

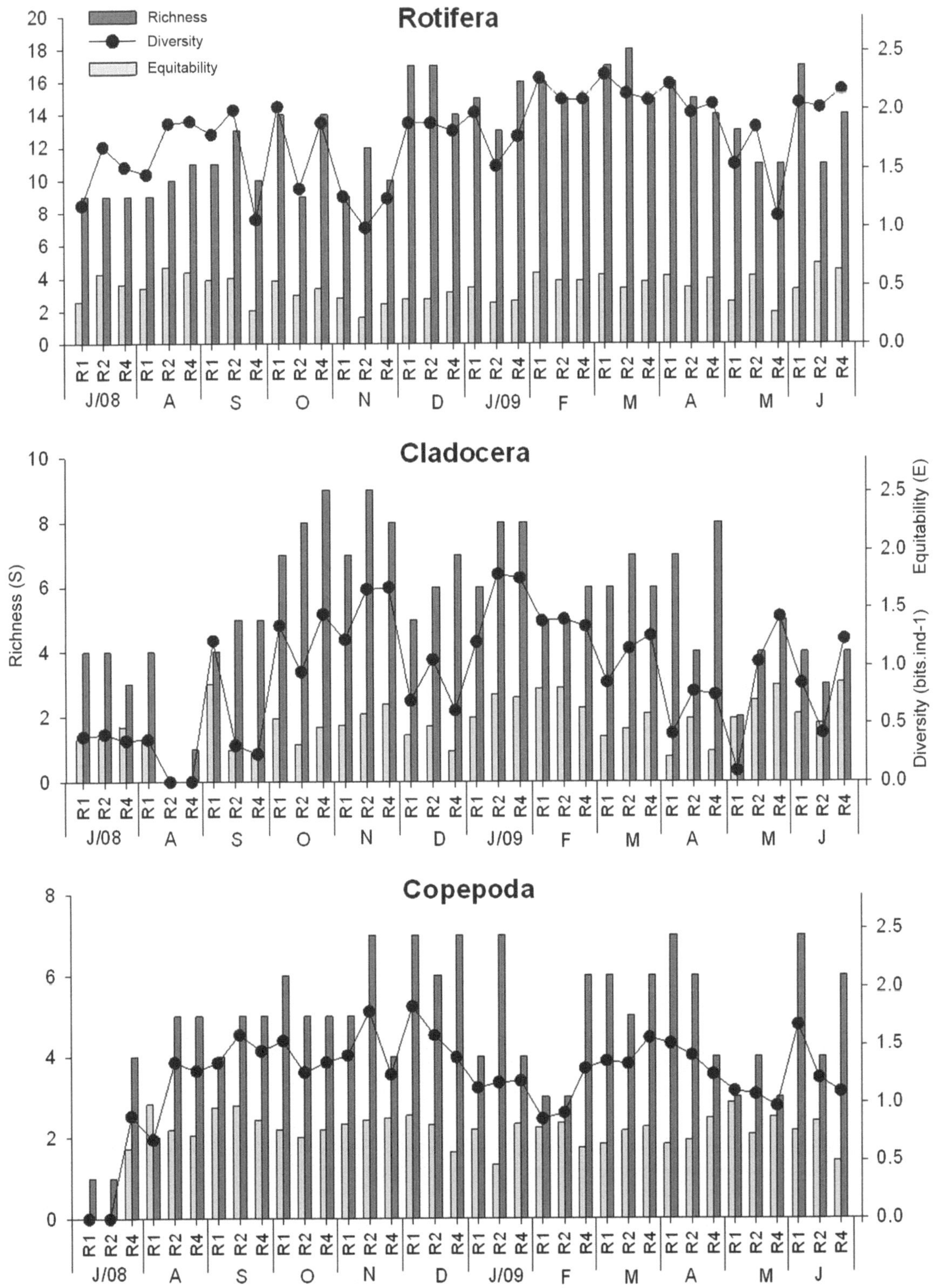

FIGURE 4 – VARIATION ACROSS SAMPLING MONTHS AND POINTS OF SOME ECOLOGICAL ATTRIBUTES (RICHNESS, EQUITABILITY, AND DIVERSITY) OF THE OBSERVED ZOOPLANKTON GROUPS

TABLE 3 – RESULTS OF ANOVA USED TO STUDY THE ABUNDANCE OF ZOOPLANCTON TAXA. DEGREES OF FREEDOM: POINTS=2, MONTHS=11. THE ABBREVIATIONS OF THE SPECIES ARE IN TABLE 1. SIGNIFICANT DIFFERENCES IN BOLD

	POINTS		MONTHS	
	F	p	F	p
Rotifera				
Colsp1	0.35	0.71	3.55	0.01
Ccoe	11.73	0.00	2.85	0.02
Cuni	4.75	0.02	1.24	0.32
Kbos	1.90	0.17	1.88	0.10
Kame	3.23	0.06	1.09	0.41
Kcoc	3.06	0.07	1.96	0.09
Klen	0.21	0.81	4.73	0.00
Pvul	2.17	0.14	0.95	0.51
Spec	0.11	0.90	1.33	0.28
Total Rotifera	6.89	0.00	1.72	0.14
Cladocera				
Bhag	0.15	0.86	1.03	0.45
Ccorc	0.75	0.48	9.96	0.00
Total Cladocera	1.38	0.27	1.52	0.19
Copepoda				
Manc	0.53	0.59	6.12	0.00
Tdec	2.97	0.07	2.52	0.03
Tmin	0.42	0.66	2.10	0.07
NCal	12.20	0.00	1.34	0.27
NCy	3.18	0.06	2.47	0.03
CCal	10.32	0.00	1.56	0.18
CCy	0.16	0.85	2.70	0.02
Total Adults	0.68	0.52	3.07	0.01
Adults Cy	0.36	0.71	4.34	0.00
Adults Ca	8.17	0.00	2.07	0.07

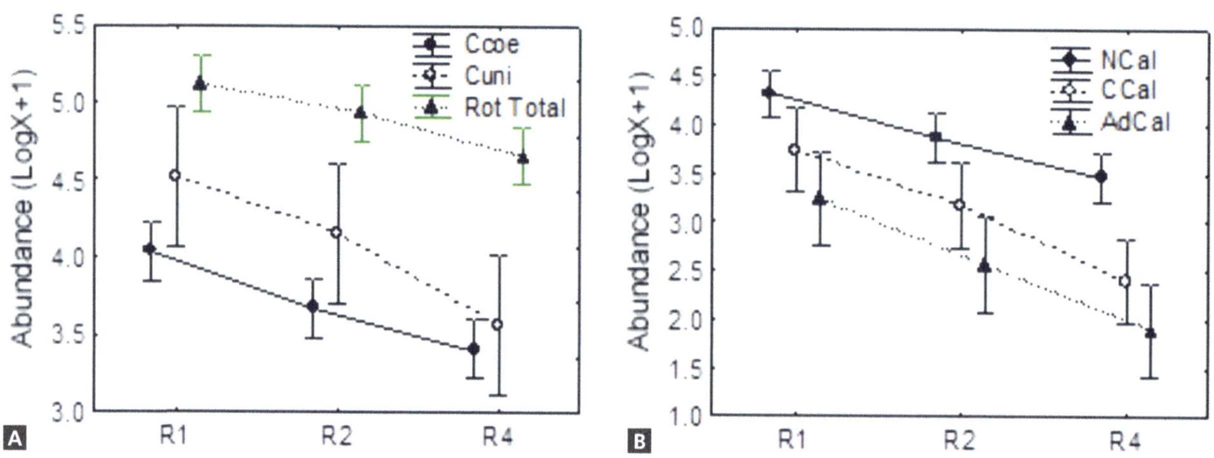

FIGURE 5 – VARIATION AMONG SAMPLING POINTS OF ABUNDANCE OF SOME TAXA OF ROTIFERA (A) AND COPEPODA (B) THAT PRESENTED P< 0.05 IN ANOVA. ABUNDANCE IN LOGARITHMIZED SCALE DUE TO SIGNIFICANT DIFFERENCES. ABBREVIATIONS ARE LISTED IN TABLE 1

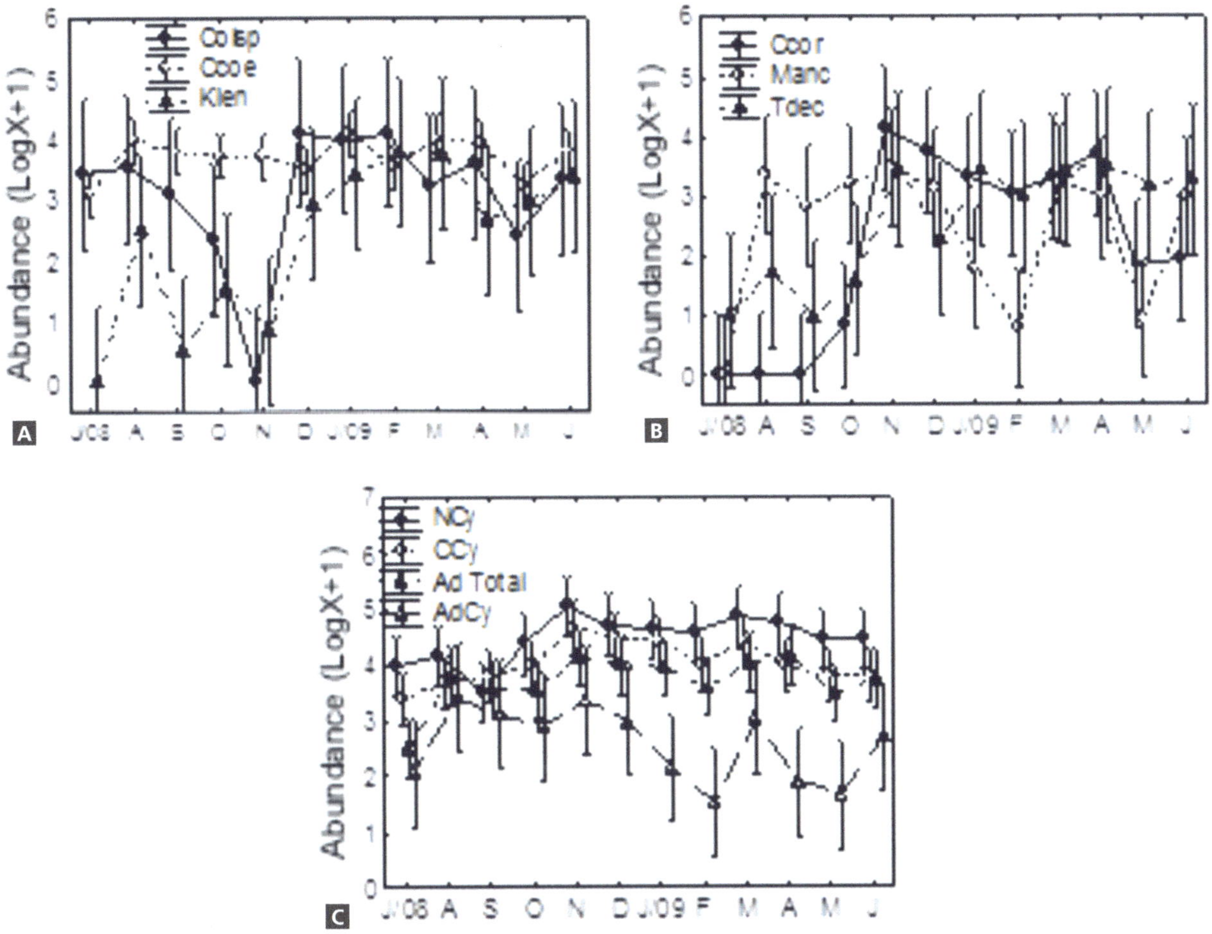

FIGURE 6 – MONTHLY VARIATION IN ABUNDANCE OF SOME TAXA OF ROTIFERA (A), CLADOCERA (B), AND COPEPODA (B AND C), DEMONSTRATED THROUGH ANOVA. ABUNDANCE IN LOGARITHMIZED SCALE DUE TO SIGNIFICANT DIFFERENCES. ABBREVIATIONS ARE LISTED IN TABLE 1

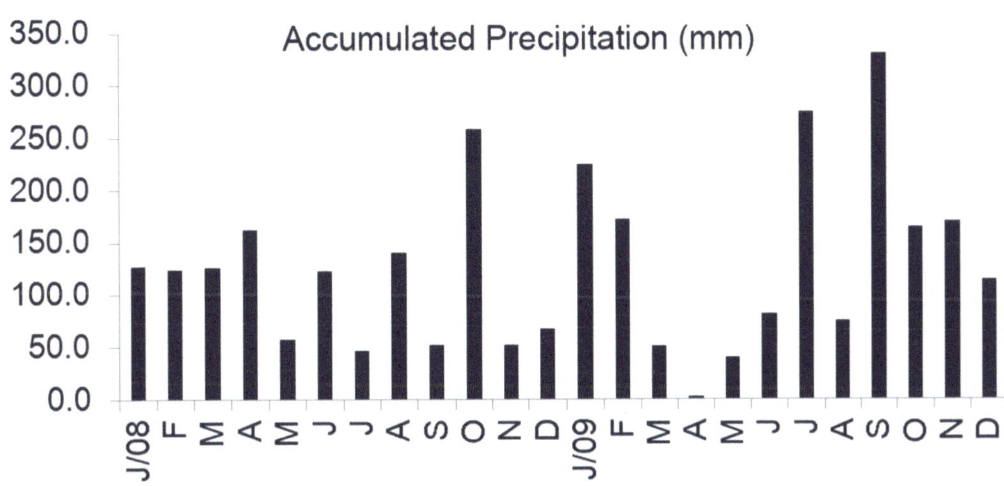

FIGURE 7 – ACCUMULATED RAINFALL IN 2008 AND 2009

The CCAs (Figure 8) explained 52% and 48% of the relation of rotifer and microcrustaceans abundance with environmental variables, respectively. Nine species of rotifers, four of cladocerans and two of copepods presented significant correlations ($p<0.05$) with these variables. In general, the CCAs showed different correlations for the zooplankton with well separated groups and more regular intervals. We found seasonal and spatial influence of different environmental variables. Highly significant correlations ($p=0.00$) occurred for the rotifers *Ascomorpha solutions*, *Collotheca* sp1, *Kellicottia bostoniensis* and *Polyarthra vulgaris*. Of the environmental variables, conductivity ($r^2=0.29$; $p=0.00$) and chlorophyll ($r^2=0.15$; $p=0.07$) showed significant correlations with the rotifers and the total phytoplankton (r^2=0.10; p=0.16). On the other hand, temperature (r^2=0.20; p=0.02), dissolved oxygen (r^2=0.21; p=0.00), transparency (measured with the Secchi disk) (r^2=0.10; p=0.16) and the total suspended solids (r^2=0.103; p=0.16), showed significant correlation with the microcrustaceans.

In CV1, the conductivity favored some species of rotifers (*P. vulgaris, Trichocerca cylindrica chatoni*, *Gastropus hyptopus*, *K. lenzi*) in autumn and winter, at points R1 and R2 (lotic and intermediate zones). Other species, such as *Plationus patulus patulus*, *Anuraeopsis navicula* and *Keratella cochlearis*, were correlated with chlorophyll in spring and summer at point R4. On the other hand, *K. bostoniensis* and *F. Opulence* positively correlate with the intermediary and lentic points in the spring and summer. However, there was no correlation of these species with any environmental variable. In the second CV, also with no environmental variable correlation, the species *P. vulgaris*, *A. saltans* and *K. bostoniensis* were associated with points R1 and R2. In contrast, *Collotheca* Sp.1 and *K. americana,* were mainly associated with point R4 and chlorophyll-a.

Among the microcrustaceans, the species of copepods of the Cyclopoida order and the cladoceans *M. minuta*, *C. reticulata*, *D. gessneri*, *C. cornuta* and *C. cornuta* var. *rigaudi* showed positive correlation with total phytoplankton, water temperature, and transparency at points R1 and R2 during the summer and R4 during the winter. In winter, *Bosmina hagmanni*, *Notodiaptomus spiniger* and *N. Incompositus* were associated with the dissolved oxygen in the lotic and intermediate points. In the second CV, *T. minutus*, *T. decipiens* and *T. inversus* at points R1 and R2 in the summer and winter, and R4 in winter, were correlated with phytoplankton. On the other hand, *D. parvula*, *C. cornuta* and *C. cornuta rigaudi*, *D. gessneri*, and *C. reticulata* were correlated with total suspended solids at points R1 and R2, and with R4 in the summer.

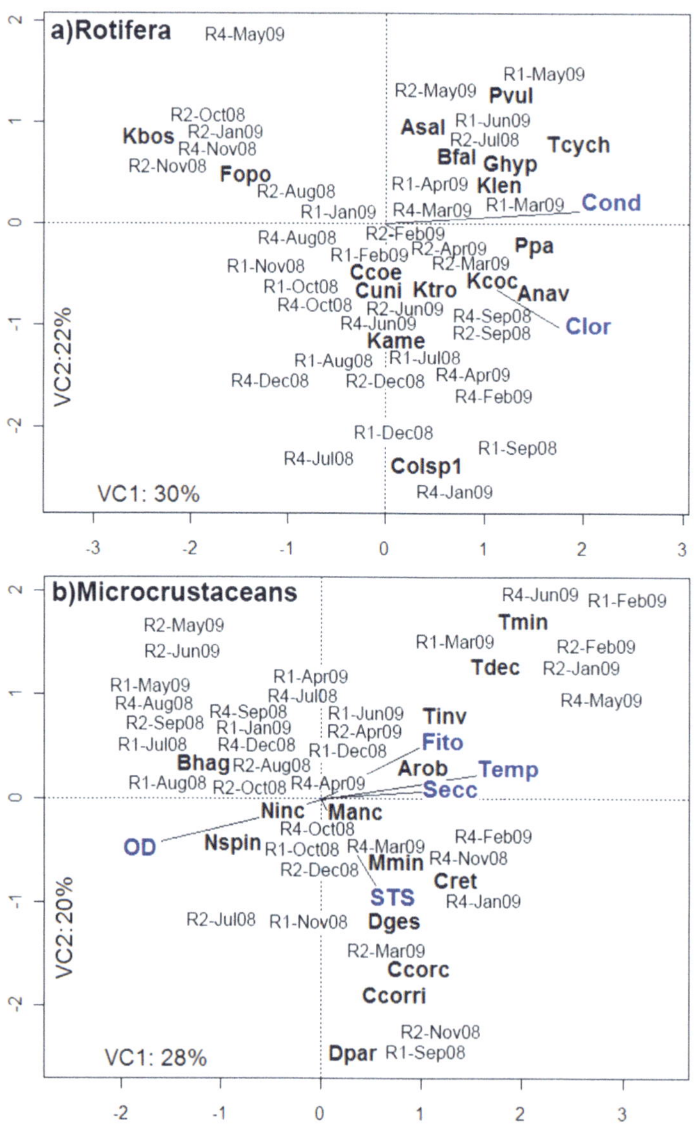

FIGURE 8 – CANONIC CORRESPONDENCE ANALYSES FOR A) ROTIFERS AND B) MICROCRUSTACEANS, WITH THE MEASURED ENVIRONMENTAL VARIABLES. ABBREVIATIONS ARE LISTED IN TABLE 1. ABBREVIATIONS OF ENVIRONMENTAL VARIABLES: COND = ELECTRIC CONDUCTIVITY; CLOR = CHLOROPHYLL-A; TEMP = WATER TEMPERATURE; SECC = TRANSPARENCY BY THE SECCHI DISK; OD = DISSOLVED OXYGEN; STS = TOTAL SUSPENDED SOLIDS; FITO = TOTAL PHYTOPLANKTON ABUNDANCE

4. DISCUSSION

In the present study, the richness of the zooplankton reached values expected for tropical and sub-tropical reservoirs. The number of species found was similar to that found in the Paraibuna and Guarapiranga Reservoirs, both located in the Southeast region of the country (CABIANCA, 1991; CALLEFI, 1994), and in the Iraí Reservoir, in the hydrographic basin of the Iguaçu River (SERAFIM-JUNIOR, *et al.* 2005). The high levels of richness observed for rotifers, regardless of the trophic state, are predictable and have been noted by several authors (SERAFIM-JUNIOR *et al.* 2010).

The zooplankton species that were recorded with high frequency and in abundance in the Rio Verde Reservoir were also recorded in other Brazilian reservoirs and are characteristic of oligo-mesotrophic environments (SENDACZ, 1999; NOGUEIRA, 2001; SERAFIM-JUNIOR *et al.*, 2005; LANSAC-TÔHA *et al.*, 2005). For the rotifers, Matsumura-Tundisi *et al.* (1990) proposed the association of *P. vulgaris* and *K. Tropica* as indicators of less eutrophic environments and *C. unicornis* and *K. cochlearis* as characteristic of more eutrophic environments. Lucinda (2003), studying the rotifers of the Tietê River Basin, suggests that the dominance of *C. unicornis* seems to be related to environments of lower eutrophication. In the Rio Verde Reservoir, *C. unicornis* was associated with *K. bostoniensis* in the spring and *P. vulgaris*, *K. cochlearis* and *K. tropica* in the summer. These associations are consistent with the studies cited above and indicate the oligo-mesotrophic state of the reservoir.

Among the microcrustaceans, the dominance of *B. hagmanni* is related to its ability to select its food, showing significant feeding flexibility by consuming several types of algae, including filamentous algae and cyanobacteria (SOMMER, 1989). This species was also the most abundant species in the Iraí Reservoir (SERAFIM-JUNIOR *et al.*, 2005; GHIDINI *et al.*, 2009). Studies in the reservoir in Barra Bonita, in São Paulo, showed that populations of *C. cornuta* var. *rigaudi*, another cladoceran species abundant in the present study, may grow in the presence of cyanobacteria (CHOUERI *et al.*, 2009). These authors demonstrated that *Anabaena spiroides exopolysaccharides* may be consumed by *C. cornuta* and promote its growth and reproductive success. In the present study, cyanobacteria algae were almost absent, as shown in Chapter 13, thus supporting the flexibility of these cladocerans species to ingest different types of algae.

Among the copepods, the *Thermocyclops* genus was the most abundant. According to Kiefer (1938), this genus has a tropical origin and is present in different types of continental aquatic environments. In the Neotropical region, this genus has a relatively low number of species, with only five species, excluding questionable reports (REID, 1985). *Thermocyclops minutes* is a common species but restricted to South America while *T. decipiens* is widely distributed. In some cases, these two species can occur together (ROCHA *et al.*, 1995; LANDA, 2007). The Calanoida *N. spiniger,* frequently recorded from in samples of the Rio Verde Basin, also occurs in the floodplain of the upper Paraná River as reported by Choueri *et al.* (2005), and this region is probably the northern boundary for the distribution of the species. Our study is the second recorded occurrence of this species in the State and the most eastern recorded to date. Data on the occurrence and dominance of this species in Brazilian reservoirs is severely lacking. The Calanoida copepods have a high degree of endemism and in Brazil most of the *Notodiaptomus* species occur within certain latitudinal bands, and few show a wide distribution from north to south (MATSUMURA-TUNDISI, 1986).

There are some restrictions in relation to the zooplankton diversity patterns in Brazilian reservoirs. Noteworthy in the present study, is the presence of "invasive" species such as *K. bostoniensis,* widely distributed in the Paraná hydrographic basins and reservoirs, and *D. parvula* recorded in the reservoir Segredo e Salto Caxias by Lopes *et al.* (1997) and Serafim-Junior (2002). According to Flössner (2000), *D. parvula* is frequently found in meso-eutrophic waters, including shallow lakes, reservoirs and meandering rivers, where it takes advantage over other rotifer species and succeeds in establishing permanent populations.

Another uncertainty that is currently being clarified is the modification of the concept that reservoirs with higher levels of eutrophication present lower diversity. In reservoirs with moderate eutrophication, such as the reservoir in Barra Bonita, SP, in the middle section of the Tietê River, Matsumura-Tundisi & Tundisi (2005) found high diversity indices of zooplankton; however, in cases of extreme eutrophication the ecological characteristics may be irregular and unstable, as indicated in some studies on the Iraí Reservoir (SERAFIM-JÚNIOR *et al.*, 2005, 2010; PERBICHE-NEVES *et al.*, 2007; GHIDINI *et al.*, 2009). In the present study, the analyzed ecological attributes showed values that were lower than those in eutrophic reservoirs, which is consistent with the tendency of less richness and diversity in oligo-mesotrophic reservoirs.

The dominance of immature forms of copepods in the zooplankton detected in almost all the sampling months is a trend that can be observed in other reservoirs and is associated with the reproduction rates of these organisms (NOGUEIRA, 2001; SERAFIM-JÚNIOR *et al.*, 2005). The abundance of nauplii and copepodites in the summer is a possible response to higher water temperatures which directly influences metabolism and reproduction. However, there is generally a greater abundance of nauplii of Cyclopoida than of Calanoida; at the copepodite stage, this trend reverses with a greater abundance of copepodites of Calanoida than Cyclopoida (NOGUEIRA, 2001; PERBICHE-NEVES *et al.*, 2007; SARTORI *et al.*, 2009).

The trends found for diversity at the sampling points differed very little from those found for composition and abundance. Regarding the richness of species, the zooplankton presented the highest values in the summer months, which is the period of greatest rainfall and highest water temperature. The means of diversity indices (H') recorded for the three groups were similar to those found in other Brazilian reservoirs (MATSUMURA-TUNDISI, 1999; NOGUEIRA, 2001). Rocha *et al.* (1999) suggest that the communities are sensitive to the reservoir system and there is likely a relationship between the residence time of water in the system and zooplankton diversity, as suggested by Perbiche-Neves & Nogueira (2010) for two reservoirs in the

Paranapanema River with contrasting systems and residence time.

The longitudinal variability observed in the present study differs partly from that proposed by Marzolf (1990), who mentions greater abundance in the intermediary zone. In the Rio Verde Reservoir, there was a decrease in abundance from point R1 towards point R4, or from the fluvial area to the lentic area. A possible explanation is that point R1 may be situated in the ecotone of the fluvial and intermediary areas, thus containing the greatest abundances.

The seasonal pattern registered for the zooplankton in the Rio Verde reservoir is consistent with those found for other Neotropical reservoirs. Lower densities frequently occur in winter months (dry season) and the higher densities in the spring and summer months (rainy season) (ARCIFA *et al.*, 1992; NOGUEIRA & MATSUMURA-TUNDISI, 1996; NOGUEIRA, 2001).

Zooplankton is highly susceptible to physicochemical alterations in water and the environment. The physicochemical factors are: temperature; oxygen concentration; luminosity; and pH. Biological factors, such as food quality and quantity, predation and parasitism, may cause changes in the structure of the community, favoring some species over others (TUNDISI & MATSUMURA-TUNDISI, 2008). Also related to these factors are variables associated with the trophic state of the aquatic environment, such as nutrients, chlorophyll-a and conductivity, which can influence the succession of zooplankton assemblages.

In the present study, the CCA showed strong correlation among *Ascomorpha saltans*, *Collotheca* sp1, *Kellicottia bostoniensis* and *Polyarthra vulgaris* and other rotifers, in general, with conductivity and chlorophyll-a. Rotifers are more susceptible to physicochemical changes due to their small size and the permeability of their tegument (NOGRADY *et al.*, 1993).

Temperature, dissolved oxygen, transparency and total solids presented significant correlations with the microcrustaceans. In the correlations between rainfall and ecological attributes of the zooplankton groups, the rainfall indices had a negative influence on the abundance of cladocerans and phytoplankton, an important food source for these organisms. The feeding of herbivorous zooplankton is a selective process that involves a set of structures and behaviors (KERFOOT & LYNCH, 1987). In tropical regions, the predictability of the phytoplankton seasonal succession and the corresponding succession of herbivorous zooplankton is complex. Physical factors such as the system's hydrology and hydrodynamics can regulate the structures of zooplankton assemblages (TUNDISI & MATUSUMURA--TUNDISI, 2008).

According to the resources used in this study, the zooplankton species of the Rio Verde Reservoir can be characterized as typical of oligo-mesotrophic environments. We found a dominance of microcrustaceans at all the sample points along the reservoir, mainly consisting of the immature forms of copepods. The highest zooplankton densities occurred in the fluvial area of the reservoir, very close to the intermediary zone. As previously mentioned, the highest densities are expected in the transition area where the drifting zooplankton positively respond to the hydraulic system and its reproductive rates maintain the large populations expected in the intermediary or transition zones of the reservoir.

5. REFERENCES

ARMENGOL, X.; BORONAT, A.; CAMACHO, A. & WURTSBAUGH, W.A. 2001. *Grazing by a dominant rotifer Conochilus unicornis Rousselet in a mountain lake: in situ measurements with synthetic microspheres*. Hydrobiologia, 446/447: 107-114.

ARCIFA, M.S.; GOMES, E.A.T. & MESCHIATTI, A.J. (1992) Composition and fluctuations of the zooplankton of tropical brazilian reservoir. *Arch. Hydrobiol., 123*(4): 479-495.

BLETTER, M. C.R.M. & BONECKER, C. C. (2007) The longitudinal distribution of zooplankton microcrustaceans biomass in three tropical reservoirs (PR, Brazil). *Acta Scientiarum (UEM)*, 29: 297-304.

CABIANCA, M.A.A. (1991) *A Represa de Paraibuna: aspectos do zooplâncton*. Master's degree dissertation, Universidade de São Paulo, SP. 92p.

CALEFFI, S. (1994) *A Represa do Guarapiranga: estudo da comunidade zooplanctônica e aspectos da eutrofização*. Master's degree dissertation. Faculdade de Saúde Pública/USP, SP. 183p.

CHOUERI, R. B.; MELÃO M. G. G.; LOMBARDI, A. T. & VIEIRA, A. A. H. (2009) Effects of cyanobacterium exopolysaccharides on life--history of *Ceriodaphnia cornuta* SARS. *J. Plankton Res.*, Vol. 31; n. 10: 1225-1233.

FLÖSSNER, D. (2000) *Die Haplopoda und Cladocera (ohne Bosminidae) Mitteleuropas*. Backhuys Publishers.

GHIDINI, A. R.; SERAFIM-JUNIOR, M.; PERBICHE-NEVES, G. & BRITO, L. (2009) Distribution of planktonic cladocerans (Crustacea: Branchiopoda) and their relationships with the limnological variables of an eutrophic reservoir (Paraná state, Brazil). *Pan-American Journal of Aquatic Sciences*, 4: 294-305.

KENNEDY, R.H. (1999) Reservoir design and operation: limnological implications and management opportunities. In: TUNDISI, J.G. & STRASKRABA, M. (Ed.). *Theoretical reservoir ecology and its applications*. São Carlos: International Institute of Ecology, p. 1-28.

KERFOOT, W. C. & LYNCH, M. (1987) Branchiopod communities: associations with planktivorous fish in space and time. In: KERFOOT, W. C. & SIH, A. *Predation. Hanover*. University Press of New England, p. 367-378.

LANDA, G.G; BARBOSA, F.A.R.; RIETZLER, A.C. & MAIA-BARBOSA, P.M. (2007) *Thermocyclops decipiens* (Kiefer, 1929) (Copepoda, Cyclopoida) as Indicator of Water Quality in the State of Minas Gerais, Brazil. *Braz. Arch. Biol. Tech.* 50: 695-705.

LANSAC-TOHA, F. A.; BONECKER, C. C. & VELHO, L. F. M. (2005) Estrutura da comunidade zooplanctônica em reservatórios do Estado do Paraná. In: Liliana Rodrigues; Sidinei Magela Thomaz; Angelo Antonio Agostinho; Luiz Carlos Gomes. (Eds). *Biocenoses em reservatórios: padrões espaciais e temporais*. São Carlos: RIMA, p. 115-127.

LUCINDA, I. *Composição de rotíferos em corpos da água da bacia do rio Tietê, São Paulo, Brasil*. Master's degree dissertation Universidade Federal de São Carlos, SP. 199p.

MARGALEF, R. (1983) *Limnología*. Omega: Barcelona. 1010pp

MARZOLF, R.G. (1990) Reservoirs as environments for zooplankton. In: THORNTON, K.W.; KIMMEL, L.B. & PAYNE, F.E. (Ed.). *Reservoir limnology: ecological perspectives*. New York: John Wiley & Sons, p. 195-208.

MATSUMURA-TUNDISI, T. (1986) Latitudinal distribution of Calanoida species in Brazil. *Rev. Bras. Biol.*, 46 (3): 527-553.

MATSUMURA-TUNDISI, T. (1999) Diversidade de zooplâncton em represas do Brasil. In: HENRY, R. (Ed.). *Ecologia de reservatórios: estrutura, função e aspectos sociais*. Botucatu: Universidade Estadual Paulista, p. 41-54.

MATSUMURA-TUNDISI, T. & TUNDISI, J. G. (2005) Plankton richness in a eutrophic reservoir (Barra Bonita Reservoir, SP, Brazil). *Hydrobiologia (The Hague), Aquatic Biodiversity II*, 542: 367-378.

NOGUEIRA, M.G. & MATSUMURA-TUNDISI, T. 1996. Limnologia de um sistema artificial raso (represa do Monjolinho – São Carlos – SP). Dinâmica das populações planctônicas. *Acta Limnol. Bras.*, 8: 149-168.

NOGUEIRA, M.G. (2001) Zooplankton composition dominance and abundance as indicators of environmental compartmentalization in Jurumirim Reservoir (Paranapanema River), São Paulo, Brazil. *Hydrobiologia,* 455:1-18.

NOGRADY, T. 1993. Rotifera. Guides to the identification of the microinvertebrates of the continental waters of the world. Vol. 1: *Biology, Ecology and Systematics*. The Hague. SPB Academic Publishing, 141p.

PANARELLI, E.; CASANOVA, S. M. C.; NOGUEIRA, M. G.; MITSUKA, P. M & HENRY, R. (2003) A comunidade zooplanctônica ao longo de gradientes longitudinais no Rio Paranapanema/Represa de Jurumirim (São Paulo, Brasil). In. Henry, R. (Ed.*).* Ecótonos nas interfaces dos ecossistemas aquáticos. São Carlos: Rima, pp. 129-160.

PERBICHE-NEVES, G. & NOGUEIRA, M. G. (2010) Multi-dimensional effects on Cladoceran (Crustacea, Anomopoda) assemblages in two cascade reservoirs in Southeast Brazil. *Lakes and Reservoirs*, 15, 139-152.

PERBICHE-NEVES, G.; SERAFIM-JUNIOR, M.; GHIDINI, A. R. & BRITO, L. (2007) Spatial and temporal distribution of Copepoda (Cyclopoida and Calanoida) of an eutrophic reservoir in the basin of upper Iguaçu River, Paraná, Brazil. *Acta Limnol. Bras.*, 19: 393-406.

PINTO-COELHO, R.M.; BEZERRA-NETO, J.F.; MORAIS-JR, C.A. (2005) Effects of eutrophication on size and biomass of crustacean zooplankton in a tropical reservoir. *Braz. J. Biol.,* 65: 325-338.

PINTO-COELHO, R.M. (1998) Effects of eutrophication on seasonal patterns of mesozooplankton in a tropical reservoir: a 4-year study in Pampulha Lake, Brazil. *Fresh. Biol.,* 40: 159-173.

SAMPAIO, E.V.; ROCHA, O.; MATSUMURA-TUNDISI, T. & TUNDISi, J.G. (2002) Composition and Abundance of Zooplankton in the Limnetic of Seven Reservoir of the Paranapanema. *River, Brazil. Braz. J. Biol.,* 62 (3), 525-545.

SARTORI, L.P.; NOGUEIRA, M. G.; HENRY, R. & MORETTO, E. M. (2009) Zooplankton fluctuations in Jurumirim Reservoir (São Paulo, Brazil): a three-year study. *Braz. J. Biol.,* 69: 1-18.

SENDACZ, S.; CALEFFI, S. & SOARES, J. S. (2006) Zooplankton biomass of reservoirs in different trophic conditions in ten State of São Paulo, Brazil. *Braz. J. Biol.,* 66: 337-350.

SERAFIM-JR., M.; LANSAC-TÔHA, F.A.; PAGGI, J.C.; VELHO, L.F.M. & ROBERTSON, B. (2003) Cladocera fauna composition in a River-Lagoon System from the Upper Paraná River floodplain, with a new record for Brazil. *Braz. J. Biol.,* 63: 349-356.

SERAFIM-JUNIOR, M.; GHIDINI, A. R.; BRITO, L. & PERBICHE--NEVES, G. (2005) Comunidade Zooplanctônica. In: Cleverson V. Andreoli; Charles Carneiro; Eduardo S. Pegorini; Marilene Maria Lora. (Eds.). *Gestão Integrada de Mananciais de Abastecimento Eutrofizados*. Curitiba: p. 411-434.

SOMMER, U. (1989) The role of competition for resources in phytoplankton succession. In: SOMMER, U. (Ed.) *Plankton Ecology: sucession in plankton communities.* Springer-Verlag. New York. 369p.

THORNTON, K.W. (1990) Perspectives on reservoir limnology. In: THORNTON, K. W.; KIMMEL, B. L. & PAYNE, F. E. (Ed.). *Reservoir Limnology: ecological perspectives.* New York: John Wiley & Sons, p. 1-13.

TUNDISI, J. G. & Matsumura-Tundisi, T. (2008*) Limnologia.* São Paulo: Oficina de Texto, 632 pp.

VAN DEN BRINK, F.W.B; VAN KATWIJK, M.M; VAN DER VELDE, G. (1994) Impact of hydrology on phyto and zooplankton community composition in floodplain lakes along the Lower Rhine and Meuse. *J. Plankton Res.,* 16: 351–373.

VELHO, L.F.M.; LANSAC-TÔHA, F.A. & BONECKER, C.C. (2005) Distribuição Longitudinal da Comunidade Zooplanctônica em Reservatórios. In: RODRIGUES, L.; THOMAZ, S.M.; AGOSTINHO, A.A. & GOMES, L.C. (Eds). *Biocenoses em Reservatórios: Padrões Espaciais e Temporais*. São Carlos: Rima, p. 129-136.

WARD, J.V. (1989) The four-dimensional natures of lotic ecosystems. *J. N. Am. Benthol. Soc.,* 8(1): 2-8.

CHAPTER 15

ICHTHYOFAUNA

Vinicius Abilhoa
Wellen Laryessa Laynes
Letícia Michalik
Letícia Lima
Luis Fernando Fávaro
Jean Ricardo Simões Vitule

ICHTHYOFAUNA

Vinícius Abilhoa
Wellen Laryessa Laynes
Letícia Michalik
Letícia Lima
Luis Fernando Fávaro
Jean Ricardo Simões Vitule

SUMMARY

The reservoirs located in the Metropolitan Region of Curitiba (RMC) are complex and diversified artificial environments and studies conducted in these reservoirs indicate that the fish community has few dominant species. The number of species registered in the Rio Verde Reservoir (11 species) corresponded to 23% of the species found in Alto Iguaçu River and 10% of the total species recorded for the basin. The highest values of catch per unit effort by individual were recorded for *Astyanax minor*, *A. bifasciatus*, *Oligosarcus longirostris* and *Geophagus brasiliensis*. The highest values of catch per unit by weight were recorded for *Hoplias*. aff. *malabaricus, A. minor*, *A. bifasciatus*, *Hypostomus commersoni*, *O. longirostris* and *Rhamdia quelen. Detritivore and piscivore fish were the most diverse and abundant groups.* Reproductive activity was higher in spring and summer, as observed for other species in the Paraná Basin. The results obtained indicate that the fish community of the Rio Verde Reservoir is not diverse and is within the patterns expected for the RMC. This reduced ichthyofauna is associated with the particular characteristics of this ecosystem, which influenced the composition and structure of the ichthyofauna during the colonization process of the Reservoir environment.

KEYWORDS

Fish assemblage, dams, trophic ecology, reproduction.

1. INTRODUCTION

The construction of a reservoir causes changes in water dynamics that directly or indirectly affecting the physical, chemical and biological attributes of the region, often leading to the formation of a new ecosystem (PETRERE JR., 1996; AGOSTINHO; GOMES, 1997). Among the impacts caused by dams, the change in the composition and structure of the ichthyofauna is one of the most important because it reduces spawning grounds and nurseries for developing juveniles, often compromising the maintenance of species in the environment and/or modifying the structure of the original community (AGOSTINHO *et al.*, 2007).

Dam construction, watercourse diversion and reservoir formation have significantly increased in recent years which have triggered responses in biological systems at all levels, from the individual to the ecosystem. Among these responses, we can highlight reproductive changes in birth and mortality rates of populations, as well as the succession of communities, changes in biodiversity, distribution patterns and even the extinction of species (CECILIO *et al.*, 1997; RODRIGUES *et al.*, 2005; AGOSTINHO *et al.*, 2007).

Fish populations subjected to the change from a fluvial to a lentic environment primarily suffer profound changes in the composition and the structure of the communities; the most affected species are migratory and rheophilic or species that prefer fluvial environments, conditions that are scarce in a dammed watercourse (AGOSTINHO *et al.*, 1992). The new environments formed by the reservoirs also lead to changes in the spatial and temporal patterns of fish communities that were previously established by seasonal cycles (AGOSTINHO; ZALEWSKI, 1996) and by associated biotic and abiotic factors (LOWE-McCONNELL, 1987; PERES NETO *et al.*, 1995).

In addition, reservoir fish populations may suffer changes caused by the use or the impacts on these water bodies, such as hydrological changes, pollution, changes in aquatic macrophyte density, species introduction, and eutrophication (AGOSTINHO *et al.*, 2007). Even when these changes are of limited duration and intensity, they can cause unpredictable effects on the environmental and physiological parameters that act on fish communities (SINDERMANN, 1979).

Understanding the role of the ichthyofauna in a modified aquatic environment is the objective of this chapter. This study significantly contributes to the ecology of small Neotropical reservoirs because fish play a significant ecological role in aquatic ecosystems, they are widely acknowledged as key-species (ecological engineers) or environmental controllers (PAINE, 1969; PAINE, 1980; JONES *et al.*, 1994), and in tropical aquatic environments these organisms are key elements and controllers of trophic webs, with a wide variety of behavioral aspects and trophic levels (WINEMILLER; JEPSEN, 1998; WOOTTON, 1990; LOWE-McCONNELL, 1999).

Based on current knowledge of the communities and aquatic environments of the many reservoirs located in the RMC, the proposal of definitive solutions for their preservation is still premature. However, since communities respond to the reorganization of aquatic ecosystems in different ways, and these responses depend on the intensity and duration of the phenomena involved, the study of the biology and behavior of a community in the face of certain environmental changes has proven critical for its preservation and management (AGOSTINHO; GOMES, 1997; VAZZOLER *et al.*, 1997). Understanding the composition, spatial and temporal monitoring, as well as the bionomics of the species is, therefore, fundamental in understanding the dynamics of these communities, effectively contributing to environmental conservation through more comprehensive management and monitoring plans (LOWE-McCONNELL, 1999).

2. RESERVOIR ICHTHYOFAUNA

The changes brought about by the damming of rivers, such as the change from a lotic to lentic or semi-lentic environment, initially result in the disappearance of strictly fluvial species and then a general rearrangement of the remaining species (LOWE-McCONNEL, 1975). The newly formed reservoir is colonized by the pre-existing ichthyofauna but as not all species are able to adapt to the new environment, this ichthyofauna is much less diverse than the original river environment (AGOSTINHO *et al.*, 1997a).

According to LOWE-McCONNEL (1975), the fish species subjected to these modifications can be divided into two basic groups. The first group is composed of rheophilic species. These species may have migratory habits, normally related to reproduction (*piracema)*, as already reported in several studies conducted on the Paraná River (AGOSTINHO, 1992). The second group consists of species adapted to lentic environments, such as deep, quiet waters and flooded regions. In theory, these species are best suited to a reservoir either because they present a broad feeding spectrum or reproductive characteristics adapted to calm water environments.

The reservoirs located in the RMC are complex and diverse artificial environments and studies conducted in these reservoirs indicate that the fish community presents few dominant species (Table 1), mostly made up of generalist/opportunist species such as *lambaris* (tetras) of the *Astyanax* genus. The ichthyofauna recorded may be divided into three types of species, depending on the original distribution: endemic species, i.e., those unique to the Iguaçu River Basin; non-endemic, native species, which also occur in other hydrographic basins; and introduced species, that occur in the environment due to accidental (aquaculture) or intentional (dam fish stocking) introduction. About 40% of the species recorded in these reservoirs are unique to the Iguaçu River hydrographic basin and this demonstrates the importance of local and regional processes in determining the structure and composition of the ichthyocenosis of these aquatic environments.

High abundance of *lambaris* of the *Astyanax* genus, was also reported for the Segredo and Foz do Areia Reservoirs (AGOSTINHO; GOMES, 1997), both in the Iguaçu River basin, in the Alagados Reservoir (LUIZ *et al.*, 2003), and in the reservoir of UHE Nova Avanhandava (CESP, 1998). These results allow us to conclude that changes that occurred in the environment as a function of dam construction, associated with a high degree of opportunism, reproductive strategy and broad feeding spectrum shown for various representatives of the genus (AGOSTINHO, GOMES, 1997; ABILHOA; AGOSTINHO, 2007), are responsible for their high number in catches because these species find favorable conditions for their development in the new environment.

3. CASE STUDY OF ICHTHYOFAUNA IN THE RIO VERDE RESERVOIR

3.1 COMPOSITION AND STRUCTURE

The survey conducted in the drainage area of the Rio Verde Basin totaled 27 fish species. From those species, 11 (41%) were considered endemic of the Iguaçu River (Table 2).

Sampling was conducted with sets of gillnets of different mesh sizes at three sampling points in the Rio Verde Reservoir, between August, 2008, and July, 2009 (Figures 1 and 2). Eleven fish species were recorded, distributed across three orders and six families, of which Characidae was the most representative, both in richness and relative abundance (Table 2).

The ichthyofauna recorded in the Rio Verde Reservoir corresponds to about 40% of the species of the Rio Verde Hydrographic Basin and 22% of the species that occur in the region called Alto Iguaçu River (ABILHOA, 2004).

Of the species recorded in the reservoir, five were considered endemic of the Iguaçu River and one introduced (Figure 5). As recorded for the Segredo Reservoir by Agostinho *et al.* (1997b) and the Iraí Reservoir (ABILHOA, 2005), the number and percentage of introduced species is low, revealing a certain failure of fish stocking.

The highest values of catch per unit effort by individu-

TABLE 1 – CHARACTERISTICS OF THE RESERVOIRS LOCATED IN THE METROPOLITAN REGION OF CURITIBA, INCLUDING DATA OBTAINED FOR THIS PROJECT FOR THE RIO VERDE RESERVOIR. THE NUMBER OF ICHTHYOFAUNA SPECIES FOR THE RESERVOIRS OF PIRAQUARA, PASSAÚNA AND IRAÍ WAS OBTAINED THOUGH MONTHLY SAMPLE COLLECTION BETWEEN JULY, 2006, AND JUNE, 2007, WITH STANDARDIZED CAPTURE EFFORT (GILLNETS)

RESERVOIRS	USE	YEAR OF CONSTRUCTION	AREA (Km^2)	NUMBER OF SPECIES (GILLNETS)
Verde	Industrial	1976	7.9	11
Piraquara	Supply	1979	3.3	8
Passaúna	Supply	1990	14	15
Iraí	Supply	2000	15	12

al were recorded for *A. minor*, *A. bifasciatus*, *O. longirostris* and *G. brasiliensis*. The highest values of catch per unit effort by weight were recorded for *H.*aff. *malabaricus*, *A. minor*, *A. bifasciatus*, *H. commersoni*, *O. longirostris* and *R. quelen*.

The species were grouped according to their consistency in the monthly samples by: accidental (frequency up to 25% in samples); accessory (frequency between 25% and 50% of the samples); and constant (frequency above 50% of the samples) (DAJOZ, 2005). The most abundant species were also considered constant in the three sampling points, although *H.* aff. *malabaricus* was constant only at point 1 (Table 3).

TABLE 2 – SPECIES SURVEYED IN THE RIO VERDE BASIN AND RESERVOIR. CATCH PER UNIT EFFORT BY NUMBER ($CPUE_{IND}$ = NUMBER OF INDIVIDUALS / 100 m^2 NET IN 24 h) AND BIOMASS ($CPUE_{WEIGHT}$ = kg/100 m^2 NET IN 24 h) WERE RECORDED ONLY FOR THE SPECIES COLLECTED IN THE RESERVOIR DURING THE STUDY. (⊗) INTRODUCED SPECIES (♦) ENDEMIC SPECIES (IGUAÇU RIVER)

SPECIES	COMMON NAME	$CPUE_{IND}$	$CPUE_{WEIGHT}$
Cypriniformes			
Cyprinidae			
Cyprinus carpio Linnaeus, 1758 ♦	Carp	-	-
Characiformes			
Characidae			
Astyanax bifasciatus Garavello & Sampaio, 2010 ♦	Red-tail tetra	178.8	3.97
Astyanax minor Garavello & Sampaio, 2010 ♦	Yellow-tail tetra	274.2	5.56
Astyanax serratus Garavello & Sampaio, 2010 ♦	Tetra	1.3	0.02
Astyanax dissimilis Garavello & Sampaio, 2010 ♦	Tetra	-	-
Bryconamericus sp. ♦	Tetra	-	-
Hyphessobrycon boulengeri (Eigenmann, 1907)	Tetra	-	-
Mimagoniates microlepis (Steindachner, 1877)	Blue tetra	-	-
Oligosarcus longirostris Menezes & Géry, 1985 ♦	Pike characin	56.9	2.57
Erythrinidae			
Hoplias aff. *malabaricus* (Bloch, 1794)	Trahira	16.7	6.63
Siluriformes			
Trichomycteridae			
Trichomycterus davisi (Haseman, 1911)	Pencil catfish	-	-
Trichomycterus castroi Pinna, 1992 ♦	Pencil catfish	-	-
Callichthyidae			
Corydoras ehrhardti Steindachner,1910	Spotted corydoras	-	-
Corydoras paleatus (Jenyns, 1842)	Spotted corydoras	5.4	0.05
Callichthys callichthys (Linnaeus, 1758)	Bubblenest catfish	-	-
Loricariidae			
Ancistrus abilhoai Bifi *et al.*, 2009 ♦	Bushymouth catfish	-	-
Hypostomus derbyi (Haseman, 1991) ♦	Armored catfish	0.4	0.07
Hypostomus commersoni (Valenciennes, 1840)	Armored catfish	7.3	3.58
Rineloricaria sp. ♦	Whiptail catfish	-	-
Heptapteridae			
Heptapterus stwearti Haseman (1911) ♦	Three-barbeled catfish	-	-
Rhamdia quelen (Quoy & Gaimard, 1824)	Silver catfish	5.8	2.27
Gymnotiformes			
Gymnotidae			
Gymnotus aff. *carapo* Linnaeus, 1758	Banded knifefish	-	-
Cyprinidontiformes			
Poeciliidae			
Phalloceros harpagos Lucinda, 2008	Speckled mosquitofish	-	-
Synbranchiformes			
Synbranchidae			
Synbranchus marmoratus Bloch, 1759	Marbled swamp eel	-	-
Perciformes			
Cichlidae			
Cichlasoma cf. *paranaense* Kullander, 1983	Banded cichlid	-	-
Geophagus brasiliensis (Quoy & Gaimard, 1824)	Pearl cichlid	49.8	1.38
Tilapia rendalli (Boulenger, 1897) ⊗	Tilapia	23.1	1.09

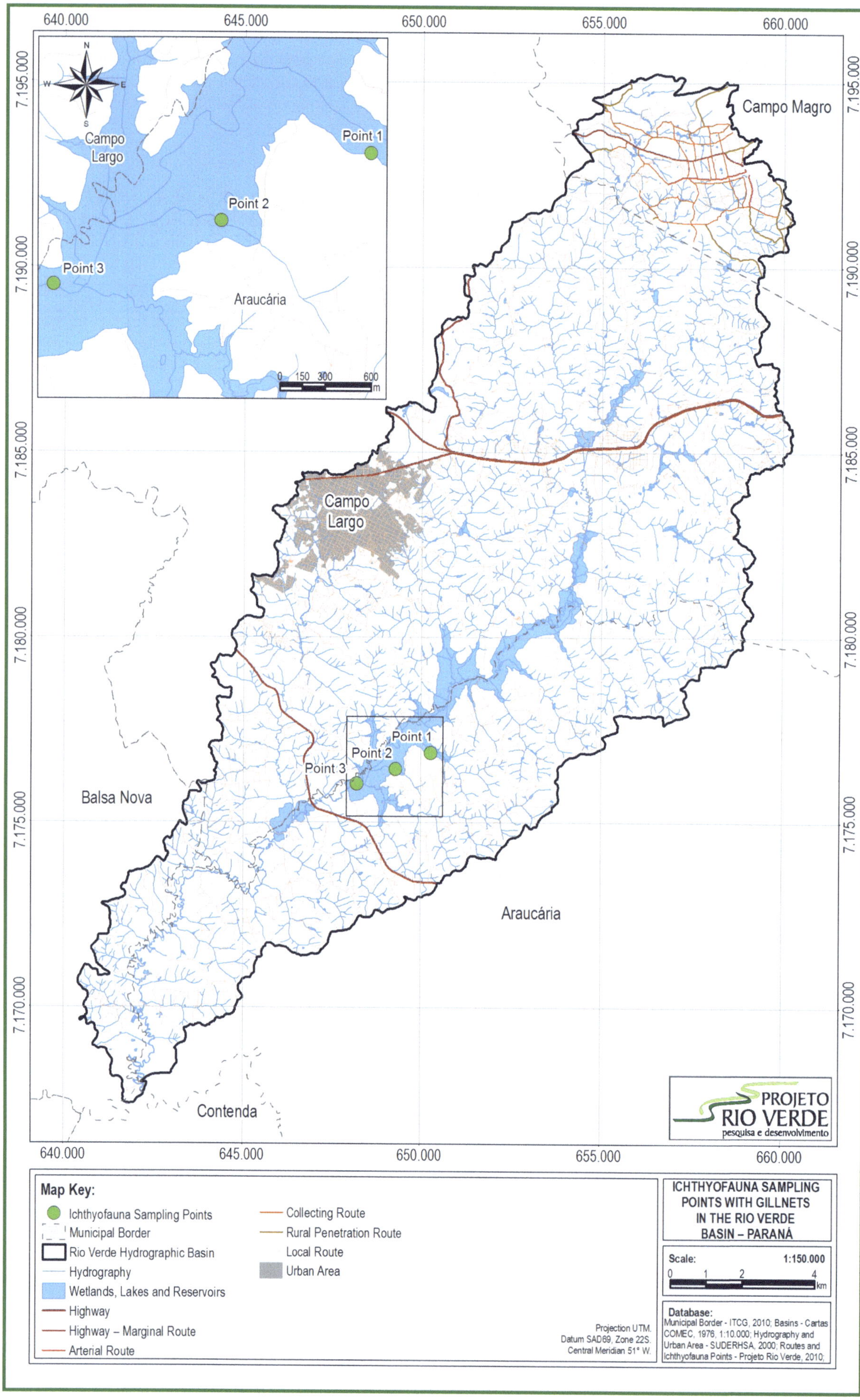

FIGURE 1 – ICHTHYOFAUNA SAMPLING POINTS IN THE RIO VERDE RESERVOIR

Capture with gillnet

Capture of H. aff. malabaricus

Capture of H. derbyi

Capture of *G. brasiliensis*

Capture of *H.* aff. *malabaricus*

Capture of *H. commersoni*

Capture of tetras (*Astyanax*)

Gillnet

Screening

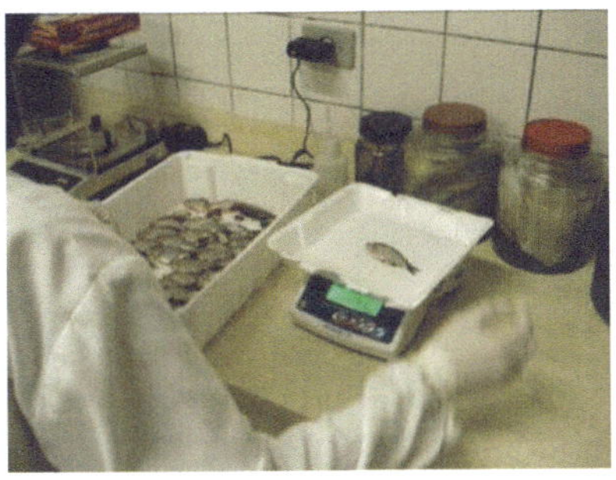
Weighing

Biometry

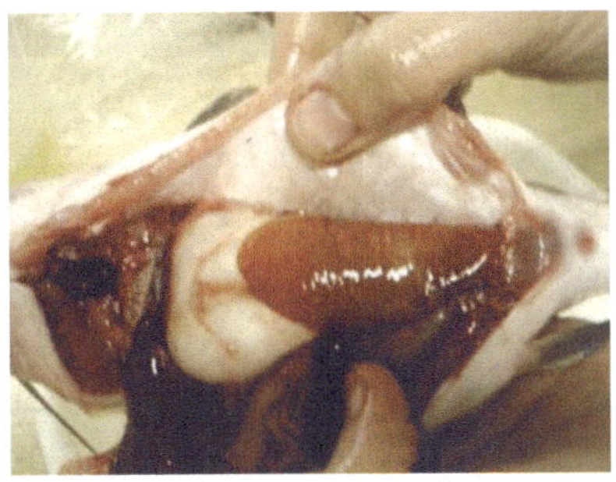
Removal of stomachs and gonads

FIGURE 2 – FIELD AND LABORATORY ACTIVITIES PERFORMED DURING THE DEVELOPMENT OF THE RIO VERDE RESERVOIR ICHTHYOFAUNA PROJECT

The highest values of abundance measured at sampling points 1, 2 and 3, in terms of catch per unit effort were recorded for *A.minor* and *A. bifasciatus*. The species *O. longirostris* and *G. brasiliensis* also presented higher values than other species except at point 3, where the capture of *Tilapia rendalli* was higher than the value recorded for *G. brasiliensis*. The highest values of catch per unit effort by weight for points 1 and 3 were recorded for *H.* aff. *malabaricus, A. minor*, *A. bifasciatus*, *H. commersoni* and *O. longirostris*. At point 2, the highest values by weight were recorded for *A. minor*, *R. quelen*, *A. bifasciatus*, *H. commersoni* and *H.* aff. *malabaricus* (Table 3).

The ichthyofauna of the study area presented the same general pattern as other reservoirs located in the Metropolitan Region of Curitiba, with few species and a high degree of endemism. About 45% of the recorded species are endemic and most of the individuals captured belong to these species (83%) (Figure 3).

Figure 4 presents a set of estimates of diversity and evenness for all three sampling points during the study period. Figure 5 presents those estimates for the monthly samplings. Diversity was estimated by the Shannon index and evenness by the Pielou index. The diversity and evenness values did not present differences across sampling points; however, when the sampling phase (month) is considered, higher values were observed in the samples collected in the summer.

To assess the structure of fish assemblages among sampling points and monthly field phases, data of catch per unit effort by individual of each species were transformed and summarized through a detrended correspondence analysis (DCA). The ordering of the results by the DCA did not reveal patterns of spatial and temporal variation through the recorded fish assemblage (Figure 6). The use of a one-factor analysis of variance on scores derived from the first axis of DCA revealed that there are no significant differences across sampling points ($F= 0.4765$; $p>0.05$) or sampling months ($F=0.7736$; $p>0.05$).

The means of total length and weight were: 12.5 ± 5.3cm and 44.6 ± 107.6 g, respectively. The intervals of length and weight ranged from 2.5 to 49.0 cm and 0.3 to 1087.0 g, respectively. This allows us to characterize the fish population of this reservoir as small to medium size.

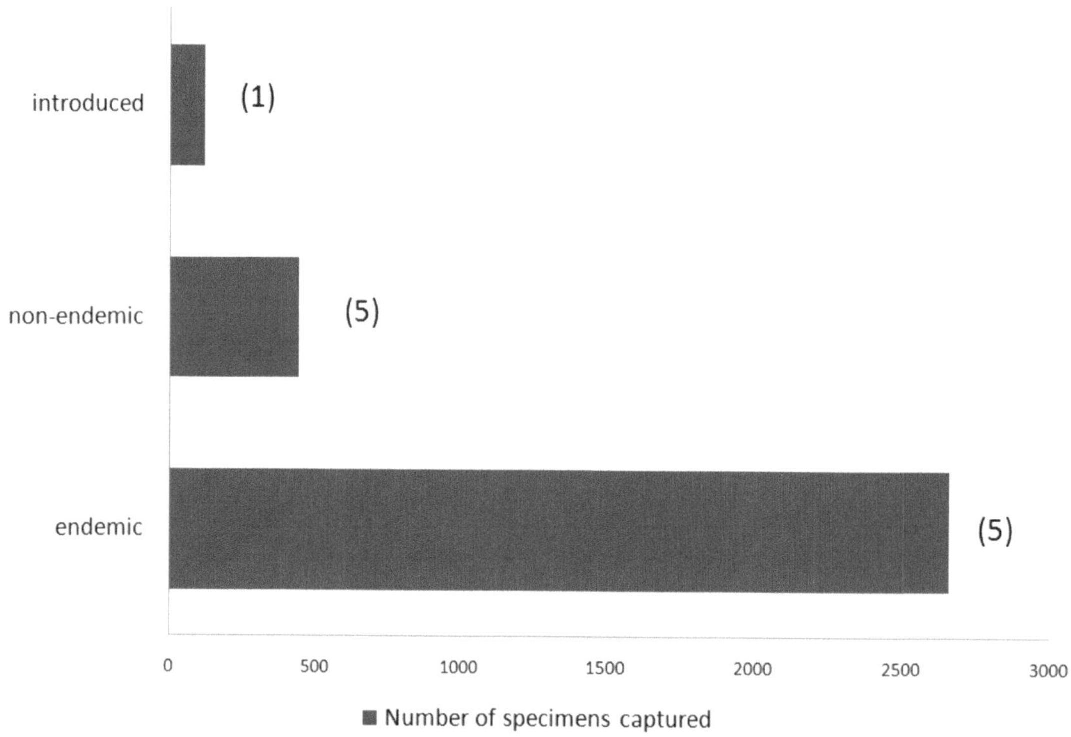

FIGURE 3 – NUMBER OF CAPTURED INDIVIDUALS OF ENDEMIC, NON-ENDEMIC AND INTRODUCED SPECIES RECORDED IN THE RIO VERDE RESERVOIR DURING THE STUDY PERIOD. THE NUMBER OF RECORDED SPECIES IS PRESENTED IN BRACKETS

SPECIES	POINT 1			POINT 2			POINT 3		
	CPUE ind	CPUE weight	C	CPUE ind	CPUE weight	C	CPUE ind	CPUE weight	C
A. bifasciatus	212	8.92	+++	193.5	8.88	+++	99.2	2.83	+++
A. minor	264.5	9.83	+++	329.5	13.87	+++	198.3	5.23	+++
A. serratus	1	0.03	+	1.5	0.04	+	1.7	0.03	+
O. longirostris	61.5	5.48	+++	55	4.71	+++	52.5	3.16	+++
H. aff. malabaricus	21.5	19.14	+++	11	6.76	++	18.3	8.56	++
C. paleatus	2	0.03	+	12	0.21	++	-	-	-
R. quelen	5	3.35	+	8.5	9.81	++	2.5	1.63	+
H. derbyi	-	-	-	1	0.34	+	-	-	-
H. commersoni	6	6.79	++	10	7.61	++	5	4.23	+
G. brasiliensis	44.5	1.77	+++	61.5	0.75	+++	39.2	0.75	+++
T. rendalli	25	1.89	++	10	1.88	++	41.7	1.89	++

TABLE 3 – CATCH PER UNIT EFFORT BY NUMBER (CPUE IND = NUMBER OF INDIVIDUALS/ 100 m^2 OF GILLNET IN 24 h) AND BIOMASS (CPUE WEIGHT = kg/100 m^2 OF GILLNET IN 24 h) AND CATCH CONSISTENCY (C) (+++ CONSTANT; ++ ACCESSORY; + ACCIDENTAL; – ABSENT) OF THE FISH SPECIES RECORDED IN THE RIO VERDE RESERVOIR DURING THE STUDY PERIOD

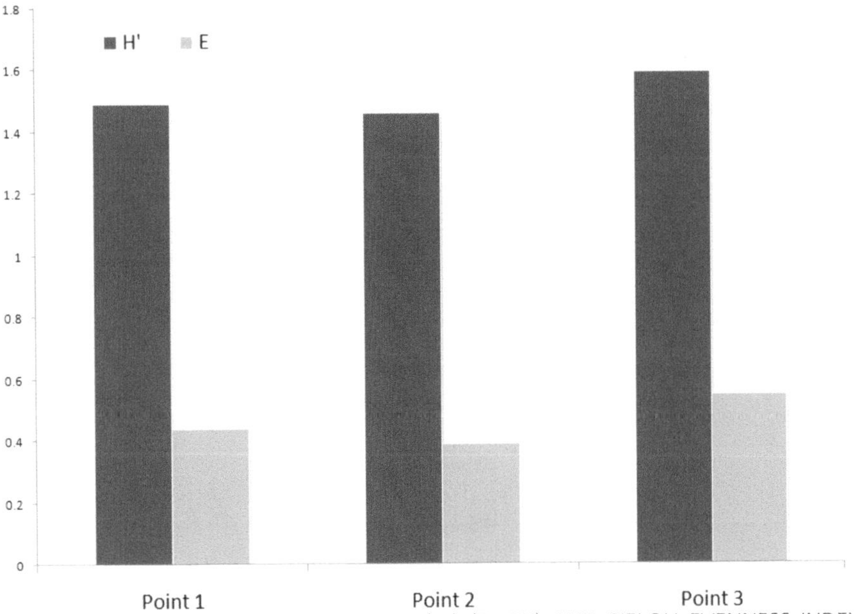

FIGURE 4 – SHANNON DIVERSITY INDEX (H') (DARK) AND PIELOU EVENNESS INDEX (LIGHT) OF THE SAMPLES COLLECTED FROM THE THREE SAMPLING POINTS IN THE RIO VERDE RESERVOIR DURING THE STUDY PERIOD

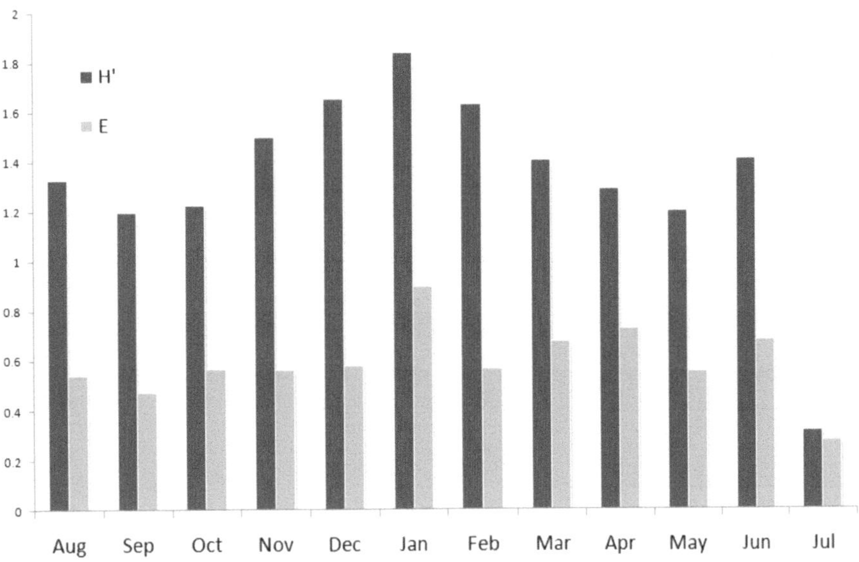

FIGURE 5 – SHANNON DIVERSITY INDEX (H') (DARK) AND PIELOU EVENNESS INDEX (LIGHT) FOR THE MONTHLY SAMPLING CONDUCTED IN THE RIO VERDE RESERVOIR DURING THE STUDY PERIOD

The relation between weight and total length was obtained for both sexes and expressed by the equation: *Weight*= *a* x *Length*^{b} (VAZZOLER, 1981), where the parameters *a* (linear coefficient) and *b* (angular coefficient/condition factor) were obtained adjusting the logarithm of the dependent and independent variables by the least squares method (Table 4). The values of angular coefficient (*b*) varied from 2.72 for *O. longirostris* to 3.54 for *R. quelen*. The value of *b* is the constant of relative growth and tends to produce values close to 3, indicating an isometric growth or with similar incremental growth rates in different parts of the body (WEATHERLEY; GILL, 1987). However, this cubic relation is not always followed and according to PAULY (1993) may be above 3 (positive allometry), or below 3 (negative allometry), indicating changes in fish development related to the environment, lack of food or even parasitism.

All specimens of *O. longirostris* (pike characin) captured in the Rio Verde Reservoir were parasitized by a large number of nematodes (Figure 7), which may have influenced the well-being of the species. It is possible that the high prevalence and intensity of the fauna parasitizing this host (paratenic or intermediate) is the result of direct consumption of larvae and pupae of Chironomidae, as the immature aquatic stages of this Diptera may act as intermediate hosts of nematodes. Furthermore, the hypothesis that the infestation may have occurred by predation of tetras of the *Astyanax* genus by the pike characin cannot be discarded due to the large amount of Chironomidae larvae in the diet of the tetra. According to Eiras (1994), piscivorous fish may be infected by predation of previously parasitized fish.

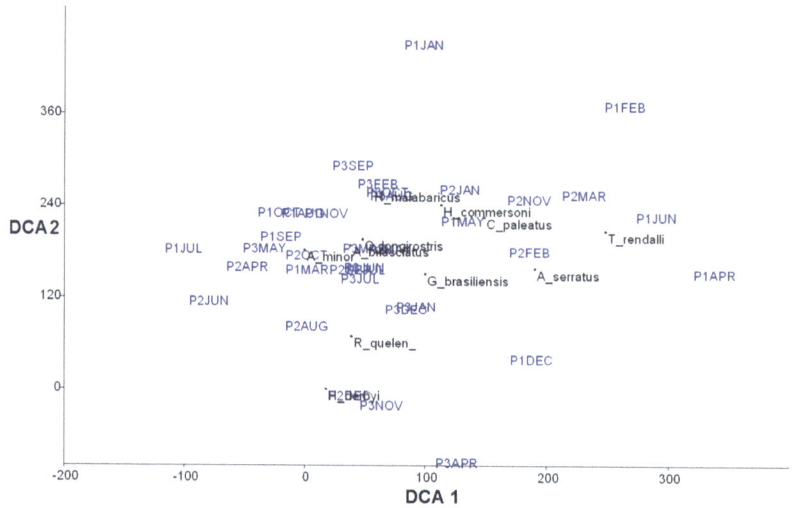

FIGURE 6 – SORTING OF FISH ASSEMBLAGES AMONG THE SAMPLING POINTS (P1 – SAMPLING POINT 1, P2 – SAMPLING POINT 2, P3 – SAMPLING POINT 3) AND THE MONTHLY FIELD PHASES (JAN., FEB., MAR., APR., MAY, JUN., JUL., AUG., SEP., OCT., NOV., DEC.) BY DETRENDED CORRESPONDENCE ANALYSIS (DCA)

CAPTURED SPECIES	TL (cm)	WEIGHT (g)	RELATION WEIGHT – LENGTH		
	(mean ± sd)	(mean ± sd)	a	b	r^2
A. bifasciatus	11.4 ± 1.15	22.2 ± 7.26	0.0058	3.35	0.88
A. minor	11.0 ± 1.02	20.3 ± 5.97	0.0064	3.34	0.94
A. serratus	10.4 ± 0.55	14.9 ± 1.59	-	-	-
O. longirostris	16.6 ± 2.27	46.6 ± 18.96	0.0213	2.72	0.92
H. aff. malabaricus	31.2 ± 7.88	452.2 ± 265.96	0.0128	2.99	0.99
C. paleatus	7.9 ± 0.36	8.4 ± 1.33	0.0188	2.94	0.77
R. quelen	31.9 ± 5.84	372.8 ± 198.61	0.0015	3.54	0.95
H. derbyi	22.0 ± 0.07	172.0 ± 0.07	-	-	-
H. commersoni	32.6 ± 11.53	450.1 ± 340.24	0.0048	3.19	0.99
G. brasiliensis	10.9 ± 3.73	32.4 ± 42.56	0.0188	2.94	0.77
T. rendalli	8.2 ± 7.41	54.7 ± 121.75	0.0147	3.14	0.99

TABLE 4 – MEAN VALUES OF TOTAL LENGTH (TL), WEIGHT AND PARAMETERS OF THE RELATION WEIGHT – TOTAL LENGTH OF THE CAPTURED SPECIES (SD – STANDARD DEVIATION, *A* – LINEAR COEFFICIENT, *B* – ANGULAR COEFFICIENT / CONDITION FACTOR, R^2 – COEFFICIENT OF DETERMINATION)

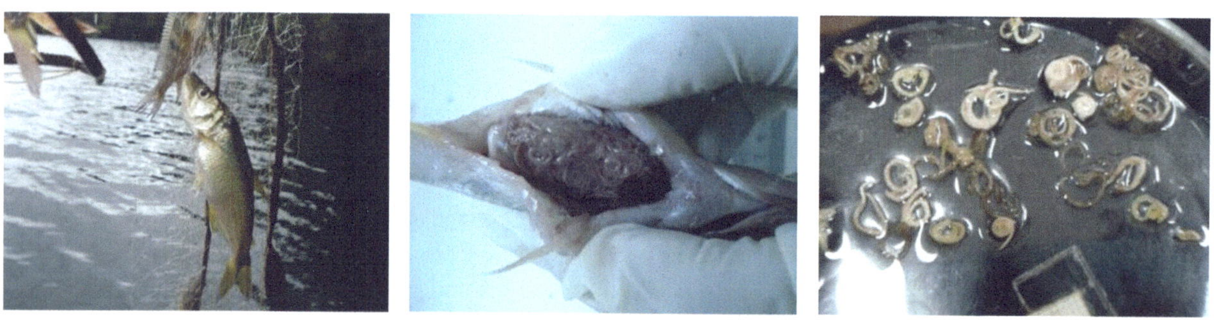

FIGURE 7 – REMOVAL AND EVALUATION OF PARASITES (NEMATODES) FOUND IN THE STOMACH OF SPECIMENS OF *O. LONGIROSTRIS* CAPTURED FROM THE RIO VERDE RESERVOIR

3.2 FEEDING

A species diet was determined through the analysis of digestive tracts. The contents were analyzed under stereoscopic microscope and the identification of the food items was performed with the aid of a specialized comparative bibliography and specialists in the area.

The items were quantified by point counting and the contribution of each item was determined by the proportion of squares occupied by the item on a checkered surface in relation to the total number of squares occupied by the content (VITULE *et al.*, 2008).

The trophic categories considered for grouping the fish species follow that proposed by HAHN *et al.* (1997): (i) herbivorous, fish that feed on higher plant material such as leaves, seeds and fruits of aquatic and terrestrial plants, besides filamentous algae; (ii) insectivorous, fish that feed on aquatic and terrestrial insects; (iii) detritivorous, fish that ingest sediment along with debris and invertebrate droppings; (iv) benthophagous, fish that explore the bottom and select their prey from the benthonic fauna; (v) ichthiophagous, also called piscivorous, fish that feed on other fish; and (vi) omnivorous, fish that indiscriminately eat substances of animal and vegetal origin.

The analysis performed allowed us to characterize the diet of the captured species according to their food preferences (Table 5).

Detritivorous and piscivorous fish were the most diverse and abundant groups. Although just one omnivorous species was identified (*A. bifasciatus*), the number of specimens and the biomass of this trophic category prevailed over the benthophagous, insectivorous and herbivorous species (Figure 8).

Although these results do not allow us to define the trophic categories unique to the species analyzed, due to the high food plasticity that most Neotropical species present (LOWE-McCONNEL, 1987), the small variety of allochthonous resources is probably due to the loss of riparian vegetation, given that the absence of this type of vegetation can reduce the supply of organic debris and materials for the shoreline of a reservoir (SMITH *et al.*, 2003).

TABLE 5 – FOOD HABITS OF THE SPECIES RECORDED IN THE RIO VERDE RESERVOIR

SPECIES	MORPHOLOGICAL CHARACTERISTIC	PREVAILING ITEM	TROPHIC CATEGORY
A. bifasciatus	Small frontal mouth, cuspid teeth	Insects/plants	Omnivorous
A. minor		Debris /insect larvae	Detritivorous
A. serratus		Insects	Insectivorous
O. longirostris	Wide mouth, developed teeth	Fish	Piscivorous
H. aff. malabaricus		Fish	Piscivorous
C. paleatus	Sub-inferior mouth, developed lips, without teeth	Insect larvae	Benthophagous
R. quelen	Wide mouth, dentigerous plates	Fish	Piscivorous
H. derbyi	Inferior oral disk, scraping teeth, long gut	Debris/sediment	Detritivorous
H. commersoni		Debris	Detritivorous
G. brasiliensis	Protractile mouth, small teeth	Insect larvae	Benthophagous
T. rendalli		Filamentous algae	Herbivorous

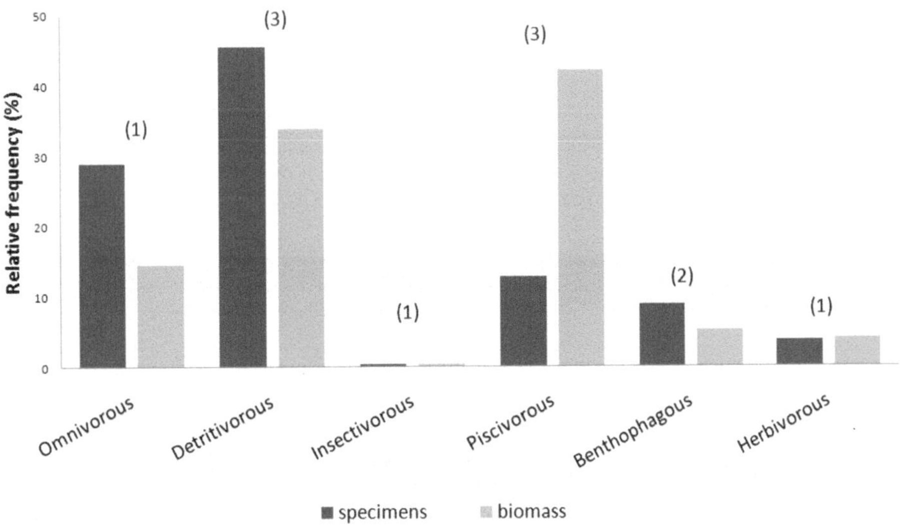

FIGURE 8 – RELATIVE FREQUENCY (%) OF THE TROPHIC CATEGORIES OF FISH RECORDED IN THE RIO VERDE RESERVOIR DURING THE STUDY PERIOD. SPECIMEN ABUNDANCE AND TOTAL BIOMASS IS INFORMED FOR EACH CATEGORY, THE NUMBER OF SPECIES IS IN BRACKETS

It is possible that the coexistence of species in the reservoir is facilitated by the observed abundance of major food resources (WOOTTON, 1990), such as debris, fish and insects, which can indicate sharing of resources (SCHOENER, 1974). The limited variety of items found in the diets may also indicate similar feeding strategies such as the ability to exploit different sections of the water column, a fact already observed for some species (ESTEVES, 1996; LUZ ; OKADA, 1999; SMITH *et al.*, 2003).

3.3 REPRODUCTION

After taking biometric data, the captured fish were ventrally sectioned to determine the sex and stage of gonadal maturation for reproduction analysis. The gonads were removed, weighed and the gonadosomatic index (GSI) was determined for each species.

The assessment of gonadal development stages was based on the scale proposed by Vazzoler (1996). The preparation of the maturity scale was based on: macroscopic characteristics, such as the size of the specimens, the occupancy of the gonads in the cavity, color, transparency, vascularization and visualization of oocytes; and microscopic characteristics, based on volume of the oocytes, affinity for dyes and the presence of membranes and their modifications.

In the present study only females had their gonads microscopically examined. This is justified because the female germline consists of larger cells than the male line, facilitating the characterization of gonadal development (WEST, 1990). Besides the production of the female gamete, females are responsible for incorporating nutrients during oogenesis for use in embryonic development, thus making the female gonadal development more complex (FÁVARO *et al.*, 2003). This analysis also allows us to determine the type of spawning practiced by the species.

According to the composition and frequency of the development phases of the ovarian follicles, the following stages of ovarian development were established: immature, maturing, mature, and spawned. Some species present a semi-spawned stage.

Reproductive activity was higher in spring and summer (Table 6), with higher frequency of females in reproductive activity between September and January. This was also observed in other species in the Paraná Basin (VAZZOLER; MENEZES, 1992; VAZZOLER *et al.*, 1997), and for the same species from the Segredo Reservoir in the middle Iguaçu River (SUZUKI; AGOSTINHO, 1997) and the Iraí Reservoir (ABILHOA, 2005).

Considering the five most abundant species, *A. minor*, *A. bifasciatus*, *O. longirostris*, *G. brasiliensis* and *H.* aff. *malabaricus*, the analysis of the female individual GSI (Figures 9, 10, 11, 12 and 13) demonstrated varying intensities of reproductive activity over the period.

The tetras, *A. minor* and *A. bifasciatus*, reached the peak of ovarian development in the months of September and October and the values of GSI slowly decreased in the subsequent months, suggesting a long reproductive period. This long period, associated with the presence of semi-spawned ovaries found in the histological analysis, allows us to characterize the spawning in stages for both species. Mature individuals were observed throughout the entire study period.

TABLE 6 – REPRODUCTIVE ACTIVITY OF THE SPECIES RECORDED IN THE SAMPLES OF RIO VERDE RESERVOIR

SPECIES		SAMPLING PERIOD (2008-2009)											
		Aug	sep	oct	nov	dec	jan	feb	mar	apr	may	jun	jul
More abundant	*Astyanax bifasciatus*												
	Astyanax minor												
	Oligosarcus longirostris												
	Geophagus brasiliensis												
	Hoplias aff. Malabaricus												
Less abundant	*Hypostomus commersoni*												
	Hypostomus derbyi												
	Rhamdia quelen												
	Tilapia rendalli												
	Astyanax serratus												
	Corydoras paleatus												

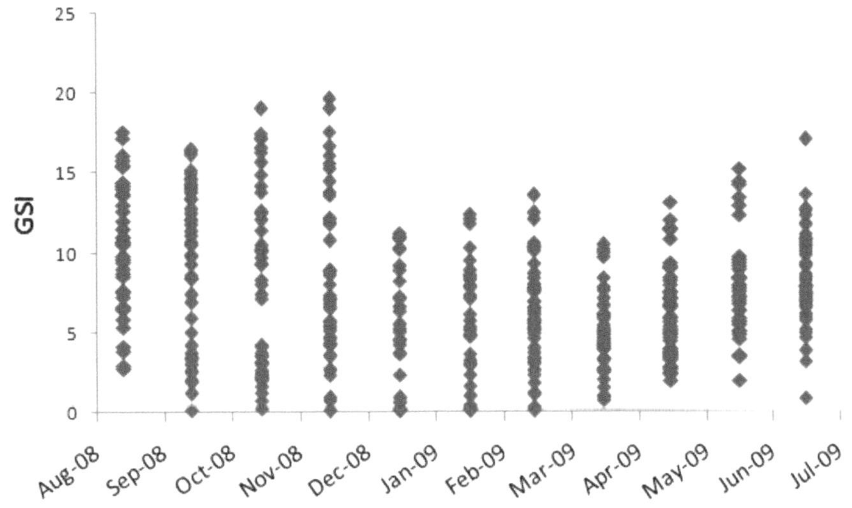

FIGURE 9 – INDIVIDUAL GONADOSOMATIC INDEX (GSI) OF *A. MINOR* FEMALES OF RIO VERDE RESERVOIR

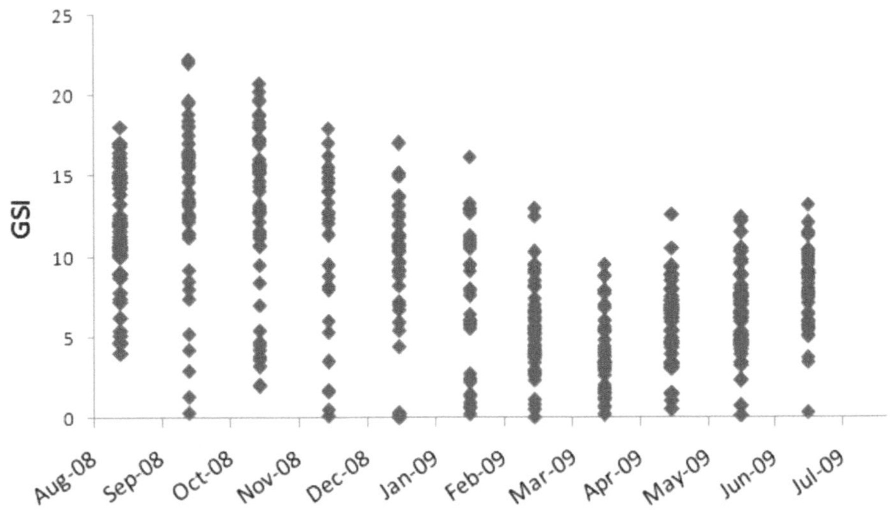

FIGURE 10 – INDIVIDUAL GONADOSOMATIC INDEX (GSI) OF *A. BIFASCIATUS* FEMALES OF RIO VERDE RESERVOIR

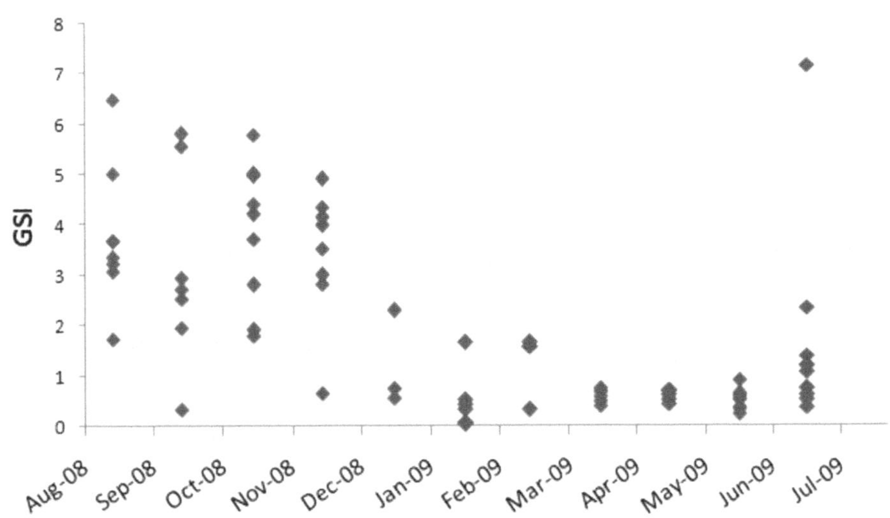

FIGURE 11 – INDIVIDUAL GONADOSOMATIC INDEX (GSI) OF *O. LONGIROSTRIS* FEMALES OF RIO VERDE RESERVOIR

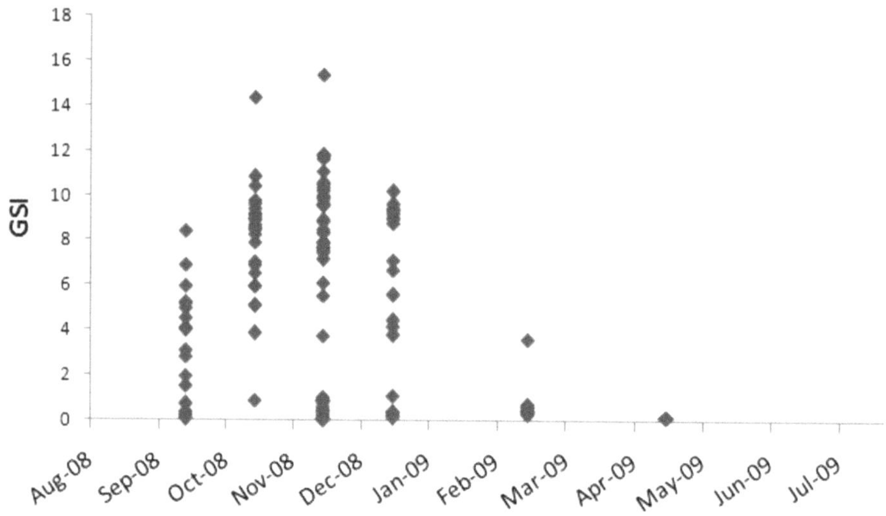

FIGURE 12 – INDIVIDUAL GONADOSOMATIC INDEX (GSI) OF *G. BRASILIENSIS* FEMALES OF RIO VERDE RESERVOIR

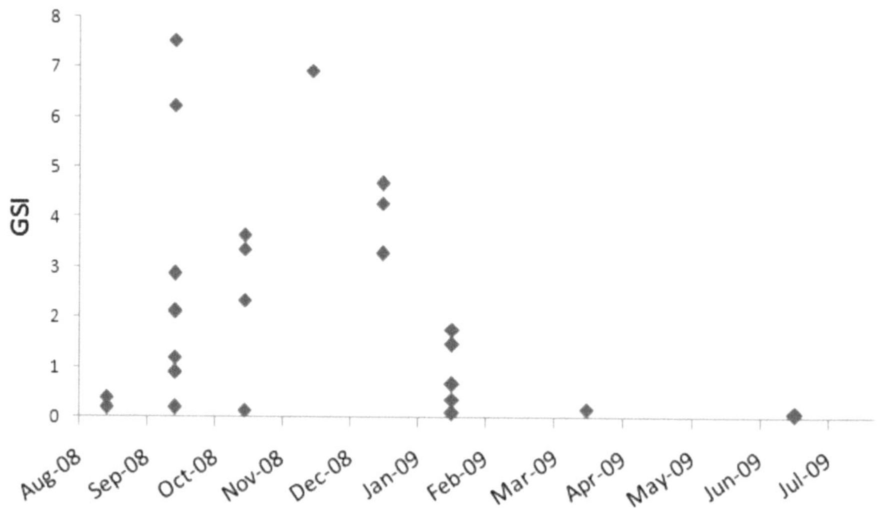

FIGURE 13 – INDIVIDUAL GONADOSOMATIC INDEX (GSI) OF *H.* AFF. *MALABARICUS* FEMALES OF RIO VERDE RESERVOIR

The species *G. brasiliensis, H.* aff. *Malabaricus,* and *O. longirostris* showed reproductive activity from the end of winter to the beginning of summer. The *O. longirostris* presented a long spawning period and a period of recovery from reproduction activity, characterized by low GSI values between February and July, at which point their gonads mature again. Semi-spawned ovaries were found in the histological analysis, proving that the species has partial spawning. Although the analyzed species have long spawning periods, *O. longirostris* has a long period of gonadal recuperation after the reproductive period. This characteristic is dissimilar to the *Astyanax* genus which always have mature individuals in the reproductive process, despite a decrease in the reproductive intensity during the annual cycle. This opportunistic tactic (WINEMILLER, 1989) is important for success in this type of environment (BAILLY *et al.*, 2005); with drastic changes in river levels and physical and chemical conditions. The maintenance of viable populations is ensured due to a long reproductive period.

Thus, as already observed in other reservoirs of the Iguaçu River, the species with greater flexibility in their reproduction strategies (spawning period and place) were more successful in colonizing the reservoir.

4. FINAL CONSIDERATIONS

The number of species recorded in the Rio Verde Reservoir (11 species) accounted for 23% of the types of species found in the Upper Iguaçu River region and 10% of all species recorded for this basin (SEVERI; CORDEIRO, 1994; GARAVELLO *et al.*, 1997; INGENITO *et al.*, 2004) (Figure 14). The reservoir ichthyofauna presented the same general pattern of the ichthyofauna of the Iguaçu River Basin and the contribution of different orders reflects the situation described for Neotropical rivers by Lowe-McConnell (1987).

The sampled ichthyofauna can be divided into three categories of species depending on their original distribution: (i) endemic species, i.e., those unique to the Iguaçu River Basin; (ii) non-endemic native species, those that also occur in other hydrographic basins; and (iii) introduced species, which occur in this environment as a function of accidental or intentional introduction (aquaculture or dam stocking). About 45% of the species recorded in the Rio Verde Reservoir are unique to the Iguaçu River Basin thus demonstrating the importance of the local and regional processes in determining the composition and structure of the ichthyocenosis of these aquatic environments.

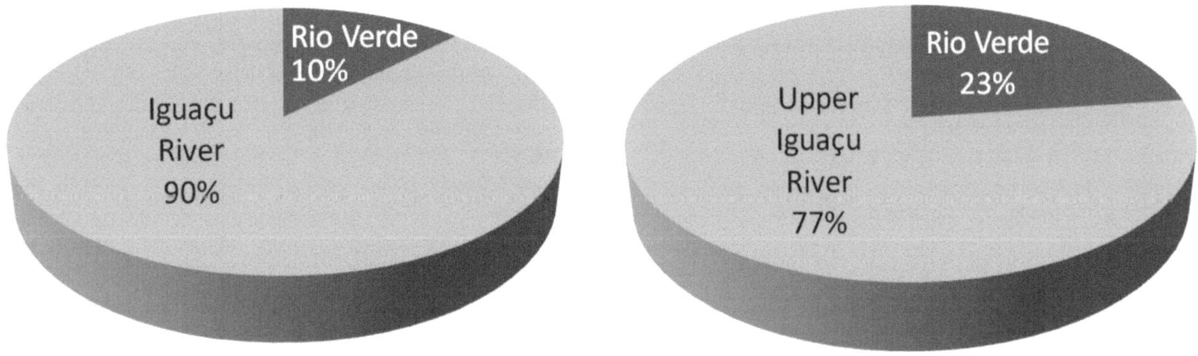

FIGURE 14 – PROPORTION OF SPECIES REGISTERED IN THE RIO VERDE RESERVOIR AND ALL THE SPECIES SURVEYED FOR THE UPPER IGUAÇU RIVER REGION AND FOR THE ENTIRE IGUAÇU RIVER BASIN

The species *Astyanax minor* (yellow-tail tetra), *Astyanax bifasciatus* (red-tail tetra), *Oligosarcus longirostris* (pike characin) and *Geophagus brasiliensis* (pearl cichlid) had markedly high and consistent catch numbers in the samples, while *Hoplias* aff. *malabaricus* (trahira) showed no significant contributions to the samples, although it was caught during most of the sampling period. The highest values of catch per unit effort by weight were recorded for *H.* aff. *malabaricus, A. minor*, *A. bifasciatus*, *H. commersoni*, *O. longirostris* and *R. quelen*. The major contribution by weight of *H.* aff. *malabaricus*, *H. commersoni* and *R. quelen* was due to the capture of larger individuals.

The results obtained indicate that the fish community of the Rio Verde Reservoir is not very diverse and with few dominant species. As such, it is within the pattern expected in the Metropolitan Region of Curitiba. This reduced ichthyofauna is associated with the particular characteristics of this ecosystem that shaped the composition and structure of the ichthyofauna during colonization of this environment

Furthermore, the analyzed data showed high catch values by individual and by weight, abundance and dominance for the tetras of the *Astyanax* genus, suggesting that the species found favorable conditions in the reservoir which enabled it to remain a dominant species.

High abundance of *A. minor* and other species of this genus was also reported for the reservoirs of Segredo and Foz do Areia (AGOSTINHO; GOMES, 1997), Alagados (LUIZ *et al*., 2003), for the UHE Nova Avanhandava (CESP, 1998), and Iraí (ABILHOA, 2005). The results obtained allow us to conclude that environmental changes due to the construction of the dam, combined with high levels of opportunism, reproductive strategy, and a broad feeding spectrum shown for the species (ABILHOA; AGOSTINHO, 2007), are responsible for their abundance in catches because the species found favorable conditions for its development in this environment.

5. REFERENCES

ABILHOA, V. (2004). *Composição, aspectos biológicos e conservação da ictiofauna do alto curso do rio Iguaçu, Região Metropolitana de Curitiba, Paraná, Brasil.* Thesis (Doctor's degree), Universidade Federal do Paraná. 84p.

ABILHOA, V. (2005). Ictiofauna. In: ANDREOLI, C. V.; CARNEIRO, C. (Eds.). *Gestão integrada de mananciais de abastecimento eutrofizados.* Curitiba, Sanepar, 2005.

ABILHOA, V., AGOSTINHO, A. A. (2007). *Aspectos biológicos de duas espécies de Astyanax (Teleostei, Characidae) em lagoas marginais do alto rio Iguaçu, Paraná, Brasil.* Estudos de Biologia, 29: 23-32.

AGOSTINHO, A. A. (1992). Manejo de recursos pesqueiros em reservatórios. pp.106-121. In: AGOSTINHO, A. A.; BENEDITO--CECÍLIO, E. *Situação atual e perspectivas da ictiologia no Brasil.* Maringá: Eduem, 127p.

AGOSTINHO, A. A.; JÚLIO JR., H. F.; BORGHETTI, J. R. (1992). Considerações sobre os impactos dos represamentos na ictiofauna e medidas para sua atenuação. Um estudo de caso: Reservatório de Itaipu. *Revista Unimar 14* (supl.): 89-107.

AGOSTINHO, A. A.; ZALEWSKI, M. (1996). *A planície alagável do Alto Rio Paraná: importância e preservação.* Maringá: Eduem, 100p.

AGOSTINHO, A. A.; GOMES, L. C. (1997). *Reservatório de Segredo: bases ecológicas para o manejo.* Maringá: Eduem: 387p.

AGOSTINHO, A. A.; JÚLIO JR., H. F.; GOMES, L. C.; BINI, L. M.; AGOSTINHO, C. S. (1997a). Composição, abundância e distribuição espaço-temporal da ictiofauna. pp.229-248. In: VAZZOLER, A. E. A. M.; AGOSTINHO, A. A.; HAHN, N. S. *A planície de inundação do alto rio Paraná: aspectos físicos, biológicos e socioeconômicos.* Maringá: Eduem, 460p.

AGOSTINHO, A. A.; BINI, L. M.; GOMES, L. C. (1997b). Ecologia de comunidades de peixes da área de influência do reservatório de Segredo. pp.97-111. In: AGOSTINHO, A. A.; GOMES, L. C. *Reservatório de Segredo: bases ecológicas para o manejo.* Maringá: Eduem, 387p.

AGOSTINHO, A. A.; GOMES, L. C.; PELICICE, F. M. (2007). *Ecologia e Manejo de recursos pesqueiros em reservatórios do Brasil.* Maringá: Eduem, 501p.

BAILLY, D.; AGOSTINHO, A. A.; SUZUKI, H.; LUIZ, E. A. (2005). Características reprodutivas de espécies de *Astyanax* e sucesso na colonização de reservatórios do Rio Iguaçu. pp. 243-252. In: RO-

DRIGUES, L.; THOMAZ, S. M.; AGOSTINHO, A. A.; GOMES, L. C. (2005). *Biocenoses em reservatórios: padrões espaciais e temporais*. São Carlos, Rima, 333p.

CECÍLIO, E. B.; AGOSTINHO, A. A., JÚLIO JR., H. F.; PAVANELLI, C. S. (1997). Colonização ictiofaunística do reservatório de Itaipu e áreas adjacentes. *Revista Brasileira de Zoologia 14*(1): 1-14.

CESP – Companhia Energética de São Paulo. (1998). *Conservação e manejo nos Reservatórios: Limnologia, Ictiologia e Pesca*. São Paulo: CESP, 166p.

DAJOZ, R. (2005). *Princípios de Ecologia*. Porto Alegre: Artmed, 519p.

EIRAS, J. C. (1994). *Elementos de ictioparasitologia*. Porto: Fundação Eng. António de Almeida, 339p.

ESTEVES, K. E. (1996). Feeding ecology of three *Astyanax* species (Characidae, Tetragonopterinae) from a floodplain lake of Mogi-Guaçu River, Paraná River Basin, Brazil. *Environmental Biology of Fishes 46:* 83-101.

FÁVARO, L. F.; LOPES, S. C. G.; SPACH, H. L. (2003). Reprodução do peixe-rei, *Atherinella brasiliensis* (Quoy ; Gaimard) (Atheriniformes, Atherinidae), em uma planície de maré adjacente à gamboa do Baguaçu, Baía de Paranaguá, Paraná, Brasil. *Revista Brasileira de Zoologia, 20*(3): 501-506.

GARAVELLO, J. C.; PAVANELLI, C.; SUZUKI, H. (1997). Caracterização da ictiofauna do rio Iguaçu. pp. 61-84. In: AGOSTINHO AA, GOMES LC (Eds.). *Reservatório de Segredo: bases ecológicas para o manejo*. Maringá: Eduem, 387p.

HAHN, N. S.; FUGI, R.; ALMEIDA, V. L. L.; RUSSO, M. R.; LOUREIRO, V. E. (1997). Dieta e atividade alimentar de peixes do Reservatório de Segredo. pp. 141-162. In: AGOSTINHO, A. A.; GOMES. L. C. Reservatório de Segredo: bases ecológicas para o manejo. Maringá: Eduem, 387p.

INGENITO, L. F. S.; DUBOC, L. F.; ABILHOA, V. (2004). *Contribuição ao conhecimento da ictiofauna da bacia do alto Rio Iguaçu, Paraná, Brasil.* Arquivos de Ciências Veterinárias e Zoologia da Unipar 7(1): 23-36.

JONES, C. G.; LAWTON, J. H.; SHACHAK, M. (1994). *Organisms as ecosystem engineers. Oikos,* 69: 373-386.

LOWE-McCONNELL, R. H. (1987). *Ecological studies in tropical fish communities.* Cambridge: Cambridge University Press, 382p.

LOWE-McCONNELL, R. H. (1999). *Estudos ecológicos de comunidades de peixes tropicais.* São Paulo: Edusp, 534p.

LUIZ, E. A.; GOMES, L. C.; AGOSTINHO, A. A.; BULLA, C. K. (2003). Influência de processos locais e regionais nas assembléias de peixes em Reservatórios do Estado do Paraná, Brasil. *Acta Scientiarum Biologial Science 25*(1): 107-114.

LUZ, K. D. G.; OKADA, E. K. (1999). Diet and dietary overlap of three sympatric fish species in lakes of the Upper Paraná River Floodplain. *Brazilian Archives of Biology and Technology, 42*(4): 441-447.

PAINE, R. T. (1969). *A note on trophic complexity and community stability.* American Naturalist, 103: 91-93.

PAINE, R. T. (1980). Food Webs: Linkage, Interaction Strength and Community Infrastructure. *Journal of Animal Ecology, 49*: 666-685.

PAULY, D. (1993). *Editorial: Fishbyte Section.* The ICLARMQ 16(2/3): 1-26.

PERES NETO, P. R.; BIZERRIL, C. R. S. F.; IGLESIAS, R. (1995). *An overview of son aspects of river ecology: a case study on fish assemblages distribution in an eastern Brazilian coastal river.* Oecologia Brasiliensis 1: 317-334.

PETRERE JR., M. (1996). *Fisheries in large tropical reservoirs in South América.* Lakes; Reservoirs Research and Management 2: 111-133.

RODRIGUES, L.; THOMAZ, S. M.; AGOSTINHO, A. A.; GOMES, L. C. (2005). *Biocenoses em reservatórios: padrões espaciais e temporais.* São Carlos, Rima, 333p.

SCHOENER, T. W. (1974). *Resource partioning in ecological communities.* Science 185: 27-39.

SEVERI, W. ; CORDEIRO, A. A. M. (1994). *Catálogo de peixes da bacia do rio Iguaçu.* Curitiba, IAP/GTZ, 128p.

SINDERMANN, C. J. (1979). *Pollution associated diseases and abnormalities of fish and sellfish: a review.* Fishery Bulletin 76(4): 717-749.

SMITH, W. S.; PEREIRA, C. C. G. F.; ESPÍNDOLA, E. L. G.; ROCHA, O. (2003). A importância da zona litoral para a disponibilidade de recursos alimentares à comunidade de peixes em reservatórios. pp. 233-248. In: HENRY, R. (Coord.). *Ecótonos nas interfaces dos ecossistemas aquáticos.* São Carlos: Rima, 349p.

SUZUKI, H. L.; AGOSTINHO, A. A. (1997). Reprodução de peixes do reservatório de Segredo. pp.162-182. In: AGOSTINHO, A. A.; GOMES, L. C. *Reservatório de Segredo: bases ecológicas para o manejo.* Maringá: Eduem, 387p.

VAZZOLER, A. E. A. M. (1981). *Manual de métodos para estudos biológicos de populações de peixes; reprodução e crescimento.* Brasília: CNPq, 108p.

VAZZOLER, A. E. A. M.; MENEZES, N. A. (1992). Síntese dos conhecimentos sobre o comportamento reprodutivo dos Characiformes da América do Sul (Teleostei, Ostariophysi). *Revista Brasileira de Biologia, 52*(4): 626-640.

VAZZOLER, A. E. A. M. (1996). *Biologia da reprodução de peixes teleósteos: teoria e prática.* Maringá: Eduem/SBI/CNPq/Nupelia, 169p.

VAZZOLER, A. E. A. M.; AGOSTINHO, A. A.; HAHN, N. S. (1997). *A planície de inundação do alto rio Paraná: aspectos físicos, biológicos e socioeconômicos.* Maringá: Eduem, 460p.

VITULE, J. R. S.; BRAGA, M. R.; ARANHA, J. M. R (2008). *Ontogenetic, spatial and temporal variations in the feeding ecology of Deuterodon langei Travassos, 1957 (Teleostei: Characidae) in a Neotropical stream from the Atlantic rainforest, southern Brazil.* Neotropical Ichthyology, 6(2): 211-222.

WEATHERLEY, A. H.; GILL, H. S. (1987). *The Biology of Fish Growth.* London: Academic Press, 443p.

WEST, G. (1990). Methods of assessing ovarian development in fishes: a review. *Australian Journal of Marine and Freshwater Research, 41*: 199-222.

WINEMILLER, K. O. (1989). *Patterns of variation in life history among South American fishes in seasonal environments.* Oecologia, 81: 225-241.

WINEMILLER, K. O.; JEPSEN, D. B. (1998). Effects of seasonality and fish movement on tropical river food webs. *Journal of Fish Biology, 53*: 267–296.

WOOTTON, R. J. (1990). *Ecology of teleost fishes.* London: Chapman and Hall Ltd, 404p.

SECTION V

SOCIO-ECONOMIC ASPECTS

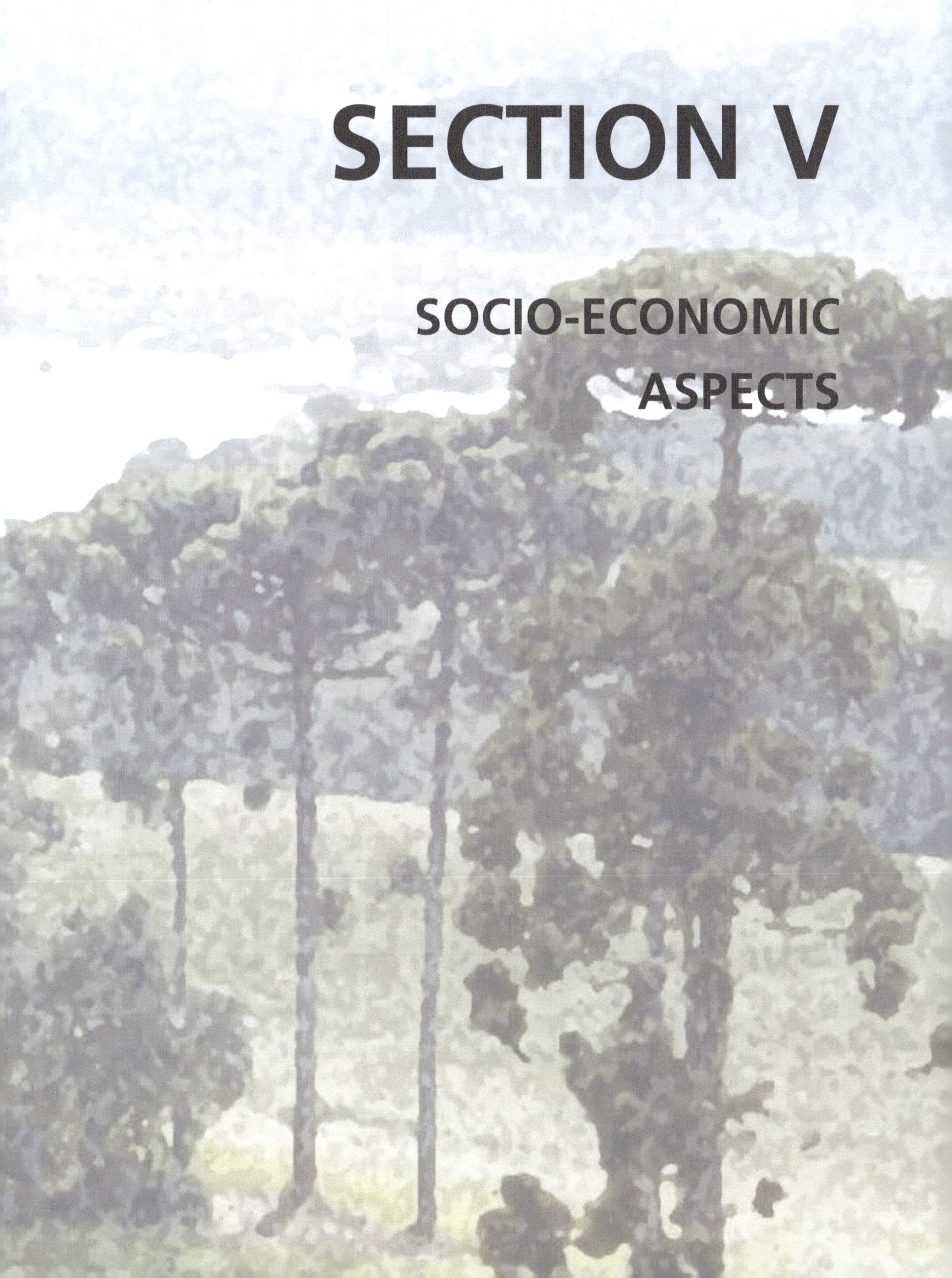

CHAPTER 16

PARTICIPATIVE METHODOLOGIES APPLIED IN THE RIO VERDE BASIN

Benno Doetzer
Fabiano A. S. Dalto
Jose Edmilson de Souza Lima
Daniel Larsen
Miguel Mansur Aisse
Edna Batistella Lopes

PARTICIPATIVE METHODOLOGIES APPLIED IN THE RIO VERDE BASIN

Benno Doetzer
Fabiano A. S. Dalto
Jose Edmilson de Souza Lima
Daniel Larsen
Miguel Mansur Aisse
Edna Batistella Lopes

SUMMARY

This chapter presents an alternative research method that has been used in communities downstream of Petrobras' Rio Verde dam, west of the Curitiba Metropolitan Region, Paraná, Brazil. This region covers three municipalities (Araucária, Campo Largo and Campo Magro) and is essentially made up of rural communities. The PRD (Participant Rural Diagnosis) method has been employed in this region. PRD involves several stages and research tools applied in collaboration with local community. The data collection methodology allows members of the local community to systematically think about their problems and possible solutions and share these with the research group. Instead of creating a 'positive' conscience, this method requires the creation of a 'dialogue of awareness' with the community. The method has been used in three communities: Colônia Cristina, located in Araucária; Colônia Dom Pedro II and Colônia Figueiredo, both in Campo Largo. Three related techniques has been employed: Participatory Mapping; Ranking of Problems and Potentials; and the Institutional Interrelationship Diagram (Venn Diagram). Questionnaires focusing on socio-economic characteristics and quality of life proved to be the most complicated step in the process. Early questionnaires proved inadequate as the community members resisted answering to some of the issues asked. Community reproach demonstrated some shortfalls of the questionnaire and the need to reformulate it. This episode illustrates one of the defining benefits of the method: interaction with the research 'object' leads to changes in the research itself. The study was carried out and concluded in July, 2010, and the collected data served as the basis for calculating the Rural Quality of Life Index for the Basin presented in this chapter

KEYWORDS

Rio Verde Basin, participant rural diagnosis, quality of life index, MEXPAR, Curitiba metropolitan region.

1. INTRODUCTION: THE NEED FOR AN INNOVATIVE RESEARCH METHODOLOGY

This chapter presents the research method that has been carried out within communities surrounding the Rio Verde. The assessment of the Rio Verde Basin, as any other field-based research, has required data on a variety of factors which contribute, or may contribute, to the degradation of local quality of life and the environment. To obtain these data, "normal" science uses positive data collection methods and data systematization according to the most accepted theory or technique until the results of the research, or a description of local conditions, are attained (DEMO, 1980; LAKATOS & MARCONI, 2007). Once the diagnosis is complete, any intervention in the object of research is simply a matter of procedure.

The application of positive science methodologies encounters great difficulties, however, when the 'object' of research involves people and their social interactions. People, as objects of research, have different interests and react differently to questions depending on who asks them and how questions are posed. Furthermore, people interact with both the research and researchers and reveal their information selectively, based on some type of strategy. As argued by Pedro Demo (1980, p.13), "Every methodological discussion carries inside a proposal, especially because it is impossible not to take a position [on an issue]." This research had, from its inception, the goal of being an agent of transformation of the local conditions. For the research to be effective in obtaining the desired information and for its proposals for intervention to be constructive and successfully transformational, engagement with, and especially engagement *of*, the communities in the research process was necessary. It would also be necessary that the community become the true protagonist of the process, allowing the stakeholders involved to find solidarity with each other and seek solutions for shared problems. The knowledge generated by the community during the research should have the ability to empower the community in processes of self-awareness and change. Such ambitious objectives could only emerge by obtaining the desired autonomy and establishing relationships of trust between researchers and community members. Relationships based on shared responsibilities in decision-making for issues with shared results (EMATER, 2006).

To conduct research within the communities surrounding the Rio Verde needed to find a methodology that differed from most conventional models of data collection and analysis. Previous experiences with the communities by local EMATER technicians and public hearings held during the macro-zoning of the local Environmental Protection Area[1], advised that the communities would be highly resistant to any sort of intervention related to the use of the region's natural resources. From the start, it was noted that this study would need a research methodology that could overcome the traumas experienced by local community members and establish a relationship based on trust and co-responsibility with researchers. Such a methodology should enable the communities to become receptive once again to information sharing and at the same time foster their engagement in the process of changing local practices in the use of water and other natural resources.

In the remainder of this Chapter we describe the choice of the research methodology, scope and focus of study, techniques, materials and objectives of the research. The decision to include this chapter was based on two reasons: a) four distinct research groups (whose specific research results will be presented in the four following chapters) were involved in this process; and b) the research activities employed an innovative methodology which was a learning experience for researchers and such a methodology may be useful to other researchers in the field who face similar difficulties in applying conventional methodologies.

2. STUDY AREA

From the beginning, the overall project to assess the Eutrophication of the Rio Verde faced the challenge of conducting an investigation which encompassed a diverse and complex array of factors that could affect the dynamics of the reproduction and distribution of water resources in the Rio Verde Basin. The interdisciplinary approach began at the inception of the research, with the establishment of thematic groups, including the group "Socioeconomics and Environmental Education." The goal of this group was to study the population's socioeconomic profile, rural sanitation, agricultural activities, risk perception and environmental education.

Immediately, this nucleus faced other issues, such as technical-operational, economic, socio-political and institutional, which would influence the chosen method of analysis. The Rio Verde Basin, located in the western part of the Curitiba Metropolitan Region (Figure 1), covers parts of the municipalities of Araucária, Campo Largo, and Campo Magro and includes mostly rural communities. The population which was the object of this research included 259 farms, generally inhabited by more than one family. As such, several problems were immediately identified regarding the methodological choices for data collection: a) the available resources were not sufficient for each group to separately reach a significant number of families in the region; b) integrating the activities of the research group was a necessity in order to obtain the desired information, otherwise the number of questionnaires, visits to community members' homes, and disruptions of their productive and routine activities would be overwhelming; c) research results would be individual and not collective constructions. Furthermore, and as mentioned above, since the establishment of the Rio Verde Environmental Protection Area (APA) and its macro-zoning, there was significant resistance from the local population to collaborate with any process that might create additional restrictions on the use of the natural resources in the basin.

Finally, according to the National Water Resources Policy "the management of water resources must be decentralized and include the participation of public authorities, users and communities" (Article 1, clause VI, National Law 9.433/97). That is, any policy recommendation emerging from this research as intended must be committed to the empowerment of the communities involved. In other words, the chosen research method considers the Rio Verde communities as agents of the research and patrons of the policies derived from it. The premise was that any proposed intervention that aimed at compliance with the law and sought to be effective in promoting the sustainable and equitable management of the resources of the Rio Verde, would have to rely on the participation of a local population that was aware of both its environment and the issues surrounding it.

Then the research approach adopted should allow interaction between the researchers and those being researched, as well as a greater involvement of the latter in conducting the research. So Participatory Research (PR) methodologies seemed fit to the task. PR is defined as "a research process in which the community takes part in the analysis of its own reality, with a view to promote social change for the benefit of the oppressed participants. Therefore, it is a research activity which is educational and guides communities to action" (GROSSI, 1981, *apud* DEMO, 1984, p. 77). According to Demo (1984, p. 78), PR consists of three steps: data gathering within the context of the action; discussion of information with the informing community to clarify problems and intentions and to develop action plans; and problem solving as developed by the stakeholders. Participatory methodologies, and more specifically the Participatory Rural Diagnosis(PRD), were thus employed in this study. PRD consists in a set of techniques and tools which allows the collection of primary or "field" information in the community as well as fostering community groups' self-analysis and self-determination (VERDEJO, 2006).

Within this methodology, a further tool was developed by EMATER – MG, also aiming rural community participation. Known as the Participatory Methodology of Rural Extension for Sustainable Development (MEXPAR), this technique is now a model for fruitful interactions with rural communities, anticipating the possible dissatisfaction felt by residents regarding the imposition of rules by governments. As mentioned above, MEXPAR takes community participation as its basic premise, favoring the use of techniques that encourage and stimulate the reflection of social groups on the relationships they establish with the physical and social environment and the elaboration of new concepts and experiences. This process creates a "certainty of the unfinished" for both researchers and the community, and shows the need for a permanent dialogue between all stakeholders involved (EMATER, 2006).

[1] See, for example, transcribed recordings of public hearings on the macro-zoning of the Rio Verde Environmental Protection Area, IAP (2002).

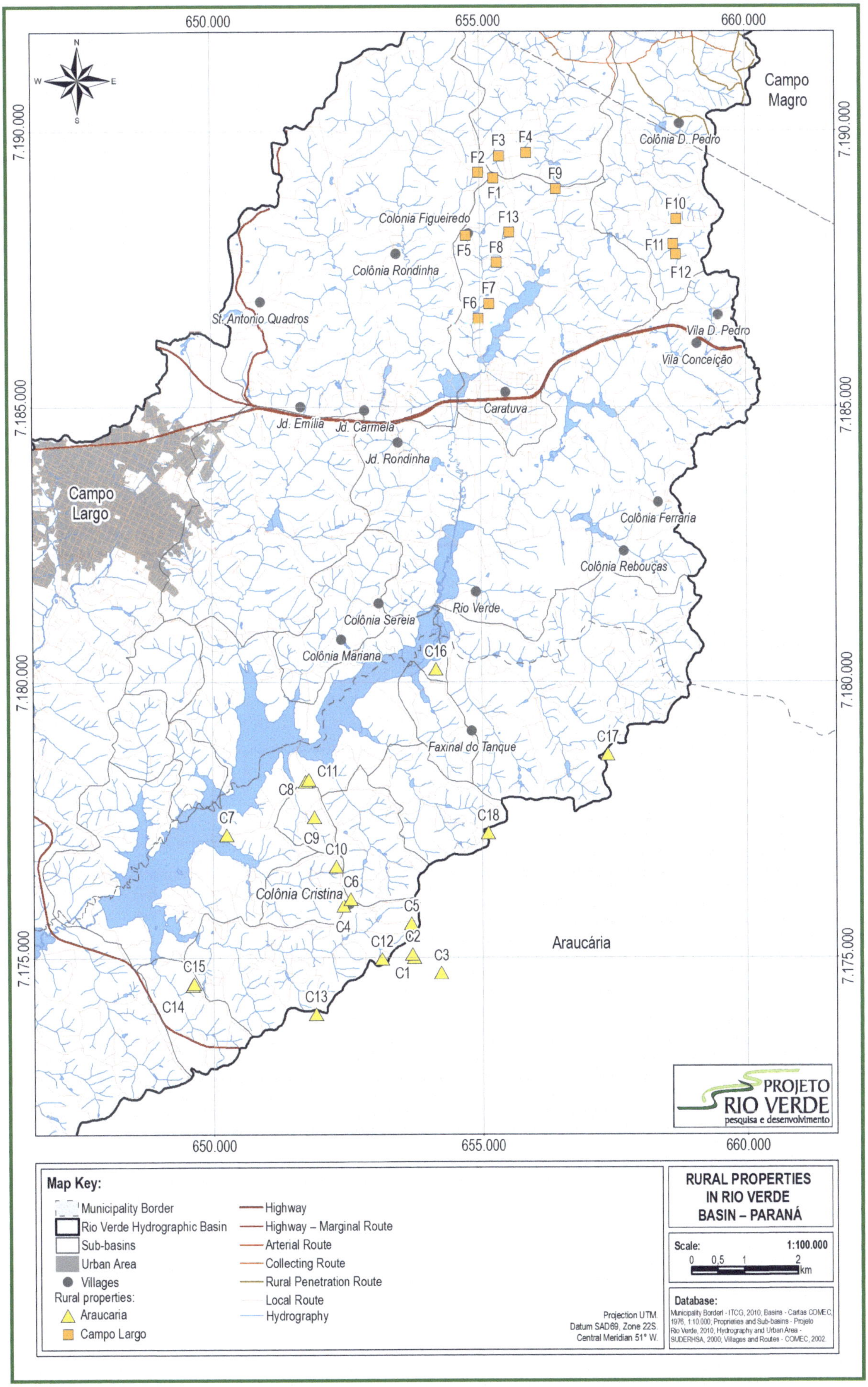

FIGURE 1 – MAP OF THE RURAL PROPERTIES VISITED TO CONDUCT QUESTIONNAIRES AND WATER AND SOIL SAMPLING LOCATIONS - RIO VERDE LAKE BASIN

3. MATERIALS AND METHODS

The Participatory Rural Diagnosis (PRD) works in several stages, using tools in conjunction with the rural community. The first step aims to gain preliminary knowledge of the community and establish ties with the community so that researchers begin to understand their habits and ways, taking into consideration that each community has its own identity. The second step consists of action planning and organization. With data gathered from the actions performed in the first step, this second phase allows build up proposals from economic, social, cultural, environmental and political standpoints, followed by the planning of necessary and possible actions to be implemented. The third step involves the planning, implementation, and monitoring of actions.

MEXPAR suggests some tools for successfully completing the participatory appraisal. In the first step, the method entails (in sequential order):

1. Historical Mapping – the community expresses itself by drawing a map of the region representing its past, its current state and how it sees its future, highlighting problems and potentials;
2. Election of Priorities – ranking problems and potentials outlined in historical maps in order of importance;
3. Frequency Hierarchy – the aforementioned problems and potentials are compared and voted on to determine priority;
4. Venn Diagram – a diagram showing the interaction of the community with other public and private institutions, based on the importance given by the community to the role these institutions play in local development;
5. Structured interview – use of a questionnaire, focused on questions regarding quality of life.

Unlike other commonly used research methods, PRD uses triangulation techniques (MARCONI & LAKATOS, 2007, p. 283) combining different methods to ensure comprehensive data collection. Such techniques may include the review of secondary data from literature, aerial photographs and satellite imagery, direct observation of events and processes, the relationships between proprieties/communities, semi-structured interviews, diagrams, maps and activity schedules.

The system of data collection must first allow the community's inhabitants to reflect systematically on their problems and possible solutions and share these with the team conducting the PRD. For that reason, the research team must understand local conditions and circumstances and attempt to analyze the problems and possible means to mitigate them.

The indicators built and data reported in this chapter were mainly obtained with the use of structured interview, which contains questions about quality of life, focusing on aforementioned factors. The questionnaire (structured interview) consisted of the following items: 1. Identification of the interviewee (subsequently withdrawn); 2. Labor/non-family work force (temporary and permanent); 3. Inventory of rural properties, current land-use, improvements, animals/traction animals, and machinery; 4. Annual revenue; 5. Quality of Life; 6. Family succession; 7. Rural sanitation; 8. Socioeconomic profile; and 9. Environmental risk perception.

4. RESULTS AND DISCUSSION

The Basin was divided into three main communities: Colônia Cristina, located in Araucária; Colônia Dom Pedro II and Colônia Figueiredo, located in the municipality of Campo Largo. The PRD meetings in the Campo Largo communities were held in Colônia Figueiredo with the participation of inhabitants from Colônia Dom Pedro II. These communities were chosen based on their central location, number of inhabitants, as well as each community's importance in the Basin.

On February 12, 13 and 17, 2009, the entire team of the Socio-Environmental research group were trained together. The training was sponsored by EMATER and involved a number of technical components from the PRD and Socio-Economic Characteristics Questionnaire.

The participants from the communities were representatives of the basin and included residents of both sexes and various age groups, and were employed in diverse economic activities and types of work. To begin the research activities, the residents were invited by local EMATER representatives. The range of participants was assured both by the invitations from EMATER and because of information spread among neighbors.

Between March 5 and March 18, 2009, three meetings were held with the community of Colônia Figueiredo, in Campo Largo, and one with Colônia Cristina, in Araucária. The intent of these meetings was to introduce the group of researchers, explain the aims of the research, assess the willingness of the community to take part in the activities, and finally, develop the PRD techniques.

The first visit took place in the annex building of the Figueiredo Church and aimed to introduce the project. From this first meeting, the difficulties in obtaining information and gaining community members' commitment to the research were made clear. The community openly declared themselves as against the activities intended by the research group, since it was believed to be associated with the public hearings on the local Environmental Protection Area. Agreement to continue the research occurred only after several meetings in which community leaders were more receptive to the research group (Figure 2). In Colônia Cristina, the first meeting was held at the São Casemiro Society, and the community debated whether or not to participate in the research during a private meeting, without the presence of the members of the research team. From that point on, the MEXPAR tools were used as described in the methods section, producing the Participatory Rural Appraisal for the community. In these communities, we applied the following methods: Participatory Mapping; Analysis of Problems and Potentials; and Institutional Interrelationship Diagram (Venn Diagram).

In both communities, the chosen methodology showed potential for increasing the awareness of the local situation, for both researchers and community itself. Young, old, men, women and children all took part in the PRD activities. The historical map drawn by the community helped teach the youth of Colônia Figueiredo how their parents and grandparents lived there 30 or 40 years ago. The talks demonstrated to the older residents that young people, their children and grandchildren, would like to keep living in the community and follow in the footsteps of their parents, as long as opportunities for education and adequate public transportation and infrastructure were available. In general, among young people there was more acceptance of change in habits and customs than among older inhabitants, but neither group accepted a loss of their culture (Figure 3).

FIGURE 2 – INTRODUCTORY AND INITIAL ASSESSMENT MEETING (A) COLÔNIA FIGUEIREDO; (B) COLÔNIA CRISTINA

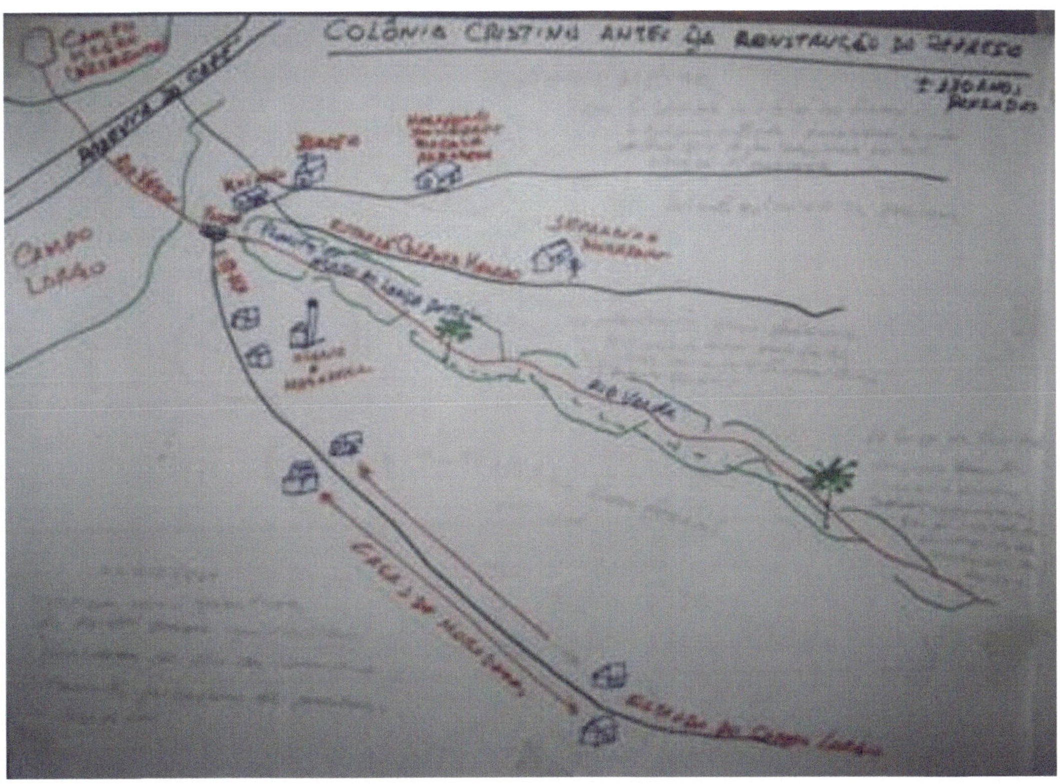

FIGURE 3 – COLÔNIA CRISTINA – HISTORICAL MAP OF THE COMMUNITY AS DRAWN BY LOCAL RESIDENTS SHOWING MAJOR LANDMARKS AND THE LANDSCAPE OF THE RIO VERDE BASIN APPROXIMATELY 30 TO 40 YEARS AGO

While external researchers observed social, economic and environmental issues of direct interest to the project, the team also learned that issues, such as negative or inexistent interaction with municipal governments and other public agencies, insufficient infrastructure and the encroaching of gated communities and the development of subdivisions, are more important for the conservation of the Rio Verde than the team initially imagined. Figure 4 is a graphic representation of the observed institutional relationships from the community's point of view.

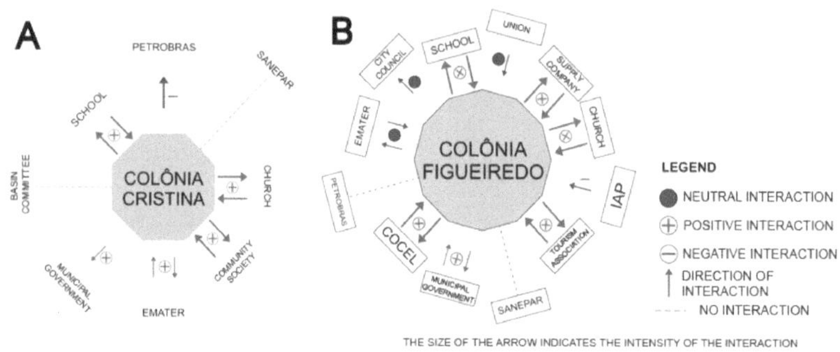

FIGURE 4 – VENN DIAGRAM – (A) COLÔNIA FIGUEIREDO; (B) COLÔNIA CRISTINA

In Colônia Cristina, the research team faced greater resistance to the project throughout. This community, which is located closer to the Reservoir than the other communities included in this study, had trouble with the policies implemented in the 1970s restricting use of the river after the construction of the Rio Verde dam. The community's distrust of public policies increased even more after the public consultations on the implementation of the local Environmental Protection Area (APA). In this community, which is more united and organized than Colônia Figueiredo, the use of PRD techniques was only made possible through methods of encouragement and participation, providing space for the community to raise their own questions and anxieties, and establishing shared points of interest with the locals, particularly those in leadership positions. It was also possible to obtain a list of problems and potentials of the communities which were ranked by the locals participating in the meetings. These are important instruments for the collaborative creation of development plans (see Tables 1 and 2).

In both communities, the use of socio-economic characteristics and quality of life questionnaires proved to be the most complicated activity. Since these questionnaires deal with personal information, they met with some resistance from the community. The foundation of this resistance was always an apprehensiveness regarding informant identification. The questionnaire also showed some incongruence with the informants' reality. Some questions were confusing, while others were apparently inappropriate, and others still were too personal in the eyes of the community members. To overcome these problems, the research team took the strategy of "offering" to take samples and conduct analyses of soil and domestic water supply while the questionnaires were given. Interaction with the community allowed the group to improve the questionnaire responses, by adapting them to local jargon and eliminating references which might identify the informant (such as headers with a blank for the respondent's name).

TABLE 1 – RANKING OF PROBLEMS AND POTENTIALS IN COLÔNIA FIGUEIREDO

PROBLEM	VOTES	POTENTIAL	VOTES
Lack of infrastructure such as post offices, roads, telecommunications and waste collection	19	Preservation with financial returns	18
Poor rural technical support	19	River conservation practices	15
Assistance from Municipal Government	13	Youth remaining in rural areas	12
Departure of youth from rural areas	10	Maintenance and diversification of agriculture	11
Subdivisions and Gated Communities	10	Increase in biodiversity with a clean river	11
-	-	Road improvements enhancing income	11

TABLE 2 – RANKING OF PROBLEMS AND POTENTIALS IN COLÔNIA CRISTINA

PROBLEM	VOTES	POTENTIALITY	VOTES
Uncertainty in agricultural production	Unanimous	Strong community identity	Unanimous
Future water availability	Unanimous	Technical qualification of producers	Unanimous
Physical insecurity	Unanimous	Production diversity	Unanimous
Flooding	Unanimous	Less aggressive production techniques	Unanimous
Urban solid waste	Unanimous	Community infrastructure	Unanimous

The structured questionnaires were distributed in the second semester of 2009. Interviews were conducted at the interviewees' property by members of the research team; some in their living room, with the whole family, others in their fields. Interviewees always provided the team with stories, some of which helped better understand the community's development. The initial goal was to make the locals as comfortable as possible so they might provide information they may not always be interested in sharing. Some information gathered during the process was initially (and erroneously) judged 'unimportant' for the research. By reading the answers to the questionnaires one might be induced to believe there were "errors" in the information provided (e.g. regarding income). However, no errors were recorded and the information given is an accurate representation of the current situation (e.g. some families have property and income in other Municipalities of Paraná State). With this strategy, the group succeeded in completing 44 questionnaires (22 in both Colônia Figueiredo and Colônia Cristina), obtaining information from over 350 people as each property in general houses more than one family. It should be noted that 31 farms (Figure 1) were visited to complete the questionnaires; the remaining questionnaires (13) were given during PRD meetings with community members.

5. RURAL LIFE QUALITY INDEX

In the last two decades there has been growing concern with the creation of sustainability indexes that serve as indicators for decision making at local, regional, national and international levels. According to Pintér *et al.* (2005), the use of indicators as parameters for assessing and monitoring a process or reality allows for the classification of complex phenomena and enables the determination of how human actions affect their surroundings. Such indicators also warn about situations at risk, predict future scenarios and provide support for political decision making. For Freudemberg (2002), a good indicator must have features such as: ease of measurement; low cost; flexibility to adapt to different local situations; identify trends; the capacity to promote interrelationships; and allow the understanding and participation of the local community.

Some authors, such as Demo (1996), show that debates surrounding a quantification of quality of life led to the need for another concept, that of human development. While the concept of quality of life can be measured based on quantitative attributes (housing, sanitation, health, schooling, transportation, etc.), human development is a wider concept that sees the human being as a citizen who aspires to participate in strategic decisions and has freedom as the ultimate goal. Thus, it is important to incorporate parameters and to devise indexes that bear a connection with "living well" (leisure, participation in social activities, entertainment) to characterize people's desirable life situations.

An important aspect is to assess the quality of life of farmers and the rural area in which they live. As such, the idea was to create an index reflecting aspects tied to the farmer's quality of life on his farm which could reveal access to basic services in the community and the municipality.

Therefore, one objective of this chapter is to show how the Rural Life Quality Index was devised as a standardized form of assessment and measurement of the well-being of the rural population, covering a variety of dimensions: socio-cultural, technical, economic, ecological and political-institutional.

One step of the participative methodology consists of a structured questionnaire with questions related to diverse areas of interest. In this project, the questionnaire included questions regarding: the population's socio-economic profile; rural community inhabitants' perception of environmental risk; agricultural and livestock activities developed in the basin; environmental education; and rural sanitation.

Unlike other steps of the participative methodology, most of the questionnaires were completed in the field, alongside the collection of water and soil samples for later analysis. Filling in the questionnaire during soil and water collection activities was a strategy to obtain information that otherwise would not likely be provided in a community generally wary of this type of approach. In meetings within the rural community, the farms participating in the assessment were listed based on their compliance with the project, i.e., only the properties whose families wished to participate were included (DOETZER *et al.*, 2010).

The assessment of quality of life was performed based on several variables and indicators (Chart 1) devised by Darolt (2000), basically representing housing conditions (appearance of the house and available appliances); sanitation (water supply, sewage system and waste disposal); transportation (vehicles); access to services (education, health and transportation); leisure (holidays); and social integration (participation in social activities).

For the assessment, all indicators are ranked between 0 and 10. Extreme values of 0 or 10 indicate very poor or excellent situations, respectively. A score of 7.0 represents the lower limit of an acceptable quality of life (between 7.0 and 10) and less favorable situations are scored below 7.0.

For all indicators, higher scores correspond to: better housing standards; better sanitation and waste treatment conditions; easier transportation and flow of production; better access to education, healthcare (doctors and dentists) and public transportation services; longer rest periods; and greater participation in social activities.

The Rural Life Quality Index is obtained by calculating the simple mean of the values in each indicator assessed. The results were organized in radar diagrams, which facilitate analyses and observation of all components of the Rural Life Quality Index.

Concerning farmers' housing, specifically their residences, the research reveals a suitable average, considering house conditions and the appliances owned by the family. For home conditions, two aspects were assessed: date of construction and predominant material, relating the two to ensure a more consistent analysis. Houses were in general well preserved: 66% of them built with masonry and within the last few years. As for available appliances, most households possess various home appliances, an indicator of satisfactory income levels for local farmers.

CHART 1 – DESCRIPTION AND FORM OF ASSESSMENT OF VARIABLES RELATED TO THE RURAL LIFE QUALITY INDEX

VARIABLES	INDICATORS	ASSESSMENT GRADE			
		(0 / 1 / 2) (poor)	(3 / 4 / 5 / 6) (acceptable)	(7 / 8 / 9) (good)	(10) (optimal)
Housing	House Appearance [1]	Poor	Acceptable	Good	Excellent
	Appliances[2]	None or up to 2 pieces	Has basic appliances	Has key appliances	Has all appliances
Sanitation	Water[3]	No access	Untreated/ not disinfected well	Source/fountain/well followed by treatment	Public network
	Sewage[4]	No treatment	Dry sump/ Cesspit	Septic Tank	Septic Tank followed by drain field
Waste[5]	Organic Waste	No treatment (disposal in rivers/on soil)	Unused	Public Collection	Recycled on the farm
	Non-organic Waste	No treatment (disposed in rivers/on soil)	Burnt or buried	Recycled without public collection	Recycled and directed to public collection
Transportation[6]	Vehicles	No vehicles or alternative means of transportation	More than one alternative means of transportation	One vehicle	More than one vehicle
Access to School Services[7]	School	No access	Service available in another municipality	Service available in municipal center	Service available locally
Access to Services	Health	No access	Service available in another municipality	Service available in municipal center	Service available locally
	Transportation	No access	Service available in another municipality	Service available in municipal center	Service available locally
Leisure[8]	Holidays and/or Rest	Never takes holidays	Takes short outings	Takes holidays every year (1 week)	Takes holidays every year (1 month)
Social Integration[9]	Social Activities[8]	Does not participate	Participates sporadically	Takes part in at least one social activity	Takes part in more than one social activity

[1]House Appearance = related to date of construction, owner's assessment (regarding the condition of the buildings) and predominant material (masonry, wood or mixed);

[2]Equipment = gas stove, wood stove, refrigerator, freezer, mixer, blender, television set, radio, stereo, computer, telephone, others; Basic equipment: gas stove, refrigerator, radio and TV; Key equipment: gas stove, refrigerator, freezer, mixer/blender, TV, radio;

[3]Water = public network with water treatment and other forms of supply (sources, fountains, wells and artesian wells);

[4]Sewage = forms of treatment (septic tank, dry sump or cesspit); Dry sump or cesspit = Fecal waste is released directly without flushing with water. Not recommended, as this may cause groundwater contamination; Septic tank = a closed, waterproof tank where solid waste is separated and processed.

[5]Waste = recycled, buried, burnt, separated and directed to public collection.

[6]Vehicles = passenger car, freight vehicle, bicycle, cart, horse, others;

[7]Services = level of access to the main services of education, healthcare (doctor and dentist) and public transportation. Availability closer to the property receives higher scores;

[8]Leisure = level of rest and entertainment, basically measured by holidays.

[9]Social activity = participation in festive or religious activities (church), organized groups or associations.

Source: Darolt (2002)

The results confirm that sanitation is a very common problem in small rural communities and particularly for more isolated properties, which are often not linked to water networks supplied by government sanitation companies. Below the acceptable average of 7.0, the indicator relating to water supply represents one of the biggest problems observed in the Rio Verde Basin. Many farmers have their own water source, generally groundwater, artesian or semi-artesian wells. In some regions (Colônia Cristina and Colônia Dom Pedro II), treated water is supplied through the public network by the municipalities of Araucária and Campo Largo. The water from the public network is demonstrably of good quality, following the standards established by Ordinance 518, 2004, of the Brazilian Ministry of Health. A study by Larsen (2010) verified that of the 41 properties that had water from their own water sources analyzed, 63% showed the presence of *Escherichia coli* (fecal coliforms). Even though they have access to water from the public network, most farmers prefer to use water from their own sources for primary consumption, including their own drinking water, food preparation and others, while the treated, chlorinated water from the public network is used for secondary purposes such as laundry and washing sidewalks, etc. When asked about their refusal to use chlorinated water, farmers claimed that they are uneasy and feel unsafe due to the taste of chlorine.

Regarding sewage, the ideal system of a septic tank followed by a drain field was found in few properties, with rudimentary systems predominating in the rural properties. Few farmers knew the difference between a septic tank and dry sumps or cesspits and there was no concern with technical standards for the construction of sewage systems. The minimum distance of 15 meters between cesspits and wells was observed in most properties, including in low-lying areas, avoiding contamination of the wells. Despite the lack of technical information for the farmers, only two properties had no sewage treatment at all, directing waste into the soil or the river.

The treatment and final disposal of organic and non-organic waste received adequate average scores. In some cases when waste is not collected, it is buried or burnt, but in most properties there is some environmental consciousness in this regard; many farmers use organic waste in their gardens and separate recyclables. Another issue not dealt with in the Rural Life Quality Index is animal waste, which is not adequately treated. This type of waste is either inappropriately disposed of on the soil or not specifically disposed of at all. The presence of animals near the water supply wells was observed and constitutes one of the main causes of the presence of coliforms in the water analyzed.

Access to basic healthcare, education and transportation services varies according to the community. Colônia Cristina, in the municipality of Araucária, has a health facility in the community center, in an area with paved roads and bus transportation to the municipal center. In Colônia Figueiredo and Colônia Dom Pedro II, the rural communities of Campo Largo, healthcare services are available in the municipality but not in the communities themselves. Furthermore, public transportation is hindered by poor road quality, preventing many families from accessing health services. Regarding education, Colônia Dom Pedro II has a private school offering Primary Education. In Colônia Cristina, access to education is affected by the distance between the school and the community.

The most negative aspect observed in the Rural Life Quality Index was the lack of opportunities for leisure and short rest periods for farmers. The vast majority of farmers reported they have no opportunity to rest, other than on Sundays, and longer periods would be desirable. Respondents report a lack of time and that the amount of work they do is necessary to provide for their families.

Lastly, the most positive point demonstrated in the research was the observation of consistent social interactions in the communities. Almost all families take part in social activities, especially those related to the church, cooperatives, labor unions, municipal councils, farmers' associations and community associations.

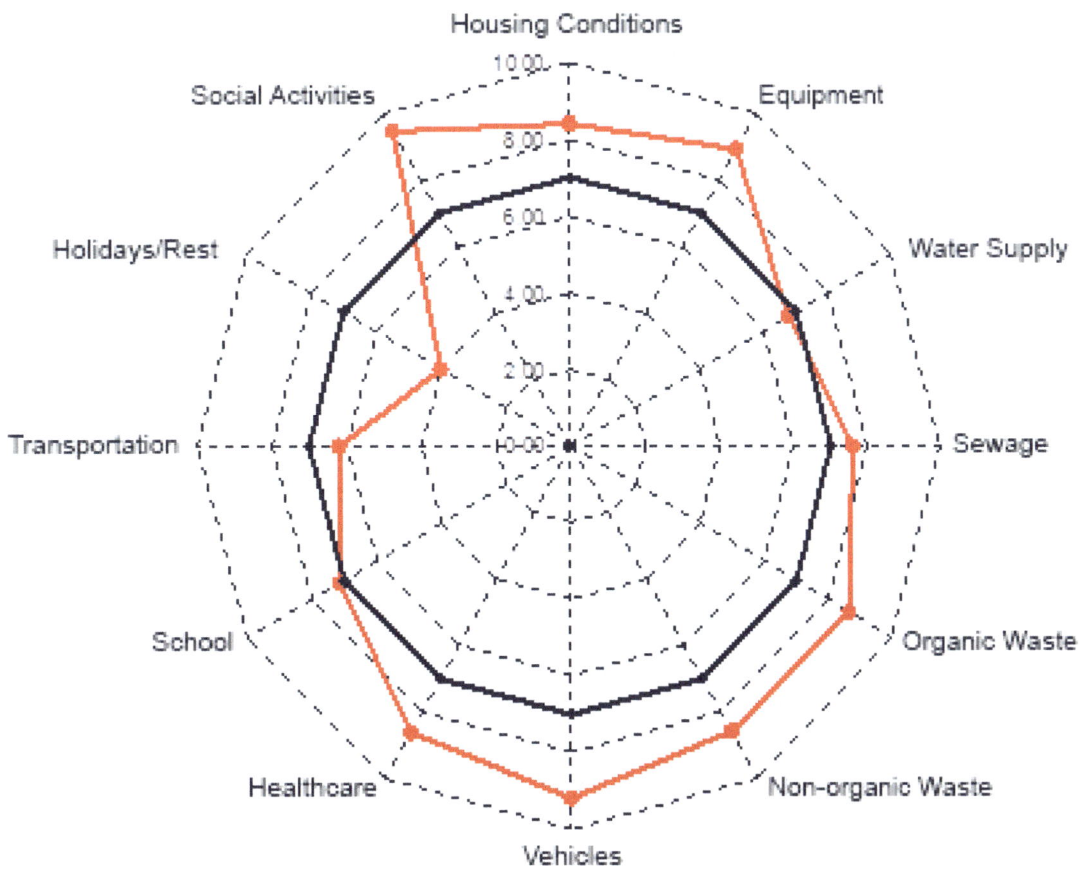

FIGURE 5 – COMPONENTS OF RURAL LIFE QUALITY INDEX FOR THE RIO VERDE BASIN

6. CONCLUSIONS AND RECOMMENDATIONS

The main objective of this study was to employ participative methodologies, with an emphasis on constructive and humanistic practical pedagogy, using techniques from the Participatory Rural Appraisal and MEXPAR, in the Rio Verde Basin, situated in the Curitiba Metropolitan Region. The study required researchers to adapt to the distinct realities found in rural communities, focusing initially on the details of the methodology and the overall project objectives, as well as emphasizing the importance of community participation in the assessment.

The initial study of communities, their customs and cultures is recommended to avoid conflict between community members and researchers, proceeding towards social transformation through the participation of different stakeholders, allowing participants involved to become collaborators.

The Rural Life Quality Index associated with participatory rural methodologies allows for the rapid assessment of the local conditions, based on a broad range of indicators, with the participation of the studied population, highlighting situations of risk and vulnerability within the studied population. The Rural Life Quality Index may be an important tool for decision makers in resource management, especially in cases when indicators are below average, serving as a guideline for the development of future actions.

Although the research indicates an above-average Rural Life Quality Index (7.8), some actions must be prioritized in the studied communities, such as: water supply, disposal of domestic sewage and animal waste, issues related to leisure, rest and entertainment of rural families. Furthermore, rural sanitation must be a priority in public policy actions related to the rural environment.

7. REFERENCES

ARAUCÀRIA, (2009) Prefeitura Municipal. *Dados do Município*.

BRASIL. *Lei Nacional n° 9.433* que institui a Política Nacional de Recursos Hídricos, cria o Sistema Nacional de Gerenciamento de Recursos Hídricos, regulamenta o inciso XIX do art. 21 da Constituição Federal e altera o art. 1° da Lei n° 8.001, de 13 de março de 1990, que modificou a Lei n° 7.990, de 28 de dezembro de 1989.1997.(establishing the National Water Resources Policy, creates the National System of Water Resources Management, regulates paragraph XIX of art. 21 of the Federal Constitution and amends art. 1 of Law 8,001, of March 13, 1990, which amended Law 7.990, of December 28, 1989.1997)CAMPO LARGO, Prefeitura Municipal. *Plano Diretor de Desenvolvimento Integrado do Município de Campo Largo – Paraná*. Campo Largo: FUNPAR, 2004.

CAMPO MAGRO, (2005) Prefeitura Municipal. *Plano Diretor de Desenvolvimento Integrado do Município de Campo Magro – PR*. Campo Magro, Ambientec – Consultoria Ambiental Ltda.

CONSILIU. (2002) *Macrozoneamento Ecológico-Econômico da APA Estadual do Rio Verde*. Curitiba, COMEC.

DAROLT, M. R. (2000) *As Dimensões da Sustentabilidade: Um estudo da agricultura orgânica na região metropolitana de Curitiba-PR*. Tese de Doutorado em Meio Ambiente e Desenvolvimento, Universidade Federal do Paraná/Paris VII. Curitiba, 310p.

DAROLT, M. R. (2002) Metodologia para Avaliação da Qualidade de Vida Rural. In: *Agricultura Orgânica: inventando o futuro*. Londrina: IAPAR, p.85-91. ISBN 85-88184-09-5.

DEMO, P. (1980) *Metodologia Científica em Ciências Sociais*. São Paulo: Atlas.

DEMO, P. (1984) *Pesquisa Participante: Mito e Realidade*. Rio de Janeiro: SENAC.

DEMO, P. (1996) *Combate à Pobreza: Desenvolvimento como oportunidade*. Campinas: Autores Associados, p.23-35.

DOETZER, B. H. W., DALTO, F. A. S., LIMA, J. E. S., LARSEN, D, AISSE, M. M., LOPES, E. Diagnóstico rural participativo aplicado a Bacia do Rio Verde, Região Metropolitana de Curitiba, PR (Brasil). (2010) In: *XXXII Congreso Interamericano de Ingeniería Sanitária y Ambiental*, 2010, Punta Cana, Rep. Dominicana. AIDIS, p.1-8

EMATER. (2006) *MEXPAR – Metodologia Participativa de Extensão Rural para o Desenvolvimento Sustentável*. Belo Horizonte, EMATER MG, 134p.

FREUDENBERG, M. (2003) *Composite indicators of country performance: a critical assessment*. Paris: OECD, 11-13 march, 32p.

GIL, A. C. (2002) *Como Elaborar Projetos de Pesquisa*. São Paulo, Atlas, 175p.

LARSEN, D. (2010) *Diagnóstico do Saneamento Rural Através de Metodologia Participativa*. Estudo de Caso: Bacia do Rio Verde, Região Metropolitana de Curitiba, Paraná. Universidade Federal do Paraná, Curitiba.

MARCONI, M.A.; LAKATO, E.M. (2007) *Metodologia Científica*. São Paulo: Atlas.

SANEPAR. *Saneamento Rural. Programa de Atuação Social*. SANEPAR, USPO. Sd.

PINTÉR, L.; HARDI,P. and BARTELMUS, P. (2005) *Sustainable Development Indicators: proposal for the a way forward*. United Nations Division for Sustainable Development, IISD. December, 42p.

VERDEJO, M. E.(2006) *Diagnóstico Rural Participativo: Guia Prático DRP*. Brasília, MDA/ Secretaria da Agricultura Familiar. 62p.

CHAPTER 17

SOCIAL AND ECONOMIC CHARACTERISTICS OF THE BASIN

Fabiano A. S. Dalto
Felipe Almeida
Noam Goldstein

SOCIAL AND ECONOMIC CHARACTERISTICS OF THE BASIN

Fabiano A. S. Dalto
Felipe Almeida
Noam Goldstein

ABSTRACT

The appropriation of good quality water resources has increasingly constituted in economic asset, that is, able to generate income for those who have access to their use. Human and economic reproduction often lead to degradation of water resources as it is not possible to build a governance system that manages water resources as a public good of common use. In this chapter, we present the results of the research on the system of governance of the Rio Verde Basin. Initially, we present the social and economic features of the region that define the context in which the economic and public demands on water are determined. Then, we provide a social and economic analysis, as well as an analysis of the institutional system that permeates the communities surrounding the Basin. Finally, we indicate the existing institutional constraints in the Basin that limit or prevent a fair and efficient administration of natural resources in the Basin.

KEY WORDS

Common resources, water resources, hydrographic basin, Rio Verde basin, governance structure.

1. INTRODUCTION: WHAT ARE THE ISSUES?

The appropriation of good quality water resources has increasingly constituted in economic asset, in other words, able to generate income for those who possess it. On the other hand, water is a crucial economic factor given its importance for the reproduction of human life. The pressure on the sources of water supply for various uses has grown with the needs of economic development, a growing population and urbanization. Andreoli *et al.* (2005) claim that "major supply issues in Brazil are related to the combination of the explosive and disorganized growth of cities, causing the accelerated degradation of water quality, and the reduction of the soil's infiltration capacity." In summary, human and economic reproduction often lead to degradation of water resources, since it is not possible to build a governance system that manages water resources as a public asset for common use.

In a recent study on environmental indicators, IPARDES (2010) estimated that water consumption in the state of Paraná totals less than 3% of the available surface water sources. The most problematic issues arise from distribution of use and quality of water resources. In Paraná, according to the same study from IPARDES, most water use is allocated to economic activities: industry (24%), agriculture (21%), livestock (12.6%) and minerals (0.4%). The remaining 42% is allocated to publicsupply. In the basin of Alto Iguaçu, where the Rio Verde Sub-Basin is located, the IPARDES study calculated that approx. 20% of surface water available is consumed, whereas 74% is used in public water supply and 26% in economic activities – industrial activities (14%) and agriculture and livestock (12%). Needless to say that the uses given to water indicate the central role it plays to the reproduction of the economic system and society. Thus, from an economic standpoint, water is not only an input in the production of wealth but also, and furthermore, an input in the economy, due to the economic and public needs that attract the interest for economic accumulation.

Considering the central role of water to social reproduction, the management of water resources has become an area of conflict of interests with great potential to lead to the degrading use and perverse distribution of water resources. Governmental and international entities have called the attention to issues related to the unequal access to good quality water. The system of governance, therefore, is the central factor in water issues (ANA, 2007a; UNDP, 2006). The various issues that have emerged regarding the governance of the water basins that affect collective goals include, but are not limited to: conflict reduction, confidence building, public education, incorporation of community knowledge and values in the decision-making process, efficient use (low costs and no waste policy) and equitable distribution of resources (Beierle, 1998). Governance means the rules (and those who define them) on use, distribution and preservation of the quantity and, especially, the quality of water resources (WALSH *et al.*, 2006; CARNEIRO *et al.*, 2005; ANA 2007a, 2007b). The regulation of the use of natural resources in its various dimensions has been addressed in the economic theory through concepts developed within the institutional approach of the governance system, that is, the system that coordinates and regulates the access and use and monitoring practices established in a given area and (or) community.

In this chapter, we present the results of the research on the governance system in the Rio Verde Basin. We begin by pointing out the methodological aspects that guided the research. We then present the social and economic characteristics of the region that define the context in which the economic and public demands on water are determined. Finally, we make an analysis of the institutional system that permeates the communities surrounding the

Basin. In this last section, we also indicate the basin institutional constraints that limit or prevent an effective and equitable administration as defined above.

2. DIGRESSIONS ON THE METHODOLOGICAL AND THEORETICAL ASPECTS THAT GUIDE THE RESEARCH

In this section, we shall present the theoretical and methodological elements that underpin the empirical analysis in the following sections. The use of water and environmental resources of a hydrographic basin necessarily involves decision coordination problems and (or) conflicts of interest. The central issue in this section is the role of institutions in establishing and solving these coordination problems and conflicts of interest. That is, we shall discuss how agents involved in the use of resources are organized to settle this kind of problems.

Common use natural resources - like water from rivers, forest reserves, fisheries, etc. - are known as resources that have some characteristics of public goods. Pure public goods have two main characteristics: i) non-exclusive use, namely the inability to build instruments to exclude potential beneficiaries of resource usage, ii) non-rivalry of use, that is, the part of the resource used by an individual will be available to others.

In the case of a river, its use is not exclusive but it is rival, which means that water resources are characterized by being common use resources. Along the river bed, there are several users, and upstream users do not suffer impacts from downstream users. On the other hand, the more downstream the resource user is, the higher the impacts from upstream users. The greater use and (or) degradation caused by upstream users, the higher costs (need for treatment due to pollution or construction of dams to increase the volume of water etc.) of downstream users.

In the economic literature, the impacts caused by the actions of some individuals over others are known as externalities. Externalities can be positive or negative, depending on whether the impact is positive or negative.[1] Hardin (1968) created the term *the tragedy of the commons* to express the dilemma of an individual regarding the collective use resources which results in effects on the wellbeing of other individuals in the community, without, however, bearing the costs or appropriating the collective benefits of his/her actions. The issue that emerges is how to make sure that, in a world in which agents interact and their decisions affect the well-being of others, deleterious actions are not encouraged while conservation actions are fostered.

The characteristics of common use goods of basin water resources present problems of governance of issues related to potential conflicts of interest and coordination problems. Institutional arrangements designed to deal with this problem have a crucial importance in the analysis. The governance of common use goods has been addressed in the economic theory through concepts developed within the institutional approach of property right systems, which coordinate and regulate the access and use and monitoring practices established in a given area and (or) community. Thus, in the next section, we shall make a brief reference to the central concepts of the institutional theory that enlighten the ways in which institutional relationships affect and evolve with the behavior of individuals and organizations. We have put greater emphasis on the most overlooked factors of the institutional literature, that is, the informal aspects of institutions, which are, in our view, the elements that give legitimacy, reliability and durability to social relations.

2.1 INSTITUTIONS AND GOVERNANCE STRUCTURES

According to Oliver Williamson (1985, 1998), the governance structure or institutions are the rules by which an ordinance of rights and duties is done in a relationship. Thus, institutions play a central role in the coordination and resolution of conflicts related to the use of natural resources by establishing the rights and duties of the entities that use such resource (COMMONS, 1931; NORTH, 1991). In addition to the rules and formal institutions that establish incentives and penalties tangible to behavior, informal institutions and habits determine individual behaviors and the effectiveness and durability of formal institutions (VEBLEN, 1899 e 1919; HODGSON, 2003 e 2007; STEIN, 1997; DUGGER, 1980).

In particular, trust and self-monitoring are key elements in the operation of any governance structure of large-scale systems and common use resources as the ones found in hydrographic basins. While some governance systems enable the commitment of individuals to solve problems through direct participation, voice and vote, others shall use less cooperative incentives as exit or boycott. In short, the rights to a common use good are socially defined and establish not only the way people interact with nature, but also, and particularly, how people relate to each other (FEENY *et al.,* 1990).

According to Elinor Ostrom & Edella Schlager (1996), five rights can be identified regarding the use of common use resources:

(I) Access: the power of access includes the right to enter a particular space and enjoy non extractive benefits such as leisure, for example;
(II) Extraction: right to obtain units or products from a resource;
(III) Management: right to regulate internal use patterns and transform the resource by making improvements;
(IV) Exclusion: right to determine who has access rights and how these rights may be transferred;
(V) Conveyance: right to sell or lease any of the rights listed above, therefore being the highest hierarchical level of power action.

The first two rights, access and extraction, are operational. The last three on the other hand – management, exclusion and conveyance, imply rights to participate in collective decisions that define the rights to be acknowledged

[1] If, for instance, users upstream are conservationists, they produce positive impacts on users downstream.

or not in the future. For example, farmers can have access to the river banks if they own properties up to a certain distance from the banks. And farmers can have a grant to use river waters for the irrigation of crops. All other rights are more encompassing in the sense that, to a greater or lesser extent, they establish the rules that shall command the two operational rights. The last three are higher power rights and are established in a broader collective context.

Individuals or entities may take five positional levels in relation to the five rights:

I) Users: individuals with operational access rights, for example, leisure (when visiting a site, but without being able to harvest);
II) Concessionaires: individuals who have the right to collectively choose to enter and extract, but are not able to negotiate such rights;
III) Claimants: individuals who have the same rights of authorized users, plus the right to manage. As such, they deal with extraction activities, including decision-making;
IV) Holders: have the rights described above, and the right to decide on exclusions. In certain locations, there also is the co-owner, who shares such liabilities;
V) Owners: inclusion of all rights described above plus the right to sell or lease their collective choice rights to define the owner of the resource. This is the highest hierarchical level of possession.

Table 1 below shows the inter-relationships between the five rights and the five positions of participants. Each participant's position implies rights with hierarchical levels of power over different resources. The more to the right on the table, the fewer the rights individuals or entities have on the resource. Property regimes can be classified into three categories (FEENY *et al.,* 1990). When private agents have full rights, property is private. When no right is clearly associated with an agent and everyone can enjoy a resource, there is a free access system. Finally, there is the common property when the rights are owned equally by a recognizable group (community, for example).

Governance structures define the holders of the rights and duties of users of common use resources. The literature that deals with coordination issues and conflicts over common use resources has listed four types of management regimes (FEENY *et al.*, 1990; BROMLEY, 1982; BROMLEY, 2000, BOWLES & GINTIS, 2002): state regulation; governance through market mechanisms, common or community governance; *laissez-faire* or free access.

In the real world, it is worth noting that the complexity of institutional governance regimes exceeds the stylized types listed above. For example, it is clear that there is no market governance system where private property prevails with no state system of governance (for example, judiciary) and no more general social system which guarantees private property rights. The same goes for the community governance of a resource that generally requires the recognition and regulation of a central authority. The community governance obviously does not exclude the private property or the market, even though it may be guided by criteria different from those that prevail in relation to market mechanisms. Finally, whatever the dominant governance structures between the ones listed above are, the state structure will somehow be present through laws, regulations and other strictly public activities. The advantages and limitations of each particular structure to inform how a specific governance structure can be established to produce both a sustainable and equitable use of water resources are presented below.

2.2 LAISSEZ-FAIRE AND MARKET MECHANISMS

The complete market freedom, i.e., the total absence of rules to limit the behavior of users/owners (*laissez-faire*), would certainly lead to the tragedy of the use of common resources. Since any economic system is somehow regulated, *laissez-faire* is only an unattainable utopia of unregulated capitalism.

Its closest substitute, the second best of liberals, is the governance system through the market, which is based on the private ownership of resources. Hypothetically, the biggest advantage of private property rights is that owners can sell or loan their shares of a common use resources. The possibility of conveyance or lending allows those who can extract more products from resources, or that value the resources more, to increase their share of the resource and thus increase the overall outcome of the products. In other words, the biggest advantage of the market solution is that, once the private property is guaranteed, the holder of rights (who pays more for the resource) can extract maximum product (allocative efficiency), or lend or sell the resource to those who do so. For this reason, it is necessary that resources have characteristics of pure private goods. Some conservationist solutions, such as payments for environmental services, promoted by multilateral institutions, are based on these principles.

The limitations of the market solution are, however, as restrictive as the assumptions underlying the efficient markets (BROMLEY, 1982 and 1985; GRAFTON, 2000; BOWLES, 2004). Due to the uncertainties involved in the natural resource management, the discount rate of the

TABLE 1 - INTER-RELATIONS OF RIGHTS PER TYPE OF PARTICIPANT

	HOLDER	PROPERTY	CLAIMANT	CONCESSIONAIRE	USER
Access	x	x	x	x	x
Extraction	x	x	x	x	
Management	x	x	x		
Exclusion	x	x			
Conveyance	x				

private owner may be higher than the rate of replenishment of the resource inventory, causing the resource to be overexploited under a private property regime (GRAFTON, 2000).

Moreover, the efficiency of the market depends on whether the markets are competitive and the available information is complete, which is strictly impossible in regard to the waters of a river. Owners who are at the river source have positional advantages over those who are downstream and it is virtually impossible for downstream owners to monitor and discern the individual contribution of each upstream owner in terms of degradation and utilization of resources. Moreover, even in competitive markets, distributive problems related to the use of resources would emerge, since the costs would be the responsibility of owners downstream. Nothing guarantees that, by taking ownership rights and preferences of individuals for granted, the distribution of initial property rights may be considered fair by participants (BROMLEY, 1982; BOWLES, 2004).[2]

In short, the system of governance of water resources based on market mechanisms can lead to three types of insurmountable problems using the own market rules: overexploitation; allocative inefficiency; and distributive injustice.

2.3 STATE REGULATION

The state system of regulation of property rights, in turn, has been required to operate where market failures are decisive. And as noted relative to common use resources – especially water resources, market failures are immeasurable. Public officials can act through laws, tax and credit incentives, administrative and fiduciary penalties to induce or compel the cooperation of private agents. The State is responsible for defining, protecting and making property rights effective. It has an obligation to provide public goods, and it can regulate the environment externalities and restrict the economic power of monopolies. In short, the state regulation and provision of common use resources tie groups subjected to its regulation in a set of mutual relations.

The state governance structure can present some obvious weaknesses. Public officials have insufficient information about the resources involved. Agents subordinated to state standards, in turn, have insufficient information about the decisions of public agents, leading to their insufficient accountability or even to the complete absence of democracy.[3] Finally, the state action assures owners the power to exclude others from having gains with resource usage. Therefore, depending on how the state action is structured, whether as a simple and more efficient market mimicry, a symbiotic and cooperative commitment, that already exists or is yet to be built can be transformed into competitive relationships connected to the position in the state; in which social conditions that could lead to collective actions - solidarity, trust and equality - are eroded. (HANNA *et al.*, 1996)

2.4 COMMUNITY GOVERNANCE

There seems to be an agreement among authors who study common use resources on the benefits of community management in relation to private property rights (AGRAWAL, 2003; AGRAWAL & GIBSON, 1999; HANNA *et al.*, 1996 WADER, 1984, BOWLES & GINTIS, 2002). International agencies for the promotion of economic development and environmental conservation also embraced the cause of decentralized administration and conditioned their resources to the decentralization of power to the local level.

Bowles & Gintis (2002) define community as "a group of people who interact directly, frequently and in multi-faceted way." They include in this definition a wide range of institutions defined by several characteristics. It is important to notice that according to these authors' definition, the connections and not the affective relations is what defines the characteristics of the community. Whether someone was born in a community (geographic) or has entered the community by choice, there will always be the costs of switching community. This ensures that community ties, once established, tend to be strong.

For the purposes of this study, the concept will be applied to a particular territory comprising the use of land for agriculture and residence whose residents make plans and take (or suffer with) governance decisions jointly. Examples of decisions made by the community are the distribution of public goods and services, the allocation of economic and environmental resources and decisions about the representation and participation in the political process.

The literature that advocates the management of common use resources by communities points the small spatial scale, bigger economic and social homogeneity and shared standards of conduct and cultural values, sometimes calling these social capital characteristics as the advantages of the community on other governance structures (BEARD & DASGUPTA, 2006; WADE, 1987; HANNA *et al.*). A small number of people sharing similar values and cultures, facilitate collective actions needed to decide on the use of common resources. When the contract is incomplete or difficult to enforce in court, the communities have advantages relative to the monitoring and obtaining of dispersed information, as well as to the imposition according to collective action by applying "rewards and punishments to members according to their conformity with or deviation from social norms" (BOWLES, 2004). Mechanisms developed by the community to coordinate and monitor behaviors generally involve: trust, reciprocity, solidarity, reputation, pride, respect, revenge and retribution among others. The fre-

---

[2] Daniel Bromley (1982, p.835) shows that even allocative efficiency (the social product) of the Paretian situation depends on the distribution when there is more that one type of product to be marketed.

[3] Authoritarian governments base their judgment on, at least, three pillars: a) political Powerp ower; b) economic power and c) power of the expert. The lack of information by those who suffer the effects of state policies falls under the third pillar. The authoritarian power or lack of accountability by the public officer is justified in the sentence "leave it with the experts" or "technicians shall decide" .

quency and repetitiveness of community interactions reinforce the mechanisms of compliance with collective rules. In short, the defining characteristics of a community have advantages in obtaining information, monitoring behavior, complying with rules and equitable distribution that private rights and state regulation could never achieve.

Not always, however, communities have been successful in managing natural common use resources sustainably.[4] Recurring problems in communities such as the difficulty to monitor the development and introduce new technologies; difficulty to respond to the growth and differentiation of market demands; tendency to exclude external members in morally repugnant form (ethnic, religious, gender); political dominance of small groups within the community, and disinterest or resistance to conservationist actions, inability to impose sanctions on actions outside their jurisdiction affecting the availability of resources among others may jeopardize the fairness, efficiency and sustainability of community governance,

In view of these evidences, several authors have tried to list a number of conditions that turn the potential benefits of the community into effective factors of efficient, equitable and sustainable governance (WADE, 1987; AGRAWAL, 2003). Based on the main studies in the area, Agrawal (2003) listed 33 factors identified in the literature as determinants of the success of a governance structure for natural common use resources. Table 2 below reproduces the four groups with 33 elements synthesized by the study of Agrawal. Naturally, the list presents some difficulties: a) the interdependence of factors contributing to the success of community governance, which makes the definition of the causality relationship difficult, b) the possibility that not all factors are present in specific cases or have varying levels of relevance according to specific cases – as it is the case study of this work, c) the large number of variables that hinders their detailed analysis, and d) the under-description of factors that may affect the success of community governance (AGRAWAL , 2003).

The difficulties mentioned above indicate that there are indeed important mediations within the communities that should be taken into account when considering each of the factors listed and their contribution to the efficiency, equity and sustainability of the community-based management of common use resources. Agrawal & Gibson (1999) argue that the community "must be examined in the context of conservation by focusing on the multiple interests and actors within communities, on how these actors influence decision making, and on the internal and external institutions that shape the decision-making process. A focus on institutions rather than 'community' is likely to be more fruitful for those interested in community-based natural resource management". This means that the selection of relevant variables in the table above and the causal relationship between them must derive from the institutional specificities of the place under analysis.

The determinant factors of the common use resource management contained in the table above will be analyzed following the methodological suggestion of Agrawal (2003) in order to take into account the elements of power and resistance involved in the governance structure of the Rio Verde Basin. We shall begin with the discussion on the sectorial distribution of production, use and ownership of the property, which ultimately determine the demands of economic activities on natural resources in the region and define the possibilities of adjustment to an environmentally sustainable and socially fair model.

2.5 THE PRODUCTIVE AND LAND STRUCTURE IN THE MUNICIPALITIES OF RIO VERDE'S APA

The income generated in the municipalities where the Rio Verde's APA is located accounts for about 7% of all income generated in the State of Paraná, and 18% of the income generated in the Metropolitan Region of Curitiba. The region has a productive structure dominated by services and industry, which together generate 95% of local income, while agriculture generates the other 5%. Economic performance, measured by the income growth in the region, was extraordinary in 2002-2007. While the annual average income grew by 7.5%, the average annual per capita income grew by 5.6%. Both rates are well above the rates of the state of Paraná and the Metropolitan Region of Curitiba

Considering the data above, we can see that the value produced in Campo Largo and Araucária is almost the same. However, aggregate data from the municipalities of Campo Largo and Araucária hide disparities between the productive structures of the two cities. Although the relative shares of different sectors are similar, in Campo Largo the income generation power compared to the one in Araucaria is much lower. Araucaria produces 6% of the domestic product of Paraná, with more emphasis on the industry that represents about 9.5% of the state's industrial production and 23% of the industrial production of the Metropolitan Region of Curitiba. The industry in Campo Largo, in turn, adds little more than 1% to the total state product. On the other hand, while agriculture in Campo Largo represents 4.5% of the income generated in that county; in Araucária, agriculture represents only 0.5% of the income generated there. This increased income generation power presented by Araucaria allows this county to enjoy a per capita income seven times greater than that of Campo Largo.

There should be no doubt as to the importance that the petrochemical industry has on those numbers as the head offices of the President Getúlio Vargas refinery of Petrobras (REPAR) is located in Araucária. It is estimated that REPAR generates between 2,000 and 2,500 direct and indirect jobs, and pays about 16% of the VAT of the State of Paraná. The industrial production of the refinery is currently the main consumer of water from Rio Verde. Geo-

[4] Robert Wade (1987), a proponent of the community solution, says that, in his study in 31 villages of southern India, he found that some communities ran their common use resources sustainably while others did not have any sustainable management in place. Wade concludes that "there can be no presumption that the collective action route (as Wade calls the communities) will generally work, any more than there can be a presumption that private property or state regulation will generally work" (p.105).

graphically speaking, the dam of the refinery from where the water resources are drained for production and cooling processes, is downstream of Rio Verde.

Thus, within our scheme, REPAR would be the economic unit which would suffer the most with negative externalities in the case of predatory use of natural resources in the region. In this case, the unregulated use of natural resources, with impacts on the quality and quantity of water, would cause the private costs of the company to grow. Likewise, granting REPAR the exclusive right to use would represent a cost of monitoring that the company probably could not or does not want to afford, which would result in the depredation and degeneration of available resources.

TABLE 2 - CRITICAL CONDITIONS FOR THE SUCCESS OF COMMUNITY GOVERNANCE OF NATURAL COMMON USE RESOURCES

1) System Features
I) Small scale
II) Well-defined limits
III) Low mobility of resources
IV) Possibility of storing resource benefits
V) Predictability
2) Group Characteristics
I) Small scale
II) Clearly defined limits
III) Sharing norms and values
IV) Past success experience – social capital
V) Appropriate leadership – young, familiar with external changes, connected to local traditional elite
VI) Interdependence among group members
VII) Heterogeneity of appropriations, homogeneity of identities and interests
VIII) Low levels of poverty
(1 and 2) Relationship between the characteristics of the system and the characteristics of the group
I) Overlap between the residential location of users and the resource location
II) High levels of dependence of the group on the system where the resource is
III) Justice in the allocation of benefits from the use of common resources
IV) Low levels of user demand
V) Gradual changes in the level of demand
3) Institutional organization
I) Simple and easy to understand rules
II) Rules of access and administration designed locally
III) Easy compliance with rules
IV) Gradual penalties
V) Availability of low cost lawsuit
VI) Accountability of monitors and other users' overseers
(1 and 3) Relationship between resource system and institutional organizations
I) Restrictions on use are suitable for regeneration of resources
4) External environment
I) Technology
a) Low cost exclusion technology
b) Time to adapt to new technologies related to the commons
II) Low level of articulation with external markets
III) Gradual change in articulation with external markets
IV) State
a) Central Government should not underestimate local authority
b) External institutions that support sanctions
c) Appropriate levels of external aid to compensate local users for conservation activities
d) Conjugated levels of appropriation, provision, obligation and governance

TABLE 1 - INCOME PER CAPITA AND SECTOR PRODUCTION STRUCTURE IN ARAUCÁRIA AND CAMPO LARGO

	2002	2007	ANNUAL AVERAGE VAR	AS A PROPORTION OF PARANÁ	AS A PROPORTION OF RMC
Araucária					
Income per Capita	R$44,846.00	R$58,592.40	6.1	5.5 times	4.1 times
Agriculture (%)	0.6	0.5	3.0	0.4	11.1
Industry (%)	50.3	43.4	4.1	9.4	23.1
Services (%)	49.1	56.1	11.9	5.3	12.6
Total			8.0	6.0	15.7
Campo Largo					
Agriculture (%)	R$7,324.0	R$8,293.4	2.6	0.8 times	0.6 times
Industry (%)	3.3	4.5	14 1	0.1	13.1
Services (%)	42.1	42.0	4.7	1.2	3.1
Total	54.7	53.5	4.2	0.7	1.6
Total			4.7	0.8	2.1

SOURCE: IBGE and Ipardes (Cadernos Municipais)

2.6 LAND TENURE AND AGRICULTURAL PRODUCTION

In the surroundings of Rio Verde, there is a variety of agricultural, leisure and even periurban villages. The property structure is typically small and, where being dedicated to production, is usually of family farming. The establishments of this kind in the region occupy 44% of the tracts of land and make up to 83.5% of the existing establishments. The average size of these properties is around 10 ha.

The vast majority of these families owns the land and employs their own family in the agricultural and livestock production. The production of foods such as corn, beans and potatoes prevails. In Araucária, these cultures represent 87% of the planted area and 88% of the amount produced. In Campo Largo, those same cultures prevail in 88% of the planted area and represent 88% of the amount produced. The other productions are onions, some fruits, tobacco (in Campo Largo) and soybeans. Livestock farming and production of derivatives is not the main or even an important activity in the region according to data available. However, although data do not register this production, there is evidence that local production of poultry for slaughter is growing in the region of the Rio Verde's APA.

TABLE 2 – RURAL POPULATION PER GENDER

		1996	2007	Var.%
Araucária	Rural population	8,036	8,563	6.6
	Male	4,174	4,376	4.8
	Female	3,862	4,072	5.4
Campo Largo	Rural population	19,225	17,233	-10.4
	Male	9,956	8,958	-10.0
	Female	9,269	8,275	-10.7

SOURCE: IBGE and Ipardes

The dynamics of the rural population in the two cities differ sharply (Table 2) Campo Largo has more than twice the rural population of Araucária. Considering the per capita income, a farmer in Araucária has an income about 1.7 times bigger than that of a producer in Campo Largo. We also note an opposed population behavior in the two counties. While Araucária saw its rural population grow in the ten years between 1996 and 2007, Campo Largo experienced a decrease of approximately 10.5% of the rural population in that same period.

Land is not only the asset of families around Rio Verde, but also their main source of income. For that reason, residents of this region are often concerned with the strictness with which environmental preservationist laws can be established. On the one hand, local producers fear that restrictions to their activities, especially with regard to the limits of extensive use of soil, impede their activities, and on the other, expect incentives that enable their production. The majority of interviewed in the communities, especially in Colônia Figueiredo, in Campo Largo, say their children will not continue to perform agricultural activities. More than 30% left or will leave the property to live elsewhere. In Araucária, the reality of producers is different. Approximately 65% of respondents say that their heirs will continue to work in agricultural activities, either on their property or in another property.

In short, there is already a clear movement of rural exodus in Colônia Figueiredo. The rural population has been decreasing, and power to generate income from agricultural activities is proving to be quite limited. In Colônia Figueiredo, economic prospects for the heirs of settlers are reduced and younger generations tend to look for housing and employment in urban areas. The sale of the properties to "chacreiros" (owners of small rural properties called "chacras") became a common financing alternative for replacement in the cities. In Araucária, the economic question does not put the same pressure on the population movement. In fact, we find cases of business expansion of some families to other regions of the state.

The theoretical and empirical literature that were reviewed in the first part of this study suggest that the problems and solutions that affect the management of common use resources result from the governance structure operating in the region. In the next section, we discuss this structure and show its limitations to account for problems that arise from the management of common use resources in the socioeconomic structure presented in Rio Verde. Going more specific, we indicate how local communities perceive the governance structure and the extent to which they seem to be willing to reduce the costs of implementation and monitoring of preservationist laws through collaboration with the public good.

TABLE 3 – QUESTIONS ON FAMILY SUCCESSION

	FIGUEIREDO	C. CRISTINA	TOTAL	PERCENTAGE
	13 quest.	31 quest.	44	
Continue to work on the property with farming activities	2	11	13	29,5
Continue on the property and work outside on other activities	6	1	7	15,9
Continue on the property and work on farming activities outside the property	0	6	6	13,6
Leave the property and continue in the countryside	1	3	4	9,1
Leave the property and move to another city	2	3	5	11,4
Sons no longer involved in the activities of the property	2	3	5	11,4
Others	0	1	1	2,3

Source: Prepared by the authors based on applied questionnaires

2.7 THE GOVERNANCE STRUCTURE OF RIO VERDE HYDROGRAPHIC BASIN

What is the current governance structure of water resources in Rio Verde (Paraná)? How was it formed and what roles its components have played? According to our theoretical framework, the answers to these questions should emerge from an analysis of the interrelationship of formal institutions that make up such an environment for decision making with the way of thinking and behavior of agents in face of these structures. For such reason, two different methodologies were applied, one related to the organizations involved with the resources of Rio Verde and the other focusing on the rural population residing on its banks.

The first empirical approach model was based on semi-structured interviews with questions involving the governance structure at stake being asked (see Chapter 16 of this book for a detailed description of the methodology and questionnaires). The second field work consisted in the development of participatory diagnosis of the populations; a survey of data that corresponds to the joint creation of information as free as possible from the vision of the researchers[5]. This empirical work tried to determine the *enforcement* of the governance structure through the ways of thinking captured in the participants' own words.

The table above highlights the results from questions asked in the region in order to assess the perception of producers related to the water resources represented by Rio Verde and the law that governs the activities in the region, the Rio Verde's APA. At a first glance, one realizes that the importance of the river in relation to productive and leisure activities (fishing, walking, bathing etc.) is low. In fact, few crops in the surveyed area use irrigation waters from Rio Verde. This data may also suggest a possible low valuation of the resource. That is, these producers probably would not take into account externalities that would result from their activities on the resource. This means that the incentives for the preservation of Rio Verde should go beyond payment for water.

In fact, during visits to communities, especially Colônia Figueiredo, the key issue found is the land use. As the properties are small on average – and the terrain is uneven in some cases, the restriction on land use on river banks means for these manufacturers a much greater economic risk than simply having to pay for water. In other words, charging for water, even if all the income was reverted to the region, does not compensate for these producers' losses of agricultural land imposed by environmental legislation. Thus, incentives to low cost adhesion to environmental laws are distorted.

In both Colonies, the laws of APA and the prospect of zoning have caused great concern in the community. Milder (2004) has done an extensive research in the annals of meetings and public hearings that occurred in 2003, when the first macro zoning of the APA was performed. Transcripts of speeches of residents show total disagreement, distrust and apprehension with the rules that were being informed by authorities during public meetings (Milder, 2004). The community remains suspicious of the process. In our survey, about 41% of the people we interviewed attended all public meetings on the macro-zoning of the APA in 2003, and 27% participated in some. The full knowledge and full agreement with the rules to be effective with APA are around 20-22%.

Part of this distrust of local producers is due to the lack of institutional integration between instances of representation of communities and public authorities who make decisions on the resources of the Rio Verde Basin. The following diagrams were drawn from community meetings. The color similar to the community represents the establishment of positive connections. The closer to the community (and with no contours), the stronger the ties of the in-

[5] Obviously, data interpretation is at the discretion of those who research, thus, there is some cognitive interference. However, if this influence is presented ex ante the field dynamic, information collected is the one that farmers deem important. Element that highlights the representativeness of such data.

stitution with the community. On the other hand, reddish coloration means negative relations from the standpoint of communities. Dashed or solid contours mean greater barriers to inter-relationship.

TABLE 4 – QUESTIONS ON RIO VERDE AND RIO VERDE'S APA

	FIGUEIREDO	C. CRISTINA	TOTAL	PERCENTAGE
What is the importance of Rio Verde to your productive activities?	13 quest.	31 quest.	44 quest.	
Essential	0	8	8	18.2
Important	1	8	9	20.5
Not important	12	9	21	47.7
What is the importance of Rio Verde to other activities?				
Essential	1	4	5	11.4
Important	3	11	14	31.8
Not important	9	15	24	54.5
Do you know the conditions of use of the natural resources established by the APA?				
Know completely	5	5	10	22.7
Know partially	6	19	25	56.8
Do not know	2	8	10	22.7
Do you agree with the conditions of use of natural resources established by the APA?				
Fully agree	3	6	9	20.5
Partially agree	4	9	13	29.5
Indifferent	0	2	2	4.5
Disagree	4	7	11	25.0
Do not know	2	6	8	18.2
Did you participate in the meetings that defined the conditions of use of the APA's natural resources?				
Fully	7	11	18	40.9
Partially	1	11	12	27.3
Did not participate	5	8	13	29.5

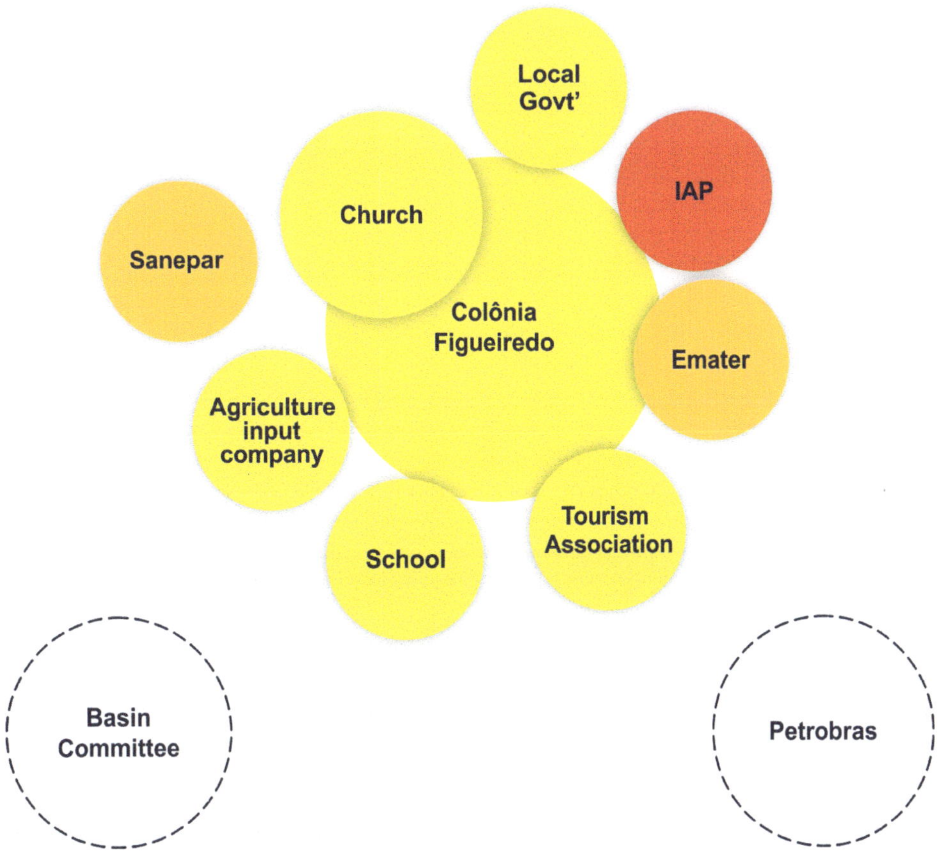

GRAPHIC 1 – DIAGRAM OF INSTITUTIONAL RELATIONS AT COLÔNIA FIGUEIREDO

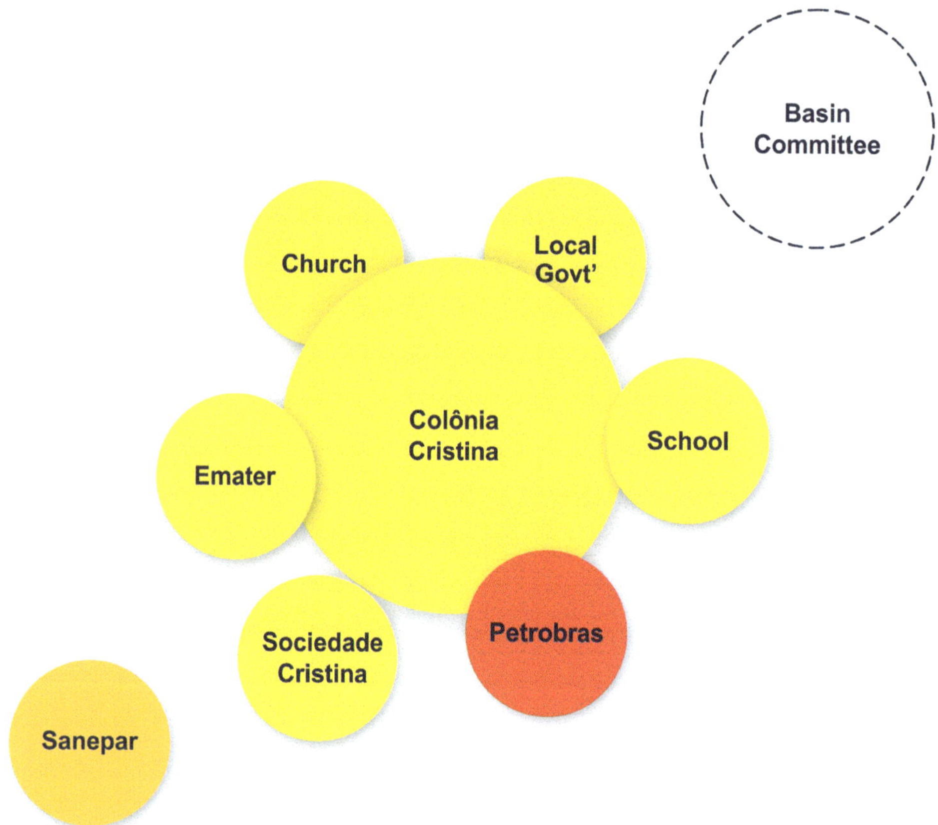

GRAPHIC 2 – DIAGRAM OF INSTITUTIONAL RELATIONS AT COLÔNIA CRISTINA

In both surveyed communities, we find a strong positive relationship with the Church, with the Local Government and with Emater. At Colônia Figueiredo, IAP is considered an institution negatively present in this community. This negative view is due to the perception that the supervisory role of IAP is performed unevenly. That is, not all offenders are equally punished. In particular, it is clear that IAP is stricter with local farmers than with urban residents, especially the residents of Curitiba. In Colônia Cristina, Petrobras is perceived as an institution with a negative relationship with the community. This perception is explained by the community as a result of past performance of Petrobras, which imposed losses of land to settlers when it created the Rio Verde dam.

However, it is noteworthy the fact that, in both communities, the Basin Committee is totally ignored. The sphere of the Basin Committee comes in white to represent the lack of mentioning by the community. Even when stimulated by researchers to position themselves in relation to the Basin Committee, community members responded that they were unaware of that entity, demonstrating the lack of presence of the comittee

The Basin Committee should be the key institution in the resolution of regulations and activities on basin resources. The Law 9,433, establishing the Brazilian National Water Resource Policy, requires that "management of water resources should be decentralized and include the participation of public authorities, users and communities" (section VI, Article 1st). This article of the law informs is need for commitment of local communities with the destination given to the natural resources they use. This commitment implies that communities must participate in the management of resources that suffer some influence of their production or other kind of work. In other words, the purpose of the Basin Committees is to engage users and communities with the integrated management of common use resources under its influence. With this engagement from communities and users, the supply of positive externalities would increase, such as sustainable land management, preservation of riparian forests, construction of sanitary infrastructure, elimination of illegal occupation, etc.

The current setting of the Alto Iguaçu Basin Committee, to which Rio Verde Basin belongs, however, is not composed by the diversity of users and communities whose actions affect the maintenance of the system. The table below shows the final configuration of the representatives on the Alto Iguaçu and Ribeira Committees, before the establishment of the Institute of Water, which replaced Sudersha as the executive body of the Paraná Basin Plans.

There is an overrepresentation of public agencies, even in those positions intended for civil society participation. There is also a large number of companies and associations that are neither users nor belong to the community. The latter are customarily captured by the first through research agreements, joint development projects, etc, increasing the effective representation of companies. Furthermore, there is a clear under-representation of communities directly affected by the water resource policies.

TABLE 5 – REPRESENTATION OF ALTO IGUAÇÚ AND RIBEIRA COMMITTEES

PUBLIC INSTITUTIONS (12)	SECTORS THAT USE WATER RESOURCES (14)	CIVIL SOCIETY ORGANIZATIONS (12)
State Secretary for Environment and Water Resources - SEMA;	Water and Sanitation Company of Paraná State - SANEPAR;	Watershed Management Council of the Metropolitan Region of Curitiba;
Superintendence of Water Resource Development and Environmental Sanitation) – SUDERHSA	Water and Sanitation Company of Paraná State - SANEPAR;	Iraí river APA Technical Chamber;
Environmental Institute of Paraná – IAP;	Water and Sanitation Company of Paraná State - SANEPAR;	Rio Passaúna's APA Technical Chamber;
Coordination of the Metropolitan Region of Curitiba – COMEC;	Energy Company of Paraná State – COPEL;	CEDEA – Center for Environmental Studies, Defense and Education;
State Secretary of General Planning and Coordination – SEPL;	Energy Company of Paraná State – COPEL;	Timoneira institute;
State Secretary of Agriculture – SEAB;	Petróleo Brasileiro S/A – PETROBRAS;	Association for the Environmental Preservation of Rio Iguaçu and Serra do Mar in Paraná;
Minerais do Paraná S/A – MINEROPAR;	Geoplan do Brasil	Federal University of Paraná – UFPR;
Local government of Piraquara City;	Grupo Votorantim;	Catholic Pontifical University of Paraná State – PUCPR;
Local government of Colombo City;	DaGranja Agroindustrial Ltda;	Federal Technological University of Paraná – UTFPR;
Local government of São José dos Pinhais City;	Trombini Industrial S/A;	Brazilian Association of Water Resources – ABRH;
Local government of Almirante Tamandaré City;	Terra Rica Indústria e Comércio de Calcáreo Ltda;	Brazilian Association of Subsurface Waters – ABAS;
Local government of Campo Magro City;	Organization of Cooperatives of the State of Paraná – OCEPAR;	Brazilian Association of Sanitary Engineering – ABES
	Agriculture Federation of the State of Paraná – FAEP;	
	Solid Waste and Urban Drainage – City of Curitiba	

3. CONCLUSION

In summary, the sustainable and equitable management of resources of Rio Verde Basin faces crucial institutional problems. A necessary condition for an efficient and equitable management is that all agents whose actions affect the maintenance of the system behave as providers of a public good. That is, users and communities should be part of the management with the same purposes that govern public administration. That would be the purpose of the Basin Committee, although as evidenced, it fails in its purpose of integrating the community as an agent for the provision of the public good. Thus, communities will continue to be seen as agents antagonistic to the management system. From the communities' point of view, management will continue to be seen as a threat to their livelihood. In particular, they see the specific regulations on resource use in the region as limiting their agricultural production. In Campo Largo, this becomes even more dramatic given that Colônia Figueiredo already undergoes a process of economic stagnation and population evasion.

Finally, even if simple propositions such as charging for water may arise as a solution, they need an operating and legitimate institutional mechanism at the community level in order to integrate it in the management of common resources. Until the management of resources in Rio Verde is deemed as a common asset and until it is organized and effectively managed so that communities feel as co-owners of the system in order to establish rules consistent with the conservationist behavior, the proposed payment for water shall continue to be inadequate and poorly located.

4. REFERENCES

AGRAWAL, A. (2003). Sustainable Governance of Common-Pool Resources: Context, Methods, and Politics. *Annual Review of Anthropology*, Vol. 32, pp. 243-262

AGRAWAL, A. e GIBSON, C. (1999). Enchantment and Disenchantment: The Role of Community in Natural Resource Conservation. *World Development* Vol. 27, N. 4, pp. 629-649

ANA (Agência Nacional das Águas) (2007a). *Cadernos de Recursos Hídricos 2: disponibilidade e demandas de recursos hídricos no Brasil.* Brasília.

ANA (Agência Nacional das Águas) (2007b). *Geo Brasil: recursos hídricos.* Relatório executivo sobre o estado e perspectivas do meio ambiente no Brasil. Brasília.

BEARD , V. A. e DASGUPTA, A.(2006). *Collective Action and Community-driven Development in Rural and Urban Indonesia Urban Studies*, Vol. 43, N. 9, 1451–1468.

BEIRLE, T. (1998). *Public Participation in Environmental Decisions: An Evaluation Framework Using Social Goals.* Resources for the Future Discussion Paper, Washington.

BOWLES, S. e GINTIS, H. (2002). Social Capital and Community Governance. *The Economic Journal,* Vol. 112, N. 483, pp. F419-F436

BOWLES, S. (2004). *Microeconomics: behavior, institutions, and evolution*. Nova York: Russel Sage Foundation.

BROMLEY, D. (1982). Land and Water Problems: An Institutional Perspective. *American Journal of Agricultural Economics*, Vol. 64, N. 5, pp. 834-844

BROMLEY, D. (2000). Property Rigths and Pricing Regimes in Water Resource Manegment. Ariel Dinar (Ed.). *In the Political Economy of Water Pricing Reforms*. Oxford university Press. pp.29-48

CARNEIRO, C.; PEGORINI, E. S., e ANDREOLI, C. (2005). Mananciais de Abastecimento Público. In: Cleverson Andreoli e Charles Carneiro (Org.). *Gestão Integrada de Mananciais de Abastecimento Eutrofizados*. Curitiba: Sanepar/Finep, pp.47-82.

COMMONS, J. (1931). Institutional economics. *American Economic Review*, vol. 21: 648-657.

CONSTANCE, P. (2005). Quem ganhou a Guerra da Àgua? BIDAmérica: *Revista do Banco Interamericano de Desenvolvimento*, publicado na web em junho de 2005. http://www.iadb.org/idbamerica/index.cfm?thisid=3539. Accessed in September 15, 2008.

DUGGER, W. (1980). "Power: an institutional framework of analysis." *Journal of Economic Issues*, vol. XV (4): 897-907.

FEENY, D., BERKES, F. McCAY, B. e ASHTON, J. (1990). The tragedy of the commons: twenty two years later. *Human Ecology*, Vol.18, N.1.

GRAFTON, R. Quentin (2000). Governance of the Commons: A Role for the State? *Land Economics*, Vol. 76, N. 4, pp. 504-517

HANNA, S. & SANDERSON, S. (1996). Human Use of the Natural Environment: An Overview of Social and Economic Dimension. In: Susan Hanna, Carl Folke, Karl-Göran Mäler (Org.). *Rights to Nature: Ecological, Economic, Cultural, and Political Principles of Institutions for the Environment*. Washington DC: Island Press, pp.35-56.

HANNA, S., FOLKE., C. & MÄLER, K-G. (1996). *Rights to Nature: Ecological, Economic, Cultural, and Political Principles of Institutions for the Environment*. Washington DC: Island Press.

HARDIN, G. (1968). The tragedy of the commons: the population problem has no technical solutions; it requires a fundamental extension in morality. *Science*, vol. 162: 1243-1248

HIRSCHMAN, A. (1970). *Exit, voice and loyalty*. Harvard: Harvard University Press.

HODGSON, G. (1998). The approach of institutional economics. *Journal of Economic Literature*, v. XXXVI, n. 1:166-192.

HODGSON, G. (2003). The hidden persuaders: institutions and individuals in economic theory. *Cambridge Journal of Economics*, 27, p. 159-175.

HODGSON, G. (2007). Institutions and individuals: interaction and evolution. *Organization Studies*, vol. 28 (1): 95-116.

IPARDES (2010). *Indicadores de Sustentabilidade Ambiental por Bacias Hidrográficas do Estado do Paraná*. Curitiba: Ipardes.

McCAY, B. J. (1996). Common and Private Concerns. In: Susan Hanna, Carl Folke, Karl-Göran Mäler (Org.). *Rights to Nature: Ecological, Economic, Cultural, and Political Principles of Institutions for the Environment*. Washington DC: Island Press, pp.111-126.

MILDER, L. I. D. (2004). *Análise do Processo de Participação Popular para a Elaboração do Macrozoneamento Ecológico-Econômico da Área de Proteção Ambiental Estadual do Rio Verde –Paraná*. Dissertação de Mestrado Pós-Graduação Em Engenharia Florestal UFPR

OSTROM, E. & SCHLAGER, E. (1996). The Formation of Property Rights. In: Susan Hanna, Carl Folke, Karl-Göran Mäler (Org.). *Rights to Nature: Ecological, Economic, Cultural, and Political Principles of Institutions for the Environment*. Washington DC: Island Press, pp.127-156.

PNUD (2006). *A água para lá da escassez: poder, pobreza e a crise mundial da água*. Nações Unidas.

RAMOS, C. & KRUSE, T. (2003). Privatização: Benefícios duvidosos, riscos concretos. *Observatório da Cidadania* 2003.

RUTHERFORD, M. (1996). *Institutions in economics: the old and the new institutionalism*. Nova York: Cambridge University Press.

SIMON, H. (1957). *Models of man: social and rational*. Nova York: John Wiley and Sons.

SIMON, H. (1978). Rationality as process and as product of though. *American Economic Review*, vol. 68 (2):1-16.

SIMON, H. (1991). Organizations and markets. *Journal of Economic Perspectives*, vol. 5 (2): 25-44.

SMITH, A. (1983) [1776]. *A riqueza das nações: investigação sobre sua natureza e causas*. São Paulo: Abril Cultural, vol.1 (Coleção Os Economistas)

STEIN, J. (1997). How institutions learn: a socio-cognitive perspective. *Journal of Economic Issues*, vol. 31(3): 729-740.

VEBLEN, T. (1899). *The Theory of the Leisure Class*. Nova York: Penguin Books.

VEBLEN, T. (1919). *The place of science in modern civilisation and other essays*. Nova York:Viking.

WADE, R. (1987). The management of common property resources: collective action as an alternative to privatisation or state regulation, *Cambridge Journal of Economics* 1987,11, 95-106.

WALSH, E.; BABAKINA, A. PENNOCK, H. SHI, Y. CHI, T. WANG, T.E. GRAEDEL. (2006). Quantitative guidelines for urban sustainability. *Technology in Society* 28, pp.45-61.

WILLIAMSON, O. (1985). *The economic institutions of capitalism*. New York: Free Press.

WILLIAMSON, O. (1998). The institutions of governance. *American Economic Review*, vol. 88(2): 75-79.

CHAPTER 18

AN ANALYSIS OF FISHING PRACTICES

Vinicius Abilhoa
Wellen Laryessa Laynes
Letícia Michalik

AN ANALYSIS OF FISHING PRACTICES

Vinícius Abilhoa
Wellen Laryessa Laynes
Letícia Michalik

SUMMARY

In large urban centers, such as the Metropolitan Region of Curitiba, amateur fishermen enjoy the practice of fishing especially in reservoirs used as water supply sources. The following analysis of fishing activities indicates that fishermen who fish around the Rio Verde Reservoir are a relatively homogeneous group in relation to gender, marital status, place of origin, materials used (rods and reels), target species (tetras), and the frequency and purpose of fishing. Fishing occurs mainly on weekends throughout the year and the most favorable months are from December to February. The fishing that is conducted around the Rio Verde Reservoir is considered recreational or for sport and the catch is not commercialized. Fishing is considered an important aspect of daily life and leisure in for most fishermen because it is a pleasant way to spend time, an enjoyable pass-time for interaction and socialization, and it also represents a complementary addition and variety to their diet.

KEYWORDS

Dams, interviews, recreational fishing, fishing resources.

1. INTRODUCTION

Research on fish harvests and fishing activities undertaken in several regions of Brazil have been insufficient, scattered and inconsistent (SANTOS *et al.*, 1995; PAIVA, 1997). Despite the fact that the practice of fishing for consumption and survival dates back to the beginning of our history (FISCHER *et al.*, 1992; PAIVA, 1997) and also considering that fishing is currently practiced for several reasons, such as leisure or sport (RIOS, 1976), the statistics available on several forms of fishing do not include historic data, thus complicating the assessment and management of this activity. According to the National Policy of Sustainable Development for Agriculture and Fisheries (Federal Law 11.959, June 29, 2009), the exploitation of fish resources can be basically divided into commercial and non-commercial fishing. Commercial fishing is seen as a source of income and can be either artisanal or industrial. Artisanal fishing is practiced by professional fishermen, who develop the activity autonomously or as part of the family economy, using their own means of production or through partnership, on land or with small vessels. Industrial fishing is conducted by individuals or companies, involving professional fishermen as employees or partners, using small, medium or large vessels. Non-commercial fishing is considered any scientific, subsistence or recreational fishing activity that is not for profit and uses devices allowed in the legislation.

According to the National Program for Development of Fishing from the Brazilian Institute for the Environment and Renewable Natural Resources (IBAMA 2008), recreational fishing in Brazil has recently experienced a significant increase, with many economic benefits, particularly in the tourism sector. Fishing, when practiced on inland waters, is strongly correlated with the social and geographic contexts in which it takes place, responding to current demographic, economic and political factors (WELCOME, 2001).

In the face of this complex scenario, an accurate assessment of the activity carried out in different ways allows for a consistent evaluation of the impacts of fishing on exploited stocks and it is imperative to include in the analysis an accurate survey of the people who take part in the practice.

2. RESERVOIR FISHING

Generally, reservoirs are colonized by the preexisting ichthyofauna, meaning that the fish species in the reservoir are comprised of the fish species that occurred in the environment before the construction of the reservoir (AGOSTINHO et al., 2007). As not all species are able to adapt to the new environment, the ichthyofauna is much less diverse than the fish populations that existed in the original river (AGOSTINHO et al., 1997). The changes produced by the creation of reservoirs, such as the change from a lotic ecosystem into a lentic ecosystem, results primarily in the disappearance of strictly river fish species and a subsequent rearrangement of the remaining species (LOWE-McCONNEL, 1975).

When introduced species (or exotic species) are present during the construction of the reservoir, the environmental disruption might allow for an increase in the population of exotic species, which are in general tougher and more tolerant than the native species (GHERARDI, 2007; VITULE et al., 2006).

The creation of reservoirs in the Iguaçu River Basin brought about important changes to the local water, limnological and ecological conditions, with both a direct and indirect influence on fish populations. Generally, there was an initial drastic impact on fish populations as the construction modified the dynamics of water and destroyed shelters, and breeding and feeding sites, which caused changes in the composition and structure of the fish populations (AGOSTINHO & GOMES, 1997; ABILHOA, 2005). The ichthyofauna of these reservoirs has the same general pattern of the Iguaçu River, with a small number of species and a high degree of endemism. A large proportion of the reported species are exclusive to this hydrographic basin and most fish caught belong to these species. This shows the importance of regional participation in procedures to assess the ichthyofauna of the area of influence of a reservoir (JACKSON & HARVEY, 1989). Only a fraction of the reported species from the reservoirs of this basin correspond to fish introduced from other hydrographic basins or from other continents, such as the black-bass *Micropterus salmoides,* transparent tetra *Charax stenopterus,* carp *Cyprinus carpio* and tilapia *Tilapia rendalli.* Although the percentage of the introduced species is modest, revealing a certain failure of fish stocking, these species may have a rapid population growth and/or cause considerable changes throughout the entire aquatic ecosystem, including an indirect influence on water quality.

In large urban centers, such as the Metropolitan Region of Curitiba, amateur fishermen mainly seek-out the activity of fishing in the reservoirs used as water supply sources. According to Agostinho *et al.* (2007), fishing on the shores of the reservoirs is an activity that happens naturally and it can have an important social role, involving both amateur and professional fishermen. Indeed, fishing in the reservoirs is considered a traditional subsistence activity, which has important environmental and social consequences (PEREIRA, 2005) and requires constant management actions.

The data available on fishing in these reservoirs are weak; fishing data can be found for the reservoirs of the Iguaçu River Basin and this information suggests that it is undertaken as an illegal activity (professional fishing) and can be a supplementary source of income (OKADA *et al.,* 1997).

3. FISHING IN THE RIO VERDE RESERVOIR: A CASE STUDY

The Rio Verde Reservoir, located in the Metropolitan Region of Curitiba, is part of the Rio Verde Environmental Protection Area (APA Rio Verde), which is occurs across parts of the municipalities of Campo Largo, Campo Magro and Araucária (PARANA, 2000). In relation to fishing, there is no historical statistical data from the region (CERGOLE *et al.,* 2005) making the diagnosis and management of the activity difficult.

Considering this scenario, it is absolutely necessary to accurately assess fishing at the site, allowing for an understanding of the interactions between the activity and the environment. The purpose of this study was to examine the fishing that takes place in the reservoir area, identify target species and describe the socioeconomic profile of the fishermen who participate in the activity in the region.

Despite the fact that this data set may have spatial and temporal restrictions, the assessment produced can be used as a guideline to establish measures of control and management on the Rio Verde Reservoir, since this activity has a relevant social role, either as a source of subsistence and/or as a leisure activity for urban community members.

3.1 METHODS

A non-probabilistic typical case sampling method was used. The goal of this method is to find a typical group in relation to the population as a whole that presents general characteristics representative of the studied population (ACKOFF, 1967). The choice of participants for the present study was made based on the willingness of the participants to help with the research. As such, fishing was considered as an activity in general, without recording data that could identify either individuals or specific communities.

The evaluation of fishing activities in the Rio Verde Reservoir was made using interviews which were conducted in two phases from November 2008 to February 2010. Interviews were conducted at the main sites where fishing takes place (Figures 1 and 2). The questionnaires used were developed from existing models in the literature (*e.g.* CAMARGO, 1998; OKADA *et al.,* 1997; SANTOS *et al.;* 1995; CEREGATO & PETRERE JR.; 2002).

In the first phase, from November 2008 to March 2009, semi-structured questionnaires were administered monthly, which collected data such as:

- interviewee's place of origin (residence);
- frequency of activity;
- ethno-species caught.

In a second phase, conducted between February 23 and 26, 2010, interviews with semi-structured and open ended questions were given with the purpose of collection information on the profile of fishermen in the region. There were six major themes in this questionnaire:

- identification of the interviewee, with questions related to age, gender, profession, marital status, education level, membership in fishery associations, and place of origin (residence);
- preservation of the region, where we sought to determine the interviewee's perspective on the preservation of the fishing area, ranking from "good" to "poor", in addition to discussing the impacts that amateur fishing could have on the local community. In considering the length of time the interviewee had been fishing, the participant was asked about the possible impacts that fish populations may have suffered in the reservoirs, such as if there had been a reduction in fish stocks and what were the possible causes;
- participation in the activity of fishing, with some questions related to the length of time they had practiced fishing in the region and if the activity was a family tradition;
- fishing equipment, with descriptions of the materials (fishing tackle) used by fishermen;
- consumption of fish, with some questions about the purpose of fishing, how often fish are consumed and the way they are preserved. This topic was used to identify the target ethno-species, asking the fishermen the type of ethno-species he liked and did not like to fish. Furthermore, information was also requested about the best time of the year for fishing, highlighting the months that

are considered "good" and "bad" for fishing. The approximate time fishermen spent practicing the activity and the estimated monthly expenses also were discussed;

- observations and suggestions for improving fishing at the site.

The results of the interviews were tabulated and quantified, aiming to develop a database of the activity carried out in the region. This database was used to assess the level of fishing activity performed in the reservoir area, described through tables and frequency distribution.

3.2 CHARACTERISTICS OF FISHING

3.2.1 General characteristics

A total of 109 questionnaires were administered in the first phase of the assessment.

According to the results, the interviewees come from the municipalities located in the area surrounding the reservoir, mostly Curitiba, Araucária and Campo Largo (Figure 3). The most frequently mentioned neighborhoods of Curitiba were Cidade Industrial (CIC), São Brás, Fazendinha, and Campo Comprido (Figure 4).

Regarding the frequency of fishing in the Rio Verde Reservoir (weekly, monthly or annually; Figure 5), the results show some degree of irregularity. Fishing is conducted as a leisure activity and in most cases it preferably occurs on the weekends.

The diversity of the ichthyofauna recorded by the fishermen represented 70% of the species identified in experimental fishing (Figure 6). According to the interviewees, fish generically identified as "catfish", "tetras" and "armored catfish" may present several types (morphotype) and the similarities are highlighted using expressions such as "it seems" and "it has the same shape as." For example, in the tetra group, some varieties were mentioned, including yellow-tail tetra (*Astyanax minor*) and red-tail tetra (*Astyanax bifasciatus*). For this reason, these species were grouped into ethno-species and referred to using the scientific identification of genus (*Astyanax*). Some names used to denominate the ethno-species may be considered polysemic or in other words they can refer to more than one species or genus, as in the case of carp (Table 1).

3.2.2 Profile of fishermen and fishing

In order to record the profile of the fishermen and fishing in the region, 30 interviews were conducted using open and semi-structured questions.

We can see that the average fisherman who practices fishing at the reservoir is: male (83.3%), married (80%), fishes together with friends or family, for leisure (73.3%). Thus, fishing in the region can be characterized as amateur. Fishermen are generally aged between 20 and 70 years old and the majority has finished secondary education (47%) (Table 2).

The practice of fishing in the Rio Verde Reservoir is not new, since the majority of interviewees (43%) had been fishing in the area for over 10 years (Figure 7).

According to 86.6% of fishermen, the place where they fish is very well preserved and they believe that fishing has no significant impact on the community in the region. When asked if the amount of fish has been decreasing over the years, the majority (63.3%) gave a negative response; however, for those who gave an affirmative response, the decrease is because of the prohibition of equipment, mainly the use of nets.

The fishing tackle used for fishing at Rio Verde Reservoir are rods and reels. The species caught are tetras (*Astyanax spp.)*, pearl cichlid *(Geophagus brasiliensis)*, tilapia (*Tilapia rendalli)*, trahira (*Hoplias aff. Malbaricus)*, pike characin *(Oligosarcus longirostris)*, carp (*Cyprinidae)*, armored catfish *(Hypostomus spp)*, marbled swamp eel *(Synbranchus marmoratus)* and catfish (*Rhamdia quelen)* (Figura 8). The most frequently caught species are tetras and pearl cichlid.

Despite the variety of species found in the region, amateur fishermen have some preferences for certain species (Figure 9). The majority (74.3%) practice the activity in the reservoir with the goal of catching tetras, which according to some interviewees, are considered as "small golden dorados" (in reference to *Salminus brasiliensis)*. Some statements made by interviewees show this preference: *"tetra is a fish that fights to get out of the water unlike other fish, besides being a tasty fish."* The data collected shows that species from the genus *Astyanax* of the region are the target species for fishing. Pearl cichlid (*G.brasiliesis*) it is the second most captured species (32%) according to the interviewees and this species is not desirable in fishing.

The fishermen practice fishing more often during January, February and December, believing these are the best months for fishing in the region (Figure 10). During these months, the frequency of fishing trips is weekly, from one to two times per week, and fishermen are at the site for an average of 12 hours. The worst months for fishing are June, July and August and there are only a few fishermen who fish at the Rio Verde Reservoir during this time of the year (Figure 11).

During the period considered good for fishing, the average number of fish caught goes from 50 to 100, with as many as 150 specimens caught. During the months that are considered poor, when fishing is undertaken the number of fish caught is less than 20.

The expenses incurred from participating in the activity range from US$ 5 to US$ 50, depending on frequency. Consumption of fish in the home occurs from two to three times per week, although the interviewees do not depend on local fishing for survival. The caught fish are usually cleaned and frozen and sometimes they are consumed on the spot.

Suggestions for improving the fishing activity are related to the fishing sites, i.e., the place where the activity is practiced at the Rio Verde Reservoir (Figure 1); the improvement of fishing piers, appropriate monitoring, and more attention to cleaning up garbage, were the most frequent complaints reported.

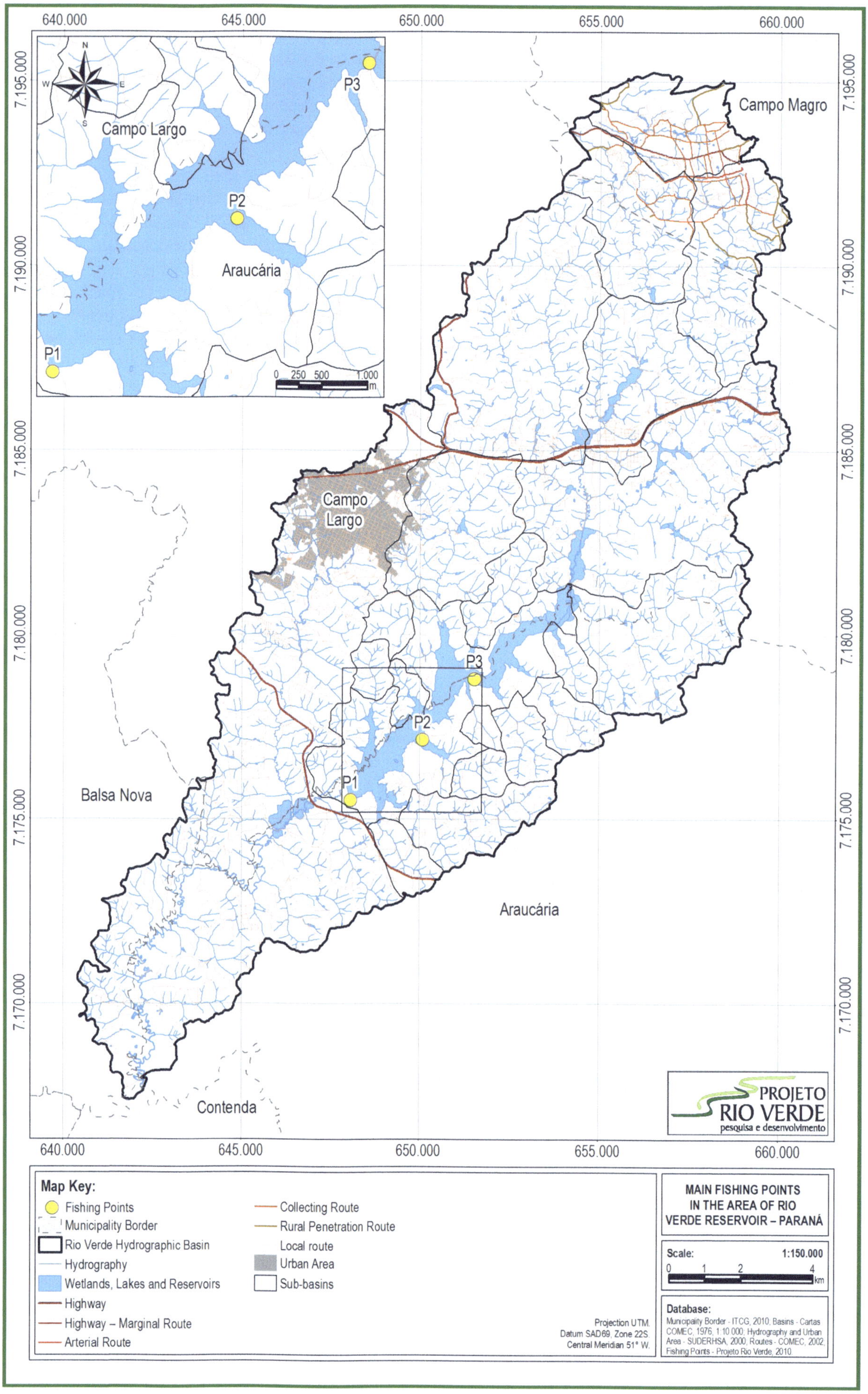

FIGURE 1 – MAJOR FISHING SITES IN THE RIO VERDE RESERVOIR

FIGURE 2 – DATA COLLECTION, CONDUCTING INTERVIEWS AND CHARACTERISTICS OF FISHING AT THE RIO VERDE RESERVOIR

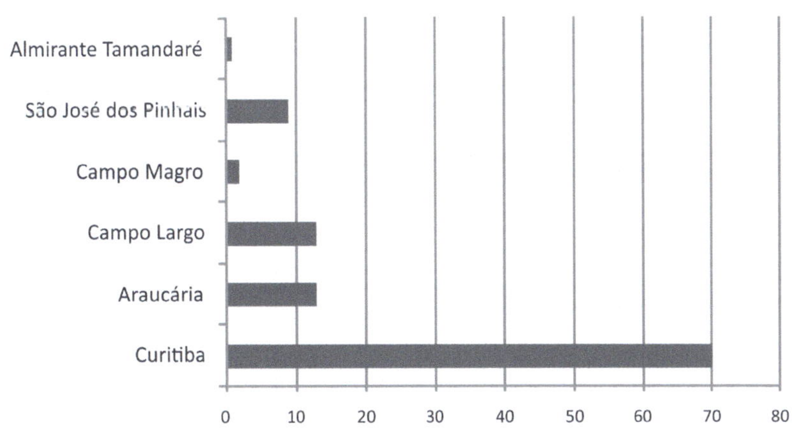

FIGURE 3 – INTERVIEWEE'S CITY OF RESIDENCE ACCORDING TO 109 INTERVIEWS CONDUCTED AT THE RIO VERDE RESERVOIR

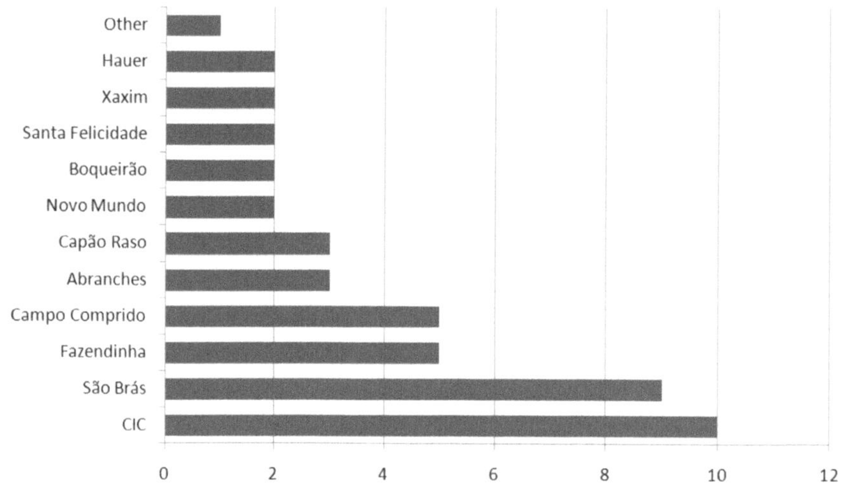

FIGURE 4 – THE NEIGHBORHOODS DISTRICTS MENTIONED BY THE 70 INTERVIEWEES WHO LIVE IN CURITIBA

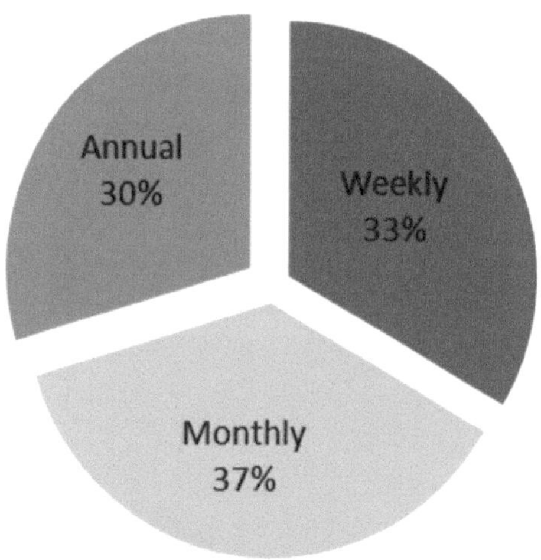

FIGURE 5 – FREQUENCY OF FISHING BASED ON 109 INTERVIEWS CONDUCTED AT THE RIO VERDE RESERVOIR

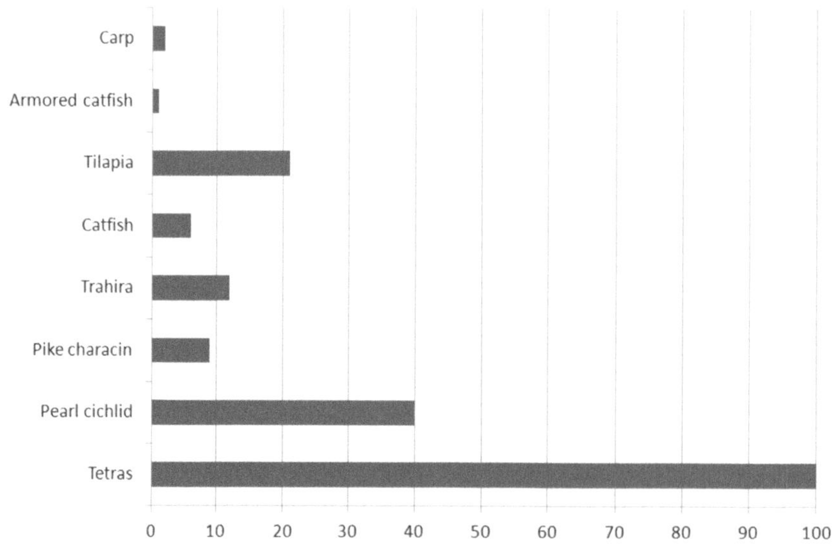

FIGURE 6 – THE ETHNO-SPECIES CAUGHT BASED ON 109 INTERVIEWS CONDUCTED AT THE RIO VERDE RESERVOIR

TABLE 1 – FISH ETHNO-SPECIES BASED ON THE PERCEPTIONS OF INTERVEWED FISHERMEN AT THE RIO VERDE RESERVOIR

ETHNOSPECIES	ORDER	FAMILY	SPECIES
Tetras	Characiformes	Characidae	*Astyanax bifasciatus*
			Astyanax minor
			Astyanax serratus
Pike characin			*Oligosarcus longirostris*
Trahira		Erythrinidae	*Hoplias aff. malabaricus*
Catfish	Siluriformes	Heptapteridae	*Rhamdia quelen*
Armored catfish		Loricariidae	*Hypostomus derbyi*
			Hypostomus commersoni
Pearl cichlid	Perciformes	Cichlidae	*Geophagus brasiliensis*
Tilapia			*Tilapia rendalli*
Carp	Cypriniformes	Cyprinidae	*Cyprinus carpio*
			Ctenopharyngodon idella
			Hypophthalmichthys nobilis

TABLE 2 – CHARACTERISTICS OF FISHERMEN BASED ON THE ANSWERS PROVIDED DURING 30 INTERVEWS, IDENTIFYING GENDER, MARITAL STATUS, AGE, LEVEL OF EDUCATION, OCCUPATION, MEMBERSHIP IN FISHING ASSOCIATIONS AND PLACE OF ORIGIN (RESIDENCE)

IDENTIFICATION	ANSWERS		
Gender	Male 83.3% (25)		Female 16.7% (5)
Marital Status	Married 80% (24)	Single 6.7% (2)	Other 13.3% (4)
Age group	< 30 years old 6.7% (2)	30 – 50 years old 20% (6)	> 50 years old 73.3% (22)
Education level	Illiterate 3.3% (1)	Unfinished Elementary School 6.7% (2)	Unfinished Secondary School 3.3% (1)
	Finished Elementary School 47% (14)		Finished Secondary School 40% (12)
Occupation	Informal 10% (3)	Employee 53.3% (16)	Retired 37% (11)
Fishing Associations	Not a member 100% (30)		
Residence	Curitiba 33.3% (10)	Campo Largo 50% (15)	Other 17% (5)

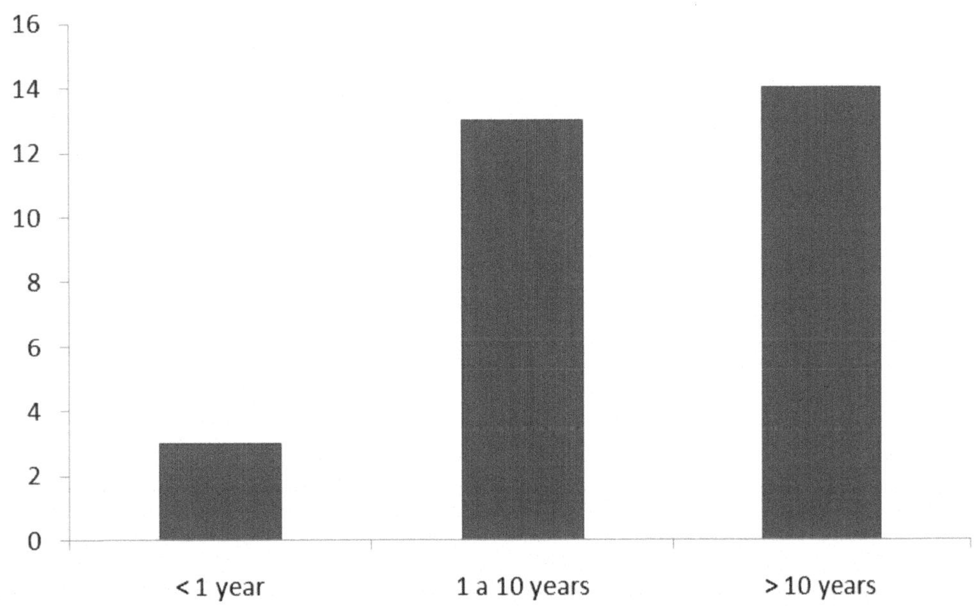

FIGURE 7 – HOW LONG THE INTERVIEWEE HAS BEEN FISHING AT THE RIO VERDE RESERVOIR

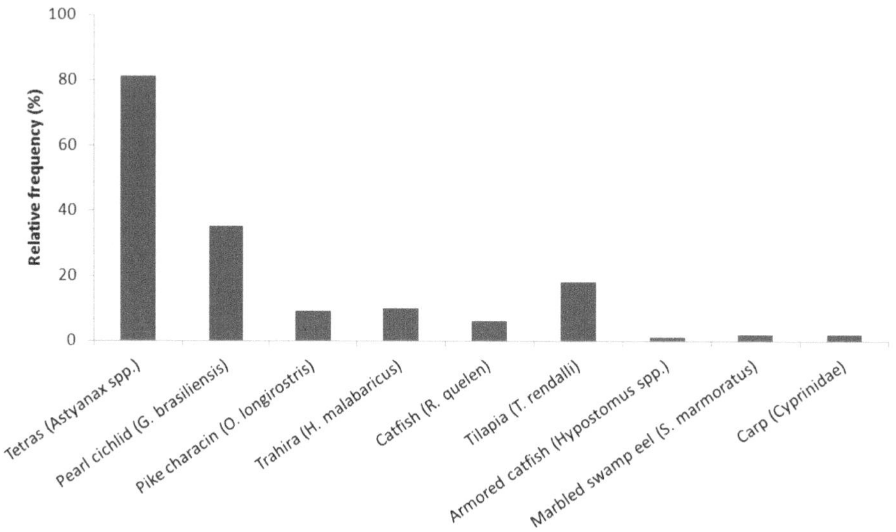

FIGURE 8 – SPECIES CAUGHT BY THE FISHERMEN AT THE RIO VERDE RESERVOIR BASED ON INTERVIEWS

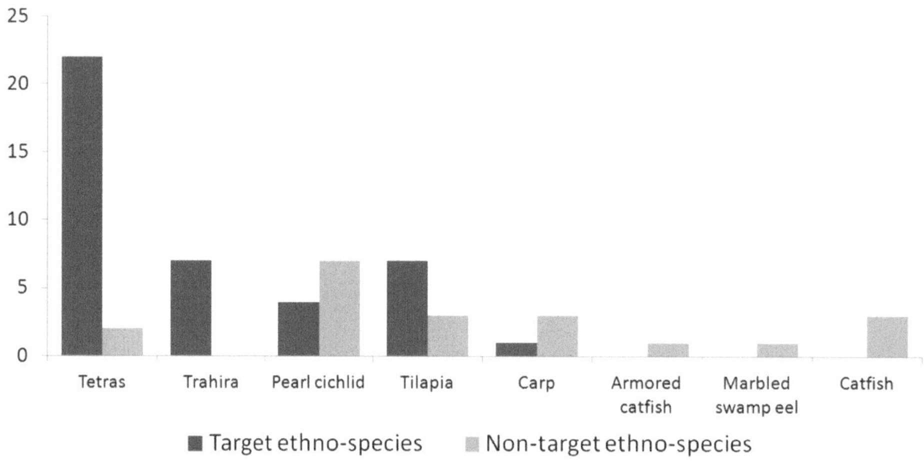

FIGURE 9 – ETHNO-SPECIES RECORDED BASED ON FISHERMEN'S PREFERENCES AT THE RIO VERDE RESERVOIR

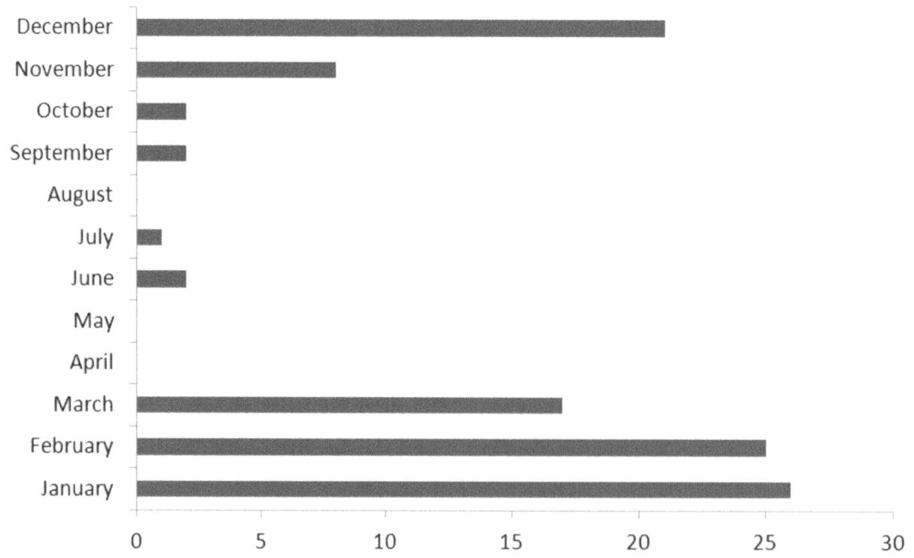

FIGURE 10 – THE BEST MONTHS FOR FISHING AND PERIOD WHEN THIS ACTIVITY IS MOST FREQUENTLY PRACTICED (PREFERENCE) AT THE RIO VERDE RESERVOIR

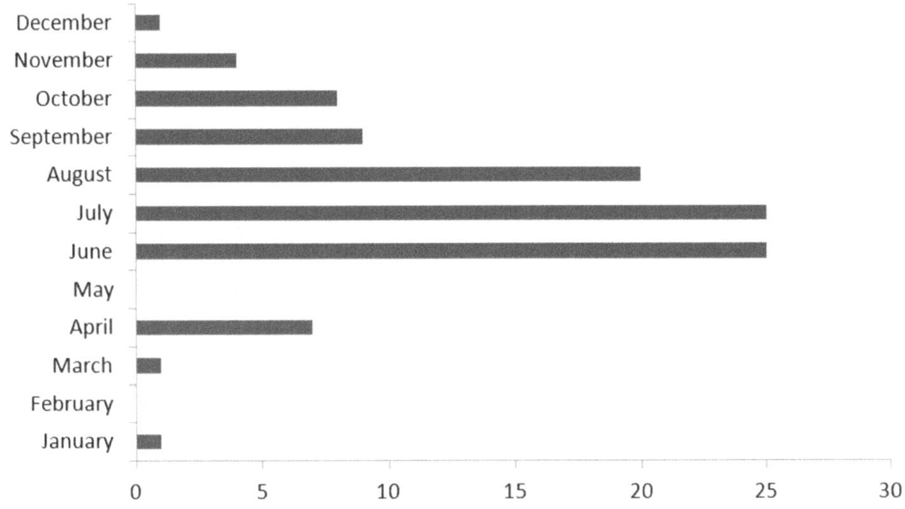

FIGURE 11 – THE WORST MONTHS FOR FISHING AT THE RIO VERDE RESERVOIR

4. CONCLUSIONS AND FINAL REMARKS

The assessment indicates that fishermen who practice fishing in the surroundings of the Rio Verde Reservoir are a relatively homogenous group in relation to gender, marital status, residence, fishing tackle used, target species, frequency and purpose. Fishing is considered an important part of the routine and *"quality of life"* of practitioners as it provides an enjoyable way to spend time, opportunities for interaction and socialization, and also represents a complement to their diet.

Regarding the frequency, fishing is practiced throughout the year but the best months are considered from December to February (*"good times for fishing"*), which corresponds to the period of greater reproductive activity of fish species in the region. Fishing is usually practiced on weekends with rods at three main points around the Rio Verde Reservoir.

Fishing is not limited by the age of its practitioners; however, it is practiced more often by amateur fishermen over 50 years of age (73.3%). This result is different from the age distribution among artisanal/professional fishermen (commercial fishing) in the Itaipu Reservoir (AGOSTINHO *et al.*, 1994), in the Segredo Reservoir (OKADA *et al.*, 1997) and in the Jupiá and Ilha Solteira Reservoirs (CEREGATO & PETRERE JR., 2002).

Another difference found in this study for this kind of fishing (amateur *versus* artisanal/professional) was the level of education of interviewees. At the Rio Verde Reservoir only one person identified themselves as illiterate (3.3% of interviewees) and a similar result was obtained by Okada *et al.* (1997), also in a reservoir in the Iguaçu River (Segredo Hydroelectric Power Plant). This level of illiteracy is significantly lower than the results from a study done with professional and artisanal fishermen (AGOSTINHO *et al.*, 1994; CAMARGO, 1998; MENDONÇA, 2000; CEREGATO & PETRERE JR., 2002).

Fishing performed in the surroundings of the Rio Verde Reservoir is considered amateur because it is practiced either for leisure or sport and the product is not for commercial purposes. In the reservoirs in the Paraná Basin, although fishing productivity is still relatively low (PETRERE & AGOSTINHO, 1993; PETRERE, 1996), artisanal fishing or fishing for commercial purposes, creates several direct and indirect jobs through the commercialization of the fish (CHAGAS, 1994), thus forming a "chain of production." Such a chain includes not only the fishermen but also the employees responsible for storage, processing, transport and distribution of fish to the consumer (SANTOS, 2005).

Fishing at the Rio Verde Reservoir resembles the activities that occur on fishing grounds and at fish farms, such as paid fish ponds (DIEGUES, 1983; CASTRO *et al.*, 2006). On the other hand, the fishing performed at the reservoir cannot be characterized as a sport activity because such sport fishing usually involves vessels, specific equipment, and practitioners are generally more well organized (PEREIRA *et al.*, 2008).

The assessment of amateur fishing performed at the Rio Verde Reservoir has provided important information in understanding the exploitation dynamics of this resource. This information can be used in developing policies for the organization and management of this environment, in order to preserve water resources and to ensure sustainable exploitation.

5. REFERENCES

ABILHOA, V. (2005). Ictiofauna. pp. 437-456. In: ANDREOLI, C. V.; CARNEIRO, C. *Gestão Integrada de Mananciais de Abastecimento Eutrofizados.* Curitiba: Ed. Gráfica Capital Ltda., 500p.

ACKOFF, R. L. (1972). *Planejamento da pesquisa social.* São Paulo: Herder/Edusp, 1967

AGOSTINHO, A. A.; GOMES, L. C.; PELICICE, F. M. (2007). *Ecologia e Manejo de Recursos Pesqueiros em Reservatórios do Brasil.* Maringá: Eduem, 501 p.

AGOSTINHO, A. A.; JÚLIO JR., H. F.; PETRERE JR., M. (1994). Itaipu reservoir (Brazil): impacts of the impoundment on the fish fauna and fisheries. pp. 171-184. In: COWX, I. G. *Rehabilitation of freshwater fisheries.* Bodmin: Fishing News Books.

AGOSTINHO, A. A.; GOMES, L. C. (1997). *Reservatório de Segre-*

do: bases ecológicas para o manejo. Maringá: Eduem, 387 p.

ATTAYDE, J. L.; OKUN, N.; BRASIL, J.; MENEZES, R. F.; MESQUITA, P. (2007) *Impactos da introdução da tilápia-do-Nilo, Oreochromis niloticus, sobre a estrutura trófica dos ecossistemas aquáticos do Bioma Caatinga*. Oecologia Brasiliensis, 11: 450-461.

BRASIL. Law 11.959 of June, 2009. Available in: http://www.planalto.gov.br/ccivil_03/_Ato2007-2010/2009/Lei/L11959.htm. Accessed in: March 22, 2010

CAMARGO, S. A. F. (1998). *Sociobiologia da gestão participativa dos pescadores comerciais do rio São Francisco, MG, Brasil*. Dissertation (Master's Degree in Resource Preservation and Management), Centro de Estudos Ambientais, UNESP, Rio Claro. 122p.

CASTRO, P. M. G.; MARUYAMA, L. S.; MENEZES, L. C. B.; MERCANTE, C. T. J. (2006). Perspectiva da atividade pesqueira no Alto Tietê: contribuição à gestão dos usos múltiplos da água. *Boletim do Instituto de Pesca*, 32(1): 1-14.

CEREGATO, S. A.; PETRERE JR., M. (2002). *Aspectos sócio-econômicos das pescarias artesanais realizadas no Complexo de Urubupungá e sua jusante no Rio Paraná*. Holos Environment, 2(1): 1-24.

CERGOLE, M. C.; ÁVILA DA SILVA, A. O; ROSSI-WONGTSCHOWSKI, C. L. B. (2005). *Análise das principais pescarias comerciais da região sudeste-sul do Brasil: dinâmica populacional das espécies em explotação*. São Paulo: Instituto Oceanográfico-USP, 176p.

CHAGAS, A. L. G. A. (1994). Pesca de águas interiores. pp. 14-23. In: *Seminário sobre fauna aquática e o setor elétrico brasileiro, reuniões temáticas preparatórias: Caderno 3 – Conservação. ELETROBRÁS*, Rio de Janeiro.

DIEGUES, A. C. S. (1983). *Pescadores, camponeses e trabalhadores do mar*. São Paulo: Ática, 287p.

FISCHER, C. F. A.; CHAGAS, A. L. G. A.; DORNELLES, L. D. C. (1992). *Pesca de águas interiores*. Brasília: Instituto Brasileiro do Meio Ambiente e dos Recursos Naturais Renováveis. 1992. 32p.

GHERARDI, F. (2007). Biological invasions in inland waters: an overview. pp. 3-25. In: GHERARDI, F. (Ed). Biological invaders in inland waters: Profiles, distribution, and threats. *Book Series Invading Nature – Springer Series in Invasion Ecology*. Vol. 2. ed. Springer.

GOVERNO DO PARANÁ. (2000). Súmula: Instituída a Área de Proteção Ambiental do Rio Verde, denominada APA do Rio Verde, localizada nos municípios de Araucária e Campo Largo, Estado do Paraná. *Diário Oficial*. Available in: <http://celepar7cta.pr.gov.br> Accessed in: May 17, 2009.

IBAMA. Instituto Brasileiro do Meio Ambiente dos Recursos Naturais Renováveis. (2008). *Guia da pesca amadora: peixes de água doce*. Available in: < http://www.ibama.gov.br/pesca-amadora > Accessed in: May 17, 2009.

JACKSON, D. A.; HARVEY, H. (1989). *Biogeographic associations in fish assemblages: local vs. regional process*. Ecology, 70: 1472-1484.

MENDONÇA, S. A. T. (2000). *Pescadores-barrageiros: perfil profissional e condições de vida no interior paulista*. Dissertação (Mestrado em Conservação e Manejo de Recursos), Centro de Estudos Ambientais, UNESP, Rio Claro. 119p.

OKADA, E. K.; GREGORIS, J.; AGOSTINHO, A. A.; GOMES, L. C. (1997). Diagnóstico da pesca profissional em dois reservatórios do rio Iguaçu. pp. 293-316. In: AGOSTINHO, A. A.; GOMES, L. C. *Reservatório de segredo: bases ecológicas para o manejo*. Maringá: Eduem, 387p.

PAIVA, M. P. (1997). *Recursos pesqueiros estuarinos e marinhos do Brasil*. Fortaleza: EUFC, 278p.

PEREIRA, J. M. A.; PETRERE JR., M.; RIBEIRO-FILHO, R. A. (2008). Angling Sport fishing in Lobo-Broa reservoir (Itirapina, SP, Brazil). *Brazilian Journal of Biology*, 68(4): 721-731.

PETRERE, M.; AGOSTINHO, A. (1993). *La pesca em el tramo brasileño del rio Parana*. FAO Informe de Pesca, 490: 52-72.

RIOS, G. S. L. (1976). *A pesca artesanal como parte do setor de subsistência*. Sua abordagem sociológica. Ciência e Cultura, 28(4): 397-406.

SANTOS, R. A.; CÂMARA, J. J. C.; CAMPOS, E. C.; VERMULM JR., H.; GIAMAS, M. T. D. (1995). Considerações sobre a pesca profissional e a produção pesqueira em águas continentais do Estado de São Paulo. *Boletim Técnico do Instituto de Pesca*, 19: 1-32p.

SANTOS, M.S. (2005). *A cadeia produtiva da pesca artesanal no estado do Pará: estudo de caso no Nordeste Paraense*. Amazônia: Ciência & Desenvolvimento, 1(1): 61-81.

VERMULM JUNIOR, H.; GIAMAS, M. T. D.; CAMPOS, E. C.; CAMARA, J. J. C.; BARBIERI, G. (2001). Avaliação da pesca extrativa em alguns rios do estado de São Paulo, no período entre 1994 e 1999. *Boletim do Instituto de Pesca de São Paulo*, 27(2): 209-217.

VITULE, J. R. S.; UMBRIA, S. C.; ARANHA, J. M. R. (2006). Introdução de Espécies, com Ênfase em Peixes de Ecossistemas Continentais. In: MONTEIRO-FILHO, E. L. A.; ARANHA, J. M. R.. *Revisões em Zoologia* – I: Volume Comemorativo dos 30 Anos do Curso de Pós-Graduação em Zoologia da Universidade Federal do Paraná. Curitiba: Secretaria do Meio Ambiente do Estado do Paraná.

WELCOMME, R. L. (2001). *Inland fisheries: ecology and management*. Oxford: FAO: Blackwell Science, 358p.

CHAPTER 19

LIVESTOCK AND AGRICULTURAL ACTIVITIES

Benno Henrique Weigert Doetzer
Edna Batistella Lopes
Oromar Berthol,
Carlos Henrique Zavelinski

LIVESTOCK AND AGRICULTURAL ACTIVITIES

Benno Henrique Weigert Doetzer
Edna Batistella Lopes
Oromar Berthol,
Carlos Henrique Zavelinski

SUMMARY

This chapter aims to survey the conditions, use, management and occupancy of the Rio Verde APA. Our focus is on understanding agricultural issues, assessing areas of production, preparing maps of land suitability, and observing techniques of soil management and farming systems, such as pesticide use, fertilization, and soil conservation, among others. With these data, we aimed to identify areas of conflict of use in order to propose strategies that better suit agricultural activities in the APA with the goal of soil management, such as more suitable land-use based on identified potential use of the land, recovery of the Legal Reserve area, and strategic development. This subproject had three phases of study. The first was a preliminary survey of the local conditions through the Participatory Rural Appraisal (PRA). With a basis in the preliminary survey, the second phase included a pedological survey, assessment of land-use suitability and identification of areas of conflict due to overuse or underuse. In this phase, critical points, such as sources of pollution, were also identified. The third phase, in consideration of the consistent data on the area of study, refers to the development of proposed adjustments to both conservation practices and land conversion systems using Pilot Plots within the area of study and a Reference Network as the monitoring methodology. As main outcomes we highlight greater diversification shown by the producers of Colônia Cristina (Araucária) in relation to Colônia Figueiredo (Campo Largo), although annual crops, especially corn, soybeans and beans still predominate. We also noted that areas with native forest cover are less than that required by current environmental legislation (Areas of Permanent Preservation and Legal Reserves). Livestock activities are not significant in the Basin considering the small amount of land used for pasture and tall grasslands. The prevailing system of cultivation is tillage with significant use of pesticides (all those interviewed mentioned that they use pesticides regularly). The soils of the region are predominantly: Nitosols, Oxisols and Inceptisols, which, due to mismanagement, show significant erosion and are today without topsoil (or have lost a significant proportion of topsoil). Although efforts have been made to control the erosion of agricultural soils, the problem has reached alarming levels. Survey data has shown that 15 to 20 tons of soil per hectare are lost yearly in areas of intense mechanized farming, as is the case in this Basin. When we cross-checked the land-use suitability data with the current usage, we developed a map of overused or conflict areas. From this, we observed that approximately 19% of the Basin area is overused and should be adapted or converted.

KEYWORDS

Soil usage, agricultural activities, environmental degradation, production systems, land-use conflict.

1. INTRODUCTION

This chapter contains the results of all the analyses completed in the 24 months of research conducted by the Livestock and Agricultural Activities sub-project. The study aimed to survey the occupation conditions, usage and management of the Rio Verde Environmental Protection Area with a focus on livestock and agriculture issues.

The Rio Verde Dam was built between 1974 and 1976 by Petrobras (Petróleo Brasileiro S.A.), with the purpose of its exclusive use for the President Getúlio Vargas Refinery. The Rio Verde's State Environmental Protection Area (APA) was established by State Decree No. 2.375, July 31, 2000, with an area of 147.56 km^2. The establishment of the APA aims to protect and conserve the quality of the environmental and natural systems, especially the quality and quantity of water for public supply, by instituting measures and instruments to manage phenomena and conflicts that arise from the various land-use activities in the area. The APA is located to the west of the Metropolitan Region of Curitiba, covering portions of the municipalities of Araucária, Campo Largo and Campo Magro, including areas upstream of the Rio Verde Dam, in the drainage basin of the Reservoir.

Changes in nature and escalating human activities have both quantitative and qualitative impacts on water resources. Historically, social development involved a change in water usage from the rural and agricultural to urban and industrial, which reflects both an increase in demand for water and in water pollution.

Each type of water usage, including abstraction and release, leads to a specific, and generally predictable, impact on the quality of water bodies. However, besides these established usages, there are several human activities that have indirect and undesirable, if not devastating, effects on the aquatic environment. Some examples are the unrestrained use of soil, deforestation, accidental (or illegal) spills of chemicals, untreated effluent discharge or toxic liquid runoff from solid waste dumps. Similarly, the excessive and uncontrolled use of fertilizers and pesticides has a long-term impact on surface and underground water resources (MEYBECK & HELMER, 1992).

Due to their characteristics, lakes and reservoirs have

somewhat of a restraining capacity, i.e., they can withstand certain levels of pollution. However, this capacity to assimilate pollutants is limited; problems may arise rapidly as a result of the entry of pollutants, which may persist in the long-term depending on the use and occupation of the drainage basin (JORGENSEN & VOLLENWEIDER, 1989). One of the main problems of agricultural activities is irrigation, which, in addition to requiring large volumes of water, is a type of consumptive use. According to Telles (2002), 98% of the water volume used in agriculture returns to the atmosphere through evapotranspiration and 2% is transformed into organic matter assimilated into the crops. Furthermore, according to the author, the use of water for livestock consumption, as well as cleanup activities related to livestock, returns liquid water to the environment but with a significant loss of quality since between 60 to 70% of the water returns in the form of urine and other wastes.

Agricultural activities are among the major non-point sources of nutrients into aquatic ecosystems, which increase eutrophication, especially those in which there are no land-use conservation practices. In such cases, the release of phosphorus is the biggest problem. Phosphorus is an essential nutrient for growing vegetables and raising cattle but when transported from agricultural areas via runoff, it can reach aquatic environments, enriching the water and triggering the process of eutrophication (SHARPLEY, 1980). Studies performed in hydrographic basins in the United States clearly demonstrate that as forests shrink and agriculture areas increase, the loss of phosphorous through runoff increases (SHARPLEY & HALVORSON, 1994). Excessive use of nitrogen, plus the high nitrogen-fixing capacity of plants, seems to be the main source of nitrate for both surface water and groundwater in agricultural areas especially when the application of nitrogen fertilizer or manure occurs before rainy periods (KIRCHMANN, 1994). The nitrogen concentration in runoff, especially in the form of nitrate, is associated with several factors, including previous soil humidity, soil properties, slope and agricultural practices, such as crop type, timing and method of planting (OWENS, 1994).

One concludes that agricultural practices aimed at reducing erosion, as well as extreme fluctuations in surface runoff, contribute substantially to the reduction of phosphorus loads from diffuse sources. A considerable amount of phosphorus from diffuse sources can be reduced with planning and generally without further costs. Considerations of the land's natural drainage when planting, maintenance of green belts around lakes and streams, careful storage of animal waste, careful application of fertilizer, and avoiding the application of fertilizer during rainy periods, are just a few measures that reduce the contribution of phosphorus to water bodies. Eutrophication is an aspect of lake productivity and similar analyses should be conducted in both natural environments and artificial reservoirs. The productivity of these environments, on one hand, is the expression of the physicochemical complexes of the hydrographic basin in which it is located. On the other hand, it depends on the internal physicochemical and biological dynamics of the elements of the lake itself. The natural eutrophication process is slow and continuous, with the nutrients brought by the rains and the surface runoff that erodes and cleans the land surface. This natural eutrophication corresponds to what one might call the "natural aging" of lakes (ESTEVES, 1998). When it does not occur naturally, i.e., when it is triggered by human action, eutrophication is called artificial, cultural or anthropogenic. The artificial eutrophication of water leads to a progressive deterioration of water quality, especially in lakes, due to the massive growth of aquatic plants, which has repercussions on the entire metabolism of the affected water body.

In their analysis in Paraná State, Bragagnolo & Pan (2001) emphasized that the main causes of erosion and pollution of water sources are illegal land occupation, inappropriate land-use and soil preparation. The invasion and occupation of land and improper planning led to narrow lots oriented towards the slope, which have strict divisions, do not consider natural water behavior, and do not permit agricultural operations on flat land (BRAGAGNOLO & PAN, 2001).

According to these authors, the location of rural roads, which are usually constructed without considering the dynamics of water flow thus turning them into drainage channels of water and sediment, has aggravated the erosion problems and caused the silting of rivers and streams.

Along with creating terraces, the restoration of riparian forests and fencing of areas of permanent preservation (APP), the use of technical strategies in micro-hydrographic basin programs seeks to increase soil vegetation cover, increase water infiltration in the soil profile, reduce superficial runoff, and control water and soil pollution (BRAGAGNOLO & PAN, 2001).

Planning is done within the framework of the micro-hydrographic basin and rural property, respecting natural dynamics. Therefore, the boundaries between farms, roads and fences cannot impede the execution of the planned activities (BRAGAGNOLO & PAN, 2001).

Usually a plan of the entire micro-basin is prepared as well as individual plans for each of the farms in the micro-basin.

The problems, guidelines, strategies and action plans of different micro-basin programs are similar. The form, the schedule, and the operationalization of activities depend on the size and abilities of the human resources available and on the involvement of the direct beneficiaries of the program, i.e., the farmers and their organizations.

2. METHODOLOGICAL CONSIDER ATIONS

2.1 USE OF PARTICIPATORY METHODOLOGIES FOR THE PREPARATION OF SUB-BASINS PLANS

Rural administration was developed in the early twentieth century, at the Universities of Agricultural Sciences, in England, and in the United States at the so-called "Land Grant College," in order to analyze the economic feasibility of agricultural techniques. It had a partial concept of rural administration and focused primarily on the area of production and control.

Application and outreach studies considered mainly the allocation of resources, accounting and financial records, with simplified accounting being the most publicized "managerial" tool.

At this early stage, rural administration was regarded as a branch of rural economics and this view still persists in many institutions.

After analyzing various approaches and definitions of Rural Administration, Lima (1982), in his article "The Goal of Rural Administration," developed the following definition: "Rural Administration is a branch of administrative science that studies the processes of rational decisions and administrative actions in rural organizations." With this concept, the author enables access to all management theories, from Taylor and Fayol's classical approach to modern theories of organizational development. Thus, concepts in the areas of finance, sales and marketing, and human resources, were added to the concept of farm administration as subjects equally relevant as production. In addition to control, it also emphasized the importance of other administrative functions: planning, organizing and directing.

For Rural Outreach in Paraná, the concept of rural administration was included from the beginning, with the foundation of the Paraná Association of Credit and Rural Assistance (*Associação de Crédito e Assistência Rural do Paraná:* ACARPA), in 1956. It was continued by the succeeding organization, Technical Assistance and Rural Extension Agency of Paraná (*Empresa de Assistência Técnica e Extensão Rural do Paraná:* EMATER–PR), whose major emphasis was seen in the 1970s and early 1980s when rural administration came to reflect specific technical structures to advise the management of farms.

Because Family Farming of Paraná covers various socioeconomic and climatic conditions, it faces the challenge of ensuring a minimum income to meet basic family needs while respecting the proper use of natural resources.

In this context, since 1998, the Reference Networks project and the Rural Extension Model of Paraná were implemented to support family farming. These tools aim to analyze and generate technical and economic references of agricultural production systems feasible for family farmers using research and development methods and family farm management tools. The foundation of the process consists of the following principles:

- A systemic approach that does not lose sight of the analyses of the system's components;
- Integrated participation between research, outreach, and farmers;
- Compliance with the technical, economic, social and environmental aspects.

A new technology that is selected with the participation of local farmers is adapted better than technologies solely recommended by experts. Moreover, when farmers are involved from the beginning of the technology generation process, their peers more easily accept the result.

A method such as this assists farmers in managing their properties and also highlights the production systems better adapted to the regional limitations/restrictions, described by them and, hence, disseminated by them. Where to place the network and what kind of production systems to select are important decisions because the use and dissemination of results depend on them. Making such choices is to bet on the future of current systems and the promotion of emerging ones.

Chambers (1992) defines the **Participatory Rural Appraisal** (PRA) as a term referring to "a set of methods and approaches that enable communities to share and analyze their perception of their living conditions, planning and acting."

The main objective is to obtain information as well as to empower local people to analyze, plan and act. This implies change in experts' attitude regarding their role in the process.

2.2 METHODOLOGY

The sub-project has three study phases. The first phase refers to the preliminary survey of the local conditions through the PRA methodology. With the preliminary assessment complete, the second phase consists of a survey of pedology, land-use suitability and identification of areas of conflict due to over- or underuse. In this stage we also identified critical points of pollution. The third phase, which is built on the data collected in the previous activities of the study, is the proposal of necessary adjustments, including both conservation practices and conversion using Pilot Plots within the study area, and implementing Reference Networks as the monitoring methodology.

The techniques used in the PRA were presented and discussed in Chapter 16. Therefore, this chapter will focus on the methods used after the completion of the PRA.

2.2.1 Identification of areas of conflict

Once the agricultural suitability of soils through pedological survey was determined, we proceeded to identify the areas of the basin that could be in conflict regarding land-use. This information is obtained through the detailed survey of the physical environment and with the aid of geoprocessing, cross-referencing the land suitability map with the land-use map. The detailed pedological survey is done through the interpretation of aerial images and data collection at the study sites. Soil analyses were performed for precise determination of soil types.

An area was considered in conflict of use if it was being used beyond its exploitation potential. Thus, a given area may be in conflict of use due to overuse (when the soil is being exploited above its potential use) or underuse (when the soil is being exploited below its potential use). An example of a conflict due to overuse is the use of an area for annual crops based on a C level management, which because of its slope limitation and/or soil, is only suitable for forestry. An example of underuse would be the opposite of the previous scenario or an area that is being used with native pastureland based on A level management that is well suited to perennial crops with a C level management. Conflicts of use could also include APPs that are being occupied by any agricultural activity. Therefore, areas with conflict of use constitute those areas where intervention should occur to maximize the use of soils that are suited for farming and to adapt agricultural areas that are better suited for permanent preservation. This form of property planning is presented in the following flowchart (Figure 1).

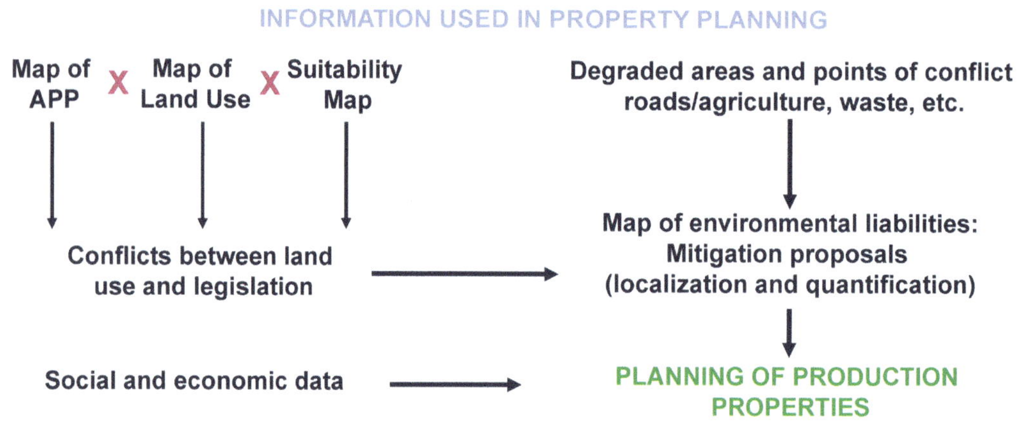

FIGURE 1 – FLOWCHART OF PROPERTY PLANNING

2.3 PROPOSING ADJUSTMENTS AND MONITORING THROUGH THE REFERENCE NETWORK SYSTEM

This process aimed to analyze and generate technical and economic references of agricultural production systems that are feasible for family farmers using research and development methods and family farm management tools. The basis of the process consists of the following principles:

- A systematic approach without losing sight of the analyses of the system components;
- Integrated participation of research, outreach and farmers;
- Compliance with technical, economic, social and environmental aspects.

- This methodology has the following objectives:
- Improve the farmers' decision-making processes;
- Analyze and plan the property's production systems;
- Improve the management process of producers;
- Improve technological processes (biological, mechanical and chemical);
- Boost farmers' economic organization processes so that they get the benefit of economies of scope and scale of production;
- Improve the process of value added to the product;
- Develop actions to minimize the risks of farming;
- Adjust family production to competitive markets;
- Adjust production and build competitive entrepreneurial and managerial skills;
- Support competitive entrepreneurial and managerial training and skill development.

In general, the methodology can be represented by Figure 2 below.

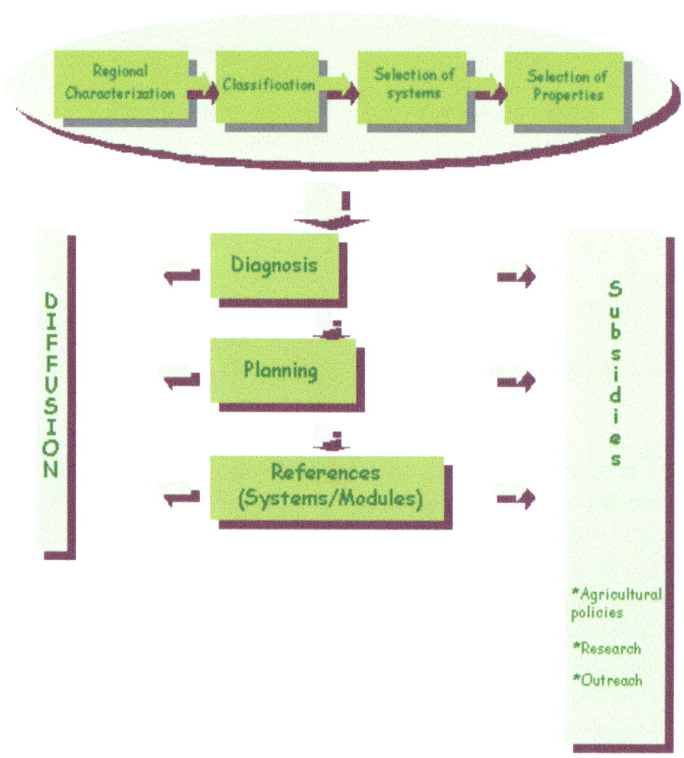

FIGURE 2 – FLOWCHART OF THE SUBPROJECT GENERAL STRATEGY

3. AGRICULTURAL AND LIVESTOCK ACTIVITIES

3.1 PREVAILING PRODUCTION SYSTEMS

In light of the analyses carried out, we found some striking differences between the opinions of communities and their current situation. Table 1 highlights some of the issues identified.

3.2 FINAL RESULTS – AGRICULTURAL AND LIVESTOCK ACTIVITIES

The results of the data collection from November 2009 to June 2010 regarding agricultural activities were tabulated and the results are discussed below.

3.2.1 Land

The preliminary results show a significant difference in relation to land-use. In the chart below, we can see that this factor is much more favorable in Colônia Cristina. That is, the average area per household is more than four times higher than Colônia Figueiredo, which suggests much more favorable agricultural production conditions. Table 2 shows the comparative averages:

3.2.2 Current use

Table 3 shows the comparative average between the two communities regarding the use of soil.

As an important result, we would like to first highlight greater diversification by farmers in Colônia Cristina, although annual crops, such as corn, soybeans and beans, are still the dominant production system. It is also noteworthy that areas with native forest cover are relatively low compared to that required in current environmental legislation (Permanent Preservation Areas and Legal Reserves).

Livestock activities are not significant in the basin given the small proportion of land occupied with pastures and tall grasslands. The prevailing cultivation system is tillage with significant use of pesticides (all participants interviewed mentioned that they use them regularly), as can be seen in Figures 3 and 4.

The soils of the region are predominantly Nitosols, Oxisols and Inceptisols (see Chapter 5), which, due to mishandling, showed significant erosion and are now without topsoil (or have lost a substantial proportion of topsoil). Potato crops contributed the most to this situation, which reached a peak in production in the 1970s and 80s.

Currently the levels of organic matter are low and the soil no longer has high natural fertility. To address this shortcoming, farmers heavily use fertilizers that are easily lost via erosion and degrade water quality.

Despite all the efforts already made in the State of Paraná to control the erosion of agricultural soils, the phenomenon has reached alarming proportions. Survey data shows that 15 to 20 tons per hectare of soil are lost each year in areas of extensively mechanized farming, such as is the case in the Rio Verde Basin.

According to zoning data, more than half of the APA has agricultural activities and most of these properties conduct their activities with the use of agrochemicals (pesticides and highly soluble fertilizers, not to mention excessive use of soil correctives). These substances have a negative impact on water quality; therefore, Resolution No. 010/88 (CONAMA) states in Article No.5 that "wherever agricultural or livestock activities occur, it must be established as Agricultural Use Zone."

In these zones, "it will be regulated or prohibited uses or practices likely to cause appreciable degradation to the environment." The first paragraph of the same article prohibits "the use of pesticides and other biocides that offer serious risks in their use, especially where its residual effects are concerned."

TABLE 1 – SUMMARY OF THE RESULTS OF THE PARTICIPATORY RURAL APPRAISAL

FINDINGS – PAST PRA	
Colônia Cristina	**Colônia Figueiredo**
• Strong community ties/family • Planted on the river bank • "everything got worse with the dam..."	• Smaller community organization • Small subsistence production • Dam did not affect land use
FINDINGS – PRESENT PRA	
Colônia Cristina	**Colônia Figueiredo**
• Strong community ties/family • Diversification • Natural resources preserved • Growth • Infrastructure • External problems	• Smaller community organization (Church) • Less diversified production • Lack of infrastructure • Revenue decrease • Urban pressure • Youth out-migration
FINDINGS – FUTURE PRA	
Colônia Cristina	**Colônia Figueiredo**
• "We want everything to stay as it is" • Recognition • Participation • Investments • Keep "rural characteristics"	• Compensation for environmental services • Diversification of production • Association/cooperatives • Infrastructure improvement

TABLE 2 – LAND TENURE

LAND	FIGUEIREDO		COL. CRISTINA	
	Area ha	%	Area ha	%
Own	8,84	47,44	34,21	41,93
Lease	8,69	46,60	44,64	54,72
Sharecropping	1,11	5,96	2,74	3,36
Ocupied	0,00	0,00	0,00	0,00
Transferred	0,00	0,00	0,00	0,00
Total	18,64	100,00	81,58	100,00

TABLE 3 – CURRENT USE OF SOIL

USO ATUAL	FIGUEIREDO	COL. CRISTINA
	%	%
Vegetable Crops	4,17	1,37
Annual Crops	70,44	56,72
Permanent Crops	0,67	21,54
Cultivated Pastures	0,89	0,74
Natural Pastures	3,24	1,56
Tall Grasslands	0,00	0,06
Area of Evironmental Preservation	3,40	4,14
Woods and Florests (Natural)	9,77	9,89
Planted Florests (Reforestation)	1,43	1,34
Fallow	0,00	0,35
Unuseable Land	3,40	0,61
Dams	0,00	0,12
Farm Headquarters	2,59	1,55
Leased to Third Parties	0,00	0,00
TOTAL	100	100

FIGURE 3 – FIELD FOR ANNUAL CROP WITH TILLAGE SYSTEM

FIGURE 4 – FUNGICIDE APPLICATION ON ANNUAL CROP

Soil management in the conventional manner causes erosion and, therefore, transportation of these agrochemicals, clay colloids and organic matter. In addition to contributing to the eutrophication of water bodies, these materials cause siltation of riverbeds, loss of fish populations, difficulties in water system treatment for human and industrial consumption, and, as a result, they lead to increases in costs.

Environmental studies contribute to a better understanding of spaces in which dynamic equilibrium prevails between competing physical, biotic and anthropogenic forces. According to Theodorovicz (2007), these studies allow us to understand how humans, with their cultural activities and procedures, interact with the environment thus triggering reactions that can alter its dynamics in both positive and negative ways. One realizes that, when and where land management practices are not implemented in a rational and integrated manner, and where there are no appropriate rules set out based on previous knowledge of the potentials and frailties of the natural physical environment, several conflicts and environmental problems ensue. These problems can have irreversible consequences requiring complex and costly solutions to society. Many of these problems, which affect quality of life, cause loss of life, and result in massive expenditures in corrective actions, could be easily avoided if some rational guidelines were followed for occupation and land-use.

The topography of the Basin is characterized by flat-bottomed valleys transitioning to gently undulated and predominantly convex tops. The slopes are predominantly of undulating steepness (3-8%) to moderately undulating (13-20%), with strong undulating (20-45%) to mountainous (45-75%) or even steep (>75%) slopes in the headwaters that form the drainage network.

It should be emphasized that a large part of the region has been significantly modified from its natural state especially through deforestation. Patches of well-preserved forest are generally restricted to areas under environmental protection. Beyond these protected forests, the remaining fragments of the Atlantic Forest (*Mata Atlântica*) have gradually been replaced by pasture or reforestation with foreign species, which progressively homogenizes the vegetation across large areas, bringing with it several negative impacts.

It should also be emphasized that, land near an important and large capital such as Curitiba, such as the Basin, many segments are under densification and land occupation pressure, including urban occupation. Urban areas in Curitiba's peripheral municipalities have been growing significantly in recent years. Some growth has occurred without planning, moving toward areas of environmental fragility and important sources of mineral and water resources.

Conflicts already exist between urbanization, agriculture and exploitation of natural resources. Some areas with potential eco-tourism that are important water sources have been or are being degraded.

Because of this complex geological history, the Rio Verde Hydrographic Basin is characterized as a region with varying geo-environmental characteristics. As such, it includes a vast array of land-use that required wide ranging behaviors in terms of responses to use and occupancy. It should be emphasized, however, that some characteristics persist throughout the Basin. Some typical regional geo-environmental aspects are therefore of fundamental importance and must be taken into account in both regional and sectorial planning.

On the issue of inputs and the use of agrochemicals, we found that most farmers have had soil analyses conducted on their propertie s with the goal of informing soil correction programs, not to ascertain soil fertility. This practice is reflected in the fertilization practices used in the basin which on most of the properties is performed without any scientific basis. This results in indiscriminate use of fertilizers without control over the quantity of nutrients applied in relation to the needs of the crop and, hence, supporting water pollution and the development of pests and diseases that require even more pesticides for their control.

Uncontrolled fertilization includes the use of minerals (formulated fertilizers) and the application of poultry manure, considered organic by the farmers and used in 95% of the surveyed properties. The fact remains that indiscriminate use of fertilizers becomes a source of contamination both through leached nutrients into groundwater and through water-soluble nutrients carried by runoff or bonded to eroded soil particles, polluting rivers, springs and groundwater.

Another aggravating factor is the direct application of poultry manure; the manure does not undergo a composting process before use. In most properties this material is left outdoors without any containment, which renders this method of fertilization a large source of contaminants to water bodies in the Basin.

Regarding the origin of this poultry manure, our assessment determined that it is purchased from independent suppliers who do not guarantee nutritional quality or even health standards; therefore, the appropriateness of fertilization using this material is unknown.

Due to the lack of control of fertilization performed in the watershed, few land owners had sufficient knowledge to accurately answer questions about the amount of inputs used. According to the data obtained, we found that the average amount of formulated fertilizer applied per hectare in one year is approximately 1.6 tons. As for the poultry manure, the average amount is about 7.8 tons/ha/year.

Some simple calculations can be performed to ascertain the amount of nutrients applied to the soil. When it comes to poultry manure, we can use average values of N=28g/Kg of compound, P=16 g/Kg and K=22 g/Kg. If we assume an average value of 7.8 tons per hectare per year, we have placed in the system an average of 218Kg of N, 125 Kg of P, and 172 Kg of K per hectare/year.

To calculate the mineral fertilizer we can build on the standard 8-20-20 formulated fertilizer, a formula used frequently in this micro-basin. If an average of 1.6 tons of fertilizer per hectare per year is used, we have 128 Kg of N, 320 Kg of P, and 320 Kg of K per hectare/year.

This quantity of nutrients is placed annually into the system. However, as no previous soil analysis has been performed, it may be applied unnecessarily.

Regarding the use of pesticides, farmers generally do not fully understand the application technique and this is aggravated by the fact that few receive technical assistance in this regard.

Among the main pesticides used in the basin, three of them stand out in terms of usage frequency among farmers, they are: GLIFOSATO®, MANZATE® and DECIS®. Other products, such as PRIMÓLIO® and LORSBAN®, were also significant but not at the same frequency as the three listed above.

An important fact noted in the Basin is that most of the technical assistance regarding pesticides is provided by private companies, in this case product vendors, and no instance of cooperatives were reported for the region.

This assistance is a direct reflection of the amount of inputs and management given to crops in the Basin. Most of the farmers do not know the reason for the use of certain products, but do so by experimentation or because they were recommended by the vendor. These data were obtained through interviews and dialogues with local farmers but were not tabulated and therefore a representative analysis of this topic was not conducted. The information is provided here to give an idea of how these farmers are being told about the use of agrochemicals in general.

Of the three most commonly used products, GLIFOSATO® is a post-emergent herbicide belonging to the chemical group of substituted glycines, classified as non-selective and systematic. It features a wide spectrum of action allowing excellent control of annual or perennial weeds of both narrow and broad leaves. The product should always be applied directly on weeds following technical guidelines and good agricultural practices. If used in this way, it will not cause any interference in the metabolism, development and productivity of crops for which it is recommended. The properties that determine the behavior of glyphosate characterize the product as being of low environmental impact, considering the range of use of this molecule. The product is degraded by microorganisms in the soil and in the water (it has a relatively short half-life, ranging from 9 to 32 days); in the soil it is strongly retained in the form of bound residues; in the water it is highly soluble and its volatility and evaporation are insignificant.

MANZATE® is a fungicide composed of mancozeb and hexamethylenetetramine. This product is highly toxic to aquatic organisms and if exposed to heat or moisture can release toxic and irritating gases. From an environmental standpoint, mancozeb, its main component, is immobile in soil, yet it degrades rapidly in the soil in about two days; it has little bioconcentration but users must be very careful that the compound does not reach water bodies due to its high toxicity to aquatic organisms.

DECIS® is an insecticide of the pyrethroid group; it is highly dangerous to the environment due to its toxicity to aquatic organisms and a high bioconcentration in fish. Its active ingredient, deltamethrin, binds itself to most soils and is degraded in up to nine days. As above, contact with aquatic environment must be prevented.

As we have seen, these three products have similar behavior regarding fixation in the soil and the relatively short half-life; however, in the case of DECIS® and MANZATE® one should make sure that the product does not come into contact with the aquatic environment.

4. CRITICAL SITUATIONS REGARDING SUSTAINABILITY OF PRODUCTION SYSTEMS

4.1 USE OF SOIL ACCORDING TO SUITABILITY

Lepsch *et al.* (1983) claim that the ability to use the land can be conceived as its adaptability for various purposes without suffering depletion by factors of wear and impoverishment through annual crops, perennials, pastures, forestry or wildlife.

The landscape classes were defined based on environmental legislation that deals with the areas of permanent preservation, the map of slope classes was developed with support of the existing digital soil map, and the map of agricultural suitability of land, from the Ministry of Agriculture (1980), were refined in more detail in the area of interest taking into account the geomorphology, hydrology and geology. The digital intersection of these thematic maps produced a map of landscape classes with their respective potential, as shown below:

a) Landscape class 1: Indicated for Permanent Preservation;
b) Landscape class 2: Suitable for Forestry exploitation under sustainable management;
c) Landscape class 3: Unfit for longer crop cycles and forestry;
d) Landscape class 4: Unsuitable for mechanized farming;
e) Landscape class 5: Unsuitable for shorter crop cycles;
f) Landscape class 6: Available for cultivation under appropriate management.

Once the agricultural suitability of the soil was determined, we proceeded to the identification of the basin areas that may be in conflict regarding land-use. This information is obtained through geoprocessing, cross-referencing the land suitability map with the map of land-use, as mentioned above.

Evaluation of agricultural suitability

For preparation of the agricultural suitability map, we used a method adapted from Ramalho Filho developed by the National Secretariat of Agricultural Planning (*Secretaria Nacional de Planejamento Agrícola*; SUPLAN) and Embrapa-Soils, previously the National Soil Survey and Conservation Service (*Serviço Nacional de Levantamento e Conservação de Solo*; SNLCS) (RAMALHO-FILHO *et al.*, 1978). This method was developed to classify soils on a regional scale and considers three management levels in its composition: Level A (agricultural practices with low technological investment); Level B (agricultural practices with average technological investment); and Level C (agricultural practices with high technological investment).

In the classification, the management level is faced with five restrictions: soil fertility, water deficiency, excess water, susceptibility to erosion, and impediments to mechanization. In this analysis, we disregarded the restrictions regarding fertility since the entire region has been used for agriculture and, therefore, some soil fertility characteristics may have changed over time. Once the cross-checking is finalized, different levels of restriction are obtained for the different management levels, ranging from "Nil" to "Very Strong." With these restriction levels, one gains an idea of the soil suitability for the level of management. The suitability classes range from "good," to "fair," "restricted," and "unsuitable."

Suitability	Type of use: Crops			Cultivated Pasture	Forest	Natural Pasture
Level of Management	A	B	C	B	B	A
Good	A	B	C	P	S	N
Fair	a	b	c	p	s	n
Restricted	(a)	(b)	(c)	(p)	(s)	(n)
Unsuitable	?	?	?	?	?	?

FIGURE 5 – CLASSIFICATION OF AGRICULTURE SUITABILITY

In short, the suitability classification is cross-referenced with data collected in the field considering different types of management.

Soil classification begins with the suitability classes. For this classification, six suitability classes are adopted: class 1 is the most suitable lands for agriculture with virtually no restrictions; while class 6 has the greatest severity of restrictions and is therefore unsuitable for agriculture and livestock, to be used exclusively for preservation.

We include Table 4 to illustrate the suitability classification:

TABLE 4 – EXAMPLE OF CLASSIFICATION OF LAND SUITABILITY

SUBGROUP	DESCRIPTION
1ABC	Lands with good suitability for management crop levels A, B and C
1ABc	Lands with good suitability for crops in management levels A and B, and fair in level C – shallow soil
1bC	Lands with good suitability for crops in management level C, fair in level B and unsuitable in level A —fertility problem
2ab(c)	Lands with fair suitability for crops in management levels A and B, and restricted in level C
2(b)c	Lands with fair suitability for crops in management level C, restricted in level B and unsuitable in level A
3(ab)	Lands with restricted suitability in levels A and B and unsuitable in level C – steep slope
4P	Lands with good suitability for planted pasture
4(p)	Lands with restricted suitability for planted pasture
5Sn	Lands with good suitability for forestry and fair for natural pasture
5s(n)	Lands with fair suitability for forestry and restricted for natural pasture
6	Lands unsuitable for agriculture

Considering the surveys performed, we developed the map shown in Figure 6 which demonstrates the suitability of land-use in the Basin.

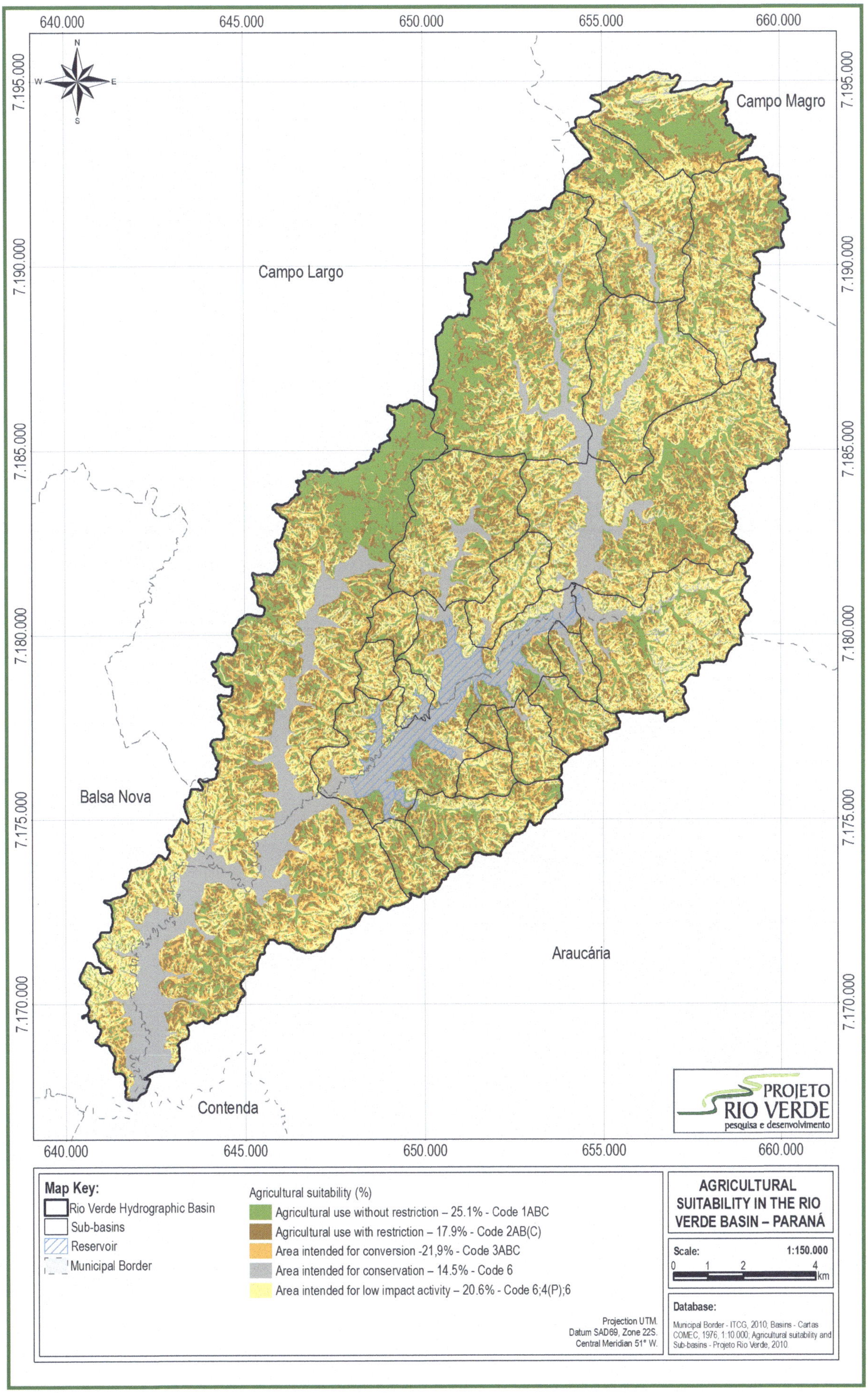

FIGURE 6 – SOIL USE SUITABILITY IN THE RIO VERDE BASIN

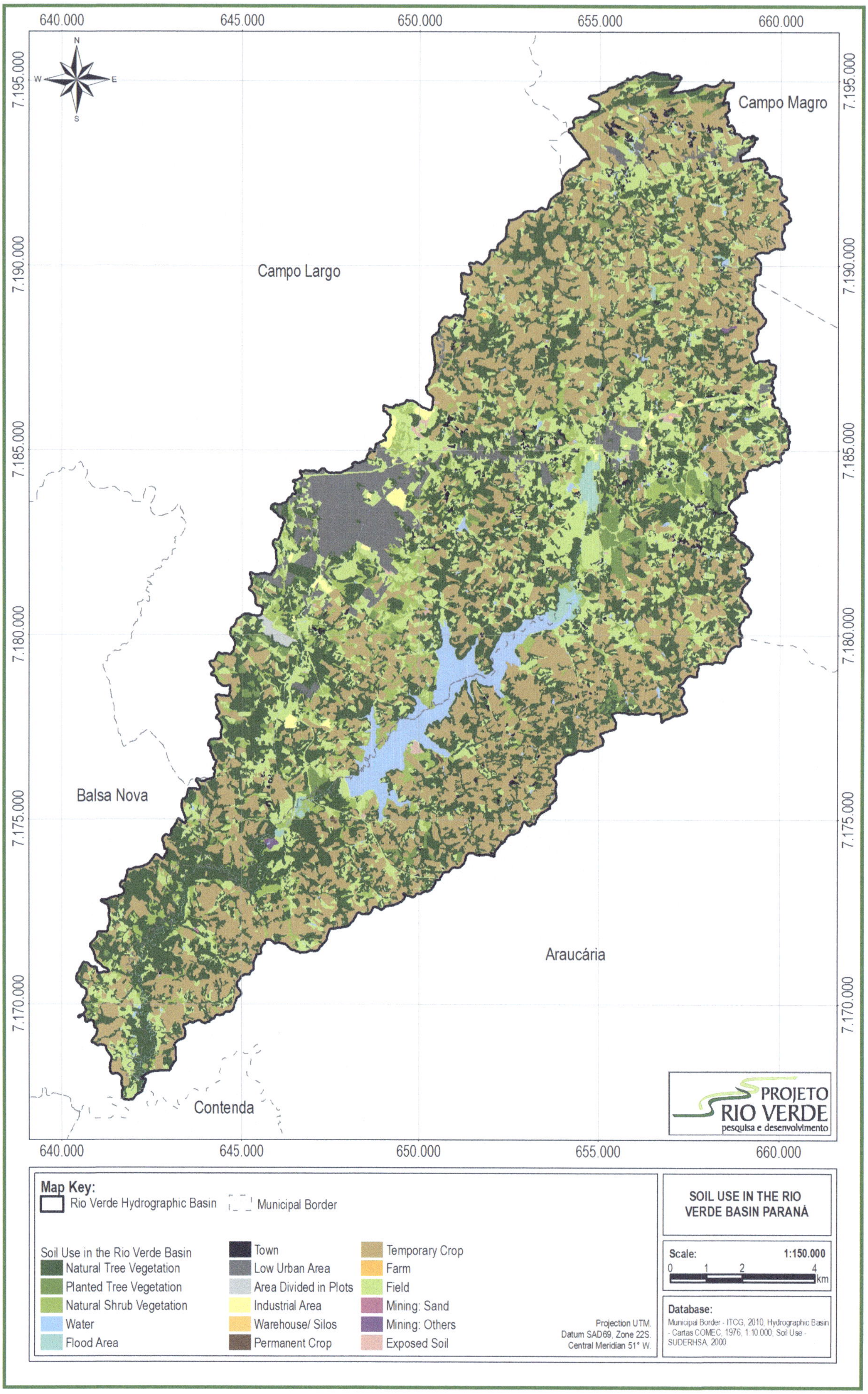

FIGURE 7 – SOIL USE IN THE RIO VERDE BASIN

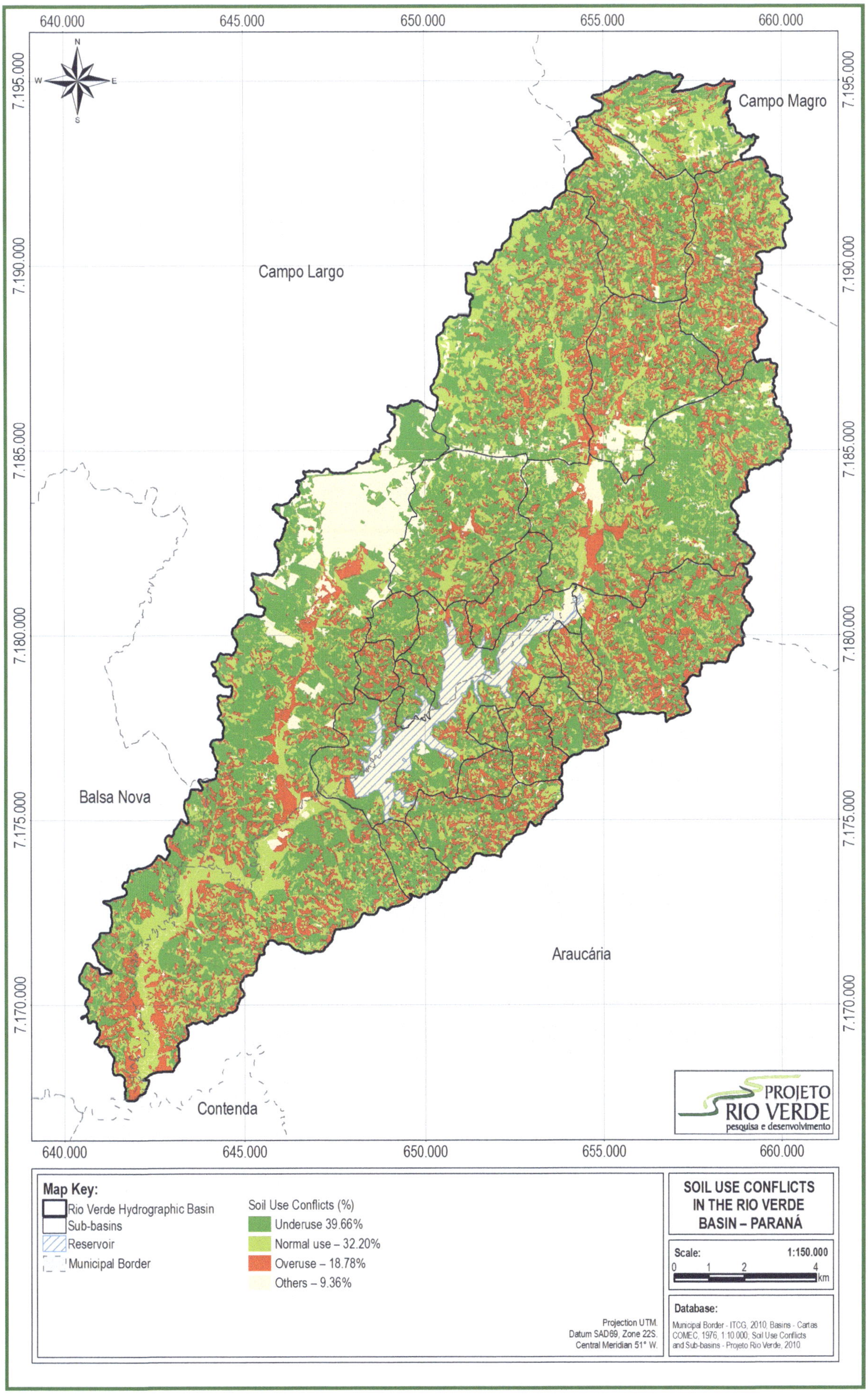

FIGURE 8 – SOIL USE CONFLICTS IN THE RIO VERDE BASIN

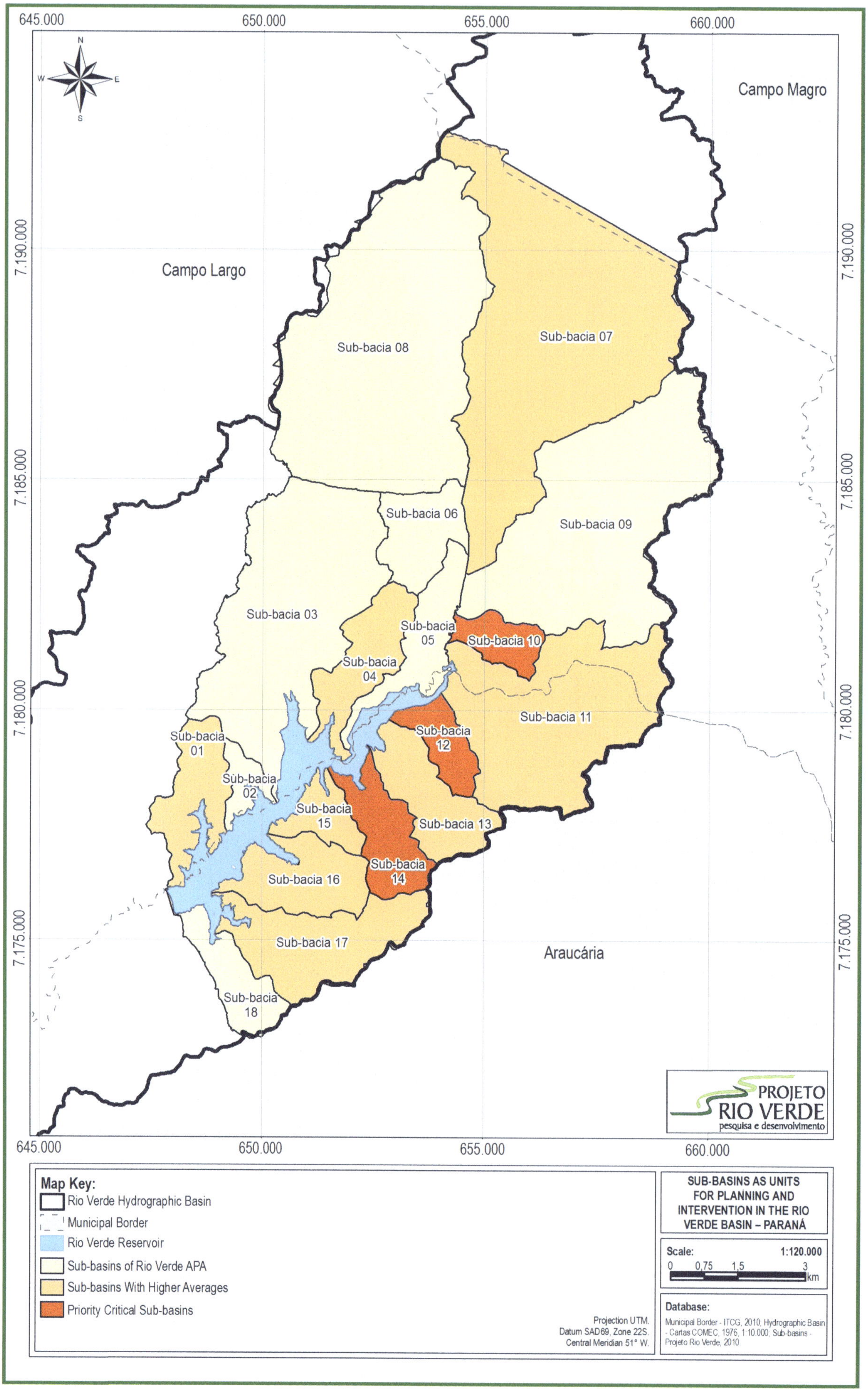

FIGURE 9 – SUB-BASINS AS UNITS FOR PLANNING AND INTERVENTION

From the map (Figure 6), we can see that approximately 43% of the basin area has good and fair suitability for agricultural use provided some measures of control and soil management are taken. For the other 57%, more drastic measures should be taken regarding production systems as these areas are more fragile.

When we verified -the suitability data versus data on current land-use, shown in Figure 7, we created the map of areas of conflict or overuse (See Figure 8). From this analysis, we noticed that approximately 19% of the basin area is overused, requiring adjustment or conversion.

With the objective of focusing intervention on the most critical areas, the Rio Verde Basin was divided into 18 sub-basins, as illustrated in Figure 9. Of these sub-basins the conflict of use areas were assessed and those most critical were selected.

If we consider the average of overused areas as 19%, we can see that 11 out of the 18 sub-basins have values higher than the average, as shown in Table 2.

TABLE 2 – OVERUSED AREA PER SUB-BASIN

SUB-BASINS	% OVERUSED AREA
12	34.32
14	30.75
10	29.85
11	26.98
1	25.68
17	25.36
7	24.61
15	24.28
13	23.09
16	22.27
4	20.99

However, in terms of total overused area, the most critical basins in which actions should be focused due to their significant contribution to the degradation of the environmental water quality of the basin are sub-basins 3, 7, 8, 9, 11 and 17, as shown Table 3.

5. PROPOSED ACTION FOR THE MICROBASIN AIMING AT SOIL MANAGEMENT AND CONSERVATION

5.1 USE AND MANAGEMENT CONSIDERING THE POTENTIAL AGRICULTURAL LAND-USE

As 22.31% of the microbasin is suitable for agricultural use based on assessments of agricultural suitability and usability, we should use these areas for more intense agriculture activities. We must keep in mind, however, that despite the suitability of these areas, they can present problems, mainly related to erosion. As such, agriculture is only possible when done in conjunction with environmental conservation techniques. We can therefore assume that in the mid- and long-term, conventional soil preparation activities should be replaced by lower-impact activities with no tillage whenever feasible.

The micro-basin includes large areas that could be used for reforestation and pasture; however, in order to be economically feasible, these activities require areas larger than the existing property sizes owned by individual farmers in the region. Therefore, we cannot suggest a radical change to these two usage systems because, despite being technically appropriate, they are economically unfeasible. An alternative for these areas is fruit production. This could be an interesting activity for the micro-basin since the analysis of the area indicates the availability of labor and quite favorable consumer markets. Obviously it is important to allocate resources for the implementation of this activity as it usually requires high initial investments.

RECOVERY OF VEGETATION COVER

The analysis of the area suggests that a significant proportion of the APP is not preserved. These areas should be recovered in order to bring the micro-basin in compliance with the requirements of current environmental legislation. In addition, the legal reserve areas should also be recovered within individual properties. In the micro-basin analysis as a whole, the quantity of areas with native forests is relatively high.

Vegetation recovery should be done with native species prioritizing technical criteria, such as the proper selection of native plants and the climate, as these determine the success of recovery efforts. In addition, native species that are favored by wildlife should also be given preference because they improve environmental quality and can be a source of income, such as honey production, that can be a meaningful alternative for the farmer.

The recovery of vegetation leads to benefits in soil and water conservation in the micro-basin that has no precise valuation, but all indications suggest that this will exist in the future and can be an excellent source of extra income for rural producers.

An alternative is the introduction of perennial native plants, fruit trees and medicinal plants, among others, for reforestation and agricultural areas replacing the traditional crops of the region. The region, including Campo Largo, Campo Magro and Araucária, offers soil and climatic characteristics that are ideal for the production of numerous species of native and cultivated medicinal plants of economic interest. For example, already cultivated in this region are *Espinheira Santa*, *yerba mate*, Surinam cherry (*pitanga*), *araçá do campo*, guava, *gabiroba*, chamomile, *quebra pedra*, burdock, *tansagem* (broadfleaf plantain), among other fruit and melliferous (plants containing nectar) trees. The production of medicinal plants makes extensive use of labor and can generate employment and income for small-scale farmers and their families. The CHAMEL industry has been in Campo Largo for 15 years, producing medicinal plants within the Curitiba Metropolitan Region. In these three municipalities there is a predominance of small-scale farmers who face difficulties in maintaining their families with the cultivation of soybeans, corn, beans and potatoes.

The production of medicinal plants and their use in the manufacture of herbal medicines are supported by the

National Policy on Integrative and Complementary Practices (*Política Nacional de Práticas Integrativas e Complementares;* PNPIC) in the Unified Health System published in the government Official Gazette, May 4, 2006, by Ordinance No.971, May 3, 2006. The PNPIC is based on the World Health Organization, which has encouraged the integrated use of Traditional Medicine, Complementary Medicine and Alternative Medicine in health systems with the techniques of modern Western medicine. In its document "WHO Traditional Medicine Strategy 2002-2005" the organization advocates the development of policies observing the requirements of safety, efficacy, quality, rational use and access. The PNPIC considers phytotherapy as a therapeutic resource characterized by the use of medicinal plants in their different pharmaceutical forms and such an approach encourages community development, solidarity and social

TABLE 3 – TYPE OF CONFLICT IN EACH SUB-BASIN

Basin	Class	Percent	Area (ha)
1	Other	0.24	0.90
	Underutilized	47.83	181.64
	Normal	26.26	99.75
	Overused	25.67	97.50
2	Class	Percent	Area (ha)
	Other	0.06	0.06
	Underutilized	54.38	59.97
	Normal	26.85	29.61
	Overused	18.71	20.63
3	Class	Percent	Area (ha)
	Other	6.97	112.90
	Underutilized	50.15	812.92
	Normal	28.36	459.67
	Overused	14.52	235.33
4	Class	Percent	Area (ha)
	Other	0.81	3.11
	Underutilized	47.47	183.40
	Normal	30.74	118.75
	Overused	20.99	81.09
5	Class	Percent	Area (ha)
	Other	11.49	38.09
	Underutilized	39.23	130.11
	Normal	36.43	120.83
	Overused	12.85	42.61
6	Class	Percent	Area (ha)
	Other	13.92	47.09
	Underutilized	44.19	149.51
	Normal	29.86	101.01
	Overused	12.03	40.69
7	Class	Percent	Area (ha)
	Other	6.66	171.75
	Underutilized	35.13	906.36
	Normal	33.60	866.86
	Overused	24.61	634.84
8	Class	Percent	Area (ha)
	Other	3.25	80.89
	Underutilized	38.62	960.95
	Normal	40.67	1012.00
	Overused	17.46	434.36
9	Class	Percent	Area (ha)
	Other	4.29	77.69
	Underutilized	53.05	960.26
	Normal	29.66	536.98
	Overused	13.00	235.33
10	Class	Percent	Area (ha)
	Other	2.11	3.79
	Underutilized	33.73	60.66
	Normal	34.31	61.71
	Overused	29.85	53.69
11	Class	Percent	Area (ha)
	Other	1.90	25.92
	Underutilized	41.51	566.55
	Normal	29.60	403.99
	Overused	26.98	368.27
12	Class	Percent	Area (ha)
	Other	0.74	1.48
	Underutilized	22.57	44.84
	Normal	42.36	84.14
	Overused	34.32	68.17
13	Class	Percent	Area (ha)
	Other	0.54	2.26
	Underutilized	38.45	160.98
	Normal	37.91	158.74
	Overused	23.10	96.70
14	Class	Percent	Area (ha)
	Other	1.79	6.12
	Underutilized	35.80	122.42
	Normal	31.66	108.26
	Overused	30.75	105.17
15	Class	Percent	Area (ha)
	Other	0.58	1.08
	Underutilized	36.11	67.62
	Normal	39.03	73.09
	Overused	24.29	45.48
16	Class	Percent	Area (ha)
	Other	2.69	11.30
	Underutilized	42.39	178.36
	Normal	32.65	137.38
	Overused	22.28	93.73
17	Class	Percent	Area (ha)
	Other	1.40	8.99
	Underutilized	37.80	242.34
	Normal	35.44	227.24
	Overused	25.36	162.63
18	Class	Percent	Area (ha)
	Other	0.99	2.82
	Underutilized	49.25	139.70
	Normal	32.62	92.54
	Overused	17.14	48.62

participation.

As such, the improvement of different stages of the supply chain of medicinal, aromatic and seasoning (culinary) plants are necessary in order to meet future demand.

Integrating environment and agriculture, social advancement, education and health through environmentally correct practices from the point of view of preservation, stimulates family farming and the use of medicinal plants to treat illnesses.

6. STRATEGIC DEVELOPMENT

In the light of the intrinsic characteristics of the micro-basin assessed in this analysis, we can see that the possible developmental actions must be focused on soil conservation and income assurance for local producers, as well as on water conservation. With this in mind, we propose some activities and strategies that could be implemented in the micro-basin:

- ✓ Technical assistance: establish and/or improve communication between farmers and institutions that provide technical assistance, such as EMATER and universities. Actions aimed at good quality and frequent technical assistance are often the key for any project to be developed in the micro-basin. As such, technical assistance would be one way to facilitate production activities and reorganize areas to reduce the impacts of agriculture on the environment;
- ✓ Erosion: processes of erosion are major limiting factors in the micro-basin; therefore, programs aimed at implementing road and soil conservation would be quite desirable for the micro-basin. Within this program we can include actions that favor areas without tillage, decreasing conventional intensive soil preparation practices, improving vegetation cover to conserve and improve the chemical quality of the soil, and crop rotation systems that not only enable improvements in other aspects of agriculture, but greatly affect soil conservation as well;
- ✓ APP and Legal Reserve: propose a program that aims to raise farmers' awareness about the need for rehabilitation of APPs and Legal Reserves in the micro-basin; this program would primarily aim at providing farmers with information regarding current and future benefits of this conservation. In a subsequent step of the project, we could include facilitating farmers' access to seedlings of native species for recovery of these areas, and these seedlings could come from the Environmental Institute of Paraná (*Instituto Ambiental do Paraná*; IAP) and the Municipalities. It should be noted that IAP already has seedling distribution programs for riparian forest restoration;
- ✓ Production systems: Existing production systems in the area can be improved by giving attention to horticulture and fruit production, as they are products that have consumer markets in close proximity. The goal of these improvements would be to optimize production and reduce cost, for example, over-fertilization generates environmental and economic problems. Another method would be to improve the study of markets and the associated activities present in the micro-basin to verify that activities, such as the more intensive fruit production in the properties, is a feasible option in the micro-basin;
- ✓ Socioeconomic: Actions to improve farmers' living conditions should always be a goal of interventions; such measures are aimed at improving the social conditions related to education, culture, relationships, sanitation, as well as assisting in farm management.

Although the proposed actions are treated separately, in practice it is difficult to apply each in isolation. To be successful, an intervention in the micro-basin should be developed out of this set of proposals. It should not become just another short-lived program with no actual gains to farmers, without sustainability over time, space, and economically.

7. REFERENCES

BRAGAGNOLO, N.; PAN, W. (2001) *A experiência de programas de manejo e conservação dos recursos naturais em microbacias hidrográficas: uma contribuição para o gerenciamento dos recursos hídricos*, Curitiba: IPARDES.

CHUEH, A.M. (2005) *Análise do Potencial de Degradação Ambiental na Bacia Hidrográfica do Rio Pequeno em São José dos Pinhais/PR, por meio do DFC - Diagnóstico Físico Conservacionista.* R. RA'E GA, Curitiba, n. 10, p. 61-71, Editora UFPR.

COORDENADORIA DE ASSISTÊNCIA TÉCNICA INTEGRAL (CATI). (2000) *Programa Estadual de Microbacias Hidrográficas.* CATI: São Paulo.

COORDENADORIA DA REGIÃO METROPOLITANA DE CURITIBA – COMEC. (2002) *Proposta de Macrozoneamento da Bacia do Rio Verde.* Curitiba: COMEC.

COORDENADORIA DA REGIÃO METROPOLITANA DE CURITIBA – COMEC. (1988) *Plano Diretor da Bacia hidrográfica: estudos geológicos.* Curitiba: COMEC, 32p.

COORDENADORIA DE ASSISTÊNCIA TÉCNICA INTEGRAL (CATI). (2005) *Incentivos ao manejo e conservação dos recursos naturais.* PEMBH - Seção Operativa – Volume III, Versão 2, Setembro, Campinas: CATI.

EHLERS, E. (1999) *Agricultura sustentável: origens e perspectivas de um novo paradigma.* 2. ed. Guaíba: Agropecuária, 157 p.

EMPRESA BRASILEIRA DE PESQUISA AGROPECUÁRIA / SERVIÇO NACIONAL DE LEVANTAMENTO E CONSERVAÇÃO DE SOLOS – EMBRAPA/SNLCS. (1984) *Levantamento de Reconhecimento de Solos do Estado do Paraná.* Curitiba: Embrapa/ SNLCS/SUDESUL/IAPAR, 791 p.

EMPRESA BRASILEIRA DE PESQUISA AGROPECUÁRIA / CENTRO NACIONAL DE PESQUISA DE SOLOS – EMBRAPA/CNPF. (1999) *Sistema brasileiro de classificação de solos.* Brasília: Embrapa/SPI e Embrapa Solos, 412 p.

FONTES, J. L. (2006) Programa de Microbacias Hidrográficas: justificativa, histórico, estratégia e resultados. In: RODRIGUES, Valdemir Antonio; BUCCI, Luis Alberto (Org.). *Manejo de microbacias hidrográficas: experiências nacionais e internacionais.* Botucatu, FEPAF, p. 200-215.

FREISCHFRESSER, V. (1999) Políticas públicas e a formação de redes conservacionistas em microbacias hidrográficas: o exemplo do

Paraná Rural. *Revista Paranaense de Desenvolvimento*. Curitiba, n. 95, jan./abr., p. 61-77.

HESPANHOL, A. N.; HESPANHOL; R. A.M. (2004) Desenvolvimento rural e poder local: a situação dos CMDRS na região de presidente prudente – SP. In: *I Seminário Internacional*, 2004, Rio Claro. CD do I Seminário Internacional. Rio Claro: IGCE-UNESP, p. 1-11.

JORGENSEN, S. E.; VOLLENWEIDER, R. A. (Ed.). (1989) *Guidelines of Lake Management*. Vol. 1. Principles of Lake Management. ILEC/UNEP. Japan, 199 p.

KIRCHMANN, H. (1994) Animal and municipal organic wastes and water quality. In: LAL, R.; STEWART, B. A. (Ed.). *Soil processes and water quality: advances in soil science. Boca Raton*. Lewis Publishers, p. 163-232.

MACHADO, M. L.S. & RODRIGUES, A. S. (1997) *A geração de tecnologia e o desenvolvimento da agricultura familiar.* Londrina, IAPAR, 80p. (IAPAR. Boletim Técnico, 57).

MEYBECK, M.; HELMER, R. (1992) An introduction to water quality. In CHAPMAN, D. (1992). *Water quality assessment.* Cambridge, University Press. 585p.

MINEROPAR. (2000) *Proposição de Expansão Urbana na Região Metropolitana de Curitiba, com Base em Levantamentos e Cadastramentos Geotécnicos.* Curitiba.

MIRANDA, M. & MIRANDA, G. M. (1995) A experiência do Instituto Agronômico do Paraná – IAPAR, na busca do desenvolvimento da agricultura familiar. In: WORKSHOP O DESENVOLVIMENTO DE UMA OUTRA AGRICULTURA: ACESSO À TERRA E A MEIOS DE PRODUÇÃO, A QUESTÃO DA FOME E A INTEGRAÇÃO SOCIAL. Anais, Cooperação Inter-Universitária Franco-Brasileira, Curitiba.

NAVARRO, Z. (2001) *Manejo de recursos naturais e desenvolvimento rural.* Porto Alegre: Programa de Pós-graduação em Desenvolvimento Rural. (Relatório preliminar).

OWENS, L. B. (1994) Impacts of soil N management on the quality of surface and subsurface water. In: LAL, R.; STEWART, B. A. (Ed.). *Soil processes and water quality: advances in soil science*. Boca Raton Lewis Publishers. p. 137-162.

PARANÁ. (1999) Secretaria de Estado da Agricultura e do Abastecimento. *Manual Operativo do Projeto Paraná 12 Meses.* Curitiba, 241p.

PASSINI, J. J. (1997) Redes de propriedades de referência. In: IAPAR. *Enfoque Sistêmico em P&D: a experiência metodológica do IAPAR.* Londrina, 152p. Ilust. (IAPAR. Circular, 97).

PASSINI, J. J. *Geração e comunicação de inovações tecnológicas para a agricultura familiar.* Curitiba, 1999. Dissertação (Mestrado) - Programa de Pós-graduação em Tecnologia, CEFET/PR, 154p.

PAYÉS, M. A. M. (1989) *Sistemas de produção predominantes no município de Rio Azul* – Paraná – uma proposta teórico-metodológica. Londrina: IAPAR, (IAPAR. Boletim técnico, 27).

PRADO, H. (1993) *Manual de Classificação de Solos do Brasil*. Jaboticabal: FUNEP, 218p.

RASCHIATORE R. A.; MOREIRA, D. A. (2006) Inovação na implementação do Programa Estadual de Microbacias Hidrográficas do Estado de São Paulo. *Gestão & Produção*, v. 13, n.° 3, p. 517-529, set.-dez.

SABANES, L. (2002) *Manejo sócio-ambiental de recursos naturais e políticas públicas: um estudo comparativo dos Projetos Paraná Rural e Microbacias.* Programa de Pós-Graduação em Desenvolvimento Rural da UFRGS, (Dissertação de Mestrado)

SHARPLEY, A. N.; HALVORSON, A. D. (1994) The management of soil phosphorus availability and its impact on surface water quality. In: LAL, R.; STEWART, B. A. *Soil process and water quality: advances in soil science.* Boca Raton: Lewis publishers, 398 p.

SHARPLEY, A. N. (1980) The enrichment of soil phosphorus in runoff sediments. *Journal of Environmental Quality*, v. 9, p. 521-526.

THEODOROVICZ, A. (2007) *Atlas geoambiental: subsídios ao planejamento territorial e à gestão ambiental da bacia hidrográfica do rio Ribeira do Iguape* / Antonio Theodorovicz, Ângela Maria de Godoy Theodorovicz. 2. ed. rev. São Paulo: CPRM.

VIEIRA, L. S.; VIEIRA, M. N. F. *Manual de Morfologia e Classificação de Solos.* 2. ed. São Paulo: Editora Agronômica Ceres Ltda. 313 p.

WORLD BANK. (1997) Land *Management III – Project São Paulo, - Environmentally and socially sustainable development* – Sector Management Unit – Latin America and the Caribbean region.

CHAPTER 20

RURAL SANITATION

Daniel Larsen
Miguel Mansur Aisse

RURAL SANITATION

Daniel Larsen
Miguel Mansur Aisse

SUMMARY

Rural sanitation aims to provide a healthy environment and use natural resources in a sustainable way. This study evaluated the rural sanitation of the Rio Verde Basin, in the Metropolitan Region of Curitiba, in collaboration with research partners from the Federal University of Paraná (UFPR) and Municipal Governments (Health Inspectors) as well as SANEPAR, the Sanitation Company of Paraná. Community-based and participatory methodologies were used with the rural population through the implementation of the Participatory Rural Appraisal (PRA). The quality of sanitation was then diagnosed in the communities, revealing levels of pollution and contamination of the water supply, identifying the factors contributing to the low quality of water, and evaluating the treatment processes and final disposal of solid and liquid waste through questionnaires. Some recommendations were made that aim to minimize the observed problems involving state and municipal stakeholders.

KEYWORDS

Rio Verde basin, participatory rural appraisal, MEXPAR, metropolitan region of Curitiba, rural sanitation, rural water supply system.

1. CONCEPTS

1.1 RURAL SANITATION

The concept of sanitation, according to the World Health Organization (WHO), is directly related to the control of all factors of the human physical environment that cause or have the potential to cause harmful effects on physical, mental and social wellbeing. In other words, sanitation is related to the state of environmental health, achieved through a set of socioeconomic measures; this implies overcoming barriers, including technological, political and managerial, that have held back the development of sanitation measures and their associated benefits, especially for the residents of rural areas, small municipalities or isolated locations (FUNASA, 2006). Moreover, rural sanitation encourages a healthy environment in rural areas, by promoting the use of natural resources in a sustainable way and thus mitigating the effects of environment degradation, in particular the contamination of freshwater springs, due to the inappropriate disposal of sanitary sewage and solid waste.

Rural communities that are located in hydrographic basins of water supply sources use water that come from artesian wells, groundwater wells and springs which are often contaminated by inadequate soil management, agriculture and livestock activities, inadequate disposal of wastes, among others activities that have developed in the region, without consideration of the environment (PILATTI, 2008).

In rural areas and in small cities with low population density, where residences are widely dispersed, the adoption of feasible and simple technologies is a common occurrence since urban methods of sanitation are almost never appropriate. The main goal of a technical sanitation program is to meet the needs of those who use it and therefore the choice of such a technology should be based on a set of economic, social, ecological and cultural guidelines that fulfill the needs of the society in question (AISSE, 2000).

The Federal Sanitation Policy, established in Law 11.445/07, considers basic sanitation as the set of services including the infrastructure and operational facilities to supply drinking water, sewage sanitation, urban sanitation and solid waste management, and drainage and management of urban storm water. In this study, emphasis was given to the first three services mentioned by Law as applied to small communities: villages, rural settlements, hamlets, and villages, as defined by the Brazilian Institute of Geography and Statistics – IBGE (BRASIL, 2007).

1.2 TECHNOLOGIES FOR RURAL SANITATION

1.2.1 Water

In rural areas, the issues related to water supply are different from urban areas with more concentrated populations. Alternative solutions to supplying water for human consumption are used that are distinct from the system found in large urban centers. In the countryside, it is common to make use of wells, springs, distribution via transportation vehicles, among other solutions used by individuals. These solutions are usually applied in areas with a dispersed population, they consider the characteristics of each place, and both the implementation of technology and the cost is assumed exclusively by the residents (FUNASA, 2006).

Two alternative solutions to water supply for human consumption include: collective solution and individual solution. The collective solution is less common in rural areas as it generally depends on a higher population concentration. The individual solution is usually used in disperse populations.

Water Supply System

Ordinance 518/04 of the Ministry of Health, defines the collective alternative solution (CAS) of water supply for human consumption as all forms of collective water supply that are different from the water supply system (WSS), including springs, community wells, distribution by transportation vehicles, and communal facilities, horizontal and vertical.

The Ministry of Health defines the WSS for human consumption as the set of civil infrastructure, material and equipment, meant for the production and piped distribution of drinking water for populations, under the responsibility of the government, even if administered by a third party under concessions and permission arrangements (BRASIL 2004; Figure 1).

Water Abstraction

Depending on the source used, water abstraction can occur in several ways, listed below and illustrated in Figure 2 (FUNASA, 2006):

- Surface abstraction (rain water)
- Collection (slope spring)
- Infiltration gallery (located at the bottom of valleys)
- Excavated well (water table)
- Tube well (groundwater)
- Direct abstraction from rivers, lakes and dams (surface watershed)

Water Treatment

Despite the general reliability of underground water quality, certain impurities can make the water unsuitable for human consumption. The levels of such impurities are defined according to parameters established by the agencies responsible, such as the Ministry of Health that indicates the standards of potability through Ordinance 518 (March 25, 2004; PILATTI & HINSCHING, 2008).

As such, water quality can be guaranteed through simple and feasible treatments. Slow filters are commonly found in small communities as an alternative to water supply treatment, mainly for the removal of turbidity. Such filters are advantageous because they do not need chemical products or sophisticated equipment. Their construction and operation are simple and there is limited production of sludge. Even with these advantages, slow filters can be inefficient during some periods throughout the year when the quality of source water significantly worsens, leading to high levels of turbidity, causing a reduction in the filtration efficiency (PATERNIANI, 2004).

Domestic ceramic filters are an alternative for small communities where there is no collective water supply as these filters have been identified as a promising and accessible technology for water treatment in residences (GUSMÃO *et al.*, 2010, citing CLASEN *et al.*, 2004). The efficiency of such filters in removing pathogenic microorganisms has been tested (especially those manufactured in Britain and the U.S.) and they have been certified for their high performance (GUSMÃO *et al.*, 2010, citing SOBSEY, 2002).

Disinfection Process

Disinfection is the last process before the water is distributed to the population. It is necessary to guarantee water quality for human consumption and to avoid waterborne diseases by inactivating the pathogenic organisms. In the rural environment, disinfection is commonly accomplished by using compounds of liquid or gas chlorine (Cl_2) considering their ease of use (BARRETO, 1984; DANIEL, 2001).

The chlorine derivatives that are usually used in small communities are sodium hypochlorite (liquid) and calcium hypochlorite (solid).

Currently, there is a tendency for *in situ* production of sodium hypochlorite and in some cases electrolysis gases are vented to the atmosphere, while in other cases all gases are used for disinfection. Industrial generators of sodium hypochlorite can produce chlorates and traces of chlorites, chlorine dioxide and ozone as by-products. The dosage used should be controlled in order to maintain chlorate levels below those suggested by the WHO (DANIEL, 2001).

Thus, we can cite the Hidrogerox (Figure 3) or Moggod (*Mixed Oxidants Gases Generated in Situ)* technology, which is a continuous system of production and dosage of oxidant solution used to disinfect the water (AISSE, 2005). In Hidrogerox, the oxidant gas is produced through the electrolysis of a saline solution of sodium chloride (cooking salt). This process also generates a solution of sodium hydroxide and hydrogen gas as by-products. The advantages of using this process are the low operational cost, the availability and low cost of raw materials, the elimination of the need for transporting, storing and handling of toxic/dangerous materials, and low levels of energy consumption (DANIEL, 2001).

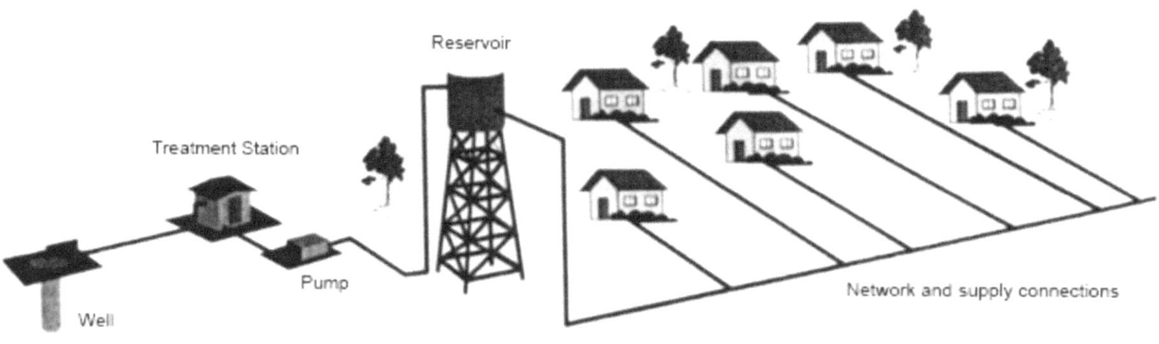

FIGURE 1 – WATER SUPPLY IN RURAL COMMUNITIES
Source: SANEPAR[1] (Sd).

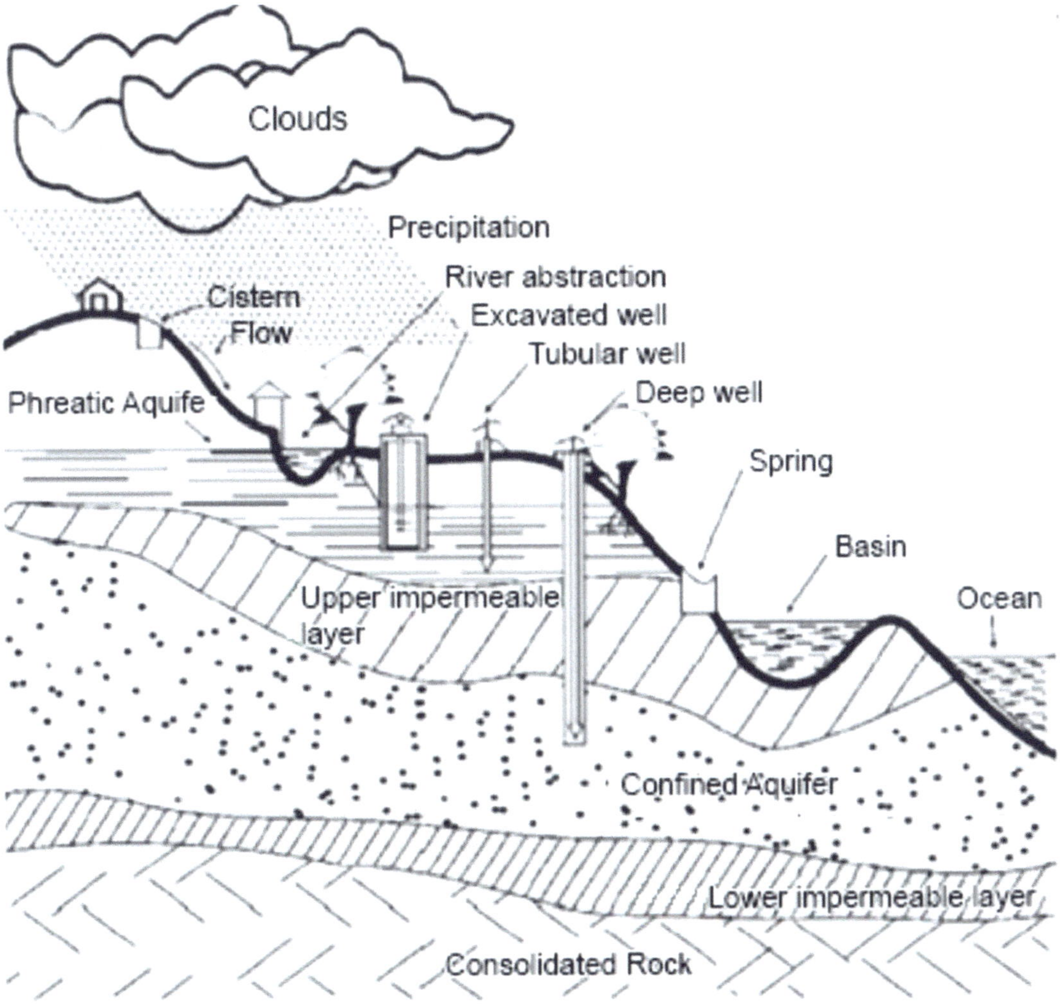

FIGURE 2 – SOLUTIONS FOR WATER ABSTRACTION IN RURAL AREAS
Source: FUNASA (2006).

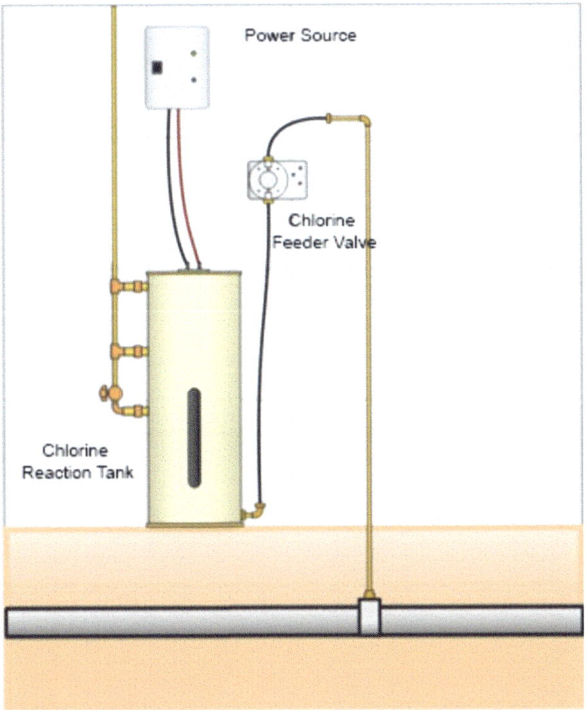

FIGURE 3 – SCHEMATIC OF THE STATIC HIDROGEROX
Source: Hidrogeron (2009).

Calcium hypochlorite is composed of the precipitated result of the dissolution of gaseous chlorine in calcium oxide and sodium hydroxide solution. It is used in the form of pellets with 60% active chlorine and can supply sufficient chlorine for disinfection for 10 to 15 hours.

Ministry of Health Ordinance 518/2004 requires that at the end of treatment (after disinfection) there must be a free residual chlorine concentration of 0.5 mg/L and a concentration at any point in the distribution network of 0.2 mg/L. It is also recommended that the maximum level at any point in the supply system is 2.0 mg/L (FUNASA, 2009).

Fluoridation Process

Considering the importance of fluorine salts in the prevention of dental cavities when given to children up to age 14, and the decreasing effectiveness of fluorine with age, the inclusion of fluoride in the public water supply has been identified as the most effective and economical way of controlling dental cavities (FUNASA, 2006).

Fluoridation of the water supply is done through dispensing devices, using sodium fluoride, sodium fluorosilicate and fluorosilicic acid (FUNASA, 2006).

The concentration of the fluoride ion is adjusted

based on average maximum daily temperatures, observed over a minimum period of one year (with five years as the recommended period of observation). The optimum concentration of fluoride is around 1.0 mg/L and the maximum concentration allowed is 1.5 mg/L, according to the Ministry of Health (FUNASA, 2009).

Water Distribution

At the end of the system, the distribution network is the set of pipelines, connections, valves and other parts designed to continuously distribute water to all users in the system. In the case of rural areas, and considering the dispersed layout of properties, the distribution network should include a main pipeline and secondary pipelines that branch off of it, with only one supply source (FUNASA, 2006).

1.2.2 Domestic Sewage

Considering the importance of sanitary sewage treatment, some fundamental goals are set that aim to: avoid the pollution of the soil and water supply sources; avoid contact of vectors with manure; encourage new hygienic habits in the population; and promote comfort and meet aesthetic needs (FUNASA, 2006).

Domestic sewage has particular characteristics depending on its origin from residences, commercial establishments, institutions or any building that has bathroom installations, laundries, and kitchens. As with any effluent, domestic sewage presents physical (solid matter, temperature, odor, color, turbidity and flow variation), chemical (organic and inorganic matter), and biological (bacteria, viruses, fungus, protozoa, algae and other indicators of pollution, such as organisms of the coliform group) characteristics that should receive adequate treatment and cannot exceed the load capacity or contaminate the environment (FUNASA, 2006). Table 1 presents typical characteristics of domestic sewage.

There are simple and economic alternatives cited in the literature related to sanitation methods in rural environments. As such, the Ministry of Health (MH), through the National Health Foundation (FUNASA), recommends the usage of septic tanks for areas of low population concentration. Such technologies are varied and simple in their conception and several forms are commonly found in rural properties, such as: toilet with a dry cesspit, which consists of an outside toilet and dry cesspit that receives only waste; toilet with a watertight cesspit, that receives waste in a tank; toilets with fermentation cesspit, comprised of two contiguous and independent chambers designed to receive waste; chemical toilet, composed of a cylindrical tank of stainless steel containing caustic soda (NaOH) with the same function as those previously mentioned; toilet with toilet bowl where there is piped water, in which the waste can be directed to a septic tank and then to a drainage pit, infiltration trench or filtration trench (FUNASA, 2006).

The septic tank, also called septic cesspit, is widespread and found in many rural properties and it is the main technology of domestic sewage treatment in the countryside (BARRETO, 1984). The tank receives the residual water, separates the solids from liquid, partially processes the organic matter, stores the solids, while the clarified liquid is finally disposed of or receives other treatment. The sedimented solids go to the bottom of the tank where an anaerobic breakdown occurs and the solids finally accumulate (AISSE, 2000).

Septic tanks are classified according to their characteristics (AISSE, 2000): number of chambers (single, two in series, two overlapped, multiple); shape (cylindrical, prismatic, rectangular); and relative placement of the chambers (in series, overlapped) (Figure 4).

TABLE 1 – TYPICAL COMPOSITION OF DOMESTIC SEWAGE

COMPONENTS	CONCENTRATION (mg/L)		
	STRONG	MEDIUM	WEAK
Total Solids	1200	720	350
Total Dissolved Solids	850	500	250
Fixed Dissolved Solids	525	300	145
Fixed Suspended Solids	75	55	20
Sedimented Solids (mL/L)	20	10	5
Biochemical oxygen demand (BOD5, 20).	400	220	110
Chemical oxygen demand (COD)	1000	500	250
Total Nitrogen	85	40	20
Organic Nitrogen	35	15	8
Ammonia Nitrogen	50	25	12
Nitrate	0	0	0
Nitrite	0	0	0
Total Phosphorous	15	8	4
Organic Phosphorous	5	3	1
Inorganic Phosphorous	10	5	3
Alkalinity	200	100	50
Oils and Greases	150	100	50
Total Coliforms (MPN/100 mL)	--	108 – 1010	--
Fecal Coliforms (MPN/100 mL)	--	106 – 10^8	--

Source: AISSE (2000) citing Metcalf & Eddy (1979) and Yanez (1984).

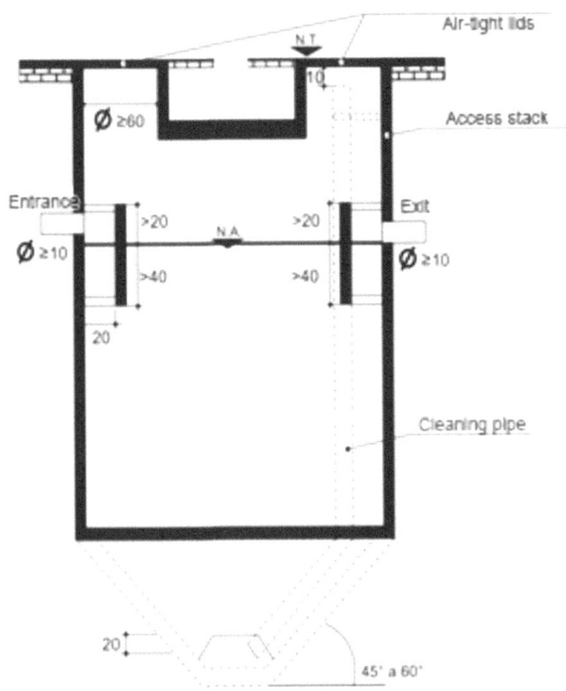

FIGURE 4 – SINGLE CHAMBER SEPTIC TANK
Source: Adapted from NBR7229 (1993)

Disposal

It is recommended that the effluent originating from the septic tank be deposited into the soil, as long as it does not negatively affect the quality of the groundwater (ABNT NBR 7229/1993). There are some more commonly used alternatives for depositing septic waste in the soil (SANEPAR[2]):

- Drainage pit: an excavated pit with walls that permit the sewage to infiltrate the soil. Usually built with hollow concrete rings, masonry, airbrick or common brick without mortar. The bottom should have a layer of gravel (Figure 5);

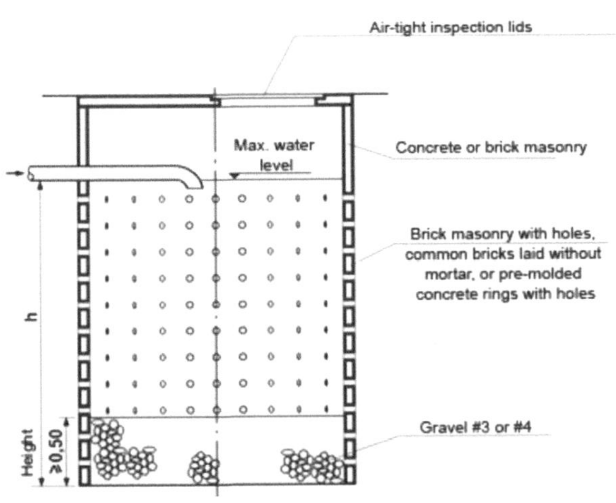

FIGURE 5 – DRAINAGE PIT
Source: Aisse (2000) citing NBR 7229 (1982)

- Infiltration trench: an excavated trench in the terrain with piping that distributes the sewage over a layer of crushed stone. After passing through the stone layer the sewage should infiltrate the soil;
- Filtration trench: a trench excavated in the terrain, occupying a large area with shallow depth. The waste passes through a perforated pipe, infiltrates through a layer of sand and is collected by another pipe below the sand layer. After the process, the sewage could be disposed of into a water body.

The choice of the alternative of deposition in the soil depends on the capacity of infiltration and the availability of space. If either of these aspects are limited, treatment of the waste should be undertaken. Another alternative is an anaerobic filter, which is basically a contact unit in which the sewage passes through a mass of biological solids contained within a reactor. The soluble organic compounds contained in the sewage come into contact with the biomass (filling) and are diffused through the surface of the biofilm or granular sludge, thus being converted into intermediate and final products, such as methane and carbon dioxide (CHERNICHARO, 2007).

The waste coming from the filter is usually sufficiently clean and has relatively low concentrations of organic matter, but is rich in minerals. It is suggested that the waste is applied to the soil for agricultural production, either by irrigation or by infiltration, provided that there is no concern related to the presence of pathogens (CHERNICHARO, 2007).

1.2.3 Solid Wastes

Solid waste is a significant factor in sanitation as it can cause extensive sanitary problems when not given the necessary treatment. The measures that can be taken to solve such problems are common aspects of sanitation that prevent and control related diseases.

The composition of solid waste can vary depending on the region, habits and customs of the community, number of inhabitants, purchasing power, seasonal variations, climate, level of development, educational level, and in rural areas the land-use activities that are conducted on the property (FUNASA, 2006).

Toxic Waste

In rural areas, packages of pesticides, or class I (dangerous) waste, are commonly found without a proper disposal destination. Such types of waste attract attention due to their potential impact on the environment. When they are not properly collected pesticide packages become a risk factor in environmental contamination that can be aggravated based on the proximity of the waste to residences and water supply sources (IBGE, 2005).

Federal Decree 4.074, January 4, 2002, determines that the users of pesticides and related products should return all packages and lids to the commercial establishments from which they were acquired or take them to collecting centers/stations within one year of the purchase date, as instructed on the labels and leaflets and included on the invoice (BRASIL, 2002). At the end of one year, if there is still product in the package that has not expired, it must be returned within six months of the date of expiration. Users should keep the return receipts of empty packages that are provided by the commercial establishments and collecting centers/stations for a period of one year after the package has been returned as this proof may be required by supervisory bodies. Containers holding dissolvable products or products that are dispersed with water should be submitted to a triple washing process or an equivalent technology, according to the guidelines outlined their labels, leaflets or information brochures.

In Paraná, Law 12.493, January 22, 1999, establishes in Article 12 that all manufacturing or commercialization companies of pesticides, their components and related products, are responsible for providing services of collection and receiving empty packages for disposal of products manufactured and (or) commercialized by them as well as products seized through enforcement actions and products deemed unfit for use, according to the conditions and criteria established by IAP (PARANÁ, 1999).

Animal Waste

Besides the issue of toxic waste, intensive agriculture and livestock activity, especially the production of swine and poultry, poses a significant contamination potential to surface and groundwater through the movement and leaching of excrement, respectively. The confinement of animals generates a concentrated production of manures that need treatment and adequate disposal. In pig farming, the use of organic fertilizer is quite common by creating dunghills that store the manure for a given time period but the use of such waste as fertilizer requires adequate planning and care in the application of the fertilizer. The excrement can reach water courses, carrying high loads of nitrogenous and thermotolerant coliforms, thus making the water unsuitable for public use (DORIGON, 2008).

Waste Treatment and Disposal

Considering all the characteristics mentioned above, waste treatment varies greatly depending on its composition and origin. FUNASA (2006) recommends that such organic waste, due to the lack of adequate collection and transportation services, should be buried in order to avoid exposure to the environment and prevent the proliferation of vectors. This waste disposal mechanism is not recommended for pesticide packages because of the risk of contaminating the soil, water bodies, affecting animals, and also the possibility of affecting nearby human populations (IBGE, 2005).

a) In the case of organic waste, composting can be of great value to the rural farmers because it is an aerobic biological process that, if properly controlled, converts organic matter into organic that already exist or were introduced into the waste mass (FUNASA, 2006). Despite the inability to commercialize the compost due to the possible presence of contaminants, as well as the negative aspect of odor, composting organic waste ensures a considerable decrease in the volume of waste produced. If well operated, this process produces compost that can be used for soil fertilization; it also recycles nutrients and is cost efficient in comparison to other forms of waste disposal that seek results and efficiency (CISAM, 2006).

Some factors are important during the composting process: the moisture levels of the waste should be between 50 and 60%; there should be constant aeration during the first 60 days (on average twice a week) thus avoiding biological activity that compromises the process; the temperature should be maintained at approximately 55°C, with a reduction of 10°C to 20°C at the end of the process. Other factors such as pH and nutrients are also controlled: the acidic pH should be maintained at 4.5 to 5.5 in the beginning of the composting and from 7.0 to 8.0 for the humidified compost; the ratio of C/N at the beginning should be in 30/1.

One of the advantages of composting is that it can be accomplished using both simple and complex technologies, as long as the waste is adequate and the biological process occurs in favorable conditions. The important consideration is that the chosen alternative is appropriate for the situation considering technical and socioeconomic factors (FERNANDES *et al.*, 1999).

According to Fernandes *et al.* (1999) composting processes can be divided into three major groups:

- Systems of plowed furrows, where the waste mixture is laid out in furrows and aeration is achieved by plowing the waste which allows for air diffusion and convection in the mass of the compound. This system can be done outdoors or in covered areas.
- System of static furrows, where the mixture to be composted is placed over a perforated pipe that injects or aspirates air into the mass of the compost. There is no mechanical turning of the furrows. This process can also be done outdoors or in covered areas.
- Finally, the closed system or biological reactors, where the waste is placed inside a closed system that controls all parameters of the process.

Other technologies employed in the rural environment consist of the temporary storing of the organic material in dunghills or biodigesters, after which the generated waste can be applied to the soil.

b) Dunghills are compared in theory to stabilization ponds, where the removal of carbonaceous material occurs with the help of facultative and strictly anaerobic bacteria that biologically transform the organic matter into a more stable product (SANTOS, 2007). For this technology, a retention time of 120 days is recommended to ensure the inactivation of pathogenic material and the stabilization of organic matter. A minimum depth of 2.5 m is also recommended (KUNZ, 2005).

c) The biodigester consists of a closed chamber in which the biomass (manure) is anaerobically fermented, producing biogas and biofertilizer. In general, the biodigester has two parts: a tank to accommodate and allow the digestion of the biomass and a gas holder for storing biogas (GASPAR, 2003). In Brazil, the model developed in India is the most common due to its functionality. It is comprised of a cylindrical body divided into a fermentation tank in two chambers with a floating gas holder, an input box, and an output box (ISHIBA, 2009). Besides the Indian model, the Chinese model is also commonly found in rural properties.

The greatest advantages of using the biodigester are the production of biogas and biofertilizers. The biogas, which is produced during the anaerobic digestion, can be used as an alternative source of energy, especially in rural properties, allowing rural residents to produce a practical and inexpensive fuel. Biofertilizer, a waste rich in humus and nutrients, can be used in the fertilization of the soil, increasing the productivity of the crops with a low cost of production (AISSE & OBLADEN, 1982).

Urban Solid Waste

In the rural environment, the reuse of urban or common solid waste is well regarded by the community, considering that in more isolated areas waste collection services provided by Municipalities are challenging. Other options used for waste disposal are the burning and the burying of waste in the rural property, preventing the accumulation of garbage and ensuring rodent and insect control.

1.3 WATER QUALITY

Pure water is not found in nature. Various substances can be found in water that were dissolved from the soil or the air and carried along the water course. Some of these substances make the water unfit for human consumption, such as substances resulting from human and industrial activity. In other cases, substances such as iron can be found in the water, which gives it an unpleasant color and taste, as well as limestone and magnesium, which make the water hard. There is a wide range of situations that can differentiate the physic-chemical and biological properties of water, even without human intervention (FUNASA, 2006).

Of the biological characteristics of the water, coliforms have been used as indicators of contamination and recent pollution of feces. However, the presence of coliforms in the water body does not necessarily relate to pathogenic organisms because the presence of such organisms require a carrier in the contributing population, while the number of coliforms (total or thermotolerant) depends only on the presence of non-sterile organic waste that is foreign to the receiving body. As such, even if there is no relation, there may be a greater possibility of the presence of pathogenic organisms in the water with a greater number of coliforms found therein. Therefore, when coliforms are found in water for human consumption treatment is needed (DI BERNARDO, 2005).

1.3.1 Waterborne Diseases

Even with the growing economic and social development of Brazil, excessive deaths related to the lack of sanitation infrastructure continue to affect the poorer regions of the country. Policies that require further investment in specific areas of the country include the implementation of preventive public health programs, which work to prevent the spread of immuno-preventable diseases, and the expansion of basic sanitation services, the absence of which is an important factor in the prevalence of deaths from infectious and parasitic diseases (IBGE, 2005).

In Brazil, it is estimated that 60% of hospital admissions are linked to the precariousness of the basic sanitation system, thereby reducing the life expectancy of the population. Other studies indicate that 90% of these diseases are due to the absence of sufficient quantity or quality of water for consumption and this situation of insufficient water quality and quantity is commonly found in Brazil (DI BERNARDO, 2005).

Within the area of sanitation, epidemiological studies can be undertaken with the goal of identifying causal factors, evaluating programs, or planning public action (HELLER, 1997). The IBGE points out that in 2007 only 54.5% of households with children had all sanitation services simultaneously and this information is relevant considering that most infant deaths are related to a lack of basic sanitation.

Under these conditions, medical-hospital action is not enough to overcome obstacles caused by the poor conditions experienced by some segments of the population. Although some actions are able to ameliorate immune deficiencies, for example by avoiding death through immunizations (vaccines), through the neutralization of bacteria (antibiotics), or through oral rehydration of children with diarrhea, many actions end up being palliative or transitory because they do not address basic factors in the manifestation of the disease. These factors are oftentimes associated with nutritional, social and environmental problems. In other words, a consistent decrease in infant mortality appears to be strongly dependent on the intervention model put in place through public policies, especially in the fields of preventive and curative medicine and basic sanitation (IBGE, 2005).

2. MATERIALS AND METHODS

2.1 Analyses of the Collective Water Supply Systems

The water quality of the implemented Water Supply Systems (WSS) was obtained initially by consulting the data from the Health Inspection Agency, when the system in question was operated by the Municipality, and from the Unit of Projects, Services and Construction Works of Curitiba, Metropolitan and Coastal Region (USPO-CT), when operated by SANEPAR. The approach (Figure 6) was divided according to the data provided by the Rural Sanitation Assessment (RSA) of SANEPAR.

Considering the number of parameters used by SANEPAR to analyze the wells under its responsibility, as compared to those used by Health Inspection Agencies, the following analyses were made:

- Physico-chemical: Color, pH, Conductivity, Turbidity, Nitrate, Chloride, Total Dissolved Solids and Fluorine;
- Heavy Metals: Iron, Arsenic, Cadmium and Lead;
- Pesticides: Glyphosate, metalaxyl, propanil, amitraz, alachlor, metolachlor, alpha-endosulfan, folpet, vinclozolin, iprodione, trifluralin, tolifluanida, Pyrazohos, chlorothalonil, aldrin, DDT 2.4, 4.4 DDT, dicofol, dieldrin, endrin, beta endosulfan sulfate, endosulfan, hexachlorobenzene, alpha-HCH, delta-HCH, heptachlor, lindane, methoxychlor, acephate, azinphos-methyl, chlorfenvinphos, cou-

mafos, demeton-S-methyl, diazinon, dichlorvos, dimethoate, disulfoton, etiona, ethoprophos, fenchlorphos, fenitrothion, fensulfotiona, fentiona, phorate, malathion, methidathion, mevinphos , monocrotophos, naled, ethyl parathion, methyl parathion, pirimiphos ethyl, phosphamidon, pirimiphos-methyl, Prophenophos, Prothiophos, terbufos, Tetrachlorvinphos, triazophos, trichlorfon, cypermethrin, deltamethrin, lambda cyhalothrin, permethrin, pyrethrins, zeta-cypermethrin, fenarimol, molinate, difenoconazole, ciproconazol, propiconazole, tebuconazole, triadimefom, triadimenol, ametrina, atrazine, simazine.

The Laboratory of Environmental Analysis of the Pontifical Catholic University of Paraná (PUC-PR) performed the physic-chemical and Iron analyses and the Center of Food Research and Processing (CEPPA), analyzed the other parameters.

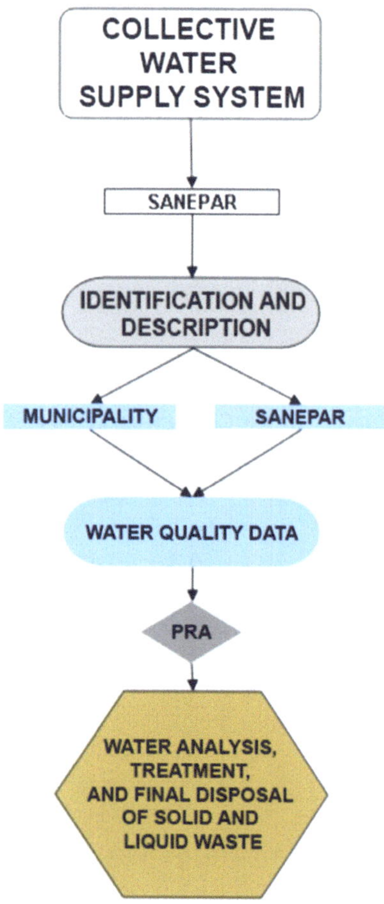

FIGURE 6 – FLOW CHART OF THE APPROACH

2.2 DIAGNOSIS OF RURAL SANITATION

2.2.1 Water

Samples were collected from properties whose owners were willing to participate in the study (see Figure 1, Chapter 16) and the sample collection was planned in meetings with the local community who participated in the Participatory Rural Appraisal (PRA). On all participating properties, samples were collected and analyses of pH, turbidity, and conductivity were performed in the field and later confirmed in the laboratory; levels of total coliform and *E. coli* were analyzed in the laboratory.

Samples were collected during the first output from the well after the faucet was disinfected with alcohol. One sample from each well was collected in sterilized bottles for later laboratory analysis of total coliform and *E. coli* at the Laboratory of Environmental Analyses, PUC-PR. For analyses of the other parameters, samples were collected in clean bottles and were conducted at the Laboratory of Environmental Engineering (LABEAM) of the Federal University of Paraná (UFPR). When measured in the field, equipment was provided by LABEAM.

Table 2 presents the methods and equipment used to assess water quality parameters.

In the field, forms were used to describe the property through detailed sketches, collection conditions, date, time and results of the analysis.

The results of the analyses were returned to the community and the Health Inspection Agency of the municipality, in order for the collected data to be a collated into a database for use in future improvement activities or maintenance of water quality according to the standards of potability established by Ordinance 518/04 of the Ministry of Health.

In the field, structured questionnaires were also undertaken with community members, which provided more information to be used in the diagnosis of the sanitary situation in the studied region. The water-related questions are presented in Table 3.

TABLE 2 – DESCRIPTION OF METHODS USED FOR WATER ANALYSIS

PARAMETER	METHOD	EQUIPMENT	BIBLIOGRAPHY	LABORATORY
pH Field	Potentiometer	WTWpH330i/SET	Standard Methods (2003)	LABEAM
pH Lab	Potentiometer	NT PHM	Standard Methods (2003)	LABEAM
Total Coliforms	Chromogenic substrate	-	Standard Methods (2003)	PUC-PR
E. coli	Chromogenic substrate	-	Standard Methods (2003)	PUC-PR
Turbidity	Turbidimetry	Portable H. 2100P	Standard Methods (2003)	LABEAM
Conductivity	Potentiometer	WTW Cond 315i Handylab LF1	Standard Methods (2003)	LABEAM

TABLE 3 – PART OF THE STRUCTURED QUESTIONNAIRE UNDERTAKEN IN RURAL COMMUNITIES REGARDING WATER

What type of water supply is used in the house where the family lives?				
Public Network		Artesian Well with electric pump		
Common well with electric pump		Common well with manual operation		
Mined water, fountain, stream, river, dam – manual operation		Mined water, fountain, stream, river, dam – mechanical operation		
What is the quality of the water?				
Very Good	Good	Reasonable	Poor	Very Poor
What are the uses of the water?				
Residential	Agricultural	Industrial	Others	
What is the water availability?				
High	Sufficient	Medium	Low	
Is disinfection conducted before the water is used?				
Yes	How?		No	

2.2.2 Domestic Sewage

The identification of the treatment systems of sanitary sewage produced in the rural properties located in the study area was done through the PRA, during the structured questionnaire (Table 4) in the questions related to quality of life.

In isolated properties where water analyses were conducted, the evaluation of the sewage treatment systems was linked to the data on water quality and the information outlined in the property sketch, thus identifying the distances between sewage disposal and the water supply well.

In the structured questionnaire, three questions were asked regarding characterization and final destination of the sewage generated in the rural property, as shown in Table 4.

2.2.3 Solid Waste

Another aspect that is considered a factor in the pollution and contamination of the water supply and consequently a factor in living conditions, is the final disposal of solid waste as well as its treatment.

In surveying the data, two PRA tools were used: historical maps that reveal possible grievances with the collection of waste or uncontrolled waste disposal, and the structured questionnaire. Questions about the production, identification, treatment and final disposal of the solid waste were raised (Table 5).

TABLE 4 – PART OF THE STRUCTURED QUESTIONNAIRE UNDERTAKEN IN RURAL COMMUNITIES REGARDING SEWAGE

What type of toilet is used where the family lives?			
Toilet inside the residence		Outside toilet, connected to the residence	
Outside toilet, like an outhouse		Other	
What is the destinati on of the waste (sewage) of the house where the family lives?			
Public system		Septic tank – closed and waterproofed tank	
Dry Cesspit – waste released directly		Piped/dumped/channeled to the river/stream/creek/terrain	
Characterization of the Sewage			
Domestic	Animal	Industrial	Others

TABLE 5 – PART OF THE STRUCTURED QUESTIONARY UNDERTAKEN IN THE RURAL COMMUNITIES REGARDING SOLID WASTE

What is the destination given to the common garbage produced on the property?				
Separation/Recycling and taken to public collection sites		Burned		
Taken to the public collection sites		Released in the terrain/river		
Buried		Others		
What is the destination given to the organic garbage produced in the property?				
Burned		Recycled (buried immediately, making compost etc.)		
Taken to public collection sites		Released in the terrain/river		
Buried		Others		
Types of Animal Manure				
Cattle	Equine	Swine	Ovine	Poultry
Manure Treatment				
Biodigester	Dunghill	Composting	Does not treat	Other
Final Destination of the Manure (Treated or not)				
Soil	Water Course		Other	
Final Destination of Pesticide Packages				
Returned	Reused	Stored	Burned	Buried

2.2.4 Waterborne Diseases

For understanding sanitary conditions, data regarding intestinal tract diseases and other possible waterborne illnesses were collected. Such information was collected from the Health Inspection Agency in each municipality, seeking seasonal data when available.

2.3 PROPOSING ALTERNATIVES FOR MINIMIZING POLLUTION AND CONTAMINATION OF THE WATER SUPPLY

Finally, the project proposed feasible alternatives for the treatment and disposal of effluents and solid waste which would reduce the impact on water and public health, as well as improve water quality of the both communal and dispersal systems, based on the literature review and the experience of the researchers involved.

3. RESULTS AND DISCUSSION

3.1. Analyses of the Collective Water Supply Systems

The water supply systems (WSS) in the Rio Verde Basin were constructed by SANEPAR through a Rural Sanitation Program aiming to improve the living conditions of the rural population. The WSS constructed by SANEPAR in the rural environment are composed of: abstraction (well)/treatment, pipeline/pumping, reservation, distribution network and individual household connections. Beyond the construction of the WSS, SANEPAR also operates some of the systems, periodically treating and monitoring water quality. In the cases where SANEPAR does not operate the system, community residents are trained by SANEPAR instructors to enable residents to operate the equipment, perform system maintenance and treat the water according to the standards of potability establish ed by Ordinance 518/04 of the Ministry of Health (SANEPAR[1]).

When obtaining information, there was significant reception to the Rural Sanitation Assessment (RSA) from SANEPAR in providing data regarding the WSS in the municipalities within the Basin and from the Secretary of Environment and Agriculture of Araucária, in which two WSS were identified: Colônia Cristina and Formigueiro. In Campo Largo, we found: Colônia Rebouças (operated by SANEPAR) and Colônia Dom Pedro II (operated by the Municipality).

Data on water quality were obtained through an existing system in Araucária (Table 6) provided by the Health Inspection Agency. In Campo Largo, the RSA/SANEPAR provided the data on quality of the Rebouças system, operated by SANEPAR itself, and the Health Inspection Agency of Campo Largo provided data on the water quality of the WSS in Dom Pedro II (Table 7).

TABLE 6 – DATA ON WATER QUALITY OF THE WSS OF COLÔNIA CRISTINA

DATE	TOTAL COLIFORMS	THERMOTOLERANT COLIFORMS	pH	CHLORINE	TURBIDITY	FLUORINE
Mar/08	Absent	Absent	6.9	-	-	-
May/08	Absent	Absent	6.5	0	0.5	-
Aug/08	Absent	Absent	6.5	0	2.8	0
Apr/09	Absent	Absent	6	0	1.3	-
Apr/09	Absent	Absent	6.7	0	1.1	-
Apr/09	Absent	Absent	6.8	0	1.3	-
Aug/09	< 1,1	< 1,1	6	0.2	1.1	-
Feb/10	25	Absent	5.5	0	0.2	0
Feb/10	Absent	Absent	6	0	0.3	0
Feb/10	Absent	Absent	6.5	0.2	0.2	0
Feb/10	Absent	Absent	6.5	0.2	0.3	0
Mar/10	Absent	Absent	6.9	0	0.4	0
Mar/10	Absent	Absent	6.8	0	0.3	0
Mar/10	Absent	Absent	6.9	0	0.3	0
Mar/10	168	Absent	6.8	0	0.2	0

Note: Coliforms in MPN/100ml; turbidity in NTU; chlorine and fluorine in mg/l.
SOURCE: Health Inspection Agency of Araucária (2010)

TABLE 7 – DATA ON WATER QUALITY OF THE WSS OF CAMPO LARGO

DATE	LOCATION	TOTAL COLIFORMS	THERMOTOLERANT COLIFORMS	pH	CHLORINE	TURBIDITY	FLUORINE
Feb/09	Colônia Dom Pedro II	< 1.0	< 1.0	-	0.5	0.21	0.23
May/10	Colônia Dom Pedro II	Absent	Absent	-	-	-	-
May/10	Colônia Dom Pedro II	Absent	Absent	-	-	-	-
Jan/08	Colônia Antônio Rebouças	Absent	Absent	6.9	5	< 0.2	0.82
Jul/08	Colônia Antônio Rebouças	Absent	Absent	6.9	3.1	0.3	0.89
Jan/09	Colônia Antônio Rebouças	Absent	Absent	6.8	4.4	< 0.2	1
Jul/09	Colônia Antônio Rebouças	Absent	Absent	6.8	4.1	0.24	0.8

Note: Coliforms in MPN/100ml; turbidity in NTU; chlorine and fluorine in mg/l.
Source: Health Inspection Agency of Campo Largo (2010)

In Araucária, the water supply well of Colônia Cristina presented a minimal level of chlorine in 21% of the samples analyzed, while chlorine was not detected in the other 79% and there was also no fluoridation of the water. Ordinance 518/04 of the Ministry of Health recommends a minimum level of 0.5mg/L of free residual chlorine after disinfection and 0.2 mg/L at any point of the distribution network. The analyses of these WSS are not conducted at only one point, but at random points throughout the network. Despite the fact that the well is believed to be chlorinated, residuals were not detected in the analysis. The municipality of Araucária requires that chlorine is in fact added to the WSS that supplies Colônia Cristina; however, the responsibility for adding the chlorine is that of the community and there has only been training by the Health Inspection Agency. We know that the population of the community does not agree with the addition of chlorine to the system which, according to farmers, gives a taste to the water. For this reason, the community prefers the private well and rejects the chlorination of the collective system.

In Campo Largo, a lack of monitoring was observed from 2007 to 2008 for the community Dom Pedro II. Monitoring resumed in 2009 but with only one analysis done in February. The well is properly disinfected and fluoride is added to the water. The analysis was done right after the disinfection and an analysis of the network was not done. In 2010, the Health Inspection Agency made another inspection, but only for assessing coliform, thus hindering a well-grounded discussion about the water quality of the WSS in question.

In Colônia Antônio Rebouças, the WSS is the responsibility of SANEPAR which conducts a semiannual analysis of metals and pesticides. Residual chlorine analyses are performed daily, while the parameters of color, turbidity, pH and total coliforms are analyzed monthly, as determined by Ordinance 518/04 of the Ministry of Health.

It should be mentioned that the WSS operated by the municipalities of Araucária and Campo Largo do not have the minimum number of samples nor the sampling frequency required by Ordinance 518/04 (Table 8).

Due to the high number of parameters used by SANEPAR, our project set out to analyze some metals, pesticides and various physico-chemical parameters in the WSS of Colônia Cristina and Colônia Dom Pedro II, as explained in Section 2. The collection was done in May 2010 with the participation of the Health Inspection Agencies of Araucária and Campo Largo. The results, presented in Table 9, correspond to what is defined in the legislation.

3.2 DIAGNOSIS OF THE RURAL SANITATION

3.2.1 Water

The isolated systems are essentially composed of excavated wells for personal consumption, especially for primary consumption, and the water is sometimes used for cleaning, irrigation, animal consumption and spraying.

The wells are excavated into the soil, lined with brick and concrete (Figure 7), have a diameter not bigger than

TABLE 8 – MINIMUM NUMBER OF SAMPLES AND MINIMUM SAMPLING FREQUENCY FOR WATER QUALITY CONTROL OF ALTERNATIVE SUPPLY SYSTEMS

PARAMETER	TYPE OF SOURCE	SAMPLES FROM THE OUTPUT OF TREATMENT	SAMPLES FROM POINT OF CONSUMPTION (1) (2)	SAMPLING FREQUENCY
Color, turbidity, pH and total coliforms	Superficial	1	1	Weekly
	Underground	1	1	Monthly
Free Residual Chlorine	Superficial or Underground	1	1	Daily

Note: (1) Per 500 inhabitants; (2) Samples must be taken from a minimum of 3 points of consumption
SOURCE: Brasil (2004)

TABLE 9 – WATER QUALITY DATA OF THE WSS OF COLÔNIA CRISTINA AND DOM PEDRO II

PHYSICAL PARAMETERS	C. CRISTINA (1)	D. PEDRO II (2)	ORDINANCE 518/04
Appearance "In natura"	Clear	Clear	-
Apparent Color (HU)	1.0	<1.0	15.0
Odor	-	-	Unobjectionable
pH	6.1	6.8	6.0 – 9.5
Turbidity (UT)	1.0	< 1.0	5.0
Electrical Conductivity (µS/cm)	71.9	282	-
Chemical Parameters			
Chloride (mg/L)	7.0	13.0	250.0
Fluoride (mg/L)	0.71	<0.1	1.5
Nitrates (mg/L)	0.10	3.6	10.0
Total Dissolved Solids (mg/L)	65.0	189.0	1000.0
Arsenic (mg/L)	< 0.01	<0.01	0.01
Cadmium (mg/L)	0.002	0.002	0.005
Lead (mg/L)	< 0.005	<0.005	0.01
Iron (mg/L)	0.1	<0.05	0.3

NOTE: (1) Water collected on August 4, 2010; (2) Water collected on April 7, 2010.

1.5m and with a depth that varies widely but is on average 16m, even in nearby properties. With the exception of one property, in which the well is not yet used, all others have an electric pump for the withdrawal of water and were constructed more than 10 years ago.

Two springs used for water consumption were found. Both were well protected and very far from the collection point.

FIGURE 7 – TYPICAL EXAMPLE OF THE WELLS FOUND AND ANALYZED IN THE RIO VERDE BASIN

Water samples were collected from the properties whose owners were willing to receive the project researchers. Thirty-nine wells and two springs were analyzed from 31 properties. The results, shown in Table 10, reveal the low water quality in most wells and some with high levels of total coliforms and *E. coli*.

The presence of coliforms confirms what was observed in the field; the sources are not being protected and the water is being significantly contaminated. Levels of *E. coli,* such as those presented in the collected samples, may cause problems to public health because they indicate the presence of feces in the water, which is a source of contamination for humans. This data can be correlated with information about diseases, such as diarrhea, for evaluating the actual sanitary conditions in the rural region of the Basin.

In an overall assessment of the water situation of the Individual Alternative Systems of the Basin (Figure 8), we notice that only the F3 well meets the standards of potability in the microbiological analyses. The other wells require treatment and disinfection to meet the standard of an absence of coliforms.

Despite the proven quality of the water of the collective water supply system, for the most part rural users prefer to use the water from private (isolated) wells for primary consumption and in some cases they use the collective system for secondary use, as shown in Figure 9. Furthermore, the rural population considers water as an abundant resource of high quality but disinfection is not yet used unanimously. Even the 48% that adheres to this process do not do so regularly.

From the responses to the questions raised in the structured questionnaire, we observed that the rural population considers the water they consume to be of good or high quality, with sufficient levels of availability or enough for residential and agricultural use, for the most part. When the water supply is through individual wells, disinfection is performed in 25% of the properties, usually with hypochlorite or chlorine added to the well.

The percentage of community members (75%) that do not disinfect the water that they consume is similar to the percentage of the samples analyzed that show *E. coli* (62%) in the water. This is one of the factors that contributes to the poor quality of individual systems. However, this section of the population disagrees with the use of chlorine.

3.2.2 Domestic Sewage

In the studied area, sewage from rural residences is comprised of grey and black water, when the toilet is inside the residence (95% of the studied properties).

Septic tanks, the system commonly used in less dense regions, was the main system of domestic sewage treatment found in the studied rural region. However, it is difficult for the local population to differentiate septic tanks from cesspits (dead well) which is a hole in the soil that is coated with concrete and closed, in which all generated sewage is released. In some properties there are two cesspits, one that receives the gray water and the other the black water. Figure 10 demonstrates the destination of mostly domestic sewage that comes from the studied residences.

The cleaning of the filled up well/cesspit is done using a cleaning truck, but there is no control or time frame. In most cases, it is concluded that the well has filled up when the sewage overflows the concrete surface.

The distance between cesspits and the water supply wells varied greatly in each property but overall they were located at least 10 meters apart. Nevertheless, domestic sewage cannot be discounted as a possible source of contamination of the wells for water consumption given the presence of coliform bacteria in most samples.

3.2.3 Solid Waste

In the results of the questionnaire, we noted a lack of knowledge related to alternatives of waste disposal, particularly for animal manure. Biodigesters were not found on the properties studied and in most cases the farmers did not have information on this technology. In the majority of cases, manure is released into the ground as fertilizer for small vegetable gardens or left in the pasture as there is a low density of animals on large areas not used for agriculture.

In the studied area, the public collection of recyclable waste is not carried out with regularity. Approximately 84% of the researched population delivers its common waste to public collection sites, while the rest burns or buries such waste. A portion of the farmers that burn or bury their common waste attributes this behavior to the lack of regularity of the collecting system.

The organic waste generated in rural residences are either collected or composted for later use. Some producers choose less common alternatives, such as burying or releasing the organic waste into the soil/river. Common waste is generally intended for public collection and in most cases is recycled prior to collection.

In rural properties, manure and pesticide package waste are generated. Animal manures are divided into bovine, equine, swine, ovine and poultry, and few properties do not have animals. In 45% of the properties, manure is not treated but rather left or released onto the soil. Releasing onto the soil is the most common destination of animal waste, with or without treatment. Many of the studied properties compost or use dunghills but no biodigester was found. Data from the treatment and destination of animal manures can be seen in Figure 12.

Most producers do not connect water quality with the presence of animals near the well. A significant proportion of the properties possesses poultry and dogs in large quantities and do not protect the well from possible water contamination from the manure of these animals.

Pesticide packages, when they are used, are always returned to the commercial establishment from which the

TABLE 10 – RESULTS OF THE MICROBIOLOGICAL PARAMETERS OF WATER QUALITY IN INDIVIDUAL SUPPLY SYSTEMS

COLLECTION POINT	DATE	TOTAL COLIFORMS MPN/100ml	E. COLI MPN/100ML
CC1a	Sept/09	1	Absent
CC1b	Sept/09	135	Absent
CC2a	Sept/09	> 2419	53
CC2b	Sept/09	816	11
CC3a	Sept/09	85	1
CC3b	Sept/09	770	180
CC4a	Sept/09	345	2
CC4b	Sept/09	345	Absent
CC5a	Oct/09	70	Absent
CC5b	Oct/09	1553	58
CC6	Oct/09	> 2419	6
CC7	Oct/09	1300	5
CC8	Nov/09	> 2419	1733
CC9	Nov/09	> 2419	1
CC10	Nov/09	649	29
CC11	Nov/09	53	Absent
CC12	Nov/09	1120	34
CC13a	Nov/09	13	Absent
CC13b	Nov/09	1553	Absent
CC14	Dec/09	1203	231
CC15	Dec/09	236	Absent
CC16	Feb/10	2419	1
CC17	Feb/10	> 2419	83
CC18a	Feb/10	816	2
CC18b	Feb/10	2419	Absent
F1	Mar/10	>23	>23
F2	Mar/10	9	1,1
F3	Mar/10	Absent	Absent
F4	Mar/10	9	7
F5a	Apr/10	187	1
F5b	Apr/10	2420	99
F6a	Apr/10	17	Absent
F6b	Apr/10	9	Absent
F7	Apr/10	291	4
F8	Apr/10	> 23	> 23
F9	Apr/10	435	29
F10	Apr/10	2	Absent
F11a	Apr/10	1	Absent
F11b	Apr/10	1300	21
F12	Apr/10	>2419	Absent
F13	Apr/10	687	1

NOTE: CC: Colônia Cristina; F: Figueiredo.

product was acquired. This happens because of the Federal Decree put in place in 2002 and instructions on the label of the product are observed. The farmers that use pesticides follow the recommendations and claim that there is supervision regarding the disposal of packages. Although they do not attribute the contamination of the Rio Verde to the use of pesticides, the interviewed farmers stated that they are aware that the excessive use of these products can harm the environment, including the water that they drink.

Among the questions raised by the PRA, waste disposal was discussed at length, especially the wastes from tourism near the reservoir. In constructing historical maps in the community of Colônia Cristina, it shows the lack of monitoring on the shores of the lake, causing uneasiness and insecurity among the local population regarding the wastes generated there.

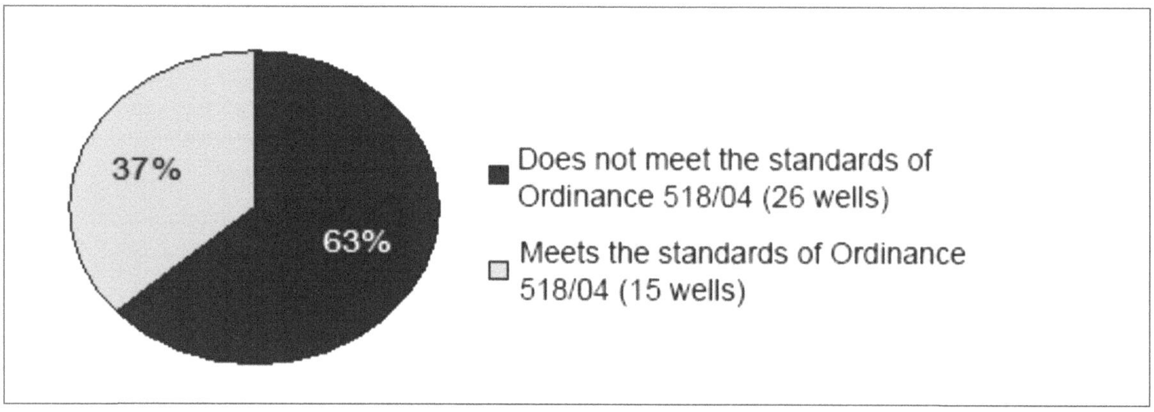

FIGURE 8 – RESULTS OF THE ANALYSES OF E. coli IN INDIVIDUAL SUPPLY SYSTEMS IN THE RIO VERDE BASIN

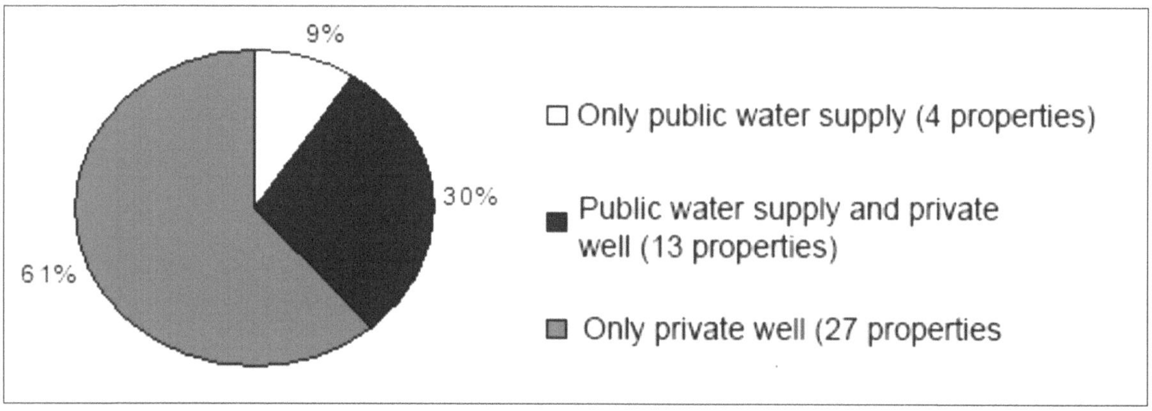

FIGURE 9 – WATER SUPPLY WHERE THE FAMILY LIVES

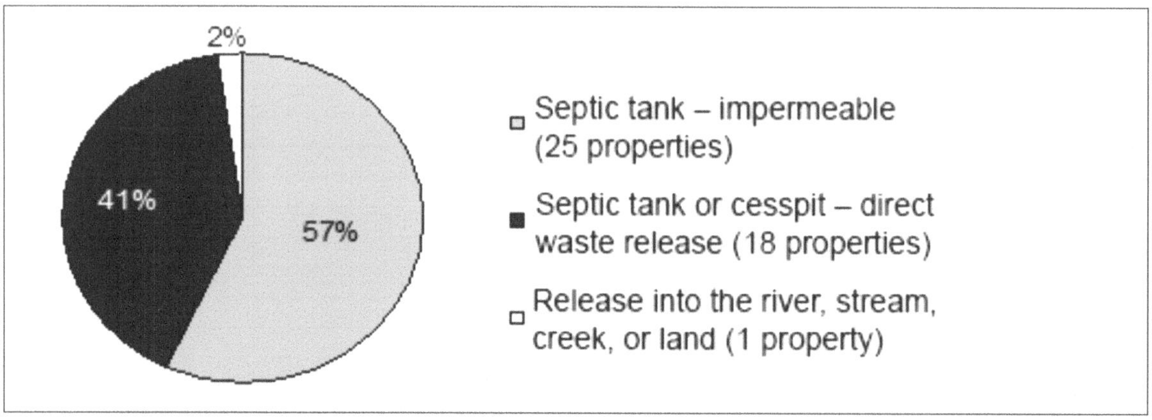

FIGURE 10 – DESTINATION OF WASTE (SEWAGE) OF THE HOUSE WHERE THE FAMILY LIVES

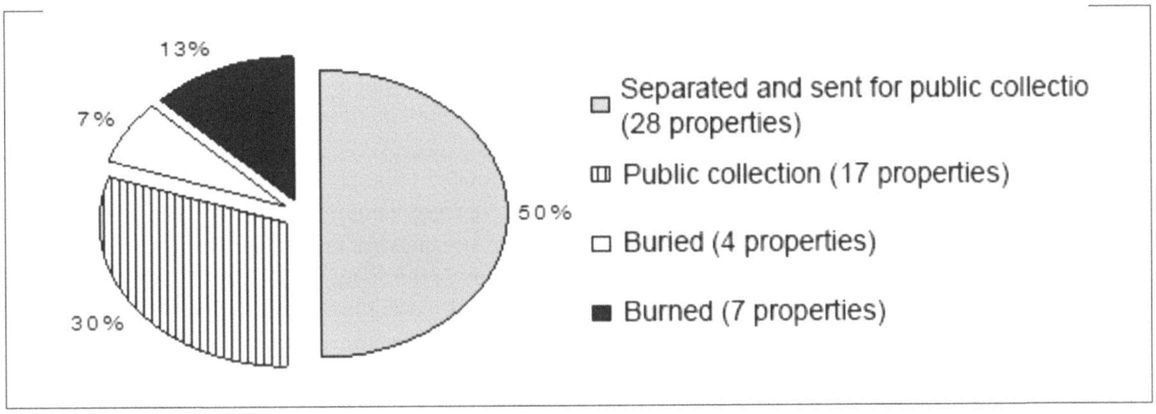

FIGURE 11 – DESTINATION OF ORGANIC WASTE PRODUCED ON THE PROPERTY

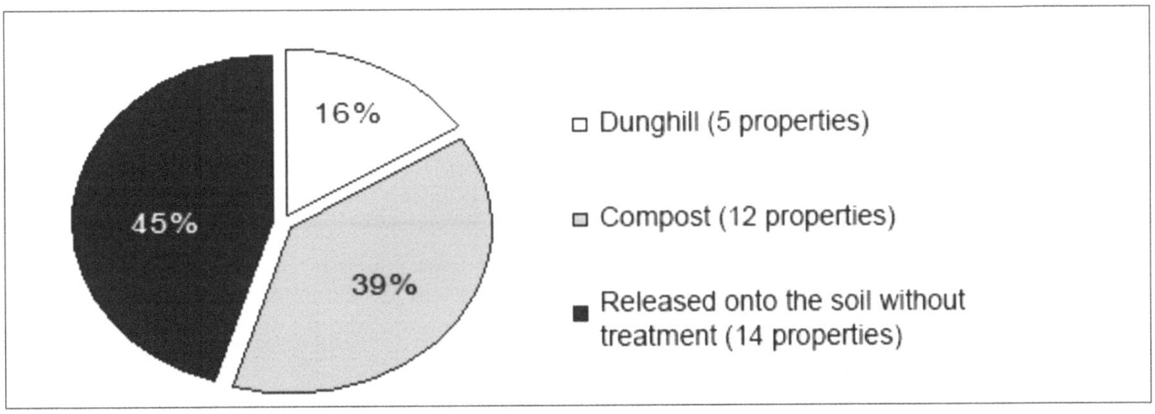

FIGURE 12 – TREATMENT AND FINAL DISPOSAL OF GENERATED MANURE

3.2.4 Waterborne Diseases

In the municipality of Araucária, which is inside the drainage basin, there are data on waterborne diseases (Table 11). Paradoxically there is a greater percentage of diarrhea cases in Colônia Cristina than in Faxinal, where, according to project data, there is no WSS. This is due to the fact that the rural population from Colônia Cristina consumes water from their own wells, in most cases without disinfection, and ignores the collective system with water that falls within the standards of potability.

In Campo Largo (Table 12), the lack of data regarding public health impedes a more thorough discussion but it can be noted that in Figueireido there were no cases of

TABLE 11 – RIO VERDE BASIN: CASES OF DIARRHEA IN THE COMMUNITIES INSIDE THE DRAINAGE BASIN IN ARAUCÁRIA, PR

COMMUNITY	YEAR	POPULATION	DIARRHEA CASES	%
Colônia Cristina	2006	444	40	9.01
Colônia Cristina	2007	440	33	7.5
Colônia Cristina	2008	428	33	7.71
Faxinal do Tanque	2006	76	1	1.32
Faxinal do Tanque	2007	65	3	4.62
Faxinal do Tanque	2008	64	2	3.12

Source: Health Inspection Agency of Araucária (2009).

TABLE 12 – CASES OF DIARRHEA IN THE COMMUNITIES INSIDE THE DRAINAGE BASIN IN CAMPO LARGO

COMMUNITY	YEAR	CASES OF DIARRHEA
Colônia Figueiredo	2008	52
Colônia Figueiredo	2009	0
Colônia A. Rebouças	2008	0
Colônia A. Rebouças	2009	25
Colônia Dom Pedro II	2008	67
Colônia Dom Pedro II	2009	155

Source: Health Inspection Agency of Campo Largo (2010).

diarrhea in 2009, unlike the communities that have chlorinated collective water supply systems. In Colônia Dom Pedro II, the cases of diarrhea increased substantially from 2008 to 2009, a number that was already high.

These data are an important tool for the Health Inspection Agencies of the municipalities in order to improve the sanitation in the more fragile communities that do not have collective systems and present high numbers of diarrhea cases. Such information can also help improve the education of the population that has access to the water supply network but does not use it.

3.3 ALTERNATIVES TO MINIMIZING THE POLLUTION AND CONTAMINATION OF THE WATER SUPPLY

Based on the results of the water analysis of the isolated supply systems, or individual alternative solutions, which show high levels of contamination, general proposed solutions can be established regardless of the region or studied community and no specific measures are recommended for each case.

From what was observed in the field, greater care with animal manures and the presence of chickens and dogs near the water supply can be considered a way of minimizing contamination. For the protection of wells, it is recommended that animals are fenced-off at least 30 meters from the well or the well is fenced to avoid the presence of animals within 30m.

Collected animal manures can be processed in composters or in biodigesters or even in dunghills, as long as they are located at a sufficient distance from the water supply well.

No specific analysis was made of each treatment system or final disposal of sewage, especially due to lack of access, information and knowledge of the proprietors about septic tanks and dry cesspits. However, a more thorough discussion with property owners is recommended regarding the systems they use, keeping cesspits 15 meters from the well and in a lower terrain that is not subject to flooding. A mound of earth can also be constructed around the pit in order to divert rainwater.

We also suggest the planting of trees around the fenced area of the well in order to avoid erosion and the construction of a brick structure with cement mortar around the well to keep it protected.

Even when taking care of the well and its potential sources of contamination, other precautions can be taken directly with the water. Common and rudimentary means of disinfecting the water in rural areas tend to be very efficient:

- The use of domestic filters that retain bacteria, particles and other substances in porous filters;
- Boiling the water for 15 to 20 minutes to destroy pathogens;
- A drop of 2.5% liquid sodium hypochlorite (bleach) in each Liter of water, that is well mixed and left for 15 minutes before consumption;
- Maintaining containers clean and sealed.

In the rural environment, some treatment processes are unusual when comparing with treatment stations for water distribution in the urban environment. Filtration is one of the recommended processes for rural areas and is used for greater purification of the water. The most common filters used on rural properties are composed of gravel and sand.

Besides filtration, chlorination is recommended for water supply systems. In the Rio Verde Basin, all collective systems use chlorine as a disinfectant but the population, even though they have access to such a network, chooses to consume water from the individual systems that in most cases do not have any type of disinfection. The recommendation is that the population consumes water from the collective systems, considering their superior quality. If there is no access to the network, disinfection can be achieved simply through the processes described above.

4. CONCLUSIONS AND RECOMMENDATIONS

Based on our study, we concluded that the WSS are made up of abstraction through wells, treatment/disinfection, pumping, reservation, distribution networks and individual household connections. The administration/operation of the WSS is performed by SANEPAR in one of the three wells located in the municipalities of Araucária and Campo Largo, and the others are responsibility of the Municipality.

The analyses of supply system water quality reveal in general the failure to comply with some parameters of Ordinance 518/04 of the Ministry of Health (for example, residual chlorine), as well as irregular monitoring in some communities carried out by the Municipality. We recommend that better planning and greater frequency of analyses of the WSS and the network is put in place and if possible we recommend an increase in the range of parameters studied, which is currently low.

The recommendation regarding the sewage and waste is that further information is made available to farmers, especially by proposing alternatives to final waste disposal and thus avoiding water contamination.

Considering the results obtained in the analyses of the isolated wells, it is of extreme importance to public health that the rural residents seek alternatives for water disinfection, protect the water sources and perform continuous analyses of the water in order for the water to comply with the standards of potability.

5. REFERENCES

ABNT. *Projetos, Construção e operação de tanques sépticos*. Rio de Janeiro, ABNT, NBR 7229/1993.

AISSE, M. M; OBLADEN, N. L. (1982) *Tratamento de Esgotos por Biodigestão Anaeróbia*. Curitiba: CNPq, ITAH/IPPUC, PUCPR.

AISSE, M. M. (2000) *Sistemas Econômicos de Tratamento de Esgotos Sanitários*. Rio de Janeiro, ABES, 19 p.

AISSE, M. M.; LOBATO, M. B.; JURGENSEN, D. (2005) *Desinfecção de Efluentes Utilizando Hipoclorito Gerado "in-loco"*. Curitiba, ABES, 13 p.

ALVAREZ, V. M. P. *A Companhia de Saneamento do Paraná: Estratégias Empresariais, Políticas Públicas e Mudanças Organizacionais*. UFPR, Sd.

ARAUCÁRIA, Prefeitura Municipal. (2009) *Dados do Município*.

ANDREOLI, C. V.; CARNEIRO, C. (Ed.). (2005) *Gestão Integrada de Mananciais de Abastecimento Eutrofizados*. Curitiba, SANEPAR, FINEP, 500p.

ANDREOLI, C. V. (Ed.). (2003) *Mananciais de Abastecimento: Planejamento e Gestão: Estudo de Caso do Altíssimo Iguaçu*. Curitiba, SANEPAR, FINEP, 494 p.

ASR. (2008) *Comunicação pessoal*. Curitiba, SANEPAR.

BARRETO, G. B. (1984) *Noções de Saneamento Rural*. Campinas, Instituto Campineiro de Ensino Agrícola, 2. ed., 54 p.

BRASIL. (2002) Decreto n.° 4.074, de 4 de janeiro de 2002. *Diário Oficial da República Federativa do Brasil*, January 08, Seção 1, p. 1.

BRASIL. (2004) Ministério da Saúde. Portaria 518. *Diário Oficial da República Federativa do Brasil n.° 59*, March 26, Seção 1, p. 266-270.

BRASIL. (2007) Casa Civil. Lei n.° 11.445, de 5 de janeiro de 2007. *Diário Oficial da República Federativa do Brasil* n. 5, January 8, Seção 1, p. 3.

CAMPO LARGO, Prefeitura Municipal. (2004) *Plano Diretor de Desenvolvimento Integrado do Município de Campo Largo – Paraná*. Campo Largo, FUNPAR.

CAMPO MAGRO, Prefeitura Municipal. (2005) *Plano Diretor de Desenvolvimento Integrado do Município de Campo Magro – PR*. Campo Magro, Ambientec – Consultoria Ambiental Ltda.

CHERNICHARO, C. A. L. (2007) *Reatores Anaeróbios. Princípios do Tratamento Biológico de Águas Residuárias*. Volume 5. 2. ed. Belo Horizonte: Departamento de Engenharia Sanitária e Ambiental – UFMG.

CONSILIU. (2002) *Macrozoneamento Ecológico-Econômico da APA Estadual do Rio Verde*. Curitiba, COMEC.

COORDENAÇÃO DA REGIÃO METROPOLITANA DE CURITIBA. (2002) *Plano Diretor Integrado* – PDI. COMEC.

COORDENAÇÃO DA REGIÃO METROPOLITANA DE CURITIBA. (2009) *Municípios da Região Metropolitana de Curitiba*. COMEC.

DANIEL, L. A. (2001) *Processos de Desinfecção e Desinfetantes Alternativos na Produção de Água Potável*. São Carlos, PROSAB.

DECRETO n.° 4.074, de 4 de janeiro de 2002. Regulamenta a Lei n.° 7.802, de 11 de julho de 1989, que dispõe sobre a pesquisa a experimentação, a produção, a embalagem e rotulagem, o transporte, o armazenamento, a comercialização, a propaganda comercial, a utilização, a importação, a exportação, o destino final dos resíduos e embalagens, o registro, a classificação, o controle, a inspeção e a fiscalização de agrotóxicos, seus componentes e afins, e dá outras providências. (Regulates Law 7.802, of July 11, 1989, which provides for research experimentation, production, packaging and labeling, transportation, storage, marketing, commercial advertising, use, import, export, the final destination of the waste and packaging, registration, classification, control, inspection and surveillance of pesticides, their components and related products, and other measures) Available in: <http://www.pr.gov.br/seab/agrotoxico/legislacao.html>. Accessed in: July, 2009.

DI BERNARDO, L.; DI BERNARDO, A. D. (2005) *Métodos e Técnicas de Tratamento de Água*. São Carlos, RIMA, v.1.

DORIGON, E. B.; STOLBERG, J.; PERDOMO, C.C. (2008) Qualidade da Água em uma Microbacia de uso Agrícola e Urbano em Xanxerê – SC. *Revista Ciências Ambientais*, Canoas, v.2, n. 2, p. 105 a 120.

EMATER. (2006) *MEXPAR – Metodologia Participativa de Extensão Rural para o Desenvolvimento Sustentável*. Belo Horizonte, EMATER MG. 134 p.

FERNANDES, F.; SILVA, S. M. C. P. (1999) *Manual Prático para Compostagem de Biossólidos*. Edição FINEP – PROSAB, Rio de Janeiro.

FUNASA. (2006) *Manual de Saneamento*. Brasília, FUNASA, 409 p.

GASPAR, R. M. B. L. (2003) *Utilização de Biodigestores em Pequenas e Médias Propriedades Rurais, com Ênfase na Agregação de Valor: Um Estudo de Caso na Região de Toledo – PR*. UFSC – PPGEP.

GIL, A. C. (2002) *Como Elaborar Projetos de Pesquisa*. São Paulo, Atlas, 175p.

GONZÁLEZ, R. G; TAYLOR, M. L; ALFARO, G. (1982) Estudio Bacteriologico del Agua de Consumo en una Comunidad Mexicana. *Bol Of Sanit Panam 93*(2).

GUSMÃO, P. T. R.; OLIVEIRA, J. W. S.; SANTOS, D. L. S. (2010) *Filtros Domésticos: Avaliação de Eficácia e Eficiência na Redução de Agentes Patogênicos*. 3° Caderno de pesquisa de engenharia de saúde pública / Fundação Nacional de Saúde, Brasília, p. 89-120.

HELLER, L. (1997) *Saneamento e Saúde*. Brasília, OPAS/OMS, 97p.

HIDROGERON DO BRASIL. *Tecnologias Avançadas em Tratamento de Água e Efluente*. Gerador de Cloro Estático. Accessed in October, 2009 at <http://www.hidrogerondobrasil.com.br/>

IBGE – Instituto Brasileiro de Geografia e Estatística. *Censo Demográfico 2000*. IBGE: Rio de Janeiro.

IBGE – Instituto Brasileiro de Geografia e Estatística. (2005) *Pesquisa de Informações Básicas Municipais – Perfil dos Municípios Brasileiros – Meio Ambiente*. Rio de Janeiro, 382p.

IBGE – Instituto Brasileiro de Geografia e Estatística. (2007) *Séries Histórias e Séries Estatísticas – Características Gerais dos Domicílios*.

IPARDES. Instituto Paranaense de Desenvolvimento Econômico e Social. (2009) *Informações Municipais*. Secretaria de Estado do Planejamento e Coordenação Geral.

KUNZ, A.; HIGARASHI, M. M.; OLIVEIRA, P. A. (2005) Tecnologias de Manejo e Tratamento de Dejetos de Suínos Estudadas no Brasil. *Cadernos de Ciência & Tecnologia*, Brasília, v. 22, n. 3. p. 651-65.

LARSEN, D. (2010) *Diagnóstico do Saneamento Rural Através de Metodologia Participativa*. Estudo de Caso: Bacia Contribuinte ao reservatório do Rio Verde, Região Metropolitana de Curitiba, PR. Universidade Federal do Paraná, Curitiba.

Lei Ordinária de Araucária – PR, n° 1547 de 14 de janeiro de 2005. Dispõe Sobre a Reestruturação da Organização Básica da Prefeitura do Município de Araucária, Revoga a Lei 1.101/97, os Artigos 10 e 20 da Lei 1.207/01, os Artigos 10, 20, Parágrafo Único do Artigo 30 e Tabela C do Anexo I, da Lei 1.304/02, e dá Outras Providências. (Features About Restructuring the Basic Organization of the Municipality of Araucaria, Repeals Law 1.101/97, the Article 10 and 20 of Law 1.207/01, Articles 10, 20, Sole Paragraph of Article 30 and Table C of Annex I, Law 1.304/02, and gives Other Arrangements). Available in <http://www.leismunicipais.com.br/cgi-local/forpgs/form_vig.pl>. Accessed in March, 2010.

MACHADO, E. S. (1998) *Introdução à História da Gestão de Recursos Hídricos no Estado do Paraná*. Simpósio Internacional sobre Gestão de Recursos Hídricos, Gramado, RS.

MARQUES, P. M.; CALZAVARA, O.; GUIMARÃES, M. F. (2001) O Projeto "Paraná 12 Meses" em um Contexto de Desenvolvimento Rural Sustentável. *Acta Scientiarum*, Maringá, 23(1): 257-261.

Ministério da Saúde. *De olho nas Metas do Milênio*. Accessed in July, 2009 no site http://portal.saude.gov.br/

OLIVEIRA, A. L. S.; FERNANDEZ, J. C. (2004) *Análise da Eficiência do Setor de Saneamento Básico no Brasil*. Fórum Banco do Nordeste de Desenvolvimento – IX Encontro Regional de Economia da ANPEC.

PARANÁ. Secretaria de Meio Ambiente e Recursos Hídricos. *Lei n° 12.493*, de 22 de janeiro de 1999. Diário Oficial do Estado do Paraná. n.° 5430 de 05 de fevereiro de 1999.

PATERNIANI, J. E. S.; CONCEIÇÃO, C. H. Z. (2004) *Eficiência da Pré-Filtração e Filtração Lenta no Tratamento de Água para Pequenas Comunidades*. Engenharia Ambiental, Espírito Santo do Pinhal, v.1, n.1, p. 017-024, jan./dez.

PHILIPPI, L. S. (2000) *Saneamento Descentralizado: Instrumento para o Desenvolvimento Sustentável*. IX SILUBESA – Simpósio Luso-brasileiro de Engenharia Sanitária e Ambiental.

PILATTI, F.; HINSCHING, M. A. O. (2008) *Saneamento Básico Rural na Bacia Hidrográfica do Manancial Alagados*. Ponta Grossa, PR: UEPG/SANEPAR.

REZENDE, S. C. (2002) *Consequências das Migrações Internas nas Políticas de Saneamento no Brasil: uma Avaliação Critica do PLANASA*. XIII Encontro da Associação Brasileira de Estudos Populacionais. Ouro Preto, MG.

SANEPAR[1]. *Saneamento Rural*. Programa de Atuação Social. SANEPAR, USPO. Sd.

SANEPAR[2]. *Projeto Unifamiliar – Construção, Operação e Manutenção das Fossas Sépticas*. SANEPAR, Sd.

SANEPAR. *Histórico da Empresa*. Accessed in September, 2009 at <www.sanepar.com.br>.

SANTOS, L. A. C. (1985) O Pensamento Sanitarista na Primeira República: Uma Ideologia de Construção da Nacionalidade. *Revista de Ciências Sociais*, Rio de Janeiro, v.28, p 123-210.

SANTOS, M. A. A.; SCHMIDT, V. C. B.; MAROSO, M. T. D. (2007) Esterqueiras: Avaliação Físico-Química e Microbiológica do Dejeto Suíno Armazenado. In: V Simpósio Internacional de Qualidade Ambiental. Eng. Agric., Jaboticabal, 2007. *Anais*... Jaboticabal: ARS Veterinária. p. 537-543.

VERDEJO, M. E. (2006) *Participatory Rural Appraisal: Guia Prático PRA*. Brasília, MDA/ Secretaria da Agricultura Familiar. 62 p.

VERTRAG PLANEJAMENTO LTDA. (2007) *Plano Diretor de Araucária*. Prefeitura do Município de Araucária. Estado do Paraná.

VIGILÂNCIA SANITÁRIA. (2010) *Comunicação Pessoal*. P. M. de Araucária.

VIGILÂNCIA SANITÁRIA. (2010) *Comunicação Pessoal*. P. M. de Campo Largo.

CHAPTER 21

THE VOICE OF THE POPULATION IN COLÔNIA CRISTINA: A STUDY ON RISK PERCEPTION

José Edmilson de Souza-Lima
Sandra Mara Maciel-Lima
Lafaiete Santos Neves
Claudia Cristina Machado
Priscila Cazarin Braga

THE VOICE OF THE POPULATION IN COLÔNIA CRISTINA: A STUDY ON RISK PERCEPTION

José Edmilson de Souza-Lima
Sandra Mara Maciel-Lima
Lafaiete Santos Neves
Claudia Cristina Machado
Priscila Cazarin Braga

SUMMARY

This chapter aims to highlight and publicize the voices of some of the human inhabitants of Colônia Cristina regarding the interaction between the human and non-human world. Located in the Rio Verde Basin, west of the metropolitan region of Curitiba, this colony is located in part of the environmental protection area (APA) of the Rio Verde. The research presented herein is qualitative and data collection was conducted using Paulo Freire's culture circle technique. We conclude that in hearing the discomfort in the voices of some interviewees, we must never forget to ask: what is the place of human inhabitants in projects targeted at areas of potential risk? Listening carefully and respecting the voices of the inhabitants of high risk areas does not imply accepting *a priori* their views as ultimate or true in absolute terms, but these voices must be taken into account.

KEYWORDS

Risk perception, risk in society, public policy, culture circle, human voices.

1. INTRODUCTION

This chapter aims to highlight the voices of some of the inhabitants of Colônia Cristina, a community located in the Rio Verde Basin, west of the Metropolitan Region of Curitiba. This community is located in part of the environmental conservation area (APA) within the Rio Verde Basin.

A tributary of the Iguaçu River, the Verde River runs through part of the state of Paraná and could meet the water needs of the Metropolitan Region of Curitiba. The quality of the water in the river, however, has prompted concern from several corporate, academic and public representatives; a concern that peaked in early 2008 which resulted in the current study.

These waters are claimed by several sectors, causing disputes over sustainable use. Therefore, understanding the importance of the voices of the inhabitants in the River Basin, their perceptions of risk regarding not only the quality of water but also the quality of life in Rio Verde, are the purposes of this chapter.

In interdisciplinary projects already completed in the Metropolitan Region of Curitiba and within the scope of the ongoing project on the eutrophication of the Rio Verde Reservoir, there was indeed room for the realization and improvement of integrative research practices. However, from the experience of research on the Iraí dam, a region previously studied, and despite its success among researchers in the natural sciences, research on indicators of water quality, soil, fish, climatic conditions, among others, were privileged. The voices of local people and the inhabitants of the communities surrounding the Iraí dam, both propagators of problem and victims, were not considered.

In comparative terms, it seems, the study developed in the Rio Verde Basin in relation to the Iraí Dam[1] has a wider focus. There is a consensus among outside consultants, coordinators and researchers involved in the project on eutrophication of the Rio Verde Basin and this consensus is directly linked with the need to take into account not only physical and biological indicators, but also and especially the perceptions of its inhabitants.

This study discusses part of the research carried out within the Research Group "Epistemology, Society and Environment" (CNPq / FAE) and aims to identify social and environmental risk perception of the inhabitants of Colônia Cristina.

2. METHODOLOGICAL PROCEDURES

This study takes a qualitative research approach and operates from the theoretical perspective that "...meaning [is] attributed by the subjects to the facts, relationships, practices and social phenomena" (DESLANDES & ASSISI, 2002). It understands the environment as a direct source of data, attempts to understand the phenomenon from the perspective of the participants, and requires direct and prolonged contact of the researcher with the participants in the research. The qualitative approach explores the intensity of the phenomena which enables a deeper understanding of groups, segments, and micro-realities, aimed at unveiling their internal and specific logic, cosmology, and vision of certain problems which are expressed in opinions, beliefs, values, relationships, attitudes and practices (MINAYO & MINAYO-GOMEZ, 2003).

[1] Additional information on the research conducted on the eutrophication processes of the Irai Dam in the Metropolitan Region of Curitiba can be found in ANDREOLI & CARNEIRO (2005).

The qualitative method, "besides unveiling social processes still poorly known [...], allows the construction of new approaches, revising and creating new concepts and categories during the investigation" (MINAYO, 2006). It should be noted that we do not seek to understand social norms but rather to understand individuals and what led them to perform certain actions. This is only possible if the subjects are able to express their own logic and rationalization of their reasons. In this scenario, one can identify beliefs more or less shared by social groups that are not static but under constant transformation (GODOI & BALSINI, 2006).

According to Deslandes & Assisi (2002), the qualitative perspective "is not intended to interpret facts but to interpret the players' interpretation of the facts, practices and conceptions". It is acknowledged that this practice will never uncover everything that is hidden or expressed in the interviews; what is produced are interpretations of reality, seen not as "truth", but as versions of truths with scientific pretensions of reality. Deslandes & Assisi claim that interpretation in qualitative research is the basis of the research action: "It will be present throughout the process (during field work to build hypotheses, in the interaction between the researcher and the researched subjects) and is the essential exercise in analysis" (DESLANDES & ASSISI, 2002).

A case study was used as a research strategy to assist in understanding complex social phenomena and preserving significant features of real-life events (Yin, 2005). The unit of analysis is the group of residents of Colônia Cristina: farmers and representatives of civil society organizations that are located throughout the colony.

Formal invitations were sent through EMATER (Paraná's Institute of Technical Assistance and Rural Outreach) and participants were those who chose to participate in the survey (sample by adherence). Meetings with community were held on April 30, 2009, May 18, 2009, May 21, 2009 and June 06, 2009. An average of 25 participants attended each meeting. All encounters were filmed, with the consent of the participants.

Data collection was conducted in two phases:

a) The first was the culture circle

This research technique, the culture circle, enabled capturing more constitutive elements and the risk perceptions of the inhabitants. The major elements of the culture circles appear in the words of Freire (1983), their main proponent:

> Rather than the teacher with strong "donor" traditions, the debate coordinator. Rather than a discursive lecture, the dialogue. Rather than the student with passive traditions, a group participant (Freire, 1983, p. 103).

During the activities, our team separated the participants from of Colônia Cristina into three subgroups and suggested that each group draw a map that represented their perceptions about the past (sub-group 1), the present (sub-group 2), and the future (sub-group 3) of Colônia Cristina. Each sub-group received cardboard and pens and was asked to perform the task with the inauguration of the PETROBRAS dam in 1975 as the initial point of reference. Thus, point zero was the date of implementation of the dam and drawings and symbols were based on the dam (Figures 1, 2 and 3).

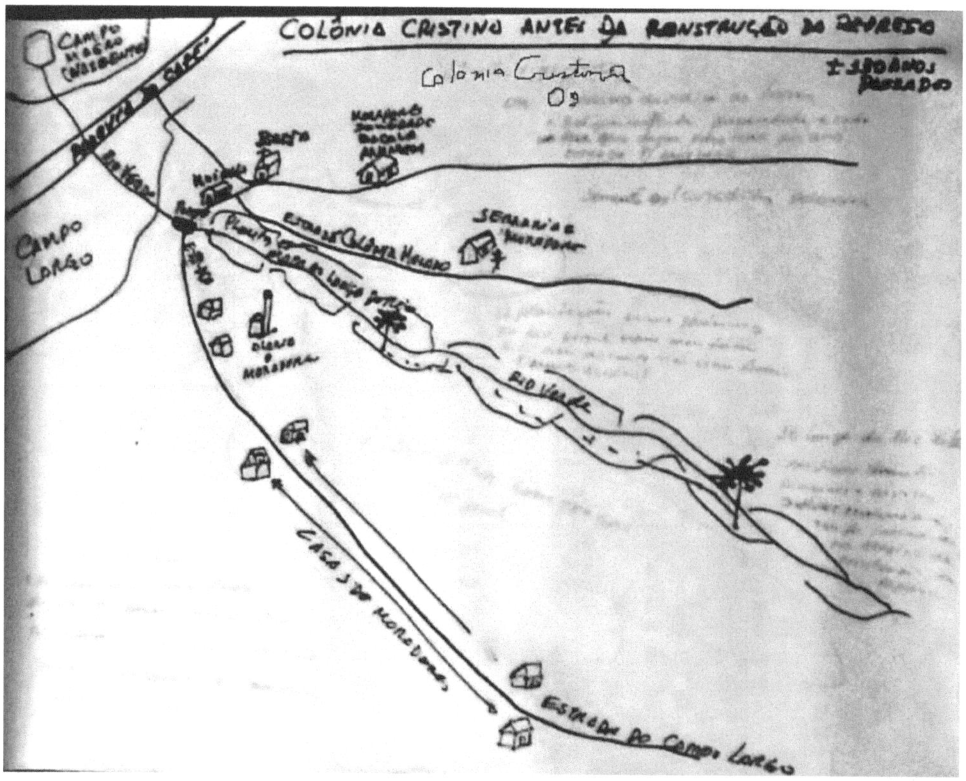

FIGURE 1 – REPRESENTATION OF COLÔNIA CRISTINA BEFORE DAM CONSTRUCTION AS DRAWN BY RESIDENTS
Source: Drawn by residents of Sub-group 1.

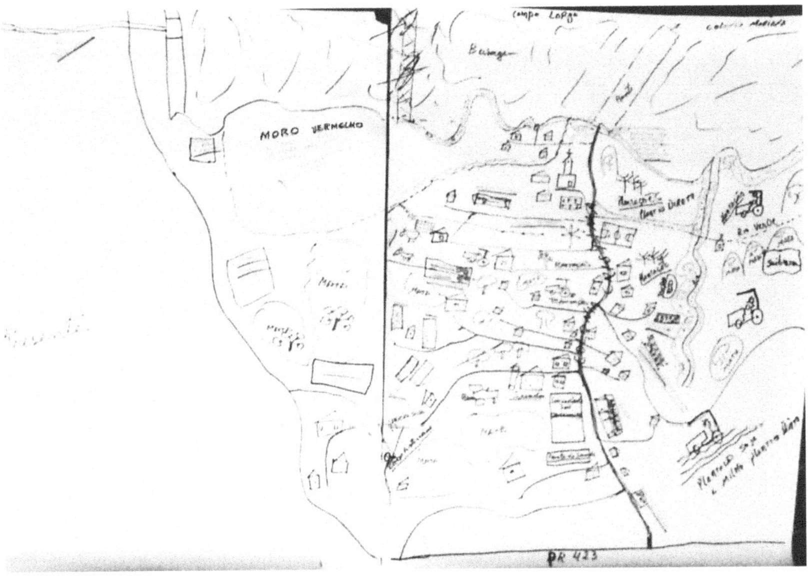

FIGURE 2 – REPRESENTATION OF PRESENT-DAY COLÔNIA CRISTINA AS DRAWN BY RESIDENTS
Source: Drawn by residents of Sub-group 2.

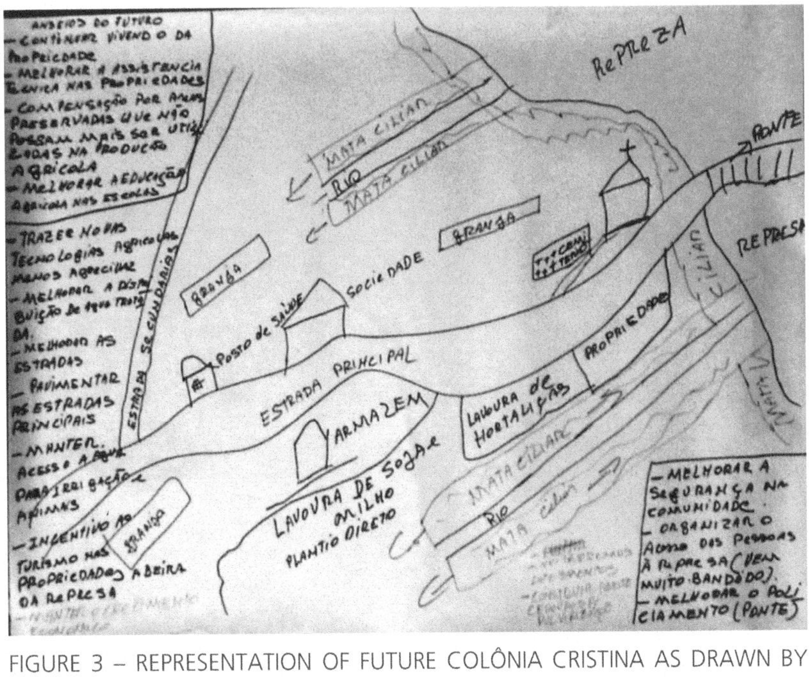

FIGURE 3 – REPRESENTATION OF FUTURE COLÔNIA CRISTINA AS DRAWN BY RESIDENTS
Source: Drawn by residents of Sub-group 3.

b) The second phase was research validation.

As agreed upon with the study participants and in order to validate the research, our team conducted a feedback session with the community to confirm that our data collection and interpretation corresponded to what the community actually thought. According to Yin (2005), this process is one of the principal elements used to validate qualitative research.

The feedback session is a process of negotiation during which the themes that the research team highlighted as important were thoroughly questioned and altered by the community. They also had the opportunity to add additional points that were not discussed during the culture circles.

Therefore, all the significant statements[2] (FREIRE, 1983) presented here are the result of feedback, as they embody the voices of the inhabitants of Colônia Cristina who chose to participate in the research.

[2] The idea of "significant statements" derives from Paulo Freire's work and it encompasses translating the deep meanings that are hidden or revealed not only through the words themselves, but especially through the spontaneous gestures of each interviewee.

3. CHARACTERISING THE BASIN: THE ENVIRONMENTAL PROTECTION AREA OF RIO VERDE AND COLÔNIA CRISTINA

A tributary of the Iguaçu River, Rio Verde is a Brazilian river that runs through part of Paraná and it has the capacity to meet the water needs of the Metropolitan Region of Curitiba.

According to Braga (2008), the Rio Verde Basin, is located in the western part of the Metropolitan Region of Curitiba (RMC) and it is located within the city limits of the three municipalities in the region: Campo Magro, Campo Largo and Araucária (see Figure 1 - Chapter 19).

The RMC has placed restrictions on the availability of water in order to meet projected demand in face of significant growth in the region. Motivated by concerns about improving and restoring the environment in the Rio Verde hydrographic basin and ensuring the potability of water for the population of the RMC, the Government of the State of Paraná decreed on July 31, 2000, that part of the Rio Verde Basin would be deemed an Environmental Protection Area (APA) of Rio Verde. The Rio Verde APA was thus established in the municipalities of Campo Largo and Araucária with 8.21% and 8.14% of their areas within the APA, respectively.

As stated in Paraná (2000) the Rio Verde APA was created by State Decree 75/2000, with the purpose of protecting and preserving the environmental systems that exist in the Basin, especially in relation to quality and quantity of water for public consumption.

According to Braga (2008), the municipality of Campo Magro, in which the springs of the Rio Verde are located, was not included in the macro-zoning because there was already zoning in place, called Territorial Planning Unit (UTP), which is not a conservation unit but does impose restrictions on land-use.

The summary of State Decree 2375 (PARANÁ, 2000) outlines restrictions, prohibitions, authorizations, responsibilities and other information of particular significance to the research that was conducted in Colônia Cristina, one of the twelve communities located within the Rio Verde APA.

Moreover, we highlight one of the 14 articles of this summary in particular because of its relevance to subsequent issues:

Art. 4° - Aiming to meet its objectives, the Rio Verde APA will have Ecological-Economic Zoning, which shall be prepared and approved within 180 (one hundred and eighty) days from the date of publication of this Decree (PARANÁ, 2000 n/p).

Located southeast of the Rio Verde APA, Colônia Cristina, which is the focus of this analysis, is one of 39 rural communities in the municipality of Araucária, as stated on the municipality's official website.

This colony was established in 1886 by Polish immigrants who settled initially on small plots of land. The community was initially called Colônia Santa Christina and it was part of the Parish of Iguaçu, which belonged to the municipality of Campo Largo. The colony became part of the municipality of Araucária in 1890. Corn, beans, vegetables, peaches and strawberries are the most important crops on their properties and in the region.

Polish cultural heritage is very present in Colônia Cristina and its surroundings. According to Brandenburg (1998) "the settler represents the foreign immigrant who freely settled in the south, on small lots or tracts, which was a process of occupation independent from larger-scale farming."

Many social and cultural practices with their origins in the beginning of the last century continue to be observed in Colônia Cristina, which seems to have a cultural heritage greater than the wealth produced on its land.

Thus, the context of the first waves of Polish immigration in Brazil and Paraná will be discussed together with the findings of the research for a deeper understanding of the cultural aspects that have been transmitted through generations. Further, we will discuss how these aspects transform the abstract theoretical perception of risk into concrete perceptions of risk in daily life.

4. PUBLICIZING THE VOICES OF "NON HUMANS"

The first question that emerges in this section is how to make public the 'voices' of non-human beings? Or, how can we make cyanobacteria, fish and riparian forest speak?

It is important to note that the voices of each of these inhabitants of the environment are mediated by the specialist or expert who is responsible for capturing and translating the information provided by of each of these inhabitants of the Basin. Despite the reservations that are made by specialists or experts[3], they are principally responsible for the growth of this new collective[4] consisting of both humans and non-humans.

The voices of non-human inhabitants can reveal the health of the Basin or trends toward situations of higher

[3] Although Sheldrake has not made direct reference to expert systems, as Giddens (1991) did, he records his distrust of them, in the following passage relativizing the supposed "objectivity" of science and expert systems that legitimate it: "many laymen are perplexed in the face of the power and apparent certainty of scientific knowledge. The same occurs with science students. Handbooks are filled with facts and supposedly unquestionable and quantitative data. Everything leads us to believe that science is highly objective. In fact, the belief in objectivity in science is an article of Faith for many people in the modern world" (SHELDRAKE, 1995, p.133).

[4] The notion of the collective is derived from Bruno Latour's approach (2004). It is a collective that brings together both humans and non-humans. For Latour, "the collective is first different from society, a term that makes us think in a poor distribution of power, then it accumulates the ancient powers of nature and society in a single place before differentiating them again into various levels of power. Although used in the singular, the term does not have the meaning of an already made unit, but of a procedure to gather the associations of humans and non-humans" (LATOUR, 2004, p. 372). Latour uses "this word only to signify a political philosophy where there are no more than two elements being attracted: one that would make the unit in the form of nature and the other that would keep the multiplicity in the form of societies. The collective means: everything and not two separate items" (LATOUR, 2004, p. 117).

risk. Thus, for example, the geochemical assessment of water quality, the effects of short-duration weather events on cyanobacteria blooms, the composition and biological aspects, and ichthyo-sanitary and histopathologic assessment of fish, create a space through which the voices, cries of despair or hope, of non-human inhabitants of the Basin can be heard.

In the case of the human inhabitants of the Basin, their voices can be interpreted by experts but they can also be publicized *in natura* or in their original form.

5. PUBLICIZING THE VOICES OF HUMANS

The goal herein is to present the interviews and the feedback given by the inhabitants of Colônia Cristina who participated in culture circles in order to make public their perceptions of risk.

5.1 IDENTIFICATION OF INHABITANT FEARS IN RELATION TO THE SOCIAL SYSTEM

One of the first reactions of the interviewed group was fear in relation to the new measures proposed in the forest code to reduce protected areas from 30 to 5 meters. Such a change appears to residents as a risk factor in their survival, as well as a major risk factor in processes of social reproduction. Moreover, it is a factor that is perceived as a risk coming from outside the Basin, derived from a wider social system.

Another reaction marked by fear was in relation to the requirement of the research that no one would be negatively affected by the research results. In several statements made by residents, phrases such as: "We would like, if possible, a written guarantee" were identified. The guarantee to which they refer is that the research conducted by our team would not pose obstacles to the various current and future forms of social reproduction of the community.

Fears about the future were evident when the community members made reference to planning. For them, "they do not have what they need to plan if there is no guarantee". This was one of the ways they expressed their fears and anxieties towards our research questions.

They also mentioned the flooded area caused by the dam: another problem coming from outside that significantly affected the community.

5.2 NOSTALGIA FOR THE PAST, REINFORCING IDENTITY AND TRADITION

The subgroup that was asked to represent the past (Figure 1) recorded images and very moving words: "people planted near the Rio Verde because of the humidity." They also made sure to explain that the most distant lands were more acidic and this prevented planting. There was preservation of the forest, despite the area needed for planting. Here the idea of conservation also appears.

Intertwined with nostalgia for the past appears one of their main fears: fear of foreign invaders. In the past, they emphasized and exalted that "the fishermen were just people living in the Colônia" and because of that there were plenty of fish and "no one was coming from outside to pollute the river."

For residents, the social fabric in the past was stronger because "the community was more united" and "people knew and visited each other." In addition, in the past there was more security because: "people went out at night and no one bothered"; "anyone left the house at two o'clock in the morning and was not afraid."

Residents believe that the dam is the main factor in the weakening of the social fabric as they are adamant that "with the coming of the dam the calm was over." The dam appears as the clear break between a virtuous and harmonious past and a present and future marked by fear and uncertainty. Box 1 summarizes the perceived nostalgia of the past as communicated by the interviewees.

BOX 1 – NOSTALGIA FOR THE PAST IN COLÔNIA CRISTINA: BEFORE THE DAM

"They all belonged to the family"
"We visited more"
"We walked at dawn"
"We did not have fear"
"The community was more united"
"The calm was over after the Dam"
"Corn was harvested ear by ear"
"Picherom – joint effort to harvest and a party at the end – a big dance"

Source: Prepared by the researchers with the guidance of the residents.

5.3 REJECTION OF THE CURRENT SITUATION

If the past was represented as a symbol of harmony, the drawing that represents the current situation (Figure 2) is the opposite. For residents, problems related to garbage, which were nonexistent before, emerges as one of the greatest concerns. This concern stems mainly from the belief that it "is not caused by the community, but by people who come from outside. Human garbage brought by fishermen and urban tourists, vandals and drug users." Everything seems to sum up to "undesired and dangerous visitors."

The idea of preservation, recorded in the drawings depicting the past, reappears in the drawings of the current situation as an indicator of what they are able to care for in their immediate environment: "around the river is all forest."

The dam is characterized as a "serious case, they [an unfriendly reference to Petrobras] took all the sand to make the dam." "Everything remains without vegetation to the present day." This area is called Red Hill.

For many people the bridge represents a symbol of progress and improvements for the entire community. For the farmers, the bridge symbolizes much of the misery of the community because "it brings major threats that put the community at risk." The bridge is represented as a gateway of the main factors that threaten the community. For example, "on Holidays and Sundays in the summer, no one can stand it there," "it is binge drinking (...). Walking on the bridge is too dangerous." Box 2 summarizes and reflects the mood of pessimism and an aversion to the current situation as expressed through the perceptions of interviewees.

BOX 2 – REJECTION OF THE CURRENT SITUATION IN COLÔNIA CRISTINA

- → **The water of the well could supply the entire Colônia, but piping is very expensive**
- → **Phone for some / Electricity 100%**
- → **Collection of organic waste and recycling**
- → **School buses and mass transportation**
- → **We have dances and entertainment of the Society**
- → **Health Center**
- → **Dirt roads that contaminate the Dam**
- → **Major problems: drugs, binge drinking, garbage, lack of security ["is very dangerous"]**
- → **Many fish and other animals/birds are always increasing in numbers**

Source: Prepared by the researchers with the guidance of the residents.

5.4 BETS ON THE FUTURE

If longing for the past and nostalgia reign while misery and misfortune prevail during the present, the future seems to be represented as the possibility of hope (Figure 3). However, it is a hope marked by fear of change that may jeopardize the traditional lifestyle of the community.

In this sense, for residents, "the perception of the future is that [the] community remains as is: planting, taking care of the farm." Their main aspirations are associated with the possibility of continuing to "go on living from the land." There seems to be no room for expectations to leave the environment in which they built their identities because one of the major goals is "to improve agricultural education in schools" in order to ensure their descendants can remain in the community.

The expectations in relation to the world outside the community is very close to what Giddens (1991) characterizes as an expert system, coordinated and conducted by professionals who can advise them in order to "improve the technical assistance in their properties." We must remember that an expert system is one in which its answers and certainties produces confidence, especially in technical rationality that sustains modern lifestyles. From this perspective, the community, which is shut-off from many innovations of modern life, is open to expert systems as they believe that expert advice would be useful and fundamental to the processes of social reproduction, as they can "bring new less aggressive agricultural technologies."

With regard to socioeconomic aspirations, it was possible to identify statements from dialogues that reveal these aspirations. The first is the acknowledgement that if wider society requires preserved areas, it should pay for it. Hence the need for "compensation for preserved areas that can no longer be used for agricultural production."

Coupled with the need for compensation, residents highlighted another clear need to "maintain economic growth" because they understand that this is the mainstay of the community. At the core of this market-oriented understanding, residents did not hide their aspirations which are directed at "encouraging tourism in properties on the edge of the dam."

As a complement, aspirations for the community were linked to the need to "improve the distribution of treated water," "improve roads", "paving the main roads", "maintain access to water for irrigation and animals" and finally "build irrigation ponds, because there is not enough water for both" came up in discussions.

In the fragments of the following significant speeches what may be called the fear of the foreign or fear of any type of invasion appears. In particular, the first invaders into the region are gated communities or subdivisions and they are perceived as follows: "We do not want subdivisions because they only bring problems." The subdivisions are inhabited by people foreign to the community in general, coming from urban areas.

As an antidote to the dramatic process of invasion, they claim that the authorities should: "improve safety in the community", "organize people's access to the dam (many thugs are coming)" and ultimately "improve policing (bridge)." Note that the prevailing idea is that if there was no invasion, the community would be happier. They indirectly link their levels of worry and anxiety to the arrival of outsiders who have no obligations to the ethnic community.

The drawing that subgroup 3 (Figure 3) created to represent the future expresses the desire of the community to keep it as it is, so that it does not change significantly. In the discussions collected "the desire of the group is that no drastic changes will take place" as one of the most prominent aspirations is "to continue living here with increased quality of life." This shows the aversion to change that compromises the longevity of the community.

However, the community is not radically opposed to any change. For them, "every project that comes to improve our lives, will be welcome." In this meaningful statement it is possible to identify indications of openness to new developments that are favorable and that do not jeopardize the conditions of social reproduction.

In the following significant dialogues it is possible to identify another very important association: that some of the initiatives introduced to the community, that are derived from expert systems, tend to reduce the uncertainty regarding the risks related to the use of pesticides on their crops. They understand that the already established practice of "annual collection of pesticide packages" is a good initiative because "each farmer has a place for storing packaging" and "this explains the reduction of pollution." For them, "once it was common, but today pesticide packages are not thrown just anywhere." Note that the guidance of expert systems in relation to the collection and "controlled" use of pesticides are basic elements in the understanding that there is no environmental risk if rules are followed appropriately, revealing that the perception of improvement is associated with trust in expert systems.

The following significant statements indicate the awareness of participation coupled with the need to ensure that the voice of the community is included in the research conducted. In their own terms residents state: "we are the smallest area, fewest people and our participation in this research will be important because of that. We are few, but we can talk, otherwise others will speak for us

(...). We are present in this research" (Box 1). The studied community, Colônia Cristina, is located in the Araucária municipality. Although they are not the smallest area (the smallest area is not Araucária with 38.95 km², but Campo Magro with 18.4 km² of area occupied by the dam), it is worth distilling another idea from these statements. Residents believe that although the municipality of Araucária participates with the fewest inhabitants, only 1,160, their voices are best represented as a function of community cohesion; as such they are relevant participants in the research. Through close contact with our research group, residents of this community realized that collaborating with the research meant making public their problems, their concerns, but also their virtues and potential.

TABLE 1 – OCCUPIED AREAS AND NUMBER OF INHABITANTS IN THE MUNICIPALITIES

MUNICIPALITY	AREA	INHABITANTS
Araucária	38.95 Km2	1,160
Campo Largo	107.98 Km2	15,807
Campo Magro	18.40 Km2	4,909
Total		21,.876

Source: Developed by a Colônia Cristina resident during the culture circle

Table 1 was prepared and submitted to the entire community during the culture circle by a resident of Colônia Cristina who participated in one of the general seminars involving all sub-projects in September 2009.

In the following statements, we can see narcissistic pleasure as the basis for sustaining the idea of preservation. "The Rio Verde water is clean because there are farms. Each driveway gives access to a farm." "The owner cares, but who comes from outsides pollutes, leaves garbage, plastic bags, plastic bottles, etc." In those two significant statements we can interpret the association between environmental preservation and "narcissistic pleasure" as described by Castoriadis (2007). Preservation of the environment is conditioned to the fact that it is assumed to be "very good." This is the materialization of narcissistic pleasure with regard to the care towards the environment. On the other hand, everything that appears as "bad" in the community comes from outside. To some extent, they are assuming that while residents, who declare themselves good and need to systematically repeat this to themselves, take care, others or the "barbarians" and "invaders" are the main polluters.

For Castoriadis, the notion of narcissistic pleasure is the "capacity of the psyche to be satisfied with the mere representation." Narcissistic pleasure is associated with a very exaggerated positive image of the community; it can be understood as an investment in oneself. This is exactly what appears in the significant statements of the interviewees as they refer to themselves with the clear objective of differentiating themselves from the "others." In their words, "they [referring to all researchers involved] will not be able to come here anyhow. We are educated, we know things. We have lived for more than 100 years here." In the collective dialogue, the materialization of Castoriadis' concept is evident because the repetition by the community, to itself, that they are educated sounds like blunt self-praise, through which they are able to construct or reconstruct the socio-cultural identity of the community. One of the consequences arising from narcissistic pleasure is the segregation of the community in relation to noise, interference and possible external influences.

Apparently, narcissistic pleasure emerges in the minds and practices of the residents of the Colônia as an antidote to all fears arising from wider society. Through narcissistic pleasure, they seem to create an identity capable of resisting imminent threats, such as environmental legislation, unemployment, macro-zoning, etc. With narcissistic pleasure the community alleviates some fears for future generations. Box 3 summarizes and translates the hopes expressed through the dialogues with the interviewees of Colônia Cristina.

BOX 3 – BETS ON THE FUTURE IN COLÔNIA CRISTINA

"That it remains as it is today"
"That it does not change drastically"
"Trucks for sewage collection"
"We want to go on living here with better quality of life"
"We don't want subdivisions"
"Each farmer takes care of his piece of land"
"It has already improved a lot, we have a special collection of pesticide packages"

Source: Prepared by the researchers with the guidance of the residents.

6. FEEDBACK ASSOCIATED WITH COMMUNITY FEARS

In this section we present the feedback from the community based on our presentation of the significant dialogues and information we had collected during the culture circles. The goal of this process was to verify whether the community agreed or not with the information we had collected. We then asked participants to establish a hierarchical order for their own significant statements. The hierarchy established by the community is presented below.

1) "We feel insecure about our future crops." "We have to ask permission to work in our own land."

These were highlighted as the biggest fears the community has. These fears are directly related to the restrictions imposed by environmental laws and to each and every factor that is an obstacle to their processes of social reproduction. They are afraid that they will not be able to continue using their own properties.

2) External waste.

"The fishermen are the ones who bring trash. We have waste collection."

"Those who leave garbage dumped in the environment are fishermen."

"The basin is also a place where corpses from Curitiba are thrown."

3) "After the dam the calm is over and hence the security."

The third greatest fear for the community was the lack of security after the construction of the dam. The bridge became the entry point of some advantages (such as garbage and sewage collection trucks, etc.) but also the point of entry of various problems, such as urban fishermen, drugs, noise and violence of all kinds.

4) After discussion, participants decided that the fourth greatest fear is associated with the availability of water.
"Sometimes we have water available, but environmental laws will not let us use it."
"In our properties, we want to have water for irrigation and for animals to drink."
"Many live near the dam and want access to the water."
5) "A lot of people are concerned about the possibility of our area being flooded."
"Petrobras paid compensation for 40 meters."
"Now IAP is demanding 100 meters to legalize the property." Here we identify an impasse, because those who wish to legalize their property must establish 100 meters of protected areas, not just the 40 m guaranteed by Petrobras.
"If we want to build a tank to create fish we risk going to jail for life."
6) In the drawing representing the present (Figure 2) the area of riparian vegetation increased.
7) The seventh biggest concern is with the law on riparian vegetation which has become more restricted thus creating difficulties for community members.
8) There is concern with the garbage brought by tourists from outside the colony.
9) The ninth is the gravel from the roads.
10) "The employees could have a little more technical training."
11) "If they [city officials of Araucária] worked with different techniques, much less gravel would go into the rivers."
12) "It is not our concern to contaminate the dam."
13) "In that devastated area that belongs to Petrobras, we could use some help from Petrobras."
14) "Since the dam was built, Petrobras never gave any support to the community."
15) "The community asked Petrobras for a solution, but had no response."
16) "The São Casemiro Society needed a machine and had no response from Petrobras."
17) "Petrobras does not even maintain the bridge."
18) "First, they [Petrobras] have to finish the Red Hill and only then can they expect our collaboration."
19) "We needed stained glass windows for the church, but we were not welcomed at Petrobras."
20) "You [Petrobras] have the dam here, but when we need support, you do not give."
21) "We cannot even talk to them [Petrobras]."

7. FEEDBACK ASSOCIATED WITH THE CONNECTION BETWEEN COLÔNIA CRISTINA AND OTHER AGENCIES

This section sets out the main relationships/links between the Colônia and other agencies or institutions present in the Basin, as shown in Figure 4.

From Figure 4 we can infer that their ties with SANEPAR, Universities and Petrobras were characterized as "nil", which means there is not any kind of interaction that in the eyes of community members favors them. On the other hand, the links between the Colônia and the school, health care providers, City Hall, the Church, EMATER, São Casemiro (neighborhood association) were characterized as "strong" or very favorable to the interests of the Colônia. They highlighted that if they were to rank these interactions, the strongest link would be with the Church. In their own terms, "everything depends on the Church."

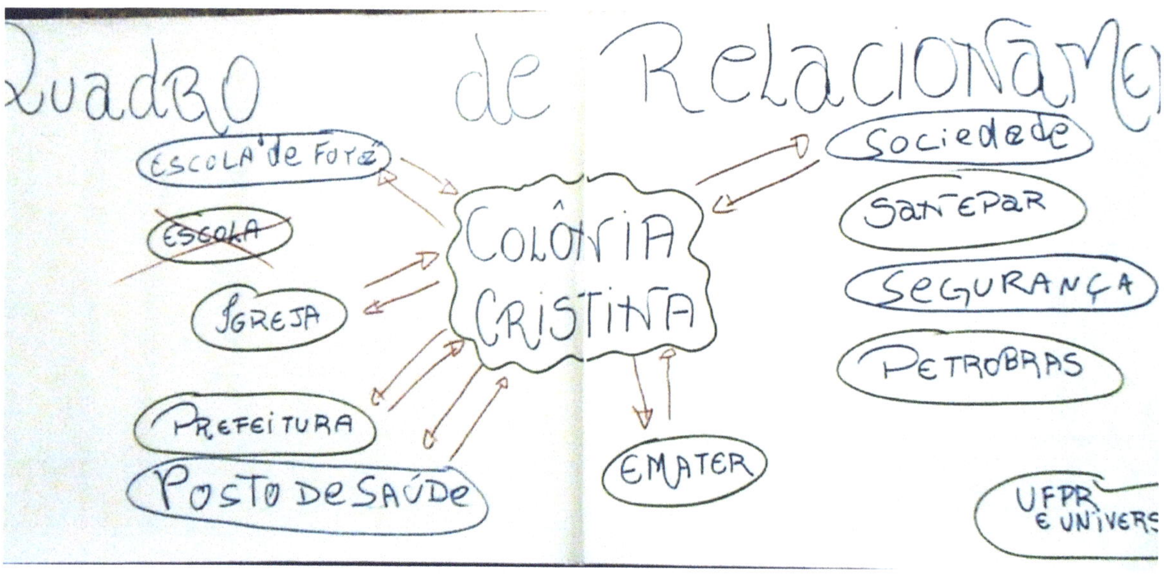

FIGURE 4 – RELATIONSHIP BETWEEN COLÔNIA CRISTINA AND INSTITUTIONS
Source: Prepared by the researchers with the guidance of the residents.

8. FEEDBACK ASSOCIATED WITH THE POTENTIAL OF COLÔNIA CRISTINA

In this section some major assets of the Colônia are described. The question asked was: what is good here that you would like to preserve? The responses are outlined in Box 4.

BOX 4 – ASSETS OF COLÔNIA CRISTINA

1) The unification of the community "one for all, all for one" (community life)
2) Good workers = high and good agricultural production
3) Sustainable community
4) Less aggressive farming techniques
5) Riparian forest
6) Roads
7) Education "It has already improved, we have a special collection for pesticide packages"
8) School Transportation
9) Health
10) Irrigation
11) Technical Assistance

Box 4 shows the perceptions that the interviewees have about themselves and these perceptions reinforce once again that they have the need to accentuate their qualities and potential in order to strengthen their sociocultural identity. The equation reported by participants, "Good workers = high agricultural production" (item 2) is an example of this. When asked about the meaning of this statement, they categorically declared themselves as "self-sustaining." This was elaborated on by stressing that "[they are] self-sustainable because they do not need outside help." "Here no one has family income assistance." Regarding the fourth item, "less aggressive farming techniques such as tillage", participants insisted on adding that "[they have] the culture of accepting what is best for the community," an indirect and laudatory reference to expert systems when they bring information relevant to their production processes which is fundamental to social reproduction.

Despite the positive perception in relation to expert systems, it seems strange that the technical assistance occupies the final position in the repertoire of assets. In conversations with some of the community members, assistance "is in last place" because they do not feel well assisted by the agencies that should or could provide this type of support.

9. FEEDBACK ON THE NEED FOR IMPROVEMENT IN COLÔNIA CRISTINA

In this section we asked: what can be improved in the community? Faced with this question, participants' reactions focused on the following point: although the assets of the community were already listed, they thought they could be improved. The responses are presented below:

1) "Keep community life." This item was one of the most discussed and strikingly reinforced by participants as the foundation for preserving the socio-cultural identity of the Colônia.
2) "Offer qualification (courses) to farmers." The desire to remain in rural life is implicit in this statement and having better farming qualifications emerges as a real possibility in realizing this desire.
3) "Establish new alternative income programs." Here, the opening-up of the Colônia to the outside world emerges. Community members were very receptive to external innovations that can be translated into practice that favor economic growth but do not jeopardize their unique way of life.
4) "Improving technical assistance."
5) "New and less aggressive technologies." Items 4 and 5 are directly related to item 3.
6) "Improving the roads." In this item, participants made sure to note that the idea of road improvement is directly linked to the imperative need to avoid contamination of the basin caused by gravel and dirt on the roads that are transported to the dam during rainfall.
7) "To improve elementary and secondary education."
8) "Improving health care."
9) "Improving public transportation."
10) "Improving irrigation."
11) "Improving school transportation."
12) "Implement agricultural education." Although this item appears in 12th position, it is directly related to items 2, 7, 9, 10 and 11, since it is very clear to participants that any proposed education will only make sense if it is related to the preservation of local social reproduction.
13) "Riparian forest (preservation)." Despite its location at the bottom of this list, community members are concerned about riparian forests. For them, this type of preservation is already a fact and their lifestyles are indicators of this preservation.

The above list of issues reveals that there are plenty of possibilities for external social actors to establish fruitful dialogues with the community. However, these dialogues must take into account some of the limitations and assets cataloged and recorded by community members.

10. RISK PERCEPTION

"Here within the APA only 300 or 400 families are involved. What will those people do if they cannot produce chicken, cattle or plant, where will they go?"

The perception, as depicted in the above statement, through discussion with a resident of the area surrounding the Rio Verde Basin, can both hide and reveal longings, anxieties, as well as hope. The statements of some interviewees seem to reveal their ability to denounce a society that on the one hand privileges and benefits some, and on the other, penalizes: "awareness has made people change (to planting organic strawberry, for example) but they also know the importance of the pesticides for budgets."

Many interviewees declared themselves victims of social institutions, such as environmental legislation, macro-zoning, the APA, etc.

"The staff of the "chacrinhas" [small leisure farms] are polluting, they do not have sewage collection and they are different people. It changed much, sidewalks were constructed and the runoff was interrupted."

For many, the risk seems to be much more social than environmental. Living within an APA, with the trend toward the area becoming high risk because of eutrophication processes, seems to be very uncomfortable.

Likewise, residents do not see their lifestyles, including the use of pesticides, as exacerbating the process of eutrophication. Strictly speaking, they " use pesticides because society asks for a specific product otherwise they certainly would not use any. You need to be aware that: they want quality and unfortunately have to use insecticides."

The voices of participants are strikingly stronger in denunciating the possible threats from wider society. One of the positive points is that a perception centered on a complaint may contain at its core elements that suggest new possibilities, new worlds. Therefore, it is imperative to try to identify in the complaints these possibilities of other worlds, of new, vital and liberating opportunities.

"There needs to be a change, I agree! But, why only the agriculture? Why in Araucária, businesses were built on springs that were buried and no one does anything about that?"

10.1 CLASSICAL ASSUMPTIONS OF PERCEPTION

Perception may be interpreted as a form of social representation (DURKHEIM, 2009). From this perspective, the assumption is that there is a reality external to the individual and that the individual is able to systematize it in light of existential, emotional and intellectual resources, which are internalized by the individual throughout their life history.

In this analytical matrix, the voices of humans could be cataloged as both constituents and constituent elements of analysis and not as living expressions of the tensions and conflicts inherent to any associative experience involving human and non-human interests. In this sense, more importance would be given to the expert responsible for capturing and translating the voices than to the voices themselves.

The analysis derived from such assumptions tends to be restricted to complaints from residents regarding the factors that threaten their social system. This understanding of perception as social representation derives from the classical tradition of sociology which is based on the separation between subject and environment. On the other hand, the understanding of perception as a possible production of the world was developed out of the first investigations into the explanatory limits of classic social representation.

10.2 NEW ASSUMPTIONS OF PERCEPTION

From another perspective, one can understand perception not as a social representation of a certain level of reality, but as the possibility of producing other worlds. Here, the assumption is not that there is an individual who thinks and that there is a level of reality that can be represented. There is not a subject and a world separate from him/her; there is an individual who, when perceiving the world, helps to produce it and therefore produces him/herself.

Castoriadis (1982) does not use the notion of perception as a synonym for representation. For him, each concept belongs to different epistemological fields. If perception belongs to a field characterized by Castoriadis as a field of "inherited thought", then representation is thought well beyond that field.

Castoriadis prefers the notion of representation because he understands that inherited thought believes in a representation apart from itself that does not exist, insisting on the separation between the subject and the environment. For him, representation is dismissed by inherited thought because, unlike perception, which is instituted, representation is instituting, rebelling and fundamentally a transgressor. If for perception the human being is a common determined entity and the same for all, then representation tends to ignore these rules.

In inherited thought, representation is reduced to an imperfect copy of objective facts or is a source of error. It is nothing more than a defective signifier of what is objectively outside the subject. Representation is the result of weakened perception.

Castoriadis rejects any association between representation and the idea of ownership, as ownership is defined as the subject itself, not as his/her property. Representation does not belong to the subject, it is the subject itself. In the situation of radical imagination, representation cannot be understood as something that produces distorted and impoverished images of the facts but as something that actually produces these realities and produces itself within them. It makes no sense to separate the real from the imaginary, since both are intrinsically linked.

Another argument that Castoriadis uses to justify his preference for the concept of representation is that in the objectivist field of perception there is no place for radical imagination. All attributes are directed at the object, shaping the worship of reality as something distinct and separate from the subject. Note that there is no place for creation, since everything is determined objectively. As in the field of perception, the fetishism of the real occupies a prominent position. In this field there is no subject and object, because the individual is also featured as an object. Thus, the relationships are not between subject and object, but among objectified things.

For Castoriadis, these hegemonic indicators of inherited thought tend to lay claim to the idea of an eternal and abstract universalism of the human representative capacity. This he categorically rejects as it overvalues instituted sacrifices, a concept that he considers fundamental in his work, the power of radical imagination, the instituting.

Like with the work of Sheldrake (2001), Castoriadis explains his reservation to the hegemonic idea of science as a science of objective facts, as the science of things. In the epistemological field of social representation, the cult of science is just a daydream because the imagination is a logical condition, but also ontological of reality. Science for him is the science of facts that is constructed and reconstructed from the potentialities of the radical imaginary. It is not a science of only subjects or objects, but the infinite possibilities of articulation between both.

In this analytical matrix, which is different from previous matrices, the voices of humans can be cataloged not only as constitutive and constituent elements of analysis, but also as living expressions of the tensions and conflicts inherent in any associative experience involving human and non-human interests. In this sense, we do not favor one level of analysis over others but we seek balance between the voices of the interviewees themselves and the discursive ability of the responsible specialist. Here, more than translating, it is essential that a dialogue is developed with the voices captured in order to distill from them new understandings of the world.

Note that here the perceptions of interviewees is not understood as a possible method of control, but as a combined responsibility in the face of social and environmental conditions of the Basin. If in the first approach the complaints were identified, here the understandings that are hidden or revealed in the voices of the human inhabitants of the Basin are documented. Therefore, if we take as a reference the second analytical perspective, the statements of some interviewees reveal or hide how they produce the world and at the same time how they are produced by it.

11. FINAL CONSIDERATIONS

In this interdisciplinary project on the eutrophication of the Rio Verde Basin, the voices of humans occupy a prominent place without necessarily overtaking the voices of the non-human inhabitants of the Basin. Taking into account the voices of the human inhabitants recognizes them not as being more important than the others, but as able to perceive themselves as protagonists of new civilizing possibilities.

Their voices, while denouncing some aggressive and abusive dimensions of a society marked by tensions and conflicts that are difficult to resolve quickly, present a desire for resistance. Resistance can be understood as an ethical constraint, a call for new dialogues between governments and communities living in areas classified as at risk.

One of the virtues of this project is the boldness and the courage to make public the voices of those communities that are rarely taken into account. On behalf of an always questionable "public interest" or "environmental" considerations, entire communities are silenced while their social fabrics are weakened or destroyed.

Finally, the discomfort in the voices of some interviewees requires us to never forget to ask where is the place of human inhabitants in projects targeted at areas that are potentially at risk. Careful listening to the voices of the inhabitants of areas deemed at risk does not imply accepting *a priori* their reasons as ultimate or truth in absolute terms, but they must be taken into account. Space must be made for community members to speak out, denounce and declare their interests; equally spaces for them to learn and listen must also be created. Perhaps from this perspective, favorable and fertile conditions for a collective understanding are being created, as suggested by Latour (2004), that consider the interests of all the inhabitants of the Basin, human and non-human alike.

12. REFERENCES

ANDREOLI, C.; CARNEIRO. C. (Ed.). (2005) *Gestão integrada de mananciais de abastecimento eutrofizados*. Curitiba: SANEPAR Finep.

BRAGA, P.C. (2008) *Turismo rural, educação ambiental e risco: um estudo da percepção dos riscos ambientais com os empreendedores de turismo rural na bacia do Rio Verde*. 2008. 68p. Trabalho de Conclusão do Curso (Especialização em Educação Ambiental, Meio Ambiente e Desenvolvimento) - Universidade Federal do Paraná. Curitiba.

BRANDENBURG, A. (1998) *Colonos: subserviência e autonomia*. Para pensar: outra agricultura. Curitiba: UFPR.

CASTORIADIS, C. (1982) *A instituição imaginária da sociedade*. 6. ed. Rio de Janeiro: Paz e terra.

CASTORIADIS, C. (2007) *Sujeito e verdade: no mundo social-histórico*. Rio de Janeiro: Civilização Brasileira.

DESLANDES, S.F.; ASSIS, S.G. de. (2002) Abordagens quantitativa e qualitativa em saúde: o diálogo das diferenças. In: MINAYO, M. C. de; DESLANDES, S.F. (Orgs.). *Caminhos do pensamento: epistemologia e método*. Rio de Janeiro: Fiocruz, p. 195-223.

DURKHEIM. (2009) É. *Sociologia e filosofia*. São Paulo: Martin Claret. (Coleção a obra-prima de cada autor).

FREIRE, P. (1983) *Educação como prática da liberdade*. 17.ed. Rio de Janeiro: Paz e Terra.

GODOI, C.K.; BALSINI, C.P.V. (2006) A pesquisa qualitativa nos estudos organizacionais brasileiros: uma análise bibliométrica. In: GODOI, C.K.; BANDEIRA-DE-MELO, R.; SILVA, A.B. da. *Pesquisa qualitativa em estudos organizacionais*. Paradigmas, estratégias e métodos. São Paulo: Saraiva. p. 89-112.

GIDDENS, A. (1991) *As conseqüências da Modernidade*. São Paulo: Edit. UNESP.

LATOUR, B. (2004) *Políticas da natureza: como fazer ciência na democracia*. Bauru: Edusc.

MINAYO, M.C. de S. (2006) *O desafio do conhecimento: pesquisa qualitativa em saúde*. 9. ed. São Paulo: Hucitec.

MINAYO, M.C. de S.; MINAYO-GÓMEZ, C. (2003) Difíceis e possíveis relações entre métodos quantitativos e qualitativos nos estudos de problemas de saúde. In: GOLDENBERG, P.; MARSIGLIA, R.M.G.; GOMES, M.H. de A. *O clássico e o novo: tendências, objetos e abordagens em ciências sociais e saúde*. Rio de Janeiro: Fiocruz. p. 117-142.

PARANÁ. Decreto nº 2375. *Súmula: Instituída a Área de Proteção Ambiental do Rio Verde, denominada APA do Rio Verde, localizada nos municípios de Araucária e Campo Largo, Estado do Paraná*. Paraná, 28/07/2000. (Decree 2375. Summary: Establishes the Rio Verde Environmental Protection Area, called Rio Verde APA, located in the municipalities of Araucaria and Campo Largo, State of Paraná, Paraná, July 28,2000) Available in<http://celepar7cta.pr.gov.br/SEEG/sumulas.nsf/fcc19094358873db03256efc00601833/c47a1e5911da8c0103256e990061e717?OpenDocument>. Accessed in: January 26, 2009.

SHELDRAKE, R. (1995) Seven Experiments That Could Change the World: a do-it-yourself guide to revolutionary science, New York, NY: Riverhead Books.

SHELDRAKE, R. (2001) *Sete experimentos que podem mudar o mundo: pode a ciência explicar o inexplicável?* São Paulo: Cultrix.

YIN, R.K. (2005) *Estudo de caso: planejamento e método*. 3.ed. Porto Alegre: Bookman.

SECTION VI

ENVIRONMENTAL EDUCATION

CHAPTER 22

ENVIRONMENTAL EDUCATION

Lucia Izabel Czerwonka Sermann
Andreia Hoffman
Janaína Kether Campos Piedade
Tania Cristina Salomão Rodini
Vânia Eloiza Cerutti

ENVIRONMENTAL EDUCATION

Lucia Izabel Czerwonka Sermann
Andreia Hoffman
Janaína Kether Campos Piedade
Tania Cristina Salomão Rodini
Vânia Eloiza Cerutt

SUMMARY

Educating the public requires knowledge and hard and complex efforts especially due to large interaction between individuals and nature. In these circumstances, education has significant potential for the future, since it represents new connections with nature, values, and ethical environmental behavior. Given this reality, environmental education takes on the task of promoting our potential to transform, since we have the strategic and decisive daily role to build a sustainable society. This chapter's aim is to describe the process of collectively creating a community environmental education program that takes a participatory, sensitizing, and mobilizing approach to promote a balance between humans and nature, with a focus on local sustainability. The focus of the research is a community called Cercadinho, located in the municipality of Campo Largo, in the Metropolitan Region of Curitiba, PR. The community is located within the environmentally protected area (APA) of the Rio Verde. It is a qualitative case study that is exploratory, descriptive, and bibliographical in nature. The information was obtained during both participatory and non-participatory observation, semi-structured questionnaires, and interviews with the Community of Practice called *Grupo Vida ao Rio Verde*. Their involvement in meetings, presentations, sensitivity workshops, ecological hikes, and the systematizing of shared knowledge, resulted in the production of an environmental education booklet. Such practices led to the organization of the stakeholders involved coming together with the shared goal of elaborating a community environmental education program.

KEY WORDS

Environmental education, communities, ethical responsibility, local sustainable development.

Educating community members brings about new knowledge that is necessary for understanding the increasingly complex social processes and environmental risks that are intensifying as a result of ecological imbalance. This imbalance is the result of the social development and also because of the inherent difficulties in the relationship between individuals and nature.

Such issues include the exaggerated use of natural resources, which create social exclusion and misery for millions of people worldwide. The uncontrollable consequences of such actions have resulted in political and academic authorities seeking solutions to this environmental crisis.

Once again, education becomes the hope for the future, since it represents the means through which society can reconnect with nature and develop environmentally ethical values and behaviors.

A world that is fairer is an attainable utopia with a profound reeducation related to the accumulation of wealth and habits of consumption. The goal of such reeducation is to ensure people act responsibly for future generations, making it possible to build a more equal society with a solid economy based on a sustainable way of life.

Faced with this reality, education, especially environmental education, has the task of promoting our transformative potential because all citizens can have a strategic and decisive role in their daily lives, becoming environmental citizens who are mobilized to build a sustainable society.

This chapter's aim is to describe the process of collectively creating a community environmental education program guided by participatory practice that seeks to reestablish the equilibrium between humans and nature, with a focus on local sustainability. The focus of the research is a community called *Cercadinho,* located in the municipality of Campo Largo, within the Metropolitan Region of Curitiba, PR, and within the Rio Verde environmental protection area. It is a qualitative case study that is exploratory, descriptive, and bibliographical. The information was obtained during meetings with both participatory and non-participatory observation, semi-structured questionnaires, and interviews performed during meetings with the community of practice *Grupo Vida ao Rio Verde* (Life for Rio Verde). These procedures resulted in meetings, presentations, sensitivity workshops, ecological hikes, systematizing of shared knowledge, and an environmental education booklet, all of which lead to the preparation of a community environmental education program.

1. EMERGENCE OF ENVIRONMENTAL EDUCATION

Imbalance in the environment has generated a worldwide crisis whereby each ecosystem across the planet experiences in some manner the negative effects of its exploitation. This is a consequence of the weak values that guide the relationship between humans and nature. This weakness has intensified over time and has become more accentuated, provoking misery, consumerism, and social and economic exclusion.

The ability of socioeconomic systems to react seems to be very slow, despite the current growing concern about environmental issues.

Based on the Gaia hypothesis of James Lovelock (2006), Earth, as a living organism, constantly attempts to maintain its equilibrium. However, humans have invariably destroyed this natural equilibrium. Therefore, the planet, which is constantly striving to restore equilibrium, is generating unpredictable events, endangering civilization itself.

Situations related to the environment and its imbalance are constantly found in the media today, such as the lack of potable water, catastrophic floods, problems with waste disposal, deforestation, global warming, and finally the contamination of air, water, and soil by countless human actions. What can be interpreted by these phenomena is that these situations represent the current environmental crisis.

Leff (2002) noted that:

> Natural catastrophic transformations occurred during different stages of geological and ecological evolution of the planet. For the first time, the current ecological crisis is not a natural transformation but one induced by the metaphysical, philosophical, ethical, scientific and technological concepts of the world. (LEFF, 2002, p. 194)

The severity of these environmental problems poses some questions for the present generation. Complex solutions are required that reconcile development with conservation/preservation and will require the participation of all.

As such, citizens have a strategic and decisive role in integrating environmental education into daily life, adopting a critical position in the face of environmental crisis and aiming to transform habits and social practices and create of an environmental citizenship that mobilizes for sustainability.

1.1 THE CONTEXT OF ENVIRONMENTAL EDUCATION

Since their occupation on the planet, humans have had an impact on their environment through the development of instruments and tools to guarantee survival. McCormick (1992) notes that 3700 years ago, the Sumerians had problems with irrigation due to the salinity of the water. In Greece, 2400 years ago, Plato regretted deforestation and the soil erosion caused by over-grazing and the cutting of trees for firewood.

In constructing vessels for the fleet of the Byzantine Empire, Venice, Genoa and other Italian states contributed to the degradation of coastal forests along the Mediterranean Sea during the ninth century AD. During the tenth century AD, the Mayan civilization began to collapse due to a significant increase in population. The problems intensified with the expansion of empires and the later advent of worldwide navigation. Inter-cultural exchange not only introduced species into new environments for economic reasons, but also accidentally created large environmental problems.

Beginning in the fifteenth century, the European colonies in America and especially the Caribbean, Central and South America, and Africa began to heavily exploit the colonized regions, thus generating a series of problems, which are visible and present until today.

The significant environmental impacts arising from the growing process of industrialization, which began during the industrial revolution and grew after World War II, stimulated wide reflection not only on the operation of industry, but also on the structure and values of contemporary society.

Based on Layrargues (2002), during the 1950s and 1960s environmental problems arose, such as air pollution in London and New York (between 1952 and 1960), as well as mercury poisoning in Minamata Bay in Japan in 1953. The risks of using DDT (Dichlorodiphenyltrichloroethane) and other pesticides that generated bio-accumulation were warning signs of environmental problems. These were situations in which the impacts of human action stretched across borders and arose as a result of significant changes in environmental processes. Beginning in the 1960s, a large number of debates and conferences were held with the intention of bringing attention to the environmental crisis, consolidating into the idea of Environmental Education.

In March 1965, the Education Conference which took place at the University of Keele, in the UK, cited for the first time the term "environmental education."

In the mid-1970s, the United Nations, through UNESCO, created UNEP (United Nations Environmental Program) demonstrating the high expectations placed on education as a means to combat environmental crises.

The UN Conference on the Human Environment (Stockholm Conference, 1972) proposed that environmental education be considered as a critical element for coping with environmental problems. As a consequence, during the event there arose the beginnings of an International Environmental Education Program, which was later founded in 1975.

In 1974, with the support of UNESCO, an Environmental Education Congress took place in Jammi, Finland, which became a consistent event.

The following year (1975), UNESCO held the First International Congress on Environmental Education in Belgrade, stressing the goals and principles of environmental education.

In the wake of the events related to the environmental crisis, The International Society for Environmental Education, the first social organization bringing together environmental educators, was created in Ohio, USA, in 1976.

In 1977, the First Intergovernmental Conference on Environmental Education, known as the Tbilisi Conference, was organized by UNESCO in cooperation with UNEP, which was considered the hallmark of the education proposal by establishing the concept. Since then, environmental education has been understood as the processes by which an individual and community build social values, knowledge, skills, attitudes and abilities for the conservation of the environment, essential for a healthy quality of life and sustainability.

In the 1990s, more precisely in 1992, the UN Conference on Environment and Development took place in Rio de Janeiro, Brazil. At this conference, known as ECO-92, "Agenda 21" was the most important commitment made between the leaders of the participating countries. It aimed to prepare the world for the challenges of the 21st century.

Parallel to ECO-92, the Global Forum took place in the same city, bringing together different ethnic groups to assess the environmental context and its degradation. This resulted in the production of the first draft of the "Earth Charter," calling for all participants to adopt its principles, both individually and on a global scale.

These two important documents propose a development model committed, above all, to preserving life on the planet by preventing harm to the environment as the best method of environmental protection. Working towards the goal of preserving life, the documents refer to alternatives that undoubtedly require intense and effective educational processes in order to re-educate society and thus promote ecological education.

In 1998, UNESCO organized the "International Conference on Environment and Society: Education and Public Awareness for Sustainability" in Thessaloniki, Greece. During the conference, it was concluded that environmental education, as well as legislation, technology and economy, would be one of the pillars of sustainability (Layrargues, 2002).

By incorporating the term 'sustainability' in the discourse of environmental preservation, a new dimension was introduced into environmental discussions which takes an interdisciplinary perspective on the various determining factors in recovering environmental equilibrium.

In addition to these international events, many other local events under the auspices of UNESCO/UNEP were held aiming to establish the principles of environmental education, as well as regional aspects based on local practice in facing environmental issues. The objective of these events was to determine ethical, methodological and conceptual advances for environmental education, aiming at social transformation.

The World Summit on Sustainable Development took place in Johannesburg, South Africa, in 2002 and was carried out simultaneously in different cities around the world. In Brazil, the event was named RIO+10, celebrating the renewal of ECO-92 held ten years earlier. The main objective of the Summit was to analyze the actions taken by countries committed to Agenda 21, as well as review concepts and principles that govern political, economic and social forces, in order to implement the provisions of these actions.

The year 2005 was crucial for Environmental Education as it began the Decade of Education for Sustainable Development (2005-2014), coordinated by UNESCO, which seeks to establish the international implementation of the principles of Agenda 21, as defined in its chapter 36. Governments were thus required to adhere to the measures necessary for the implementation of what was proposed for the decade in their educational plans and strategies.

More recently, in December 2009, the 15th session of the Conference of the Parties of the United Nations Framework Convention on Climate Change (UNFCCC) took place in Copenhagen, Denmark. The meeting was considered the most important in recent history of the multilateral environmental agreements and was aimed at establishing a treaty to replace the Kyoto Protocol, in place from 2008 to 2012. However, the overall event generated frustration because the agreement was insufficient to solve the most critical environmental issues, such as stabilizing atmospheric temperature, protecting vulnerable communities, and ensuring sustainable development in developing countries. These issues, in treaty form, were postponed until the next meeting in Mexico.

1.2 ENVIRONMENTAL EDUCATION IN BRAZIL

As an initiative by environmental authorities, environmental education was established in Brazil in the 1970s and 1980s following recognition of the need to manage the world's natural resources more sustainably (LIMA, 2009).

The institutionalization of environmental education in the Brazilian Federal Government began in 1973 with the establishment of the Special Secretary of the Environment (SEMA) within the executive branch. It was the first Brazilian official entity oriented towards integrated management of the environment (DIAS, 1992; BRAZIL, 2005). Originally conceived as a pollution control agency, it established the program of ecological research and preservation stations with strict standards.

As part of its duties, this entity helped to educate Brazilian people about the proper use of natural resources for environmental conservation. It also took on the responsibility of training human resources in order to make the first steps towards civil awareness on environmental issues. Initial projects were aimed at inserting environmental education into school curriculums at the elementary and middle school levels (BRAZIL, 2005).

In June 1992, the city of Rio de Janeiro was the setting for two important events: Rio-92 (or ECO-92) sponsored by the UN, which was the largest international gathering of all time, and the Global Forum 92, attended by more than 10,000 participants from non-governmental agencies. REBEA, the Brazilian Network for Environmental Education, was created at this event which is comprised of NGO's, educators and diverse institutions related to education. REBEA was involved with the promotion of the First Environmental Education Workshop and with the drafting of the Environmental Education Treaty. It initially held five regional environmental education forums predominantly in the Southeast of the country, but has since taken on a national mandate. The last REBEA forum was held in 2006 and it was the first international forum as it had representation from across Ibero-America (CARVALHO, 2008).

Also taking place at ECO-92 was the UNFCCC (United Nations Framework Convention on Climate Change), which aimed at stabilizing atmospheric concentrations of greenhouse gases and reducing each country's emissions to 1990 levels. In 1996, the Education Ministry organized the National Curriculum Parameters (PCN) that included environmental education as a nation-wide theme in the Brazilian educational curriculum.

At the First National Conference on Environmental Education of Brasilia (CNEA) in 1997, the Brasilia Declaration for Environmental Education was developed and later presented in Thessaloniki, Greece. This document recognizes that the vision of education and public awareness was enriched and reinforced by international conferences and action plans developed at these conferences should be implemented by national governments and civil society, including NGOs, businesses and educational communities, the UN and other international organizations.

Established in Law 9.795/1999 and regulated by Decree 4,281, June 25, 2002, the National Policy for Environmental Education (PNEA) describes environmental education as a process through which the individual and the collective build social values, knowledge, skills, attitudes and competencies for the conservation of the environment (Article 1. Law 9.795/1999). This reinforces the collective responsibility of its implementation, its basic principles, objectives, and strategies. This law provides a roadmap for the practice and regulation of environmental education and recommends the Ministry of Education (MEC) and the Ministry of the Environment (MMA) as the managing bodies of the policy.

After the enactment of the National Environmental Education Policy (PNEA), the Ministry of Education, the General Coordination of Environmental Education (CGEA), the Ministry of Environment, and the Directorate of Environmental Education (DEA) enforced PNEA, inserting environmental education into Brazilian public policy.

This law also establishes criteria and standards for environmental education both in formal education, in public and private schools, as well as informal education, through actions aimed at raising community awareness and education on environmental issues and the participation of community organizations in the protection of the environment (Art. 13, Law 9.795/1999).

Chart 1 indicates the main environmental education public policies put into place in Brazil since the 1980s:

The trajectory described herein contributed to achieving the goal of environmental education which is to stimulate citizen awareness, prepare citizens to make sound decisions in today's social/environmental reality, and commit to the well-being of each individual and society, both locally and globally.

CHART 1 – MAIN ENVIRONMENTAL EDUCATION PUBLIC POLICIES ESTABLISHED IN BRAZIL SINCE 1980

YEAR	PUBLIC POLICY
1984	Creation of National Environmental Education Program (PRONEA).
1988	Inclusion of Environmental Education as a right for all and a responsibility of the State, in the Environmental chapter of the Constitution.
1992	Creation of Environmental Education Centers by the Brazilian Environmental Institute and Renewable Natural Resources (IBAMA) and Environmental Education Centers by the Ministry of Education (MEC).
1994	Creation of the National Environmental Education Program (PRONEA) by MEC and the Ministry of Environment (MMA).
1997	Development of MEC Department of Elementary Education curriculum parameters, including "environment" as one of the transversal themes.
1999	Approval of the National Policy for Environmental Education by Law nº 9795.
2001	Implementation of the program "Parameters in Action: Environmental Education at School" by MEC
2002	Regulation of the National Policy for Environmental Education (Law 9795) by Decree 4281
2003	Creation of the Governing Body of the National Policy on Environmental Education, uniting MEC and MMA

SOURCE: CARVALHO, I. C. M. Educação Ambiental: a formação do sujeito ecológico. São Paulo, Cortez, 2. Ed 2006.

Thus Environmental Education takes on an interdisciplinary dimension, leading to active participation, enhancing pedagogical action, creating a process of continuing education and lifelong learning. It is not necessarily a formalized discipline as its interaction with other disciplines provides an awareness regarding the environment, greatly aiding in the development of citizenship.

The needs and demands in Environmental Education can be divided into two categories: Formal Education and Informal Education.

Formal or School Environmental Education has as reference the Ministry of Education National Curriculum Parameters and the Law of Guidelines and Frameworks (LDB). This occurs within the school system, aiming to educate citizens to observe and understand reality, pose questions in light of world challenges, be concerned with collective outcomes, and focus on the quality of life and development of citizenship, guided by principles of solidarity, ethics, health, respect for nature and life, cultural diversity and responsibility.

Informal Environmental Education works through public awareness campaigns that aim to influence actions and attitudes leading to knowledge and understanding of environmental issues, preservation of natural resources, prevention of environmental disasters and correcting actions that are threatening to the quality of life on Earth.

Thus formed, Environmental Education in Brazil becomes part of a citizen's rights, similar to the fundamental rights and duties linked to the constitution. Therefore the social relationship of environmental education becomes evident in the duty of the community to protect future generations through responsible, ethical, and effective actions in order to achieve a sustainable society.

2. LOCAL SUSTAINABLE DEVELOPMENT AND ETHICS OF RESPONSIBILITY

Currently, studies on development analyze the production process, prioritizing the transformation of its structure and adapting it to the generation of wealth, technical progress, economic growth, industrialization, and the modernization of society.

Meanwhile, the notion of local development combines the specific location with its associated variables: wind, humidity, temperature, rivers, terrain, and biological and social factors that affect the area. According to research, local development is realized through people and their relationships in which local and public institutions act to organize society.

Thus, we can say that local development is related not only to economic growth but also to improving quality of life and conservation of the environment. These three interrelated and interdependent factors determine the conditions for sustainable development.

Therefore, local sustainable development is "the process of social change and increasing opportunities of society, aligning, in time and space, economic growth and efficiency, environmental conservation, quality of life and social equity, based on a clear commitment for the future and solidarity between generations" (Buarque, 1994).

Largely related to experiences of different groups or communities, such development is not restricted to specific problems, such as unemployment or health care, but to the whole set of problems that interconnect and influence each other. Everything has to do with everything. Each of the social actors has their own role to contributing to local development (Buarque, 1999).

For Avila (2003), local development is created principally through solidarity and participation; it is based on respect for the identities of the local population and each person, group or individual, including an appreciation for multicultural society.

Thus, one can say that the globalized world becomes a mutually dependent local world, since the global influences directly upon the local, and the local directly on the global.

Such influences and mutual dependencies call for a world in which humans learn to live harmoniously in their habitat because environmental issues permeate all human activities. As such, it calls for the reevaluation of all patterns and established models, in turn bringing about change and transformation on both the personal and institutional level.

This setting calls for new educational proposals that emphasize respect for cultural and individual diversity, as well as biodiversity.

Thus, a process of local sustainable development is possible that incorporates a new understanding of the world, one that includes knowledge, theories and practical skills that are the foundation of every culture, the strength of locality, the shared territory identified by a social and community conscience, the essence of which is the very lived and collective history (LE BOURLEGAT, 2000).

This is the scenario for the emergence of a proposal for Environmental Education that will become effective in local sustainable development, considering that its practice stresses the importance of the environment, quality of life, and especially an ethical coexistence that is guided by the principle of responsibility.

2.1 ETHICS OF RESPONSIBILITY

Contemporary science created a divide between knowledge and ethics causing serious consequences generated by the extraordinary powers of death and the manipulation raised by scientific progress (MORIN, 2005b, p.25). As a result, we are faced with an ethical crisis.

According to Morin (2005), this crisis requires rewiring the individual, society and the human species to regenerate responsibility and solidarity, essential foundations of ethics, thus leading to the achievement of a community ethic, an ethic of humanity or, as the author defines it, a planetary ethic.

Jonas (2006) notes that this crisis is enhanced by modern technology, which threatens not only the destiny and the physical survival of humans, but also poses a threat to the integrity of the natural world.

By proposing a new ethic focused on human thought and behavior that overcomes technological necessity, reviving consciousness, the individual and freedom, the author points to a new subject that does not endanger the continuity of humanity on Earth.

From the viewpoint of the author, the vulnerability of nature, which is extremely critical and caused by the technical intervention of humanity, gave rise to environmental science and ecology, which modifies human self-understanding and understanding of the nature of human action. He points out that new types and limits of human action require an ethic of foresight and responsibility consistent with these limits and the situations with which humans must cope (Jonas, 2006).

Therefore, it is necessary to reestablish social responsibility, or human solidarity as defined by Morin (2005b), reconnecting knowledge and building the collective consciousness, thus bringing about a new ethics of responsibility.

When one thinks about who should take on the actions of social responsibility, we first tend to follow historical practice and look to the State. We point to the government as being responsible for ensuring or promoting citizenship and meeting social needs.

However, according to Guerreiro (2009), the limitations of public policies always gives rise to new players: community organizations that reinforce the process of social democracy and guarantee social rights, occupying spaces forgotten by public powers and thereby contributing to the local quality of life.

Thus, it is possible to understand that the community is also an environment of citizenship, commitment and involvement with a social cause. This is the essential nature of local development and human solidarity and as such seeks creative solutions to social problems generated in various segments of society.

By linking social responsibility, human solidarity and local development, a project of intervention and technological development is necessary that is capable of enhancing the sustainability of the local community and promoting qualitative changes in the lives of people in situations of social risk.

Environmental Education emerges an alternative solution, retrieving its essence of education in the social and interpersonal sense, in the pursuit of individual, collective, environmental and ethical welfare. This is realized by the people in the community who decide to intervene in the current reality to overcome socio-environmental problems. In doing so, communities are formed that are able to learn collectively through the sharing of knowledge.

3. COMMUNITY OF PRACTICE AS A PATH TO ENVIRONMENTAL EDUCATION

The expansion of community networks has inspired a new way of acting, living and belonging, both in personal relationships and at work, giving meaning and creating a new way of living. Thus, from the theoretical perspective of

Capra (2005), we find that increasingly we live in all kinds of networks, whether virtual or physical, creating new social structures, expanding areas of operation, and disseminating knowledge within and outside these organizations.

Network systems in communities, whether virtual, physical or both, as disseminators of knowledge have been the subject of both organizational and scientific interest, awakening in the scientific community the need to understand how these systems influence the process of collective learning within a community.

In this context, "community" cited here should be perceived not only in terms of a locality, but as:

> The entity to which people belong, larger than kinship relations, but more immediate than the abstraction we call 'society.' It is the arena in which people acquire their most fundamental and substantial experience of social life outside the boundaries of their homes. (COHEN, 1985, p.15)

Thus, the community is understood as a space of observation and reflection that enables one to learn social experiences of everyday life.

In this particular case, a local community with common interests focused on certain goals is characterized as being a community of practice, where learning occurs through collaboration and cooperation.

The dissemination of knowledge in this community happens through social interaction, by acting as a facilitator of learning and providing incentives to members of a group with the purpose of encouraging their participation in the effort to learn together.

From this perspective, communities will contribute to the sharing of practices by socializing learning and disseminating knowledge.

3.1 COMMUNITIES OF PRACTICE

In 1991, Etienne Wenger, Jean Lave, Richard McDermott and William Snyder developed studies on the management of organizational knowledge. Their research focused on the learning relationships within organizations that initiated new practices and procedures, which in turn helped them learn how to transform organizations into communities of practice.

Based on their research, the authors defined a Community of Practice as:

> Groups of people who share a concern, a set of problems, or a passion about a topic, and who deepen their knowledge and expertise in this area by interacting on an ongoing basis... These people don't necessarily work together every day, but they meet because they find value in their interactions. (WENGER, MCDERMOTT & SNYDER, 2002)

Thus, the process of sharing ideas, reaching goals and achieving common objectives also promotes social relations, making groups cohesive and multidisciplinary. The interaction and diversity that permeate contributes to knowledge building since people spend time together, sharing ideas and testing their limits. This process facilitates creativity and promotes a sense of trust and belonging.

Accumulated knowledge thus adds value to spontaneous and rewarding learning that takes place collectively. Members feel self-motivated to learn how to build together what interests them. Through the satisfaction of sharing, the individual member becomes recognized and valued for what he/she knows.

Wenger (2004) explains the importance of becoming aware of belonging to a community of practice, or the awareness of its members to develop a sense of belonging, as this perception will enable them to adopt strategic measures for the acquisition and dissemination of knowledge.

Mobilization only around common interests, without having a collective goal to fulfill, does not constitute a community of practice but rather a collaborative working group. In Chart 2, we show the differences between communities of practice and other collective organizations (based on WENGER & SNYDER, 2002).

In this context, it is clear that Communities of Practice are characterized by skills and expertise, as well as by the importance of social interaction that unites community members around interests and objectives chosen and carried out.

Therefore, an environment of trust and respect should be developed where members can exercise their autonomy and present and share their ideas.

When the practice of knowledge sharing evolves, it becomes a collective product and it is integrated into the work of the people organizing the knowledge: every com-

CHART 2 – A SUMMARIZED COMPARISON BETWEEN COMMUNITIES OF PRACTICE AND OTHER COLLECTIVE ORGANIZATIONS

	WHAT IS THE OBJECTIVE?	WHO PARTICIPATES?	WHAT KEEPS THEM UNITED?	HOW LONG DOES IT LAST?
Communities of Practice	Develop participants' skills; generate exchange and knowledge	Participants who choose to	"Passion", commitment, and identification with the specialized knowledge of the group	As long as there is interest in maintaining the group
Formal Working Group	Develop a product or perform a service	Any person who presents himself to the manager of the group	Requirement of the work and common goals	Until the next re-organization
Project Team	Perform a determined task	Employees chosen by senior managers	Goals important to the project	Until the end of the project
Informal Network	Collect and transmit company information	Friends and relations within the organization	Mutual needs	As long as people are motivated to maintain contact

SOURCE: Adapted from Wenger & Snyder (2002)

munity has a specific way to make its practice visible in order to develop and share knowledge (WENGER, MCDERMOTT & SNYDER, 2002).

In this sense, community actions occur through a practical and collective activity intermediated by dialogue. Thus, one can say that the "community activity is that which gives visibility to these actions that are realized through cooperation and dialogue, the development of social and personal awareness, and the building of community responsibility" (GÓIS, 2005).

In this network of social interactions, it is clear that three core elements are necessary: domain, community, and practice. These represent different aspects of participation that motivate people to be part of a community (WENGER, MCDERMOTT & SNYDER, 2002).

Based on Wenger (2004), the elements of Domain, Community and Practice will establish and develop a group, making it a community where participants are taught and learn to share knowledge via networks, thus socializing the knowledge.

In a community of practice it is important that all members realize that the domain, like the knowledge, encompassing everything that involves the community, becomes the responsibility of all. This is extremely important because decisions are collective.

In this context, as a community of practice grows, other roles emerge and are adapted to meet the needs of the community.

In the case of leadership, Wenger (2004) suggests the appointment of an external coordinator with extensive knowledge of the group and its members, as well as the domain of the community, i.e., the issues that mobilize the group and lead to action.

The coordinator and the group elect a core of approximately ten people who will motivate the community of practice, keeping it alive, enabling its language and supporting its history.

After the formation of the core, each core member must enlist two more people to assist in various activities. Thus, the allocation of tasks for each supporting person must be consistent with the domain (specialty) of each (i.e., communication, computing, distributing materials, etc.).

Over time, the relations established by members within the community will build the language of the community, creating the space in which issues of practice can be discussed.

Thus, when entering into a community of practice, an individual gains acceptance by others and becomes autonomous while at the same time engaging in collective thinking and action. Mutual support strengthens the unity and sense of belonging, making this a supportive community.

For the community of practice to share knowledge and learn to build knowledge, it needs to make use of the important strategy known as the Tree of Knowledge, the result of the theory of Pierre Lévy and Michel Authier.

Through interactive maps, the Tree of Knowledge gives visibility to different types of knowledge, aiding in building the process of learning to know. When mapping the knowledge of members of the community of practice through the construction of the Tree, this strategy represents what and how much people know and how the interactions occur. By giving value to knowledge, it mediates, organizes, and systemizes the knowledge. Thus, learning is achieved through cultural and social relations.

4. COLLECTIVE CONSTRUCTION OF A PROGRAM OF ENVIRONMENTAL EDUCATION

The focus of this study was the Rio Verde Basin and the circumstances that led to its eutrophication, especially in the community of Cercadinho, Campo Largo, which is located within the environmental protection area (APA) of the river. This study originated from the Interdisciplinary Project on Water Eutrophication in the Rio Verde Reservoir, under the auspices of Petrobras/UN REPAR (President Getúlio Vargas Refinery), beginning in 2008. Specifically it was part of the subproject Development of an Environmental Education Program for the Community Surrounding the Rio Verde Basin.

The Rio Verde is a tributary of the Iguaçu River that covers a part of Paraná State and it has significant potential for meeting the water demands of most of the Metropolitan Region of Curitiba (RMC). Its socio-environmental management requires actions emanating from public, private and civil society in order to avoid eutrophication. In this sense, it is worth mentioning the creation in 2000 of the Rio Verde APA, located across the municipalities of Campo Largo and Araucária, in order to protect the river, as can be seen in Figure 1.

In Campo Largo and within the APA, there is a community that was established along the Rio Verde. This community, called Cercadinho, is an atypical neighborhood because it is situated in the rural context but has urban characteristics and it is threatening the sustainability of the Rio Verde.

Considering this, the community has become the focus of and reason for developing an environmental education program.

4.1 THE CONTEXT OF CERCADINHO

Evidence indicates that in the mid-nineteenth to early twentieth centuries, the region was a large farm owned by the family of the Honorable Mariano de Almeida Torres. Over the years, sections of the property were sold to farming families, mainly of Polish and Italian descent, who formed the foundation of some of the colonial communities in the region.

This led to the formation of the Cercadinho neighborhood, along the old road to Campo Largo, known at the time as "The Mato Grosso Road." In the past, cattle herders stopped in the community along their journey to rest and trade with the locals and settlers (Figure 2).

As the cattle herding traffic was rather intense and economically important at the time, large enclosures (*Cercados,* in Portuguese) were constructed to shelter, protect and care for the animals (three of them still exist today), giving rise to the place name now known as "Cercadinho."

Over the years, the Cercadinho neighborhood grew and today is home to more than 3000 inhabitants who are mostly employed in the twelve regional industries, or work in Curitiba or Campo Largo.

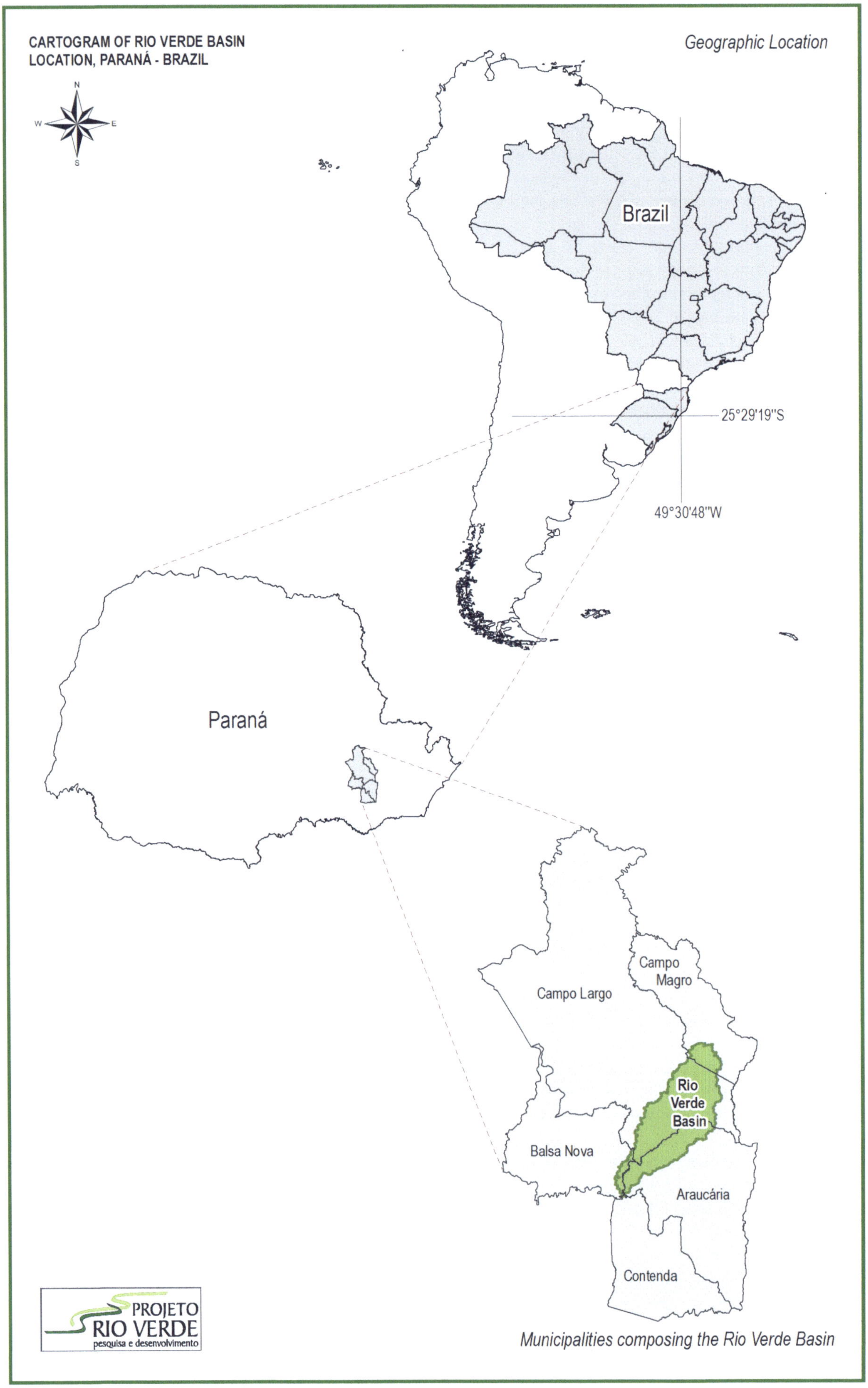

FIGURE 1 – LOCATION OF THE RIO VERDE BASIN IN THE METROPOLITAN REGION OF CURITIBA

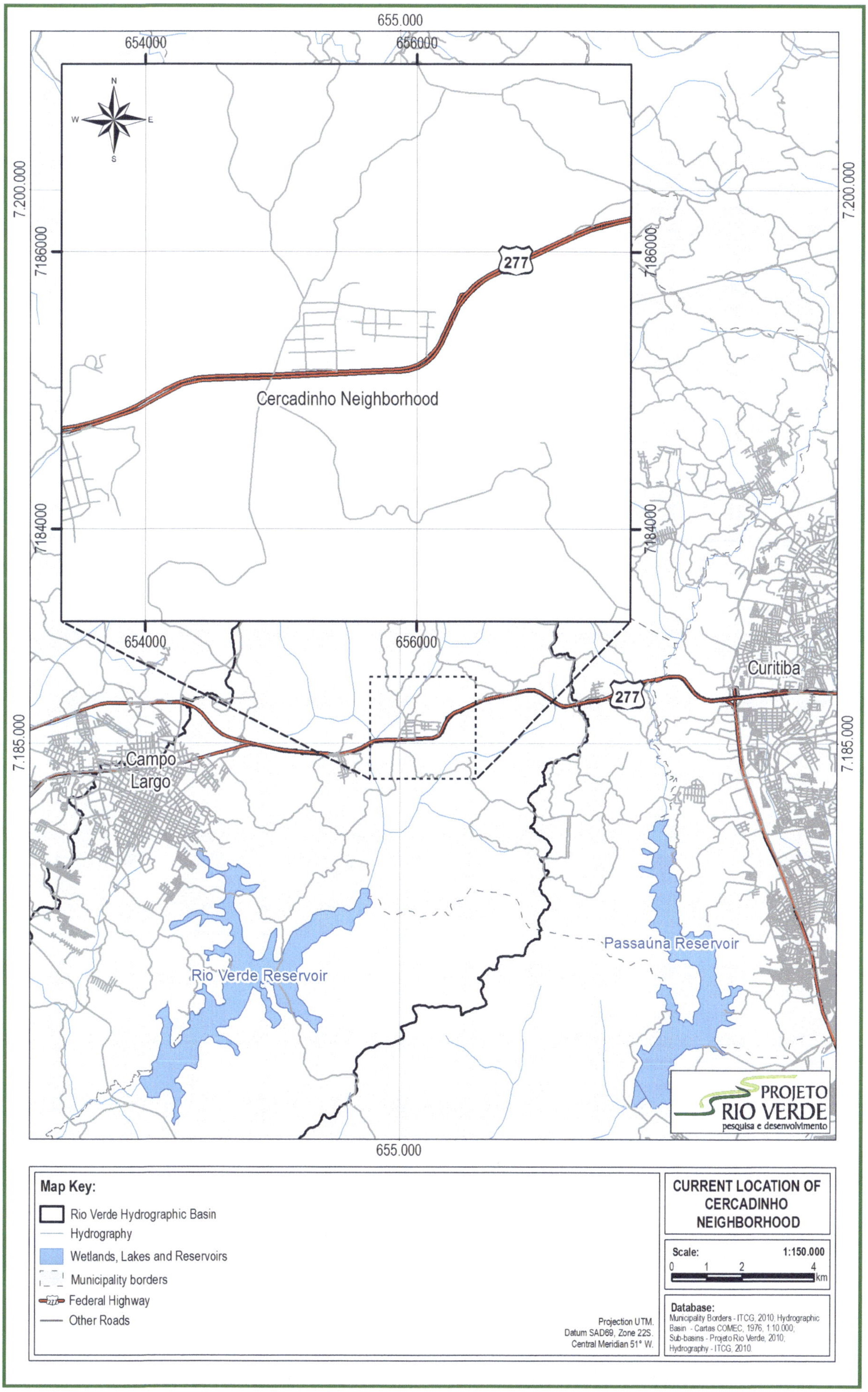

FIGURE 2 – CURRENT LOCATION OF CERCADINHO NEIGHBORHOOD. SOURCE: GOOGLE MAPS, 2010

Today, some streets are paved and urban transportation is available to various locations within the RMC. The community also has schools, markets and a Basic Health Center (UBS) (Figure 3).

FIGURE 3 – CURRENT VIEW OF THE CERCADINHO COMMUNITY

4.2 CHARACTERISTICS OF THE STUDY'S SOCIAL STAKEHOLDERS

The key social stakeholders in this study are the management team of the Municipal School 'Vereador Jose Andreassa', members of the school's Parent, Teacher and Employee Association, teachers, local community leaders, neighborhood merchants, residents and representatives of municipal organizations.

4.3 PROCEDURES USED TO DEVELOP THE STUDY

The procedures discussed below outline the steps taken to complete the qualitative, exploratory, descriptive and bibliographic case study which enabled the collective development of an environmental education program.

To support the research, the following studies were consulted: Wenger (2000) on Communities of Practice; Lévy (1998), on mapping of collective knowledge using the Knowledge Tree; Morin's various works on Complexity Theory; Jonas (2006), on the Ethics of Responsibility; and Leff (2003), on the focus of environmental education. Other studies were also consulted from scientific journals, monographs, dissertations, and theses.

All these sources contributed to the development of a community of practice that is capable of building an environmental education proposal guided by ethics of responsibility.

To carry out the project, 39 meetings and three events involving the community were conducted. The first, "River Day", occurred in November 2009, which involved an ecological walk with about 200 participants to clean the banks of the Rio Verde in Cercadinho. The second event was organized around "Water Day" and was held in March 2010. It included a week of activities involving students of Vereador Jose Andreassa Municipal School and the José Ribas Vidal State School. A Community Environmental Education Booklet, made collectively by the community of practice *Grupo Vida ao Rio Verde*, was distributed at the opening ceremony , which was attended by the Municipal Secretary of the Environment. The event attracted about 150 people from the community.

The third event was held in May 2010 and was called "Evening Pastel" (a *pastel* being a Brazilian fried snack). This event was promoted in partnership with the Parent, Teacher and Employee Association of Vereador Jose Andreassa Municipal School, to mobilize parents, students and the local community. The community of practice *Grupo Vida ao Rio Verde* held a lecture on local sustainability and alternatives for the community to reestablish lost practices. These activities intended to help local residents act sustainably through techniques such as making soap from the oil used for frying *pastels*, a way of demonstrating sustainable reuse of waste. The development of community gardens and composting was also demonstrated, prompting those present to develop a new Project called "I Take Care of My Garbage."

In order to collect data, five semi-structured interviews, two semi-structured questionnaires, 25 participatory observation meetings, and 14 non-participatory observation meetings took place and were documented. In addition, to verify the quality of the recorded statements and observations, five tape recordings were taken at different times throughout data collection.

The first questionnaire consisted of 22 questions (including open, closed and multiple-choice) relating to the environment surrounding the Rio Verde. The questionnaire was aimed at understanding the perceptions of those involved in the study and how the community perceives its relationship with the river. We also assessed the knowledge of neighborhood residents about environmental issues in the region around the Rio Verde, the environmentally protected area, and the Rio Verde basin.

The second structured questionnaire included five open-ended questions and four multiple-choice questions. The objective was a group self-assessment on the collective learning process and knowledge sharing developed with the community of practice *Grupo Vida ao Rio Verde*.

To map the perceptions of the social stakeholders in relation to the knowledge about the community, five semi-structured interviews were held because "they are appropriate when the researcher wants to grasp an understanding of the world of the interviewee and the elaborations that he uses to support his opinions and beliefs" (GODÓI, FLAG-DE-MELLO & SILVA, 2007). The participants were selected based on the number of years they had lived in the neighborhood; the older residents of Cercadinho were selected to participate in this aspect of the study.

The *daily routine notebook* is a daily diary used to record the observations, perceptions and behaviors of the individuals involved in the study and compare them with the empirical procedures and personal data recorded by the researcher.

In this case study, the daily routine notebook was the medium used to record the perceptions of those involved, as observed during meetings.

To analyze the data collected through the different techniques used, five interpretive axis were built, based on the theoretical foundations of the study. For each axis we developed analysis parameters and behavioral indicators that enabled interpretation of the perceptions of the participants involved.

Axis 1: Socio-Environmental Crisis

- Environmental Imbalance - described as a result of the actions of man in nature.
- Factors of socio-environmental crisis - characterized by demographic increase, consumption, poor quality of life.
- Environmental Education - understood as a condition for acquiring habits to care for the environment in which they live.

Axis 2: Complexity for Sustainability

- Complex thought - characterized by a thought devoid of certainties and truths; diversity of ideas; thought reeducation.
- Quality of life - described as a desire for a healthy environment.

Axis 3: Communities of Practice

- Develops members' skills - described as experience and participation that contributes to personal growth.
- Generates Knowledge - is characterized by the sharing of collective knowledge.
- Mutually defined identities (interdependent) establishing their own language - described as the mutual construction of identity and identification with the expertise of the group.
- Mutual assistance in problem solving - characterized by mutual support, appreciation of each other's opinions, shared humor.
- Conflict as a learning opportunity for the community - characterized by position negotiation.
- Commitment - described as commitment and collective engagement, marked by the presence, participation and responsibility of participants.
- Central nucleus/leadership - characterized by cooperative and collaborative leadership among members.
- External coordination - promotes development, guides and motivates the core group.

Axis 4: Knowledge Trees

- Self-organization and relative self-sufficiency - develops own regulatory mechanisms.
- Democracy of knowledge - characterized by the recognition of collective knowledge.
- Free exchange of knowledge between individuals - characterized by the willingness to learn and teach.

Axis 5: Ethics based on Hans Jonas

- Presence of Techniques - understood as the initial necessity to seek survival and quality of life.
- Ethics Crisis and Change of Imperatives - the individual cannot risk human life.
- Responsibility - characterized by an awareness of attitudes and consequences.

4.4 CASE STUDY: CERCADINHO IN THE MUNICIPALITY OF CAMPO LARGO

The aforementioned contextualization of the Cercadinho case study in Campo Largo, RMC, permitted the analysis of the procedures that mobilized the local community to build a Community Environmental Education Program. The findings resulting from the experience are described below.

4.4.1 Understanding the reality

In order to identify the social stakeholders' perceptions and their understanding of environmental issues in the region surrounding the Rio Verde and their relationship with the river, they were asked to contemplate their views on their way of life.

To identify their levels of knowledge during the meetings, the participants were asked to complete a semi-structured questionnaire consisting of 22 questions (open-ended, closed and multiple-choice) related to environmental issues in the Rio Verde region, organized as follows:

- Six open-ended questions to identify the importance of the Rio Verde for the community;
- Eight direct, closed questions aimed at identifying the perceptions regarding the quality of life of both the participant and the rest of the community;
- Three indirect, closed questions related to the participant's perceptions in relation to the quality of life in the region as well as the importance of the Rio Verde in their lives;
- Five multiple choice questions to identify the importance of the Rio Verde in the residents' activities, as well as their perceptions regarding their quality of life and quality of the community involved in this study.

The respondents are a group of 18 people, including managers of the municipal schools from the Rio Verde area in the municipality of Campo Largo, the board members of the Parent, Teachers and Employees Associations of the same schools, representatives from the Municipal Secretaries of Education and Environment, the president of the neighborhood association, businessmen, and citizens from the district.

When asked about the importance of the Rio Verde for their productive activities, 89% of respondents reported that the Rio Verde is important or very important for productive activities, highlighting the value of the river in every mentioned sector; 11% did not respond to the question.

Regarding the importance of the Rio Verde in other activities (leisure, household, etc.), 94% of the respondents indicated that it was important or very important, demonstrating the high degree of awareness with regard to their relationship with the river.

The rates 38% and 56%, as indicated by the participants, strengthen the hypothesis that they have a clear understanding of the role of the river for the community.

It became necessary to identify how the community could contribute to the improvement and conservation of the Rio Verde. This index reveals that 78% of those involved were aware of how they could better conserve the river and its surroundings. Participants noted that they should not throw garbage into the river and that they should conserve riparian vegetation.

This understanding was crucial for the realization of the proposed actions, in particular the ecological walk on River Day for the purpose of removing of garbage from the river banks. It was also played a critical role after River Day, enabling the construction of the environmental education proposal.

As for the actions needed to develop a community environmental education program and readiness to act and intervene, the stakeholders indicated possible programs such as exchanging trash for food (23%), replanting of riparian vegetation (33%), and implementing forms of surveillance (44%); the latter showing little direct influence on the river, but identifying ways to bring about improvement in the community.

A fundamental aspect of the study was identifying the stakeholders' knowledge about the restrictions on use (incentives, prohibitions, regulations, etc.) of natural resources that currently exist within the environmentally protected area (APA) of Rio Verde. The majority of respondents (85%) indicated they were not familiar with the APA and its objectives. This data is concerning but it represents an important indicator for the research because it was possible to question the respondents about their knowledge of information provided in relation to the City's Strategic Plan for Rio Verde's water resources.

Regarding the awareness of the Strategic Plan, the responses indicate the very limited awareness about the subject (6% were familiar with it, whereas 94% were not). This allows us to consider that when people are not privy to key decisions and episodes in relation to community well-being, there is very little citizen engagement, demonstrating that awareness is a necessity.

Considering the responses, we opted to identify the degree of participation of individual players in the meetings that defined the restrictions of use, the incentives, prohibitions, and regulations of natural resources, established in the City's Strategic Plan. Corroborating previous findings, the participants reported that a minority (6%) attended the meetings that defined the Strategic Plan. This emphasizes the low levels of participation and citizen engagement.

Based on the data collected, it was possible to assess the perceptions of the participants involved and conclude that although the participants demonstrated understanding of the importance of the Rio Verde and awareness of how to intervene in order to help in its preservation, the participants did not emphasize citizen activism as a means to achieve these ends.

4.4.2 Building a Community of Practice

A community of practice (COP) is created through the sharing of knowledge; sharing what you know and listening to what others know is critical.

Guided by this vision, the stakeholders involved in the case study were sensitized to the process of learning and the development of knowledge within the community. Thus, the project development across the 39 meetings demonstrates the interaction and knowledge exchange of the group, confirming the experience of building a program of community environmental education, through the principles of Communities of Practice and the Tree of Knowledge.

Stakeholders who willingly volunteered to participate in the experiment came to define themselves as a group called *Grupo Vida ao Rio Verde*. During the third meeting, they held democratic elections and agreed to develop the proposal:

> *"Folks, let's vote for the name of our community of practice. Think about it and write it down ... then we'll choose the name that fits best, okay?"*

The process was centered on dialogue, the free exchange of knowledge, and knowledge construction. Thus they began to cooperate and learn by experiencing that openness to others' knowledge and the recognition of each person's knowledge, allows for the autonomy to express one's thoughts and actions, thus constructing identity. This principle can be perceived at different times throughout the meetings:

> *"I know the history of the neighborhood; I've lived here for 37 years. Before, down here was an enclosure where they raised goats, there were a lot of fences and so became known as Cercadinho"*
> *"Yes, but it wasn't only that ... there was a Pole who planted back here on a great big plantation ... then he sold it to some people that ended up parceling off this part of his property... "*
> *"Wow, really? I never heard that story ... then what was the Pole's story, do you know?*
> *"Look, I don't think it was anything like that ... the story goes that the region belonged to Commander Mariano Torres. And it was his family that parceled it off and sold the property."*

Thus, access to other's knowledge reveals the desire to learn about the reality into which the subject is participating, contributing to the participant's own personal growth.

As the history of the community was revealed, identities, a sense of belonging, and appreciation of knowledge were constructed, developing the participants' competencies and resulting in growth of community knowledge.

In an interview, one resident, age 77, who resided for more than 60 years in the district, reported:

> *" This here was a vast stretch of land; it was such a forest that we were scared of wild beasts. Some of them attacked us and a pack of them destroyed the crops.*
> *"We had practically no neighbors ... they were far away and most were from the colonies in the region. There was no electricity; drinking water from the tap wasn't even heard of ... no streets ... and today there is even asphalt."*

To which another participant added:

> *"Hey guys, I already told you ... here in the Rio Verde, I fished all kinds of fish .... Lambari, Piaba, and even Dog's Head ..."*
> *"What? Dog's Head? What kind of fish is that?*
> *"Folks, we need to find out about this fish, I'm curious ..."*

It was observed that knowledge, when shared, brings new information to the discussion and renews old knowledge. Accordingly, at some point the interest in the history of the neighborhood was revived:

> "Before, the bus passed only at 9 am and at 2pm."
> *"Markets like we have today didn't exist back then, at that time there only those little groceries .... General stores, you know? We only ate meat on Sundays, Christmas, New Year's and Easter."*

Due to the rapid growth of the neighborhood, the Cerdadinho staekholders, both old and young, have reported concerns about violence.

They recall the past with nostalgia, when they could leave clothes, tools and other objects outside and the objects remained where they were; as there was no crime, everyone's personal property was respected. Today, crime is a constant problem in the region:

> *"Nowadays, by the time you walk to the front door, some crook is already slipping in your back door."*

Another time it was noted the importance of mobilizing to action through shared knowledge, as seen through the statement of one of the participants:

> *"I do not know much, I can hardly write, but I'm a good electrician and I understand mechanics too. I can teach the kids that spend their time on the street to start enjoying tinkering in the workshop ..."*

When a member of the community demonstrates his knowledge and shows initiative in sharing his skills, and others do the same, this results in the need to attract and integrate new members.

The members of the COP decided to seek the participation of religious leaders in the community:

> *"I have a proposal, I don't know if you agree ... but we're missing the religious leaders of the community ... should we call the priest and pastors?"*
> *"So if the pastors all get to know us and see what we are doing here, then they go back to their churches and spread the word... that will get a lot of people interested in what we are doing here!"*

It was found that understanding, social interaction and knowledge exchange between members of the community, through dialogue, allowed them to create their own ways of communicating and creating identity:

> *"Folks, we need to get everything straight ... but each in his own way ... this way we'll achieve our final goal, and one more thing, we need to learn to all talk the same language..."*
> *"We had to start working more during the daytime with them, right? What do you think?"*
> *"Yes, good idea....but wouldn't it be better to talk a little first, to find out what we each know about the environmentally protected area and the history of Cercadinho, before we go talk to others?"*

The group began making its decisions based on the general consensus:

> *"OK people, so everyone agrees to divide up the meeting time next week?"*
> *"Whoever agrees, raise your hand."*

It was also observed that at different moments of interaction members supported each other in seeking solutions to problems, making suggestions while at the same time accepting constructive criticism, valuing the opinions of each member.

The positive relationships formed enabled the members to participate in discussions and make decisions together. The willingness to listen and help each other made the community more united in pursuit of their interests. However, not all decisions were made calmly, sometimes generating conflict between COP members.

At one meeting, during the delegation of tasks for the River Day activities, there was some tension regarding who would do what.

The President of the COP asks:

> *"Attention everyone ... The spring at the water tank, we can leave that to D. and C.?"*

Both answered:

> *"What?? In the middle of nowhere?? Two women?? Why not send two men?"*

It was observed that negotiating the tasks helped to lead to consensus:

> *"Okay, okay, you're right, we have to think about the kids and their safety ..."*
> *"Yes, it's better for them to go to the springs than to the banks of the river ..."*

It was observed that negotiating was always for the benefit of the group, not personal choice, when making decisions about the River Day activities.

In this context, dialogue was established as a means of managing conflict. These moments were not considered as victory or defeat but as a learning experience for the entire community, generating consensus:

> *"At the next meeting I can arrive early, organize the room and arrange the snack table."*
> *"Don't worry; we'll do all the survey of the springs surrounding Cercadinho."*
> *The project coordinator asks:*
> *"Wow, you guys will have time for that? It's a lot of work and it might take a long time."*
> *"Don't worry, Professor. We'll split up on Sunday and walk through the woods. If each one picks an area ... we can manage!"*

It was observed that by seeking to do things together, participants were experiencing a strong sense of belonging. This sentiment was voiced repeatedly demonstrating their satisfaction in residing in Cercadinho and therefore the need to work towards improving both the community and the environmental quality of the Rio Verde.

> *"I want this community idea to work ... if it were up to me ... the people of Curitiba need to know that good and honest people live here in this neighborhood."*
> *"It's important that we write the history of how the neighborhood was created ... and this should even be taught to children in school."*
> *"I was born here; my father grew up here and caught a lot of fish in that river there ... Someday, I'd like my kids to be able to do the same thing."*
> *"I came here when I was 17. I like it here and want to improve the quality of the neighborhood. We have to have a representative who is from here, from our community."*

The collaborative and cooperative leadership happened through encouragement, motivation and commitment, mobilizing people and delegating duties. Examples of this are shown below:

> *"Shouldn't we exchange ideas first, finding out what everyone knows about the environmentally protected area and the history of Cercadinho, before going to talk to others?"*
> *"Now we need to strengthen the group with information, right? All information stemming from facts is very good!"*
> *"Hey folks, it's in our own interest to get organized. We can't wait around for something that may never come... we live here and it is our responsibility."*
> *"Look, I'll knock on doors and deliver leaflets to advertise River Day, I'll be working..."*
> *"I can put the brochures on gas cylinders."*
> *"I'll get two posters that the kids will do with the teachers and hang them in my butcher shop."*
> *"Let's discuss the theme with the university students after the lecture."*

The above examples show us that the establishment of a COP owes its initiation to external coordination but as it grows, it begins to stand alone.

In the beginning, the external coordinators brought the group together and guided and motivated the core participants in "managing" the COP. As such, it began to represent the means through which the Environmental Education Program was to be built.

4.4.3 Building the Environmental Education Program

The formation of the community of practice allowed us to systematize the findings emerging from the study at different times. This enabled the group to identify the human actions leading to environmental imbalance and understand that instead of transferring the responsibility for the solutions to others, each one should contribute to the solution.

In the meetings, the participants shared their perceptions about this local reality:

> *"There's a pig farm there in Campo Magro and the workers were throwing their manure into the river."*
> *"Our neighbor took her trash and threw it there on the side of the drainage ditch; then L. took it all and dumped rubble in the mouth of the ditch, everything from the wheelbarrow."*
> *"So I said, this is not right, because acting like this shows no conscience, no common sense."*
> *"If there's a flood, the water will have nowhere to drain. There's human waste in the drainage ditches and also rocks the size of a VW Beetle tire in there. When it floods I can already imagine someone running to blame the mayor; but we can't blame him."*
> *"We already suggested cleaning the river but even the neighbors who had a meter of water in their flooded houses backed off."*
> *"It's funny this mindset. Just this week I worked on a plumbing job. A man called me; he was building two houses and throwing the sewage from both directly into the river."*
> *"Last summer a child died of leptospirosis because he bathed in that river."*

Through comments reported during the meetings of the COP *Grupo Vida ao Rio Verde*, it was revealed that the increase in the local population and increased consumerism interfered with the environment in which they live. Also, listening to the words of some, neglect of natural resources and increase in consumerism was noted:

> *"If you look at the ravine, the mud and the stench are unbearable. 28-30 years ago, my mother washed clothes there, and we swam in the river as children, now if you want to get sick, just swim in this river."*
> *"The children once bathed in the river, but today our grandchildren only see the river as ugly and dirty."*
> *"I can save on some things, but I won't give up taking a long shower ... over half an hour."*
> *"Today there is asphalt in the neighborhood, buses and other improvements. A long time ago we had none of this, but at least the river didn't flood and the water was clean."*

Based on these findings, the group was realizing the need to seek ways to educate themselves and the community in order to improve their environment. Thus, we studied the concepts of environmental education and alternative activities that could be undertaken by the group while also demonstrating the need for more knowledge on the subject.

> *" We need a lecture on pollution... littering on the riverbanks ... look, if you were to take a walk by the river, you would be horrified because it has a lot of bottles, plastic bags, garbage, tires ... it's terrible.... it's very bad on the riverbanks ... and when the river rises it will bring all the garbage with it."*
> *" I thought that environmental education was just about cleaning the river and making a vegetable garden, but now I see that it's more than that; it's the way we do things and learn and always do....for the rest of our lives."*

The social actors indicated the need to reeducate people so that they understand the complex environment around them, showing that they accept the idea of relearning:

> *"Before you explained things to us, I never thought of the way that everything is linked together... Now when I see a bird eating a moth, I realize that nature provides everything!"*
>
> *"The issue of reeducation is very important to make people aware of the problems that happen here."*

The desire for a healthy environment became a constant theme during meetings favoring the collective construction of the Environmental Education Program.

The information collected about the views of the group at this stage show the stakeholders' awareness about the possibility of changes in the neighborhood, through the proposed Environmental Education Program.

> *"God willing, four years from now I do not want to see anyone drinking bottled water, but good, clean tap water."*
>
> *"I was born here, my father grew up here and caught a lot of fish in that river there ... Someday I want my kids to do the same thing."*
>
> *"I came here when I was 17. I like it here and want to improve the quality of the neighborhood. We have to have a representative who is from here, from our community.."*
>
> *"I want my children to grow up in a better world, so I will do everything in my power to make this happen."*

At this stage of the process, some aspects were considered fundamental to improving the quality of life of the community:

> *"The sewage system is essential, especially for those who live further down, because there if you dig a 50cm deep hole, you already hit water, so someone can dig a dead well but in reality the well is alive, you know? It's sad."*
>
> *"Here there is no stationery store, beauty salon, bakery, pharmacy, clothing stores or gift shop; if you want something like this you have to go to the center or even to Curitiba."*

The desire for a better quality of life, shown as imperative by the group, promoted the understanding that nature is vulnerable and that people can not put humanity at risk.

Such learning sparked the awareness of a necessity for new ethical attitudes within the community:

> *"I always say that unity means strength and disunity weakens. I need to give my opinion on a few points: we are lacking a social reeducation; and another thing is individualism, people are thinking only about themselves, not about society; in society your goal is to help the community so if we reeducate people to think about society instead of thinking only of themselves, we would have a better neighborhood."*
>
> *"I do not want to be misunderstood and say that the people are ignorant, but what we lack are leaders to provide the information so we can reach our ultimate goal."*

This awareness led to action, motivating the group to write a document that depicted a community practice, guided by responsibility, solidarity, ethics and education.

It was also considered necessary to disclose the activities and proposed activities of the COP. For this, it was decided to create a publicly accessible blog hosted on <www.vidarioverde.blogspot.com>. The blog is interactive so that each participant in the group can share their knowledge.

As theory and practice began to give way to reality, the collective wisdom transformed into knowledge shaped what the group called the 'Primer on Environmental Education for the Cercadinho neighborhood' and gave rise to the Environmental Education Program which incorporated the lessons learned and the desire to improve life along the river.

The results of this study demonstrate a practice of collective participation and cooperative learning. With the guidance of researchers and coordinators from outside the COP, the group that went through this process developed a text in which they described Cercadinho, the rediscovery of its history, and its occupation and location in the environmentally protected area of the Rio Verde. When conceptualizing local sustainability they developed a set of rules and norms aimed at continuous action for preserving the Rio Verde and its surroundings. The document created by the group was called 'Booklet on Environmental Education: Living alongside the Rio Verde'.

This process was considered by the group as a starting point for creating a program of environmental education for the whole community.

5. FINAL CONSIDERATIONS

This case study allowed us to adopt practices that are dedicated to environmental preservation working towards sustainability and a better quality of life locally.

In light of the emergence of Environmental Education and its historical trajectory, we sought to involve the community in bringing about changes to the way they live and interact with the Rio Verde.

Therefore, the goals of the meetings with the COP *Grupo Vida ao Rio Verde* from Cercadinho was to understand the relationships between the people in this community and the place in which they live, and to collectively build knowledge about the environment in order to promote the development of an Environmental Education Program

Thus, it was possible to find alternative means of providing information from Environmental Education to the stakeholders involved, so that they could seek balance in the relationship with their environment and local sustainability.

As a result, Environmental Education came to be considered as an instrument of change and ethical transformation, with the possibility of restoring balance between people, the environment and the community.

This finding was observed in the discussion of the social actors when they shared what they had learned. Also, by mobilizing themselves they were able to promote activities such as River Day and Water Week.

Community involvement with the Rio Verde Eutrophication project was significant because by implementing the principles of the COP as a strategy for the construction of collective knowledge, the community learned how to coexist with the river.

The day-to-day interviews, questionnaires, meetings and events of the COP portrayed the perceptions of the participants and demonstrated that the stakeholders involved, when mobilized to meet their own needs, were engaged and showed readiness to intervene.

Collective learning permeated the group's actions. The responsibility and commitment to each other promoted a sense of belonging and group identity.

Despite the contradictions and conflicts arising from local circumstances considering their history and culture, community members showed the maturity to deal with their differences.

Confronting the natural resistance to a collective learning process helped the group discover the value of dialogue, responsibility and ethics, and led to the production of a document representing their concerns and provided a glimpse into the ways life along the Rio Verde can be improved.

Respecting the river and recognizing its importance and role in the community became a personal responsibility and a reason to intervene and bring about changes in the community.

Implementing knowledge and transforming it into actions that, in turn, become life-long habits represented the greatest challenge for a community that was willing to learn how to learn.

Finally, respect for culture and local history and the mobilization of people and local authorities have demonstrated that it is possible to build a successful environmental education program.

6. REFERENCES

ÁVILA, V. F. *Educação escolar e desenvolvimento local: realidade e abstração no currículo.* Brasília: Plano, 2003.

BRASIL. IBGE. *Perfil dos Municípios Brasileiros; Meio Ambiente 2002.* Rio de Janeiro: IBGE, 2005.

BRASIL. *Lei n.° 9.795 de 27 de abril de 1999.* Dispõe sobre a educação ambiental, institui a Política Nacional de Educação Ambiental e dá outras providências.(Law 9725, from April27,1995 Provides for environmental education, establishing the National Environmental Education Policy and other provisions)

____. ProNEA – Programa Nacional de Educação Ambiental. Ministério do Meio Ambiente, Diretoria de Educação Ambiental; Ministério da Educação, *Coordenação Geral de Educação Ambiental.* 3. ed. Brasília: MMA, DF, 2005.

BUARQUE, S. C. *Metodologia de planejamento do desenvolvimento local e municipal sustentável.* Material para orientação técnica e treinamento de multiplicadores e técnicos em planejamento local e municipal. Brasília, DF: IICA, 1999.

CAPRA, F. *Conexões Ocultas.* São Paulo: Pensamento – Cultrix, 2005.

CARVALHO, I. S. M. *Salto para o futuro. Educação Ambiental no Brasil.* Ano XVIII boletim 01 – March, 2008. p.14.

COHEN, A. *The Symbolic Construction of Community.* Londres: Routledge, 1985.

DIAS, G. F.. *Educação ambiental: princípios e práticas.* São Paulo: Gaia, 1992.

GÓIS, C. W. L. *Psicologia comunitária: atividade e consciência.* Fortaleza: Publicações Instituto Paulo Freire de Estudos Psicossociais, 2005.

GOOGLE MAPS. Available in <http://maps.google.com.br/> Accessed in June, 2010.

GUERREIRO, E. P. *Responsabilidade social: a solidariedade humana para o desenvolvimento local.* Available in: <http://www.gtsi.com.br>. Accessed in:Sept 16, 2009.

JONAS, H. *O princípio responsabilidade: ensaio de uma ética para a civilização tecnológica.* Rio de Janeiro: PUC-Rio, 2006.

LAYRARGUES, P. P. Muito prazer, sou a educação ambiental, seu novo objeto de estudo sociológico. *Anais do I Encontro da Associação Nacional de Pós-graduação e Pesquisa em ambiente e sociedade.* Indaiatuba-SP, 2002. Available in: <www.anppas.org.br/.../Philippe%20Pomier%20Layrargues.pdf>. Accessed in May 23, 2010.

LAYRARGUES, P.P. Educação no processo de gestão ambiental. In: SIMPÓSIO SUL BRASILEIRO DE EDUCAÇÃO AMBIENTAL, 1. *Simpósio Gaúcho de Educação Ambiental, Semana Alto Uruguai do Meio Ambiente.* Erechim: RS Anais. Erechim: EdiFAPES, 2002.

LE BOURLEGAT C. A.. *Sustentabilidade Local – IV Simpósio Sobre Recursos Naturais e Socioeconômicos do Pantanal* – Corumbá/MS – 23 a 26 de novembro de 2004 (SIMPAN 2004 – Sustentabilidade Regional)

LEFF, E. *Epistemologia Ambiental.* Tradução de Sandra Valenzuela. 2. ed. São Paulo: Cortez, 2002.

LÉVY, P; AUTHIER, M. *As árvores de conhecimentos.* 2. ed. São Paulo. Editora Escuta, 2000.

LÉVY, P. *Inteligência Coletiva.* São Paulo. Edições Loyola, 1998.

LIMA, G. F. C. Educação ambiental crítica: do socioambientalismo às sociedades sustentáveis. *Educação e Pesquisa.* São Paulo, jan./abr. 2009 vol. 35 n.° 1. ISSN 1517-9702

LOVELOCK, J. *A vingança de Gaia.* Tradução de Ivo Korytowski. Rio de Janeiro: Intrínseca, 2006.

McCORMICK, J. *Rumo ao Paraíso: a história do movimento ambientalista.* Rio de Janeiro: Relume-Dumará, l992.

MORIN, E. *Introdução ao pensamento complexo.* Sulina, 2005.

Tratado de Educação Ambiental para Sociedades Sustentáveis e Responsabilidade Global-1992. Available in: <http://www.rebea.org.br/vquemsomos.php?cod=908>. Accessed in August 10,2005.

WENGER, E. *Comunidades de prática, 2004.* Available in: <www.ewenger.com>. Accessed in: March 23, 2009.

WENGER, E.; McDERMOTT, R.; SNYDER, W. M. *Cultivating communities of practice: a guide to managing Knowledge.* Harvard Business School Press, 2002.

SECTION VII

MANAGEMENT TOOLS

CHAPTER 23

ADVANCED OXIDATION PROCESSES

Patricio Peralta-Zamora

ADVANCED OXIDATION PROCESSES

Patricio Peralta-Zamora

SUMMARY

The main objective of the present study was to verify the capability of some advanced oxidation processes, such as heterogeneous photocatalysis (TiO_2/UV and ZnO/UV) and its electrochemically assisted version (photoelectrochemical system), photolysis in the presence of H_2O_2 (UV/H_2O_2 system), and Fenton processes (Fe^{2+}/H_2O_2 and Fe^{2+}/H_2O_2/ UV), in relation to the remediation of waters contaminated by cyanotoxins. In laboratory experiments, practically all the processes evaluated produced an efficient degradation of the substrates under study (2-MIB, geosmin and microcystin). Within this context, attention should be given to the heterogeneous photocatalysis process, which quickly degraded microcystin in aqueous solution, with an almost complete removal in less than 5min. When applied in a continuous mode, the photocatalysis process was equally efficient, obtaining values of residual microcystine below the limit re commended by the World Health Organization (1g L^{-1}).

KEYWORDS

Advanced oxidation processes, remediation,cyanotoxins.

1. INTRODUCTION

Poor quality water is currently one of the most serious problems worldwide, second only to unequal income distribution. According to the latest Human Development Report (2006), nearly two million children die each year around the world, mainly due to diseases associated with poor sanitation. In Brazil, 47.8% of municipalities still lack systems of sewage collection and only 20.2% of the existing sewage collection systems treat the waste. In the majority of cases, sewage is discharged *in natura* into rivers, the soil or the sea, compromising water quality and encouraging eutrophication.

Under these conditions, certain populations of aquatic microorganisms grow excessively, such as cyanobacteria, leading to the phenomenon of algae blooms. These microorganisms are the main producers of compounds that give taste and odor to water, such as geosmin and 2-methylisoborneol (2-MIB). They are also the main producers of hepatotoxic and neurotoxic substances, such as microcystins, saxitoxins and cylindrospermopsins.

In general, conventional treatment procedures are ineffective in removing the contaminants mentioned above. As such, new treatment technologies have been developed that employ physical processes based on adsorption and chemical processes using oxidants, such as chlorine and ozone. In recent years, advanced oxidation processes have emerged as a promising treatment alternative as they degrade numerous resistant pollutants. Usually, the radicals that are created in this type of reaction, mainly hydroxyl radicals, enable a complete degradation of organic substrates, with high rates of mineralization. Although promising, few studies involving the degradation of 2-MIB, geosmin and cyanotoxins by oxidation processes have been assessed in the literature.

The objective of the present study is to verify the effectiveness of some homogeneous and heterogeneous advanced oxidation processes in the degradation of aqueous solutions containing geosmin, 2-MIB, and microcystin LR.

2. LITERATURE REVIEW

2.1. EUTROPHICATION AND WATER QUALITY

The poor quality of water is currently one of the most severe problems worldwide and it is a particular problem for developing countries. According to data presented in the 2006 Human Development Report (UNDP, 2006), at least one in every two people in developing countries do not have access to good quality water and a lack of sanitation is considered among the leading causes of childhood mortality. In Brazil, the degradation of water quality has created unsustainable situations in the largest metropolitan regions and domestic effluents are the main source of pollution in aquatic ecosystems. According to relatively recent research, 47.8% of Brazilian municipalities still lack sewage collection systems and only 20.2% of the existing systems treat the sewage (IBGE, 2000). In other municipalities, sewage is discharged *in natura* into rivers, the soil or the sea, jeopardizing water quality and causing eutrophication.

Eutrophication is a phenomenon of complex causes and effects; but it can be characterized as the acceleration of a natural biological production process in rivers, lakes and reservoirs, caused by an increase in the level of nutrients, such as nitrogen and phosphorus. In natural water bodies two types of eutrophication can be observed: natural and anthropogenic (RAST & THORNTON, 1996). Natural eutrophication depends only on local geology and the characteristics of the surrounding environment. Anthropogenic eutrophication is associated with human activities which usually contribute heavily to the acceleration of the natural process. Some water bodies are naturally eutrophic; however, in most cases an excess in nutrients is of anthropogenic origin, especially related to urban sewage and agricultural and industrial effluents.

The eutrophication of water bodies results in countless adverse effects and in general the greater the increase in nutrients, the poorer the water quality. One of the main problems related to the quality of eutrophicated water is the proliferation of cyanobacteria (or blue algae) called blooms (AZEVEDO *et al.*, 2002). The excessive growth of these microorganisms has various impacts, including: ecological, related to the loss of biodiversity; aesthetic, associated with the production of compounds that change the odor and taste of the water; economic, mainly associated with an increase in the cost of water treatment; and human health, mainly because of toxin production.

The most frequently produced compounds are geosmin (*trans*-1,10-dimethyl-*trans*-9-decalol) and 2-methylisoborneol (1,2,7,7-tetramethyl-exo-bicyclo-[2,2,1]-heptan-2--ol or 2-MIB), the chemical structure of which are presented in Figure 1. These substances have been identified as key agents of change in the organoleptic properties of water, giving an earthy and musty odor to aquatic environments. Since humans detect these compounds even at very low levels (6-10 ng L^{-1} for geosmin and 2-20 ng L^{-1} for 2-MIB), their presence has become a serious problem for sanitation companies (CHORUS & BARTRAM, 1999; DI BERNARDO, 1995). Toxicological studies involving geosmin and 2-MIB are still limited. Dionigi *et al.* (1993), however, demonstrated that these two compounds induce no mutagenic responses in tests using *Salmonella typhimurium,* even when testing concentrations that are six-fold higher than the detection limit of each of the compounds.

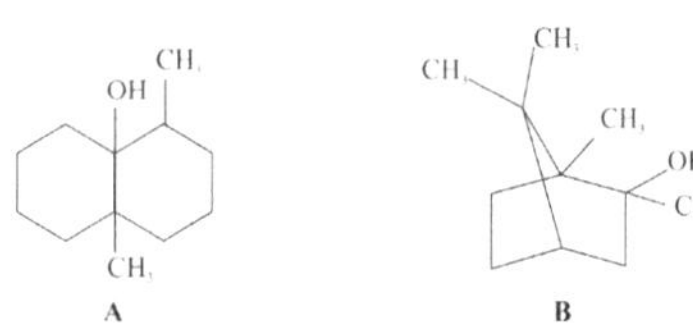

FIGURE 1 – CHEMICAL STRUCTURE OF GEOSMIN (A) AND 2-METHYLISOBORNEOL (B)

In addition, some species of cyanobacteria produce toxins, so-called cyanotoxins. Although the causes of this production are not yet well understood, it seems that these products have a protective function against herbivory, as occurs with some metabolites of vascular plants (CARMICHAEL, 1992). In Brazil, studies have confirmed the occurrence of toxic strains of cyanobacteria in water bodies (public water supply reservoirs, artificial lakes, brackish lagoons, and rivers) across several states, such as São Paulo, Rio de Janeiro, Minas Gerais, Pará, Paraná, Rio Grande do Sul, Bahia, Pernambuco, and the Federal District (FUNASA, 2003; YUNES *et al.*, 1996).

According to Sant'Anna and Azevedo (2000), the occurrence of 20 species of potentially toxic cyanobacteria, from 14 genera, has been observed in different Brazilian aquatic environments. According to the authors, *Microcystis aeruginosa* is the most widely distributed in Brazil, while *Anabaena* is the genus with the greatest number of potentially toxic species (*A. circinalis, A. flos-aquae, A. planctonica, A. solitaria* and *A. spiroides*). In addition, over the last decade a significant increase in the occurrence of the species *Cylindrospermopsis raciborskii* was observed in different regions of Brazil (BOUVY *et al.*, 1999; CONTE *et al.*, 2000; HUSZAR *et al.*, 2000).

Cyanotoxins are released into the environment when the cells are broken down through the natural biological process of cell lysis or by artificial induction of cell destruction during treatment processes. The latter is of particular interest in the production of drinkable water. The most common type of poisoning involving cyanobacteria is caused by hepatotoxins, which have a slower rate of action and can cause death within a span of a few hours to a few days. Microcystins produced by the genera *Microcystis, Nodularia, Anabaena, Oscillatoria, Planktothrix* and *Nostoc,* are among the main hepatotoxic cyanotoxins (HAIDER *et al.*, 2003; CARMICHAEL, 1994).

Microcystins are low molecular weight, cyclic heptapeptides, with a generic structure represented by cyclo-(--D-Ala^1-X^2- D -MeAsp^3-Y^4-Adda^5- D –Glu^6-Mdha^7-), where X and Y are variable L-amino acids (RINEHART *et al.*, 1988). The Adda portion was determined as one of the segments responsible for the biological activity of hepatotoxins (HARADA *et al.*, 1990; NISHIWAKI-MATUSUSHIMA *et al.*, 1992). At least 70 different types of microcystins have been found in cyanobacterial blooms or cultures in the laboratory. However, the variant most frequently found in environmental samples, is one of the most toxic, microcystin-LR, which contains L-leucine and L-arginine in variable positions (Figure 2) (SIVONEN, 1998; CARMICHAEL, 1994). The toxicity of these microcystins in laboratory animal tests was measured as LD_{50} (p.i.) between 25 and 150mg kg^{-1} body weight and between 5,000 and 10,900µg kg^{-1} body weight by oral administration (CHORUS & BARTRAM, 1999).

Based on oral toxicity studies at sub-chronic levels performed on mice (FAWELL *et al.*, 1994) and pigs (FALCONER *et al.*, 1994), the tolerable daily intake (TDI) for microcystin-LR was established as 0.04µg kg^{-1} body weight (CHORUS & BARTRAM, 1999). From this value, a maximum acceptable limit of 1µg L^{-1} of microcystins in water for human consumption was established by the World Health Organization (WHO) and incorporated into the addendum of the Standard of Treated Water Quality published in 1998 (WHO, 1998). Brazilian legislation adopted the same parameters established by the WHO requiring weekly analysis of microcystin levels whenever the number of cells (cyanobacteria) exceeds 20,000 cells mL^{-1} at the abstraction point. The maximum allowed level must be carefully controlled for public water supply, hemodialysis clinics, and in the manufacturing of injectable drugs (MS, 2004).

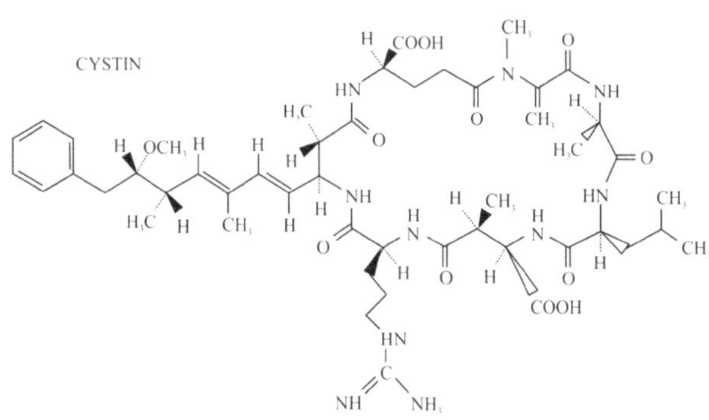

FIGURE 2 – CHEMICAL STRUCTURE OF MICROCYSTIN-LR

2.2 TREATMENT OF WATER CONTAMINATED BY CYANOTOXINS

For a long time, the use of algicides, such as copper sulfate, was considered the most effective procedure to control the growth of cyanobacteria in water reservoirs and minimize toxin production (McGUIRE & GASTON, 1988; CHORUS & BARTRAM, 1999). However, some cases have demonstrated that this process often exacerbates the problem, since it promotes the release of greater amounts of toxins and selects for species that are increasingly resistant to treatment (IZAGUIRRE, 1992). For this reason, Brazilian legislation prohibits the use of algicides in water sources when the density of cyanobacteria exceeds 20,000 cells/mL.

In general, conventional water treatments have proven inadequate for the complete removal of dissolved geosmin, 2-MIB and microcystins (LAWTON & ROBERTSON, 1999; ELLIS & KORTH, 1993; HIMBERG *et al.*, 1989; KEIJOLA *et al.*, 1988). Because they are tertiary alcohols, both geosmin and 2-MIB are very resistant to oxidation by chloride, chloramine and chlorine dioxide. In general, the efficiency of oxidation greatly depends on the type and concentration of the chlorine compound and the pH of the medium (HAIDER *et al.*, 2003; HOFFMANN, 1976; KEIJOLA *et al.*, 1988; HIMBERG *et al.*, 1989). In addition, the excess of chlorine agents in the medium can result in a residual chlorine dosage above the recommended levels and as such the water becomes unsuitable for consumption.

On the other hand, microcystins are very stable and resistant to chemical hydrolysis and to oxidation in neutral pH, due to their cyclic peptide structure. Furthermore, microcystins maintain their toxicity even after boiling. Although microcystins are resistant to many eukaryotic and bacterial peptidases, they are susceptible to degradation by some bacteria naturally found in rivers and reservoirs. Bacteria capable of degrading microcystins have been found in several aquatic ecosystems and sewage effluents (CHORUS & BARTRAM, 1999). This process can lead to the degradation of 90% of all the microcystins within 2 to 10 days, mainly depending on initial toxin concentration and water temperature.

Early studies involving microcystin degradation by chlorine indicated that the process was inefficient (HOFFMANN, 1976; KEIJOLA *et al.*, 1988; HIMBERG *et al.*, 1989). Furthermore, and as previously mentioned, the indiscriminate use of chlorine may worsen the situation in the case of a bloom. When in contact with chlorine, many organic compounds released by cyanobacteria can created chloroform and other potentially toxic organochlorines (HOEHN *et al.*, 1980).

Physical processes focusing on the elimination of volatile species (e.g. *air stripping*) are not effective for contaminants such as geosmin and 2-MIB due to the relatively low volatility of these compounds.

To solve this problem, mainly in relation to cyanotoxins, many treatment techniques have been assessed in recent years. Among the most common are the oxidation methods involving ozone and adsorption, based on the use of activated charcoal and biological processes (MEUNIER *et al.*, 2006; ELLIS & KORTH, 1993; KOCH *et al.*, 1992; LALEZARY-CRAIG *et al.*, 1988; IZAGUIRRE *et al.*, 1982).

The use of activated charcoal grew significantly in Europe and North America over the last three decades, due to the high removal capacity of soluble organic matter. Powdered activated charcoal (PAC) and granular activated charcoal (GAC) have been used in many water treatment facilities around the world (GLAZE *et al.*, 1990; LAMBERT *et al.*, 1996; COOK *et al.*, 2001). Unfortunately, because this system is not destructive, it inconveniently entails additional expenses in the disposal and treatment of residual sorbants (COOK & NEWCOMBE, 2002).

Filtration processes using membranes, such as micro- and ultra-filtration, have shown positive results in removing geosmin, 2-MIB and cyanotoxins, with 98% efficiency (CHOW *et al.*, 1997). However, the high cost of the equipment precludes its application on a large scale and it is restricted to the treatment of water used in hemodialysis or parenteral nutrition clinics (LAWTON & ROBERTSON, 1999).

The use of ozone proved to be very efficient in the removal of 2-MIB and geosmin as well as microcystins, surpassing chlorine, hydrogen peroxide, and potassium permanganate (ROSITANO *et al.*, 1998). Ozonation followed by filtration with GAC, proved to be the most efficient treatment for the removal of geosmin and 2-MIB (95% removal efficiency) (ANDO *et al.*, 1992). However, despite the high oxidizing power of ozone, its use has disadvantages, including the high cost of production and the non-selective character of the reactions (LAWTON & ROBERTSON, 1999). Ozone reacts with natural organic compounds present in the water, producing molecules with low molecular weight. These molecules, although more biodegradable, create a biologically unstable environment in water supply systems (GOEL *et al.*, 1995). In addition, an excess of natural organic matter in the water to be treated compromises the efficiency of the process, which operates best with small loads (GLAZE *et al.*, 1990; ELLIS & KORTH, 1993).

Gajdek *el al.* (2004) verified that UV radiation was effective in degrading microcystin-LR, using doses in the order of 20,000mWs/cm^2. Although effective, this alternative becomes impractical for treatment stations, since the radiation doses required significantly exceed those typically used in disinfection processes (approximately 30mWcm^{-2}). To overcome this inconvenience, recent studies have investigated the feasibility of Advanced Oxidation Processes (AOPs).

2.3 ADVANCED OXIDATION PROCESSES (AOPS)

By definition, AOPs are different reactive systems in which the hydroxyl radical (•OH) acts as the principle oxidizing agent. It is a high oxidative power species (E° = 2.8V) that must be produced *in situ* and that allows complete mineralization of many chemical species of environmental significance in a relatively short period of time (ANDREOZZI *et al.*, 1999).

Hydroxyl radicals may be generated by reactions using strong oxidants, such as ozone (O_3) and hydrogen peroxide (H_2O_2), semiconductors, such as titanium dioxide

(TiO_2) and zinc oxide (ZnO), and ultraviolet radiation (UV) (MANSILLA *et al.*, 1997).

Depending on the structure of the organic contaminant, the degradation process may involve hydrogen abstraction reactions, electrophilic addition in substances containing unsaturations and aromatic rings, electronic transference and radical-radical reactions (LEGRINI *et al.*, 1993).

When compared to other remediation processes, the main advantages of AOPs are their high mineralization capacity of resistant species, the possibility of integration with other treatment systems, and the possibility of *in situ* applications (TEIXEIRA & JARDIM, 2004).

AOPs are divided into homogeneous and heterogeneous systems, in which hydroxyl radicals are generated with or without radiation (NOGUEIRA & JARDIM, 1998).

2.3.1 Heterogeneous Photocatalysis

Within the scope of AOPs, heterogeneous photocatalysis may be considered a universal technique from which most of the concepts involved in advanced oxidation processes were derived. Moreover, recent literature on the application of these processes is quite abundant, mainly in relation to environmental decontamination (NOGUEIRA & JARDIM, 1998; LAWTON *et al.*, 2003a.)

Since 1972, when Honda and Fujishima described the decomposition of water on a TiO_2 electrode, heterogeneous photocatalysis has been used with great success in the remediation of a wide variety of contaminants, including aromatic hydrocarbons, organochlorine compounds, phenols and chlorophenols, herbicides, surfactants and dyes, among others (HERMANN *et al.*, 1993; LEGRINI *et al.*, 1993; NOGUEIRA & JARDIM, 1998; ANDREOZZI *et al.*, 1999; LAWTON *et al.*, 2003a).

Heterogeneous photocatalysis involves the activation of a semiconductor (usually TiO_2) by sunlight or artificial light. A semiconductor is characterized by valence bands (VB) and conduction bands (CB) and the region between them is called the band gap (Figure 3). Photocatalysis occurs through the absorption of a photon with energy equal to or greater than the band gap energy of the semiconductor (SC) used. This allows an electron to jump from the valence band (VB) to the conduction band (CB) and results in a gap in the valence band (GRELA *et al.*, 2001, Equation 1).The photogenerated species can participate in redox reactions with various chemical species, since the gap in the valence band is strongly oxidizing and the electron in the conduction band is moderately reducing. In aqueous systems, the most important constituents are water, molecular oxygen and other dissolved species, and any possible contaminants to be removed from the system. The gaps can react directly with the adsorbed pollutant (RXad) in the semiconductor (Equation 2); however, typically the reaction takes place in an aqueous medium promoting the oxidation of water molecules (Equation 3) or surface hydroxyl groups (Equation 4) by the photogenerated gap. Consequently, they form hydroxyl radicals (• OH) on the surface of the catalyst (GRELA *et al.*, 2001).

$$SC + hv \rightarrow e^-_{(BC)} + h^+_{(BV)} \quad (1)$$
$$SC(h^+) + RX_{ad} \rightarrow SC + RX_{ad}^{+\bullet} \quad (2)$$
$$SC(h^+) + H_2O_{ad} \rightarrow SC + {}^\bullet OH_{ad} + H^+ \quad (3)$$
$$SC(h^+) + OH^-_{ad} \rightarrow SC + {}^\bullet OH_{ad} \quad (4)$$

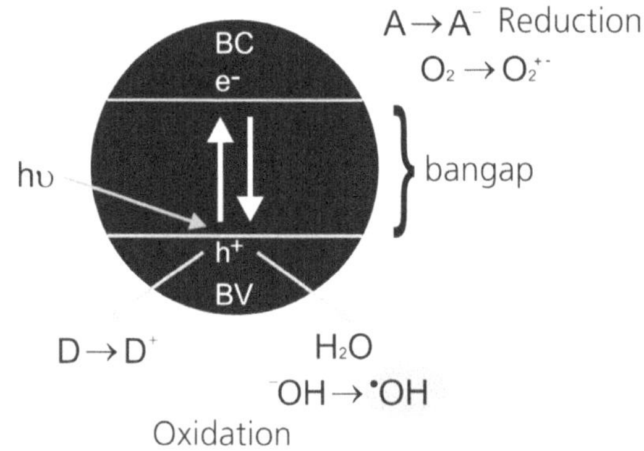

FIGURE 3 – SCHEMATIC REPRESENTATION OF THE PRINCIPLES OF HETEROGENEOUS PHOTOCATALYSIS. A: ELECTRON ACCEPTOR SPECIES; D: ELECTRON DONOR SPECIES (SOURCE: ADAPTED FROM KUNZ *et al.*, 2002).

Usually, the photocatalysts used in this type of process are semiconductor oxides, such as TiO_2, ZnO, CdS, ZnS, and Fe_2O_3, used in the form of a fine suspension. Among the semiconductors, TiO_2 is the most widely used due to its non-toxicity, photostability, low cost, and chemical stability over a wide pH range (BUTTERFIELD *et al.*, 1997; NOGUEIRA & JARDIM, 1998).

Unfortunately, there are some disadvantages associated with the use of the heterogeneous photocatalysis process, including: the need for artificial sources of radiation, which ends up increasing the cost of the process; the low quantum yield of the process for high concentrations of the photocatalyst due to the low penetration of radiation into a medium containing a suspension of fine opaque particles; the need for electron sequestering agents (typically oxygen) in order to favor the charge separation process and increase the lifetime of the electron-gap pair; and the difficulty in removing the photocatalysts at the end of the process (HERMANN *et al.*, 1993).

To circumvent the drawbacks outlined above, which greatly hinder the development of systems for large-scale operation (ZHU *et al.*, 2000; ZAMORA *et al.*, 1997), several alternatives have been proposed, including the use of solar radiation (MALATO *et al.*, 1998), the use of immobilized semiconductors (PARRA *et al.*, 2004), and the use of photoelectrochemical processes (PALOMBARI *et al.*, 2002).

2.3.2 Photoelectrochemical process

The potential of heterogeneous photocatalysis using TiO_2 in the detoxification of water was first explored using the oxide in suspension; however, the process requires a long settling time before the catalyst can be removed from the purified water. To eliminate this problem, proposals using the immobilized catalyst have attracted great interest, most notably photoelectrochemical processes in which the catalyst is immobilized on electrodes (HOIGNE & BADER, 1983; WALNUT GARDEN, 1998; SANTANA *et al.*, 2003).

In this type of process, TiO_2 is used in the form of a thin film supported on a conductive material (usually titanium). Besides the photochemical process, made possible by the use of artificial sources of radiation, an electrochemical component is incorporated that provides an external potential. In theory, the photocatalytic process is carried out in a conventional manner while applying an external potential that favors the charge separation process and thus the generation of hydroxyl radicals. The combination of the two processes has shown a synergistic effect in which the degradation rates observed are up to an order of magnitude higher than the sum of those resulting from the application of the processes individually (PELEGRINI *et al.*, 1999; PELEGRINI *et al.*, 2000).

The main reactions involved in a generic photoelectrolytic process are shown in Figure 4. In the electrolysis, the discharge of water molecules on the surface of the metal oxide anode (MOx) forms hydroxyl radicals (•OH) which are physically adsorbed (a). The incidence of radiation (*hν*) on the surface of the semiconductor film leads to the formation of the electron/gap pair with the movement of electrons to the conduction band, as shown in (b). The gap formed (h^+) makes the anodic water discharge possible according to the process shown in (c). The hydroxyl radicals adsorbed on the oxide can be transformed into higher oxides by promoting gradual oxidation, as shown in Figure 4B, processes (d) and (e), or they may oxidize organic compounds (R) directly, as shown in process (f). Equations (e) and (f) occur with simultaneous oxygen evolution (BERTAZZOLI & PELEGRINI, 2002).

The photoelectrochemical process has proven to be efficient in treating various matrices, such as landfill leachate (TAUCHERT *et al.*, 2006; BERTAZZOLI & PELEGRINI, 2002), paper industry effluents (BERTAZZOLI & PELEGRINI, 2002; PELEGRINI *et al.*, 2000), leather industry effluents (RODRIGUES *et al.*, 2008), reactive textile dyes (CATANHO *et al.*, 2006; CARNEIRO *et al.*, 2004), and chlorophenols (YANG *et al.*, 2006; WALDNER *et al.*, 2003).

2.3.3 UV/ H_2O_2 System

The UV/H_2O_2 process is one of the most well established AOPs, having been successfully used in the removal of various contaminants present in industrial waste water and effluents, including: organochlorine compounds (GLAZE *et al.*, 1995), organophosphate pesticides (BADAWY *et al.*, 2006), aromatic pesticides (AARON & OTURAN, 2001), phenols and chlorophenols (MOMANI *et al.*, 2004; PERA-TITUS *et al.*, 2004; ESPLUGAS *et al.*, 2002), dyes (MURUGANANDHAM & SWAMINATHAN, 2004; KURBUS *et al.*, 2003; ALATON *et al.*, 2002), and polycyclic aromatic hydrocarbons (PAHs) (AN & CARRAWAY, 2002).

The mechanism is characterized by homolytic cleavage of one molecule of hydrogen peroxide by UV radiation at wavelengths shorter than 280 nm, forming two hydroxyl radicals (Equation 5):

$$H_2O_2 + h\nu \rightarrow 2\ {}^{\bullet}OH \qquad (5)$$

The use of this process offers many advantages, mainly because H_2O_2 is an commercially affordable oxidant that is thermally stable and can be stored at the site with due care. As it has very high water solubility, there are no mass transfer problems associated with gases, such as is the case with ozone for example.

The efficiency of hydroxyl radical generation is dependent on the pH, concentration of substrate to be treated, and H_2O_2. In general, an excess of peroxide in the medium can promote radical-radical combination, again producing H_2O_2 and reducing reaction efficiency (ANDREOZZI *et al.*, 1999).

2.3.4 Fenton Process (Fe^{2+}/H_2O_2)

The Fenton reaction was discovered approximately 100 years ago but its application as an oxidation process for destroying toxic organic compounds was not utilized until the 1960s (HUANG *et al.*, 1993). Currently, the process is successfully used in the remediation of various types of industrial effluents (NEYENS & BAEYENS, 2003).

Although the generation of ${}^{\bullet}OH$ radicals involves a complex sequence of reactions, the process may be summarized as the generation of hydroxyl radicals from the decomposition of H_2O_2, catalyzed by ferrous ions (Fe^{2+}) in aqueous medium (Equation 6):

$$Fe^{2+} + H_2O_2 \rightarrow Fe^{3+} + {}^{\bullet}OH + OH^- \qquad (6)$$

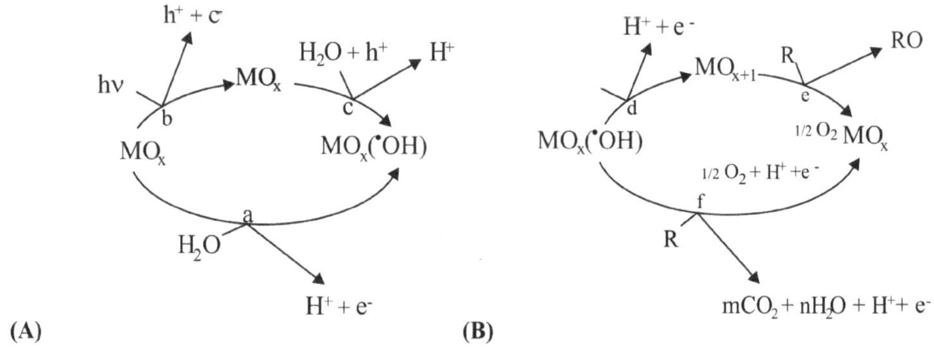

FIGURE 4 – SCHEMATIC REPRESENTATION OF OXIDATION REACTIONS IN THE PHOTOELECTROCHEMICAL PROCESS (SOURCE: BERTAZZOLI & PELEGRINI, 2002)

The Fenton process is strongly dependent on pH and the presence of H^{+} ions is required for the decomposition of H_2O_2. Previous studies have shown that pH levels near 3 are usually optimal for the occurrence of these reactions (NEYENS & BAEYENS, 2003).

The Fenton reaction began to be applied in the oxidation of organic contaminants in water, wastewater and soil after almost a century since it was first studied (NOGUEIRA *et al.*, 2007). The Fenton process has been successfully used in the treatment of water and wastewater containing phenols and chlorophenols (PEAR-TITUS *et al.*, 2004; ESPLUGAS *et al.*, 2002), organophosphate pesticides (Badawy *et al.*, 2006), hydrocarbons derived from oil (MATER *et al.*, 2007; TIBURTIUS *et al.*, 2005), landfill leachate (LOPEZ *et al.*, 2004), effluents from industrial food production (AHMADI *et al.*, 2005), and the textile industries (PEREZ *et al.*, 2002).

Features such as its operational simplicity, homogeneous character and low cost of the reactants involved, make the Fenton process a good alternative for water treatment on a large scale. On the other hand, the limitations of pH and the formation of colloidal ferric hydroxide precipitates represent the greatest difficulties in applying this method.

The oxidation of organic compounds under UV radiation in the presence of ferric ions in an acid medium was verified in the 1950s, when it was postulated that the electron transfer initiated by radiation resulted in the generation of $^{\bullet}OH$ (NOGUEIRA *et al.*, 2007). The Fenton process can be associated with UV-B radiation (280 to 320nm), UV-A (320 to 400nm), and VIS (400 to 800nm) and this is called the Photo-Fenton system. With the presence of radiation, regeneration of Fe^{2+} species occurs that reacts with the H_2O_2 in the medium (Eq. 6), thus closing the catalytic cycle and producing two hydroxyl radicals for each initially decomposed mole of H_2O_2 (Equation 7).

$$Fe^{3+} + H_2O + \text{Radiation} \rightarrow Fe^{2+} + H^{+} + {}^{\cdot}OH \qquad (7)$$

Like the Fenton process, there are limitations in using the Photo-Fenton process on large-scale as the process requires a very narrow pH range between 2.5 and 3.0 (PIGNATELLO, 1992). Above this value, the ferric ions form precipitates, drastically reducing its interaction with H_2O_2 and hence the production of $^{\bullet}OH$ radicals. At pH below 2.5 the degradation rate also decreases despite the fact that Fe species remain soluble. This is because high concentrations of H^{+} can sequestrate $^{\bullet}OH$ radicals (PIGNATELLO, 1992).

The effect of radiation on the degradation of an organic contaminant was first studied by Pignatello *et al.* (1992) who observed a significant increase in the degradation of the herbicide 2.4-dichlorophenoxyacetic acid (2.4-D) under irradiation in the presence of H_2O_2 and Fe^{3+}. Subsequent studies demonstrated that the Photo-Fenton process is efficient in degrading many toxic compounds present in water and wastewater, such as phenols and chlorophenols (POULOPOULOS *et al.*, 2008; MOMANI *et al.*, 2004; GERNJAK *et al.*, 2003), pesticides (BADAWY *et al.*, 2006; MALATO *et al.*, 2000; HUSTON & PIGNATELLO, 1999), textile dyes (PETERNEL *et al.*, 2007), organophosphate insecticides (EVGENIDOU *et al.*, 2007), and landfill leachate (PRIMO *et al.*, 2007; MORAIS & ZAMORA, 2005).

In the last 20 years, several solar photocatalytic reactors for water treatment have been developed and tested. The degradation efficiency of different classes of toxic organic compounds has made the Photo-Fenton process using sunlight quite attractive (RODRIGUEZ *et al.*, 2005; KOSITZI *et al.*, 2004; EMILIO *et al.*, 2002). Studies range from the use of different iron complexes (APLIN *et al.*, 2001) to the construction of reactors that allow for a more efficient use of solar radiation (RODRIGUEZ *et al.*, 2005; MORAES *et al.*, 2004).

2.3.5 Application of advanced processes for the remediation of contaminated waters

Despite promising results in the treatment of various kinds of pollutants, few studies on the degradation of 2-MIB, geosmin, and microcystin-LR using AOPs have been reported in the literature.

Glaze *et al.* (1990) evaluated the effectiveness of direct ultraviolet radiation (photolysis) and of the UV/ H_2O_2 process in the removal of 2-MIB and geosmin in samples of untreated and treated water. The application of 1000 $mJcm^{-2}$ of UV radiation degraded 25-50% of the initial concentration of the compounds; with the addition of hydrogen peroxide to the system, degradation reached values greater than 70%. Rosenfeldt *et al.* (2005) showed that the UV/ H_2O_2 process requires less than 5 kW/h of UV light to oxidize 90% of the initial concentration of 2-MIB and geosmin in untreated water corresponding to a cost that is lower than US$0.35 for each 3,700L.

The combined use of ozone and hydrogen peroxide, the peroxon process, removed about 90% of 2-MIB and geosmin in up to 12 minutes (KOCH *et al.*, 1992). Pilot studies showed an 80-90% degradation when peroxon (O_3/ H_2O_2) was used (KOCH *et al.*, 1992), while the heterogeneous photocatalysis using TiO_2 was able to completely degrade them after 60 minutes of treatment (LAWTON *et al.*, 2003a). Collivignarelli & Sorlini (2004) observed a complete removal of both geosmin and 2-MIB after use of ozone with UV radiation (O_3/UV).

In relation to microcystins, chlorination and ozonation were effective in the degradation of dissolved microcystins but the costs can be prohibitive, particularly since microcystin contamination is typically seasonal and unpredictable (LAWTON & ROBERTSON, 1999). Systems with TiO_2/UV, UV/H_2O_2 and TiO_2/UV/H_2O_2 have proven extremely effective in removing microcystin in water, with complete degradation after 10-35 minutes of treatment (QUIAO *et al.*, 2005; LEE *et al.*, 2004; LAWTON *et al.*, 2003b; SHEPARD *et al.*, 2002; CORNISH *et al.*, 2000; FEITZ *et al.*, 1999; LAWTON *et al.*, 1999; SHEPARD *et al.*, 2002).

Using the Fenton process, the same result was obtained after 30-180 minutes (GADJEK *et al.*, 2001; BANDALA *et al.*, 2004). However, with the Photo-Fenton process, the same removal rate was obtained after 40 minutes of treatment (BANDALA *et al.*, 2004).

3. OBJECTIVES

The main objective of this study is to verify the ability of some advanced oxidation processes in the remediation of water contaminated by cyanotoxins. The processes tested include: heterogeneous photocatalysis (TiO_2/UV and ZnO/UV) and its electrochemically assisted version (photoelectrochemical system); photolysis in the presence of H_2O_2 (UV/ H_2O_2 system); and Fenton processes (Fe^{2+}/ H_2O_2 and Fe^{2+}/ H_2O_2/ UV).

4. EXPERIMENTAL DESIGN

4.1 REAGENTS

Camphor solutions (pharmaceutical grade, 98%) were prepared with deionized water at a concentration of 50mg L^{-1}. Solutions of 2-MIB and geosmin were prepared in deionized water using Supelco (Sigma-Aldrich) standards at a concentration of 50mg L^{-1}. A stock solution of Microcystin-LR (500mg L^{-1}) was prepared in deionized water based on the Abraxis standard.

In the process of heterogeneous photocatalysis we used: TiO_2 (Degussa P25), ZnO (Merse), and commercial grade oxygen from the White Martins Company.

For the Fenton and UV/H_2O_2 processes, aqueous solutions of H_2O_2 (10% w/v), prepared from a 50% (w/v) stock solution (Peróxidos do Brasil Ltda.), and ferrous sulfate ($FeSO_4.7H_2O$, Isofar) were used. The other reagents used were all of analytical grade.

4.2 METHODOLOGY

4.2.1 Heterogeneous Photocatalysis and UV/H_2O_2

Experiments were conducted in a borosilicate glass reactor of 300 mL capacity, fitted with water cooling, magnetic stirring and oxygenation system (Figure 5). The UV radiation was provided by a mercury arc lamp of 125W (without the bulb shield), inserted into the solution through a quartz bulb. In the degradation experiments we used 250 mL of the aqueous solution of the substrate of interest in optimized pH values, with optimized quantities of the semiconductor (TiO_2 or ZnO) and hydrogen peroxide. In photocatalytic processes, oxygen was bubbled into the solution at a flow rate of approximately 50mL min^{-1}.

4.2.2 Fenton and Photo-Fenton Process

The experiments were conducted in the same reactor described above (Figure 5). 250 mL samples had their pH adjusted to previously determined and optimized values and the addition of hydrogen peroxide and Fe^{2+} solution was in previously determined concentrations. In the case of Photo-Fenton processes, two sources of radiation were used: artificial UV light (with quartz and glass bulb) and sunlight.

The experiments with sunlight were conducted first on a bench-top photochemical reactor with a 250mL capacity, equipped with magnetic stirring and an opening in the top. Solar radiation was concentrated using a parabolic solar collector coated with aluminum (Figure 6). In this reactor, 250 mL samples of camphor solution were added to optimized quantities of hydrogen peroxide and Fe^{2+} solution. The studies were conducted in the city of Curitiba (latitude 25°25'N and longitude 49° 16'E), on clear days with minimal cloud cover, between 11:00 and 14:00. Under these conditions, the average intensity of UVA radiation was 2.2 to 3.8 mW cm^{-2} (measurement performed with CosmoLUX-UVATEST® 3000 radiometer).

Degradation tests were also carried out in a continuous treatment unit (Figure 7), with four reactors of 2.6 liter capacity. UV radiation was provided by a 125W mercury vapor lamp (without original bulb) inserted at the top of each reactor and protected by a glass bulb. The residue was fed in ascending flow with the aid of a peristaltic pump. The process was performed using a 50L sample of microcystin-LR in a concentration of 5µg L^{-1}, 12.5 g of photocatalyst (TiO_2 and ZnO) and natural pH.

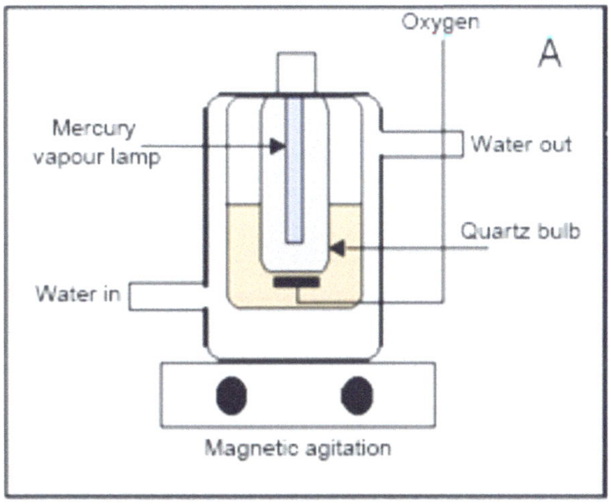

FIGURE 5 – SCHEMATIC REPRESENTATION OF THE PHOTOCHEMICAL REACTOR WITH ARTIFICIAL RADIATION

FIGURE 6 – PHOTO OF PHOTOCHEMICAL REACTOR OPERATED WITH SOLAR RADIATION

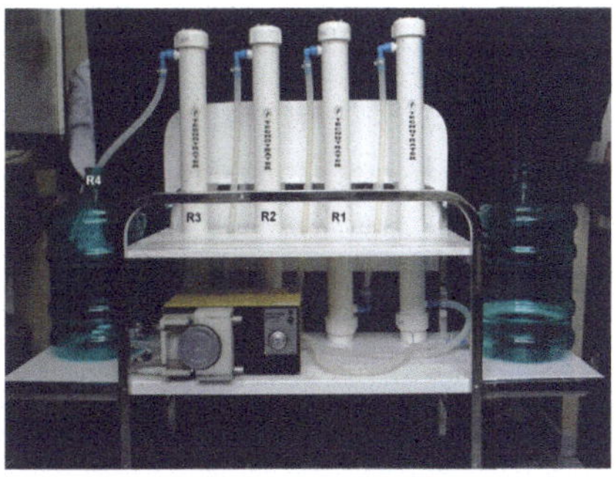

FIGURE 7 – PHOTO OF CONTINUOUS PHOTOCHEMICAL REACTOR

4.2.3 ELECTROCHEMICAL PROCESS

Experiments were performed in a 750mL borosilicate reactor (Figure 8), equipped with water cooling and magnetic stirring. The UV radiation was provided by a mercury vapor lamp (125W Philips). The current was maintained with the aid of a voltage source EMG 18134, 30V. The anode was manufactured from a titanium cylinder plate ($138cm^2$) covered with titanium and ruthenium oxides in the following proportions (w / w) 70% TiO_2 / 30% RuO_2. As a cathode, an expanded titanium grid was used. The distance between electrodes was kept at 1cm. Analytical grade Na_2SO_4 and NaCl were used as electrolytes.

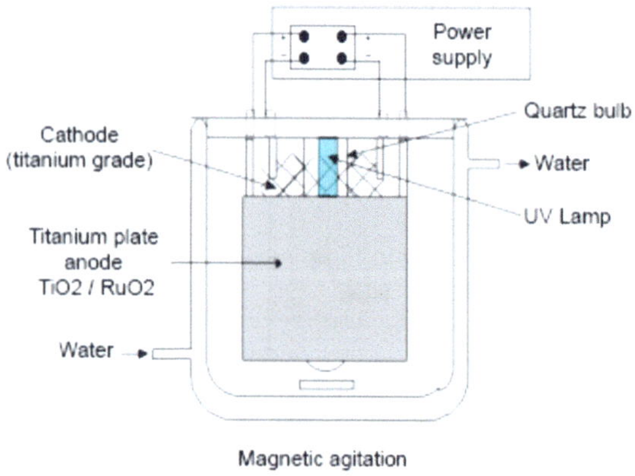

FIGURE 8 – SCHEMATIC REPRESENTATION OF THE ELECTROCHEMICALLY ASSISTED PHOTOCATALYTIC REACTOR

4.3 ANALYTICAL CONTROL

The degradation of camphor, 2-MIB and geosmin was monitored by gas chromatography after conventional liquid-liquid extraction, assisted by the salting-out effect (NOZAL *et al.*, 2002). The chromatographic analyses were performed on a Shimadzu (Model 14B) gas chromatograph equipped with a flame ionization detector (FID) and a Shimadzu model CRGA area integrator. A DB-Wax (J & W Scientific) capillary column (30m x 0.25mm) was used for all analyses. The working range was set between 5 and 50 mg L^{-1} for camphor and from 5 to 15mg L^{-1} for 2-MIB and geosmin. The preparation of the analytical curves allowed for the determination of linear correlation coefficients of 0.991, 0.981, and 0.989 for camphor, 2-MIB, and geosmin, respectively, and average coefficients of variation of 2.5%, 1.9% and 0.5%, respectively.

The degradation of microcystin-LR was monitored by enzyme-linked immunosorbent assay (ELISA ENVIROLOGIX Kit), based on the reaction between microcystin and specific monoclonal antibodies, a procedure that has a linear response range between 0.16 and 2.5mg L^{-1}. Controls were also performed by liquid chromatography coupled with mass spectrometric detection using an Agilent chromatography system consisting of a quaternary pump, degasser and Autosampler coupled to an API 4000 (Applied Biosystems) triple quadropole mass spectrometer operated in positive mode by electrospray ionization (ESI). The determination was made possible by external calibration, using aqueous standards of microcystin-LR in the concentration range between 1.0 and 10mg L^{-1}. The quantification limit of the method was estimated at 1mg L^{-1}.

5. RESULTS

5.1 CAMPHOR DEGRADATION

In order to optimize each advanced oxidation process proposed in the study, preliminary studies focused on the degradation of aqueous solutions of camphor, a chemical species selected as a standard substrate due to its structural similarity to 2-MIB.

All processes were optimized by factorial experiment design and subsequently applied in conditions of maximum efficiency of degradation. The results, summarized in Figure 9, indicate that virtually all processes showed high degradation efficiency, allowing complete removal of camphor in reaction times of less than 30 minutes. Within this context, the Photo-Fenton processes, assisted by UV-C, UV-A and solar radiation, should be highlighted as they achieved complete removal of the standard substrate in reaction times of less than 15min.

From this, the kinetic parameters presented in Table 1 were calculated, which confirmed the high degradation efficiency of the processes under study, with half-lives between 1 to 12 minutes.

TABLE 1 – KINETIC PARAMETERS OF THE ADVANCED PROCESSES APPLIED TO CAMPHOR DEGRADATION IN AQUEOUS SOLUTION

PROCESS	CANPHOR DEGRADATION	
	K_a (min^{-1})	$T_{1/2}$ (min)
Photo-Fenton (UV-C)	0,506	1,37
Photo-Fenton (UV-A)	0,238	2,91
UV/H_2O_2	0,178	3,88
Photoelectrochemical (NaCl)	0,132	5,24
UV/TiO_2	0,121	5,74
Photo-Fenton solar	0,099	7,00
ZnO/UV	0,089	7,82
Photoelectrochemical (Na_2SO_4)	0,058	11,97
Fenton	0,012	56,86

5.2 GEOSMIN AND 2-MIB DEGRADATION

For studies involving degradation of geosmin and 2-MIB only the Photo-Fenton processes assisted by UV-A and solar radiation were selected. Although they showed high degradation efficiency, processes involving UV-C were discarded. This was primarily due to the costs associated with the use of UV-C radiation, which usually precludes its implementation in the treatment of large volumes of water.

The degradation efficiency of the processes tested can be calculated from the data shown in Figure 10 and the kinetic parameters shown in Table 2. For both the process assisted by artificial radiation and by solar radiation, degradation rates above 80% were observed for both ge-

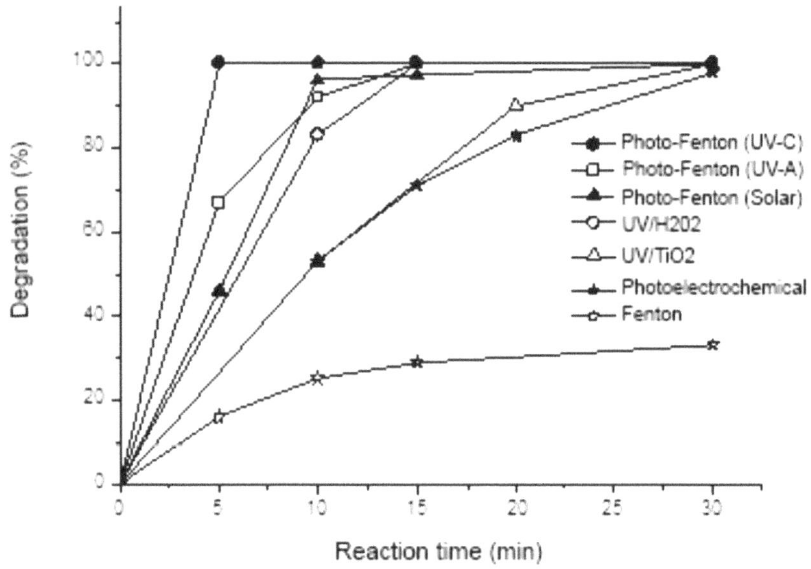

FIGURE 9 – CAMPHOR DEGRADATION USING THE ADVANCED OXIDATION PROCESSES UNDER STUDY

osmin and 2-MIB, in up to 60 minutes. In the Photo-Fenton process, radiation promotes the photoreduction of ferric ions and the formation of a further hydroxyl radical equivalent, with the concomitant regeneration of ferrous ion, a process that allows the Fenton reaction to sustain itself as long as there is H_2O_2 in the reaction medium (data not shown).

In general, solar radiation-assisted processes perform slightly poorer than artificial radiation due to the low photon efficiency of the system. In the study involving geosmin, however, we observe a significant improvement in the rate of degradation in the final treatment times, probably due to the effect of temperature. In the system assisted by solar radiation, the lack of refrigeration meant that temperatures rose significantly, up to 40°C. Under these conditions, the generation of hydroxyl radicals in Fenton processes is favored, especially when using low concentrations of ferrous ion.

The results show the feasibility of Fenton systems assisted by solar radiation. It should be noted, however, that the efficiency of this system is greatly influenced by the type of irradiation used; in this study they were extremely variable. At the time this study was conducted, the intensity of UV-A showed peaks of 40W m^{-2}, alternating in the order of 10W $m^{-2 \text{ at minimum}}$. As a result, one can expect that efficiency would increase significantly by improving the insolation conditions.

TABLE 2 – KINETIC PARAMETERS OF GEOSMIN AND 2-MIB DEGRADATION BY PHOTO-FENTON PROCESSES (GEOSMIN AND 2-MIB: 50mg L^{-1}, 200 mL; Fe^{2+}: 10 mg L^{-1}; H_2O_2: 75 mg L^{-1}; pH: 3, TEMPERATURE: 25°C± 2)

PROCESS	GEOSMIN		2-MIB	
	k_a (min^{-1})	$t_{1/2}$ (min)	k_a (min^{-1})	$t_{1/2}$ (min)
Photo-Fenton (UVA)	$5.83x10^{-2}$	11.9	$6.22x10^{-2}$	11.1
Photo-Fenton (Solar)	$5.12x10^{-2}$	13.6	$4.24x10^{-2}$	16.6

5.3 DEGRADATION OF MICROCYSTIN-LR

Initially, two processes of photocatalytic degradation of microcystin were tested: one involving titanium dioxide and the other zinc oxide. For titanium dioxide, the results indicate a strong primary adsorption, which reduces the microcystin concentration by approximately 80% during the first minutes of contact (Figure 11). Subsequently, the photocatalytic degradation induces a rapid degradation of microcystin in solution, reducing its concentration values close to the limit of quantification (about 0.3µg L^{-1}) after only 2 minutes of treatment.

The high adsorption capacity of titanium dioxide of aqueous solutions of microcystin is a well-documented fact, in some cases removing in the order of 50 to 80% (LAWTON *et.al.*, 2003).

Degradation processes involving the use of zinc oxide are quite different, with little preliminary adsorption. In this case, the photocatalytic degradation can be easily observed, which reaches values greater than 95% during the first minutes of the reaction (Figure 11).

In both cases, chromatographic monitoring allowed us to observe the appearance of degradation intermediaries. In the case of titanium dioxide, these species have low intensity peaks, while for zinc oxide significantly greater intensity is observed.

Due to the high capacity of degradation observed in preliminary tests, the main operating variables involved in the photocatalysis process were optimized by factorial experiment design. This study involved the following variables: pH and weight of photocatalyst, each of which was investigated based on the levels shown in Table 3. Due to the high adsorption capacity presented by TiO_2 in the preliminary study, the microcystin concentration was raised to 50µg L^{-1}, so as to allow observation of the photocatalytic degradation process.

The results presented as geometric representation diagrams in Figures 12 and 13 show an insignificant influence of the photocatalyst weight on the degradation ability of the photocatalytic process. In turn, the pH has a negative effect (-5) on the degradation ability of TiO_2-assisted system (Figure 12) and positive (3.5) in ZnO-assisted systems (Figure 13), suggesting a favoring of the degradation process under different conditions of acidity.

In general, the effect of pH relates to the modification of the surface properties of the semiconductor and the chemical form of the target compound. Thus, the photocatalysis process mediated byTiO_2 tends to be more efficient in an acidic medium, typically with a pH between 3 and 5.

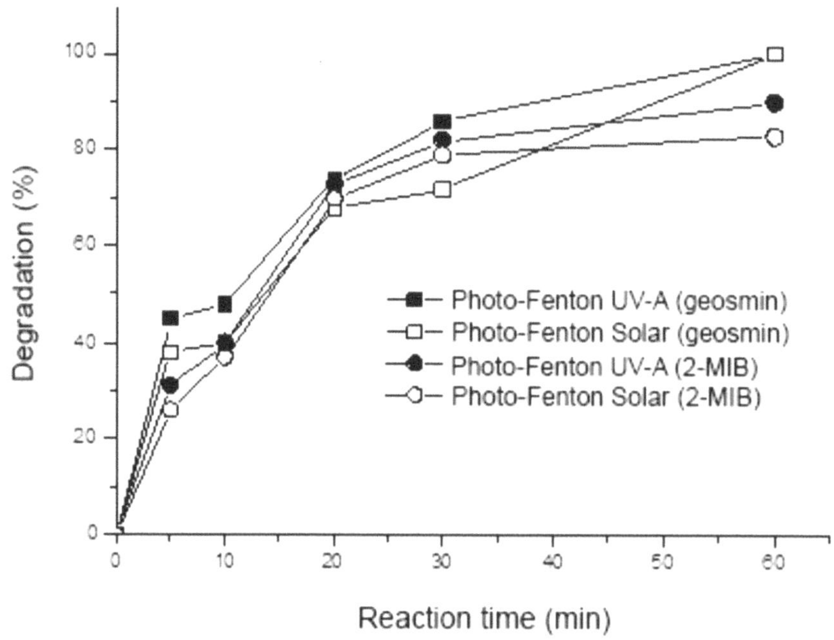

FIGURE 10 – REDUCTION OF PERCENTAGES OF GEOSMIN AND 2-MIB DURING PHOTOCHEMICAL TREATMENT BY PHOTO-FENTON PROCESSES (GEOSMIN AND 2-MIB: 50 mg L^{-1}, 200 mL; Fe^{2+}: 10 mg L^{-1}; H_2O_2: 75 mg L^{-1}; pH: 3).

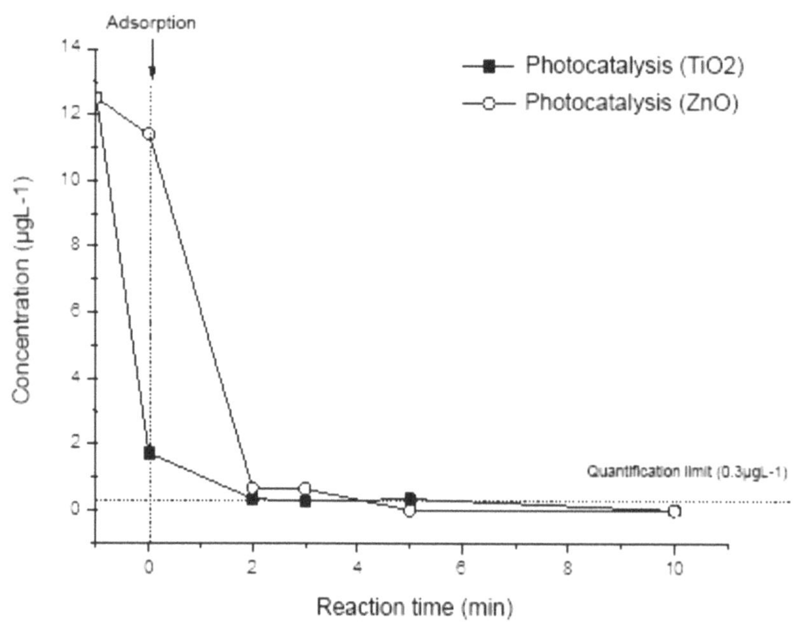

FIGURE 11 – MODIFICATION OF MICROCYSTIN CONCENTRATION DURING PHOTOCATALYTIC DEGRADATION IN THE PRESENCE OF TITANIUM DIOXIDE AND ZINC OXIDE (MICROCYSTIN: 12.5 µg L^{-1}; VOLUME: 250 mL; SEMICONDUCTOR: 50 mg; pH: 6.0 (ZnO) AND 4.0 (TiO_2).

TABLE 3 – EXPERIMENTAL CONDITIONS FOR THE OPTIMIZATION OF THE HETEROGENEOUS PHOTOCATALYSIS (MICROCYSTIN: 50µg L^{-1}; VOLUME: 200mL; REACTION TIME: 1 min.; CONTROL: ELISA)

VARIABLES/LEVEL	+	0	-	
pH MASS TiO_2 (mg)	4 50	6 100	8 150	
EXPERIMENT	pH	Mass	%DEGRADATION	
			TiO_2	ZnO
1	-	-	96	89
2	+	-	89	91
3	-	+	95	90
4	+	+	95	95
5	0	0	95+/-0,24	94+/-1,3

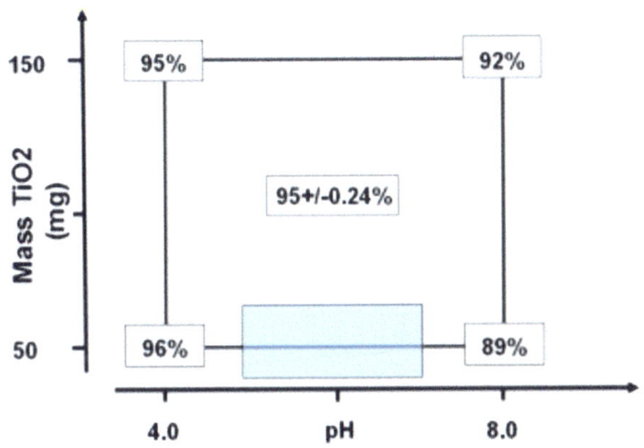

FIGURE 12 – GEOMETRIC REPRESENTATION OF THE FACTORIAL DESIGN USED FOR THE OPTIMIZATION OF HETEROGENEOUS PHOTOCATALYSIS (TiO_2). (MAIN EFFCTS: pH: -5.0+/-0.24; WEIGHT: 1.0+/-0.24; pH X WEIGHT: 2.0 +/- 0.24).

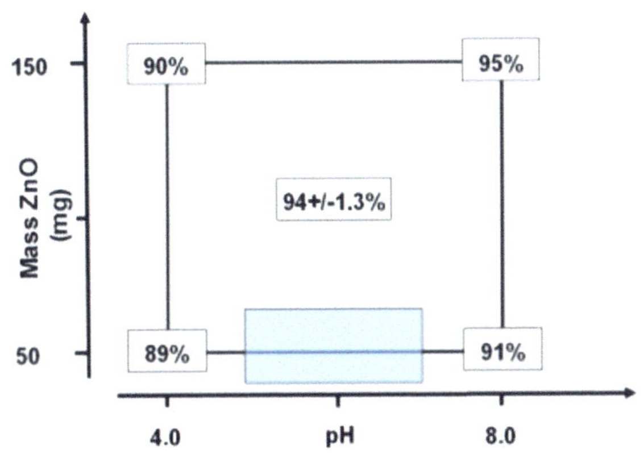

FIGURE 13 – GEOMETRIC REPRESENTATION OF THE FACTORIAL DESIGN USED FOR THE OPTIMIZATION OF HETEROGENEOUS PHOTOCATALYSIS (ZnO). (MAIN EFFECTS: pH: 3.5+/-1.3; WEIGHT: 2.5+/-1.3; pH X WEIGHT 1.5+/-1.3).

Although the calculated effects indicate that pH values produce greater efficiency, the percentage differences observed between the different conditions are relatively small. Therefore, the conditions selected for subsequent studies correspond to the natural pH of the aqueous solution of microcystin (approximately 6) and 50mg of photocatalyst. Under these conditions, both photocatalysts degrade more than 90% of the initial concentration of microcystin in only 1 minute of treatment.

Using the experimental conditions previously discussed, the degradation of aqueous solutions of microcystin (50µg $L^{-1)}$ was evaluated for a period of up to 10 minutes (Figure 14). In TiO_2-assisted processes (Figure 14A), there is a progressive adsorption of the substrate on the photocatalyst, which results in the removal of approximately 60% of the initial amount of microcystin in 10 minutes of treatment. In processes which involve the use of ZnO (Figure 14B), adsorption is much less favored, which means removal rates are in the order of 30% with the same reaction time.

The process of photolysis (UV radiation) induces a rapid degradation of the model substrate, which results in an almost complete removal in reaction times of 5 minutes. The high photosensitivity of microcystin-LR was reported by Tsuji *et al.* (1995), who demonstrated the occurrence of conformational changes in the Microcystin-LR molecule, when subjected to ultraviolet radiation. This involves a slight modification of the adda group leading to the generation of a new form of microcystin (6(Z)-Adda--microcystin-LR), which has no mutagenic or carcinogenic potential.

As seen in preliminary studies, the photocatalytic process allowed the rapid degradation of the microcystin molecule, removing about 90% in 5 minutes of treatment.

Due to the important effect caused by ultraviolet radiation, further degradation studies were carried out in the presence of lower energy radiation. To this end, the mercury vapor lamp was protected by a Pyrex glass bulb, which allows the passage of radiation with a wavelength greater than 320 nm (UV-A). Under these conditions, the photolysis process becomes less intense, while the photocatalysis process maintains its high degradation efficiency (Figure 15).

Proceeding with the studies of degradation by heterogeneous photocatalysis, we sought to expand the range of treatment by using a continuous reactor with a treatment capacity of approximately 1 L min^{-1}. The assays were performed with 50 L of an aqueous solution of microcystin-LR (5µg $L^{-1)}$ at a pH of about 6, using 12.5 g of TiO_2 or ZnO.

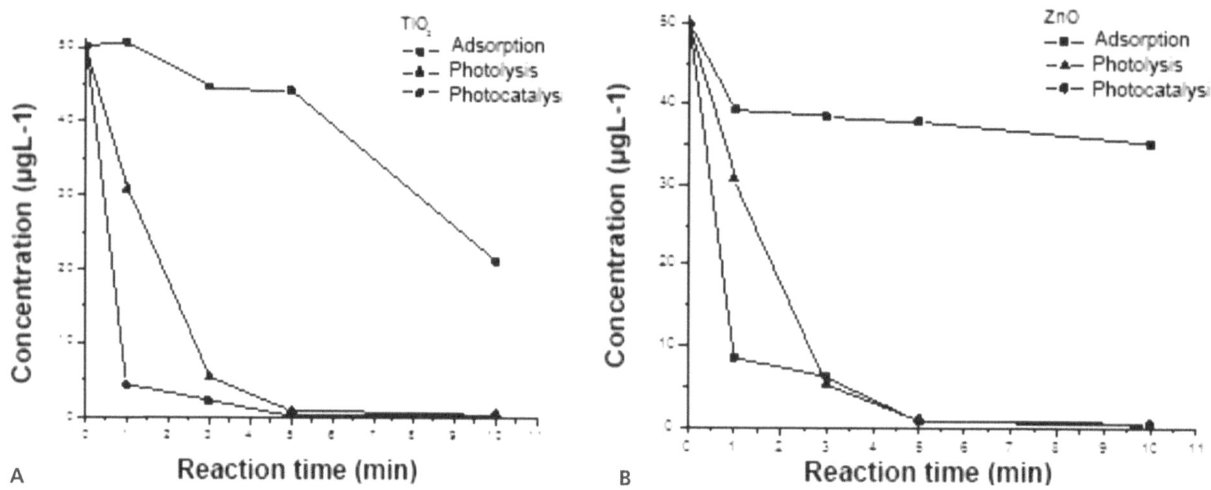

FIGURE 14 – VARIATION OF THE CONCENTRATION OF MICROCYSTIN-LR DURING THE PHOTOCATALYTIC TREATMENT ASSISTED BY (A) TiO_2 AND (B) ZnO (RADIATION UV-C).

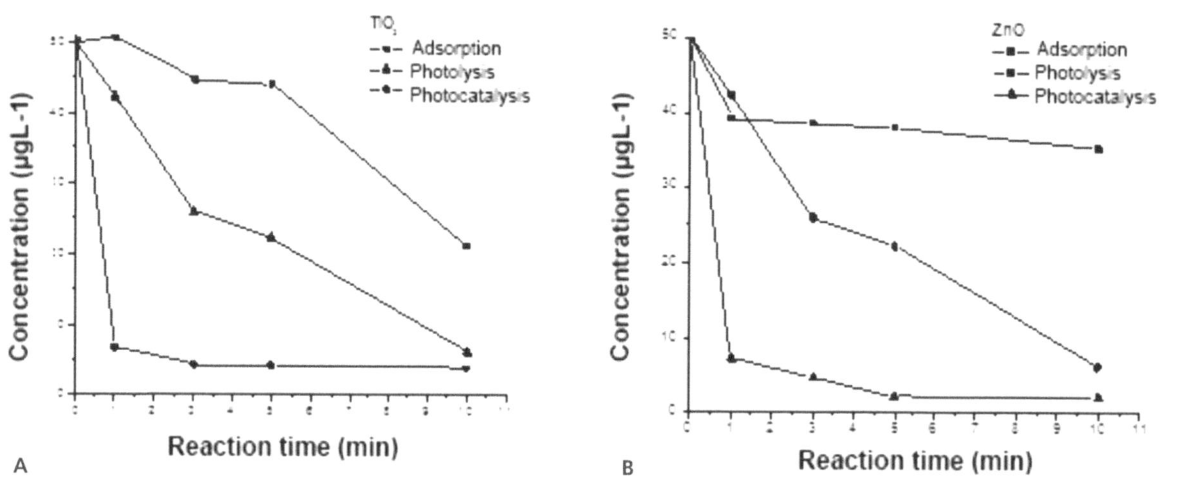

FIGURE 15 – VARIATION OF THE CONCENTRATION OF MICROCYSTIN-LR DURING THE PHOTOCATALYTIC TREATMENT ASSISTED BY (A) TiO_2 AND (B) ZnO (RADIATION UV-A)

Aliquots were taken from each of the four reactors at regular intervals, analyzing the content of residual microcystin by ELISA and LC-MS. For the system UV/TiO_2 (Figure 16), the results indicate a high degradation capacity in continuous mode, which results in residual concentrations of microcystin below the limit recommended by the WHO (1 μgL^{-1}) in all reactors and at all times sampled. In reactor 4, which corresponds to the highest retention time (approximately 10 minutes) we found concentrations below the limit of detection, set at approximately 0.1 $\mu g\ L^{-1}$.

The degradation efficiency of the system UV/ZnO (Figure 17) was slightly lower, resulting in residual concentrations below the limit stipulated by the WHO only in the final reactor.

These results suggest a significant potential of the photocatalytic continuous process for the treatment of waters contaminated with this type of substrate.

Due to the high degradation efficiency presented by the photocatalytic process, further degradation studies were performed using photocatalytic processes assisted by electrochemical processes (photoelectrochemical process). Because of the difficulties encountered initially in monitoring the processes of microcystin degradation, all preliminary studies aimed at optimizing the photoelectrochemical process were conducted with blue dye QR 19, a standard dye easily detectable in spectrophotometric monitoring.

Initially, the effect of relevant experimental variables was investigated by factorial experiment design, using the discoloration of the standard dye as an analytical response. The study involved the following variables: pH, concentration of electrolyte, and current density (J), each of which was studied at the levels shown in Table 4.

Sodium sulphate and chloride solutions were used as electrolyte. Due to the low relative efficiency when using sodium sulfate (discoloration in the order of 80% with reaction times of 40 minutes), only the results involving the use of sodium chloride are shown and discussed below.

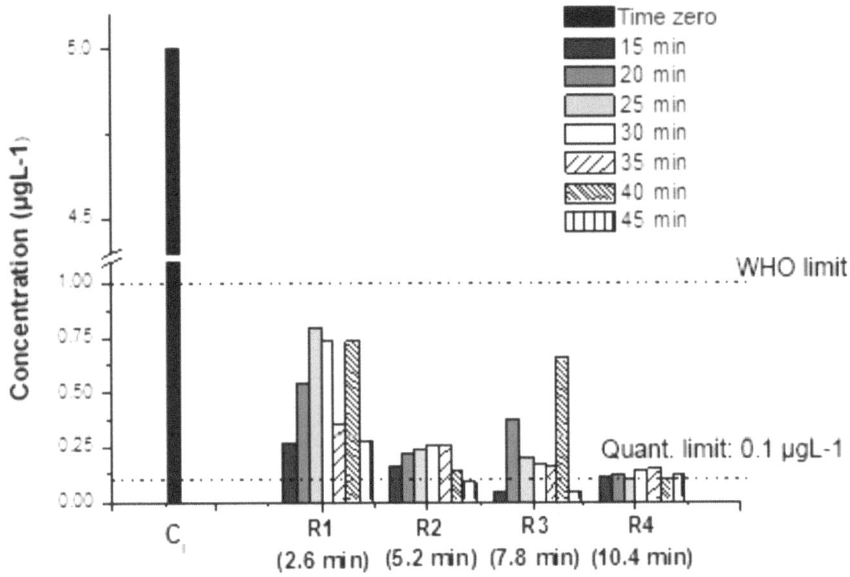

FIGURE 16 – CONCENTRATION OF RESIDUAL MICROCYSTIN-LR DURING CONTINUOUS PHOTOCATALYTIC TREATMENT OF AQUEOUS SOLUTIONS OF MICROCYSTIN-LR (UV/ TiO_2 SYSTEM).

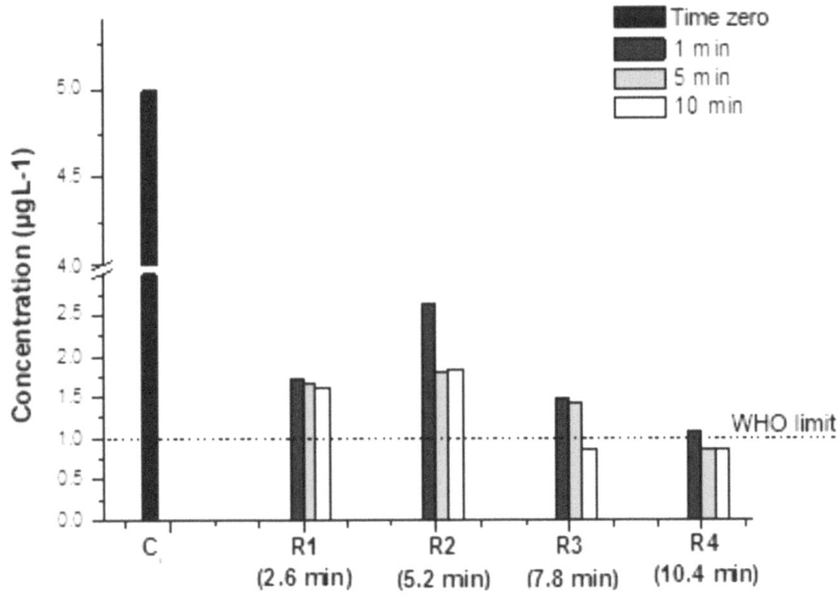

FIGURE 17 – CONCENTRATION OF RESIDUAL MICROCYSTIN-LR DURING CONTINUOUS PHOTOCATALYTIC TREATMENT OF AQUEOUS SOLUTIONS OF MICROCYSTIN-LR (UV/ZnO SYSTEM).

The use of sodium chloride increased significantly the degradation reaction rate, allowing almost complete removal of color in 1 to 2 minutes. This finding is very well illustrated in the spectra sequence shown in Figure 18, which corresponds to the degradation study performed under the conditions of the central point. As can be seen in this figure, the color is removed to a large extent in the first minute of the reaction and is practically unnoticeable in the aliquots collected at minute two. Therefore, it was not possible to assess the kinetic parameters and our study was restricted to the assessment of the degradation rate observed in one minute.

The results of factorial experiment design, expressed as percentage of discoloration at 1 minute of reaction time, are shown in Figure 19. This representation as well as the effects, indicate that the efficiency of the degradation process is significantly enhanced by increasing the concentration of the electrolyte and the current density (main effects of +47 and +39 percentage points, respectively), which is consistent with the literature (PELEGRINI *et al.*, 1999). In turn, the effect of pH is negative (main effect -9 percentage points), which implies favoring degradation process at lower pH values.

The favoring of the degradation process for higher

concentrations of electrolyte and higher current values is consistent with indirect degradation reactions mediated by active chlorine species that arise by electrochemical decomposition of chloride ion, that can form species with high oxidation potential, such as hypochlorous acid (E° = 1.5 V) or hypochlorite ion (E° = 0.89 V) (SCIALDONE *et al.*, 2009):

$$2Cl^- \rightarrow Cl_2 + 2e^-$$
$$Cl_2 + H_2O \rightarrow HClO + H^+ + Cl^-$$
$$HClO \leftrightarrow H^+ + ClO^-$$

TABLE 4 – EXPERIMENTAL CONDITIONS FOR THE OPTIMIZATION OF THE PHOTOELECTROCHEMICAL PROCESS IN THE DEGRADATION OF THE MODEL DYE (DYE: BLUE QR 19, 50mg L^{-1}, 750mL)

VARIÁBLES/LEVELS	(-)	0	(+)
pH	4	6	8
[NaCl] (mol L-1)	0.01	0.03	0.05
J (mA cm-2)	5	10	15
EXPERIMENT	**pH**	**[Na_2SO_4]**	**J**
1	-	-	-
2	+	-	-
3	-	+	-
4	+	+	-
5	-	-	+
6	+	-	+
7	-	+	+
8	+	+	+
*9	0	0	0
*10	0	0	0
*11	0	0	0

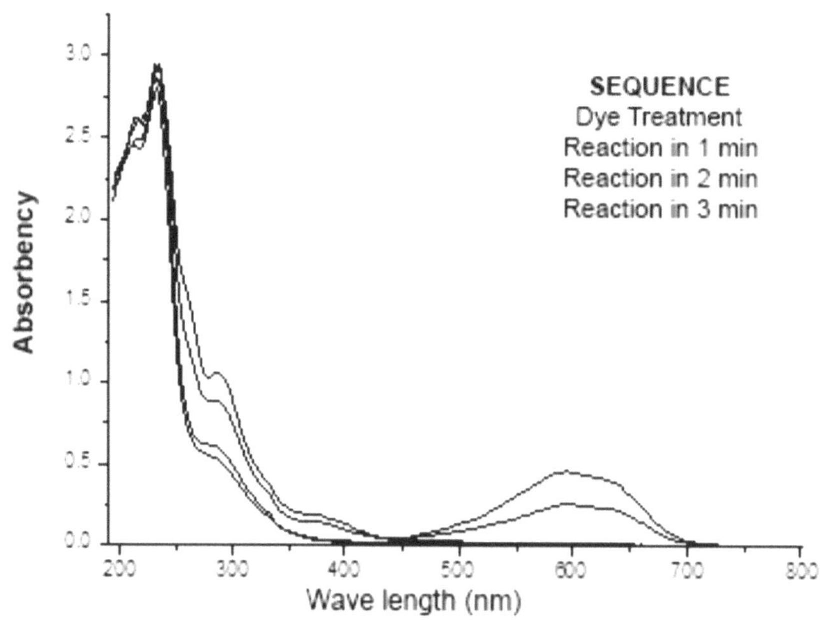

FIGURE 18 – SEQUENCE OF SPECTRA REGISTERED DURING THE PHOTOELECTROCHEMICAL DEGRADATION OF MODEL DYE, IN THE PRESENCE OF SODIUM CHLORIDE

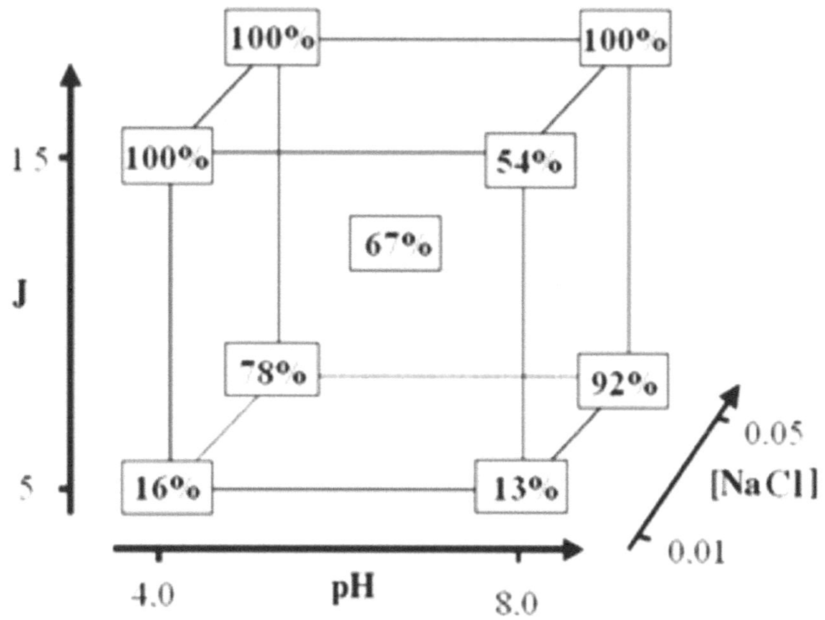

FIGURE 19 – GEOMETRIC REPRESENTATION OF FACTORIAL DESIGN USED TO OPTIMIZE THE PHOTOELECTROCHEMICAL PROCESS (NaCl). (MAIN EFFECTS: pH: -9; J: 39; [NaCl]: 47; AVERAGE DEVIATION: 2).

Generally, four concomitant processes contribute to the degradation observed in the application of a photoelectrochemical process. These are: photolysis (isolated presence of radiation); heterogeneous photocatalysis (electrode containing titanium dioxide subjected to radiation); electrolysis (electrodes subjected to external potential); and complete photoelectrochemical process. To evaluate each of these contributions, degradation tests were performed at the natural pH of the dye solution (approximately 6), using the lower concentration of NaCl (0.01 mol L^{-1}) and the lowest current density (5 mA cm^{-2})

The results (Figure 20) indicate that the photolysis process induces significant changes in the dye molecule, allowing discoloration in the order of 30% after 60 minutes. With the same reaction time, the process of heterogeneous photocatalysis allows discoloration in the order of 50%, which is relatively low, especially taking into account the isolated effect of radiation.

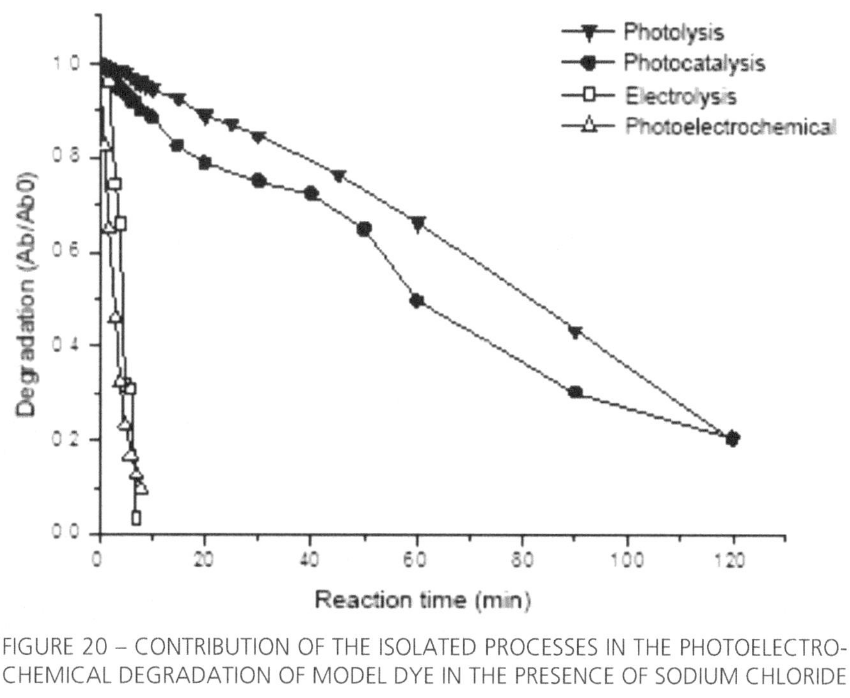

FIGURE 20 – CONTRIBUTION OF THE ISOLATED PROCESSES IN THE PHOTOELECTROCHEMICAL DEGRADATION OF MODEL DYE IN THE PRESENCE OF SODIUM CHLORIDE

In the presence of sodium chloride the electrochemical process is manifested intensely, greatly exceeding the associated processes. Additionally, the extreme similarity observed in the behavior of the dye in electrochemical and photoelectrochemical processes suggests a lesser contribution of radiation-assisted processes, suggesting again an indirect degradation mediated by chlorine oxidants.

To verify this hypothesis, both the consumption of chloride ion and the emergence of active chlorine were monitored (results not shown). To avoid interference of the dye with degradation by-products, these experiments were performed in aqueous solution containing chloride ion in a concentration of 0.01 mol L^{-1}. The results show a rapid consumption of chloride ion and maintenance of concentrations equivalent to 50% of the initial concentration until the longest times assessed (90 min.). Hypochlorite is produced concomitantly, reaching concentrations between 3.0×10^{-4} and 3.5×10^{-4} mol L^{-1} in the initial minutes of reaction, which corresponds to a small portion of the chloride decomposed initially. The hypochlorite is consumed during the degradation process, reaching values close to zero at the end of the reaction (90 min).

In order to investigate the importance of the reactions mediated by chlorine, the dye solution was subjected to chemical oxidation in the presence of hypochlorite concentration equivalent to that generated electrochemically (3.2×10^{-4} mol L^{-1}). In these conditions (Figure 21) the chromophore of the dye molecule is effectively degraded, allowing discoloration in the order of 70% in 120 minutes of treatment. In the presence of ultraviolet radiation (Figure 22), the degradation process is much faster, almost completely removing color in 4 minutes of treatment.

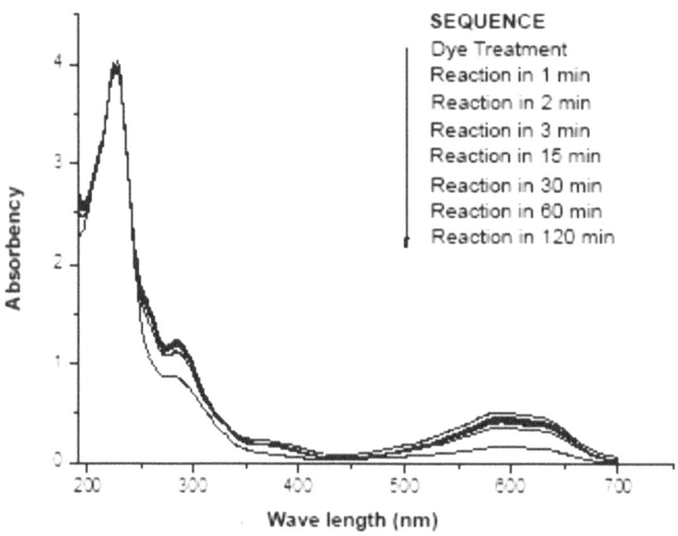

FIGURE 21 – MODIFICATION OF THE SPECTRUM PROFILE OF DYE SAMPLES SUBMITTED TO REACTION WITH HYPOCHLORITE (3.2×10^{-4})

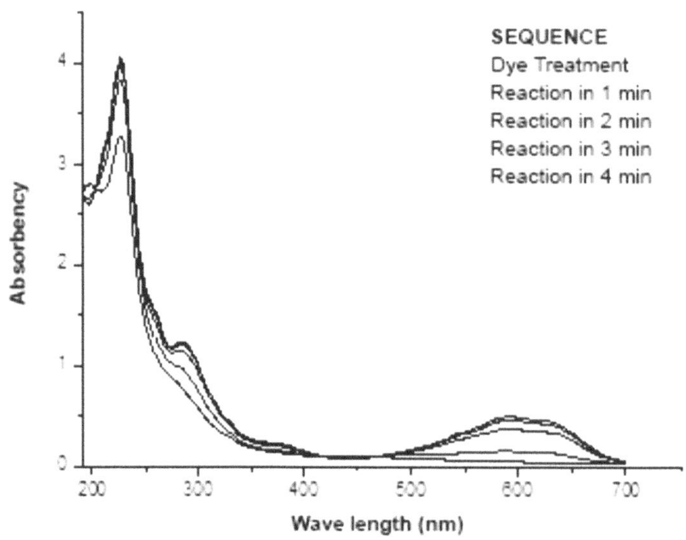

FIGURE 22 – MODIFICATION OF THE SPECTRUM PROFILE OF DYE SAMPLES SUBMITTED TO REACTION WITH HYPOCHLORITE AND UV RADIATION

The spectral behavior in the presence of light suggests the occurrence of reactions in the photochemical stage. Thus, the electrochemical process leads to the generation of strong oxidants such as hypochlorite which can be photochemically converted into radical species of chlorine, with higher oxidizing potential.

Using the previously optimized conditions (pH: 5 to 6, J: 5mA/cm^2 and [NaCl] 0.01 mol L^{-1}) photoelectrochemical experiments were carried out aiming at the degradation of microcystin-LR (50μgL^{-1}) in aqueous solution. Preliminary results are promising; a degradation rate of about 90% of the substrate is observed within reaction times of about 1 minute (Figure 23). Under these conditions, the isolated effect of hypochlorite (3.2 x 10^{-4}mol L^{-1}) and the combined action of hypochlorite and ultraviolet radiation allows degradation rates that are fairly similar (60% in 1min treatment), but significantly different from those observed in the photoelectrochemical process. Although further tests are being conducted to explain these differences, it is possible to make the preliminary assumption that the electrochemical component of the process favors the occurrence of side reactions, which contribute to the degradation process.

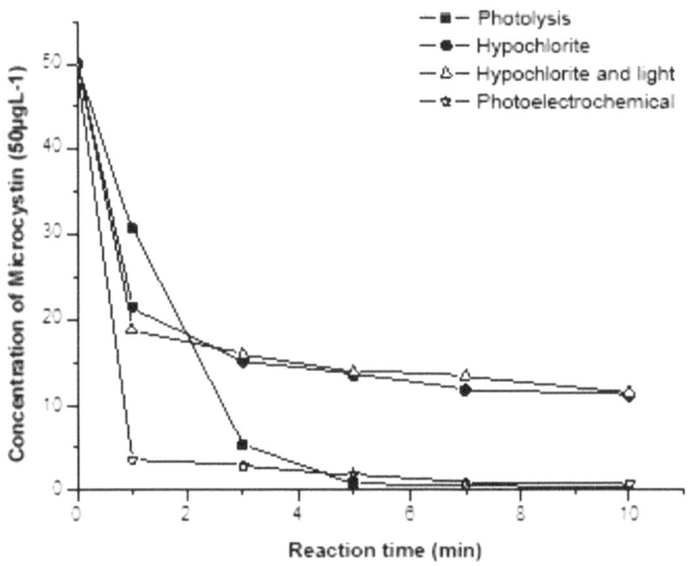

FIGURE 23 – EVOLUTION OF MICROCYSTIN-LR CONCENTRATION DURING THE PHOTOELECTROCHEMICAL PROCESS AND IN THE REACTIONS WITH HYPOCHLORITE AND HYPOCHLORITE/UV RADIATION

It is noteworthy, that there are no reports in the literature on the degradation of microcystin by photoelectrochemical processes. In studies of electrooxidation of microcystin (40 μgL^{-1}), involving the use of the electrode Ti/RuO_2 and a current density of 5mA/cm^2, the complete degradation of the substrate is only observed after 140 minutes (LIANG *et al.*, 2008). This result demonstrates the favorable effect of radiation in such processes.

6. CONCLUSIONS

Except for the Fenton process, all the advanced oxidation processes assessed present efficient camphor degradation in aqueous solution, allowing its complete degradation in reaction times of less than 30 minutes. Although more efficient, the processes assisted by UV-C radiation must be carefully considered, mainly due to the high cost of treatment systems developed using quartz. Thus, the most promising processes for application on a large scale are the Photo-Fenton processes, assisted by solar or UV-A radiation.

The use of these processes allows efficient degradation of 2-MIB and geosmin in aqueous solution, allowing the removal of more than 80% of the initial concentration in 60 minutes of treatment. Despite showing lower performance when compared to the other systems evaluated, the Photo-Fenton process mediated by solar radiation appeared as a promising alternative for treating water contaminated with geosmin and 2-MIB, mainly due to the economic and environmental advantages.

The process of heterogeneous photocatalysis rapidly degrades microcystin in aqueous solution, allowing for its virtually complete removal in reaction times of less than 5 minutes. When applied in continuous mode, the process of photocatalysis proves equally efficient, reducing the concentration of residual microcystin to values below the limit recommended by the World Health Organization (1 mg L^{-1}).

Because of the high photosensitivity of microcystin, significant degradation is observed during photolysis processes. However, this procedure is usually associated with only small conformational changes, which, having a reversible character, may revert to the starting compound.

Photoelectrochemical processes applied in the presence of sodium sulfate have low degradation efficiency of the model dye. On the other hand, processes applied in the presence of sodium chloride enable rapid degradation which causes the virtually complete discoloration of the dye solution after only 1 minute of treatment. This feature as well as the observation of the significant formation of hypochlorite, suggests the existence of indirect oxidative mechanisms which are enhanced by concomitant use of UV radiation.

When applied to microcystin degradation studies, the photoelectrochemical process has high degradation efficiency, removing approximately 90% in reaction times of 1 minute.

7. REFERENCES

AARON, J.J.; OTURAN, M.A. (2001) New photochemical and electrochemical methods for the degradation of pesticides in aqueous media. Environmental applications. *Turish Journal of Chemistry*, 25, 509-520.

AHMADI, M.; VAHABZADEH, F.; BONAKDARPOUR, B.; MOFARRAH, E.; MEHRANIAN, M. (2005) Application of the central composite design and response surface methodology to the advanced treatment of olive oil processing wastewater using Fenton's peroxidation. *Journal of Hazardous Materials*, 123, 1-3, 187-195.

ALATON, I.A.; BALCIOGLU, I.A.; BAHNEMANN, D.W. (2002) Advanced oxidation of a reactive dyebath effluent: comparison of O_3 H_2O_2/UV-C and TiO_2/UV-A processes. *Water Research*, 36:1143-1154.

AN, Y.J.; CARRAWAY, E.R. (2002) PAH degradation by UV/H_2O_2 in perfluorinated surfactant solutions. *Water Research*, 36, 1, 309-314.

ANDO, A.; MIWA, M.; KAJINO, M.; TATSUMI, S. (1992) Removal of musty-odorous compounds in water and retained in algal cells through water purification processes. *Water Science and Technology*, 25, 2, 299-306.

ANDREOZZI, R.; CAPRIO, V.; INSOLA, A.; MAROTTA, R. (1999) Advanced oxidation processes (AOP) for water purification and recovery. *Catalysis Today*, 53, 51-59.

APLIN, R.; FEITZ, A.J.; WAITE, T.D. (2001) Effect of Fe(III)-ligand properties on effectiveness of modified photo-Fenton processes. *Water Science and Technology*, 44, 5, 23–30.

AZEVEDO, S.M.F.O.; CARMICHAEL, W.W.; JOCHIMSEN, K.L.R.; LAU, S.; SHAW, G.R.; EAGLESHAM, G.K. (2002) Human intoxication by microcystins during renal dialysis treatment in Caruaru-Brazil. *Toxicology*, 181-182, 27, 441-446.

BADAWY, M.I.; GHALY, M.Y.; GAD-ALLAH, T.A. (2006) Advanced oxidation processes for the removal of organophosphorus pesticides from wastewater. *Desalination*, 194, 1-3, 166-175.

BANDALA, E.R.; MARTÍNEZ, D.; MARTÍNEZ, E.; DIONYSIOU, D.D. (2004) Degradation of microcystin-LR toxin by Fenton and Photo-Fenton processes. *Toxicon*, 43 (7), 829-832.

BERTAZZOLI R., PELEGRINI R. (2002) Descoloração e degradação de poluentes orgânicos em soluções aquosas através do processo fotoeletroquímico. *Química Nova*, 25, 3, 477-482.

BOUVY, M.; MOLICA, R.; OLIVEIRA, S.; MARINHO, M.; BEKER, B. (1999) Dynamics of a toxic cyanobacterial bloom (*Cylindrospermopsis raciborskii*) in a shallow reservoir in the semi-arid region of northeast Brazil. *Aquatic Microbial Ecology*, 20, 285-297.

BUTTERFIELD, I.M.; CHRISTENSEN, P.A.; HAMNETT, A.; SHAW, K.E; WALKER, G.M.; WALKER, S.A.; HOWARTH, C.R. (1997) Applied studies on immobilized titanium dioxide films as catalysts for the photoelectrochemical detoxification of water. *J. Appl. Electrochem.*, 27, 4, 385-395.

CARMICHAEL, W.W. (1992) Cyanobacteria Secondary Metabolites – The Cyanotoxins. *J. Appl. Bacteriol.*, 72, 445.

CARMICHAEL, W.W. (1994) Toxins of Cyanobaceria. *Scientific American*, 270, 1, 7-86.

CARNEIRO, P.A.; OSUGI, M.E.; SENE, J.J.; ANDERSON, M.A.; ZANONI, M.V.B. (2004) Evaluation of color removal and degradation of a reactive textile azo dye on nanoporous TiO_2 thin-film electrodes. *Electrochimica Acta*, 49, 22-23, 3807-3820.

CATANHO, M.; MALPASS, G.R.P.; MOTHEO, A.J. (2006) Photoelectrochemical treatment of the dye reactive red 198 using DSA® electrodes. *Applied Catalysis B: Environmental*, 62, 3-4, 193-20 0.

CHORUS, I.; BARTRAM, J. (Eds.). *Toxic Cyanobacteria in Water*. E & FN Spon: London and New York, 1999.

CHOW, C.W.K.; PANGLISCH, S.; HOUSE, J.; DRIKAS, M.; BURCH, M. D.; GIMBEL, R. (1997) A study of membrane filtration for the removal of cyanobacterial cells. *Journal of Water SRT – Aqua*, 46, 6, 324-334.

COLLIVIGNARELLI, C; SORLINI, S. (2004) AOPs with ozone and UV radiation in drinking water: contaminants removal and effects on disinfection byproducts formation. *Water Sci Technol.*, 49(4):51-6.

CONTE, S. M.; RABELO, I. M. M.; GIORDANI, A. T.; DEWES, W. (2000) Ocorrência de Cylindrospermopsis raciborskii nas bacias hidrograficas dos rios dos Sinos e Caí, RS-Brasil. In: *Anais do XXVII Congresso Interamericano de Engenharia Sanitária e Ambiental*, Porto Alegre, Brasil.

COOK D.; NEWCOMBE G. (2002) Removal of microcystin variants with powdered activated carbon. *Water Science & Technology: Water Supply 2*(5/6), 201-207.

COOK, D.; NEWCOMBE, G.; SZTAJNBOK, P. (2001) The Application of powdered activated carbon for MIB and geosmin removal: predicting PAC doses in four raw waters. *Water Research*, 35, 5, 1325-33.

CORNISH, B.J.P.A.; LAWTON, L.A.; ROBERTSON, P.K.J. (2000) Hydrogen peroxide enhanced photocatalytic oxidation of microcystin-LR using titanium dioxide. *Applied Catalysis B: Environmental*, 25, 1, 59-67.

Di BERNARDO, L. (1995) *Algas e suas influências na qualidade das águas e nas tecnologias de tratamento*. ABES: Rio de Janeiro.

DIONIGI, C.P.; LAWLAO, J.E.; McFARLAND; JOHNSEN, P.B. (1993) Evaluation of the effects of geosmin and 2-methylisoborneol on the histidine dependence of TA98 and TA100 *Salmonella typhimurian tester trains. Water Resources*, 27, 1615-8.

ELLIS, J; KORTH, W. (1993) Removal of geosmin and methylisoborneol from drinking-water by adsorption on ultrastable zeolite-y. *Water Research*, 27, 4, 535-539.

EMILIO, C.A.; JARDIM, W.F.; LITTER, M.L.; MANSILLA, H.D. (2002) EDTA destruction using the solar ferrioxalate advanced oxidation technology (AOT): Comparison with solar photo-Fenton treatment. *Journal of Photochemistry and Photobiology A: Chemistry*, 151, 1-3, 121-127.

ESPLUGAS, S; GIMENEZ, J; CONTRERAS, S; PASCUAL, E; RODRIGUEZ, M. (2002) Comparison of different advanced oxidation processes for phenol degradation. *Water Res.*, 36, (4), 1034-42.

EVGENIDOU, E. KONSTANTINOU, I.; FYTIANOS, K.; POULIOS, I. (2007) Oxidation of two organophosphorous insecticides by the photo-assisted Fenton reaction. *Water Research*, 41, 9, 2015-2027.

FALCONER, I.R.; BURCH, M.D.; STEFFENSEN, D.A.; CHOICE, M.; COVERDALE, O.R. (1994) Toxicity of the blue-green alga (cyanobacterium) *Microcystis aeruginosa* water to growing pigs, as an animal model for human injury and risk assessment. *Environmental Toxicology and Water Quality Journal*, 9, 131-9.

FAWELL, J.K.; JAMES, C.P.; JAMES, H.A. (1994) *Toxins from blue-green algae: Toxicological assessment of microcystin-LR and a method for its determination in water.* Medmenham: Water Research Centre.

FEITZ, A.J.; WAITE, T.D.; JONES, G.J.; BOYDEN, B.H.; Orr, P.T. (1999) Photocatalytic degradation of the blue green algal toxin microcystin-LR in a natural organic-aqueous matrix. *Environ. Sci. Technol.* 33, 243-249.

FUNASA- Fundação Nacional de Saúde (2003). *Cianobactérias Tóxicas na Água para Consumo Humano na Saúde Pública e Processos de Remoção em Água para Consumo Humano.* Available in http://bvsms.saude.gov.br/bvs/publicacoes. Accessed in January, 2008.

GAJDEK, P.; LECHOWSKI, Z.; BOCHNIA, T.; KEPCZYNSKI, M. (2001) Decomposition of microcystin-LR by Fenton oxidation. *Toxicon*, 39, 1575-9.

GAJDEK P, BOBER B, MEJ E, BIALCZYK J. (2004) Sensitised decomposition of microcystin-LR using UV radiation. *Journal of Photochemistry and Photobiology B-Biology* 76 (1-3): 103-106.

GERNJAK, W.; KRUTZLER, T.; GLASER, A.; MALATO, S.; CACERES, J.; BAUER, R.; FERNÁNDEZ-ALBA, A.R. (2003) Photo-Fenton treatment of water containing natural phenolic pollutants. *Chemosphere*, 50, 1, 71-78.

GLAZE, W.H.; SCHEP, R.; CHAUNCEY, W.; RUTH, E.C.; ZARNOCH, J.J.; AIETA, E.M.; TATE, C.H.; McGUIRE, M.J. (1990) Evaluation oxidants for the removal of model taste and odor compounds from a municipal water supply. *JAWWA*, 82 (5), 79-84.

GLAZE, W.H.; LAY, Y.; KANG, J. (1995) Advanced Oxidation Processes. A kinetic model for oxidation of 1,2-dibromo-3-chloropropane in water by combination of hydrogen peroxide and UV radiation. *Ind. Eng. Chem. Res.*, 34, 2314-23.

GOEL, S.; HOZALSKI, R.M.; BOUWER, E.J. (1995) Biodegradation of NOM – effect of NOM source and ozone dose. *Journal American Water Works Association*, 87 (1): 90-105.

GRELA, M. A.; LOEB, B.; RESTREPO, G. M.; LAGORIO M. G.; SAN ROMÁN, E. (2001) *Los mecanismos de destrucción de contaminantes orgánicos*. In: *Eliminación de Contaminantes por Fotocatálisis Heterogénea*. La Plata: Digital Graf.

HAIDER, S.; NAITHANI, V.; VISWANATHAN, P.N.; KAKKAR, P. (2003) Cyanobacterial toxins: a growing environmental concern. *Chemosphere*, 52, 1-21.

HARADA, K.I.; OGAWA, K.; MATSUURA, K.; MURATA, H.; SUZUKI, M.; WATANABE, M.F.; ITEZONO, Y.; NAKAYAMNA, N. (1990) Strutural determination of geometrical-isomers of microcystins LR and RR from cyanobacteria by two-dimensional NMR spectroscopic techniques. *Chemical Research Toxicology*, 3, 473-81.

HERMANN J. M., GUILLARD C., PICHAT P. (1993) Heterogeneous photocatalysis: an emerging technology for water treatment. *Catalysis Today*, 17, 7-20.

HIMBERG, K; KEIJOLA, A.M.; HIISVIRTA, L; SIVONEN, K.; PYYSALO, H. (1989) The effect of water-treatment processes on the removal of hepatotoxins from microcystis and *Oscillatoria* cyanobacteria – a laboratory study. *Water Research*, 23, 979.

HOFFMAN, J.R.H. (1976) Removal of *Microcystis* toxins in water purification processes. *WATER SA*, 2, 2, 58-60.

HOIGNE, J; BADER, H. (1983) Rate constants of reactions of ozone with organic and inorganic-compounds in water.1. Non-dissociating organic-compounds. *Water Research*, 17, 2, 173-183.

HUANG, C. P.; DONG, C.; TANG, Z. (1993) Advanced chemical oxidation: its present role and potential future in hazardous waste treatment. *Waste Management*, 13, 361-377.

HUSTON, P.L.; PIGNATELLO, J.J. (1999) Degradation of selected pesticide active ingredients and commercial formulations in water by the photo-assisted Fenton reaction. *Water Research*, 33, 5, 1238-1246.

HUSZAR, V.L.M.; SILVA, L.H.S.; MARINHO, M.; DOMINGOS, P.; SANT'ANNA, C.L. (2000) Cyanoprokaryote assemblages in eight productive tropical Brazilian waters. *Hydrobiologia*, 424, 67-77.

IBGE – Instituto Brasileiro de Geografia e Estatística. *Pesquisa Nacional de Saneamento Básico*, 2000. Available in www.ibge.gov.br. Acessado em 2008.

IZAGUIRRE, G.; HWANG, C.J.; KRASNER, S.W.; McGUIRE, M.J. (1982) Geosmin and 2-methylisoborneol from cyanobacteria in three water supply systems. *Appl. Envir. Microbiol.*, 43, 708-14.

IZAGUIRRE, G. (1992) A copper-tolerant *Phormidium* species from lake Mathews, California, that produces 2-methylisoborneol and geosmina. *Water Science and Technology*, 25, 2, 217-223.

KEIJOLA, A.M.; HIMBERG, K.; ESALA, A.L.; SIVONEN, K.; HIISVIRTA, L. (1988) Removal of cyanobacterial toxins in water-treatment processes – laboratory and pilot-scale experiments. *Toxic. Assess.*, 3, 643.

KOCH, B.; FERGUSON, D.W.; GRAMITH, J.T.; DALE, M.S. (1992) Control of 2-methylisoborneol and geosmina by ozone and peroxone: a pilot study. *Water Science and Technology*, 25, 291-98.

KOSITZI, M.; POULIOS, I.; MALATO, S.; CÁCERES, J.; CAMPOS, A. (2004) Solar photocatalytic treatment of synthetic municipal wastewater. *Water Research*, 38, 5, 1147-1154.

KUNZ, A.; PERALTA-ZAMORA, P. G.; MORAES, S. G.; DURÁN, N. (2002) Novas tendências no tratamento de efluentes têxteis, *Química Nova*, 25, 1, 78-82.

KURBUS, T.; SLOKAR, Y.M.; MARECHAL, A.M.; VONCINA, D.B. (2003) The use of experimental design for the evaluation of the influence of variables on the H_2O_2/UV treatment of model textile wastewater. *Dyes and Pigments*, 58, 171-178.

LALEZARY-CRAIG, S; PIRBAZARI, M; DALE, M.S.; TANAKA, T.S.; McGUIRE, M.J. (1988) Optimizing the removal of geosmin and 2-methylisoborneol by powdered activated carbon. *Journal of the American Water Works Association JAWWA*, 80, 3, 73-80.

LAMBERT, T.W.; HOLMES, C.F.B.; HRUDEY, S.E. (1996) Adsorption of microcystin-LR by activated carbon and removal in full scale water treatment. *Water Research*, 30, 6, 1411-1422.

LAWTON, L.A.; ROBERTSON, P.K.J. (1999) Physico-chemical treatment methods for the removal of microcystins (cyanobacterial hepatotoxins) from potable waters. *Chem.Soc. Ver.* 28, 217-24.

LAWTON, L.A.; ROBERTSON, P.K.J.; ROBERTSON, R.F.; BRUCE, F.G. (2003a) The destruction of 2-methylisoborneol and geosmin using titanium dioxide photocatalysis. *Applied Catalysis B: Environmental*, 44, 1, 9-13.

LAWTON, LA; ROBERTSON, PKJ; CORNISH, BJPA; MARR, IL; JASPARS, M. (2003b) Processes influencing surface interaction and photocatalytic destruction of microcystins on titanium dioxide photocatalysts. *Journal of Catalysis*, 213 (1): 109-113.

LEE, D.K., KIM, S.C., CHO, I.C., KIM, S.J., KIM, S.W. (2004) Photocatalytic oxidation of microcystin-LR in a fluidized bed reactor having TiO_2-coated activated carbon. *Separation and Purification Technology* 34 (1-3): 59-66.

LEGRINI, O.; OLIVEROS, E.; BRAUN, A. M. (1993) Photochemical processes for water-treatment. *Chem. Rev.* 93, 671-698, 1993.

LIANG, W.; Qu, J.; WANG, K.; WANG, J.; LIU, H.; LEI, P. (2008) Electrochemical degradation of cyanobacterial toxin Microcystin-LR using Ti/RuO_2 electrodes in a continuous tubular reactor. *Environmental engineering science* 25, 635-641.

LOPEZ, A.; PAGANO, M.; VOLPE, A.; Di PINTO, A.C. (2004) Fenton's pre-treatment of mature landfill leachate. *Chemosphere*, 54, 7, 1005-1010.

MALATO S.; BLANCO J.; RICHTER B.; BRAUN B.; MALDONADO M.I. (1998) Enhancement of rate of solar photocatalytic mineralization of organic pollutants by inorganic oxidizing species. *Applied Catalysis B:Environmental*, 17, 347-356.

MALATO S.; BLANCO J.; RICHTER B.; MALDONADO M.I. (2000) Optimization of pre-industrial solar photocatalytic mineralization of commercial pesticides Application to pesticide container recycling. *Applied Catalysis B: Environmental*, 25, 1, 31-38.

MANSILLA, H.D.; YEBER, M.C.; FREER, J.; RODRIGUEZ, J.; BAEZA, J. (1997) Homogeneous and heterogeneous advanced oxidation of a bleaching effluent from the pulp and paper industry. *Water Science and Technology*, 35, 4, 273-278.

MATER, L.; ROSA, E.V.C.; BERTO, J.; CORRÊA, A.X.R.; SCHWINGEL, P.R.; RADETSKI, C.M. (2007) A simple methodology to evaluate influence of H_2O_2 and Fe^{2+} concentrations on the mineralization and biodegradability of organic compounds in water and soil contaminated with crude petroleum. *Journal of Hazardous Materials*, 149, 379-386.

McGUIRE, M.J.; GASTON, J.M. (1988) Overview of technology for controlling off-flavors in drinking water. *Water Science and Technology*, 20, 8/9, 215-228.

MEUNIER, L., CANONICA, S., von GUNTEN, U. (2006) Implications of sequential use of UV and ozone for drinking water quality. *Water Research* 40 (9): 1864-1876.

MOMANI, F.A.; SANS, C.; ESPLUGAS, S. (2004) A comparative study of the advanced oxidation of 2,4-dichlorophenol. *Journal of Hazardous Materials,* 107, 3, 123-129.

MORAES, J.E.F.; QUINA, F.H.; NASCIMENTO, C.A.O.; SILVA, D.N.E; CHIAVONE-FILHO, O. (2004) Treatment of saline wastewater contaminated with hydrocarbons by the photo-fenton process. *Env. Sci. Tech.*, 38, 4, 1183-1187.

MORAIS, J.L.; ZAMORA, P.P. (2005) Use of advanced oxidation processes to improve the biodegradability of mature landfill leachates. *Journal of Hazardous Materials***,** 123, 1-3, 181-186.

MS- MINISTÉRIO DA SAÚDE, Brasil. Portaria n.º 518 de março de 2004.

MURUGANANDHAM, M; SWAMINATHAN, M. (2004) Solar photocatalytic degradation of a reactive azo dye in TiO2-suspension. *Solar Energy Materials and Solar Cells,* 81, 4, 439-457.

NEYENS, E.; BAEYENS, J. (2003) A review of classic Fenton's peroxidation as an advanced oxidation technique. *Journal of Hazardous Materials*, B98, 33-50.

NISHIWAKI-MATSUSHIMA, R.; OHTA, T.; NISHIWAKI, S.; SUGUNUMA, M.; KOHYAMA, K.; ISHIKAWA, T.; CARMICHAEL, W.W.; FUJIKI, H. (1992) Liver tumor promotion by the cyanobacterial cyclic peptide toxin microcystin-LR. *Journal Cancer Research Clinical Oncology,* 118, 420-4.

NOGUEIRA, R.F.P.; JARDIM, W.F. (1998) A fotocatálise heterogênea e sua aplicação ambiental. *Química Nova*, 21, 1, 69-72.

NOGUEIRA, R.F.P.; TROVÓ, A.G.; SILVA, M.R.A.; VILLA, R.D. (2007) Fundamentos e aplicações ambientais dos processos Fenton e Foto-Fenton. *Quimica Nova,* 30, 2, 400-408.

NOZAL, M.J.; BERNAL, J.L.; JIMENEZ, J.J.; GONZALEZ, M.J.; HIGES, M. (2002) Extraction of thymol, eucalyptol, menthol, and camphor residues from honey and beeswax – Determination by gas chromatography with flame ionization detection. *Journal of Chromatography A*, 954, 1, 207-215.

PALOMBARI R., RANCHELLA M., ROL C., SEBASTIANI G. V. (2002) Oxidative photoelectrochemical technology with Ti/TiO_2 anodes. *Solar Energy Materials & Solar Cells*, 71, 359-368.

PARRA, S.; SARRIA, V.; MALATO, S.; PÉRINGER, P.; PULGARIN, C. (2002) Photochemical versus coupled photochemical–biological flow system for the treatment of two biorecalcitrant herbicides: metobromuron and isoproturon. *Applied Catalysis B: Environmental*, 27, 3, 153-168.

PELEGRINI, R.; PERALTA-ZAMORA, P.; ANDRADE, A. R.; REYES, J.; DURÁN, N. (1999) Eletrochemically assisted photocatalytic degradation of reactive dyes. *Applied Catalysis B:Environmental*, 22, 83-90.

PELEGRINI, R.; REYES, J.; DURAN, N.; PERALTA-ZAMORA, P.; De ANDRADE, A. R. (2000) Photoelectrochemical degradation of lignin. *Journal of Applied Electrochemistry*, 30, 953-8.

PERA-TITUS, M.; GARCÍA-MOLINA, V.; BAÑOS, M.A.; GIMENEZ, J.; ESPLUGAS, S. (2004) Degradation of chlorophenols by means of advanced oxidation processes: a general review. *Applied Catalysis B: Environmental*, 47, 4, 219-256.

PÉREZ, M.; TORRADES, F.; DOMENECH, X.; PERAL, J. (2002) Fenton and photo-Fenton oxidation of textile effluents. *Water Research*, 36: 2703-2710.

PETERNEL, I.T.; KOPRIVANAC, N.; LONCARIC BOZIC, A.M.; KUSIC, H.M. (2007) Comparative study of UV/TiO_2, UV/ZnO and photo-Fenton processes for the organic reactive dye degradation in aqueous solution. *Journal of Hazardous Materials*, 148, 477-484.

PIGNATELLO, J.J. (1992) Dark and photoassisted Fe^{3+} catalyzed degradation of chlorophenox herbicides by hydrogen peroxide. *Environmental Science and Technology,* 26, 944-51.

PNUD – Programa das Nações Unidas para o Desenvolvimento. *Relatório do Desenvolvimento Humano*, (2006). Available in www.pnud.org.br/rdh. Accessed in January, 2008.

POULOPOULOS, S.G.; NIKOLAKI, M.; KARAMPETSOS, D.; PHILIPPOPOULOS, C.J. (2008) Photochemical treatment of 2-chlorophenol aqueous solutions using ultraviolet radiation, hydrogen peroxide and photo-Fenton reaction. *Journal of Hazardous Materials*, 153, 582-587.

PRIMO, O.; RIVERO, M.J.; ORTIZ, I. (2007) Photo-Fenton process as an efficient alternative to the treatment of landfill leachates. *Journal of Hazardous Materials*, no prelo. Available in www.sciencedirect.com. Accessed in September, 2007.

QIAO, R.P.; LI, N.; QI, X.H.; WANG, Q.S.; ZHUANG, Y.Y. (2005) Degradation of microcystin-RR by UV radiation in the presence of hydrogen peroxide. *Toxicon*, 45, 745-752.

RAST, W.; THORNTON, J. A. (1996) Trends in eutrophication research and control, *Hydrological Processes*, 10, 295-313.

RINEHART, K.L.; HARADA, K.; NAMIKOSHI, M.; CHEN, C.; HARVIS, C.A. (1988) Nodularin, microcystin and the configuration of Adda. *Journal of American Chemical Society*, 110, 8557-8.

RODRIGUES, M.A.S.; AMADO, F.D.R.; XAVIER, J.L.N.; STREIT, K.F.; BERNARDES, A.M.; FERREIRA, J.Z. (2008) Application of photoelectrochemical–electrodialysis treatment for the recovery and reuse of water from tannery effluents. *Journal of Cleaner Production*, 16, 5, 605-611.

RODRÍGUEZ, M.; MALATO, S.; PULGARIN, C.; CONTRERAS, S.; CURCÓ, D.; GIMÉNEZ, J.; ESPLUGAS, S. (2005) Optimizing the solar photo-Fenton process in the treatment of contaminated water. Determination of intrinsic kinetic constants for scale-up. *Solar Energy,* 79, 360–368.

ROSENFELDT, E.J.; MELCHER, B.; LINDEN, K.G. (2005) UV and UV/H_2O_2 treatment of methylisoborneol (MIB) and geosmin in water. *Journal Of Water Supply Research And Technology-Aqua* 54 (7), 423-434.

ROSITANO, J.; NICHOLSON, B.C.; PIERONNE, P. (1998) Destruction of cyanobacterial toxins by ozone. *Ozone: Science & Engineering,* 20, 223-238.

SANT'ANNA, C.L.; AZEVEDO, M.T.P. (2000) Contribution to the knowledge of potentially toxic Cyanobacteria from Brazil. *Nova Hedwigia*, 71, 359-385.

SANTANA, H.; BONANCÊA, C.E.; TAKASHIMA, K. (2003) Photoelectrochemical degradation of atrazina on titanium dioxide: Effect of different experimental parameters. *Química Nova,* 26, 6, 807-811.

SCIALDONE, O.; RANDAZZO, S.; GALIA, A.; SILVESTRI, G. (2009) Electrochemical oxidation of organics in water: Role of operative parameters in the absence and in the presence of NaCl, *Water Research*, 43, 2260-2271.

SHEPARD, G.S.; STOCKENSTRÖM, S.; de VILLIERS, D.; ENGELBRETCH, W.J.; SYDENHAM, E.W., WESSELS, G.F.S. (2002) Degradation of microcystin toxins in a falling film photocatalytic reactor with immobilized titanium dioxide catalyst. *Water Research* 36 (1), 140-146.

SIVONEN, K., 1998. Toxins produced by cyanobacteria. In: Miraglia, M.; Van Egmond, H.; Brera, C.; Gilbert, J. (Eds.), *Mycotoxins and Phycotoxins – Development in Chemistry, Toxicology and Food Safety*, EUA, 547-568.

TAUCHERT, E., SCHNEIDER, S., MORAES, J. L., PERALTA-ZAMORA, P. (2006) Photochemically-assisted electrochemical degradationof landfill leachate, *Chemosphere*, 64 (9): 1458-1463.

TEIXEIRA, C.P.A.B.; JARDIM, W.F. (2004) *Processos Oxidativos Avançados: Conceitos teóricos*. Vol. 5. Disponivel em http://lqa.iqm.unicamp.br/cadernos/caderno3.pdf. Acessado em Janeiro 2008.

TIBURTIUS, E. R. L.; PERALTA-ZAMORA, P.; EMMEL, A. (2005) Treatment of gasoline-contaminated waters by advanced oxidation processes. *Journal of Hazardous Materials*, 126, 86–90.

TSUJI, K.; NAITO, S.; KONDO, F.; ISHIKAWA, N.; WATANABE, M.F.; SUZUKI, M.; HARADA, K.I. (1993) Stability of microcystins from cyanobacteria: effect of light on decomposition and isomerization. *Environmental Science and Technology,* 28, 173-7.

WALDNER, G; POURMODJIB, M; BAUER, R; NEUMANN-SPALLART, M. (2003) Photoelectrocatalytic degradation of 4-chlorophenol and oxalic acid on titanium dioxide electrodes. *Chemosphere*,

50, (8), 989-98.

WHO: WORLD HEALTH ORGANIZATION. *Guidelines for Drinking water Quality, Second Edition, Addendum to Volume 2, Health Criteria and other supporting information.* World Health Organization, Geneva, 1998.

YANG, S.; LIU, Y.; SUN, C. (2006) *Preparation of anatase TiO_2/Ti nanotube-like electrodes and their high photoelectrocatalytic activity for the degradation of PCP in aqueous solution.* Applied Catalysis A: General, 301, 2, 284-291.

YUNES, J. S.; SALOMON, P. S.; MATTHIENSEN, A.; BEATTIE, K. A.; RAGGETT, S. L.; CODD, G. A. (1996) Toxic blooms of cyanobacteria in the Patos Lagoon Estuary, Southern Brazil. *J. Aq. Ecos. Health.*, 5, 223-229.

ZAMORA, P. P.; ESPÓSITO, E.; REYES, J.; DURÁN, N. (1997) Remediação de efluentes derivados da indústria de papel e celulose: Tratamento biológico e fotocatalítico. *Química Nova*, 20, 2, 186-189.

ZHU, C.; WANG, L.; KONG, L.; YANG, X.; WANG, L.; ZHENG, S.; CHEN, F.; MAI-ZHI, F.; ZONG, H. (2000) Photocatalytic degradation of azo dyes by supported TiO_2 + UV in aqueous solution. *Chemosphere*, 41, 303-309.

CHAPTER 24

ENVIRONMENTAL ZONING OF THE PROTECTION AREAS IN THE RIO VERDE BASIN

Eduardo Felga Gobbi
Enéas Souza Machado
Ednei Bueno do Nascimento

ENVIRONMENTAL ZONING OF THE PROTECTION AREAS IN THE RIO VERDE BASIN

Eduardo Felga Gobbi
Enéas Souza Machado
Ednei Bueno do Nascimento

SUMMARY

The Environmental Protection Area of the Rio Verde, called the Rio Verde APA (*Area de Proteção Ambiental)*, was established in 2000 and covers the municipalities of Campo Largo and Araucária; the Rio Verde Basin encompasses a larger area, including part of the municipality of Campo Magro in the upper basin. In Campo Magro, the management tool used is the Territorial Planning Unit of Campo Magro. Due to its location and to interests related to water quality, the hydrographic basin that drains its waters into the Rio Verde Reservoir requires a protection strategy. This strategy, in essence, involves planning of land-use in the basin and mechanisms of monitoring and control. This chapter analyzes the relevant legislation that aims to ensure good water quality of this important hydrographic basin with particular interest to issues related to land-use and occupation. Of special interest is the zoning of land-use in the portions of the municipalities that comprise the drainage area of the reservoir that are present in the master plans as well as the zoning of the Rio Verde APA. The chapter also presents experiences of managing water sources through Technical Support Boards (*Câmaras do Apoio Técnico*-CATs*)*.

KEYWORDS

Environmental zoning, basin management, public water supply, Rio Verde APA.

1. INTRODUCTION

The management of any territory in Brazil -- municipality, hydrographic basin, geographical area, and so forth -- must consider a rather sophisticated and comprehensive legal framework. This framework was developed over the last three decades and relied on the cooperation of various interest groups. As such, it has different perspectives in its constitution, which implies a complex effort of analysis and articulation of the laws.

The hydrographic basin that drains its waters into the Rio Verde Reservoir, because of its location and the interests already outlined in the introductory chapter of this book, requires a strategy for protection. Essentially, this strategy involves planning land-use in the basin and implementing monitoring and control mechanisms. Federal, State, and Municipal laws burden the territory of the basin. Of special interest is the zoning of land-use in the sections of the municipalities that are located within the drainage area of the reservoir which are addressed in the master plans as well as the zoning of the Rio Verde APA.

Water source management in the Metropolitan Region of Curitiba relies on two distinct tools: the Territorial Planning Unit (*Unidade Territorial de Planejamento-* UTP*)* and the Environmental Protection Area (APA). In cases where water sources have very fragile environments, for example, water sources that have reservoirs or areas of replenishment of the Karst aquifer, the instrument used is an APA. In cases where there are municipal centers in the area of influence, the instrument used is the UTP. The Rio Verde Basin integrates these two instruments since the head of the basin houses part of the urban area of the municipality of Campo Magro, with the Rio Verde Reservoir downstream.

This chapter provides an analysis of the relevant legislations that aim to ensure good water quality of this important hydrographic basin with particular interest in issues related to the use and occupation of land and the experience developed in the Metropolitan Region of Curitiba with the Technical Support Boards *(*CATs) as units of integrated and participatory management.

2. THE EVOLUTION OF THE LEGAL FRAMEWORK

The starting point for the analysis of interest for this study was Law 6.938/81 from the National Policy for the Environment (*Política Nacional do Meio Ambiente)*. In article nine, Law 6.938/81 defines the instruments of the National Policy for the Environment. Two of these instruments are of particular interest to this chapter: environmental zoning and the creation of territorial spaces specially protected by the federal, state, and municipal governments, such as environmental protection areas of relevant ecological interest and extractive reserves. In the specific case of the State of Paraná, Law 6.938/81 influenced an important development of the legal framework, aiming to protect the environment and resource areas.

a) Concept of Water Supply Source Basin

Of particular interest for this study, we must mention Law 8.935 from 1989. This law established the concept of "Water Source Basin," as a basin destined as a source for public water supply or the hydrographic basin area located upstream of an existing or forecasted dam built to collect water for public consumption. It is important to remember that, at the time the Law was enacted, the rate of population growth in the Curitiba Metropolitan Region (RMC) was very high. For public authorities, this caused a concern in relation to securing sources for future water supply. This Law was the seed of what we call today Areas of Interest of Water Supply Sources for the Curitiba Metropolitan Region (PARANÁ 2008).

Law 8.935 / 1989 outlines the minimum requirements for water originating from source basins intended for public consumption. It provides that rivers intended for public supply must meet the requirements within the framework for Class 2, from former CONAMA Resolution 20/86. It prohibits the installation of a number of industrial activities, hospital facilities, landfills, and land subdivisions with high demographic density. Polluting industries and landfills must submit proposals for sewage and waste treatment, anticipating if possible, their disposal in another basin. It also mentions the alternative of transferring polluting activities to another area outside the basin on land that is going to be expropriated by the government. The lots already approved should have their domestic sewage collected, treated, and released outside of the basin. The law allows farming and reforestation within the basin, provided a supervisory body monitors pesticide use and the misuse of land that can lead to erosion.

Because the law has never been regulated, a number of issues remain ambiguous. For instance, basins with a large drainage area cannot accommodate "highly polluting" industries, even if there is a rigorous waste treatment and the remaining loads are diluted or sedimented up to the point of abstraction. The economic feasibility of basin reversal for the effluents is not taken into account. The expropriation of garbage dumps or landfills by the Government does not consider the economic capacity, either at municipal or state level. Indeed, law 8.935 / 1989 has restricted applicability and should have been revised as its applicability was not regulated.

b) State Environmental Protection Area of Passaúna

Inspired by the aforementioned conditions of Law 6.938 / 81, among other laws, Decree 458 / 1991was set up to create the State Environmental Protection Area (APA) of Passaúna and its respective Ecological-Economic Zoning located in the municipalities of Almirante Tamandaré, Araucária, Campo Largo, and Curitiba, Paraná. The establishment of the State APA of Passaúna aimed to protect and conserve environmental quality and the natural systems found there, especially the quality and quantity of water for public supply, by setting up measures to manage all phenomena and their conflicts arising from varied and incompatible usage in the Passaúna River Hydrographic Basin.

Along this same line, the APA in the Iraí River Basin, called the State APA of Iraí, was established by Decree 1.753 in 1996 and its Ecological-Economic Zoning was set up by Decree 2.200 of 2000, while the APA of the Rio Verde, described in detail later in this chapter, was established through Decree 2.375 / 2000.

When establishing these APAs, the legislation provided for the creation of the Technical Support Boards (CAT). The Technical Support Boards created for the various APAs had the function of implementing the activities of management, zoning, and monitoring for the APAs and provide advice on other issues when requested by government departments.

The institutionalization of the Technical Support Boards began with COMEC (Coordination of Curitiba Metropolitan Region), with support from the Institute of Land, Cartography and Forest (*Instituto de Terras e Cartografia e Florestas*-ITCF) and the Superintendent of Water Resources and Environment (*Superintendência dos Recursos Hídricos e Meio Ambiente*-SUREHMA). Thus, the Technical Support Boards were legalized by regulations established by COMEC and all advisors must be appointed as legal representatives by their institutions.

The creation of Technical Support Boards allows for the representation of various entities with legal abilities to participate in territorial management. Its membership is comprised of representatives from the following agencies and organizations: Forest Police Battalion (*Batalhão da Policia Florestal*-BPFlo); Coordination of Curitiba Metropolitan Region (COMEC); Department of Highways of the State of Paraná (*Departamento de Estradas de Rodagem do Estado do Paraná*- DER); Technical Assistance and Rural Extension Agency of Paraná (*Empresa Paranaense de Assistência Técnica e Extensão Rural—Paraná*-EMATER/PR); ITCF; State Public Prosecutor through specialized prosecutors; municipalities; SANEPAR (*Companhia de Saneamento do Paraná*); SUREHMA; and a representative appointed by environmental groups.

Territorial Planning Units (UTPs) are territorial areas that are under occupation pressure and are located in urban areas of the municipalities in the areas of interest for source protection. Overall, they are designed to make the transition between the already established urban areas with the areas of greatest environmental restriction, such as the APAs, and/or rural areas.

In the urban agglomeration there are five UTPs regulated by state laws: the UTPs of Pinhais, Guarituba, Itaqui, Quatro Barras, and Campo Magro. Among the areas of interest in protecting water sources, these units were considered areas with high occupation pressure given the proximity to the metropolitan hub and road accessibility. The planning of UTPs was performed based on already defined current conditions, with high density plots of land that had been approved in previous decades being used as guiding parameters. Thus a maximum density of one dwelling every 2000m^2 on average was established, taking into account the capacity of sewage and drainage treatment systems.

Decree 1.611/99 established the Campo Magro UTP in order to ensure appropriate environmental conditions for the preservation of water sources through the preservation and restoration of the natural and anthropogenic environment with effective control over degradation processes and environmental pollution.

This decree regulates three types of intervention areas:

I – Areas with Restricted Occupation—those of interest in relation to preservation, aiming at promoting the restoration and conservation of natural resources, ensuring the maintenance of biodiversity and ecosystem conservation;

II – Areas of Targeted Occupation—those targeted for land subdivisions, urban occupation, and transition zones between rural and urban areas, that are subject to occupation pressure requiring public authorities' intervention in order to minimize the contaminating effects on water sources;

III – Areas of Consolidated Urbanization—areas of interest for the consolidation of urban occupation, improving and recuperating environmental conditions.

c) Ecological ICMS Tax

With the creation of source interest areas and APAs in the context of what was defined in the 1981 National Policy for the Environment, many municipalities where these source and conservation areas were implemented began to experience serious constraints in relation to conventional economic development. Many economic activities were restricted or eliminated in these areas. This fact has led many municipalities to suffer a reduction in economic growth rates with consequential loss of tax revenue.

A way of minimizing potential problems related to loss of revenue due to restrictions on the establishment of polluting activities, Complementary Law 59/91 set an allocation of 5% of the ICMS Tax from the State of Paraná as a way of compensating municipalities. Therefore, municipalities that house environmental conservation units, or are directly influenced by them, or those with public water supply sources, began to receive a portion of their ICMS Tax, creating the Ecological ICMS.

The portion that each municipality is entitled to receive is defined by technical criteria related to the size and quality of the water source, the type and the condition of the Conservation Units. The amount received annually by the municipality may vary up or down based on regular assessments of water sources and Conservation Units made by government State agencies.

The Paraná law for the Ecological ICMS was pioneering in the country and several states followed with similar programs. It is noteworthy that some municipalities receive more than 50% of their ICMS revenues from the Ecological ICMS (PARANÁ, 2010b). It is also important to note that the Rio Verde Reservoir, as a water source, allows the affected municipalities to receive tax revenues in the form of the Ecological ICMS.

d) Policy and Management of Water Resources

In the past three decades, due to economic development, industrialization, population growth, and increasing population concentration in urban areas, Brazil experienced its first problems related to conflicts over water use. This led to the inclusion of the national system of water resources management as a prominent theme in item XIX of Article 21 of the 1988 Constitution. As a consolidation of this constitutional provision, the National Policy for Water Resources was established leading to the National Water Resources Management System that was set up in 1997 by Federal Law 9.433. In the State of Paraná, the State Policy on Water Resources was established and the State System of Water Resources Management was set up in 1999, by State Law 12.726.

This chapter will not provide a specific analysis of these laws; however, it is appropriate to emphasize some important points of their more direct interfaces with water source protection. Two instruments are worth mentioning that are quite relevant to the protection of water sources: the Basin Plan and the Framework for the Classification of Water Bodies.

According to Law 12.726, the Basin Plan must contain, at minimum, the following items that have direct relation to water source areas:

- analysis of alternative scenarios of population growth, development of productive activities and modifications of land-use patterns;
- balance between availability and future demand of water resources in quantity and quality with identification of potential conflicts;
- goals to rationalize use, suitability of supply, improve quality of available water resources, protection and development of aquatic ecosystems;
- measures to be taken, programs to be developed and projects to be implemented to meet planned targets;
- division of watercourses along stretches of rivers indicating the grantable flow in each stretch;
- priorities for granting of water resource use rights;
- proposals for the establishment of areas subjected to use constraint to protect water resources and aquatic ecosystems.

With regard to classification, this should target water quality standards compatible with the uses to which they are intended, supporting the process of granting licenses for water resource use rights. The classes are those set forth by CONAMA Resolution 357 / 2005.

We furthermore draw attention to some general guidelines presented in both Law 9.433 and Law 12.726, which show the interdependency among the various legal instruments. These guidelines are essential to ensure water availability in quantity and quality from water sources.

- integration of water resources management and environmental management;
- joint management of the water resources with various sectors, including users and regional, state, and national planning;
- integrated management of water resources with land-use and flood control.

The Rio Verde is located in the area composed of the Upper Iguaçu basins and the Upper Ribeira tributaries. Currently the Basin Plan is being drafted and has built-in the watercourse classification of these basins. This classification prioritizes the recovery of water quality of public water supply source basins, allocating resources for collection, transportation, and domestic sewage treatment systems. The rivers considered current and future water sources receive infrastructure to ensure quality compatible with the current classification, which is Class 2.

The State Water Resources Plan (AGUASPARANÁ, 2011), one of the instruments of the State Policy of Water Resources, was recently completed by the Water Institute of Paraná, and approved at the December 2009 plenary session of the State Water Resources Council. The State Plan has a strategic vision of water resources, offers guidelines for the sound implementation of the management system over four years. In general, the Plan assesses demands and availability of water resources, develops scenarios, and proposes a robust strategy for the implementation of the management system. There is no specific reference to public water supply sources; this issue was left to the more specific scope of the Basin Plans.

e) Integrated System of Water Source Management and Protection for the Metropolitan Region of Curitiba

The National Policy of Water Resources has a strong focus on water availability and on the proposal of processes aimed at managing potential conflicts over water use. In Paraná, particularly within the Metropolitan Region of Curitiba, there was the opportunity for a specific legislation aimed at land-use that was complementary to the Management System of Water Resources. Thus, the Integrated System of Water Source Management and Protection of the Metropolitan Region of Curitiba – SIGPROM was enacted on July 31, 1998.

This law is a clear improvement over the previous State law regarding source protection. The law creates the Source Management Council of the Metropolitan Region of Curitiba, a joint committee with advisory, deliberative, and normative authority. The Council is chaired and administered by COMEC and its members are, namely: COMEC itself, IAP, SANEPAR, AGUASPARANÁ, the Mayors of four municipalities and alternates, a representative of educational institutions, a representative of the construction and real estate industries, and a representative of NGOs.

The law also creates the Territorial Planning Units (UTP), that include the sub-basins containing water sources of interest to the RMC, to facilitate planning and combining municipalities with specific features to be developed jointly. The UTPs of Pinhais, Guarituba, Itaqui, Quatro Barras, and Campo Magro, have already been created and are regulated by State law. The planning of the UTPs is carried out based on current conditions, that is, in areas already occupied.

The law also proposes the Environmental Protection Plan and Territorial Reorganization (PPART), establishing deadlines and goals for interventions in the areas requiring water source protection. The PPART should be included as a separate section in the Water Resources Plan. Finally, the law creates the Environmental Preservation Fund for the Metropolitan Region of Curitiba to meet the objectives of the system created by law.

Decree 3411 / 2008 defines Areas of Interest of Water Sources for Public Supply for the Metropolitan Region of Curitiba. The hydrographic basin of the Rio Verde Reservoir is included as a water supply source.

f) Statute of the City

Item XX of Article 21 of the Constitution, 1988, defines that the Union must establish guidelines for urban development including housing, basic sanitation and urban transport. To determine guidelines for urban development, directives were set by the Urban Policy, in 2001, by Federal Law 10.257, also known as the Statute of the City. While we will not go into an in-depth analysis of the Statute of the City, one of the instruments of this law, "city planning," must be highlighted. We would like to draw attention to the master plan and the environmental zoning as they are of fundamental importance to territorial organization. Thus, another tool must be analyzed in the context of ensuring environmental quality in the Rio Verde Reservoir basin, which ultimately will ensure the quality of its waters.

g) National Policy for Basic Sanitation

Law 11.445 of 2007 establishes national guidelines for basic sanitation and for the federal policy of basic sanitation, as stated in item XX of Article 21 in the 1988 Federal Constitution. The provision of public services for basic sanitation must be based on a plan developed by the municipality and shall include at least:

- I – diagnosis of the situation and its impact on living conditions, using health, epidemiology, environment (including hydrological), and socioeconomic indicators, pinpointing causes of detected deficiencies;
- II – goals for the short-, mid-, and long-term with the aim of achieving universal access to services, allowing gradual and progressive solutions and ensuring compatibility with other sectorial plans;
- III – programs, projects and actions necessary to attain goals and objectives in a manner consistent with their respective multi-year planning and other related government plans, identifying potential funding sources;
- IV – actions for emergencies and contingencies; and
- V – mechanisms and procedures for systematic evaluation of the effectiveness and efficiency of programmed actions.

Thus, we have another important tool to ensure the environmental quality of the Rio Verde Reservoir Basin.

3. RIO VERDE APA AND ITS ZONING

As mentioned earlier in this chapter, the Environmental Protection Area of the Rio Verde, called Rio Verde APA, was established by Decree 2.375 in 2000. The Ecologic-Economic Zoning (EEZ) of the Rio Verde APA was proposed by the COMEC and by the *Consiliu* Company in 2002. This EEZ has never been sanctioned. In 2009, the company Vertrag (VERTRAG, 2010) was commissioned to review the EEZ and the work was coordinated by the Municipality of Campo Largo, with technical supervision by COMEC, and contributions from other municipalities within the Rio

Verde Basin (Campo Magro and Araucária). State Decree 6.171/2010 regulated this version of the zoning.

It is important to note that the research that led to this book, was critical in the review of the EEZ, as can be seen in the introductory chapter of the work completed by Vertrag: "Also contributed to this study are the state entities involved in the issue of water resources and environmental protection areas – IAP, SANAPAR, Public Prosecutor, MINEROPAR, SUDERHSA, and EMATER. Besides the mentioned entities, PETROBRAS and FUNPAR, which simultaneously developed a study on the characteristics of the Rio Verde Basin and the risks of eutrophication in the reservoir of the same name decisively contributed to the development of this study." It should be stressed, however, that the interpretation of the information that led to zoning are the sole responsibility of Vertrag since some of the proposals were not based on data made available by this research project.

Regarding this proposed zoning, the Water Sources Management Council of the Metropolitan Region of Curitiba approved a set of adjustments proposed by the Municipality of Campo Largo, on March 31, 2011, in order to improve the document presented based on internal discussions and consultations held by the Municipality.

The APA of the Rio Verde covers parts of the municipalities of Campo Largo and Araucária, and the Rio Verde Basin covers a larger area also including a portion of the municipality of Campo Magro in the upper basin. In this section, the UTP of Campo Magro is in force and defined according to the SIGPROM discussed above (PARANÁ 1998). Thus, it is important to understand that in the hydrographic basin of the Rio Verde Reservoir, we have the Campo Magro UTP in its upper stretch and in the remainder of the basin the APA of Rio Verde. In the EEZ review developed by Vertrag, we note that it was agreed among the institutions that participated in the meetings held during the study, that the Campo Magro UTP would be incorporated into the Rio Verde APA and would abide by the same general land-use and occupancy guidelines.

The study developed by Vertrag included a detailed analysis of the municipalities' zoning, assessing the compatibility among these and the zoning proposed for the Rio Verde APA (also considering the UTP of Campo Magro, as explained). The results of the study established the following zones in the Rio Verde APA:

a) Consolidated Urban Areas

Medium density urban areas already consolidated (for which relocation processes are considered unfeasible) or in the process of consolidation. These areas are considered a priority to receive sanitation and environmental recovery programs.

Consolidated Urban Zone

Objective: increase density of areas already consolidated according to the availability of existing infrastructure networks or the expected deployment and/or expansion.

Characteristics: Comprises areas that are divided up and occupied regularly or irregularly, with consolidated urbanization or in the process of consolidation, with urban infrastructure, especially relating to rainwater drainage and intensive sanitation systems, to harmonize use with the preservation objectives of the APA.

b) Targeted Occupation Areas

These are areas under occupation pressure that have physical and territorial characteristics suitable for urban occupation, requiring public intervention in order to minimize the impacts of pollution on the water source.

Corridor of Special Use

Objective: organize land-use along highways BR 277 and PR 510 located within the APA in order to harness the industrial-logistical potential of the highways and establish control and parameters for the deployment of commercial activities and services and/or non-polluting industries, seeking to minimize their impact on the APA.

Characteristics: comprises a mid-range strip of land of 300 meters – respecting physical limits, especially water bodies – along BR 277 and PR 510 starting from the road verge along the highway established by DER-PR.

Special Corridor for Tourism Use

Objective: encourage rural tourism and activities related to rural-metropolitan tourism and stimulate the preservation of structures of historical and cultural value.

Characteristics: access routes to rural communities, in which rural-metropolitan tourism activities already occur, restricted to properties facing towards the highway, within a range of 100 meters from the end of the road on both sides. As a measure to implement this corridor, buildings and/or façades are listed for historical preservation by the Historic and Artistic Heritage Trusteeship of Paraná given the importance of colonization, architectural style, and cultural identity.

Targeted Occupation Zone I

Objective: guide land-use and promote the minimization of the impacts of pollution on water sources in areas pressured by land subdivision and urban occupation.

Characteristics: consists of the area south of highway BR 277 toward the dam and part of Rincão/Cachoeira, amenable to low density occupation and with a strong tendency toward occupation. Encompasses the transition between more intensive urban occupation areas, especially in the western portion of the APA, bordering the urban area of Campo Largo, and restricted occupancy areas and/or rural areas.

c) Restricted Occupancy Areas

These are areas identified for environmental preservation as outlined by federal and state legislation that must be maintained or have their natural characteristics restored, ensuring the maintenance of biodiversity and the conservation of ecosystems.

Yerba Mate Park Zone

Objective: protect the area of the Yerba Mate Historical Park.

Characteristics: corresponds to the Yerba Mate

Historical Park, listed for preservation in 1984 by IPHAN, under the responsibility of the Cultural Agency of the State Government. It has restricted use and infrastructure dedicated to maintaining environmental quality for the ecosystems involved.

Dam Zone

Objective: limit activities at the dam to ensure safety and maintenance of water quality.

Characteristics: comprises the Rio Verde Dam and its floodplains.

Dam Preservation Zone

Objective: ensure water flow through the restoration and maintenance of the area of permanent preservation (APP).

Characteristics: refers to a strip of 100 meters of vegetation cover around the Rio Verde Dam, which must be preserved and/or recovered in areas where vegetation has been removed or degraded.

Valley Bottom Preservation Zone

Objective: protect water courses, ensure water quality, as well as create corridors of biodiversity in order to maintain the balance of the ecosystem.

Characteristics: includes areas of significant environmental fragility, namely: areas of permanent preservation set out by the Forest Code, floodplains (set out by SUDERHSA, 2000), and wetlands and their immediate surroundings, as established by the Joint Resolution IBAMA/SEMA/IAP 05/2008.

Wildlife Conservation Zone

Objective: retain forested areas in intermediate and advanced stages of preservation, in order to provide relevant sites for local wildlife, as well as ensure the maintenance and balance of the ecosystem.

Characteristics: corresponds to forested areas in intermediate and advanced stages of conservation with areas greater than $1000m^2$.

d) Rural Areas

Rural areas are considered to be spaces reserved for agroforestry activities.

Agricultural and Livestock Use Zone

Objective: guide the use of land for rural activities in compliance with the conservation practices outlined by the Rio Verde APA.

Characteristics: comprise areas with land-use geared toward agriculture, livestock, forestry, fisheries, vegetable and fruit produce, diverse cultivation, and animal husbandry without urban occupation and no tendency toward the development of this type of occupation. Uses or practices likely to cause appreciable degradation to the environment are prohibited; therefore, the use of pesticides and other biocides that present serious risks in their use, including their residual capacity, are prohibited. According to CONAMA 10/88, IBAMA will list the classes of pesticides allowed for use in APAs.

Although the final zoning was the result of a detailed study carried out by Vertrag, we believe that a new revision is necessary. This is related to the fact that the revision by Vertrag took place at a time when the research projects that led to this book were still in progress. Thus, we believe that with the final results of the current research project, a further revision may add some important implications for more effective zoning. We also recommend a better legal framework for incorporating the UTP of Campo Magro into the Rio Verde APA.

4. COLLECTIVE MANAGEMENT OF ENVIRONMENTAL PROTECTION AREAS – EXPERIENCES OF THE TECHNICAL SUPPORT BOARD

Territorial management in public water supply sources is quite complex, for it affects land from different municipalities. It is responsible for the mediation of public, private, environmental, and economic interests that involve highly valued urban and rural areas. It must strictly follow environmental regulations, as it involves safeguarding the availability, quantity and quality of one of society's most important resources: water.

This section discusses the experiences in the Curitiba Metropolitan Region, which adopted Environmental Protection Areas (APAs) as legal instruments for restricting use in more vulnerable water source areas, those with reservoirs and karst aquifer replenishment areas. Technical Support Boards (CATs) were a strategy adopted as paradigms of integrated management that were responsible for zoning compliance and enforcement of environmental legislation in these areas.

The establishment of a program for territorial management, such as the Technical Support Boards, was set when the Environmental Protection Areas were created in the 1980s. APAs are environmental conservation units with an area defined by State decree involving more than one municipality. The Technical Support Boards are a mechanism to establish management practices that integrate the various government sectors that have sectorial duties affecting the area and agencies responsible for environmental and territorial management. We should also note that these territorial units accommodate various economic interests, in both urban and rural areas.

The first experience with this type of territorial management was in the Passaúna River APA, in 1982, and continued until 2003 with the Iraí River APA. These areas integrate regions of public water supply sources and the use of the APA is based on the regulations of CONAMA. As defined in the National System of Conservation Units (*Sistema Nacional das Unidades de Conservação*-SNUC), the APAs are Conservation Units outside of the public domain and the management of these units should be based on the development of a model that is consistent with the purpose of these areas. It is up to the government to create Conservation Units at the municipal, state, or federal level.

The creation of the Technical Support Boards is the responsibility of the institutions that create the APAs. In Paraná, this procedure began with COMEC, with support from ITCF and SUREHMA (the organization responsible for environmental and water resources management in the State of Paraná in the 1980s). Thus, the Technical Support Boards were legalized by regulations created by COMEC. This process of territorial management had the legal framework supported by the functions and powers of COMEC. Their institutions officially appointed all directors as legal representatives.

Therefore, its constitution required that representatives of the entities on the Technical Support Boards had the legal ability to act in territorial management. Thus, there was representation from local government (the Municipalities), the environmental authority (forest protection, wildlife), ITCF, the water resources and environment entity (SUREHMA), enforcement (Forestry Battalion, currently Green Force), Technical Assistance and Rural Outreach entities (EMATER), SANEPAR, COMEC, representative of the Public Prosecutor for the Environment, and a Civil Society Organization operating in the geographic area of the Environmental Protection Area.

The main guiding document used by the Technical Support Board is the Ecological-Economic Zoning Decree developed in accordance with the regulations of CONAMA. The applicable Federal laws, CONAMA Regulations, State Legislation, and Municipal Legislation were the set of regulations that were considered when the Zoning was developed. As water source areas in some cases cover urban and rural areas, we should take into account that the enforcement of the legislation is variable. Oftentimes the state creates APAs through decrees without the necessary Zoning preparation, which is necessary to guide actions for effective implementation. For this reason, in these cases the Technical Support Board guides their actions based on existing sectorial legislation.

An important definition of the features of Technical Support Board was to act as a forum for resolution, finding solutions where the Ecological-Economic Zoning was deficient or did not enforce the land-use policy. In cases of disagreement regarding the interpretation of legal mechanisms, the NGOs participating in the Technical Support Board in partnership with the Public Prosecutors Office promoted Civil Action as a means of exerting pressure in relation to certain issues. This has occurred countless times in Technical Support Board management.

The main commission of the Technical Support Board is reviewing economic licensing processes that have been submitted to the Municipalities or other government institutions. The Technical Support Board also monitors sectorial activities that are in progress in the APA, such as inspections, environmental education, and sectorial cooperation.

Bylaws establishing standards of operations and management are required for the operations of the Technical Support Board. The bylaws define the Presidency, the Executive Secretariat, as well as the regulatory functions of its administrative management. This process of collaborative building of bylaws is characterized as a way of establishing everyone's commitment to the management work. The bylaws must be built in a participatory and democratic way so as to instill greater individual commitment. It is a process of becoming aware of the importance of establishing ground rules in a working group, thus developing a strategic step toward the consolidation of these working groups.

The Executive Secretariat is an important forum, as it provides the necessary momentum behind the completion of the work of the Technical Support Board. It has the duties to organize the Meeting Agenda, call participants, record the meeting minutes, and update the records of the Board discussions. The registration and organization of these files are important as oftentimes these decisions are discussed and ratified in court. The working Agenda organizes the demands of processes coming from municipalities and from institutions that are part of the Board. This demand is only forwarded to the Board in cases where the Zoning application is unclear, or when there is an appeal on the part of the entrepreneur regarding the decision made by the municipality. The environmental authority may also request a cross-referencing from the Board, when the statement needs a broader assessment from the inter-institutional collegiate bodies.

The President is responsible for planning and moderating the work schedule. He or she allots the time, organizes presentations, promotes discussions, and leads the resolutions and referrals of scheduled topics. Thus, the discussion of an issue scheduled for the Meeting is allocated according to its importance and complexity. Time management is critical to the efficiency of Technical Support Board and usually follows the following steps:

a) Presentation of the situation-problem or scheduled process, discussion among advisers with the organization of demonstrations, drawing up proposals, and voting. After these steps, the decision is made and submitted to a voting by all participants;
b) Decision-making occurs through consensus or by vote. The technical and legal arguments are voiced by the Advisor followed by the debate mediated by the President who manages the discussion. If there is consensus, the deliberation yields a Resolution, otherwise, after the debate ends, the different propositions are elaborated and voted on. In this case, each board member is entitled to one vote.

The Technical Support Board is required to act by means of Resolutions, which are duly published and registered. These Resolutions are an instrument of decision for the Board and is characterized as a simplified tool that concisely delivers an option on the petition submitted to the Technical Support Board.

For administrative norms of the Technical Support Board meetings, it is important to set a meeting place and start and end time of the meeting. The meetings of the Technical Support Board are open to interested parties, but they are scheduled. The Advisors, who are institutional representatives, usually attend the meetings, and members and alternates participate as well. Schedules are strictly observed respecting the required number of Advisors that ensure the *quorum* defined by the rules; thus, presence, absence, and tardiness are recorded in the minutes.

For a more efficient management process, it is important that the Technical Support Board promote a training process for advisors involving topics that are relevant to management. In this case, the technical discussions and excursions to the area of interest are important so that all advisors are aware of the problems to be addressed.

Establishing Technical Support Boards as instruments of Territorial Management for a geographical area deemed as an Environmental Protection Area is important as it ensures and guarantees that territorial administration is consistent with the planned Zoning. This territorial management process, along with effective participation of Board Members, who relate decisions to their own institutions, increases the process efficiency. The main outcome of this work is to avoid the implementation of projects that do not meet the regulations or suggest mediations that promote the proper use of these areas, such as:

a) Suspension of the establishment of Gas Stations along Highways, in the case of BR 277, in Curitiba, and of PR 092 in Campo Magro;
b) Deployment of consistent monitoring processes in the Landfill Station *Lamenha Pequena*, in Almirante Tamandaré;
c) Implementation of technical activities of Rural Outreach with farmers, geared toward soil conservation, reduction of pesticide use, Rural Tourism activities, riparian reforestation along reservoirs, and the suitability of rural roads.
d) Inclusion of local requests from the communities for various improvements in sanitation, garbage collection, land-use supervision and environmental protection, public safety, and the activities of farmers, in the planning processes of the various institutions that make up the Board;
e) Disclosure of Collective Management Plan for the people living within the APA, demonstrating the seriousness and commitment to the Zoning;
f) Strict adhesion to the principles proposed in the Ecologic-Economic Zoning. This Zoning has long-term applicability if it is upgraded, properly administered, and implemented;
g) Issuing permits for major events in the area of the former Castelo Branco Country Fair Park;
h) Monitoring of cemeteries;
i) Deployment of actions for sustainable local development through rural tourism and ecotourism;
j) Encouraging the adoption of farming systems with less environmental impact, like Organic Production and the Agroecology System. Today the Metropolitan Region of Curitiba is a region with success with these production systems.

The transparency of the activities of the Board is provided by the effective participation of all board members and by the management mechanisms. To increase their effectiveness, the continuous training of the advisors in the field of environmental sciences is important and requires constant monitoring. The Network of institutional coordination is a challenge, as operational and bureaucratic issues are constantly interfering. As the result of this analysis, we can state:

a) The work practice of Technical Support Boards stimulated the training of technicians in environmental science where multi- and inter-disciplinary methodologies were constant practice. The whole group discussed topics such as agriculture, land-use, inspection, and environmental law. The discussion was not restricted to the advisors of each specific area;
b) The publication of the acts of the Technical Support Boards and the opening of the meeting to public participation ensured the proper level of transparency;
c) Training Group Managers to discuss and implement a single protocol for the submission of projects requesting environmental licensing.

It is important that the territorial management of APAs, through the Technical Support Boards, have legal support, as these actions may be questioned by the various interests vying for the use of these areas.

a) Establishment of legal groundwork for formalizing various land management tools such as By-laws, Resolutions, and the production of ATAs.
b) The provision of expert opinions on certain topics.

All the work of territorial management must be understood as in the interest of the public, that goes beyond economic or localized political interests. As a final reflection, the dynamics established, the formation of a team of technicians with experience in territorial management, the domain of the work strategies, and the improvement of the regional action plans are important outcomes obtained from this collective work.

5. CONCLUSIONS

As mentioned in the introduction to this chapter, the management of any territory in Brazil, be it a municipality, hydrographic basin, geographic region, etc., relies on a sophisticated and comprehensive legal framework. The various legal instruments of interest that are applicable to the study area have been addressed throughout this chapter.

In the area under study, there currently exists: Master Plans at the municipal level; APAs, their zoning and the National System of Conservation Units in the regional level; Basin Plans at the hydrographic basin level; State Plan of Water Resources at the state level; National Plan of Water Resources at the national level; and the prospective Municipal Sanitation Plans (not yet in existence in any of the municipalities of the basin) at the municipal or regional level.

These plans intersect with the Regional Development Plan for Paraná (PARANÁ, 2006), the State Ecologic-Economic Zoning (in preparation), with the State Agricultural Zoning, among others. Currently many studies are being developed at several scales. As an example we can mention a few of interest that are currently underway, including: more comprehensive strategies for biodiversity protection;

payment for ecosystem services; building of ecological corridors; dam safety; risks of extreme events; climate change, and so forth.

The great challenge we have is to bring all this research together. There is a need to seek as much consistency as possible among all these tools. Another relevant issue is that municipalities should be able to have sufficient numbers of qualified human resources to deal with such complexity. Many practical and relevant issues were found during the development of the research project which strongly influences the environmental quality in the basin. Some suggestions that the results of this study point out are: improve liaising between water resources policy and basic sanitation; improve the links between water resources policy and land-use and occupancy; define monitoring programs with indicators that can be regularly measured and used to improve public policy; define a permanent supervision policy of land-use and occupancy and encourage appropriate and compatible activities in hydrographic microbasins. Accordingly, several researchers of the project have developed, in a joint effort, a preventive action plan for the Rio Verde Basin. This plan is presented in the next chapter (Chapter 25).

6. REFERENCES

AGUASPARANÁ. (2011) *Plano Estadual de Recursos Hídricos*.

BRASIL. 1981. *Lei n.° 6.938 da Política Nacional do Meio Ambiente*.

BRASIL. (1988) *Constituição Federal*.

BRASIL. (1997) *Lei Federal n.° 9.433*. Institui a Política Nacional de Recursos Hídricos.(Establishes de National Policy of Water Resources)

BRASIL. (2001) *Lei Federal n.° 10.257*. Estabelece diretrizes gerais da política urbana.(Establishes general guidances of urban policy)

BRASIL. (2007) *Lei n.° 11.445*. Estabelece as diretrizes nacionais para o saneamento básico.(Establishes national guidance of basic sanitation)

PARANÁ. (1989) *Lei n.° 8.935*. Dispõe sobre requisitos mínimos para as águas provenientes de bacias manancials destinadas a abastecimento publico.(Provides on minimum requirements for water from source basins and designed to public supply)

PARANÁ. (1991a) *Decreto n.° 458*. Cria a Área de Proteção Ambiental Estadual do Passaúna.(Creates the State Environmental Protection Area of Passauna)

PARANÁ. (1991b) *Lei Complementar 59*. Dispõe sobre a repartição de 5% do VAT. Provides on the destination oF 5% VAT)

PARANÁ. (1998) *Lei Estadual n.° 12.248*. Cria o Sistema Integrado de Gestão e Proteção dos Mananciais da RMC.(Creates the Integrated System of Water Source Management and Protection)

PARANÁ. (1999) *Lei Estadual n.° 12.726*. Institui a Política Estadual de Recursos Hídricos.(Establishes the Water ResourceS State Policy)

PARANÁ. (2000) *Decreto n.° 2.375*. Cria a Área de Proteção Ambiental do Rio Verde.(Creates the Rio Verde Environmental Protection Area)

PARANÁ. (2006) *PRDE – Planos Regionais e Desenvolvimento Estratégico*. SEDU.

PARANÁ. (2008) *Decreto n.° 3.411*. Declara as Áreas de Interesse de Mananciais de Abastecimento Público para a Região Metropolitana de Curitiba.(Declares the Areas of Interest of Water Sources for Public Supply of the Curitiba Metropolitan Region)

PARANÁ. (2010a) *Decreto n.° 6.171*. Estabelece o Zoneamento Ecológico Econômico da Área de Proteção Ambiental do Rio Verde.(Establishes the Ecologic Economic Zoning of the Rio Verde Environmental Protection Area)

PARANÁ. (2010b) Secretaria da Fazenda. *Tabela de municípios com participação no índice ambiental*.

VERTRAG. (2010) *Revisão do Zoneamento Econômico Ecológico da APA do Rio Verde*.

CHAPTER 25

PREVENTIVE ACTION PLAN FOR THE RIO VERDE BASIN

Ana Sylvia Zeny
André V. L. Bittencourt
Benno Doetzer
Carlos V. Roderjan
Charles Carneiro
Cleverson V. Andreoli
Cynara L. N. Cunha
Eduardo F. Gobbi
Fabiano Dalto
Kelly C. Gutseit
Lucia I. C. Sermann
Miguel M. Aisse
Patrício P. Zamora
Ricardo Godoi

PREVENTIVE ACTION PLAN FOR THE RIO VERDE BASIN

Ana Sylvia Zeny
André V. L. Bittencourt
Benno Doetzer
Carlos V. Roderjan
Charles Carneiro
Cleverson V. Andreoli
Cynara L. N. Cunha
Eduardo F. Gobbi
Fabiano Dalto
Kelly C. Gutseit
Lucia I. C. Sermann
Miguel M. Aisse
Patrício P. Zamora
Ricardo Godo

SUMMARY

This chapter presents the Preventive Action Plan for the Rio Verde Basin, which was developed through the joint efforts of the project's researchers and coordinators and based on the data presented in this book. The goal is to propose mid- and long-term actions that allow for efficient management of the Rio Verde Basin in order to prevent eutrophication of the Rio Verde Reservoir. The plan was structured with a basis in the principles and guidelines set forth by the World Lake Vision (WLV) and International Lake Basin Management (ILBM) for sustainable use of the lakes/reservoirs and their hydrographic basins. The plan is divided into five sections: Knowledge and Information, Policy, Technology, Information and Finances, each one containing guidelines for preventive action in the basin.

KEYWORDS

Reservoir basin, action plan, management, Rio Verde.

INTRODUCTION

The Preventive Action Plan is presented as an instrument of planning and management of both the water and the land within the Rio Verde Hydrographic Basin. The Basin is located in the State of Paraná, covering parts of the municipalities of Campo Magro, Campo Largo and Araucária, and it is part of the Alto Iguaçu Basin.

The plan seeks the preservation, conservation, reasonable and multiple uses of the natural and scenic resources of the region through harmonious integration of the needs of local communities, compliance with regulations established by environmental agencies, and activities carried out in the region, including those related to the use of water: with the Rio Verde Reservoir an important source of water.

The idea is to have an action plan that is justified and based on the data resulting from the research and surveys presented herein. It should be noted that in this type of research, all projects involved necessarily have guidelines of action for the basin as main outcomes. Therefore, the action plan outlines a set of mid- and long-term activities that allow for efficient management and can be replicated in other river basins.

What differentiates this new form of investigation is the proposal of a preventive management system, since the water of the Rio Verde Reservoir is currently in a condition of good quality. The emphasis on preventive action allows the development of actions before problems arise, thus promoting a plan that is slightly different from traditional management plans that are based on mitigation and compensation.

The elaboration of the Preventive Action Plan was based on the assumptions set forth by World Lake Vision (WLV) and International Lake Basin Management (ILBM) for sustainable use of lakes/reservoirs and their hydrographic basins, both platform driven by International Lake Committee Foundation (ILEC) and United Nations Environment Programme - International Environmental Technology Centre (UNEP-IETC).

WLV works with seven guiding principles:

1. A harmonious relationship between humans and nature is essential for sustainable use of water resources;
2. The hydrographic basin is the environmental scale that should be used to guide planning and management actions for the sustainable use of lakes/reservoirs;
3. A long-term preventive approach aimed at preventing the causes of the degradation of the lakes/reservoirs is essential;
4. Development of policies and decision making for the management of lakes/reservoirs should be based on scientific data and reliable information;

5. The management of lakes/reservoirs for sustainable use requires solving resource use conflicts, taking into consideration the needs of current and future generations, and the needs of the natural environment;
6. Citizens and concerned parties should be encouraged to participate in identifying and solving critical problems related to lakes/reservoirs;
7. Good governance that is based on justice, transparency and empowers all interested parties is essential to ensure the sustainable use of lakes/reservoirs.

While, ILBM platform works with six pillars as guideline for action plans:

1. Information (science): considers that scientific knowledge and public perception regarding the management of the lake/reservoir can differ and therefore the generation of knowledge and the dissemination of this knowledge are crucial.
2. Policies: political tools should be developed in order to facilitate joint social actions for the sustainable use of the river basin/reservoir.
3. Technologies: new technologies can significantly improve lacustrine environments, although sometimes with only localized and short-term effects.
4. Participation: all interested parties should have an appropriate role in the decision making process of the management in order to achieve sustainable management.
5. Finances: financial resources should come from all interested parties that financially benefit from the direct or indirect use of the lake/reservoir.
6. Institutions: a management system with an appropriate organizational scheme that helps to ensure sustainable benefits for the people that use the resources of the basin of the lake/reservoir.

The elaboration of this Rio Verde preventative action plan followed the following steps:

- Based on the concepts and guidelines of the WLV and ILMB, a preliminary structure was developed, containing the main organizing principles of the present Action Plan;
- The researchers responsible for the specific sub-projects elaborated preliminary proposals within their fields of study and presented them in a specific seminar;
- The coordination team worked on these proposals, homogenized their presentation, and grouped the proposals that dealt with related themes into a more consistent proposal;
- The proposed structure was then presented to the researchers at meetings of the thematic nuclei, which then enabled elaboration of each project within a uniform structure that orients the action plan.

The result of this joint effort between the researchers and the coordinators is discussed further below. The Preventive Action Plan of the Rio Verde Basin has five overarching themes:

I) Knowledge and Information
This section presents proposals for monitoring vegetation cover, water quality in the basin and in the reservoir, and chemical and mineralogical characteristics of the reservoir sediment.
II) Policies
This section presents strategies that support forest recovery in areas of permanent preservation, promotes improvements in the sanitary conditions of the population in the reservoir basin, as well as presents proposals for the planning, use and management of rural land.
III) Technology
This section proposes advanced technology developments for the remediation of water contaminated by toxic and recalcitrant substrates.
IV) Participation
This section presents a proposal for Environmental Education programs in the Rio Verde Reservoir Basin.
V) Finances
In this section we present a proposed social support program that aims to promote the participation of the communities around the reservoir in the administration of available natural resources.

Since this is a proposal of continuous monitoring, the deadlines indicated refer to the beginning of the activities after approval and transfer of funds to the institutions responsible.

Reports and annual evaluations will be submitted no later than 60 days after the completion of the results of each year of work. These reports will be presented in digital and in print to Petrobras.

I. INFORMATION (SCIENCE)

1. VEGETATION COVER MONITORING

1.1 INTRODUCTION

The transformation of the landscape is a natural consequence of the expansion of human occupation, ever since humans learned to domesticate plants and animals. However, the exponential increase of the world's population, combined with the unrestrained pressure in the interest of wealth accumulation (capital), is leading to the depletion of the capacity, balance and functionalities of natural systems.

In countless situations around the world, the limits have already been surpassed. Catastrophes and the suffering of a great number of people are frequently reported in the media. While this is not yet the case in the Rio Verde Basin, more than half of its surface has been completely transformed by anthropogenic activity.

Even within the locations determined in the environmental legislation as areas of permanent preservation

(APPs), 47.40% are occupied with inappropriate land-use activities, such as agriculture and livestock, reforestation, urbanization and mining, as shown in the map of natural vegetation cover and land-use completed in 2009.

Although an intense program of vegetation recovery within these areas is being proposed, through the planting of native arboreal, pioneer and fast growing species, pressures will persist within the remaining areas. Since this is a cultural issue, strongly rooted in the public consciousness since colonization, the monitoring of these processes is necessary in order to avoid the collapse of the ecosystem and, consequently, preserve the quality of life for its current and future inhabitants.

1.2 OBJECTIVE

Implement a monitoring program of land-use in the Rio Verde Basin, Metropolitan Region of Curitiba, PR, based on the vegetation cover and land-use map completed in 2009 (Chart 1).

1.3 PARTICIPATING INSTITUTIONS

- Specific Objective 1: Municipal administration (logistical support), Petrobras (resources), outsourced support, UFPR or Catholic Pontifical University of Paraná - PUC-PR (implementation).
- Specific Objective 2: Outsourced company, UFPR, or PUC-PR (implementation).

2. MONITORING OF WATER QUALITY

2.1 INTRODUCTION

A great challenge for urban development in the Metropolitan Region of Curitiba (RMC) is directly associated with the problems related to water resources: loss of water quality, restrictions in availability, and increases in demand. On the issue of water availability, since the RMC is located near the headwaters of the river, the river flow is low, requiring the construction of large volume reservoirs to ensure public water supply. As a consequence, the reservoirs of the region are extensive and have a long residence time, which favors the accumulation of significant loads of pollutants and nutrients.

The direct and indirect discharge of polluting loads into the water courses contributes greatly to the increase in nutrient concentrations that reach the reservoir and as such are a determining factor in the eutrophication process. Industrial discharges and releases, regular or not, as well as agriculture and livestock activities are the main polluting sources of water courses.

In degraded environments, or environments in the process of becoming degraded, that receive high quantities of nutrients (especially phosphorus and nitrogen), an intense growth of microalgae may occur, a phenomenon called blooms. Some of these blooms may be harmful or producers of powerful toxins, which reduces the quality of water, making treatment difficult and expensive.

For the nutrient and pollutant loads to be understood, it is necessary to develop a specific monitoring plan that enables a comprehension of the relationship between the history of the occupation of the drainage basin and its effects on the reservoir. Through monitoring, we can create a strategy for the collection, treatment and storage of data and their correlation with environmental models that can be used as management and decision-making tools. In this context, the monitoring of physical, chemical and biological parameters over long periods of time is essential.

The long-term data series can be used in environmental models that simulate water quality parameters. Such models have been used to monitor and study the impacts caused by the discharge of effluents into water bodies.

Besides monitoring the concentrations and loads of nutrients in the tributaries, the monitoring of the phytoplankton community is extremely important. It can serve as a final indicator of the water quality in the reservoir.

The Rio Verde Reservoir will be used as a source of public water supply in the near future. This reservoir can be classified as mesotrophic, with relatively good water quality. In 2005, the Rio Verde Reservoir experienced a bloom of *Cylindrospermopsis raciborskii*, a potentially toxic cyanobacteria species. This species produces cylindrospermopsin, which affects the liver and kidneys, and PSP that affects neuromuscular systems. Between 2005 and 2009 there was no record of another bloom of that microalgae species in the Rio Verde Reservoir. Although the occurrence of such bacteria is low, the potential for other blooms does exist. As such, it is essential to maintain the water quality in the reservoir. Monitoring is therefore a primary tool that researchers and managers can use to evaluate the evolution of the quality of the water destined for human consumption.

CHART 1 - SPECIFIC OBJECTIVES, ACTIVITIES, INDICATORS, GOALS AND DEADLINES PROPOSED FOR THE VEGETATION COVER MONITORING PLAN

SPECIFIC OBJECTIVES	ACTIVITY	INDICATOR	GOAL	DEADLINE FOR START UP
1. Implement Infrastructure	Establish cartographic database	Vegetation Cover and Land-use Map, 2009	Elaborated Map	1 month
	Define monitoring frequency	Monitoring Plan	Elaborated monitoring plan	1 month
	Acquire satellite images	Satellite Images	Satellite images acquired every year	End of each year
2. Implement Monitoring	Analysis of acquired image(s)	Updated Vegetation Cover and Land-use Map	Map updated every year	90 days after acquisition of corresponding images

The results obtained through monitoring will allow us to adopt control measures for possible problems in the management of the drainage basin; it will also allow us to continuously validate existing environmental models. In this sense, the role of monitoring water quality parameters becomes important; by monitoring variations in these parameters it is possible to determine the impact of the variations on the water quality of the reservoir. In the event of changes in load variation tendencies, it is possible to use models to predict the impacts on the reservoir.

The monitoring program should be implemented based on two issues: the quantity and quality of the water. Quantity refers to flow measurement in the main tributaries. Quality refers to the sampling efforts that will be carried out every two months in order to assess physical and chemical parameters in both the tributaries and the reservoir.

Meetings with the Environmental Institute of Paraná (IAP), the institution that monitors the reservoir every semester, will be scheduled to establish common protocols of sample collection, preservation and analyses, share results, and, ideally, establish a collaborative agreement for the mutual exchange of data.

The obtained results will be evaluated to ensure compliance with the standards for the classification of water bodies, as defined by CONAMA Resolution 357/05 for surface waters.

Partial reports will be developed based on the data obtained from each sampling episode. Each year, reports will be produced that consolidates the data obtained in the bimonthly reports and the results obtained through the environmental modeling.

Twenty-three sampling points will be used, as shown in Table 1 and Figure 1. This sampling effort is essential to have a broad and detailed monitoring program that is able to efficiently respond to identified impacts and act quickly to mitigate them.

During the interdisciplinary Research Project on Water Eutrophication of the Rio Verde/Petrobras Reservoir, two fixed monitoring stations were installed: i) a meteorological station installed near the dam (Latitude 25°31´36,83´´S and Longitude 49°31´39,07´´) that provides information on the wind speed and direction, air temperature, solar radiation, humidity, and rainfall; ii) and another, in Rio Verde, called the HARAS station (Latitude 25°27' 23" and Longitude 49°27' 43"), that measures the flow of the Rio Verde, rainfall and wind direction and speed.

These stations will be maintained and will provide meteorological data every 15 minutes throughout the monitoring period. In the case of the HARAS station, the flow of the main channel of the Rio Verde will be measured continuously. The HARAS will also provide information on the flow of the other tributaries through the "midsection method" with the help of a gaging reel. The measurements will be made concurrently with the sample collection for physicochemical analysis. At point F5, located downstream of the reservoir, the measurement will be performed with the use of a boat.

The monitoring of the tributaries of the Rio Verde Reservoir basin will be done every two months at the sampling points presented in Table 1 and Figure 1. The samplings will be carried out so as to consider periods of significant rainfall and drought. During the cropping season between September and November, during which the majority of soil movement occurs and agrochemicals are applied, the sampling efforts will be intensified, with planned monthly frequency.

The surface water will be collected in the tributaries. The parameters to be analyzed in these water samples are presented in Table 2. The variables that will be measured in the field are: conductivity, turbidity, pH, dissolved oxygen concentration, and water temperature.

The monitoring of within reservoir will be done every two months with collection at points R1, R4 and R5 (Figure 1). The parameters to be analyzed are presented in Table 3. Some parameters will be sampled at three depths in the water column (marked with an asterisk in the Table 3). The water temperature, pH and the dissolved oxygen will be measured in the field every 0.5 meters with the aid of a Multiparameter probe. The other parameters will be sampled only at the surface.

As part of the Rio Verde project, environmental modeling of the Rio Verde Reservoir was developed, with calibration for a water quality model. The hydrodynamic model and the water quality model used are part of the Environmental Hydrodynamics Database System called SisBAHIA®, developed by the Coastal and Oceanographic Engineering Area of the Oceanic Engineering of COPPE/UFRJ. The water quality model developed for the Rio Verde Reservoir considers the following parameters: temperature, dissolved oxygen, biochemical oxygen demand, ammonia nitrogen, nitrate nitrogen, chlorophyll-a, organic phosphorus, inorganic phosphors and zooplankton. During this period, the measurements of these parameters were taken at five points in the reservoir. The results of temperature and other parameters modeled by SisBAHIA® were compared with field data from the reservoir meteorological station and were shown to be consistent. The data collected in the tributaries were used in defining boundary conditions of the two models. In the monitoring phase, the collected data will be used to generate and calibrate the environmental models.

2.2 OBJECTIVES

The main objective is to obtain a long-term data series, necessary to verify and maintain water quality and quantity levels in the Rio Verde Reservoir, that are affected by land-use within the hydrographic basin. This will allow us to identify and interpret variations, detecting potentially degrading tendencies regarding the quality and quantity of water through environmental models (Chart 2).

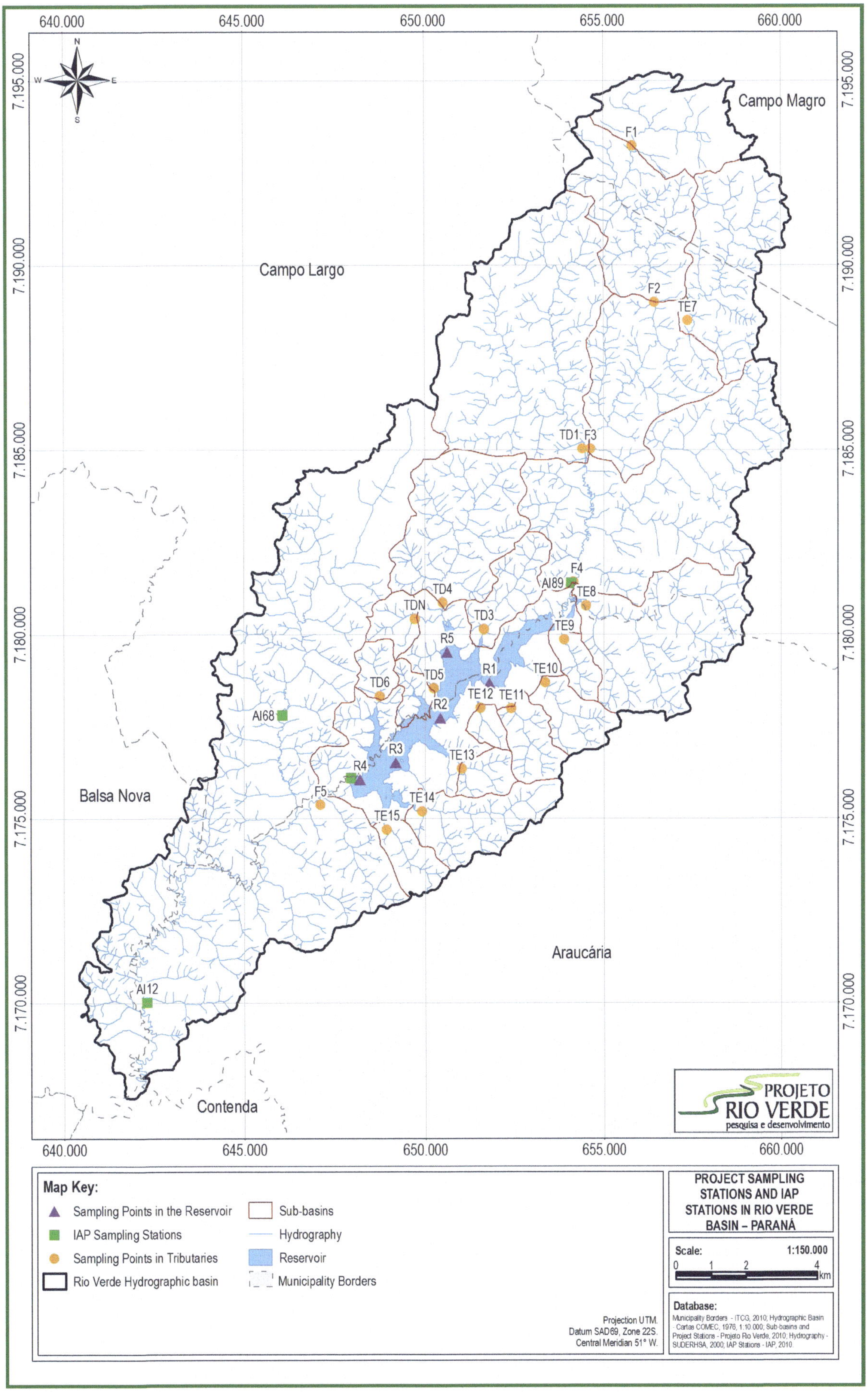

FIGURE 1 – MAP OF SAMPLING AND MONITORING POINTS. THE PERMANENT COLLECTING STATIONS ARE CIRCLED IN RED.

TABLE 1 – POINTS TO BE SAMPLED IN THE RIO VERDE RESERVOIR BASIN

CODE	POINT NAME	UTM COORDINATES	
F1	Main Fluvial Channel 1	22 J 0655824	7193242
F2	Main Fluvial Channel 2	22 J 0656445	7193242
F3	Main Fluvial Channel 3	22 J 0654646	7185027
F4	Main Fluvial Channel 4	22 J 0654152	7181423
F5	Main Fluvial Channel 5	22 J 0647124	7175377
TD1	Right Riverbank Tributary 1	22 J 0654407	7185035
TD3	Right Riverbank Tributary 3	22 J 0651670	7180151
TD4	Right Riverbank Tributary 4	22 J 0650516	7180868
TD5	Right Riverbank Tributary 5	22 J 0650256	7178540
TD6	Right Riverbank Tributary 6	22 J 0648772	7178335
TDN	Right Riverbank Tributary	22 J 0649740	7180427
TE7	Left Riverbank Tributary 7	22 J 0657369	7188510
TE8	Left Riverbank Tributary 8	22 J 0654516	7180781
TE9	Left Riverbank Tributary 9	22 J 0653904	7179869
TE10	Left Riverbank Tributary 10	22 J 0653375	7178699
TE11	Left Riverbank Tributary 11	22 J 0652424	7177998
TE12	Left Riverbank Tributary 12	22 J 0651562	7178017
TE13	Left Riverbank Tributary 13	22 J 0651044	7176363
TE14	Left Riverbank Tributary 14	22 J 0649926	7175208
TE15	Left Riverbank Tributary 15	22 J 0648965	7174716
R1	Reservoir – Headwater 1m	22 J 0651812	7178676
	Reservoir – Headwater Bottom		
R4	Reservoir – Dam 1m	22 J 0648214	7176058
	Reservoir – Dam Deep Euphotic Zone		
	Reservoir – Dam Bottom		
R5	Reservoir – Backwater 1m	22 J 0650625	7179505
	Reservoir – Backwater Bottom		

TABLE 2 – PARAMETERS TO BE ANALYZED IN WATER SAMPLES COLLECTED IN THE TRIBUTARIES OF THE RIO VERDE RESERVOIR BASIN AND THEIR QUANTIFICATION LIMITS

GENERAL PARAMETERS	UNITS	QUANTIFICATION LIMIT
Pesticides	μg/L	0.001
Alkalinity	mg/L	3.0
Bromide	mg/L	0.018
Calcium	mg/L	0.007
Total Organic Carbon – TOC	mg/l	0.3
Chloride	mg/L	3.0
Thermotolerant Coliforms	NMP/100mL	1.0
Electrical Conductivity	μS/cm	1.0
Apparent Color	UC	< 2.5
Biochemical Oxygen Demand	mg/L O_2	1.0
Chemical Oxygen Demand	mg/L O_2	5.0
Fluoride	mg/L	0.02
Total Phosphorus	mg/L P	0.002
PAH's	μg/L	0.01
Magnesium	ml/L	0.005
Sedimentable Material	ml/L	0.1
Metals and non-metals	mg/L	Variable
Nitrate	mgNO_3(N) /L	0.002
Nitrite	mgNO_2(N) /L	0.001
Ammonia Nitrogen	mgN/L	0.002
Organic Nitrogen	mg/L	0.002
Total Nitrogen	mg/L	0.002

Oxalate	mg/L	0.009
Dissolved oxygen	mg/L	0.01
PCB's	µg/L	0.001
pH	-	1 -- 13
Potassium	mg/L	0.06
Sodium	mg/L	0.06
Dissolved Solids	mg/L	≥ 1
Particulate Solids	mg/L	≥ 1
Sedimentable Solids	mL/L	≥ 0.1
Fixed Suspended Solids	mg/L	1.0
Total Suspended Solids	mg/L	1.0
Volatile Suspended Solids	mg/L	1.0
Total Solids	mg/L	1.0
Fixed Total Solids	mg/L	1.0
Total Volatile Solids	mg/L	1.0
Sulfate	mg/L	10
Turbidity	NTU	1.0

TABLE 3 – PARAMETERS TO BE ANALYZED IN WATER SAMPLES COLLECTED IN THE RESERVOIR AND THEIR LIMITS OF QUANTIFICATION. PARAMETERS MARKED WITH ASTERISKS (*) INDICATE PARAMETERS THAT WILL BE SAMPLED AT THREE DEPTHS IN THE WATER COLUMN

Parameters	Units	Quantification Limit
Alkalinity	mg/L	3.0
Bromide	mg/L	0.018
Calcium	mg/L	0.007
Total Organic Carbon *	mg/l	0.3
Chloride	mg/L	3.0
Chlorophyll-a *	mg/l	0.001
Conductivity (Electric) *	µS/cm	1.0
Phytoplankton Count *	Cells/mL	> 1
Chemical Oxygen Demand	mg/L O2	5.0
Fluoride	mg/L	0.02
Reactive Phosphorus *	mg/l P	0.002
Total Phosphorus *	mg/l P	0.002
Phytoplankton Identification *	-	-
Magnesium	mg/L	0.005
Suspended Particulate Material *		
Metals	mg/L	Variable
Nitrate *	mgNO3 (N) /L	0.002
Nitrite *	mgNO2 (N) /L	0.002
Ammonia Nitrogen *	mgN/L	0.002
Organic Nitrogen *	mg/L	0.002
Total Nitrogen *	mg/L	0.002
Oxalate	mg/L	0.009
Dissolved Oxygen	mg/L	0.01
pH	-	1 – 13
Potassium	mg/L	0.06
Silicate *	mg/l	0.001
Sodium	mg/L	0.06
Sulfate	mg/L	10
Water temperature	°C	0°C a 30°C (resolution 0.1°C)
Air Temperature	°C	-10°C a 50°C (resolution 0.1°C)
Transparency – Secchi Disk	m	0 to 20
Turbidity *	NTU	1.0

CHART 2 - SPECIFIC OBJECTIVES, ACTIVITIES, INDICATORS, GOALS AND DEADLINES FOR THE WATER QUALITY MONITORING PLAN

SPECIFIC OBJECTIVES	ACTIVITY	INDICATOR	GOAL	DEADLINE FOR BEGINNING OF ACTIVITY
Monitor quality and quantity of water in the tributaries of the Rio Verde reservoir	Data collection in the tributaries in the Rio Verde Reservoir	Parameters of water quality and flow	Comply with CONAMA Resolution 375/05 and reduce risk of waterborne diseases	1 month
Monitor water quality and phytoplankton community in the Rio Verde Reservoir	Data collection in the Rio Verde Reservoir	Parameters of water quality and phytoplankton density and composition	Comply with CONAMA Resolution 375/05. Appropriate use of water resources and protection of public health among the population that use the water from the reservoir	1 month
Obtain knowledge of the mechanisms associated with hydrodynamic circulation and dispersion of pollutants in reservoirs through SisBAHIA®	Simulate and study the hydrodynamics in the reservoir	No. of simulations done	One annual simulation	1 month after the first data collection
	Simulate and study the transport of water quality parameters in the reservoir	No. of simulations done	One annual simulation	1 month after the first data collection

2.3 PARTICIPATING INSTITUTIONS

- Pontifical Catholic University of Parana (PUC-PR)
- Federal University of Paraná (UFPR)
  Environmental Engineering Department
  Geology Department, Laboratory of Hydrogeological Research (LPH)
  Department of Botany

3. MONITORING OF CHEMICAL AND MINERALOGICAL ASPECTS OF THE RESERVOIR SEDIMENT

3.1 INTRODUCTION

The quality of water, particularly in reservoirs, is naturally maintained through the water level and the settlement and suspension of particulate matter in the water column, which play an important role in aquatic ecosystems. Suspended particles attenuate the entry of light into water bodies and react with the medium. Along with the chemical characteristics of the environment, such reactions also affect the concentration and stoichiometry of the ionic and metallic macro-constituents of the system.

The dynamics of suspension and sedimentation of the particles can also be triggers of the eutrophication process. This is one of the main concerns when the reservoir is destined to supply water for human consumption. Eutrophication can be natural or artificial. When it is natural, it is a slow and continuous process that results in the input of nutrients from rainfall and surface water runoff, in the form of particles or not, that alter the physicochemical characteristics of the system. When eutrophication is anthropogenic in origin, it is normally associated with the discharge of domestic and industrial effluents, agriculture, or the disposal of garbage on the banks of aquatic ecosystems. However, when it comes to a water body situated in a region with high levels of human occupation, such as the Rio Verde Basin, it is difficult to distinguish between factors independent of human occupation and those arising directly or indirectly from it.

One of the main nutrients, phosphorus (P), can be found in the sediment associated with complex salts of calcium, iron, aluminum and organic species; it can also be found adsorbed on the surface of minerals. Phosphorus is particularly important in the cell growth of algae, so much so that its excess can lead to blooms. Currently, much attention has been given to the process of adsorption/desorption of P in natural sediment and pure minerals.

Laboratory experiments with Fe(OH) showed that the increase of the pH produces a release of P adsorbed in iron complexes because of competition with OH- ions. However, there is little information on the effect of pH in the processes of P sorption in natural sediments.

The "natural" content of exchangeable phosphorus, as well as sorption characteristics, are related to sediment composition, such as active iron, aluminum, and organic matter. The sediment composition, on the other hand, has been linked to the chemistry of the reservoir, such as hardness, alkalinity and pH, among other parameters.

When the "natural" phosphorus content is significant, the adsorption/desorption process cannot be disregarded. When the sediment that contains phosphorus comes in contact with water, the P is exchanged with water at the sediment-water interface until a dynamic balance is reached. The concentrations in the sediment and in the water are stabilized when the exchanges, from sediment to water and vice versa, equalize.

Sediment from three points in the reservoir will be sampled each semester every two years. The points to be sampled are the same as those used in monitoring of water quality: points R1, R4 and R5 (Figure 1). The parameters to be analyzed are pH, redox potential and a set of 53 periodic table elements.

3.2 OBJECTIVES

Monitor the chemical and mineralogical aspects of the sediment of the Rio Verde Reservoir that contribute to water quality of the water body (Chart 3).

3.3 PARTICIPATING INSTITUTIONS

- Federal University of Parana (UFPR)
  Geology department, Laboratory of Hydrogeological Researches (LPH)
  Environmental Engineering Department

II POLICIES

1. REFORESTATION IN AREAS OF PERMANENT PRESERVATION

1.1 INTRODUCTION

Among the environmental impacts caused by the expansion of human occupation, the most serious and troubling is, without a doubt, the degradation of water sources. However, what is more worrying is that humans have expressed awareness of the protective function of vegetation for natural and artificial water bodies throughout human history. However, humans have ignored the facts and used the soil as it suited them. Now, solutions are sought in order to revert the situation, starting with the most obvious: the restoration of the protective vegetation around springs and headwaters, and on the banks of rivers, lakes and reservoirs.

Considering this context, the surface of the Rio Verde basin was mapped. The different forms of land-use were identified and quantified and the locations where vegetation restoration and readjustment are required were identified.

1.2 OBJECTIVE

Promote the restoration of vegetation on the banks of the water bodies in the Rio Verde Basin where human activities do not comply with environmental regulations (Chart 4).

1.3 PARTICIPATING INSTITUTIONS

- Specific objective 1: Municipal Governments and Environmental Institute of Paraná (IAP) (institutional support)
- Specific objectives 2 and 3: Municipal Governments and Petrobras
- Specific objective 4: Municipal Governments and property owners

CHART 3 - SPECIFIC OBJECTIVES, ACTIVITIES, INDICATORS, GOALS AND DEADLINES PROPOSED FOR THE SEDIMENT MONITORING PLAN.

SPECIFIC OBJECTIVES	ACTIVITY	INDICATOR	GOAL	DEADLINE FOR START UP
1 – Monitor the elements in the reservoir sediment	Collect sediment from 3 sampling points	No. of sediment collection episodes	Perform 1 episode each semester every 2 years at 3 sampling points	2 months
	Analyze the mineralogical and chemical composition	No. of samples analyzed in the laboratory (for 53 elements)	Analyze semiannually 3 sediment samples (mineralogy and elemental composition)	1 week after collection

CHART 4 - SPECIFIC OBJECTIVES, ACTIVITIES, INDICATORS, GOALS AND DEADLINES FOR THE REFORESTATION PLAN

SPECIFIC OBJECTIVES	ACTIVITY	INDICATOR	GOAL	DEADLINE FOR START UP
1 – Locate and Educate property owners	Locate non-compliant APPs	Vegetation cover and land-use map from 2009 with identified non-compliant APPs	Register and map non-compliant properties	1 month
	Team training	Number of trained teams	2 trained teams	1 month
	Visit the registered property owners	Number of visits to property owners	All registered owners visited	1 month
2 – Establish infrastructure	Construction of the Forest plant nursery	Forest plant nursery	Nursery built	2 months
3 – Seedling Production	Locate bracatingais	Locate bracatingais on the Vegetation Cover and Land-Use Map, 2009	Bracatingais identified	1 month
	Seed gathering (Mimosa scabrella)	Number of gathered seeds	3,250,000 gathered seeds by maturation period	2 months
	Maturation of the seeds of the Mimosa scabrella (summer)	No. of produced seedlings	Production of 3,191,400 seedlings by maturation period	4 months
4- Planting	Distribute seedlings in properties	No. of properties participating in the Project	Seedling distribution to all participating land owners	1 year
	Instruct and train land owners	No. of land owners that acquire environmental awareness	All participants visited	2 years
	Monitor plantation	No. of properties visited quarterly	All participants visited quarterly	1.5 years

2. RURAL SANITATION

2.1 INTRODUCTION

In hydrographic basins of water supply sources, it is common for rural communities to consume water that comes from rudimentary sources, rarely found in the urban environment, such as artesian and groundwater wells and springs. Such sources can be contaminated due to deficiencies in rural sanitation, especially with inadequate management of animal manure and the lack of technical assistance for treatment and final disposal of sewage.

The use of sanitation in rural areas promotes environmental health using natural resources in a sustainable way, thus reversing environmental degradation and negative impacts on public health.

The rural communities located in the Rio Verde Basin consume water from two different sources: Water Supply System (WSS), constructed by SANEPAR, and operated by SANEPAR or the Municipality; and private sources for individual consumption, such as shallow wells and springs, located within the rural property.

Through the water analyses of the private sources, conducted at the end of 2009 and beginning of 2010, we observed that more than 63% of the samples presented *Escherichia coli*. Ordinance 518, 2004, from the Ministry of Health, which defined the standards of water potability, does not permit the presence of *E. coli*.

The high percentage of contaminated samples found in the study requires a detailed analysis of the sanitary conditions of each property. It is known that animal manure, as well as the poor location, construction and operation of septic tanks and dry cesspits can be responsible for the high *E. coli* levels.

Lastly, the assessed rural population is resistant to the use of chlorinated water. Despite receiving good quality water from the WSS, many community members prefer to consume water from their own sources, of which approximately 70% of the studied cases do not have any type of disinfection or treatment.

The preventive action plan aimed at rural sanitation in the Rio Verde Basin is justified by the results discussed in the assessment of sanitation in the region. The analysis demonstrated that the poor management of animal waste and the lack of technical information (assistance) regarding the treatment and final disposal of domestic sewage has an impact on the water used for personal consumption, and therefore on public health.

2.2 OBJECTIVES

Promote the improvement of sanitary conditions in the rural area of the Rio Verde Basin (Chart 5).

2.3 PARTICIPATING INSTITUTIONS

Some institutions and public bodies stand out in relation to sanitation. In the municipality, its administration, represented by the Health Inspection agencies, are responsible for water quality in the collective Water Supply Systems of Colônia Cristina, located in Araucária, and Colônia Dom Pedro II, in Campo Largo.

Similarly, SANEPAR, represented by the Unit of Services, Projects and Operation – Curitiba, Metropolitan and Coastal Region (USPO-CT), is responsible for the operation of the WSS in Colônia Antônio Rebouças in Campo Largo.

The Municipal administration can also assist rural producers regarding the treatment and final disposal of domestic sewage, guiding project development and review of the current sewage system, to ensure limited impacts on the quality of the supplied water.

Due to the established relationships with rural producers, local EMATER offices can lead measures regarding the use of animal manure, with the aid of the Secretariats of Agriculture within each Municipality.

It is worth mentioning that considering the experience of the researchers, the team responsible for the sub-project (DHS/UFPR) assessing rural sanitation in the Rio Verde Basin will be able to participate in the initiatives to achieve these goals.

3. URBAN SANITATION

3.1 INTRODUCTION

When domestic sewage is released into rivers or lakes without proper treatment the nutrient supply increases which can lead to the acceleration of the eutrophication process.

To manage and control the trophic state of the water in a reservoir, the most recommended strategy is the adoption of preventive measures. These measures are usually less expensive and more effective, as they are aimed at the reduction of nutrient input in the water bodies, or rather, a direct interference in external nutrient sources, among them being domestic sewage.

In the drainage area of the Rio Verde Reservoir, there are some urban communities that do not have a system for the collection, treatment and final disposal of generated sewage. Thus, it is up to SANEPAR to develop, implement, and manage a system aimed at controlling nutrient input directly to the reservoir or through its tributaries. In Figure 2, we outline the sanitary sewage system and sanitation works planned for the region.

3.2 OBJECTIVES

Manage and control the input of loads to the Rio Verde Reservoir and its tributaries, through the implementation of systems of collection, treatment and final disposal of domestic sewage being emitted from the urban agglomerations in the drainage basin of the reservoir (Chart 6).

3.3 PARTICIPATING INSTITUTIONS

- SANEPAR – Water and Sanitation Company of Paraná State
- Municipalities

CHART 5 - SPECIFIC OBJECTIVES, ACTIVITIES, INDICATORS, GOALS AND DEADLINES FOR THE RURAL SANITATION PLAN

SPECIFIC OBJECTIVES	ACTIVITY	INDICATOR	GOAL	DEADLINE FOR START UP
Water Supply Systems (WSS)	Control of water quality in the WSS of Colônia Dom Pedro II	Sampling frequency	Comply with ordinance 518/04 of the Ministry of Health	1 month
		Sampling Parameters		
	Control of water quality in the WSS of Colônia Antônio Rebouças	Sampling frequency	Accomplished (Available in SQA SANEPAR);	
		Sampling Parameters		
	Control of water quality in the WSS of Colônia Cristina	Sampling frequency	Comply with ordinance 518/04 of the Ministry of Health (ongoing)	1 month
		Sampling Parameters		
	Implementation of WSS in the Colônia Figueiredo	WSS in Colônia Figueiredo	WSS project developed and approved	1st year of project
			Construction of the WSS (depends on community acceptance)	2 years
			WSS in operation	3 years
	Fluoridation in all WSSs	Fluorine (mg/L)	Comply with ordinance 518/04 of the Ministry of Health (1.5 mg/L) (difficult to implement; suggest an alternative approach)	2 years
Individual Water Supply, Sanitary Sewage and Solid Waste	Environmental Education	Participation of the communities from Colônia Cristina, Figueiredo and Dom Pedro II	Elaboration of Tools for Awareness and Instruction (Preparation of Primer)	6 months
			Implementation of Tools for Awareness and Instruction (Primer)	10 months
Personal Water Supply	Control of Water Quality in Personal Systems	Absence of Thermotolerant Coliforms	Water chlorination and other practices	1 year
Sanitary Sewage	Elaborate project, implement and start operation of Septic Tank and Supplemental Treatment	Standard Project	Revision of Septic Tanks and Supplementary Treatments	6 months
			Financing	2 years
		No. of sanitary sewage systems installed	30 sanitary sewage systems installed	3 years
		No. of operating sanitary sewage systems	Annual cleaning of septic tanks	3 years
Solid Waste	Produce, stabilize and use of manures	Production of animal manure on the properties	Field surveys in 50 properties	6 months
		Demonstration of animal manure stabilization	Demonstration concluded	2 years
		Demonstration of animal manure use	Demonstration concluded	2 years
	Selective collection of urban solid waste	Expansion of selective waste collection	Implementation of collection and final disposal throughout 100% of the rural area	1 year
	Conventional collection of urban solid waste	Expansion of conventional waste collection	Implementation of conventional collection operating in 100% of the rural area	1 year

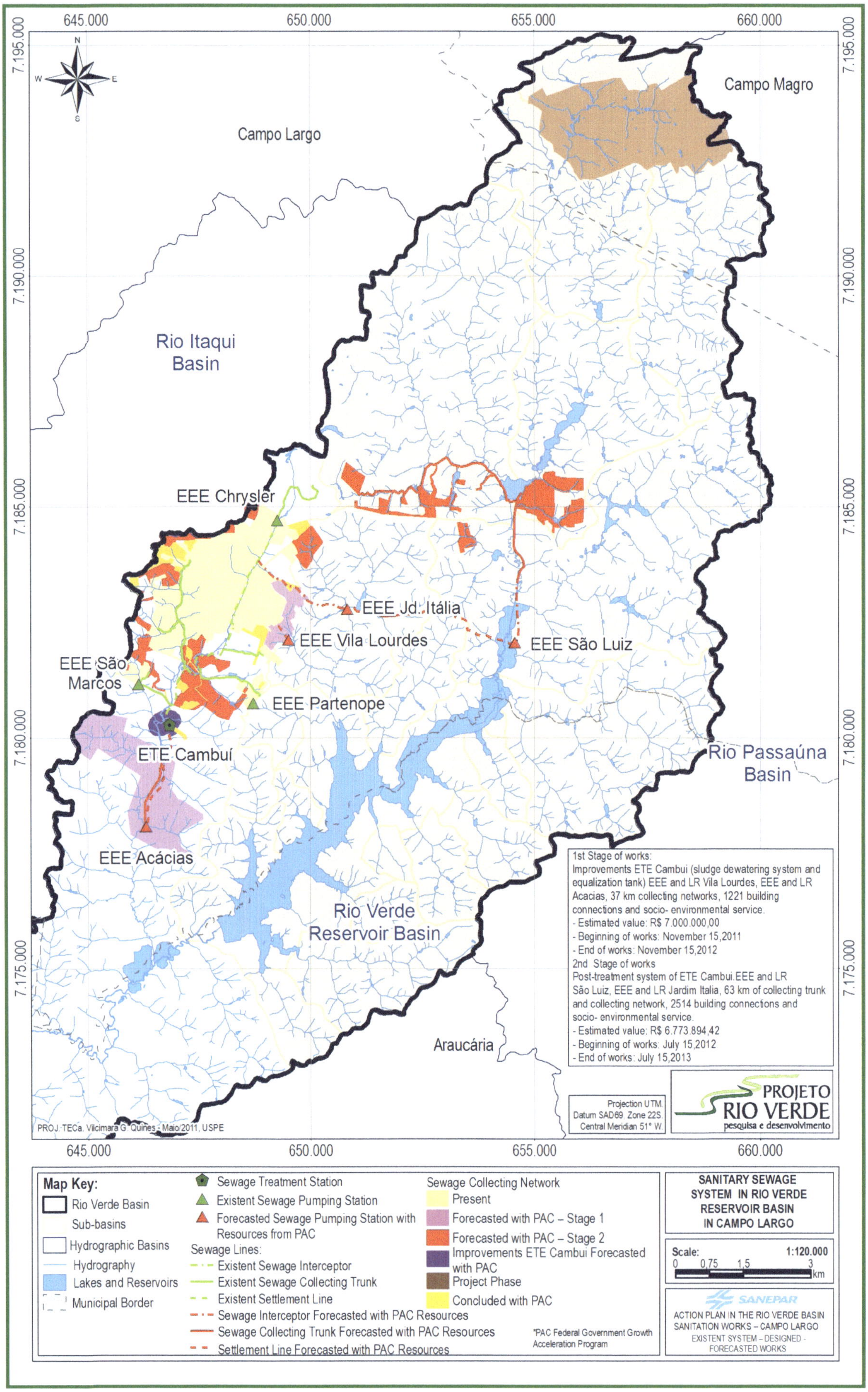

FIGURE 2 – SYSTEM OF SANITARY SEWAGE IN THE BASIN OF THE RIO VERDE RESERVOIR AND SANITATION WORKS PLANNED FOR THE REGION

CHART 6 - SPECIFIC OBJECTIVES, ACTIVITIES, INDICATORS, GOALS AND DEADLINES OF THE URBAN SANITATION PLAN

SPECIFIC OBJECTIVES	ACTIVITY	INDICATOR	GOAL	DEADLINE FOR START UP
Expand the coverage of the collection, treatment and disposal system of domestic sewage	Develop project plan for the expansion of the Cambuí Sewage System	Improvement project of Cambuí Sewage System	Project approved	November, 2011
	Implement construction work to expand the system - Phase 1	Number of Kms of collection networks	37 km of collection networks and 1221 residential connections implemented	One year from the beginning of construction
		Sewage pumping stations (Vila Lourdes and Acácias)		
		Improvements in the Cambuí Sewage System	Improvements (sludge dewatering system and equalization tank) implemented in Sewage System	
	Operate expanded system	Operator Training	Operator trained	1 month (before the end of construction)
		Monitoring Plan	Monitoring Plan Implemented	2 months (before the end of construction)
	Implement construction work to expand the system - Phase 2	Number of Kms of sewage collection networks and trunk collectors	63 km of collection networks and trunk collectors, and 2514 residential connections implemented	1 year after start of Phase 1
		Interceptors Rincão and Cercadinho	Interceptors implemented	
		Sewage pumping stations (São Luiz, Jd Italia)	Sewage pumping stations implemented	
		Improvements in the Cambuí Sewage System	Improvements in the implemented Sewage System (post-treatment flotation system)	
	Operate expanded system	Training of Operator	Operator trained	1 month (before the end of construction)
		Revise monitoring plan	Monitoring plan revised and implemented	2 months (before the end of construction)

4. PLANNING, USE AND MANAGEMENT OF RURAL LAND

4.1 INTRODUCTION

Changes to nature and the expansion of human activities have both quantitative and qualitative consequences for water resources.

The artificial eutrophication of water bodies leads to a progressive degradation of water quality, especially in lakes, due to the massive growth of aquatic plants, which affects the metabolism of entire water ecosystem.

Among the most influential non-point sources of nutrients for the aquatic ecosystem, which increase eutrophication, are agricultural areas, especially those where there are no land-use conservation techniques in place.

Therefore, possible actions should be focused on soil conservation while safeguarding farmers' income, as well as water conservation.

4.2 OBJECTIVES

Monitor the agriculture and livestock activities (land-use) in the regions of influence on the draining basins of the Rio Verde Reservoir. Monitoring will consider planning, use and management of the land, based on Ecologic-Economic Zoning, and follow-up on the proposed suitability in conflict areas, reconciling agricultural production with sustainable use of natural resources (Chart 7).

4.3 PARTICIPATING INSTITUTIONS

- EMBRAPA – Brazilian Agricultural Research Corporation
- EMATER – Technical Assistance and Rural Extension Institute of Paraná
- SEAB – State Secretary of Agriculture and Water Supply, through the Program of Integrated Environmental Management in Micro basins (PGAIM)
- SANEPAR
- Petrobras
- Municipal Administrations
- Farmers/resident associations
- Private enterprise

CHART 7 - SPECIFIC OBJECTIVES, ACTIVITIES, INDICATORS, GOALS AND DEADLINES PROPOSED FOR THE RURAL LAND--USE AND MANAGEMENT PLAN

SPECIFIC OBJECTIVES	ACTIVITY	INDICATOR	GOAL	DEADLINE FOR START UP
Survey of land-use conflict areas	Perform flight at low altitude: 200 meters, range of 2 km to each side of the flight line	Images of the area of interest	Images of the area of interest obtained	6 months
	Compile and analyze data-base	Vegetation Cover and Current Land-use Map	Location of areas with land-use conflicts	3 months after obtaining images
Elaborate proposal of joint intervention	Elaborate a Plan of Integrated Management in the basin	Management plan	Management plan developed	3 months
Implement monitoring of the Management Plan	Perform flight at low altitude: 200 meters, range of 2 km to each side of the flight line	Images of the area of interest	Images of the area of interest obtained each year	At the end of each year
	Compile and analyze the database	Vegetation Cover and Current Land-use Map	Updated Map of Vegetation Cover and Current Land-use	3 months after obtaining images
Provide training regarding the sustainable use of natural resources	Detailed survey of agricultural capability in the basin	Map of agricultural capability	Accomplished	
	Evaluate the viability and impacts of activities that are being developed on the properties	No. of evaluated properties	200 properties with the most critical areas of conflict	1 year
	Elaborate proposal for the conversion of production systems	Conversion plan of production systems	Conversion plan of production systems elaborated	6 months
Implement collaborative environmental education	See "Environmental Education Program – CONVIVERDE"			

III TECHNOLOGY

1. CONTINUOUS ADVANCED OXIDATIVE PROCESSES IN THE TREATMENT OF WATERS CONTAMINATED BY CYANOTOXINS AND OTHER SIGNIFICANT MICROPOLLUTANTS

1.1 INTRODUCTION

In the last several years, the process of environmental degradation has reached global proportions, demonstrating the planet's vulnerability. As such, rigorous policies of environmental protection have been implemented, with the application of more restrictive legislation and more rigorous inspection being critical. Unfortunately, the polluting tendencies of industrial processes and domestic waste continue to acutely manifest itself due to difficulties in the modification of productive processes and minimization of the volume of domestic waste.

Given the inability to reduce waste emission at the source, the use of treatment procedures becomes absolutely essential. Unfortunately, the available technologies do not satisfy the current needs of waste treatment and contaminated mediums, leading to a very significant number of unaddressed issues. In this context, we can highlight the contamination of natural and artificial water bodies by toxin- producing cyanobacteria, as well as other relevant micro-pollutants that are emerging (pharmaceutical residues, estrogens, etc.)

Due to the high polluting potential and acknowledged toxicity of the above mentioned chemical species, much effort has been devoted to the establishment of treatments that aim at the remediation of contaminated mediums. Unfortunately, the conventional treatment processes, which are usually based on biological processes, present limited efficiency in degrading contaminants due to the limited biodegradability of these substrates. Therefore, the search for alternatives in the remediation of contaminated water becomes absolutely essential.

In the past several years, advanced oxidation processes (AOPs) have emerged as an excellent alternative for waste treatment, mainly due to their high efficiency of degradation of resistant substrates. The process, based on the production of highly oxidizing hydroxyl radicals (HO.), allows the quick and indiscriminate degradation of a wide variety of organic compounds, and, often, their complete mineralization.

Among the AOPs, the Fenton process and its photo--assisted versions have shown to be very promising, especially due to their high degradation efficiency of notably recalcitrant substrates, such as: cyanotoxins, pharmaceuticals and estrogens.

The Fenton process is based on the production of hydroxyl radicals from the reaction between ferrous ions and hydrogen peroxide (Equation 1). When assisted by ultraviolet or visible radiation, the system presents even higher degradation efficiency, due to the fact that the photo reduction of ferric ions leads to the formation of additional amounts of hydroxyl radicals (Equation 2). In this case, besides the generation of two moles of hydroxyl radical per mole of initially decomposed peroxide, a catalytic cycle that maximizes the efficiency of the system is closed (NEYENS & BAEYENS, 2003).

$$Fe^{2+} + H_2O_2 \rightarrow Fe^{3+} + OH^- + OH^\cdot \quad (1)$$
$$Fe^{3+} + H_2O + h\nu \rightarrow Fe^{2+} + H^+ + OH^\cdot \quad (2)$$

When processed in solution, the Fenton system proves very efficient in the degradation of many organic substances. However, the need to remove the sludge generated by the precipitation of iron oxides and the strict limits of pH, which is required to prevent precipitation, are factors which impair its applicability on a large scale. The use of immobilized forms of iron seems to be a promising alternative to work around these drawbacks. Several studies have reported the maintenance of process efficiency in systems using immobilized forms of iron.

Although many studies have been developed that demonstrate the high degradation capacity of advanced oxidation processes, few technologically feasible proposals have been suggested. In this sense, the use of Fenton systems is quite advantageous because of operational simplicity, low cost and high degradation efficiency of numerous recalcitrant substrates. Additionally, the use of immobilized forms of iron can make the development of a continuous treatment system feasible and applicable on a large scale.

1.2 OBJECTIVES

Develop an advanced technology focused on the remediation of aquatic environmental matrices contaminated by toxic and recalcitrant substrates.

Among the proposed remediation processes, the Fenton-type oxidative systems stand out. By employing immobilized forms of iron II or III, they can enable the development of continuous treatment systems (Chart 8).

1.3 PARTICIPATING INSTITUTION

- Federal University of Paraná (UFPR)
- Chemistry Department – Laboratory of Environmental Chemistry and Materials

REFERENCES

NEYENS, E.; BAEYENS, J., 2003. A review of classic Fenton's peroxidation as an advanced oxidation technique. *Journal of Hazardous Materials*, B98, 33-50.

CHART 8 - SPECIFIC OBJECTIVES, ACTIVITIES, INDICATORS, GOALS AND DEADLINES FOR THE TREATMENT OF CONTAMINATED WATER PLAN THROUGH ADVANCED OXIDATIVE PROCESSES

SPECIFIC OBJECTIVES	ACTIVITY	INDICATOR	GOAL	DEADLINE FOR START UP
Evaluate the potential of immobilized forms of iron in the degradation of standard substrates (microcystin, pharmaceuticals and estrogens) in aqueous solution in the laboratory. Conduct a comparative study involving Fenton systems in homogeneous phase	Preparation and characterization of immobilized forms of iron (II and III), using supports such as calcium alginate and chitosan	No. of prepared and charactered materials	Preparation and characterization of two supported photocatalysts	2 months
	Studies of degradation of standard substrates	No. of processes per substrate	3 processes	
	Selection of processes with best performance	Evaluation report	Finalized evaluation report	1 year
Study the remediation of contaminated waters	Development, construction and characterization of a large continuous treatment unit	No. of constructed units	1 constructed unit	2 years
	Complete studies on the remediation of contaminated water using the new unit	No. of conducted tests	12 conducted tests	3 years

IV PARTICIPATION

1. ENVIRONMENTAL EDUCATION PROGRAMS

1.1 INTRODUCTION

CONVIVERDE is based on participative methodologies, in which community action is the axis of the educational program, providing activities that contribute to the reconciliation of humans with their environment, with a particular focus on recognizing water as essential for the quality of life locally.

The degradation of the Rio Verde Basin is a local crisis that requires intervention actions that seek to rebalance this basin and its surroundings.

1.2 OBJECTIVES

The ECOEDUCATION program, through CONVIVERDE, proposes to strengthen the guidance and organization of the stakeholders in a collective process of rebuilding the social, cultural, economic and environmental conditions of the region, through the planning and implementation of solutions for the problems identified in the Rio Verde Basin, aiming to change the current situation (Chart 9). The specific objectives are:

- Organize, strengthen and advise the local stakeholders;
- Plan and implement intervention activities to address environmental problems in the Rio Verde Basin;
- Monitor the collective process of rebuilding the social, cultural, economic and environmental conditions of the region.

1.3PARTICIPANT INSTITUTIONS

- Local Municipalities
- Residents' Association
- Commercial/Trade Associations
- Petrobras
- Universities involved in the programs of environmental education

CHART 9 - SPECIFIC OBJECTIVES, ACTIVITIES, INDICATORS, GOALS AND DEADLINES PROPOSED FOR THE ENVIRONMENTAL EDUCATION PROGRAM

<table>
<tr><th>SPECIFIC OBJECTIVES</th><th>ACTIVITY</th><th>INDICATOR</th><th>GOAL</th><th>DEADLINE FOR START UP</th></tr>
<tr><td rowspan="11">Organize, strengthen and guide local stakeholders</td><td>Contact the municipalities</td><td rowspan="4">Partnership with municipalities and city council</td><td rowspan="4">Establish partnerships with the three municipalities and city councils involved</td><td>1 month</td></tr>
<tr><td>Define timeline of meetings</td><td rowspan="3">2 months</td></tr>
<tr><td>Produce the formal partnership agreement</td></tr>
<tr><td>Formalize the partnership</td></tr>
<tr><td>Contact the local trade associations</td><td rowspan="3">No. of contacted local businesses</td><td rowspan="3">Establish contact with 60% of local entrepreneurs</td><td>2 months</td></tr>
<tr><td>Define timeline of meetings together with trade associations</td><td rowspan="2">4 months</td></tr>
<tr><td>Present the ECOEDUCATION/ CONVIVERDE proposal</td></tr>
<tr><td>Define timeline of local community meetings</td><td rowspan="4">No. of mobilized stakeholders</td><td rowspan="4">Mobilize local and regional stakeholders to support the actions of CONVIVERDE</td><td rowspan="4">1 year</td></tr>
<tr><td>Present the ECOEDUCATION/ CONVIVERDE proposal</td></tr>
<tr><td>Develop local stakeholder groups</td></tr>
<tr><td>Integrate the stakeholders into the working groups of CONVIVERDE</td></tr>
<tr><td rowspan="8">Implement CONVIVERDE center</td><td>Define the location within the neighborhood to build the CONVIVERDE center</td><td rowspan="4">Center implementation</td><td rowspan="4">Implementation of the CONVIVERDE activity center in the Cercadinho neighborhood</td><td rowspan="2">2 years</td></tr>
<tr><td>Define the physical space for the development of activities</td></tr>
<tr><td>Implement CONVIVERDE</td><td>2.5 years</td></tr>
<tr><td>Establish goals for the development of activities</td><td>3 years</td></tr>
<tr><td>Set location in Campo Magro and Araucária for the construction of the CONVIVERDE centers</td><td rowspan="4">Center implementation</td><td rowspan="4">Implementation of CONVIVERDE center in Campo Magro and Araucária</td><td rowspan="2">4 years</td></tr>
<tr><td>Define the physical space for the development of activities</td></tr>
<tr><td>Implementation of CONVIVERDE center</td><td>4 years</td></tr>
<tr><td>Establish goals for the development of activities</td><td>4 years</td></tr>
<tr><td rowspan="6">Chose and train local management teams for the centers</td><td>Elect management teams for each center</td><td rowspan="2">No. of inducted management teams</td><td rowspan="2">3 inducted local management teams, one in each center</td><td rowspan="2">4.5 years</td></tr>
<tr><td>Induct management teams</td></tr>
<tr><td>Define the syllabus and prepare training courses</td><td rowspan="4">No. of trained management teams</td><td rowspan="4">3 trained local management teams, one in each center</td><td rowspan="4">5 years</td></tr>
<tr><td>Create teaching materials</td></tr>
<tr><td>Prepare the schedule of training courses</td></tr>
<tr><td>Conduct training courses for management teams</td></tr>
<tr><td rowspan="3">Integrate the Centers and prepare collaborative program</td><td>Meetings for integration and sharing of information and contents of the training courses and other activities to be developed by the Program</td><td rowspan="2">Action Program</td><td rowspan="2">Action Program developed</td><td rowspan="2">5 years</td></tr>
<tr><td>Adapt the Program activities according to the current situation of each municipality</td></tr>
<tr><td>Perform training courses for projet champions and public mobilization</td><td>No. of trained projet champions</td><td>10 per center</td><td>5.5 years</td></tr>
</table>

<table>
<tr><td rowspan="5">Implement the Training Program of Environmental Educators for the Sustainability of Rio Verde and APA</td><td>Define syllabus and prepare training courses</td><td rowspan="2">Training course</td><td rowspan="2">Training course prepared</td><td rowspan="3">5years</td></tr>
<tr><td>Create teaching material</td></tr>
<tr><td>Prepare the training schedule calendar</td><td>Scheduling of training courses</td><td>Schedule of training courses established</td></tr>
<tr><td>Identify and pre-register personnel to be trained</td><td rowspan="2">No. of trained environmental educators</td><td rowspan="2">10 trained environmental educators per center</td><td rowspan="2">5.5 years</td></tr>
<tr><td>Conduct training courses</td></tr>
<tr><td rowspan="8">Realize activities of CONVIVERDE</td><td>Conduct workshops "Getting to know the Rio Verde"</td><td>No. of workshops conducted</td><td>1 workshop per center every 6 months</td><td rowspan="8">6 years</td></tr>
<tr><td>Conduct collaborative efforts involving all centers</td><td>No. of collaborative efforts realized</td><td>1 collaborative effort every 6 months</td></tr>
<tr><td>Publicize on the radio and in local newspaper</td><td>No. of media publications</td><td>1 published every 2 months by the media</td></tr>
<tr><td>Promote educational activities</td><td>No. of educational activities promoted</td><td>1 educational activity per center every 3 months</td></tr>
<tr><td>Implement the Theatre Group "Getting to know the Rio Verde"</td><td>No. of presentations of the Theatre Group</td><td>1 presentation every 6 months</td></tr>
<tr><td>Develop Mobilization Campaign involving all centers</td><td>No. of campaigns performed</td><td>1 campaign performed every 6 months</td></tr>
<tr><td>Create website</td><td>Website ready</td><td>100 hits per month</td></tr>
<tr><td rowspan="4">Monitor the collective process of rebuilding of the socio-cultural, economic and environmental conditions of the Basin</td><td>Conduct evaluation seminars</td><td>No. of evaluation seminars</td><td>4 evaluation seminars per year (1 per quarter)</td><td rowspan="4">6.5 years</td></tr>
<tr><td>Meetings with the management teams of the centers</td><td>No. of meetings of the management teams of the centers</td><td>6 meetings per center per year (one meeting every two months per center)</td></tr>
<tr><td>Preparation of reports</td><td>Bimonthly reports</td><td>Bimonthly reports prepared and delivered</td></tr>
<tr><td>Assessment of events / activities</td><td>Questionnaires filled out at every event / activity</td><td>Evaluations conducted at the end of each event</td></tr>
</table>

V FINANCES

1. SOCIAL ASSISTANCE PROGRAMS

1.1 INTRODUCTION

The Rio Verde Basin is composed of social micro-systems that require varied, but integrated, programs.

The proposal that follows takes into account such diversity and the need for differentiated intervention according to local requirements. The actions are based on two fundamental pillars:

- The proposed actions seek to reconcile: economic development through the creation of jobs in sectors that contribute to environmental preservation based on environmental education, the construction of health infrastructure, and protection of riparian forests and river slopes; the diversification of existent agricultural and non-agricultural activities in the region based on family properties with crops that expand riparian forests; and replace activities that currently do not meet legal requirements.
- The administration of resources in the Rio Verde Basin needs to be broad in order to develop the various aspects of the action plan such as planning, technology, education, financing, with intensive community participation in the process. It also needs to be based on the participation of the local community in an organized and feasible way to maintain continuous action plans that are democratically accepted related to the use of natural resources. Therefore, the proposal is to create a Project Management Committee for the Rio Verde Basin that includes representatives from public bodies and from the community, which will manage the developed projects.

1.2 OBJECTIVES

Establish a governance structure that promotes the effective participation of communities in the Rio Verde Basin in the administration of its available natural resources.

Create an administrative body that promotes the sustainable and equitable use of the resources and that increases economic and environmental security of the region. This Management Committee will be responsible for the organization and financing of the activities that promote harmonious interaction between economic activities and environmental preservation, and that increase the awareness and participation of local communities.

Promote employment and income generation within communities. This way, the proposed activities only involve the use of community residents of Colônia Cristina and Colônia Figueiredo/Cercadinho.

Chart 10 presents proposals for income and job creation considering the need for environmental preservation.

1.3 PARTICIPATING INSTITUTIONS

Partnerships required for training, financing and activity management: EMATER; IAP; Petrobras; SANEPAR; Municipal Administrations; Community Associations; Federal Government.

1.4 FUNDING SOURCES

PRONAF, ATER Programs, Minimum Price Guarantee Program;

Petrobras Environmental Program, Petrobras Cultural Program, Petrobras Development and Citizenship Program

Tax Incentives from State and Municipal Governments.

CHART 10 - SPECIFIC OBJECTIVES, ACTIVITIES, INDICATORS, GOALS AND DEADLINES PROPOSED FOR THE SOCIAL SUPPORT PROGRAM

SPECIFIC OBJECTIVES	ACTIVITY	INDICATOR	GOAL	DEADLINE FOR START UP
Develop a Production System of Plants and Phytotherapeutic Drugs (Only in Colônia Figueiredo/ Cercadinho)	Reforesting of 10,000ha for production of medicinal plants	Number of ha of reforesting using medicinal species	2,000 ha in the first year 3,000 in the second year 2,000 in the third year 3,000 in the fifth year	1 year
	Creation of an Incubator Laboratory for production of products derived of medicinal plants	Incubator Laboratory	Incubator Laboratory in operation	3 years
	Implementation of a Cooperative of Commercialization of Medicinal Plants and derivatives	Cooperative in operation	Sales of phytotherapeutic medicines, teas, cosmetics, candies, soaps, etc.	5 years
	Training of local workforce	Number of people trained	50 trained people	1 year 1 year
	Job generation in the Laboratory and the Cooperative	Number of qualified jobs generated	100 qualified jobs generated including; Pharmacists; Agricultural technicians; Sales promoters; Accountants; Administrators; Secretaries.	
Environmental Services, Health and Social Infrastructure	Recruitment and training of community members as agents	Number of environmental monitors	Training of 10 monitors every 6 months	1 year
	Training and employment of labor for construction	Number of qualified jobs	Formation of 20 qualified jobs, per semester, for construction and maintenance of health and social infrastructure	2 years
	Maintenance of health and social infrastructure	Number of km of roads with permanent monitoring; Percentage of coverage of health infrastructure;	30 km of roads with permanent monitoring; 100% of coverage of health infrastructure	2 years
	Training and employment of labor for community service	Number of community workers trained	Formation of 10 agents every 6 months	6 months
Tourist Services	Expansion of Colônia Polonesa Tourist Route	Expanded route including Campo Largo and Araucária	10 new tourist stops (museum, church, cave, Rio Verde) and 2 new guesthouses and restaurants	3 years
	Incentives to the Tourism Association of Campo Largo	Number of jobs created	40 new direct and indirect jobs to support the Tourism Association of Campo Largo	3 years
	Training of employed labor	Number of people trained by the Tourism Association of Campo Largo	40 travel agents working in various areas: commercialization of local products; guides; stewards of the local culture, maintenance of historic buildings, etc.	3 years
	Creation of Ecological Seal of Approval for the Colônia Polonesa Route	Certification of products produced according to the environmental laws of the APA	All products in the region certified by Ecological Seal of Approval	4 years
	Creation of Scheduled Daily Tour on Tourist Route	Number of people participating in the tours	10 people per day in the first year 20 people daily in the second year 40 people daily in the third year	4 years
	Implementation of point of sale of colonial products	Point of sale	Point of sale set up and the realization of a Fair of Certified Rio Verde Basin Products	4 years

IWA Publishing's authorised EU representative for General Product Safety Regulations is Diane D'Arras, 15 rue Duret, 75116 Paris, France, e-mail: safety@iwap.co.uk.

Printed and bound by CPI Group (UK) Ltd, Croydon, CR0 4YY
16/06/2026
02140391-0001